Water Science and Technology Library

Volume 103

The aim of the *Water Science and Technology Library* is to provide a forum for dissemination of the state-of-the-art of topics of current interest in the area of water science and technology. This is accomplished through publication of reference books and monographs, authored or edited. Occasionally also proceedings volumes are accepted for publication in the series. *Water Science and Technology Library* encompasses a wide range of topics dealing with science as well as socio-economic aspects of water, environment, and ecology. Both the water quantity and quality issues are relevant and are embraced by *Water Science and Technology Library*. The emphasis may be on either the scientific content, or techniques of solution, or both. There is increasing emphasis these days on processes and *Water Science and Technology Library* is committed to promoting this emphasis by publishing books emphasizing scientific discussions of physical, chemical, and/or biological aspects of water resources. Likewise, current or emerging solution techniques receive high priority. Interdisciplinary coverage is encouraged. Case studies contributing to our knowledge of water science and technology are also embraced by the series. Innovative ideas and novel techniques are of particular interest.

Comments or suggestions for future volumes are welcomed.

Vijay P. Singh, Department of Biological and Agricultural Engineering & Zachry Department of Civil and Environment Engineering, Texas A&M University, USA Email: vsingh@tamu.edu

All contributions to an edited volume should undergo standard peer review to ensure high scientific quality, while monographs should also be reviewed by at least two experts in the field.

Manuscripts that have undergone successful review should then be prepared according to the Publisher's guidelines manuscripts: https://www.springer.com/gp/authors-editors/book-authors-editors/book-manuscript-guidelines

More information about this series at https://link.springer.com/bookseries/6689

Ashish Pandey · V. M. Chowdary ·
Mukunda Dev Behera · V. P. Singh
Editors

Geospatial Technologies for Land and Water Resources Management

Editors
Ashish Pandey
Department of Water Resources
Development and Management
Indian Institute of Technology Roorkee
Roorkee, Uttarakhand, India

Mukunda Dev Behera
Centre for Oceans, Rivers, Atmosphere
and Land Sciences (CORAL)
Indian Institute of Technology Kharagpur
Kharagpur, West Bengal, India

V. M. Chowdary
Department of Agriculture, Cooperation
and Farmers Welfare, Ministry
of Agriculture and Farmers Welfare,
Government of India
Mahalanobis National Crop Forecast
Centre (MNCFC)
Delhi, India

V. P. Singh
Texas A&M University
College Station, TX, USA

ISSN 0921-092X ISSN 1872-4663 (electronic)
Water Science and Technology Library
ISBN 978-3-030-90481-4 ISBN 978-3-030-90479-1 (eBook)
https://doi.org/10.1007/978-3-030-90479-1

This Springer imprint is published by the registered company Springer Nature Switzerland AG
The registered company address is: Gewerbestrasse 11, 6330 Cham, Switzerland

Contents

Chapter 1
Overview of Geospatial Technologies for Land and Water Resources Management

Ashish Pandey, Gagandeep Singh, V. M. Chowdary, Mukunda Dev Behera, A. Jaya Prakash, and V. P. Singh

Abstract Land and water resources management are essential for the future sustainability of the environment. The studies on land and water resources require basic geo-referenced data, such as land use-land cover (LULC), soil maps, and digital elevation models (DEMs) for capturing the spatio-temporal variations of thematic layers. These data can be easily obtained from remote sensing images and limited ground truth. Hydro-meteorological data, such as precipitation, air, land surface temperature, solar radiation, evapotranspiration, soil moisture, river and lakes water levels, river discharge, and terrestrial water storage, can also be derived from remote sensing as well as from point-based ground instruments. Then, studies can be carried out at various spatio-temporal scales.

1.1 Introduction to Geospatial Technology

Remote Sensing, Geographic Information Systems (GIS), and Global Positioning System (GPS) form a revolutionary combination often referred to as Geospatial Technologies. Geospatial Technologies is the most powerful and transformational modern-day technologies used extensively to address real-time problems on the

A. Pandey (✉) · G. Singh
Department of Water Resources Development and Management, Indian Institute of Technology Roorkee, Roorkee, Uttarakhand 247 667, India
e-mail: ashish.pandey@wr.iitr.ac.in

V. M. Chowdary
Mahalanobis National Crop Forecast Centre (MNCFC), Department of Agriculture, Cooperation & Farmers Welfare, Ministry of Agriculture & Farmers Welfare, Government of India, New Delhi, India

M. D. Behera · A. J. Prakash
Indian Institute of Technology Kharagpur, Kharagpur, West Bengal, India

V. P. Singh
Department of Biological & Agricultural Engineering and Zachry Department of Civil & Environmental Engineering, Texas A&M University, College Station, TX 77843-2117, USA

A. Pandey et al. (eds.), *Geospatial Technologies for Land and Water Resources Management*, Water Science and Technology Library 103,
https://doi.org/10.1007/978-3-030-90479-1_1

earth's surface. The conjunctive use of remote sensing and GIS has proved to be highly effective to analyze diverse phenomena on the earth's surface (Davis et al. 1991; Lo et al. 1997; Huggel et al. 2003; Kaab et al. 2005; Pandey et al. 2007; Patel and Srivastava, 2013; Calera et al. 2017; Chae et al. 2017; Borrelli et al. 2017). The capability of satellites and sensors for earth observations through numerous spectral bands has enhanced the umbrella of applications manifold. Analysis of land and water resources using vast volumes of data demands a robust database management system. GIS serves as a perfect platform for storing, managing, and analyzing voluminous spatial and non-spatial data (Chang 2008). It provides a robust computing environment and platform for re-scaling models and supports handling complex data-method relationships (Pandey et al. 2016b). Groot (1989) defined geospatial technology or geoinformatics as "the science and technology dealing with the structure and character of spatial information, its capture, its classification and qualification, its storage, processing, portrayal, and dissemination, including the infrastructure necessary to secure optimal use of this information." Various applications of this technology can be broadly categorized into two significant domains, namely land resources and water resources. These two domains cover many applications in natural resources management, where geospatial technology serves as a very effective decision-making tool in these applications. This technology is being extensively used for effective and sustainable planning, management, and development of natural resources (Verbyla 1995).

Land resources form the core of sustainable existence and development in critical challenges, like agriculture, food production, poverty, and climate change impacts (Muller and Munroe 2014). Issues like improving agricultural production, soil conservation, deforestation, land degradation, and climate change require repeated observations of the nature, extent, and spatial variations of the earth surface with a high spatial resolution (Buchanan et al. 2008; Pandey et al. 2011; Yang et al. 2013; Calvao and Pessoa 2015; Huang et al. 2018; Pandey and Palmate 2018; Pandey et al. 2021a). Rapid geospatial technology advancements have revolutionized land resources mapping, monitoring, and management (Velmurugan and Carlos 2009). This technology also facilitates the generation of time-series databases enabling the scientists and researchers to derive meaningful results, recommendations, and action plans for the decision-makers at various implementation levels.

Water, the most precious natural resource, experiences immense pressure due to overexploitation to satisfy the ever-growing population's needs (Wang et al. 2021). Moreover, factors like urbanization, globalization, infrastructural developments, and climate change have posed a massive threat to the limited freshwater resources available on earth (Chapagain and Hoekstra 2008; Giacomoni et al. 2013; Nair et al. 2013). Geospatial technology plays an instrumental role in analyzing, modeling, and simulating water quality, water availability, water supply management, floods, and droughts under various climate change scenarios. There are numerous applications of this technology addressing sustainable water resources management viz. assessment of groundwater recharge potential; integrated watershed management and development (Pandey et al. 2004); design flood estimation (Sharma et al. 2021);

flood modeling (Patro et al. 2009); flood inundation and hazard mapping (Singh and Pandey 2021); sediment dynamic modeling (Pandey et al. 2016b).

Remote sensing forms the most integral component of the geospatial technology serving the purpose of a data source. Remote sensing has a unique capability of observing the earth's surface in numerous spectral bands covering different wavelength ranges (Lillesand et al. 2015). Optical remote sensing uses visible, near-infrared, and short-wave infrared sensors to form images of the earth's surface by detecting the solar radiation reflected from targets on the ground (Lillesand et al. 2015). Different materials reflect and absorb differently at different wavelengths. Thus, the targets can be differentiated by their spectral reflectance signatures in remotely sensed images.

There are few open source satellites that provide solutions to geospatial technologies with easier access to the user. Remote sensing satellite sensors gather information from space and generate a large number of datasets that are difficult to manage and analyze using software packages or applications that may require significant time and labor. The cloud computing systems, such as Amazon Web Services (AWS) and Google Earth Engine (GEE), have been developed to address this issue. Although cloud computing platforms and other emerging technologies have demonstrated their significant potential for monitoring land and water resources management, they have not been appropriately examined and deployed for RS applications until recently. Users can access various data sets on those platforms without having to download anything. Both GEE and AWS offer similar features, such as automatic parallel processing and a fast computational platform for successfully dealing with substantial data processing or time-series analysis in a quick interval.

1.2 Cutting Edge—Techniques and Applications of Geospatial Technologies in Land and Water Resources Management

Various types of geospatial technologies have been made accessible to end users in recent years for use in a variety of applications in land and water and other emerging applications.

1. **Remote Sensing**—High-resolution satellite imagery is acquired from space using a camera or sensor platforms mounted to the spacecraft. There were fewer high-resolution satellite images with centimeter resolution accuracy needed for monitoring in many applications, to meet human requirements and study the earth's climate.
2. **Geographic Information Systems (GIS)**—An application or software package for analyzing or mapping satellite data and performing additional operations, such as geo-referencing and geocoding, if the particular location of the earth's surface is known. The model can then be used to do various analyses through the use of different techniques.

3. **Global Positioning System (GPS)**—The discipline of earth monitoring has grown significantly in recent years. It has three basic components: the space segment, the control segment, and the user segment. It is a cutting edge technology capable of providing greater accuracy, less than a millimeter or meter. In the application to land and water resources, the most important requirement is to gather the geographical coordinates of any object present on the earth's surface and gain information from the object features with geographical data, which was acquired in real time and directly from the field at a reasonable cost.
4. **Internet Mapping Technologies**—Cloud computing platforms, such as Google Earth Engine, Microsoft Virtual Earth, Amazon Web Services, as well as other web features, are gradually improving how geographical data is analyzed and disseminated. With the availability of many modern technologies to users and other agencies, began analyzing data for satellite photos without prior experience or any pre-processing processes. By comparison, traditional GIS procedures are limited to highly skilled individuals for analyzing satellite data and mapping data for a variety of applications. As a result, internet mapping offers more opportunities to users who are willing to invest efforts in complex algorithms.

There are numerous uses for land and water resources, such as rainfall, land cover, snow cover extent, surface water extent, soil moisture, and hydrological cycle. All of these application parameters are quantified using various approaches including satellite data. Surface water bodies can be identified using remote sensing techniques; meteorological variables, such as temperature and precipitation can be estimated; hydrological state variables, such as soil moisture and land surface features can be estimated; and fluxes, such as evapotranspiration can be estimated. Availability of different sensors which directly gather information from land water bodies provides significant information in modeling algorithms. Moreover, it can be applied to crop inventory and forecasts; drought and flood damage assessment; and land use monitoring and management. Today, India is one of the major providers of earth observation data in the world in a variety of spatial, spectral, and temporal resolutions, meeting the needs of many applications of relevance to national development.

Based on multiple spectral bands used in the imaging process, optical remote sensing systems are categorized into three basic groups viz. panchromatic (single band), multispectral and hyperspectral systems. Table 1.1 offers many Indian and global panchromatic and multispectral satellite data products extensively utilized to address land and water resources management challenges. Table 1.2 shows a list of hyperspectral satellite data products.

Microwave remote sensing is very popular in the research community to map and monitor water resources primarily because of the capability of microwaves to accurately detect water (Ulaby 1977; Engman 1991) due to its all-weather ability. Synthetic Aperture Radar (SAR) has been one of the most prominently used microwave remote sensing data products to address water-related applications. Table 1.3 presents a list of SAR and other satellite data products available in the microwave region of the electromagnetic spectrum (Brisco et al. 2013; Singh and Pandey 2021).

Table 1.1 List of optical remote sensing (panchromatic and multispectral) data products

S. No.	Satellite mission	Sensor	Spatial resolution (m)	Temporal resolution	Manufacturer	Data available from
1	Cartosat-3	Pan	0.28	5 days revisit	ISRO	15-Feb-2020
2	Cartosat-3	MX	1.12	4/5 days	ISRO	15-Feb-2020
3	Cartosat-2 series	Pan	0.65	4/5 days	ISRO	01-Aug-2016
4	Cartosat-2 series	MX	1.6	4/5 days	ISRO	01-Aug-2016
5	Cartosat-2B	PAN	1	4/5 days	ISRO	13-Jul-2010
6	Cartosat-2A	PAN	1	4/5 days	ISRO	29-Apr-2008
7	Cartosat-2	PAN	1	4 days	ISRO	14-Apr-2007
8	Cartosat-1	PAN-F	2.5	5 days	ISRO	08-May-2005 to 31-Jan-2019
9	Cartosat-1	PAN-A	2.5	5 days	ISRO	08-May-2005 to 31-Jan-2019
10	Cartosat-1	Stereo	2.5	5 days	ISRO	08-May-2005 to 31-Jan-2019
11	Cartosat-1	Widemono	2.5	5 days	ISRO	27-May-2005 to 31-Jan-2019
12	Resourcesat-2A	AWIFS	56	2–3 days	ISRO	06-Jan-2017
13	Resourcesat-2A	Liss-3	23.5	12–13 days	ISRO	06-Jan-2017
14	Resourcesat-2A	Liss-4-FMX	5.8	25–26 days	ISRO	06-Jan-2017
15	Resourcesat-2A	Liss-4-SMX	5.8	2–3 days	ISRO	15-Dec-2016 to 17-May-2017
16	Resourcesat-2	AWiFS	56	2–3 days	ISRO	30-Sep-2011
17	Resourcesat-2	Liss-3	23.5	2–3 days	ISRO	30-Sep-2011
18	Resourcesat-2	Liss-4-FMX	5.8	2–3 days	ISRO	30-Sep-2011
19	Resourcesat-2	Liss-4-SMX	5.8	2–3 days	ISRO	28-Sep-2017
20	Resourcesat-1	AWiFS	56	5 days	ISRO	07-Dec-2003
21	Resourcesat-1	Liss-3	23.5	5 days	ISRO	07-Dec-2003
22	Resourcesat-1	Liss4-SMX	5.8	5 days	ISRO	11-Dec-2003
23	Oceansat-2	OCM	360	2 days	ISRO	01-Jan-2010
24	Landsat 7	ETM+	30	16-day	NASA/USGS	1999 to present
25	Landsat 8	OLI/TIRS	30	16-day	NASA/USGS	2013 to present

(continued)

Table 1.1 (continued)

S. No.	Satellite mission	Sensor	Spatial resolution (m)	Temporal resolution	Manufacturer	Data available from
26	Sentinel-2A and 2B	MSI	10, 20, 60	10 and 5 days	ESA	2015 to present

Dataset source for downloading, Cartosat series: https://bhoonidhi.nrsc.gov.in/ Resourcesat series: https://bhoonidhi.nrsc.gov.in/ and https://glovis.usgs.gov/ Landsat series: https://earthexplorer.usgs.gov/ Sentinel Series: https://scihub.copernicus.eu/

Table 1.2 List of optical remote sensing (hyperspectral) data products

S. No.	Satellite	Sensor type	Resolution		Organization
			Spectral (nm)	Temporal (day)	
1	EO-1	Hyperion	10	16–30	NASA
2	Shenzhou-8	Tiangong-1 hyperspectral imager (HSI)	10 (VNIR) 23 (SWIR)	–	Chinese academy of science physics
3	PRISMA	PRISMA	10	14–7	Agenzia Spaziale Italiana
4	HISUI	HISUI	30	2–60	Japanese ministry of economy, trade, and industry
5	EnMAP HSI	EnMAP	30	27 (VZA $\geq$ 5°)	GFZ-DLR
6	SHALOM	Improved multi-purpose satellite-II	10	4 (VZA $\geq$ 30°)	ASI-ISA
7	HyspIRI	HyspIRI	30 (60)	5–16	JPL-NASA

Dataset source for downloading, S. No. 1—http://earthexplorer.usgs.gov/ 2—http://www.msadc.cn/sy/ 3—http://prisma-i.it/index.php/en/ 4—https://oceancolor.gsfc.nasa.gov/data/hico/ 5—https://earth.esa.int/eogateway/catalog/proba-chris-level-1a 6—https://www.nrsc.gov.in/EOP_irsdata_Products_Hyperspectral

1.3 Methodology Development in Land Resources Management

Geospatial technologies play a pivotal role in monitoring and managing land resources. One of the most widely exploited applications is Digital Terrain Modeling (DTM), which characterizes the topography of any area using digital elevation models (DEMs) (Zhou et al. 2007). DEM products of different spatial resolutions are extensively used for topographic mapping, relief mapping, and terrain analysis (Yang et al. 2011). They also serve as a preliminary input in various hydrological studies (Nagaveni et al. 2019; Himanshu et al. 2015). A list of several DEM products available for use is presented in Table 1.4.

Table 1.3 List of microwave remote sensing data products

S. No.	Satellite	Sensor	Resolution	Data available from	Organization
1	Risat-1 C-band SAR	FRS 1	3 m	01-Jul-2012 to 30-Sep-2016	ISRO
2	Oscar	Ku-band	50 km 25 km	09-Feb-2010 to 01-Mar-2014	AMSAT
3	Scatsat-1	Ku-band scatterometer	50 km 25 km	26-May-2017 onwards Release of SCATSAT-1 V1.1.4 data products	ISRO
4	SARAL	Ka-band radar altimeter	–	01-Nov-2014 onwards	CNES, ISRO
5	Sentinel-1	SAR C-band (center frequency: 5.405 GHz)	10 m		ESA
6	TerraSAR-X	X	HR (5–20 m) VHR (0–5 m)	2007 onwards	ISC Kosmotras
7	Radarsat-2	C	3–300 m	2007 onwards	MDA
8	ALOS-2	L	10 m	2014 onwards	JAXA

Dataset source for downloading, S. No. 1–4 https://bhoonidhi.nrsc.gov.in/ 5—https://scihub.copernicus.eu/ 6—https://earth.esa.int/ 7—https://tpm-ds.eo.esa.int/oads/access/collection/Radarsat-2 8—https://www.jaxa.jp/projects/sat/alos2/index_j.html

Table 1.4 List of DEM data products

S. No.	Product	Spatial resolution (m)	Developing agency	Accuracy
1	SRTM (shuttle radar topographic mission)	30, 90	NASA	RMSE ~10 m
2	CartoSAT	30	ISRO	Approx. 8 m
3	ASTER GDEM	30	NASA, METI	RMSE 2–3 m
4	ALOS PALSAR	12.5 and 30	JAXA and Japan resources observation system organization (JAROS)	RMSE of 4.6 m and 4.9 m
5	ESA—ACE-2	90, 270, 1 and 10	ESA	Extremely high

Dataset source for downloading, S. No. 1—https://dwtkns.com/srtm30m/ 2—https://search.earthdata.nasa.gov/search 3—https://search.earthdata.nasa.gov/search 4—https://search.asf.alaska.edu/#/ 5—https://sedac.ciesin.columbia.edu/data/set/dedc-ace-v2/data-download

DEMs are processed and analyzed in a GIS environment to derive numerous indices, which enable understanding various environmental processes (Gajbhiye et al. 2015; Rao et al. 2019). Additionally, DEMs are also extensively used in morphometric characterization of watersheds (Wang et al. 2010). Parameters like slope, aspect, contours, curvature are effectively derived from the DEMs (Gajbhiye et al. 2014).

Soil resources mapping is another major application under the gamut of land resources management powered by geospatial technologies. The satellite image interpretation and image classification techniques are employed to identify and map different land uses and vegetation types (Robertson and King 2011). Remote sensing and GIS are effectively used for crop mapping, inventory, and management (Wardlow et al. 2007). This domain features serve many purposes, such as crop acreage estimation, condition assessment, yield forecasting, cropping system analysis, and precision farming. Crop type mapping, acreage, and condition assessment are mainly carried out using image interpretation and digital image processing, wherein the spectral response of crop types is analyzed. The variations in the signatures of different wavelength bands help with discrimination among additional features (Foerster et al. 2012).

Additionally, image classification supported with ground truth information helps generate land use maps spatially. Medium and high-resolution time-series satellite data are beneficial for discriminating and monitoring various crops periodically. Assessment and monitoring of droughts are one of the most critical food security issues of concern globally (Swain et al. 2021). Significantly, agrarian countries are primarily dependent on their agricultural production, which is a significant economic driver. Climate change and water availability pose a substantial threat to the world's agricultural sector (Tarquis et al. 2010).

Geospatial technology has extensive scope for drought monitoring and assessment. Satellite remote sensing enables the monitoring of crops at various growth stages. Additionally, remote sensing data is used to compute spectral indices such as normalized difference vegetation index (NDVI) and normalized difference water index (NDWI), which provide essential inputs for drought assessment and monitoring (Pandey et al. 2010).

Soil erosion is a serious problem that poses a threat to agricultural land and infrastructure globally. One of the most popular methods used for soil erosion assessment and soil loss estimation is the Universal Soil Loss Equation (USLE) (Pandey et al. 2009b). This method involves the computation of rainfall erosivity factor (R), soil erodibility factor (K), topographic factor (LS), crop management factor (C), and conservation supporting practice factor (P). GIS provides a platform to prepare and analyze the spatial layers of each of these factors to estimate the average annual soil loss rate (Dabral et al. 2008). Subsequently, researchers across the globe have employed the Revised Universal Soil Loss Equation (RUSLE) to assess the soil loss status (Pandey et al. 2021b).

Land use/land cover data is a standard input used in sediment yield modeling (Pandey et al. 2007, 2009a, b). Satellite data is being very efficiently used in reservoir sedimentation assessment studies (Pandey et al. 2016a). The change in water spread area is assessed using satellite image processing at different times using indices like NDVI and NDWI, and deposition of sediments is evaluated. Consequently, loss in the live storage of reservoirs due to sedimentation is estimated (Jain et al. 2002).

1.4 Methodology Development in Water Resources Management

Water resources can be undoubtedly argued to be the most benefitted domain from the advent of geospatial technologies. These advanced technologies play a key role in conducting hydrological studies for rainfall estimation, soil moisture estimation and modeling, streamflow estimation, rainfall-runoff modeling, rainfall forecasting, water balance modeling, hydrological modeling, hydraulic and hydrodynamic modeling (Milewski et al. 2009; Singh et al. 2015, 2019; Himanshu et al. 2017, 2021; Jaiswal et al. 2020). Application of remote sensing and GIS in water resources also extends in identifying suitable sites for soil and water conservation structures, sediment yield modeling, reservoir sedimentation, watershed characterization, and management plan (Pandey et al. 2011; Pandey et al. 2016b; Dayal et al. 2021).

Satellite data for rainfall estimation has been one of the most popular applications, especially in the data-scarce regions or lack of adequate ground-based instrumentation for measuring rainfall. Numerous operational satellite-based rainfall products provide rainfall estimates at various spatial and temporal resolutions (Table 1.5). Numerous studies have been carried out to evaluate the performance of these data products before and after bias correction and were used in many hydrological studies (Behrangi et al. 2011; Himanshu et al. 2018).

Soil moisture estimation using remote sensing data is another rapidly evolving application in the water resources domain (Srivastava et al. 2009; Singh et al. 2015). Soil moisture is a crucial parameter used in various hydrological, land surface modeling, and meteorological studies (Albergel et al. 2013; Wanders et al. 2014). Interestingly, satellite-derived soil moisture products are also used to monitor and predict natural disaster events (Abelen et al. 2015). Additionally, these products also find application in climate variability studies (Loew et al. 2013).

The microwave band of the electromagnetic spectrum is exclusively used for soil moisture estimation. Table 1.6 presents a list of remote sensing-based soil moisture products available for use. Apart from the advantages of all-weather and day-night coverage, passive microwave sensors provide soil moisture estimation capability with good temporal resolution. In contrast, active microwave sensors provide finer, more satisfactory spatial resolutions (Singh et al. 2015).

The majority of the water resources management projects or research, especially at small and medium scales, are carried out at the watershed level (Sivapalan 2003). At this level, the analysis demands operational tools for simulating various processes

Table 1.5 List of satellite-based rainfall data products

S. No.	Product	Spatial resolution	Temporal resolution	Data available from	Developing agency
1	IMERG (early run, late run, and final run)	0.1	30 min, 1 day	2000–	NASA
2	GSMaP	0.1	1 h, 1 day, 1 month	2000–	JAXA
3	CHIRPS	0.05, 0.25	1 day, 5 day, 1 month	1981–	USAID, NASA, and NOAA
4	CMORPH	0.05, 0.25	30 min, 1 h, 1 day	1998–	National weather service climate prediction center (CPC)
5	PERSIANN	0.25	1 h, 1 day, 1 month, 1 year	1983–	Center for hydrometeorology and remote sensing (CHRS)
6	MSWEP	0.1	3 h	1979–2016	CMWF, NASA, and NOAA
7	SM2RAIN-ASCAT	0.1	1 day	2007–2020	–
8	SM2RAIN-CCI	0.25	1 day	1998–2015	–
9	GPM + SM2RAIN	0.25	1 day	2007–2018	–

Dataset source for downloading, S. No. 1—https://disc.gsfc.nasa.gov/dtasets/GPM_3IMERGHHE_06/summary?keywords=%22IMERG%20Early%22 2—ftp://hokusai.eorc.jaxa.jp/ 3—https://data.chc.ucsb.edu/products/CHIRPS-2.0/ 4—https://www.ncei.noaa.gov/data/cmorph-high-resolution-global-precipitation-estimates/access/ 5—https://chrsdata.eng.uci.edu/ 6—https://gwadi.org/multi-source-weighted-ensemble-precipitation-mswep 7–9 https://zenodo.org/record/3854817#.YC3BV3Uza5w

and interactions associated with water resources (Hingray et al. 2014). Therefore, watershed modeling becomes essential to understand and analyze the interactions between nature, climate, and human interventions. The distributed models employed for watershed modeling are data-intensive, and in data-scarce areas, geospatial technology plays a prominent role in addressing data gaps (Stisen et al. 2008). The topography data is one of the essential datasets in any watershed modeling assignment. The most widely available source of topographic data is open source DEMs. Advanced data capture techniques, such as Light Detection and Ranging (LiDAR), are being deployed to gather higher-accuracy terrain information. Table 1.7 lists a few LiDAR datasets exclusively available for the USA.

Climate data specifically, temperature, relative humidity, solar radiation, and wind speed, are the primary inputs required to analyze the hydrology of any watershed. All these parameters are being monitored repeatedly using various satellite sensors. Additionally, the National Center for Environmental Prediction (NCEP) provides

Table 1.6 List of satellite-based soil moisture data products

S. No.	Product	Spatial resolution	Temporal resolution	Data available from	Agency
1	ESA CCI soil moisture	0.25	1 day	1978–2020	European space agency's (ESA)
2	ASCAT soil moisture	12.5 km, 25 km	1–2 day	2007–	EUMETSAT H-SAF
3	SMAP L3	9 km, 36 km	1 day	2015–	NASA (NSIDC DAAC)
4	SMOS L2	15 km	1–3 days	2010–	ESA
5	SMOS CATDS soil moisture	25 km	1 day	2010–	ESA
6	SMOS BEC soil moisture	25 km	1 day	2010–	ESA
7	AMSR-2	50 km	1–3 days	2012–	Japan Aerospace Exploration Agency (JAXA)

Dataset source for downloading, S. No. 1—https://esa-soilmoisture-cci.org/data 2—https://hsaf.meteoam.it/Products/ProductsList?type=soil_moisture 3—https://smap.jpl.nasa.gov/data/ 4—http://www.catds.fr/Products/Available-products-from-CPDC 5—https://smos-diss.eo.esa.int/oads/access/ 6—http://bec.icm.csic.es/land-datasets/ 7—https://suzaku.eorc.jaxa.jp/GCOM_W/data/data_w_product-2.html

Table 1.7 List of LiDAR data products

S. No.	Product	Region	Accuracy
1	Open topography	USA	–
2	U.S. Interagency elevation inventory	USA	–
3	NOAA digital coast	USA	–
4	NEON open data portal	USA	–

Dataset source for downloading, S. No. 1–https://portal.opentopography.org/dataCatalog 2—https://coast.noaa.gov/digitalcoast/data/inventory.html 3—https://www.coast.noaa.gov/dataviewer/#/lidar/search/ 4—https://data.neonscience.org/data-products/explore

the Climate Forecast System Reanalysis (CFSR) data in a gridded format to be conveniently used for watershed modeling applications (Fadil and Bouchti 2020).

Satellite altimetry is a unique application of geospatial technology in water resources management. Altimetry provides a means to monitor the water level of rivers and reservoirs using satellite observations. Moreover, repeated observations allow evaluation of change in water storage in reservoirs and overcome the limitation of the spare in-situ network of gauge stations. The water levels from altimetry can also be used to calibrate and validate hydrological and hydrodynamic models (Thakur et al. 2021). Table 1.8 presents a list of some radar altimetry data products.

Table 1.8 List of basic characteristics of radar altimetry data products

S. no.	Mission	Equator track distance (km)	Band	Frequency (GHz)
1	GEOSAT	163	Ku	13.5
2	ERS-1/2	80	Ku	13.8
3	TOPEX/POSEIDON Jason-1/2/3 Sentinel-6	315	Ku/C	13.6/5.3
4	GFO	163	Ku	13.5
5	ENVISAT	163	Ku/S	13.6/3.2
6	CryoSat-2	7	Ku	13.6
7	HY-2A/2B	90	Ku/C	13.6/5.3
8	SARAL/ALTIKA	90	Ka	35
9	Sentinel -3A	104	Ku-band and C-bands	23.8
10	Sentinel- 3B	52	S band, X band	36.5

Dataset source for downloading, S. No. 1—https://science.nasa.gov/missions/geosat 2—https://aviso-data-center.cnes.fr/ 3—https://earth.esa.int/eogateway 4—https://aviso-data-center.cnes.fr/ 5—https://earth.esa.int/eogateway 6–8 https://aviso-data-center.cnes.fr/ 9,10—https://scihub.copernicus.eu

1.5 Conclusions

The application of geospatial technologies for land use-land cover analysis and mapping, digital terrain modeling, soil resource inventory, crop monitoring, and mapping, estimation of evapotranspiration, soil moisture measurement, morphometric parameter analysis, drought monitoring, soil erosion modeling, watershed management, agricultural land use planning, water quality assessment, reservoir sedimentation, flood mapping, monitoring reservoir/lake water levels, river discharge, and spatial modeling have revolutionized the assessment, mapping, and monitoring of land and water resources. The case studies provided in this book will serve as a valuable resource for scientists and researchers involved in planning and managing land and water resources sustainably.

This book offers an overview of geospatial technologies in land and water resources management. It consists of four main sections: land use land cover dynamics, agricultural water management, water resources assessment and modeling, and natural disasters. From leading institutions, such as the IITs and ISRO, the authors have shared their experiences and offered case studies to provide insights into the application of geospatial technologies for land and water resources management.

References

Abelen S, Seitz F, Abarca-del-Rio R, Güntner A (2015) Droughts and floods in the La Plata basin in soil moisture data and GRACE. Remote Sens 7(6):7324–7349

Albergel C, Dorigo W, Balsamo G, Muñoz-Sabater J, de Rosnay P, Isaksen L, Brocca L, De Jeu R, Wagner W (2013) Monitoring multi-decadal satellite earth observation of soil moisture products through land surface reanalyses. Remote Sens Environ 138:77–89

Behrangi A, Khakbaz B, Jaw TC, AghaKouchak A, Hsu K, Sorooshian S (2011) Hydrologic evaluation of satellite precipitation products over a mid-size basin. J Hydrol 397(3–4):225–237

Borrelli P, Robinson DA, Fleischer LR, Lugato E, Ballabio C, Alewell C, Meusburger K, Modugno S, Schütt B, Ferro V, Bagarello V (2017) An assessment of the global impact of 21st century land use change on soil erosion. Nat Commun 8(1):1–13

Brisco B, Schmitt A, Murnaghan K, Kaya S, Roth A (2013) SAR polarimetric change detection for flooded vegetation. Int J Digit Earth 6(2):103–114

Buchanan GM, Butchart SH, Dutson G, Pilgrim JD, Steininger MK, Bishop KD, Mayaux P (2008) Using remote sensing to inform conservation status assessment: estimates of recent deforestation rates on New Britain and the impacts upon endemic birds. Biol Cons 141(1):56–66

Calera A, Campos I, Osann A, D'Urso G, Menenti M (2017) Remote sensing for crop water management: from ET modelling to services for the end users. Sensors 17(5):1104

Calvao T, Pessoa MF (2015) Remote sensing in food production—a review. Emirates J Food Agric 27(2(SI)):138–151

Chae BG, Park HJ, Catani F, Simoni A, Berti M (2017) Landslide prediction, monitoring and early warning: a concise review of state-of-the-art. Geosci J 21(6):1033–1070

Chang KT (2008) Introduction to geographic information systems, vol 4. McGraw-Hill, Boston

Chapagain AK, Hoekstra AY (2008) The global component of freshwater demand and supply: an assessment of virtual water flows between nations as a result of trade in agricultural and industrial products. Water Int 33(1):19–32

Dabral PP, Baithuri N, Pandey A (2008) Soil erosion assessment in a hilly catchment of North Eastern India using USLE, GIS and remote sensing. Water Resour Manage 22(12):1783–1798

Davis F, Quattrochi D, Ridd M, Lam N, Walsh SJ, Michaelsen JC, Franklin J, Stow DA, Johannsen CJ, Johnston CA (1991) Environmental analysis using integrated GIS and remotely sensed data—some research needs and priorities. Photogramm Eng Remote Sens 57(6):689–697

Dayal D, Gupta PK, Pandey A (2021) Streamflow estimation using satellite-retrieved water fluxes and machine learning technique over monsoon-dominated catchments of India. Hydrol Sci J 66(4):656–671

Engman ET (1991) Applications of microwave remote sensing of soil moisture for water resources and agriculture. Remote Sens Environ 35(2–3):213–226

Fadil A, El Bouchti A (2020) Global data for watershed modeling: the case of data scarcity areas. In: Geospatial Technology. Springer, Cham, pp 1–14

Foerster S, Kaden K, Foerster M, Itzerott S (2012) Crop type mapping using spectral–temporal profiles and phenological information. Comput Electron Agric 89:30–40

Gajbhiye S, Mishra SK, Pandey A (2014) Prioritizing erosion-prone area through morphometric analysis: an RS and GIS perspective. Appl Water Sci 4(1):51–61

Gajbhiye S, Mishra SK, Pandey A (2015) Simplified sediment yield index model incorporating parameter curve number. Arab J Geosci 8(4):1993–2004

Giacomoni MH, Kanta L, Zechman EM (2013) Complex adaptive systems approach to simulate the sustainability of water resources and urbanization. J Water Resour Plan Manag 139(5):554–564

Groot R (1989) Meeting Educational Requirements in Geomatics. ITC J 1:1–4

Himanshu SK, Pandey A, Shrestha P (2017) Application of SWAT in an Indian river basin for modeling runoff, sediment and water balance. Environ Earth Sci 76:3. https://doi.org/10.1007/s12665-016-6316-8

Himanshu SK, Pandey A, Dayal D (May 2018) Evaluation of satellite-based precipitation estimates over an agricultural watershed of India. In: World Environmental and Water Resources

Congress 2018: watershed management, irrigation and drainage, and water resources planning and management. American Society of Civil Engineers, Reston, VA, pp 308–320
Himanshu SK, Pandey A, Dayal D (2021) Assessment of multiple satellite-based precipitation estimates over Muneru watershed of India. In: Water management and water governance. Springer, Cham, pp 61–78
Himanshu SK, Pandey A, Palmate SS (2015) Derivation of Nash model parameters from geomorphological instantaneous unit hydrograph for a Himalayan river using ASTER DEM. In: Proceedings of international conference on structural architectural and civil engineering, Dubai
Hingray B, Picouet C, Musy A (2014) Hydrology: a science for engineers. CRC Press
Huang Y, Chen ZX, Tao YU, Huang XZ, Gu XF (2018) Agricultural remote sensing big data: management and applications. J Integr Agric 17(9):1915–1931
Huggel C, Kääb A, Haeberli W, Krummenacher B (2003) Regional-scale GIS-models for assessment of hazards from glacier lake outbursts: evaluation and application in the Swiss Alps. Nat Hazard 3(6):647–662
Jain SK, Singh P, Seth SM (2002) Assessment of sedimentation in Bhakra reservoir in the western Himalayan region using remotely sensed data. Hydrol Sci J 47(2):203–212. https://doi.org/10.1080/02626660209492924
Jaiswal RK, Yadav RN, Lohani AK et al (2020) Water balance modeling of Tandula (India) reservoir catchment using SWAT. Arab J Geosci 13:148
Kaab A, Huggel C, Fischer L, Guex S, Paul F, Roer I, Salzmann N, Schlaefli S, Schmutz K, Schneider D, Strozzi T (2005) Remote sensing of glacier-and permafrost-related hazards in high mountains: an overview. Nat Hazard 5(4):527–554
Lillesand T, Kiefer RW, Chipman J (2015) Remote sensing and image interpretation. Wiley
Lo CP, Quattrochi DA, Luvall JC (1997) Application of high-resolution thermal infrared remote sensing and GIS to assess the urban heat island effect. Int J Remote Sens 18(2):287–304
Loew A, Stacke T, Dorigo W, Jeu RD, Hagemann S (2013) Potential and limitations of multidecadal satellite soil moisture observations for selected climate model evaluation studies. Hydrol Earth Syst Sci 17(9):3523–3542
Milewski A, Sultan M, Yan E, Becker R, Abdeldayem A, Soliman F, Gelil KA (2009) A remote sensing solution for estimating runoff and recharge in arid environments. J Hydrol 373(1–2):1–14. https://doi.org/10.1016/j.jhydrol.2009.04.002
Muller D, Munroe DK (2014) Current and future challenges in land-use science. J Land Use Sci 9(2):133–142. https://doi.org/10.1080/1747423X.2014.883731
Nagaveni C, Kumar KP, Ravibabu MV (2019) Evaluation of TanDEMx and SRTM DEM on watershed simulated runoff estimation. J Earth Syst Sci 128(1):1–11
Nair RS, Bharat DA, Nair MG (2013) Impact of climate change on water availability: case study of a small coastal town in India. J Water Clim Change 4(2):146–159
Pandey A, Palmate SS (2018) Assessments of spatial land cover dynamic hotspots employing MODIS time-series datasets in the Ken river basin of Central India. Arab J Geosci 11(17):1–8
Pandey A, Bishal KC, Kalura P, Chowdary VM, Jha CS, Cerdà A (2021a) A soil water assessment tool (SWAT) modeling approach to prioritize soil conservation management in river basin critical areas coupled with future climate scenario analysis. Air, Soil Water Res 14:11786221211021396
Pandey A, Chaube UC, Mishra SK, Kumar D (2016a) Assessment of reservoir sedimentation using remote sensing and recommendations for desilting Patratu reservoir, India. Hydrol Sci J 61(4):711–718
Pandey A, Chowdary VM, Mal BC (2004) Morphological analysis and watershed management using GIS. Hydrol J (India) 27(3–4):71–84
Pandey A, Chowdary VM, Mal BC (2007) Identification of critical erosion prone areas in the small agricultural watershed using USLE, GIS and remote sensing. Water Resour Manage 21(4):729–746
Pandey A, Chowdary VM, Mal BC (2009a) Sediment yield modelling of an agricultural watershed using MUSLE, remote sensing and GIS. Paddy Water Environ 7(2):105–113

Pandey A, Chowdary VM, Mal BC, Dabral PP (2011) Remote sensing and GIS for identification of suitable sites for soil and water conservation structures. Land Degrad Dev 22(3):359–372
Pandey A, Gautam AK, Chowdary VM, Jha CS, Cerdà A (2021b) Uncertainty assessment in soil erosion modeling using RUSLE, multisource and multiresolution DEMs. J Indian Soc Remote Sens 49(7):1689–1707
Pandey A, Himanshu SK, Mishra SK, Singh VP (2016b) Physically based soil erosion and sediment yield models revisited. CATENA 147:595–620
Pandey A, Mathur A, Mishra SK, Mal BC (2009b) Soil erosion modeling of a Himalayan watershed using RS and GIS. Environ Earth Sci 59(2):399–410
Pandey RP, Pandey A, Galkate RV, Byun HR, Mal BC (2010) Integrating hydro-meteorological and physiographic factors for assessment of vulnerability to drought. Water Resour Manage 24(15):4199–4217
Patel DP, Srivastava PK (2013) Flood hazards mitigation analysis using remote sensing and GIS: correspondence with town planning scheme. Water Resour Manage 27(7):2353–2368
Patro S, Chatterjee C, Mohanty S, Singh R, Raghuwanshi NS (2009) Flood inundation modeling using MIKE FLOOD and remote sensing data. J Indian Soc Remote Sens 37(1):107–118
Rao KD, Alladi S, Singh A (2019) An integrated approach in developing flood vulnerability index of India using spatial multi-criteria evaluation technique. Curr Sci 117(1):80
Robertson L, King DJ (2011) Comparison of pixel-and object-based classification in land cover change mapping. Int J Remote Sens 32(6):1505–1529
Sharma I, Mishra SK, Pandey A (2021) A simple procedure for design flood estimation incorporating duration and return period of design rainfall. Arab J Geosci 14(13):1–15
Singh G, Pandey A (2021) Mapping Punjab flood using multi-temporal open-access synthetic aperture radar data in Google earth engine. In: Hydrological extremes. Springer, Cham, pp 75–85
Singh G, Srivastava HS, Mesapam S, Patel P (2015) Passive microwave remote sensing of soil moisture: a step-by-step detailed methodology using AMSR-E data over Indian sub-continent. Int J Adv Remote Sens GIS 4(1):1045–1063
Singh G, Srivastava HS, Mesapam S, Patel P (2019) An attempt to investigate change in crop acreage with soil moisture variations derived from passive microwave data. World Environmental and Water Resources Congress 2019: watershed management, irrigation and drainage, and water resources planning and management. American Society of Civil Engineers, Reston, VA, pp 83–90
Sivapalan M (2003) Process complexity at hillslope scale, process simplicity at watershed scale: is there a connection? In: EGS-AGU-EUG joint assembly, p 7973
Srivastava HS, Patel P, Sharma Y, Navalgund RR (2009) Large-area soil moisture estimation using multi-incidence-angle RADARSAT-1 SAR data. IEEE Trans Geosci Remote Sens 47(8):2528–2535
Stisen S, Jensen KH, Sandholt I, Grimes DI (2008) A remote sensing driven distributed hydrological model of the Senegal river basin. J Hydrol 354(1–4):131–148
Swain S, Mishra SK, Pandey A (2021) A detailed assessment of meteorological drought characteristics using simplified rainfall index over Narmada river basin, India. Environ Earth Sci 80(6):1–15
Tarquis A, Gobin A, Semenov MA (2010) Preface. Clim Res 44:1–2. https://doi.org/10.3354//cr00942
Thakur PK, Garg V, Kalura P, Agrawal B, Sharma V, Mohapatra M, Kalia M, Aggarwal SP, Calmant S, Ghosh S, Dhote PR (2021) Water level status of Indian reservoirs: a synoptic view from altimeter observations. Adv Space Res 68(2):619–640
Ulaby FT (1977) Microwave remote sensing of hydrologic parameters
Velmurugan A, Carlos GG (2009) Soil resource assessment and mapping using remote sensing and GIS. J Indian Soc Remote Sens 37(3):511–525
Verbyla DL (1995) Satellite remote sensing of natural resources, vol 4. CRC Press
Wanders N, Bierkens MF, de Jong SM, de Roo A, Karssenberg D (2014) The benefits of using remotely sensed soil moisture in parameter identification of large-scale hydrological models. Water Resour Res 50(8):6874–6891

Wang D, Hubacek K, Shan Y, Gerbens-Leenes W, Liu J (2021) A review of water stress and water footprint accounting. Water 13(2):201

Wang D, Laffan SW, Liu Y, Wu L (2010) Morphometric characterisation of landform from DEMs. Int J Geogr Inf Sci 24(2):305–326

Wardlow BD, Egbert SL, Kastens JH (2007) Analysis of time-series MODIS 250 m vegetation index data for crop classification in the US central great plains. Remote Sens Environ 108(3):290–310

Yang J, Gong P, Fu R, Zhang M, Chen J, Liang S, Xu B, Shi J, Dickinson R (2013) The role of satellite remote sensing in climate change studies. Nat Clim Chang 3(10):875–883

Yang L, Meng X, Zhang X (2011) SRTM DEM and its application advances. Int J Remote Sens 32(14):3875–3896

Zhou H, Sun J, Turk G, Rehg JM (2007) Terrain synthesis from digital elevation models. IEEE Trans Visual Comput Graphics 13(4):834–848

Chapter 2
Hybrid Approach for Land Use and Forest Cover Classification in Sikkim Himalaya

Mukunda Dev Behera, Narpati Sharma, Neeti, V. M. Chowdhary, and D. G. Shrestha

Abstract Land use and forest cover (LUFC) classification from satellite data in mountainous terrain offers challenge due to varied topography and complexities owing to different illumination conditions. Digital classification following supervised and/or unsupervised techniques in combination with/without ancillary information often does not provide acceptable level of accuracy. This chapter formulates and applies a hybrid approach for LUFC classification using moderate resolution satellite data. Both 'elimination' and 'fishing' approaches were used to classify the state of Sikkim into seventeen categories. The classification accuracy was estimated at 94.87% at 1:50,000 scale, which is suitable for utilization in further studies such as surface hydrological and energy fluxes. Further, the digital elevation model was utilized to derive the topographic units at 1000 m elevation steps, slope and aspect and their distribution across the seventeen LUFC classes. The distribution of various LUFC classes across different elevation, slope and aspects offers useful information for ecosystem planning and management.

M. D. Behera (✉)
CORAL, Indian Institute of Technology Kharagpur, Kharagpur, West Bengal, India
e-mail: mdbehera@coral.iitkgp.ac.in

N. Sharma · D. G. Shrestha
Department of Science and Technology, Gangtok, Sikkim, India

Neeti
TERI School of Advance Studies, Delhi, India
e-mail: neeti@terisas.ac.in

V. M. Chowdhary
Regional Remote Sensing Centre-North, ISRO, New Delhi, India

A. Pandey et al. (eds.), *Geospatial Technologies for Land and Water Resources Management*, Water Science and Technology Library 103,
https://doi.org/10.1007/978-3-030-90479-1_2

2.1 Introduction

Land cover reflects the biotic and abiotic influence of an area and therefore is a prerequisite for any land use planning activities. The conventional ground-based approaches adopted for land use and land cover mapping and monitoring are often time- and labor-intensive and therefore is least preferred. In recent years, satellite remote sensing has proven its importance beyond doubt in generation of accurate land use and forest cover maps for studying change dynamics at various temporal scales (Rao et al. 1996). Remote sensing helps in deriving spatially continuous and persistent thematic maps, representing the Earth's surface (Banskota et al. 2014). In the state of Sikkim, where many places are inaccessible and most of the LUFC features are dynamic, this satellite-based digital classification technique perhaps offers the only method of thematic map preparation.

The state of Sikkim comprises of a mountainous terrain with varied topography. Satellite data interpretation for this area is very tough due to hill shadows and varied illumination conditions. A classified map with acceptable level of accuracy is really a challenging task. There exists no detailed study on LUFC mapping for Sikkim state except a few studies on forest cover and density mapping with limited classes. The Forest Survey of India (FSI) is involved in mapping forest cover of the state at various scales using different classification methods, wherein the classification accuracy is not appropriately taken care. In 1994, with the joint collaboration of forest department, state government of Sikkim and RRSSC-ISRO, forest cover of the Sikkim state was mapped using IRS 1A LISS-II satellite data following digital interpretation technique (Sharma et al. 2015). This was the first comprehensive study on forest type and density mapping in Sikkim state that reported 43.2% forest cover. Image classification in mountainous terrain for preparation of LUFC maps has provided better results with digital techniques. Behera et al. (2000a, b) reported difficulty in vegetation stratification and mapping due to the presence of shadow and varied illumination conditions in mountain state of Arunachal Pradesh. They also suggested that using ratio-based vegetation indices the effect of shadow can be minimized, while the illumination conditions can be addressed by incorporating the aspect map in to classification exercise (Behera et al. 2001). Using a hybrid technique, which is an integration of supervised and unsupervised classification and incorporation of human knowledge base; the LUFC classification can be achieved with relatively higher accuracy (Roy and Behera 2005).

The increased population pressure and anthropogenic activities have led to the degradation of forests and its environment in the state. The rise in human population has led to proportionate increase in demand for land for various activities such as construction of houses and roads, agricultural expansion, urbanization and other developmental activities such as industries and infrastructure. The fuel wood demand has gone up to cater to the growing rural population for cooking and other uses, putting additional pressure on the forest resources. Further, increase in livestock population has increased the burden on forests day by day. The forest area has been shrinking in order to meet the land-diversion requirements to various sectors. Thus, the forests are only restricted to the hilly and undulating terrain in the form of protected areas. The steady increase in vehicular traffic and emission thereof has led to rise in air pollution both in urban and in peri-urban areas. Thus, there is an emerging need to overcome the deteriorating situation and to contribute to the conservation and management of remaining forest resources for achieving sustainable development goals. The first step towards it is to take stock of land use and forest cover status. Land use planning and management are very important for optimal utilization of existing land resources (Armenteras et al. 2013), wherein satellite-based images provide information on the extent and quality of land resources.

2.2 Study Area

The state of Sikkim in the eastern Himalaya, India is located in the west of the Mount Kanchenjunga (8588 m height). The state is divided into four districts, of which North district is the largest one and is located on the upper part (Fig. 2.1). The West, South and East districts form the lower part of the state. The state has some good natural forests that are mostly inaccessible and unexploited, apart from alpine barren lands. The Himalayan state has unique collection of fauna, birds, mountain peaks and glaciers, lakes and rivers. The state accommodates about 4000 varieties of flowering plant species including the symbolic Rhododendrons and Orchids in the temperate/alpine and tropical forests, respectively.

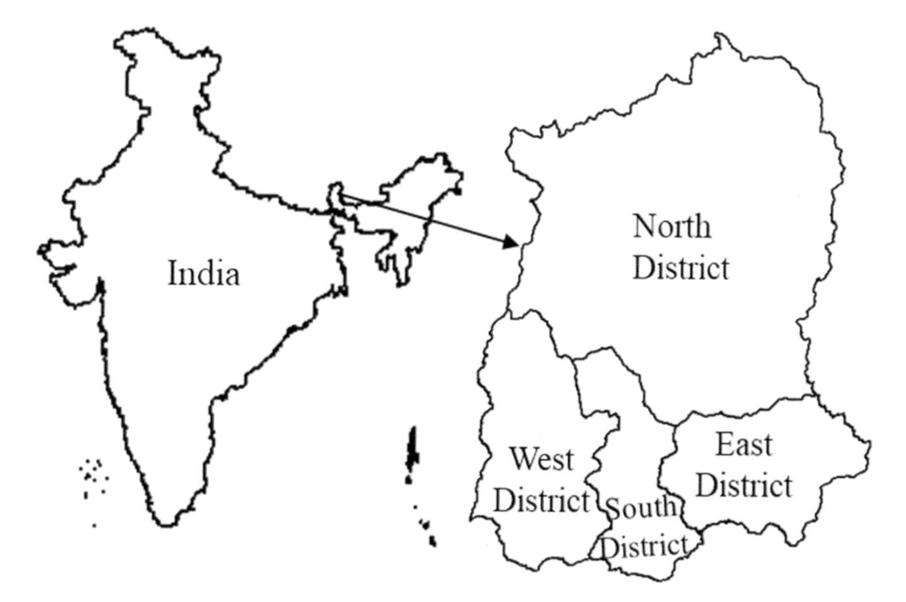

Fig. 2.1 Location Map of Sikkim state, comprises of four districts

It is a sparsely populated state with around 15% of the geographical area suitable for agriculture. The cultivation of large cardamom has grown into establishing a dominant agroforestry system in the state and is regarded as the most promising activity both ecologically and economically. The large cardamom-based agroforestry system contributes to moisture balance in microclimate by regulating the precipitation partitioning and adjustment of high moisture maintenance.

However, the land use and land cover of the state has come under severe pressure as >75% of the state's land area has been brought under either one or the other form of conservation and management measures by the state administration. Therefore, the remaining land is too stressed to accommodate the needs of the growing human and cattle population as well as various developmental activities. It is therefore the need of the hour to understand the LUFC distribution pattern over the state for sustainable development and planning.

2.3 Methodology

2.3.1 Data Used

The spring or early winter is the preferred season of a year for satellite data pick for the state, as there exists minimum cloud cover. A cloud-free IRS-1D LISS-III (path 107, row 51; 80% south shift) scene was acquired for the present study (Fig. 2.2a).

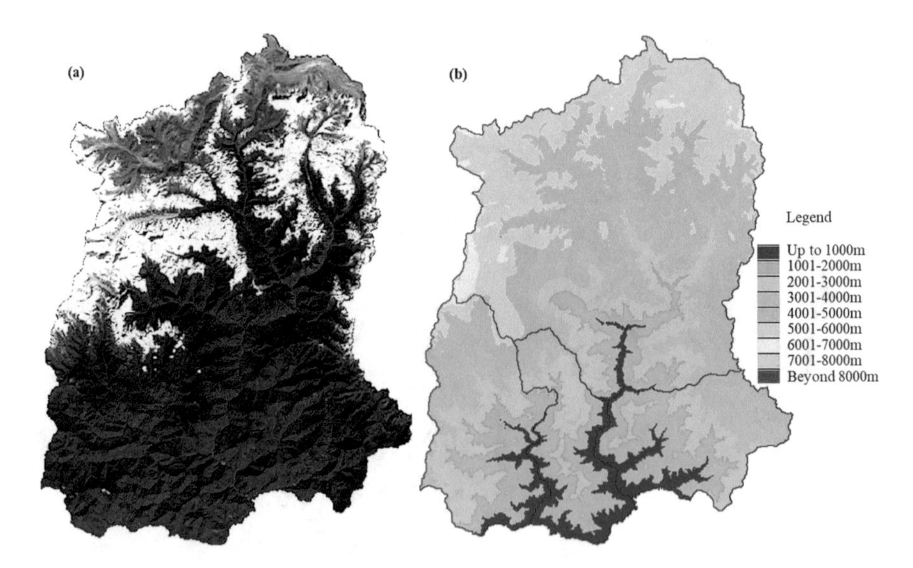

Fig. 2.2 **a** Standard False Color Composite Image (IRS 1D LISS-III) of Sikkim state; **b** elevation image of Sikkim state at 1000 m interval prepared using SRTM data

This data has spatial (23.5 m), spectral (operates in green, red, near infrared and short-wave infra-red portions of electro-magnetic spectrum) with radiometric (8-bits) and temporal (21-days) resolutions.

The major objective of this study was to map the status of LUFC of the state using hybrid digital classification approach that could yield optimum level of accuracy. Land availability across various topographical units such as elevation, slope and aspect were analyzed. SRTM elevation data was used to generate the topographical maps and for three-dimensional visualization of land surface (Fig. 2.2b). Satellite images integrated with the topographical data help in better classification of the land surface area.

2.3.2 *Field Visit and Reconnaissance*

An interpreter needs to have on-field *apriori* information for utilization in classification exercise using satellite data (Behera et al. 2001). This helps in correlating the field-based variations with the satellite-based signatures. During the reconnaissance survey, the major LUFC classes were checked across the elevation, slope and aspects gradients and along the roads, drainage and valleys. On-field observations on distribution of vegetation types and their floristic composition; various land use classes and their different neighbors; discrimination between agricultural and natural vegetation, etc., were noted. Information on the past/present status of forest cover, causes of land use changes, influence of any episodic or disturbance events such as landslide, forest fire, melting of glacier lakes, etc., were gathered from the locals and forest officials through personal interactions and informal interviews for utilization in the classification.

2.3.3 *Design of Classification Scheme*

An appropriate classification scheme was prepared that followed sequence of levels I, II and III categories (Table 2.1). The land use and forest cover map was prepared using a combination of unsupervised, supervised techniques and multi-criteria (Fig. 2.3). Available Survey of India (SOI) topographical sheets were used for geo-referencing of satellite data at 1:50,000 scale using the 'nearest neighbor resampling' algorithm that preserves the original pixel values of the image data (Fig. 2.3).

Further, the scenes were geometrically corrected using ground controlled points (GCPs) with second-order polynomial statistics and confirmed to a root-mean-square (RMS) error of <1.06 pixels. Due to extremely dissected terrain conditions, allocation of GCPs was a problem during geo-referencing (Behera et al. 2005). The state administrative boundary was used to extract the study area and the satellite image (Fig. 2.2a). Further, the images were re-projected to UTM projection with WGS 84

Table 2.1 Classification scheme for land use and forest cover mapping

1.0 Forest
1.1 Evergreen/semi-evergreen
1.1.1 Dense/closed
1.1.2 Open
1.1.3 Scrub
1.1.4 Forest blanks
1.2 Forest plantations
1.3 Crop land in forest
2.0 Grassland
3.0 Agriculture
3.1 Crop land
3.2 Fallow
3.3 Plantation
3.3.1 Tea
4.0 Built-up
1.1 Towns/cities (urban)
1.1.1. Residential
1.1.2. Mixed built-up land
5.0 Wastelands
5.1 Barren/rocky land
6.0 Water bodies
6.1 River
6.2 Lakes
7.0 Snow covered
7.1 Perennial
7.2 Glacial area

datum (north zone 45). Reconnaissance survey offers opportunity to obtain and visualize the information pertaining to the existing field conditions and assessment, the accessibility of the area, the pattern and distribution of various land cover elements (Behera et al. 2014). Major land use and forest cover classes were identified on FCC for preparation of interpretation keys that were further used for classification (Fig. 2.3).

2.3.4 Normalized Difference Vegetation Index

Various indices are available with respect to the reflectance of different features on Earth surface. Level–I separation between vegetation and non-vegetation classes

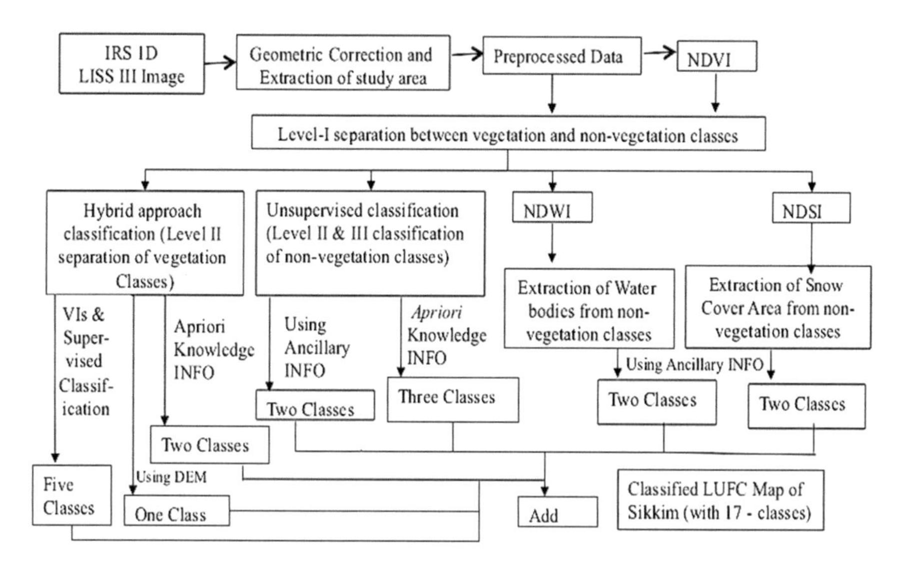

Fig. 2.3 Methodology flow chart for land use and forest cover classification for Sikkim state

was done using *Normalized Difference Vegetation Index* (NDVI) as it is a measure of the vegetation density (Behera et al. 2001). Due to high reflectance of vegetation in near infrared (NIR) region compared to red, the positive value represents vegetation, whereas zero and negative values represent other classes (Boschetti et al. 2010). Further, the images pertaining to vegetation and non-vegetation classes were classified separately by alternately zeroing of the masked-in and masked-out pixels (Fig. 2.3).

2.3.5 *Normalized Difference Water Index and Normalized Difference Snow Index*

The water reflectance at green and absorption in NIR wavelength regions and high reflectance of other features in NIR wavelengths compared to green were exploited to create a *Normalized Difference Water Index* [NDWI = (Green − NIR)/(Green + NIR)]. NDWI is an index similar to NDVI and was attempted to delineate water bodies from non-vegetation classes. The water features get positive value in NDWI, and other terrestrial features have zero and negative values (Fig. 2.3). Further, the cluster of pixels pertaining to water body was subtracted from the parent image and binary FCC was generated for utilization in further classification. Ancillary information was used for the Level-II classification of water bodies. Drainage of fifth and sixth order was digitized from Survey of India (SOI) topographical maps. A buffer of 250 m was generated on either sides of drainage channels and was used to extract river falling inside the buffer on the basis of spectral reflectance value.

The remaining water body clusters were classified as lakes. Similarly, *Normalized Difference Snow Index,* NDSI was used to depict snow cover area in the image. The positive values showed snow cover; those were further extracted using NDSI. The glacial map was overlaid over the lake pixels to pick the moraine lakes (Fig. 2.3). Thus, glacial and perennial water body areas were separated from each other.

2.3.6 Image Clustering and Classification

The spectral clusters from the satellite image were identified using ISODATA (Iterative Self-Organizing Data Analysis) algorithm (Vanderee and Ehrlich 1995). We found 60 as the optimum number after empirical trials with 20, 40, 60 and 80, etc., wherein the dimension of spectral variability of LISS III (3 bands) was reduced by 88.27%. Further, the convergence value was fixed at 0.99 so that the data processing would stop when >99% pixels belong to a cluster within a given number of iterations is achieved. Here, the maximum number of iterations was fixed at 21 to allow the utility to stop on reaching the pre-defined convergence threshold. This clustering technique was utilized to derive all the land use and forest cover classes (Fig. 2.3).

The residential area was separated from mixed built up land and wasteland based on field observation and *apriori* knowledge (Behera et al. 2001). ISOData unsupervised classification was carried out to separate between wasteland and mixed built up area. Field observations were used to demarcate between wasteland and built up class. Built-up clusters were masked out from image, and the binary FCC containing impure grassland was further classified using ancillary information. Logical rules were defined to do further level of classification like proximity to agriculture land was considered as fallow (Sarkar et al. 2014). The forest mask provided by the State Forest Department was used to dissociate the forest blank class from wasteland class.

The tea plantation was separated from the image containing vegetation classes by *apriori* knowledge of field observation. DEM was used for differentiating scrub from other vegetation (Pasha et al. 2020). It is known that state is dominated by scrub above approximately 3000 m altitudes and is known as *Tree line.* The scrub thus obtained was impure due to mixing of open forest at the altitudinal boundary, and there is no sharp separation between the two. Principal component analysis was done to separate pure scrub and open forest from each other. The open forest thus separated was combined with the other vegetation classes. The image was further subjected to hybrid classification where NDVI and VI were used as bands for supervised classification (maximum likelihood) to separate dense forest, open forest, grassland and agriculture (Behera et al. 2001). Sikkim state has 88% notified forest area (FSI 2003). The forest mask over agricultural land provided the ‘crop land in forest’ class. All the derived classes were further added together to get final LUFC classified image (Li and Feng 2016).

2.3.7 Accuracy Assessment

Accuracy assessment is the comparison of true data on geographic locations to evaluate the classification accuracy of various land use and land cover classes (Roy et al. 2015). It is not feasible to test every pixel of a classified image. Therefore, a set of reference pixels is generally used. Reference pixels are usually the well-known geographic points on a classified image. Here, stratified random sampling technique was used to select 195 points to avoid any bias. The overall accuracy and kappa coefficients were calculated. Producer's and user's accuracy were also generated for each individual class. The producer's accuracy indicates the probability of a reference pixel being correctly classified and is related to omission error, while the user's accuracy indicates the probability of a pixel classified on the image actually represents that category on the ground and measures the commission error (Behera et al. 2000a, b). Kappa coefficient expresses the proportionate reduction in error generated by a classification process compared with the error of completely random classification.

2.4 Results

2.4.1 Land Use and Forest Cover Classification

The state of Sikkim was classified into seventeen categories using digital classification technique (Fig. 2.4), with total land area of 7096.22 km^2. The classification accuracy was estimated at 94.87%, which is acceptable and suitable for many applications including integration with other thematic maps in GIS domain (Behera et al. 2018a). The area estimates were made for each land cover category (Table 2.2). All the vegetation classes were described as per their occurrence, spectral response on satellite data and distribution owing to elevation, slope and aspects. It may be observed that the vegetation distribution is restricted to the southern part of the state owing to elevation. The hybrid approach following multi-criteria was used for the classification, which is an amalgamation of unsupervised clustering, cluster labeling following multi-criteria (Fig. 2.3).

The tea plantation was delineated on ground and later on projected onto the classified map. The boundary of tea plantation was delineated on ground and projected onto the classified map later on. Only 4 towns could be picked, and few mixed built-up areas were delineated owing to minimum mappable area criteria. All the four district headquarters were delineated using *apriori* knowledge in combination with spectral signatures. Due to extremely dissected terrain conditions, allocation of GCPs was a problem during geo-referencing. The classification exercise took a longer time since it involved many steps. All the LUFC classes were described with respect to their occurrence, spectral response on satellite image, classification accuracy, occurrence zone with respect to altitude, slope and aspect.

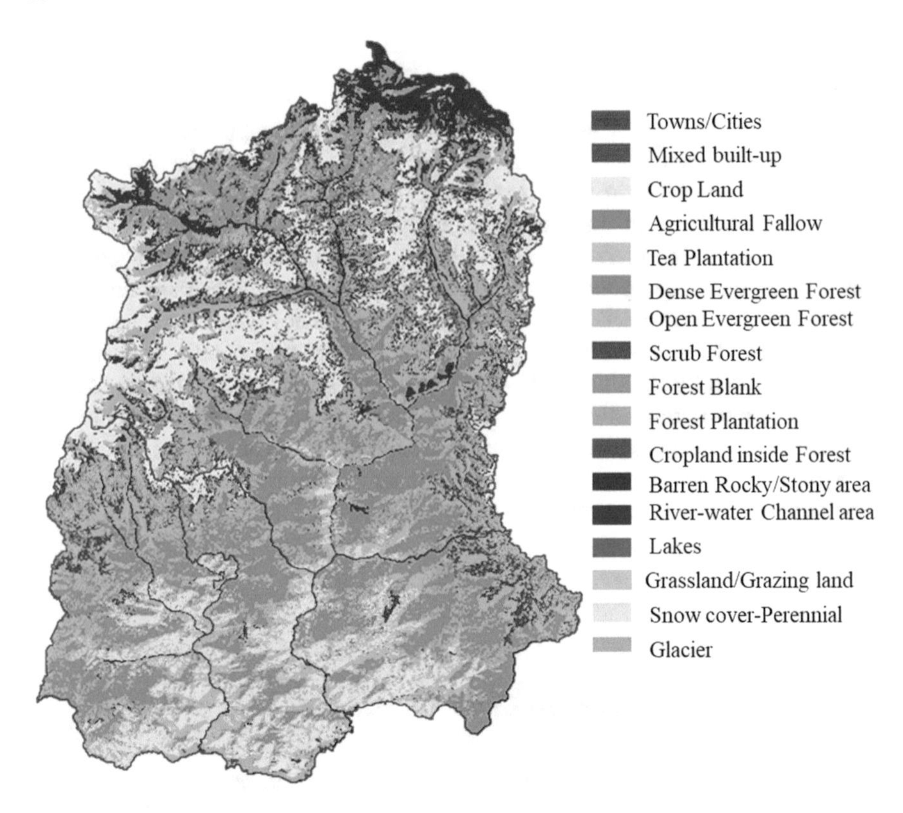

Fig. 2.4 Classified land use and forest cover map of Sikkim state

Forest/Vegetation: Dense forests occupy 18,101 km^2 (25.5%) of the state's area, and majority of them were found distributed between 1000 and 3000 m altitude (Tables 2.2 and 2.4). This category could be identified with good accuracy (κ = 0.95). The dense forests imparted a dull red tone with smooth texture on the satellite image and were found distributed in all aspects throughout, but restricted between 10° and 50° slope (Table 2.6). Open forests are found distributed over 16.3% (1158 km^2) of the state's geographical area and could be mapped with good kappa accuracy (κ = 0.95) (Table 2.3, 2.4 and 2.5).

In many places at lower elevations, these open forests were found forming a buffer layer around dense forests, while in higher elevations, the open forests are sandwiched between the dense forests and scrublands (Fig. 2.4). These forests demonstrated pink to bright red color on satellite FCC, but deviations were also found. Open forests showed universal distribution in all slopes and aspects but were restricted up to 5000 m elevations. The scrubs occupied only 382.2 km^2 (5.67%) land area of the state and could be classified with good accuracy (κ = 0.93). *Rhododendron* species and some seasonal scrubs form this category and were distributed throughout the upper breadth of the state. They reflect a bright to brick red color on standard FCC.

Table 2.2 Area statistics for LUFC classes as derived using hybrid classification technique along with accuracy level

Forest/land cover class	Area		Classification Accuracy (Overall—94.87%)		
	In km^2	In % age	Producers	Users	Kappa
Dense forest	1810.90	25.52	92.31%	96.00%	0.9538
Open forest	1158.27	16.32	95.24%	95.24%	0.9466
Scrub	382.22	5.39	93.33%	93.33%	0.9278
Forest blank	901.34	12.70	93.33%	73.68%	0.7149
Forest plantation	28.93	0.41	100.00%	100.00%	1
Crop land in forest	12.38	0.17	100.00%	50.00%	0.4974
Grassland	187.52	2.64	92.86%	100.00%	1
Crop land	524.30	7.39	100.00%	100.00%	1
Agricultural-fallow	86.42	1.22	100.00%	100.00%	1
Tea plantation	2.17	0.03	**	**	**
Residential (Town)	6.19	0.09	**	**	**
Mixed built-up	1.20	0.02	**	**	**
Barren/rocky area	649.49	9.15	85.00%	100.00%	1
River	65.60	0.92	100.00%	100.00%	1
Lake	19.76	0.28	88.89%	100.00%	1
Snow cover	1062.82	14.98	92.86%	100.00%	1
Glacial area	196.72	2.77	100.00%	100.00%	1
Total	7096.22	100.00	Overall kappa statistics—0.944		

** Not estimated

They were found distributed in all aspects but were restricted up to 5000 m elevations and a maximum of 60° slope (Fig. 2.5a; Tables 2.5 and 2.6).

The non-forest lands present inside the notified forest boundary were classified as 'forest blank.' This category was found in all aspects but restricted between 3000 and 5000 m elevation and found distributed mainly up to 50° slope. Forest blanks share 901.3 km^2 (12.70%) land area of the state and could be classified with good accuracy ($\kappa = 0.72$). Forest plantations occupy 28.9 km^2 (0.4%) of the state and were found distributed in three districts of Sikkim except North district. They were restricted up to 1000 m elevations and were classified with 100% accuracy. Most of the forest plantations were found in the eastern and southern aspects, preferably up to 40° slope. The croplands inside notified forestland were mapped to judge the amount of intrusion. Only 12.4 km^2 (0.17%) of state's area could be classified under this category and were primarily distributed inside open forests, thereby leaving a fragmented appearance. The croplands inside the notified forestland were scattered up to 50° slopes and primarily found in the southern and eastern aspects (Table 2.7).

Grasslands, which are found distributed mainly in higher elevations beyond 3000 m, are seasonal in nature due to excess cold weather conditions and are

Table 2.3 Error matrix for land use/land cover classification

	Crop land	Agricultural fallow	Dense forest	Open forest	Scrub	Forest blank	Forest plantation	Crop land in forest	Barren/Rocky area	River	Lake	Grass land	Snow cover	Glacial area	Row total
Crop land	16														16
Agricultural fallow		10													10
Dense forest			24	1											25
Open forest			1	20											21
Scrub					14	1									15
Forest blank					1	14			3			1			19
Forest plantation							4								4
Crop land in forest			1					1							2
Barren/rocky area									17						17
River										12					12
Lake											8				8
Grassland												13			13
Snow cover													20		20
Glacial area											1			12	13
Column total	16	10	26	21	15	15	4	1	20	12	9	14	20	12	195

Table 2.4 Physiographic distribution of Sikkim state at 1000 m elevation steps

1000 m elevation step			Slope-wise			Aspect		
	Area in in km^2	Area in %	(in °C)	Area in km^2	in %	Type wise	Area in km^2	Area in %
Up to 1000 m	437.11	6.16	1–5	315.17	4.44	N	701.10	9.88
1001–2000 m	1309.65	18.46	6–10	568.71	8.01	NE	877.82	12.37
2001–3000 m	1136.01	16.01	11–20	1716.37	24.19	E	1013.85	14.29
3001–4000 m	1046.75	14.75	21–30	2245.92	31.65	SE	1062.89	14.98
4001–5000 m	1619.06	22.82	31–40	1651.37	23.27	S	988.86	13.94
5001–6000 m	1415.81	19.95	41–50	531.96	7.50	SW	925.19	13.04
6001–7000 m	127.55	1.80	51–60	64.35	0.91	W	789.64	11.13
7001–8588 m	4.19	0.06	61–68	2.38	0.03	NW	736.87	10.38

composed mostly of seasonal herbs and grasses. This vegetation was continuous in nature but have a shorter snow-free period. They are distributed over 187.5 km^2 area (2.64%) and could be mapped with 100% accuracy. They occur between the scrubland and snow-covered areas and imparted a typical pink to light red tinge on FCC. Grasslands demonstrated a global distribution in all slopes and aspects of the state (Fig. 2.5a; Tables 2.6 and 2.7).

Agricultural land includes the present cropland and agricultural fallow land. Permanent agriculture is practiced across the river channels. The cropland and fallow land shared 524.3 km^2 (7.4%) and 86.42 (1.22%), respectively, and could be classified with 100% accuracy.

They were found distributed in all aspects but restricted up to 3000 m elevations and 50° slopes (Fig. 2.5b; Table 2.7). Tea plantation was found at 'Temi' in the South district between 1000 and 2000 m elevations. This class was present in the north-eastern aspect and restricted between 10° and 40° slopes. The area of tea plantation accounts to 2.17 km^2 (0.03%).

Topological distributions: The arrangement of the state's land area according to 1000 m elevation steps demonstrated maximum land area of 1619 km^2 (22.8%) between 4001 and 5000 m, followed by 1415.8 km^2 (20%) and 1309.65 km^2 (18.5%) for 5001–6000 m and 1001–2000 m elevation step, respectively (Table 2.4). Only 4.2 km^2 land area is present above 7000 m altitude. The slope of the state has been divided into eight categories that varied between 0° and 68°. Maximum land area of 2245.9 km^2 (31.7%) falls in 21°–30° slope category of the state followed by 1716.4 km^2 (24.2%) and 1651.4 km^2 (23.3%) for 11–20° and 31–40° slope categories, respectively (Table 2.4). Only 2.38 km^2 (0.03%) land area of the state falls under 61°–68° slope category. The land cover distribution of the state was divided into eight aspects of which the southeastern aspect occupied maximum land area of 1062.9 km^2 (15%) of the state followed by 1013.9 km^2 (14.3%) and 988.9 km^2 (13.9%) for the eastern and southern aspects, respectively (Table 2.4). Minimum area of 701.1 km^2

Table 2.5 Distribution of LUFC classes in nine-elevation steps of 1000 m

Vegetation class	Up to 1000	1001–2000	2001–3000	3001–4000	4001–5000	5001–6000	6001–7000	7001–8000	8001–8588
Dense forest	26.69	560.8	878.2	354	–	–	–	–	–
Open forest	132.6	355.3	216.3	355	94.82	3.94			
Scrub	0.38	0.73	6.82	221	148.7	4.20			
Grassland				14.3	148.0	24.83	0.39		
Forest plantation	28.54	0.39							
Tea plantation		2.17							
Cropland in forest	3.96	2.69	5.68	0.04	0.01				
Forest blank				48.1	531	314	8.52	0.02	
Town/cities	0.9756	4.84	0.3701						
Barren/rocky area	1.38	3.17	8.39	15.8	223	384	14.2	0.28	
River	22.35	10.0	7.07	12.8	12.9	0.56			
Lakes		0.12		0.69	7.07	11.9			
Snow cover			0.976	27.2	403	539	88.4	3.38	0.03
Glacial area				0.01	49.6	134	13.9	0.39	0.05
Mixed built-up land	0.72	0.484	0.003						
Crop land	184.6	323.91	15.82						
Agricultural fallow	33.72	51.10	1.60						

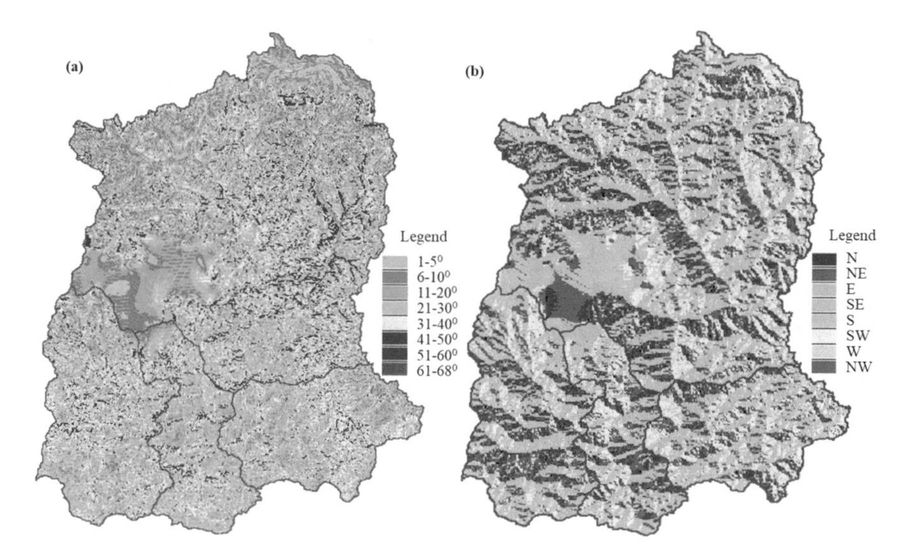

Fig. 2.5 **a** Slope and **b** aspect images of Sikkim state

Table 2.6 Distribution of land use/land cover classes in eight slope classes

Vegetation class	0–5°	5–10°	10–20°	20–30°	30–40°	40–50°	50–60°	60–68°
Dense forest	14.27	55.51	368.68	679.00	532.01	146.48	14.66	0.28
Open forest	15.32	46.27	237.43	404.69	333.05	110.25	10.94	0.32
Scrub	11.35	28.02	94.54	116.31	94.36	32.45	4.80	0.37
Grassland	7.90	18.71	49.59	57.08	40.00	12.69	1.47	0.07
Forest plantation	0.64	1.28	6.14	13.84	6.08	0.92	0.02	0.00
Tea plantation	0.00	0.00	1.16	0.95	0.06	0.00	0.00	0.00
Cropland in forest	0.35	0.56	2.44	4.30	3.41	1.23	0.09	0.00
Forest blank	71.58	113.84	243.15	236.52	164.17	61.85	9.83	0.40
Town/cities	0.41	0.39	3.02	1.86	0.47	0.05	0.00	0.00
Barren/rocky area	80.04	97.63	176.63	158.80	102.88	36.28	5.04	0.19
River	5.85	6.47	14.88	18.29	14.34	4.81	0.66	0.00
Lakes	4.18	3.01	5.46	4.29	2.10	0.70	0.02	0.00
Snow cover	57.75	137.50	309.82	276.12	192.40	76.78	11.87	0.60
Glacial area	37.66	39.12	58.63	32.93	20.01	6.83	1.00	0.02
Mixed built-up land	0.06	0.18	0.38	0.34	0.22	0.02	0.00	0.00
Crop land	6.34	16.82	120.53	213.53	128.95	36.86	3.61	0.11
Agricultural fallow	1.73	4.46	25.04	35.20	16.13	3.89	0.33	0.02

Table 2.7 Distribution of LUFC classes across eight aspect classes

Vegetation class	N	NE	E	SE	S	SW	W	NW
Dense forest	254.9	226.7	185.5	164.7	173.7	219.2	292	293.7
Open forest	85.82	133.9	178.8	201.4	195.8	168.9	108	85.81
Scrub	29.27	50.05	48.69	59.72	58.33	59.84	41.2	35.16
Grassland	17.25	21.00	18.62	19.18	27.80	39.19	26.4	18.11
Forest plantation	0.25	1.44	4.39	8.50	8.26	4.04	1.60	0.44
Tea plantation	0.16	1.66	0.30	0.03	0.00	0.00	0.00	0.01
Cropland in forest	0.36	1.26	2.58	3.51	3.04	0.89	0.56	0.19
Forest blank	92.63	113.0	120.6	127.1	127.0	123.3	99.4	98.36
Town/cities	0.117	0.542	0.616	0.832	0.683	1.283	1.85	0.308
Barren/rocky area	60.92	77.42	99.48	111.4	93.22	81.76	60.1	65.12
River	2.926	1.65	2.98	5.917	13.18	18.78	13.3	7.05
Lakes	1.505	1.033	1.104	1.081	2.715	6.063	4.19	2.056
Snow cover	101.5	158.0	197.1	182.0	142.6	106.3	88.7	86.46
Glacial area	23.71	26.9	34.20	32.86	23.02	20.07	16.2	19.93
Mixed built-up land	0.043	0.009	0.059	0.343	0.246	0.205	0.17	0.126
Crop land	15.9	51.469	112	136.56	109.11	62.629	23.62	13.003
Agricultural fallow	13.62	10.969	5.8917	7.3985	9.8281	13.054	13.93	11.729

(9.9%) was found in the northern aspect. The area estimates of different LUFC classes as per elevation, slope and aspect are provided in Tables 2.5, 2.6 and 2.7.

2.5 Discussion and Conclusions

As per the classification, the residential township of the four district headquarters occupied only 6.2 km^2 followed by the mixed built-up lands of 1.2 km^2 area. The villages are sparsely distributed and thus could not be delineated owing to minimum mappable area (MMA) criteria. However, some settlements could be delineated as mixed built-up area (1.2 km^2) using *apriori* knowledge. Barren and rocky areas occupied 650 km^2 (9.15%) of the state's land area and could be classified with optimum accuracy. They are found distributed up to 6000 m altitudes across all slope and aspect categories (Fig. 2.5; Table 2.5 and 2.6). River water channels and lakes were classified with good accuracy and occupied 65.6 km^2 (0.92%) and 19.76 km^2 (0.28%) land area of the state. Majority of the river water channel flowed between 10° and 40° slope as per the classified LUFC map. Interestingly, the lakes are found between 3000 and 6000 m elevation and distributed across all slope and aspect categories (Tables 2.6 and 2.7). Upper reaches of the state remained under permanent/frozen snow cover. Snow cover and glacial area contributed to 1062.82 km^2 (14.98%) and 196.72 km^2 (2.77%) of the state's land cover, respectively. They could be classified with absolute accuracy. The lower distribution limit for snow cover and glacial area was 300 m and 4000 m, respectively (Fig. 2.5; Table 2.7). The topographic complexity can be judged with the distribution of LUFC areas across different elevations, slope and aspect categories in an area of 7096.22 km^2. The results demonstrated that high level of classification accuracy for all LUFC classes could be achieved using multi-criteria (Behera et al. 2018a). Any classified map with such level of accuracy is useful for various applications such as surface hydrological and energy flus-based studies (Behera et al. 2018b).

Conservation Plans: Existing forests and forestlands can be protected and efforts should be made to improve their net ecosystem productivity. To minimize soil erosion and check siltation, it is prescribed that efforts should be put to increase the natural vegetation cover along high slope and adjoining the water bodies (Lele et al. 2008). There should be strict enforcement against the diversion of productive forest and agricultural lands. The conversion of forestlands for non-forest purposes should be strictly restricted. Appropriate technological interventions and management practices be implemented to increase land productivity to cater the population, thereby reducing any extra burden on the shrinking forest resources. Various plantation and afforestation schemes should be in place to cater to the fuel wood-based energy requirements of the rural poor and to meet the fodder requirement of the cattle. This may as well be combined with the agroforestry and other horticultural or economic plantations activities. Further, for afforestation schemes, the possibility of execution of carbon credit under REDD$^+$ program may be considered.

Non-timber forest produce is another means of revenue generation without harming the forest resources that may be fully employed for generating sustenance for the villagers located adjoining the forests. In terms of planning horticultural crop plantation, the high elevation and undulating areas may be preferred over relatively less slope and flat lands that are otherwise more suitable for agricultural crops. Crop diversification and multi-cropping may be targeted in very suitable agricultural lands adjoining to the valley areas. Further, diversification of production and mixed farming should be encouraged along with other occupations such as livestock and poultry, fishery, sericulture and apiculture that could cater to round the year employment. Such planning of land use and forest cover resources adds to better land use management practices. This chapter very well exemplifies the utility of satellite data for accurate mapping of land use and forest cover resources of the state and informs their distribution across different physiographic units.

References

Armenteras D, Rodriguez N, Retana J (2013) Landscape dynamics in northwestern Amazonia: an assessment of pastures, fire and illicit crops as drivers of tropical deforestation. PLoS ONE 8:e54310. https://doi.org/10.1371/journal.pone.0054310

Banskota A, Kayastha N, Falkowski MJ, Wulder MA, Froese RE, White, JC (2014) Forest monitoring using Landsat time series data: a review. Can J Remote Sens 40:362–384. https://doi.org/10.1080/07038992.2014.987376

Behera MD, Jeganathan C, Srivastava S, Kushwaha SPS, Roy PS (2000) Utility of GPS in classification accuracy assessment. Curr Sci 79(12):1996–1700

Behera MD, Kushwaha SPS, Roy PS (2001) Forest vegetation characterization & mapping using IRS-1C satellite images in eastern Himalayan region. Geocarto Int 16(3):53–62

Behera MD, Neeti JA, Jayaraman V (2005) Analysis of land cover change and misregistration error in North district of Sikkim, India. Int J Geoinform 1(4):43–58

Behera MD, Patidar N, Chitale VS, Behera N, Gupta D, Matin S, Tare V, Panda SN, Sen DJ (2014) Increase of agricultural patch contiguity over past three decades in Ganga River Basin, India. Curr Sci 107(3):502–515

Behera MD, Gupta AK, Barik SK, Das P, Panda RM (2018a) Use of satellite remote sensing as a monitoring tool for land and water resources development activities in an Indian tropical site. Environ Monit Assess 190:401. https://doi.org/10.1007/s10661-018-6770-8

Behera MD, Tripathi P, Das P, Srivastava SK, Roy PS, Joshi C, Behera PR, Deka J, Kumar P, Khan ML (2018b) Remote sensing based deforestation analysis in Mahanadi and Brahmaputra river basin in India since 1985. J Environ Manag 206:1192–1203

Behera MD, Srivastava S, Kushwaha SPS, Roy PS (2000) Stratification and mapping of *Taxus baccata* L. bearing forests in Tale valley using remote sensing and GIS. Curr Sci 78(8):1008–1013

Boschetti M, Stroppiana D, Brivio PA (2010) Mapping burned areas in a Mediterranean environment using soft integration of spectral indices from high-resolution satellite images. Earth Interact 14:1–20

Das P, Mudi S, Behera MD, Barik SK, Mishra DR, Roy PS (2021) Automated mapping for long-term analysis of shifting cultivation in Northeast India. Remote Sens 13(6):1066

FSI (2003) India State of Forest Report, 2003. Forest Survey of India, Dehradun

Lawrence D, Radel C, Tully K, Schmook B, Schneider L (2010) Untangling a decline in tropical forest resilience: constraints on the sustainability of shifting cultivation across the globe. Biotropica 42:21–30

Lele N, Joshi PK, Agrawal SP (2008) Assessing forest fragmentation in northeastern region (NER) of India using landscape matrices. Ecol Ind 8:657–663. https://doi.org/10.1016/j.ecolind.2007.10.002

Li P, Feng Z (2016) Extent and area of Swidden in Montane Mainland Southeast Asia: estimation by multi-step thresholds with Landsat-8 OLI data. Remote Sens 8:44. https://doi.org/10.3390/rs8010044

Pasha SV, Behera MD, Mahawar SK, Barik SK, Joshi SR (2020) Assessment of shifting cultivation fallows in Northeastern India using Landsat imageries. Trop Ecol 1:1–11. https://doi.org/10.1007/s42965-020-00062-0

Rao DP, Gautam NC, Nagaraja R, Ram MP (1996) IRS-1C applications in land use planning. Curr Sci 70(7):575–581

Roy PS, Behera MD (2005) Rapid assessment of biological richness in a part of Eastern Himalaya: an integrated three-tier approach. Curr Sci 88(2):250–257

Roy PS, Behera MD, Murthy MSR, Roy A, Singh S, Kushwaha SPS, Jha CS, Sudhakar S, Joshi PK, Reddy CS (2015) New vegetation type map of India prepared using satellite remote sensing: comparison with global vegetation maps and utilities. Int J Appl Earth Obs Geoinf 39:142–59. https://doi.org/10.1016/j.jag.2015.03.003

Sarkar A, Ghosh A, Banik P (2014) Multi-criteria land evaluation for suitability analysis of wheat: a case study of a watershed in eastern plateau region, India. Geo-Spatial Inform Sci 17:119–128. https://doi.org/10.1080/10095020.2013.774106
Sharma N, Das AP, Shrestha DG (2015) Land use and Land cover mapping of East District of Sikkim using IRS P6 satellite imagery. Pleione 9(1):193–200. ISSN: 0973-9467

Chapter 3
Appraisal of Land Use/Land Cover Change Over Tehri Catchment Using Remote Sensing and GIS

Sabyasachi Swain, Surendra Kumar Mishra, and Ashish Pandey

Abstract The Himalayan reservoirs have immense significance from the point of view of water resources planning and management. However, natural and anthropogenic changes and their effects upon these reservoirs are often not explored, mainly due to limitations of data availability. This chapter presents an appraisal of land use/land cover (LULC) changes over the Tehri catchment located at the lower Himalayan region, using remote sensing and geographic information system (GIS). The imageries are collected for different years, i.e., 2008, 2014, and 2020 from the Landsat 5, Landsat 8, and Sentinel 2 satellites, respectively, with the objective of deriving information on different LULC classes. Following a supervised classification, the catchment area is divided into eight classes, viz. open forest, dense forest, water bodies, shrubland, agricultural land, settlements, barren land, and snow covers. The accuracy of classification is assessed with respect to the Google Earth images and ground truth verification. A comparison between the areal coverage of the LULC classes was analyzed for temporal LULC change detection over the catchment. Comparing 2008 and 2020, it is clear that the dense forests and barren land have decreased. On the other hand, an increase in the open forests, water bodies, shrubland, snow, and settlement is observed. The accuracy assessment results confirm that the LULC changes reported in this study are justifiably accurate and utilizable for further applications. The results reported in this study may be helpful to frame solutions to hydrological problems of the Tehri catchment. Moreover, this study highlights the usefulness of remote sensing and GIS in hydrological applications, even in mountainous catchments.

S. Swain (✉) · S. K. Mishra · A. Pandey
Department of Water Resources Development and Management, Indian Institute of Technology Roorkee, Roorkee, Uttarakhand 247 667, India
e-mail: sswain@wr.iitr.ac.in

S. K. Mishra
e-mail: s.mishra@wr.iitr.ac.in

A. Pandey
e-mail: ashish.pandey@wr.iitr.ac.in

A. Pandey et al. (eds.), *Geospatial Technologies for Land and Water Resources Management*, Water Science and Technology Library 103,
https://doi.org/10.1007/978-3-030-90479-1_3

3.1 Introduction

In India, the continuous demand for economic growth coupled with the population explosion has resulted in substantial land use/land cover (LULC) changes in the past century (Tian et al. 2014). The ramifications of LULC changes include influencing ecosystem services, altering hydrological components, triggering and intensifying the natural hazards, complicating the hydroclimatic predictions, affecting quantity and quality of available water resources, etc. (Aadhar et al. 2019; Astuti et al. 2019; Bahita et al. 2021; Chen et al. 2020; Hengade and Eldho 2016; Saputra and Lee 2019; Sharma et al. 2020; Singh et al. 2020; Swain et al. 2018, 2019a, b; Talukdar et al. 2020; Tripathi et al. 2020). The impacts of droughts are more severe, where the LULC is mostly dedicated to agriculture (Swain et al. 2017, 2020a, 2021a, b, b). An improved understanding of the LULC and climatic changes of a particular area can be pivotal for effective policy framing, specifically in water resources planning and management (Anand et al. 2018; Himanshu et al. 2018, 2019; Dayal et al. 2019, 2021; Guptha et al. 2021; Kalura et al. 2021; Sahoo et al. 2021). Due to all these reasons, a detailed assessment of LULC change patterns has become necessary, which is typically carried out by analyzing historical LULC changes through multi-temporal remotely sensed images. Several research works have been carried out in the last few years to investigate the LULC changes, their future predictions, and consequential effects (Dutta et al. 2019; Liping et al. 2018; Palmate et al. 2017a, b; Pandey and Khare 2017; Pandey and Palmate 2018; Rimal et al. 2017; Rwanga and Ndambuki 2017; Singh et al. 2018; Tran et al. 2017).

The Himalayan catchments have immense significance from the point of view of water resources planning and management (Singh and Pandey 2021; Swain et al. 2021c). However, the natural and anthropogenic changes and their effects upon these reservoirs are often not explored, mainly due to limitations of data availability. With the advancement of remote sensing and geospatial technologies, the LULC information of these catchments have become easily accessible. Recently, Mishra et al. (2020) used Landsat 5 and Sentinel 2A for supervised LULC classification over the Rani Khola watershed located in the Sikkim Himalaya, India. They reported a series of complicated changes in LULC over the watershed during 1988–2017. Therefore, this study aims to carry out a detailed assessment of LULC changes over a Himalayan catchment considering multi-temporal satellite images and supervised classification. Further, it is wise to cross-check the LULC classification of the recent period by ground truth verification. In this regard, the details of the study area, methodology, results and discussion, ground-truthing information, and the conclusions derived from this study are presented in the subsequent sections.

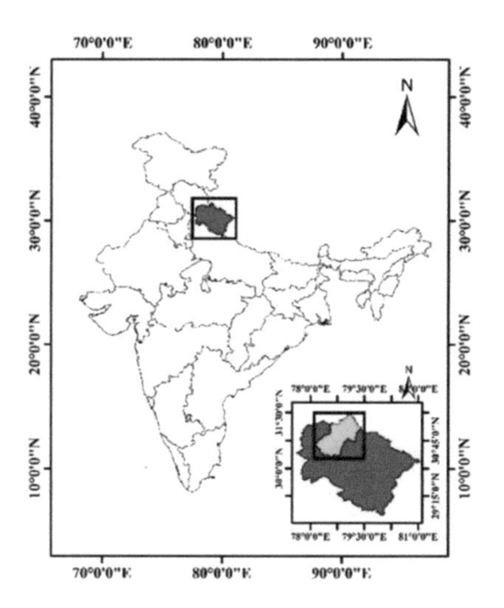

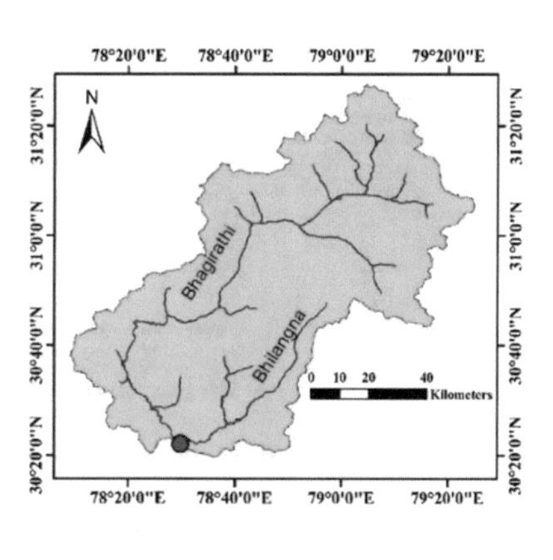

Fig. 3.1 Location map of the study area

3.2 Study Area and Data

The Tehri catchment located in the state of Uttarakhand, India, is considered as the study area. The catchment covers an area of 7295 km^2. The catchment lies in the lower Himalayan region and thus is associated with very steep slopes. This is the main reason for the very high velocity of flow, which consequently leads to mass erosion. The location of the study area is shown in Fig. 3.1. The maximum and minimum temperatures over the catchment are 36 °C in summer and 0 °C in winter, respectively. A good amount of rainfall is received all over the catchment, though there are remarkable spatial variations (Kumar and Anbalagan 2015; Rautela et al. 2002).

The satellite-based imageries were collected from the website of United States Geological Survey (USGS) EarthExplorer. While the image for 2008 was taken from Landsat 5, the images for 2014 and 2020 were taken from Landsat 8 and Sentinel 2 satellites, respectively.

3.3 Methodology

The extraction of LULC information from the imageries is carried out by remote sensing and GIS techniques. The two softwares, viz. ERDAS IMAGINE and ArcGIS are widely used to carry out different image processing and geospatial operations, which were also used in this study. The overall methodology adopted for LULC classification and change detection is presented in Fig. 3.2. First of all, the satellite images for different years are collected and their preprocessing is carried out. The area of

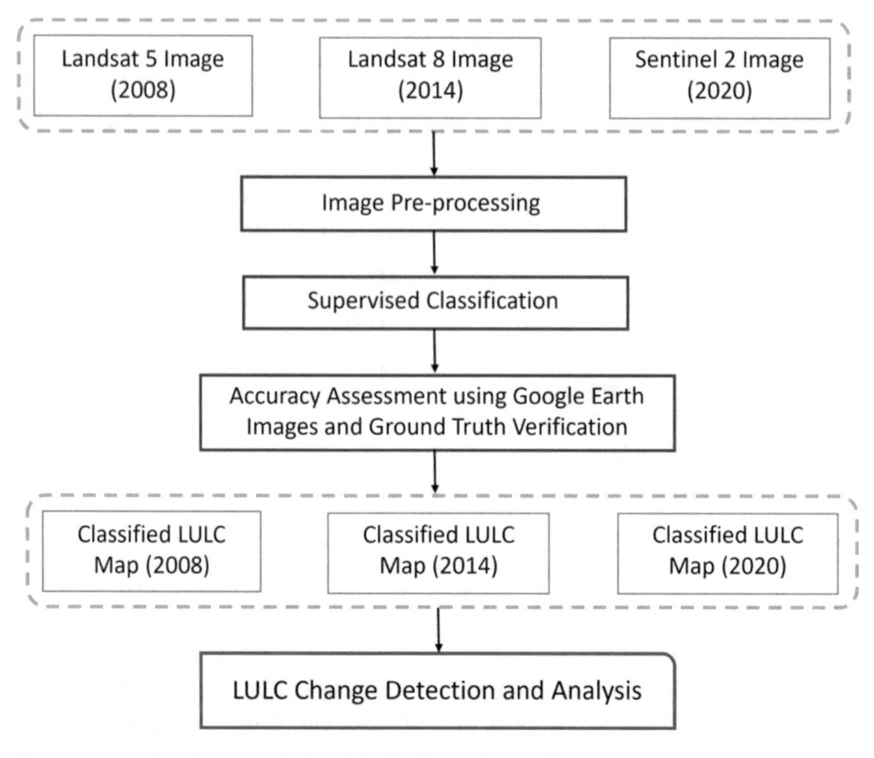

Fig. 3.2 Overall methodology for analyzing the multi-temporal LULC changes

interest may not fall under a single satellite image, and the collected data may be available in different file formats or projection systems. Therefore, stacking, mosaicing, and adjusting the coordinate systems, etc., were performed using ERDAS IMAGINE and ArcGIS 10.2.4 softwares. Moreover, for better interpretation of the imageries, false-color composites, contrast stretching, and image enhancement operations were also carried out.

The next step is the supervised classification, where the sample pixels in an image representing particular classes are selected by the user based on his/her knowledge. These are called the input classes or the training sites. The classification of all the remaining pixels can be carried out using these training sites through an image processing software. The reflectance of each pixel is the core of the image classification. For a particular class, the higher the number of training sites, the better is the precision of the classification. Therefore, LULC classification is based on the concept of segmenting the spectral domain into distinct ground cover classes. In this study, the study area is divided into eight different LULC classes, viz. open forest, dense forest, water bodies, shrubland, agricultural land, settlements, barren land, and snow covers.

The next step is the accuracy assessment, whose purpose is to validate the classification results. This justifies the utility of the classified maps for further applications.

Table 3.1 Measures for accuracy assessment of classification

Accuracy measures	Formula
Producer accuracy (A_P)	$\frac{x_{ij}}{x_j}$
User accuracy (A_U)	$\frac{x_{ij}}{x_i}$
Overall accuracy (A_O)	$\frac{1}{N}\sum_{i=j=1}^{n} x_{ij}$
Kappa coefficient (K_C)	$\frac{N\times\sum_{i=j=1}^{n} x_{ij}-\sum_{i=j=1}^{n}(x_i\times x_j)}{N^2-\sum_{i=j=1}^{n}(x_i\times x_j)}$

x_{ij} = value of ith row and jth column, x_j = sum of all values in jth column, x_i = sum of all values in ith row, N = total number of reference points, n = total number of rows/columns

This can be achieved by ground truth verification in terms of field visits. However, it is practically infeasible to collect information on the entire study area through field visits. Moreover, ground-truthing of the LULC for past years is almost impossible. Therefore, Google Earth images for a particular period can be used as a reference for validating the classifications. Using Google Earth as the reference is convenient and requires minimal cost. Nevertheless, for assessing the classification accuracy relevant to the recent period, it is always wise to conduct a field visit to some portions of the study area.

In this study, 400 random points from various classes were taken across the LULC maps. Considering their corresponding points from Google Earth image or ground-truthing information, a confusion matric is prepared. The producer accuracy (A_P) and the user accuracy (A_U) are calculated for each LULC class, whereas the overall classification accuracy (A_O) and the Kappa coefficient (K_C) are calculated to assess the LULC classification of the entire area. A_P, A_U, and A_O are expressed in percentage with a range from 0 to 100. On the other hand, K_C ranges from 0 to 1. The formula for these accuracy measures is provided in Table 3.1. The detailed procedure of accuracy assessment may be referred from literature (Manandhar et al. 2009; Rwanga and Ndambuki 2017; Sarkar 2018). Following the accuracy assessment, the spatiotemporal LULC changes are detected and analyzed.

3.4 Results and Discussion

The multi-temporal remotely sensed imageries were used for the detailed LULC classification. The classified maps of the years for 2008, 2014, and 2020 are presented in Fig. 3.3. The spatial variation of LULC classes (open forest, dense forest, water bodies, shrubland, agricultural land, settlements, barren land, and snow covers) in different years can be visualized clearly.

The areal coverage details of the individual classes in 2008, 2014, and 2020 are presented in Table 3.2. For all three years, it can be observed that the dense forest is the most dominant LULC class over the catchment, followed by shrubland and barren land. A significant portion of the catchment is covered by snow, which is

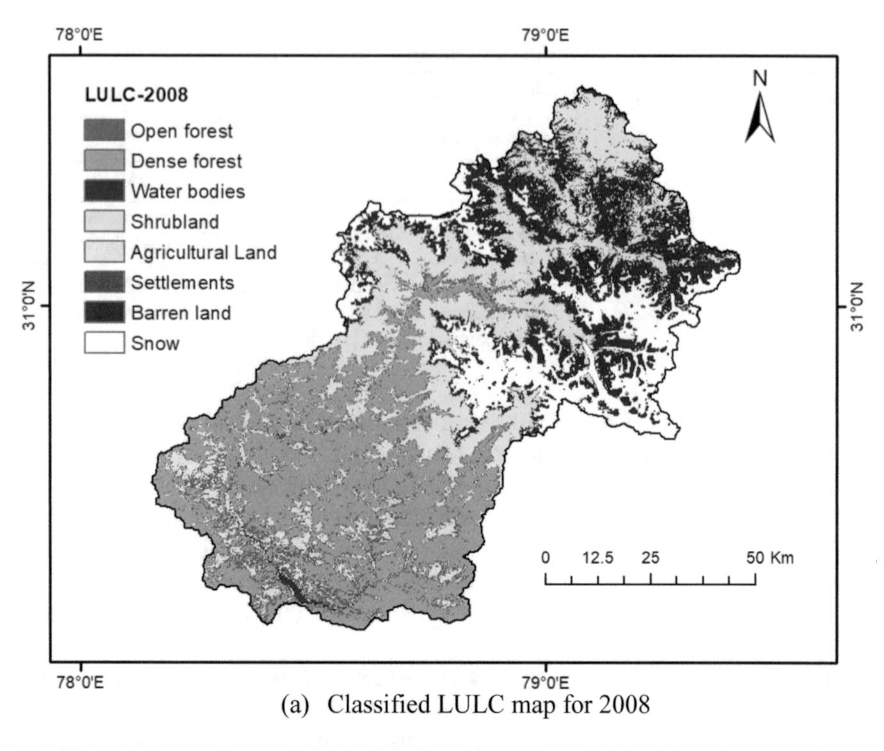

(a) Classified LULC map for 2008

Fig. 3.3 Multi-temporal supervised classification of LULC over Tehri catchment

inherent in the Himalayan conditions. The settlement constitutes the least portions of the catchment among all the classes.

From Table 3.2, the temporal changes in individual LULC classes over the Tehri catchment can be noticed. The percentage of catchment area under each of these classes during 2008, 2014, and 2020 is also presented. The dense forest has decreased by nearly 60 km^2 from 2008 to 2014, whereas there is no change between 2014 and 2020. On the other hand, there is a clear increase in open forests from 2008 to 2014. Hence, it can be fairly inferred that the canopy density has reduced over the years. As a result, the dense forests have been converted to open forests. An increase in the settlement is also observed. These may be attributed to anthropogenic activities, resulting in aggravated soil erosion. No appreciable change in agricultural land is observed between 2008, 2014, and 2020. There is a decrease in barren lands over the years. The shrubland has witnessed a remarkable increase from 2008 to 2014. Similarly, there is a clear increase in the snow covers from 2014 to 2020. The LULC changes were drastic from 2008 to 2014 over most of the classes, whereas there is hardly any change in LULC classes from 2014 to 2020, excluding barren land and snow (Table 3.2).

The results of the accuracy assessment are presented in Table 3.3. Considering the Google Earth images and the information collected during the field visits, the

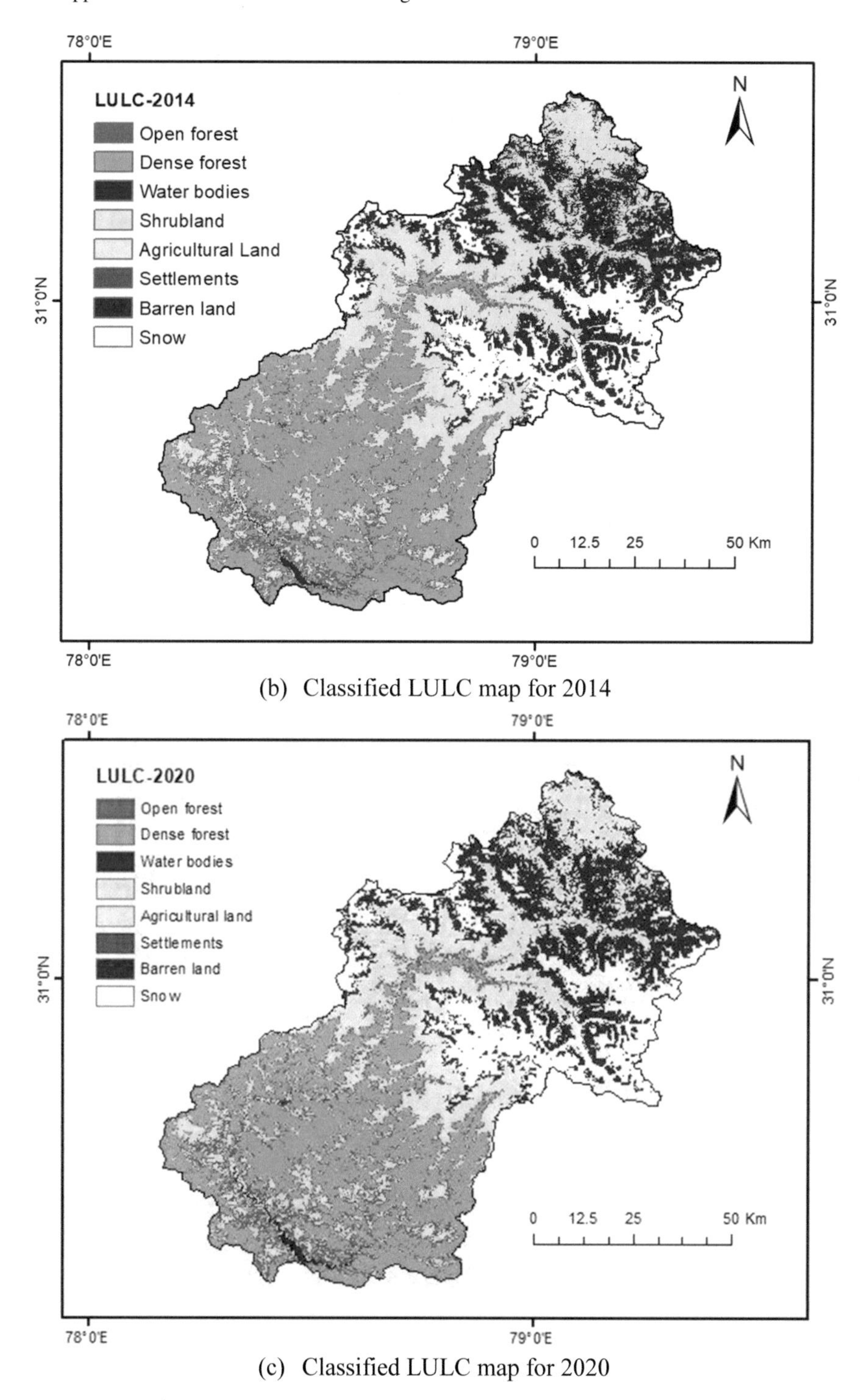

(b) Classified LULC map for 2014

(c) Classified LULC map for 2020

Fig. 3.3 (continued)

Table 3.2 Areal coverage details of LULC classes over the Tehri catchment

Sl. No.	LULC class	2008		2014		2020	
		Area (km^2)	% Area	Area (km^2)	% Area	Area (km^2)	% Area
1	Open forest	648.5	8.89	713.5	9.78	712.7	9.77
2	Dense forest	2197.3	30.12	2140.4	29.34	2139.6	29.33
3	Water bodies	23.3	0.32	27.0	0.37	35.0	0.48
4	Shrubland	1691.7	23.19	1780.0	24.4	1773.4	24.31
5	Agricultural land	176.5	2.42	172.9	2.37	171.4	2.35
6	Settlement	4.4	0.06	5.1	0.07	5.8	0.08
7	Barren land	1591.0	21.81	1486.7	20.38	1434.2	19.66
8	Snow	962.2	13.19	968.8	13.28	1022.0	14.01
Total		7295	100	7295	100	7295	100

Table 3.3 Accuracy assessment results of LULC classification

LULC class	Accuracy (%)	2008	2014	2020
Open forest	Producer	88.2	85.3	94.4
	User	84.0	88.9	90.9
Dense Forest	Producer	95.7	87.3	96.0
	User	94.2	89.2	92.5
Water Bodies	Producer	100.0	97.9	100.0
	User	100.0	96.0	97.4
Shrubland	Producer	86.3	81.7	85.0
	User	80.0	83.3	86.4
Agricultural Land	Producer	85.0	78.3	92.0
	User	86.7	82.0	90.0
Settlement	Producer	84.0	82.4	91.7
	User	84.0	87.5	93.3
Barrenland	Producer	73.3	78.8	80.0
	User	78.0	74.5	86.0
Snow	Producer	82.2	83.7	86.2
	User	79.2	80.0	84.0
Overall classification accuracy (%)		83.6	82.5	88.9
Kappa coefficient		0.821	0.803	0.873

accuracy measures were estimated. The A_P and A_U values for individual classes are quite encouraging (Table 3.3). The results of the accuracy assessment reflect a precise identification of LULC classes over the catchment for all three periods. The overall classification accuracy (Kappa coefficient) for 2008, 2014, and 2020 are found to be 83.6 (0.821), 82.5 (0.803), and 88.9 (0.873), respectively. These high values of A_O and K_C confirm that the LULC changes reported in this study can be justifiably regarded as accurate and, hence, are utilizable for further applications.

3.5 Ground Truth Verification

It is always wise to cross-check the supervised detailed LULC classification by ground truth verification. Therefore, a field visit was made to some portions of the catchment area to collect the land use/land cover observations along with their appropriate location details so that it would be helpful for proper validation and accuracy assessment. Moreover, it aimed to obtain relevant information from the local people regarding the causes of LULC changes, other hydrological problems over the study area (particularly soil erosion), and steps taken to combat those issues. This can be very helpful in preparing an effective catchment area treatment plan. Thus, a field visit was made to accomplish the aforementioned objectives. The details of the locations are provided in Fig. 3.4 and Table 3.4.

Due to the constraints of cost and time, only a portion of the catchment was covered during the field visit. The photographs collected during the field visits along with their location details were useful for the accuracy assessment of the LULC classification pertaining to 2020. A few photographs are presented in Fig. 3.5. As the catchment is prone to soil erosion, several protection measures were adopted, which is evident from some of the photographs (Fig. 3.5).

3.6 Conclusion

The detailed supervised LULC classification for the Tehri catchment is carried out using the Landsat 5, Landsat 8, and Sentinel 2 data pertaining to 2008, 2014, and 2020, respectively. The various land covers that the catchment area is classified into are water bodies, agricultural land, dense forest, open forest, shrubland, settlement, barren land, and snow. The LULC changes are found to be drastic from 2008 to 2014 over most of the classes, whereas no appreciable changes in classes are found from 2014 to 2020 except for snow and barren lands. Comparing 2008 and 2020, an increase in the open forests, water bodies, shrubland, snow, and settlement is observed, whereas a decrease in dense forests and barren land is noticed. The accuracy assessment results confirm that the LULC changes reported in this study are justifiably accurate and utilizable for further applications.

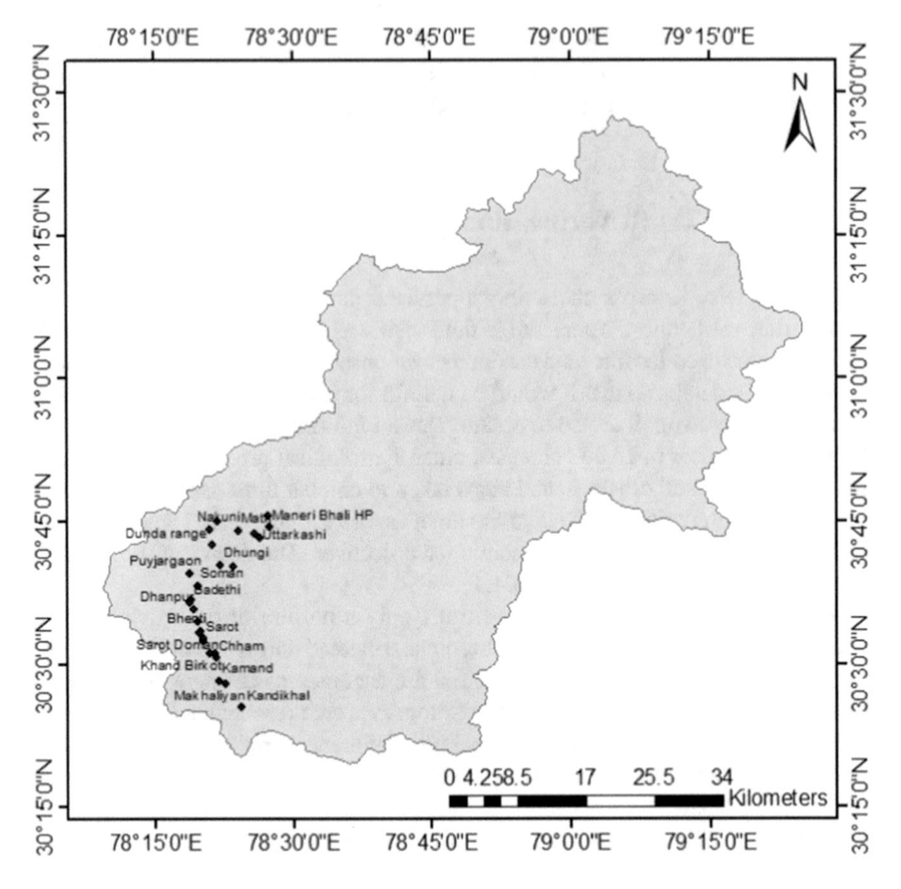

Fig. 3.4 Locations covered during the field visit

Table 3.4 Location details of the points covered during the field visit

S. No.	Name	Latitude (°N)	Longitude (°E)
1	Kandikhal	30.424	78.408
2	Makhaliyan	30.466	78.378
3	Kamand	30.471	78.365
4	Chham	30.511	78.360
5	Khand Birkot	30.518	78.349
6	Manjaruwal Village	30.519	78.359
7	Unial Village	30.540	78.336
8	Sarot	30.545	78.337
9	Sarot Doman	30.554	78.329
10	Bhenti	30.558	78.332
11	Chinyalisaur	30.574	78.328
12	Dhanpur	30.597	78.319
13	Badethi	30.608	78.314
14	Pujar Village	30.611	78.315
15	Dharashu Band	30.613	78.316
16	Soman	30.637	78.327
17	Puyjargaon	30.659	78.314
18	Singuni	30.671	78.392
19	Dhungi	30.674	78.369
20	Dunda range	30.709	78.353
21	Uttarkashi	30.721	78.440
22	Maneri Bhali	30.730	78.431
23	Matli	30.734	78.401
24	Nakuni	30.737	78.350
25	Maneri Bhali HP	30.740	78.458
26	Genwala	30.751	78.364
27	Assi Ganga Sangam	30.760	78.456

Fig. 3.5 Photographs of some locations in the Tehri catchment collected during the field visit

Acknowledgements This study is a part of the project THD-1176-WRC sponsored by the Tehri Hydro Development Corporation India Limited (THDCIL). The authors are thankful to THDCIL for the same. The resources and facilities provided by the Department of Water Resources Development and Management (WRD&M), IIT Roorkee, are acknowledged. The support provided by Lingaraj Dhal and Praveen Kalura during the study is also thankfully acknowledged.

References

Aadhar S, Swain S, Rath DR (2019) Application and performance assessment of SWAT hydrological model over Kharun river basin, Chhattisgarh, India. In: World Environmental and Water Resources Congress 2019: watershed management, irrigation and drainage, and water resources planning and management. American Society of Civil Engineers, pp 272–280

Anand J, Gosain AK, Khosa R (2018) Prediction of land use changes based on land change modeler and attribution of changes in the water balance of ganga basin to land use change using the SWAT model. Sci Total Environ 644:503–519

Astuti IS, Sahoo K, Milewski A, Mishra DR (2019) Impact of land use land cover (LULC) change on surface runoff in an increasingly urbanized tropical watershed. Water Resour Manage 33(12):4087–4103

Bahita TA, Swain S, Dayal D, Jha PK, Pandey A (2021) Water quality assessment of upper ganga canal for human drinking. In: Climate impacts on water resources in India. Springer, Cham, pp 371–392

Chen Q, Chen H, Zhang J, Hou Y, Shen M, Chen J, Xu C (2020) Impacts of climate change and LULC change on runoff in the Jinsha River Basin. J Geog Sci 30(1):85–102

Dayal D, Gupta PK, Pandey A (2021) Streamflow estimation using satellite-retrieved water fluxes and machine learning technique over monsoon-dominated catchments of India. Hydrol Sci J 66(4):656–671

Dayal D, Swain S, Gautam AK, Palmate SS, Pandey A, Mishra SK (2019) Development of ARIMA model for monthly rainfall forecasting over an Indian River Basin. In: World Environmental and Water Resources Congress 2019: watershed management, irrigation and drainage, and water resources planning and management. American Society of Civil Engineers, pp 264–271

Dutta D, Rahman A, Paul SK, Kundu A (2019) Changing pattern of urban landscape and its effect on land surface temperature in and around Delhi. Environ Monit Assess 191(9):1–15

Guptha GC, Swain S, Al-Ansari N, Taloor AK, Dayal D (2021) Evaluation of an urban drainage system and its resilience using remote sensing and GIS. Remote Sens Appl: Soc Environ 23100601. https://doi.org/10.1016/j.rsase.2021.100601

Hengade N, Eldho TI (2016) Assessment of LULC and climate change on the hydrology of Ashti Catchment, India using VIC model. J Earth Syst Sci 125(8):1623–1634

Himanshu SK, Pandey A, Patil A (2018) Hydrologic evaluation of the TMPA-3B42V7 precipitation data set over an agricultural watershed using the SWAT model. J Hydrol Eng 23(4):05018003. https://doi.org/10.1061/(ASCE)HE.1943-5584.0001629

Himanshu SK, Pandey A, Yadav B, Gupta A (2019) Evaluation of best management practices for sediment and nutrient loss control using SWAT model. Soil Tillage Res 19242–19258. https://doi.org/10.1016/j.still.2019.04.016

Kalura P, Pandey A, Chowdary VM, Raju PV (2021) Assessment of hydrological drought vulnerability using geospatial techniques in the Tons River Basin India. J Ind Soc Rem Sens. https://doi.org/10.1007/s12524-021-01413-7

Kumar R, Anbalagan R (2015) Landslide susceptibility zonation in part of Tehri reservoir region using frequency ratio, fuzzy logic and GIS. J Earth Syst Sci 124(2):431–448

Liping C, Yujun S, Saeed S (2018) Monitoring and predicting land use and land cover changes using remote sensing and GIS techniques—a case study of a hilly area, Jiangle, China. PloS One 13(7)

Manandhar R, Odeh IO, Ancev T (2009) Improving the accuracy of land use and land cover classification of Landsat data using post-classification enhancement. Remote Sens 1(3):330–344

Mishra PK, Rai A, Rai SC (2020) Land use and land cover change detection using geospatial techniques in the Sikkim Himalaya, India. Egypt J Remote Sens Space Sci 23(2):133–143

Palmate SS, Pandey A, Mishra SK (2017) Modelling spatiotemporal land dynamics for a trans-boundary river basin using integrated cellular automata and Markov Chain approach. Appl Geogr 82:11–23

Palmate SS, Pandey A, Kumar D, Pandey RP, Mishra SK (2017) Climate change impact on forest cover and vegetation in Betwa Basin, India. Appl Water Sci 7(1):103–114

Pandey A, Palmate SS (2018) Assessments of spatial land cover dynamic hotspots employing MODIS time-series datasets in the Ken River Basin of Central India. Arab J Geosci 11(17):1–8

Pandey BK, Khare D (2017) Analyzing and modeling of a large river basin dynamics applying integrated cellular automata and Markov model. Environ Earth Sci 76(22):1–12

Rautela P, Rakshit R, Jha VK, Gupta RK, Munshi A (2002) GIS and remote sensing-based study of the reservoir-induced land-use/land-cover changes in the catchment of Tehri dam in Garhwal Himalaya, Uttaranchal (India). Curr Sci 308–311

Rimal B, Zhang L, Keshtkar H, Wang N, Lin Y (2017) Monitoring and modeling of spatiotemporal urban expansion and land-use/land-cover change using integrated Markov chain cellular automata model. ISPRS Int J Geo Inf 6(9):288

Rwanga SS, Ndambuki JM (2017) Accuracy assessment of land use/land cover classification using remote sensing and GIS. Int J Geosci 8(04):611

Sahoo S, Swain S, Goswami A, Sharma R, Pateriya B (2021) Assessment of trends and multi-decadal changes in groundwater level in parts of the Malwa region Punjab India. Groundwater Sustain Dev 14100644. https://doi.org/10.1016/j.gsd.2021.100644

Saputra MH, Lee HS (2019) Prediction of land use and land cover changes for north Sumatra, Indonesia, using an artificial-neural-network-based cellular automaton. Sustainability 11(11):3024

Sarkar A (2018) Accuracy assessment and analysis of land use land cover change using geoinformatics technique in Raniganj Coalfield Area, India. Int J Environ Sci Nat Resour 11(1):25–34

Sharma I, Mishra SK, Pandey A, Kumre SK, Swain S (2020) Determination and verification of antecedent soil moisture using soil conservation service curve number method under various land uses by employing the data of small indian experimental farms. In: Watershed management 2020. American Society of Civil Engineers, pp 141–150

Singh SK, Laari PB, Mustak SK, Srivastava PK, Szabó S (2018) Modelling of land use land cover change using earth observation data-sets of Tons River Basin, Madhya Pradesh, India. Geocarto Int 33(11):1202–1222

Singh S, Bhardwaj A, Verma VK (2020) Remote sensing and GIS based analysis of temporal land use/land cover and water quality changes in Harike wetland ecosystem, Punjab, India. J Environ Manag 262:110355

Singh G, Pandey A (2021) Evaluation of classification algorithms for land use land cover mapping in the snow-fed Alaknanda River Basin of the Northwest Himalayan Region. Appl Geomat. https://doi.org/10.1007/s12518-021-00401-3

Swain S, Dayal D, Pandey A, Mishra SK (2019a) Trend analysis of precipitation and temperature for Bilaspur District, Chhattisgarh, India. In: World Environmental and Water Resources Congress 2019: groundwater, sustainability, hydro-climate/climate change, and environmental engineering. American Society of Civil Engineers, pp 193–204

Swain S, Mishra SK, Pandey A (2019b) Spatiotemporal characterization of meteorological droughts and its linkage with environmental flow conditions. In: AGU fall meeting abstracts (AGUFM 2019), H13O–1959

Swain S, Mishra SK, Pandey A (2020a) Assessment of meteorological droughts over Hoshangabad district, India. In: IOP conference series: earth and environmental science, vol 491, p 012012. IOP Publishing

Swain S, Mishra SK, Pandey A (2021a) A detailed assessment of meteorological drought characteristics using simplified rainfall index over Narmada River Basin, India. Environ Earth Sci 80:221

Swain S, Mishra SK, Pandey A, Dayal D (2021b) Identification of meteorological extreme years over central division of odisha using an index-based approach. In: Hydrological extremes. Springer, Cham, pp 161–174

Swain S, Mishra SK, Pandey A (2021c) Assessing contributions of intensity-based rainfall classes to annual rainfall and wet days over Tehri Catchment, India. In: Advances in Water Resources and Transportation Engineering. Springer, Singapore, pp 113–121. https://doi.org/10.1007/978-981-16-1303-6_9

Swain S, Patel P, Nandi S (2017) Application of SPI, EDI and PNPI using MSWEP precipitation data over Marathwada, India. In: 2017 IEEE international geoscience and remote sensing symposium (IGARSS). IEEE, pp 5505–5507

Swain S, Sharma I, Mishra SK, Pandey A, Amrit K, Nikam V (2020b) A framework for managing irrigation water requirements under climatic uncertainties over Beed District, Maharashtra, India. In: World Environmental and Water Resources Congress 2020: water resources planning and management and irrigation and drainage. American Sociiety of Civil Engineers, pp 1–8

Swain S, Verma MK, Verma MK (2018) Streamflow estimation using SWAT model over Seonath river basin, Chhattisgarh, India. In: Hydrologic modeling. Springer, Singapore, pp 659–665

Talukdar S, Singha P, Mahato S, Pal S, Liou YA, Rahman A (2020) Land-use land-cover classification by machine learning classifiers for satellite observations—a review. Rem Sens 12(7):1135

Tian H, Banger K, Bo T, Dadhwal VK (2014) History of land use in India during 1880–2010: large-scale land transformations reconstructed from satellite data and historical archives. Global Planet Change 121:78–88

Tran DX, Pla F, Latorre-Carmona P, Myint SW, Caetano M, Kieu HV (2017) Characterizing the relationship between land use land cover change and land surface temperature. ISPRS J Photogramm Remote Sens 124:119–132

Tripathi G, Pandey AC, Parida BR, Kumar A (2020) Flood inundation mapping and impact assessment using multi-temporal optical and SAR satellite data: a case study of 2017 Flood in Darbhanga district, Bihar, India. Water Resour Manag 34(6):1871–1892

Chapter 4
Land Use Land Cover Change Detection of the Tons River Basin Using Remote Sensing and GIS

Praveen Kalura, Ashish Pandey, V. M. Chowdary, and P. V. Raju

Abstract Land use and land cover are the two separate but related concepts that are predominant characteristics of the land. In general, land cover is defined as the observed surface cover on the ground, such as vegetation, water bodies, barren land, or man-made features, while land use refers to the purpose for which land is being used. Land use and land cover (LULC) change have a significant impact on water resources. Hence, it has become one of the critical components in most studies related to water resources. Remote sensing (RS) and Geographical Information System (GIS) techniques are extensively used to detect the location of changes, type of changes, and quantification of changes in LULC. In this study, an attempt is made to detect a change in LULC in the past three decades (1985–2015) for the Tons river basin, a sub-basin of the Ganges river basin. The supervised classification method was employed to classify the satellite images of the years 1985, 1995, 2005, and 2015 to study the change in LULC. The Maximum Likelihood Classification (MLC) method was used for the classification. Landsat 5 Thematic Mapper (TM) and Landsat 8 Operational Land Imager (OLI) datasets were used to prepare LULC maps. Seven LULC class types are observed in the basin, namely agricultural land, barren land, built-up area, dense forest, open forest, shrubland, and water. The LULC study showed agricultural land is the dominant class in the basin, followed by forest land, the second dominant class. An increase in built-up area from 45.73 km^2 (1985) to 105.41 km^2 in 2015 shows rapid urbanization. Some change in water body class was also seen, which

P. Kalura (✉) · A. Pandey
Department of Water Resources Development and Management, Indian Institute of Technology Roorkee, Roorkee, Uttarakhand, India
e-mail: p_kalura@wr.iitr.ac.in

A. Pandey
e-mail: ashish.pandey@wr.iitr.ac.in

V. M. Chowdary
Regional Remote Sensing Centre—NRSC-N, New Delhi, India
e-mail: chowdary_vm@nrsc.gov.in

P. V. Raju
National Remote Sensing Centre, ISRO, Hyderabad, India
e-mail: raju_pv@nrsc.gov.in

A. Pandey et al. (eds.), *Geospatial Technologies for Land and Water Resources Management*, Water Science and Technology Library 103,
https://doi.org/10.1007/978-3-030-90479-1_4

increased from 235.24 km^2 in 1985 to 255.96 km^2 in 1995, it showed a decrease in the area up to 211.38 km^2 in 2015. There was an increase in open forest area, and no significant change was observed for agricultural land. The overall classification accuracy varies from 85.4 to 90.6%, and the kappa coefficient value varies from 0.83 to 0.89 shows the satisfactory classification of the satellite-derived LULC maps.

4.1 Introduction

Land use defines the purpose for which the land is being used, and land cover refers to the type of cover that exists over land. Natural factors and anthropogenic activities can cause land use land cover (LULC) changes over the decades, and recently the speed of change has increased due to rapid urbanization and growing population. So, knowledge about past, current, and future LULC is a prerequisite for better land use planning by decision makers (Palmate et al. 2017a, b; Berihun et al. 2019; Chamling and Bera 2020).

Worldwide LULC change in recent decades is a major cause of concern because it along with other anthropogenic activities have resulted in an increased number of natural disasters (Swain et al. 2017, 2018, 2021a; b; Tripathi et al. 2020; Guptha et al.2021; Kalura et al. 2021). LULC change significantly influence on ecosystem services of the area. (Arowolo et al. 2018; Singh et al. 2020; Talukdar et al. 2020). LULC change affects the hydrological cycle components, such as Evapotranspiration and water yield, thus offering quantitative information to decision makers about the framing of the policy regarding land and water resources management (Anand et al. 2018; Dayal et al. 2019, 2021). Change in LULC is the major source of uncertainty in predicting the water balance of a basin. (Saputra and Lee 2019; Chen et al. 2020). LULC change is one of the major driving forces that impact water quantity and quality (Hengade and Eldho 2016; Himanshu et al. 2018, 2019; Swain et al. 2020a, b; Bahita et al. 2021; Pandey et al. 2021). Therefore, an evaluation of historical LULC change is critical to better understand the present situation and predict future impact of LULC changes on water resources.

Multi-temporal satellite imagery datasets derived LULC maps can be used to identify the pattern of urban expansion. (Rwanga and Ndambuki 2017; Dutta et al. 2019). LULC change has a significant relationship with the Land surface temperature, which can be used to understand the Urban heat island effect in the context of urbanization (Tran et al. 2017; Rizvi et al. 2020). Change in LULC due to rapid urbanization can cause reduction in agricultural area which in turn impact soil organic carbon sequestration (Barakat et al. 2021).

The satellite remote sensing technique provides multi-temporal data, which is also cost-effective and helps in monitoring LULC change, while GIS provides a platform for data storage, analysis, and display of data required for the detection of LULC change (Thakur et al. 2021). Multi-spectral satellite data from various missions such as Landsat, Sentinel, and Cartosat can be used to prepare LULC maps. High-resolution Sentinel 2A data have also been used for LULC classification in a Himalayan watershed (Mishra et al. 2020). The problem with optical data is that it cannot be used to prepare multi-temporal LULC maps for the area with frequent cloud cover. Microwave remote sensing-based Synthetic Aperture Radar (SAR) can play a crucial role to minimize this limitation (Parida and Mandal 2020; Prudente et al. 2020). Optical data uses physical, chemical, and biological properties of targets, whereas Microwave data uses texture, structure, and dielectric properties for image classification (Pereira et al. 2018). Recently, many studies have shown satisfactory performance of fusion of optical and microwave remote sensing data for LULC mapping (Clerici et al. 2017; Thanh and Kappas 2017; Khan et al. 2020).

The process of grouping pixels of imagery with similar characteristics is termed image classification. Various classification algorithms have been developed to detect LULC change, among which the most widely used technique is Maximum Likelihood classifier as it requires less time as compared to others (Shivakumar and Rajashekararadhya 2018; Chamling and Bera 2020; Singh and Pandey 2021). Application of machine-learning algorithms for image classification is now becoming popular among the research community (Maxwell et al. 2018; Mirici et al. 2018; Talukdar et al. 2020). Support Vector Machine (Abdi 2020; Jozdani et al. 2019; Karimi et al. 2019), Random Forest (Nguyen et al. 2020), Neural network classifiers, and decision tree classifiers (Yang et al. 2017) are some of the Machine-learning algorithms to prepare LULC maps. It is also important to model future LULC for better planning, for which, state of the art Cellular Automata- Markov Chain (CA-MARKOV) is one of the most popular techniques (Rimal et al. 2017; Palmate et al. 2017a; Liping et al. 2018; Singh et al. 2018).

In this study, we have conducted the quantitative assessment of the LULC change in the Tons river basin using multi-temporal satellite data.

4.2 Material and Methods

4.2.1 Study Area

The Tons river is a tributary of the Ganga river, which originates at Tamakund in the Kaimur Range in Madhya Pradesh at an elevation of 610 m. The geographical extent of the basin lies between 80° 20′ E to 83° 25′ E longitudes and 23° 57′ N to 25° 20′ N latitudes. Belan River is the major tributary of this river. The total catchment area is approximately more than 17,441 km^2, out of which 12,165 km^2 (70%) lies in MP, and the remaining area 5276 km^2 (30%) lies in UP. The location map of the study area is shown in Fig. 4.1. The tons river basin covers ten districts, out of which three major sections are Satna, Rewa, and Allahabad. About 90% of the rainfall is received during the monsoon season, with average annual rainfall between 930 to 1116 mm per year. The maximum and minimum temperature is 41 °C in summer and 8 °C in winter.

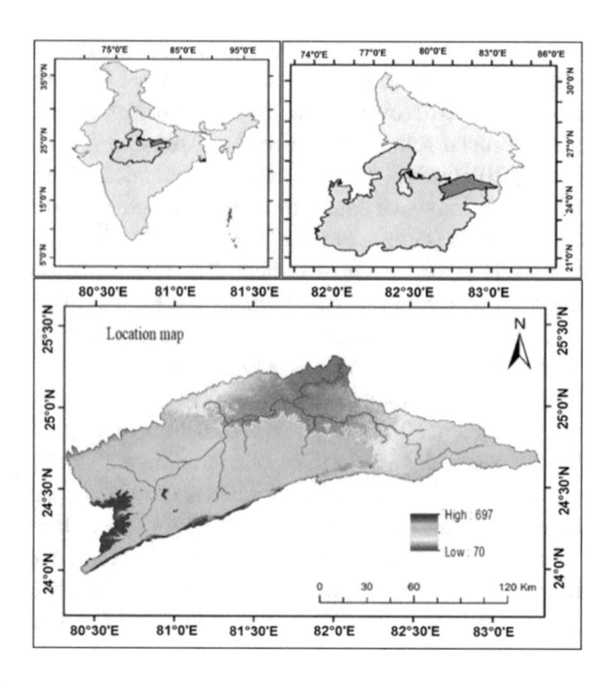

Fig. 4.1 Location map of the study area

4.2.2 *Data Used*

Landsat imagery was downloaded from the earth explorer portal (https://earthexplorer.usgs.gov/). Dates were selected based on data quality, data availability. Total of twelve Landsat imageries was downloaded for the years 1985, 1995, 2005, and 2015. All the images have the same spatial resolution (30 m), but different sensors and satellites sensed data at different times of the year. These data were used for the generation of LULC maps fed into an image processing software.

4.2.3 *Methodology*

One of the most important data sources for the preparation of LULC mapping is remote sensing satellite data. In this study, remote sensing and GIS techniques were used to extract spatial information of the Tons basin using satellite data of different spatial and temporal resolutions. Image classification is carried out to identify the features occurring in an image that are actually present on the ground. Image processing satellite data and analysis of interpreted maps were carried out using ArcGIS 10.2.4 version software packages. The detailed methodology flowchart is shown in Fig. 4.2.

Satellite data was available in different raster file format layers, and the study area falls under more than one satellite imagery. Therefore, these layers were stacked, and

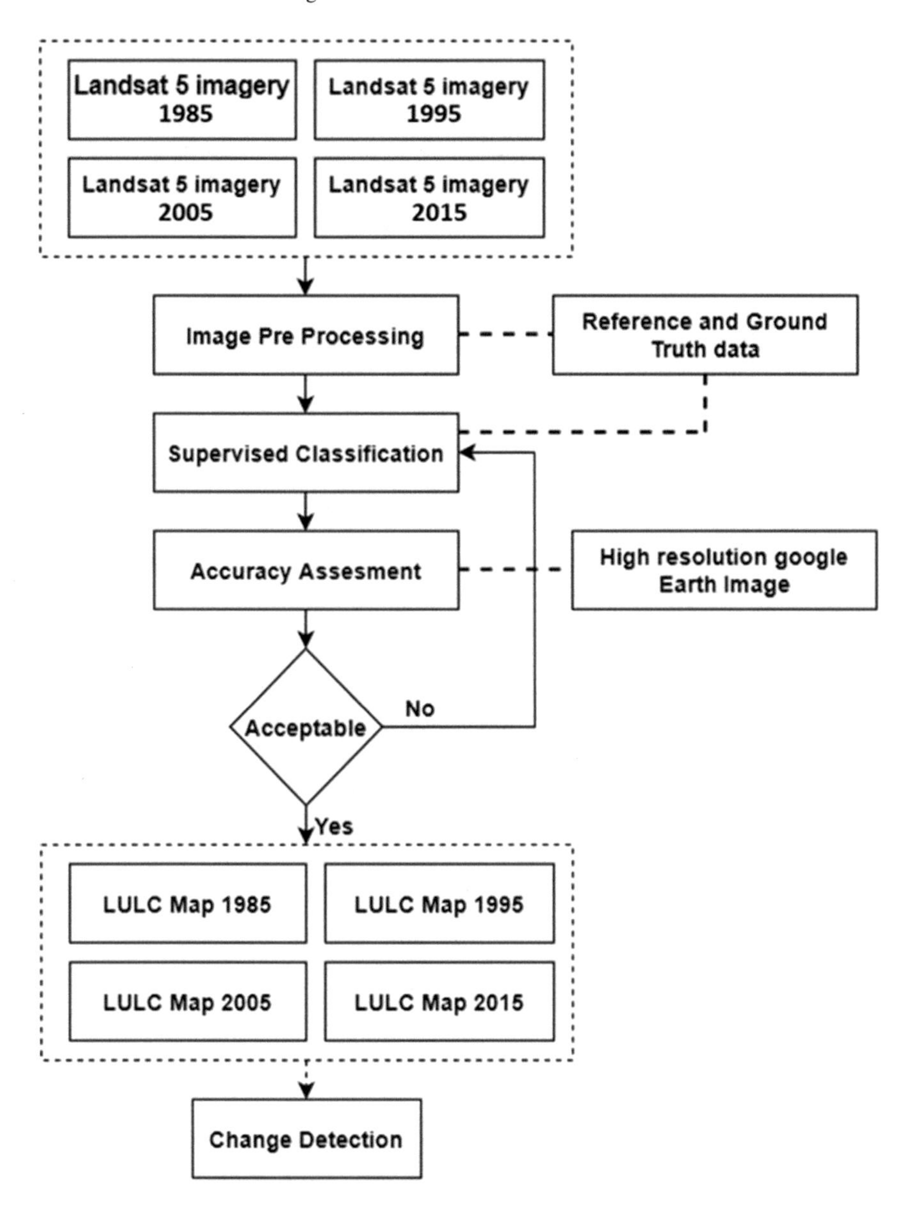

Fig. 4.2 Methodology flowchart for LULC change detection

then mosaicking of the stacked imagery was carried out using a GIS package. Further, image enhancement, contrast stretching, and false-color composites were carried out to make easy visual interpretation and understanding of the satellite imagery. Satellite imagery data follows some processes: radiometric and geometric corrections, image segmentation, image enhancement, and classification using spectral and spatial information. Geometric and radiometric corrections of satellite imagery were also carried out to improve image accuracy for further analysis. An image has been classified by determining the reflectance for each pixel. In this study, an image classification categorizes seven LULC classes, namely dense forest, waterbody, settlement, agriculture area, and barren land. In this study, a supervised classification method was used to prepare the Tons basin's LULC map.

Accuracy assessment is essential to understand accurate and valid results of the classified imagery, without which the LULC map is simply an untested hypothesis. Accuracy assessment of LULC map is required to evaluate its suitability for the application. The classified LULC classes are Agricultural land, Barren land, Built-up, dense forest, open forest, scrubland, and water. In this method, some points are selected and verified through ground truth or high-resolution images. In this study, 350 stratified random accuracy assessment points were generated across the LULC map. It is calculated in terms of overall classification accuracy (%) and Kappa coefficient.

Based on inter-transitions calculated by confusion matrix between reference data and classified imagery data, several statistical measures, namely users accuracy (Eq. 4.1), producers accuracy (Eq. 4.2), overall accuracy (Eq. 4.3). In this study, these statistical terms have been used for accuracy assessment.

Kappa coefficient (k_c) was estimated for all satellite-derived LULC maps using Eq. 4.4 Higher percentage of users' accuracy/producers' accuracy/overall accuracy indicates the more precise classification of satellite imagery. Kappa coefficient value ranges from 0 to 1. The high value of the Kappa coefficient indicates more accuracy in the LULC map, and the low value indicates less accuracy.

$$\text{User Accuracy} = \frac{x_{ij}}{x_i} \tag{4.1}$$

$$\text{Producer Accuracy} = \frac{x_{ij}}{x_j} \tag{4.2}$$

$$\text{Overall Accuracy} = \frac{1}{N}\sum_{i=1}^{r} x_{ii} \tag{4.3}$$

$$\text{Kappa Coefficient } (K_c) = \frac{N \times \sum_{i=1}^{r} x_{ij} - \sum_{i=1}^{r} (x_i \times x_j)}{N^2 - \sum_{i=1}^{r} (x_i \times x_j)} \tag{4.4}$$

where, N = Total Reference points, r = number of rows in error matrix x_{ij} = value in row i column j; x_i = total of row i; x_j = total of column j.

4.3 Result and Discussion

In the present study, LULC analysis was carried out using multi-temporal remote sensing data. LULC change analysis of the basin shows the rapid urbanization in the basin for the past four decades. The accuracy assessment results of the LULC maps indicate that the land use change has been extracted accurately. Figure 4.3 show the spatial representation of the LULC pattern for each year. LULC distribution results show that agricultural land is the primary dominant LULC class followed by open forest, the second dominant LULC class. There is not a significant change in the area of agricultural land cover. Table 4.1 gives details about the total area in km^2 and the percentage for each LULC class.

The barren land decreased from 432.23 km^2 in 1985 to 393.55 km^2 in 2015. This decrease is probably due to the increase in the built-up area, which increased from 45.73 km^2 in 1985 to 105.41 Km^2 in 2015 dense forest area decreased from 2848.71 km^2 in 1985 to 2744.28 km^2 in 2015, and scrubland area also fell from 1235.80 km^2 in 1985 to 1126.14 km^2 in 1985. This decrease in the areal extent of dense forest and scrubland may be due to conversion into open forest areas, which increased from 121.70 km^2 in 1985 to 369.64 km^2 in 2015. The total areal extent of water bodies increased from 235.24 km^2 in 1985 to 255.96 km^2 in 1995, and then it decreased to 211.38 km^2 in 2015. This initial increase may be due to a large number of water storage structures developed during the 1990s and further decrease may be due to the drying up of seasonal water bodies. The average rate of change in LULC areas in the Tons basin for 1985, 1995, 2005, and 2015 is given in Table 4.1.

The average rate of increase for the built-up area was 0.47% during 1985–1995 and 0.48% during 2005–2015. The rate of change of Open forest area during 1985–1995 was 0.614%, and during 2005–2015 it was 0.821. There was no significant rate of change in other LULC classes. Table 4.2 gives the detail about the average rate of change in percentage for each LULC class. Figure 4.4 Shows Overall changes in various LULC classes in percentage (1985–2015).

Table 4.2 shows the overall classification accuracy and Kappa coefficient that was calculated for the LULC map of the years 1985, 1995, 2005, and 2015. Overall classification accuracy was found to be 85.4, 85.1, 87.1, and 90.6%, and Kappa coefficient was found to be 0.83, 0.83, 0.85, and 0.89 for the years 1985, 1995, 2005, and 2015 respectively.

The accuracy assessment results of the classified imageries of the years 1985, 1995, 2005, and 2015 indicate that the land use changes have been accurately identified and extracted during the classification, which is also confirmed by the overall accuracy Kappa indices.

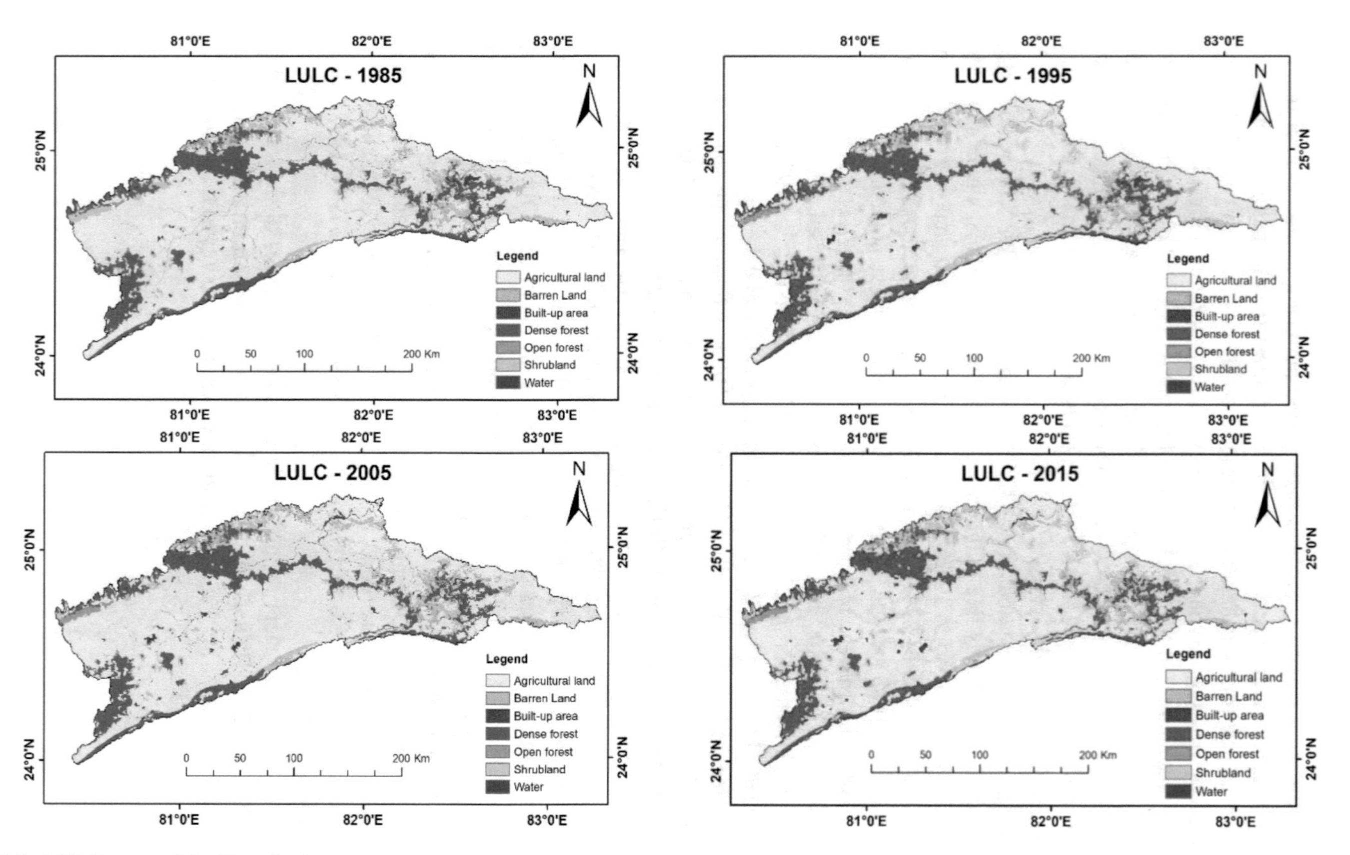

Fig. 4.3 LULC maps of the Tons Basin

Table 4.1 Area (%) under land use/land cover classification of the Tons River Basin

Class		1985	1995	2005	2015
Agricultural land	km^2	12,522.3	12,520.1	12,525.6	12,491.3
	%	71.8	71.8	71.8	71.6
Barren Land	km^2	432.2	423.6	433.7	393.6
	%	2.5	2.4	2.5	2.3
Built-up area	km^2	45.7	67.4	84.5	105.4
	%	0.3	0.4	0.5	0.6
Dense forest	km^2	2848.7	2830.9	2786.0	2744.3
	%	16.3	16.2	16.0	15.7
Open forest	km^2	121.7	196.5	203.0	369.6
	%	0.7	1.1	1.2	2.1
Shrubland	km^2	1235.8	1147.3	1168.0	1126.1
	%	7.1	6.6	6.7	6.5
Water	km^2	235.2	256.0	241.0	211.4
	%	1.3	1.5	1.4	1.2

Table 4.2 Accuracy assessment results of LULC classification

	Accuracy (%)	1985	1995	2005	2015
Agricultural land	Producer	80.0	85.7	80.8	82.1
	User	88.0	84.0	84.0	92.0
Barren land	Producer	70.2	72.9	74.5	84.3
	User	80.0	86.0	82.0	86.0
Built-up area	Producer	84.4	85.7	88.9	92.2
	User	76.0	82.4	80.0	94.0
Dense forest	Producer	95.8	93.5	95.7	97.9
	User	92.0	91.5	90.0	94.0
Open forest	Producer	89.6	85.4	90.2	93.8
	User	86.0	82.0	92.0	90.0
Scrubland	Producer	85.7	80.4	86.3	87.5
	User	84.0	80.4	88.0	84.0
Water	Producer	97.9	97.9	95.9	97.9
	User	93.9	92.0	94.0	94.0
Overall classification accuracy (%)		85.4	85.1	87.1	90.6
Kappa coefficient		0.83	0.83	0.85	0.89

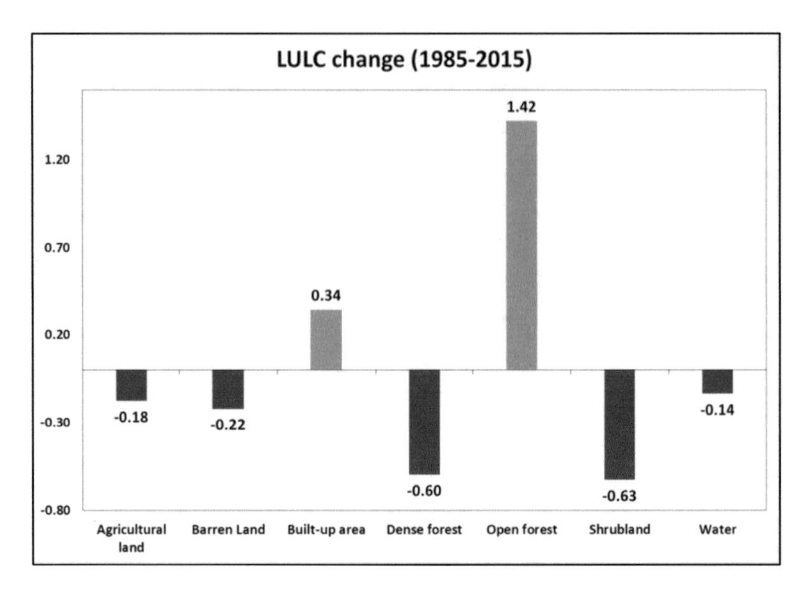

Fig. 4.4 Overall change of LULC (1985–1995)

4.4 Conclusion

The changes in LULC of a basin have a predominant effect on the hydrological response of the basin. In this study, we have analyzed the LULC change of Tons river basin for 40 years from 1985 to 2015. A considerable change was noted for the built-up area (0.34%) and open forest area (1.42%). Changes in other classes were not that significant. The increase in the built-up area can be correlated with the rapid urbanization in recent years. This decrease in the agricultural area is mainly due to its conversion into the built-up area. Dense forest and shrubland are now converted to open forest.

Acknowledgements The authors are thankful to the Indian space research organization (ISRO) for providing financial support during the study period. We are also grateful to the Department of Water Resources Development and Management (WRD&M), IIT Roorkee, for providing resources to conduct the research work.

References

Abdi AM (2020) Land cover and land use classification performance of machine learning algorithms in a boreal landscape using Sentinel-2 data. Gisci Rem Sens 57(1):1–20

Anand J, Gosain AK, Khosa R (2018) Prediction of land use changes based on land change modeler and attribution of changes in the water balance of Ganga basin to land use change using the SWAT model. Sci Total Environ 644:503–519

Arowolo AO, Deng X, Olatunji OA, Obayelu AE (2018) Assessing changes in the value of ecosystem services in response to land-use/land-cover dynamics in Nigeria. Sci Total Environ 636:597–609
Bahita TA, Swain S, Dayal D, Jha PK, Pandey A (2021) Water quality assessment of upper Ganga canal for human drinking. In: Climate impacts on water resources in India. Springer, Cham, pp 371–392
Barakat A, Khellouk R, Touhami F (2021) Detection of urban LULC changes and its effect on soil organic carbon stocks: a case study of Béni Mellal City (Morocco). J Sediment Environ 1–13
Berihun ML, Tsunekawa A, Haregeweyn N, Meshesha DT, Adgo E, Tsubo M, Yibeltal M et al (2019) Exploring land use/land cover changes, drivers and their implications in contrasting agro-ecological environments of Ethiopia. Land Use Policy 87:104052
Chamling M, Bera B (2020) Spatio-temporal patterns of land use/land cover change in the Bhutan–Bengal foothill region between 1987 and 2019: study towards geospatial applications and policy making. Earth Syst Environ 1–14
Chen Q, Chen H, Zhang J, Hou Y, Shen M, Chen J, Xu C (2020) Impacts of climate change and LULC change on runoff in the Jinsha River Basin. J Geog Sci 30(1):85–102
Clerici N, Valbuena Calderón CA, Posada JM (2017) Fusion of Sentinel-1A and Sentinel-2A data for land cover mapping: a case study in the lower Magdalena region Colombia. J Maps 13(2):718–726
Dayal D, Gupta PK, Pandey A (2021) Streamflow estimation using satellite-retrieved water fluxes and machine learning technique over monsoon-dominated catchments of India. Hydrol Sci J 66(4):656–671
Dayal D, Swain S, Gautam AK, Palmate SS, Pandey A, Mishra SK (2019) Development of ARIMA model for monthly rainfall forecasting over an Indian River Basin. In: World Environmental and Water Resources Congress 2019: watershed management, irrigation and drainage, and water resources planning and management. American Society of Civil Engineers, pp 264–271
Dutta D, Rahman A, Paul SK, Kundu A (2019) Changing pattern of urban landscape and its effect on land surface temperature in and around Delhi. Environ Monit Assess 191(9):1–15
Guptha GC, Swain S, Al-Ansari N, Taloor AK, Dayal D (2021) Evaluation of an urban drainage system and its resilience using remote sensing and GIS. Remote Sens Appl Soc Environ 23100601. https://doi.org/10.1016/j.rsase.2021.100601
Hengade N, Eldho TI (2016) Assessment of LULC and climate change on the hydrology of Ashti Catchment, India using VIC model. J Earth Syst Sci 125(8):1623–1634
Himanshu SK, Pandey A, Patil A (2018) Hydrologic evaluation of the TMPA-3B42V7 precipitation data set over an agricultural watershed using the SWAT model. J Hydrol Eng 23(4) 05018003. https://doi.org/10.1061/(ASCE)HE.1943-5584.0001629
Himanshu SK, Pandey A, Yadav B, Gupta A (2019) Evaluation of best management practices for sediment and nutrient loss control using SWAT model. Soil Tillage Res 19242–19258. https://doi.org/10.1016/j.still.2019.04.016
Jozdani SE, Johnson BA, Chen D (2019) Comparing deep neural networks, ensemble classifiers, and support vector machine algorithms for object-based urban land use/land cover classification. Rem Sens 11(14):1713
Kalura P, Pandey A, Chowdary VM, Raju PV (2021) Raju assessment of hydrological drought vulnerability using geospatial techniques in the tons river basin India. J Ind Soc Remote Sens https://doi.org/10.1007/s12524-021-01413-7
Karimi F, Sultana S, Babakan AS, Suthaharan S (2019) An enhanced support vector machine model for urban expansion prediction. Comput Environ Urban Syst 75:61–75
Khan A, Govil H, Kumar G, Dave R (2020) Synergistic use of Sentinel-1 and Sentinel-2 for improved LULC mapping with special reference to bad land class: a case study for Yamuna River floodplain, India. Spat Inf Res 1–13
Liping C, Yujun S, Saeed S (2018) Monitoring and predicting land use and land cover changes using remote sensing and GIS techniques—a case study of a hilly area, Jiangle, China. PloS One 13(7)
Maxwell AE, Warner TA, Fang F (2018) Implementation of machine-learning classification in remote sensing: an applied review. Int J Remote Sens 39(9):2784–2817

Mirici ME, Berberoglu S, Akin A, Satir O (2018) Land use/cover change modelling in a mediterranean rural landscape using multi-layer perceptron and markov chain (mlp-mc). Appl Ecol Environ Res 16:467–486

Mishra PK, Rai A, Rai SC (2020) Land use and land cover change detection using geospatial techniques in the Sikkim Himalaya, India. Egypt J Rem Sens Space Sci 23(2):133–143

Nguyen LH, Joshi DR, Clay DE, Henebry GM (2020) Characterizing land cover/land use from multiple years of Landsat and MODIS time series: a novel approach using land surface phenology modeling and random forest classifier. Rem Sens Environ 238:111017

Palmate SS, Pandey A, Mishra SK (2017a) Modelling spatiotemporal land dynamics for a trans-boundary river basin using integrated cellular automata and Markov Chain approach. Appl Geogr 82:11–23

Palmate SS, Pandey A, Kumar D, Pandey RP, Mishra SK (2017b) Climate change impact on forest cover and vegetation in Betwa Basin, India. Appl Water Sci 7(1):103–114

Pandey A, Bishal KC, Kalura P, Chowdary VM, Jha CS, Cerdà A (2021) A soil water assessment tool (SWAT) modeling approach to prioritize soil conservation management in river basin critical areas coupled with future climate scenario analysis. Air Soil Wat Res 14:1–17

Parida BR, Mandal SP (2020) Polarimetric decomposition methods for LULC mapping using ALOS L-band PolSAR data in Western parts of Mizoram, Northeast India. SN Appl Sci 2:1–15

Pereira LO, Freitas CC, Sant SJ, Reis MS (2018) Evaluation of optical and radar images integration methods for LULC classification in Amazon region. IEEE J Sel Top Appl Earth Observ Rem Sens 11(9):3062–3074

Prudente VHR, Sanches ID, Adami M, Skakun S, Oldoni LV, Xaud HAM, Zhang Y et al (2020) SAR data for land use land cover classification in a tropical region with frequent cloud cover. In: IGARSS 2020–2020 IEEE international geoscience and remote sensing symposium. IEEE, pp 4100–4103

Rimal B, Zhang L, Keshtkar H, Wang N, Lin Y (2017) Monitoring and modeling of spatiotemporal urban expansion and land-use/land-cover change using integrated Markov chain cellular automata model. ISPRS Int J Geo Inf 6(9):288

Rizvi SH, Fatima H, Iqbal MJ, Alam K (2020) The effect of urbanization on the intensification of SUHIs: analysis by LULC on Karachi. J Atmos Sol-Terr Phys 207:105374.

Rwanga SS, Ndambuki JM (2017) Accuracy assessment of land use/land cover classification using remote sensing and GIS. Int J Geosci 8(04):611

Saputra MH, Lee HS (2019) Prediction of land use and land cover changes for north Sumatra, Indonesia, using an artificial-neural-network-based cellular automaton. Sustainability 11(11):3024

Shivakumar BR, Rajashekararadhya SV (2018) Investigation on land cover mapping capability of maximum likelihood classifier: a case study on North Canara, India. Procedia Comput Sci 143:579–586

Singh SK, Laari PB, Mustak SK, Srivastava PK, Szabó S (2018) Modelling of land use land cover change using earth observation data-sets of Tons River Basin, Madhya Pradesh, India. Geocarto Int 33(11):1202–1222

Singh S, Bhardwaj A, Verma VK (2020) Remote sensing and GIS based analysis of temporal land use/land cover and water quality changes in Harike wetland ecosystem, Punjab, India. J Environ Manag 262:110355

Singh G, Pandey A (2021) Evaluation of classification algorithms for land use land cover mapping in the snow-fed Alaknanda River Basin of the Northwest Himalayan Region. Appl Geomatics https://doi.org/10.1007/s12518-021-00401-3

Swain SS, Mishra A, Sahoo B, Chatterjee C (2020a) Water scarcity-risk assessment in data-scarce river basins under decadal climate change using a hydrological modelling approach. J Hydrol 590:125260

Swain S, Mishra SK, Pandey A (2020b) Assessment of meteorological droughts over Hoshangabad district, India. In: IOP conference series: earth and environmental science, vol 491, p 012012. IOP Publishing

Swain S, Mishra SK, Pandey A (2021) A detailed assessment of meteorological drought characteristics using simplified rainfall index over Narmada River Basin, India. Environ Earth Sci 80:221

Swain S, Mishra SK, Pandey A, Dayal D (2021b) Identification of meteorological extreme years over central division of odisha using an index-based approach. In: Hydrological extremes. Springer, Cham, pp 161–174

Swain S, Patel P, Nandi S (2017) Application of SPI, EDI and PNPI using MSWEP precipitation data over Marathwada, India. In: 2017 IEEE international geoscience and remote sensing symposium (IGARSS). IEEE, pp 5505–5507

Swain S, Verma MK, Verma MK (2018) Streamflow estimation using SWAT model over Seonath river basin, Chhattisgarh, India. In: Hydrologic modeling. Springer, Singapore, pp 659–665

Talukdar S, Singha P, Mahato S, Pal S, Liou YA, Rahman A (2020) Land-use land-cover classification by machine learning classifiers for satellite observations—a review. Rem Sens 12(7):1135

Thakur PK, Garg V, Kalura P, Agrawal B, Sharma V, Mohapatra M, Chauhan P et al (2021) Water level status of Indian reservoirs: a synoptic view from altimeter observations. Adv Space Res 68(2):619–640

Thanh Noi P, Kappas M (2017) Comparison of random forest, k-nearest neighbor, and support vector machine classifiers for land cover classification using Sentinel-2 imagery. Sensors 18(1)

Tran DX, Pla F, Latorre-Carmona P, Myint SW, Caetano M, Kieu HV (2017) Characterizing the relationship between land use land cover change and land surface temperature. ISPRS J Photogramm Remote Sens 124:119–132

Tripathi G, Pandey AC, Parida BR, Kumar A (2020) Flood inundation mapping and impact assessment using multi-temporal optical and SAR satellite data: a case study of 2017 Flood in Darbhanga district, Bihar, India. Water Resour Manage 34(6):1871–1892

Yang C, Wu G, Ding K, Shi T, Li Q, Wang J (2017) Improving land use/land cover classification by integrating pixel unmixing and decision tree methods. Rem Sens 9(12):1222

Chapter 5
Modeling Landscape Level Forest Disturbance-Conservation Implications

Mukunda Dev Behera

Abstract Increasingly forest land is diverted to different land uses leading to various levels of disturbances in a landscape. Any disturbance could affect the structure and functions of a landscape, their inherent properties and interactions, thereby could lead to temporal or irreversible changes. In this chapter, the spatial distribution of various forests and non-forest patches were combined with road and settlement proximity zones using remote sensing and GIS tools to generate disturbance index (DI) of a landscape, by adopting landscape ecological principles. Various landscape ecological matrices such as forest fragmentation, interspersion, juxtaposition patchiness, and porosity, were analyzed using spatial analysis. Field sampling data on species richness from 862 plots (nested quadrats of 20×20 m^2) were analyzed to adjudge the correlation between different *DI* levels and their diversity content, and interestingly, higher species richness was observed for lower *DI* levels. The study was selected is the northeastern India of the eastern Himalaya accommodating Arunachal Pradesh, Assam, and Meghalaya states. *DI* demonstrated progressive disturbance in the forest structure and composition. Meghalaya state has better reflected the decreasing pattern of *DI* with species richness and its endemic subset. Disturbance Index, a landscape-based model proved to have well captured the patterns and processes, thereby advocating wider application and replication for conservation planning.

5.1 Introduction

Disturbance is responsible for triggering forest fragmentation, edge-interactions, and species migration in a landscape. The landscape level disturbance can be perceived with respect to natural and anthropogenic disturbances. Understanding the spatial configuration and the forces prevailing over a landscape helps study the possible

M. D. Behera (✉)
Spatial Analysis and Modeling Lab., Centre for Oceans, Rivers, Atmosphere and Land Sciences (CORAL), Indian Institute of Technology Kharagpur, Kharagpur, India
e-mail: mdbehera@coral.iitkgp.ac.in

A. Pandey et al. (eds.), *Geospatial Technologies for Land and Water Resources Management*, Water Science and Technology Library 103,
https://doi.org/10.1007/978-3-030-90479-1_5

impacts (Moloney and Levin 1996). The human being has been altering the frequency and extent of various disturbances at a faster face and greater extent (Barik and Behera 2020). In absence of strict land use and land cover policy enforcement, the settlements and transportation networks such as roads and rails add to the existing sources of biotic disturbances. The disturbances can be visualized across multiple scales and geographical space and can be studied using patchiness or interspersion matrices (Murthy et al. 2016).

Disturbances could change the homogeneity or heterogeneity of elementary patches in a landscape, thereby defining renewed or new functionalities. Any disturbance may alter the homo- or heterogeneous structural organization of patches. Individual-level disturbances may lead to homogeneous conditions at higher scale, while at lower scale multiple disturbances lead to patch heterogeneity. Any alternation in the spatial configuration of a landscape, therefore could be a reflection of the influencing disturbances itself (Das et al. 2017). Further, continuous or periodic disturbances could lead to propagation of the impacts or changes across the landscapes (Behera et al. 2018), thereby posing cascading challenges for management (Das et al. 2021). Therefore, the landscape disturbances must be studied in terms of their nature and sources including periodicity of occurrence and scale, for appropriate management and restoration.

The response of biological diversity to landscape disturbance could vary from individual species to community across a changing disturbance regime (Chitale et al. 2020). The response of species assemblages to any disturbance may demonstrate varied time-lag and reorganization periods. This may lead to alternation of established and matured ecosystems or creation of new ecosystems with renewed species assemblages with tendency of accommodating more invasive and faster growing species. Therefore, the disturbances alter the species assemblage pattern, either by loss or establishment of new species owing to the suitability of the changing environment. Alternations in disturbance regimes and their impact on biotic interactions could modify or redefine the structure and function of biodiversity in a landscape (Behera et al. 2018). However, it may be hard to estimate any quantitative or qualitative change in lieu of complexity and cascading effects of the interactions. Changes in the types, intensity, and frequency of disturbances could affect the composition of biodiversity and primary production within an ecosystem (Mahanand et al. 2021). The impact of climate change with respect to increased extreme weather events also has greater impact on landscape.

Landscape is often regarded as the best spatial scale to assess the responses as it can reflect ecosystem level alternations where the patches are the individual units. A landscape accommodates different homogeneous patches that can be studied to understand their spatial arrangement, shape and size, and neighborhood effect using various landscape matrices (Turner et al. 1989). The arrangement of patches over space such as the number of a particular patch in an area, the shape, and size of the patches, the configuration of patches, can be considered to understand different communities and species level interactions in a natural landscape (Lidicker 1995). The landscape matrices adopt a geographical window approach that could be overlapping or non-overlapping in nature (Dillworth et al. 1994). Fragmentation risks patch level disturbance having implications such as edge effects and alternations in energy balance in a landscape (Nilsson and Grelsson 1995). Therefore, disturbance induces alternations in spatial arrangement of patches and their functional characteristics including biodiversity and primary productivity.

Remote sensing images can be interpreted to define the spatial arrangement of different patches, and are therefore potentially used to analyze disturbances across time and scale. The patch or class wise information can be transferred and linked to characterize biodiversity structure and function using statistical and modeling approaches. The Geographical Information System (GIS) helps in integration of the satellite derived spatial maps such as fragmentation, patchiness, porosity, interspersion, and juxtaposition for estimating landscape disturbance, thereby useful in management decision-making. The Global Positioning System (GPS) provides information on the geographic locations of different landscape elements, thereby helpful in locating the sample plots for mapping, classification accuracy assessment, spatio-statistical analysis, and model validation (Behera et al. 2000).

Indian national level forest cover assessment is done biannually using visual and digital interpretation techniques by Forest Survey of India (FSI; SFR 2019). In a maiden attempt, Roy et al. (2015) that have prepared a nation-wide vegetation type map of India using satellite data, with 100 classes such as mixed, gregarious, locale specific, degradational and scrub formations along with plantations. Vegetation type map serves as an intermediate input for calculation of disturbance index (*DI*). Here, the vegetation type map was generated using satellite data that was further utilized for generating a *DI* image along with other information using a customized package, Bio_CAP (Behera et al. 2005).

5.2 Study Site

The Arunachal Pradesh, Assam, and Meghalaya states are located in the eastern Himalaya, and categorized under the Himalaya-East-Himalaya biogeographic zone (Rodgers and Panwar 1988), were selected as study sites (Fig. 5.2a; Table 5.1). The three states accommodate diverse tropical ecosystems comprising of mixed wet and dry evergreen, and deciduous broadleaved forests. The study site experiences very high humidity thereby supporting diverse forests with several primitive species. The three states vary as per climate and topographic conditions as well as human activities.

The state of Arunachal Pradesh accommodates a wide range of ecosystems such as grasslands, dense evergreen broadleaved and needle leaf forests, alpine scrubs, open forests adjoining to settlements, and road networks, and disturbed secondary formations (Roy and Behera 2005). The state of Meghalaya conglomerates several undulating hills in east-west orientation, thereby accommodating mosaic of valleys and hillocks. The flora of Meghalaya has been reported as the richest in India with a dominant number of angiospermic plant species. Recently large-scale deforestation have been reported from the forests of Assam and Arunachal Pradesh (Pasha et al. 2020). The driving forces for forest loss and conversion in the region are deforestation

Table 5.1 Characteristics of the study site, three states in northeastern India

Parameters	Arunachal Pradesh	Assam	Meghalaya
Total geographic area (km^2)	83,743	78,438	22,429
Forest area (km^2) (%)	68,019 (81.22)	27,827 (35.48)	16,839 (75.08)
Forest types; Champion and Seth (1968)	Tropical, sub-tropical temperate, sub-alpine alpine scrub	Tropical	Tropical, sub tropical
Altitude	50–7000 m	Up to 1000 m	Up to 1600 m
Physiognomy	Mountainous, complex terrain	Mainly plain land, partly mountainous	Mountainous, complex terrain
Dominant land use/cover	Forest	Agriculture	Forest
Single factor hitched to forest conservation	Complex terrain	Protected forests	Sacred groves[a]
Driving force for forest loss	*Jhumming* and deforestation	Deforestation	*Jhumming*
Hotspot category	Himalaya	Partly in Himalaya and Indo-Burma	Indo-Burma

[a] The scared groves of Meghalaya are a very rich storehouse of vegetation wealth. They form isolated patches of vegetation that remain untouched due to the religious beliefs and myths of the tribal people. Many endangered species have been reported from the scared groves

and shifting cultivation activities (Table 1). In past few decades, the forest loss due to slash-and-burn agricultural practices (also called jhumming or shifting cultivation) has slowed down (Das et al. 2020). The jhumming is a process of slash-and-burn agriculture, wherein the farmers clear a patch of forest for cultivation and leave it fallow for certain number of years. Deforestation has a strong relationship with the proximity to roads and habitations. The rise in population growth, the shifting cultivation activities, and industrialization in the form of plywood factories are some of the major proximate factors responsible for deforestation and forest degradation in northeastern India (Khan et al. 1997).

5.3 Methodology

Forest fragmentation can be defined as the number of forest patches to the non-forest patches in a landscape (Monmonier 1974), whereas patchiness is estimated as the density and number of patches of all types occurring in a landscape (Romme 1982). The higher fragmentation and patchiness are indications of landscape heterogeneity, thus associated with disturbance. The measure of interspersion and juxtaposition in a landscape indicate dissimilar and proximal neighborhoods of various patches respectively (Lyon 1983). Therefore, higher interspersion indicates greater disturbance while lower juxtaposition indicates higher disturbance in a landscape. The porosity can be calculated to realize the disturbance within a native or original vegetation type in a defined region (Forman and Godron 1986). For the northeastern region, the porosity was judged for all the primary forests. The species richness information from 862 field plots provided inputs for juxtaposition calculation and correlating between different *DI* levels and their biodiversity content (Fig. 5.1).

The landscape level disturbance Index (*DI*) was assessed as cumulative function of spatial and social attributes using a customized package Bio_CAP (Fig. 5.1). IRS LISS-III satellite data were used for classification of forest vegetation types using a pre-defined classification scheme (Roy et al. 2012). The simplified forest fragmentation map was utilized to generate a *DI* image along with a few other landscape matrices such as patchiness, porosity, juxtaposition and interspersion, and settlement and road buffers. Stratified random sampling following probability proportionate to the size (PPS) was practiced for field sampling and linking in-situ forest composition. A total of 862 field plots were laid in Arunachal Pradesh (405), Assam (324), and Meghalaya (133) respectively, those accounted for 0.002–0.005% of the natural vegetation area. The measurements for all tree and liana species (20′20 m), shrubs (5′5 m), and herbs (1′1 m) were taken within each sample plot (Roy et al. 2012). The location and extent of the natural vegetation types and their landscape matrices were integrated with the social attributes such as road and settlement buffer, by assigning proportionate weights (Behera et al. 2005) to derive *DI* (Eq. 5.1; Fig. 5.3). The *DI* was scaled at 10-levels and an area estimate was done (Fig. 5.4).

$$\text{Disturbance Index}(DI) = f\int \{\text{disturbance buffer}(B), \text{interspersion}(I), \text{juxtaposition}(J), \text{fragmentation}(F), \text{patchiness}(P), \text{and porosity}(PO)\} \quad (5.1)$$

$$DI = \{\text{Wti}(F) + \text{Wti}(P) + \text{Wti}(I) + \text{Wti}(PO) + \text{Wti}(J) + \text{Wti}(B)\}$$

where, $\text{Wti}(t = 0.1 - 1.0)$ indicates proportionate weight.

For validation of *DI*, the field plots were segregated into 10-categories using their field gathered GPS coordinates (Fig. 5.1c). Each group of sample plots pertaining to different levels of *DI* were enumerated further to study the distributional pattern of the life forms such as tree, shrub, herb, and liana species, along with the endemic species and mean tree BA. The regression fitness curves were plotted to study the agreement between modeled output (i.e., *DI*) and field conditions (species diversity; Fig. 5.5).

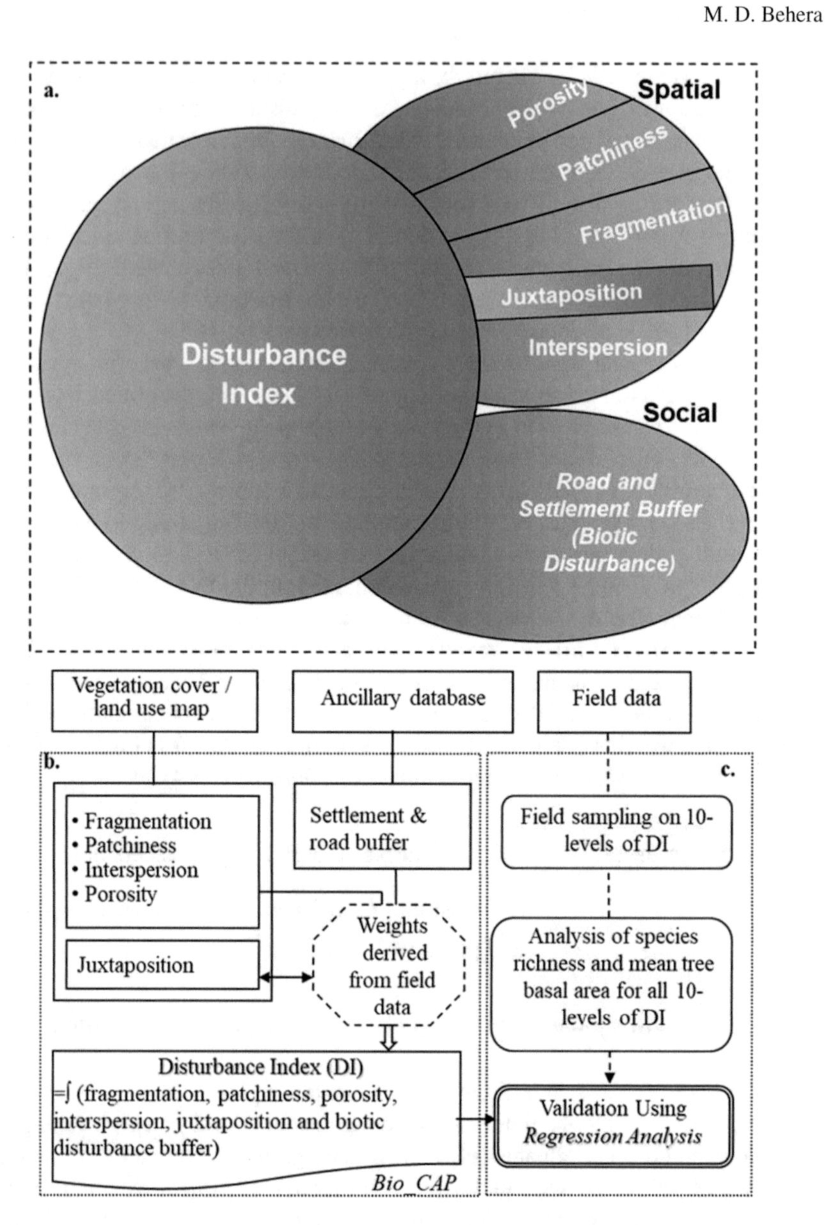

Fig. 5.1 **a** Disturbance index (*DI*) is estimated as a cumulative property of social and spatial attributes in a landscape; paradigm for **b** Assessment of forest disturbance regimes using remote sensing, GIS, and GPS tools (Modified from Behera et al. 2005) and **c** Validation of *DI* using field sampling plots

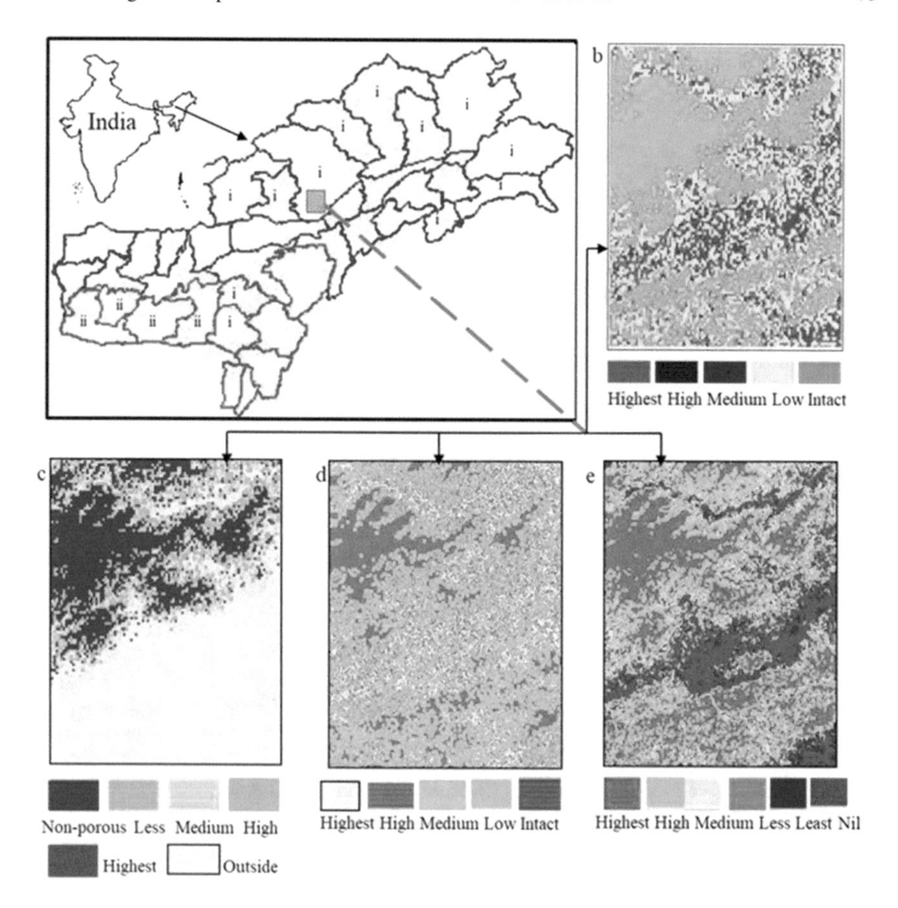

Fig. 5.2 **a** location map of (i) Arunachal Pradesh, (ii) Meghalaya and Assam, falling two hotspots; Arunachal Pradesh and upper part of Assam belong to Himalaya hotspot [shown in blue color]; Meghalaya and southern part of Assam belong to Indo-Burma hotspot. Depiction of **b** Fragmentation image, **c** Porosity image of sub-tropical evergreen forest, **d** Interspersion and **e** Juxtaposition image for part of the study area

5.3.1 Biodiversity Characterization Package (Bio_CAP)

Bio_CAP is a semi-expert package, developed for uniformal use in the 'Biodiversity Characterization at landscape level' project (Roy et al. 2012). The package takes a vegetation type map as the primary input and calculates other landscape parameters as grid data. Bio_CAP was customized using pre-defined codes for different classes of vegetation type map such as for fragmentation estimate, all the class codes from 1 to 150 are automatically considered as forest, and from 151 to 255 picked as non-forest classes. This package was used for estimating the *DI* that provided 10-levels within forest vegetation patches (Behera et al. 2005). The non-forest areas and unclassified

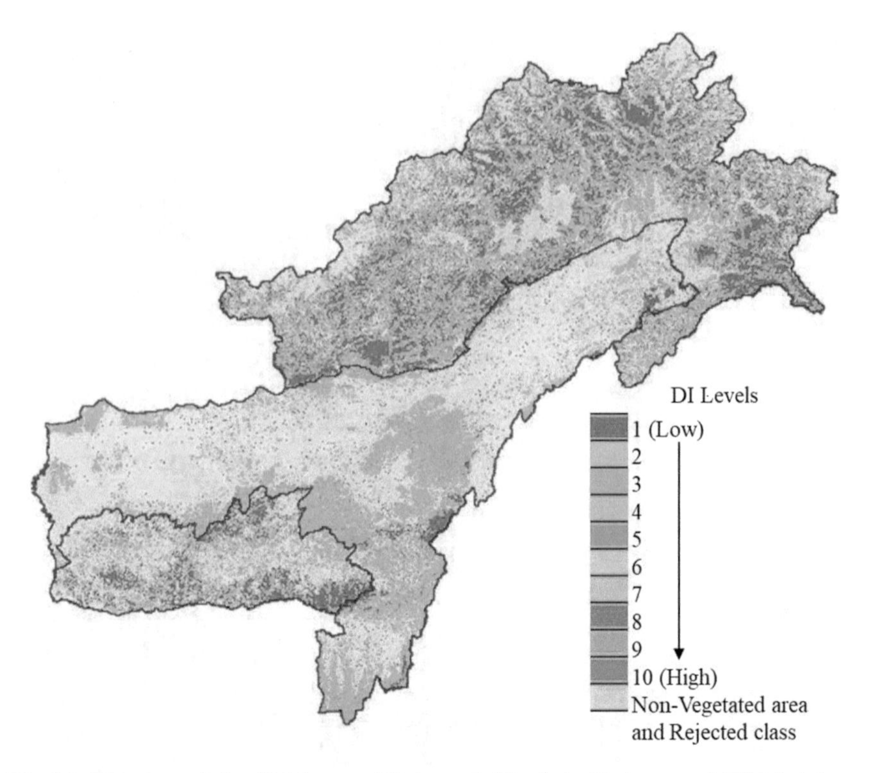

Fig. 5.3 Disturbance Index (*DI*) image of (i) Arunachal Pradesh, (ii) Assam and (iii) Meghalaya; Various *DI*-levels range from 1 to 10

areas such as cloud, shadow, etc., were not considered for landscape analysis and *DI* estimate (Fig. 5.3).

5.4 Results

5.4.1 Landscape Analysis

The forest type map generated from satellite imagery using a hybrid classification approach offered the requisite input spatial layer for analysis of landscape indices (Roy and Behera 2005). The tropical semi-evergreen forests are distributed in the regions of Arunachal Pradesh and Assam state border areas. However, the forests are exposed to large-scale exploitation and deforestation owing to their easy accessibility with respect to adjacency to the road network and simpler topography. The sub-tropical evergreen forests are found distributed in the entire study site and are the most affected by jhumming, due to their proximity to human settlements. The degraded

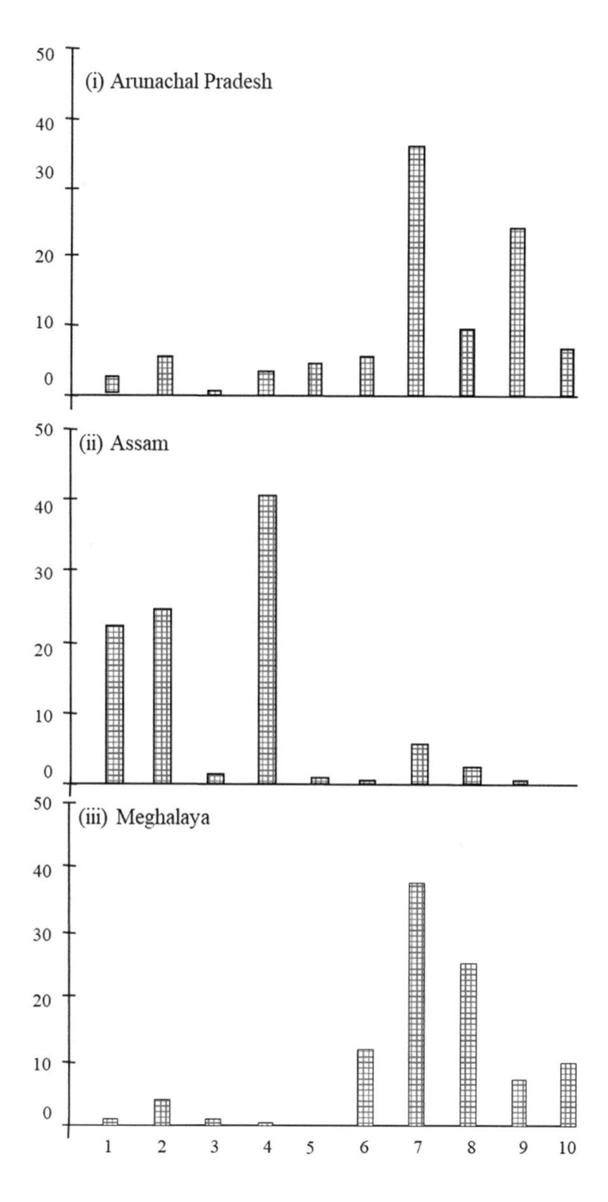

Fig. 5.4 Percentage area (Y-axis) distribution in (i) Arunachal Pradesh (ii) Assam and (iii) Meghalaya; X-axis represents different *DI* levels. Note that severity of disturbance varies from level 1 to 10

forests are originated through various activities such as jhumming, landslide, fire, etc. The secondary forests or the abandoned jhum lands are found dominated by varieties of scrubs, herbs, grasses, bamboos, weeds, etc., which receive higher direct sunlight and do faster primary production. The temperate and sub-tropical evergreen forests accommodate thorny bamboos forming an impenetrable canopy.

The higher level of forest fragmentation was observed adjacent to settlement and road networks owing to resource use by the local people (Fig. 5.1b). The patchiness image demonstrated higher homogeneity for the state of Arunachal Pradesh

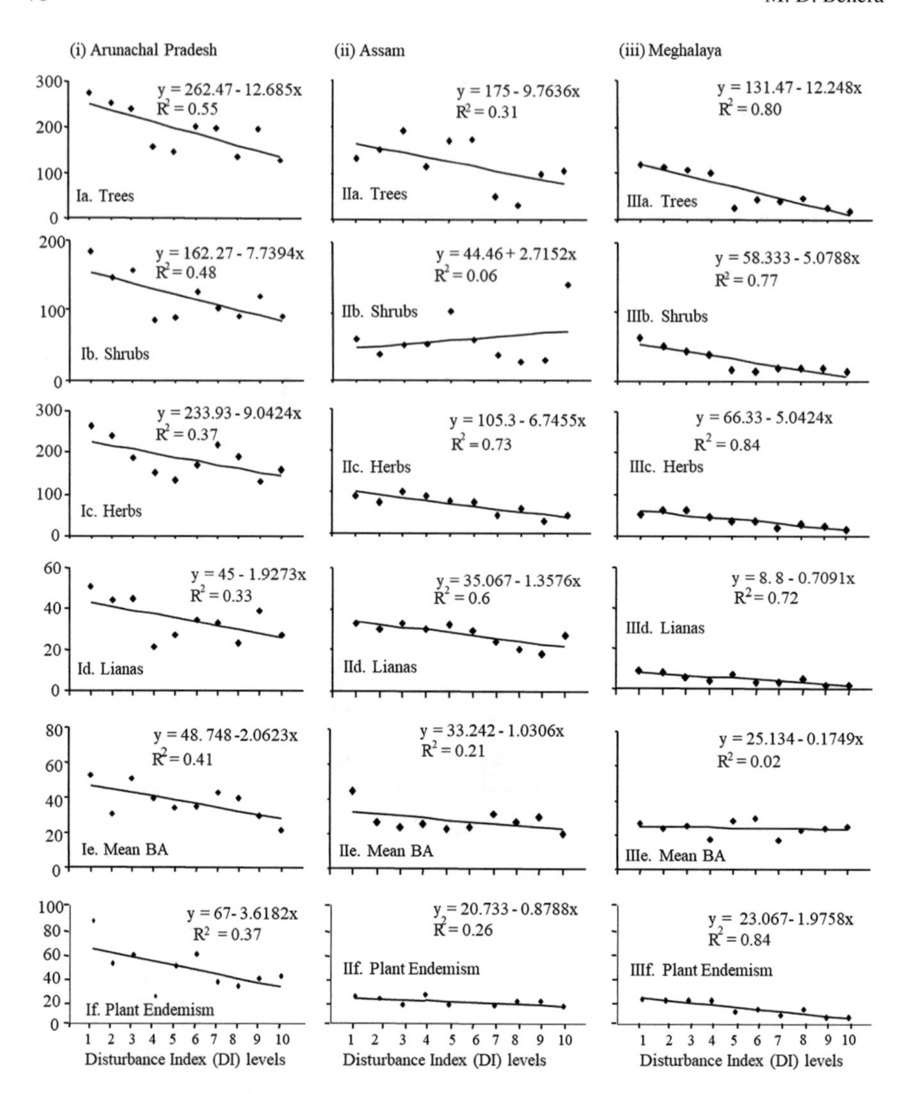

Fig. 5.5 Distribution of various life forms such as **a** tree, **b** shrub, **c** herb, **d** liana species, along with **e** mean BA and **f** plant endemism across various disturbance indexes (*DI*) levels for (I) Arunachal Pradesh, (II) Assam, and (III) Meghalaya; Y-axis represents the number of species. *DI*-level is plotted from 1 to 10 along *X*-axis. All the regression curves have shown decreasing trend along increased levels of *DI*, except the shrubs in Assam state

and higher heterogeneity for the state of Assam, whereas Meghalaya showed an intermediate level of patchiness. The medium level of patchiness is an indication of landscape disturbance. Porosity was calculated for both the tropical and sub-tropical forests that are semi-evergreen and evergreen in nature respectively (Fig. 5.1c). Both the forests were found to be relatively porous with very less patches in an intact state. Since any sort of disturbance start affecting the edges, intermediate level of porosity was found across the edge transitions, which could be attributed to degradation and deforestation activities. 80% area was found under different degrees of interspersion in the landscape, pointing to low patch diversity (Fig. 5.1d). This indicated that the dispersal ability of the central class is low or reduced. The juxtaposition image using the adjacency criteria of central pixel with the neighboring pixels was regrouped under five categories (Behera et al. 2005). The juxtaposition image demonstrated greater adjacency or neighborhood effects between the tropical semi-evergreen and sub-tropical evergreen forests owing to their nature of occurrence (Fig. 5.1e). Therefore, juxtaposition is found as a critical indicator revealing the patch interactions in a landscape.

5.4.2 *Disturbance Index (*DI*) Model*

The *DI* image reflects various levels of disturbance prevailing in the states (Fig. 5.3). Overall, the *DI* image demonstrates that the forests adjoining to human settlements and road networks are highly disturbed in the northeaster landscape (Fig. 5.3). The non-forest classes, snow, and cloud were not considered for estimating disturbance as they are relatively inert from anthropogenic point of view. The highest level of disturbance was found adjoining the agricultural area and natural vegetation. The forests with low level of interspersion and higher juxtaposition demonstrated the lowest level of disturbance on the *DI* image owing to their remoteness from habitations.

Many of the reserve forests are found having very low *DI* value that owes to the protection effort by the state forest departments. In Arunachal Pradesh, maximum area was found under low levels *DI* since most of the forest fragmentation, patchiness, porosity, and interspersion values were low too. The *DI* image of Assam shows highly disturbed patches in North Cachar district (Fig. 5.3). The evergreen and semi-evergreen forest belts of Karbi Anglong and North Cachar hill districts are found the worst affected. 20.87% area was disturbance-free and the man-made landscapes comprising of 52.11% area was not considered for *DI* analysis in Assam state. The disturbance in Meghalaya state was the highest in Garo hills followed by Khasi and Jaintia hills. It was observed that the level of disturbance was high in the areas accessible to human beings. The inaccessible areas with complex topography demonstrated lower *DI* values for the state of Meghalaya.

The percentage area for different levels of *DI* was plotted in Fig. 5.4. 36.6% forest area of the state of Arunachal Pradesh demonstrated medium (level-6) *DI* (Fig. 5.4a). In Assam, 40.73% forest area demonstrated intermediate (level-4) level of disturbance (Fig. 5.5b). Similarly, the state of Meghalaya accommodates 38.06% forest area under medium level of *DI* (level-6). These figures clearly indicate that more of the forest areas in Arunachal Pradesh and Meghalaya states suffer from low degree of disturbance (Fig. 5.4). However, for Assam, more of the forest areas experience high degree of disturbance.

5.4.3 *Validation of* DI *Model*

All the 405, 324, and 133 plots were splitted into 10-levels corresponding to their locations on *DI* image for Arunachal Pradesh, Assam, and Meghalaya respectively (Table 5.2). The species richness demonstrated S-shaped pattern across the *DI* levels and R^2 value decreases gradually from trees to lianas (i.e., 0.55, 0.48, 0.37, and 0.33, whereas the mean tree BA resulted average R^2 value of 0.41 (Fig. 5.5). In Assam state, the trees, herbs, lianas, and mean tree BA showed an overall decrease in species richness; whereas the shrubs demonstrated a reverse trend across *DI* levels ($R^2 =$ 0.06; Fig. 5.5). The deviation in case of shrubs could be explained by new formations of weed species at intermediate levels of disturbances. Herbs demonstrated a very good fit with $R^2 = 0.73$, followed by lianas with ($R^2 = 0.6$; Fig. 5.5).

The state of Meghalaya has shown a significant decrease in species richness for all life forms with high degree of correlation (i.e., trees, $R^2 = 0.80$; shrubs, $R^2 = 0.77$; herbs, $R^2 = 0.84$ and lianas, $R^2 = 0.73$). Plant endemism demonstrated a decreasing pattern along *DI*, with the highest regression fitness for the state of Meghalaya, followed by Arunachal Pradesh and Assam states (Fig. 5.5f). The mean tree BA demonstrated an *S*-shaped curve with maxima at intermediate level of *DI*.

5.5 Discussion and Conclusions

Satellite images provided forest vegetation cover map for the states with reasonably good accuracy for generating the disturbance index (*DI*). The distribution of various forest types across elevation was first attempted by Kaul and Haridasan (1987) in the state of Arunachal Pradesh. Of the 16 major forest types classified by Champion and Seth (1968), 13 types and 54-ecologically stable formations were observed in northeastern India (Roy et al. 2012). Landscape level patch analysis very well revealed the ecological patterns and processes and thereby is important in determining diversity and distribution of species assemblages. The increase in the level of forest fragmentation leads to reduction of patch size and the edge effect. The forest fragmentation, patchiness, and porosity images uniformly demonstrated lower values for dense and intact forests that are mostly located on difficult topography, remotely from the settlements and road networks. 41% area reported under intact or low levels of interspersion, well implies medium level of interaction among and between the patches. Juxtaposition, which is based on the relative adjacency of patches, demonstrates low *DI* values for intact patches in the landscape. The *DI* image reflects high disturbance level for the areas having more biotic access and vice versa, indicating that the *DI* has efficiently incorporated the spatial phenomena. The juxtaposition image has relevance in wildlife studies as some fauna just needs one habitat type in contrast to many by other groups. Therefore, patches with low forest fragmentation and high juxtaposition may be given prior attention for management.

Table 5.2 Distribution of life forms along various levels of *DI* (increases from 1 to 10) for Arunachal Pradesh, Assam, and Meghalaya

Levels	Trees	Shrubs	Herbs	Lianas	Endemism
A. Arunachal Pradesh					
1	275	185	265	51	87
2	254	147	240	44	51
3	240	157	188	45	58
4	157	86	152	21	23
5	146	90	133	27	48
6	201	127	168	34	59
7	197	103	219	33	35
8	134	91	189	23	32
9	196	120	130	39	38
10	127	91	158	27	40
B. Assam					
1	133	60	91	33	22
2	151	39	76	30	20
3	191	50	101	33	14
4	114	53	88	30	24
5	169	100	79	32	14
6	172	58	74	29	7
7	49	37	43	24	13
8	30	28	60	20	17
9	98	30	28	18	17
10	106	139	43	27	11
C. Meghalaya					
1	120	64	53	9	20
2	114	51	63	8	19
3	107	43	63	6	19
4	102	39	44	4	19
5	26	18	35	7	9
6	44	15	36	3	11
7	39	20	22	3	6
8	47	19	29	5	11
9	25	20	24	2	4
10	17	15	17	2	4

Disturbance regimes could vary across a given landscape as a function of biotic interactions, proximity to settlement and transport networks, and complex topography (Turner 1989). The landscape level disturbance offers fundamental inputs to the understanding of changes in various ecological structures and functions (Pickett and White 1985). It was found that the pattern of disturbance in the landscape has influenced the species diversity of the region (Pickett and Thompson 1978). These disturbance activities such as jhumming and deforestation have changed the disturbance regimes and altered the ecosystem processes through habitat loss and forest fragmentation in northeastern India.

The deforestation level was found higher in low elevation regions of Arunachal Pradesh that have a direct relationship with settlement and transport networks (Khan et al. 1997). Shukla and Rao (1993) have attributed to jhum cultivation and industrial activities as the dominant proximate factors for landscape disturbance in Arunachal Pradesh. Assam is located in the Brahmaputra valley that is dominated with permanent cultivation, while the border regions with Arunachal Pradesh has lot of tea gardens owing to sloppy terrain. The shifting cultivation practices primarily attribute to biodiversity decline in Meghalaya (Roy and Tomar 2000). In spatial context, any pixel in *DI* model is the resultant function of fragmentation, patchiness, porosity, interspersion, juxtaposition, and road and settlement buffer (Roy and Tomar 2000). Few studies were done in India that establish relationships between the landscape level disturbance and the biological richness (Pandey and Shukla 1999), which could be the next step forward.

In general, the regression fitness curves demonstrated clear decreasing trend among the field-derived species richness, mean tree BA, and endemism with *DI* (Fig. 5.5). The tree species richness pattern across *DI* levels decreased prominently from the state of Meghalaya with greater degree of regression fitness ($R^2 = 0.80$), followed by Arunachal Pradesh ($R^2 = 0.55$) and Assam ($R^2 = 0.31$). Shrub richness pattern along *DI*-levels also showed greater agreement in Meghalaya ($R^2 = 0.77$), followed by Arunachal Pradesh ($R^2 = 0.48$) and a reverse trend was seen for the state of Assam ($R^2 = 0.06$). Greater degree of agreement was observed in Meghalaya for herbs and lianas, followed by Assam and Arunachal Pradesh (Fig. 5.5). The endemic subsets of plant species demonstrated the highest fit ($R^2 = 0.84$) for Meghalaya followed by Arunachal Pradesh and Assam (Fig. 5.5). In contrast, Arunachal Pradesh has demonstrated good agreement of mean tree BA with high R^2 value (0.42), followed by Assam ($R^2 = 0.21$) and Meghalaya ($R^2 = 0.02$; Fig. 5.5). Each field sample plot represents different spatial and ecological properties that are reflections of fine scale local variations. The regression fitness curves well demonstrate the relationship between various landscape structural indices and their species diversity content. Such regression fitness gives the first-hand assurance that intact patches with lower *DI* values hold higher species richness in a landscape.

The states of Arunachal Pradesh and Meghalaya accommodate higher of species endemism, while the state of Assam has more agricultural lands (Table 5.1). Almost all of the forests are found in difficult terrain, and therefore Assam state has less forest land than Arunachal Pradesh and Meghalaya states.

Therefore, utilization of field data forms an absolute basis to validate *DI* model. Here, the statistical tests depend very much on the types of data used; and therefore are simple and preferred for first-hand correlation studies. The models of ecological and biological systems facilitate understanding the possible consequences of human action, that might not be possible otherwise. Similarly, the Ganga River basin that accommodates 40% of India's population is experiencing expansions at the periphery of agricultural and forest lands due to a significant increase in built-up area (Patidar and Keshari 2020).

The rate of plant endemism declined from tropics to alpine region in northeastern India owing to greater variation in microclimates with complex terrain and remoteness from biotic interferences (Heywood and Watson 1995). These findings partially agree with Myers's (1988) reporting that tropics could harbor higher number of endemic species than non-tropics. Behera and Kushwaha (2007) reported a decrease in species richness and an increase in species endemism across the elevation gradients in Arunachal Pradesh. Further, Behera et al. (2002) reporter higher endemism in tropics than other regions that may be explained with higher plant functional activities. This study clearly indicated that species endemism decreases along the disturbance regimes in the eastern Himalaya. The potential of satellite remote sensing along with the kindred technologies such as GIS and GPS, for landscape level patch characterization, is well demonstrated in the study.

Acknowledgments The Author thanks authorities of Spatial Analysis and Modelling (SAM) Laboratory, Centre for Oceans, Rivers, Atmosphere and Land Sciences (CORAL) at Indian Institute of Technology Kharagpur for providing facilities for preparation of the chapter.

References

Barik SK, Behera MD (2020) Studies on ecosystem function and dynamics in Indian sub-continent and emerging applications of Satellite remote sensing technique. Tropical Ecol 61(1):1–4

Behera MD, Kushwaha SPS (2007) An analysis of altitudinal behavior of tree species in Subansiri district. Eastern Himalaya, Biodivers Conserv 16:1851–1865

Behera MD, Jeganathan C, Srivastava S, Kushwaha SPS, Roy PS (2000) Utility of GPS in classification accuracy assessment. Current Sci 79(12):1996–1700

Behera MD, Kushwaha SPS, Roy PS, Srivastava S, Singh TP, Dubey RC (2002) Comparing structure and composition of coniferous forests in Subansiri district, Arunachal Pradesh. Current Sci 82(1):70–76

Behera MD, Kushwaha SPS, Roy PS (2005) Geo-spatial modeling for rapid biodiversity assessment in Eastern Himalayan region. For Ecol Manag 207:363–384

Behera MD, Tripathi P, Das P, Srivastava SK, Roy PS, Joshi C, Behera PR, Deka J, Kumar P, Khan ML, Tripathi OP, Dash T, Krishnamurthy YVN (2018) Remote sensing based deforestation analysis in Mahanadi and Brahmaputra river basin in India since 1985. J Environ Manage 206:1192–1203. https://doi.org/10.1016/j.jenvman.2017.10.015

Champion HG, Seth SK (1968) A revised survey of forest types of India, Manager of Publications, Government of India, New Delhi

Chitale VS, Behera MD, Roy PS (2020) Congruence of endemism among four global biodiversity hotspots in India. Curr Sci 118(1):9

Das P, Behera MD, Pal S, Chowdary VM, Behera PR, Singh TP (2020) Studying land use dynamics using decadal satellite images and Dyna-CLUE model in the Mahanadi river basin, India. Environ Monit Assess 191(3):804. https://doi.org/10.1007/s10661-019-7698-3

Das P, Mudi S, Behera MD*, Barik SK, Mishra DR, Roy PS (2021) Automated mapping for long-term analysis of shifting cultivation in Northeast India. Remote Sens 13(6):1066

Das P, Behera MD, Patidar N, Sahoo B, Tripathi P, Behera PR, Srivastava SK, Roy PS, Thakur P, Agrawal SP (Oct 2017) "Changes in evapotranspiration, runoff and baseflow with LULC change in eastern Indian river basins during 1985–2005 using variable infiltration capacity approach". 38th Asian conference on remote sensing—space applications: touching human lives, ACRS 2017

Dillworth ME, Whistler JL, Merchant JW (1994) Measuring landscape structure using geographic and geometric windows. Photogram Eng Remote Sens 60:1215–1224

Forman R, Godron M (1986) Landscape Ecology. John Wiley & Sons, New York

Heywood VH, Watson RT (1995) Global biodiversity assessment. Cambridge University Press, New York

Kaul RN, Haridasan K (1987) Forest types of Arunachal Pradesh – A preliminary study. J Econ Taxonomic Bot 9:379–389

Khan ML, Menon S, Bawa KS (1997) Effectiveness of protective area network in biodiversity conservation: a vase study of Meghalaya state. Biodiver Conserv 6:853–868

Li X, Gong P, Liang L (2015) A 30-year (1984–2013) record of annual urban dynamics of Beijing city derived from landsat data. Remote Sens Environ 166:78–90. https://doi.org/10.1016/j.rse.2015.06.007

Lidicker WZ (1995) Landscape approaches in mammalian ecology and conservation, University of Minnesota Press, Minneapolis, Minnesota

Lyon JG (1983) Landsat derived landcover classifications for locating potential nesting habitat. Photogrammetric Eng Remote Sens 49(2):245–250

Mahanand S, Behera MD, Roy PS, Kumar P, Barik SK, Srivastava PK (2021) Satellite based fraction of absorbed photosynthetically active radiation is congruent with plant diversity in India. Remote Sens 13(2):159

Moloney KA, Levin SA (1996) The effects of disturbance architecture on landscape-level population dynamics. Ecology 77:375–394

Monmonier MS (1974) Measure of pattern complexity for chor opleth maps. American Cartographer 1(2):159–169

Murthy MSR, Das P, Behera MD (2016) Road accessibility, population proximity and temperature increase are major drivers of forest cover change in Hindu Kush Himalayan region: study using Geo-Informatics approach. Curr Sci 111(7):1599–1602

Myers N (1988) Threatened biotas: 'hotspots' in tropical forestry. Environmentalist 8:1–20

Pandey SK, Shukla RP (1999) Plant diversity and community patterns along the disturbance gradient in plantation forests of sal (Shorea robusta Garten.). Curr Sci 77(6):814–818

Pasha SV, Behera MD, Mahawar SK, Barik SK, Joshi SR (2020) Assessment of shifting cultivation fallows in Northeastern India using Landsat imageries. Tropical Ecol 61(1):65–75

Patidar N, Keshari AK (2020) A rule-based spectral unmixing algorithm for extracting annual time series of sub-pixel impervious surface fraction. Int J Remote Sens 41(10):3970–3992. https://doi.org/10.1080/01431161.2019.1711243

Pickett STA, Thompson JN (1978) Patch dynamics and the design of nature reserves. Biol Conserv 13:27–37

Pickett STA, White PS (1985) The ecology of natural disturbance and patch dynamics. Academic Press, London

Romme W, Knight DH (1982) Landscape diversity: the concept applied to Yellowstone National Park. Bioscience 32:664–670

Roy PS, Tomar S (2000) Biodiversity characterization at land scape level using geospatial modelling technique. Biol Conser 95:95–109

Roy PS, Behera MD (2005) Rapid assessment of biological richness in a part of Eastern Himalaya: an integrated three-tier approach. Current Sci 88(2):250–257

Roy PS, Kushwaha SPS, Murthy MSR, Roy A, Kushwaha D, Reddy CS, Behera MD, Mathur VB, Padalia H, Saran S, Singh S, Jha CS, Porwal MC (eds) (2012) Biodiversity characterisation at landscape level: national assessment, Indian Institute of Remote Sensing, Dehradun, India, pp 140, ISBN 81-901418-8-0

Roy PS, Roy A, Joshi PK, Kale MP, Srivastava VK, Srivastava SK, Dwevidi RS, Joshi C, Behera MD, Meiyappan P, Sharma Y, Jain AK, Singh JS, Palchowdhuri Y, Ramachandran RM, Pinjarla B, Chakravarthi V, Babu N, Gowsalya MS, Kushwaha D (2015) Development of decadal (1985–1995–2005) land use and land cover database for India. Remote Sens 7(3):2401–2430. https://doi.org/10.3390/rs70302401

The State of India Forest Report (2019) Forest Survey of India, DehraDun

Turner MG (1989) Landscape ecology: the effect of pattern on process. Ann Rev Ecol Syst 20:171–197

Turner MG, Gardener RH, Dale VH, O'Neill RV (1993) A revised concept of landscape equilibrium: disturbance and stability on scaled landscapes. Landsc Ecol 8(2):13–227

Wagner PD, Kumar S, Schneider K (2013) An assessment of land use change impacts on the water resources of the Mula and Mutha rivers catchment upstream of Pune, India. Hydrol Earth Syst Sci 17:2233–2246. https://doi.org/10.5194/hess-17-2233-2013

Chapter 6
Spatiotemporal Dynamics of Land and Vegetation Cover in Cold Desert Region of the Ladakh Himalaya

Mukunda Dev Behera, Viswas Sudhir Chitale, Shafique Matin, Girish S. Pujar, Akhtar H. Malik, and Seikh Vazeed Pasha

Abstract The Himalayan ecosystems have characteristics land and vegetation distribution pattern owing to its varied complexities in topography, seasonality, changing climate and socioeconomic interventions. The comprehensive mapping of land and vegetation cover in the Himalayas has always been a great challenge to the cartographers and remote sensing scientists. The focus of this chapter is to demonstrate a practical approach to map and understand land and vegetation cover distribution, and their dynamics over interval of three decades using earth observation data. The study provides an insight to characterize the vegetation pattern across an elevation gradient using geospatial techniques in a test site of Kargil district in the Ladakh Union Territory, India. Two set of images during August–November 1975 and 2005 were used in classification that provided high classification accuracy of >90% (overall, and 0.86 kappa), as per field correspondence. This spatial analysis has indicated that LULC demonstrated significant changes during 1975–2005. It was observed that the barren lands and the snow cover areas together contributed to nearly 80% of the total area in 1975, whereas in 2005, they contributed to nearly 60% area of the district. Variation in elevation range owing to distribution of the vegetation classes was realized for the eastern and western aspects. The separation of classes with a sharp boundary between two adjoining classes is absolutely impossible in nature. Therefore, some

M. D. Behera (✉) · V. S. Chitale · S. Matin · S. V. Pasha
CORAL, Indian Institute of Technology (IIT), Kharagpur, West Bengal, India
e-mail: mdbehera@coral.iitkgp.ac.in

V. S. Chitale
e-mail: vishwas.chitale@icimod.org

V. S. Chitale
International Centre for Integrated Mountain Development, Kathmandu, Nepal

S. Matin
Environmental Protection Agency, Dublin, Ireland

G. S. Pujar · S. V. Pasha
National Remote Sensing Centre, ISRO, Balanagar, Hyderabad, India

A. H. Malik
KASH, Kashmir University, Srinagar, India

A. Pandey et al. (eds.), *Geospatial Technologies for Land and Water Resources Management*, Water Science and Technology Library 103,
https://doi.org/10.1007/978-3-030-90479-1_6

misclassification was noticed at the ecotone region between two spectrally similar classes such as agroforest-scrubs, scrubs-pastures. Another noticeable observation is the mapping of water body pixels in 2005 (58.88 Km^2) in contrast to 1975, which may well be attributed to the low resolution (60 m) of satellite data used for 1975, and rise in the extent of water bodies such as alpine lakes, and more water flow through river channels owing to snow and glacier melting. The geospatial database integrated with field data plays an important role toward land and vegetation cover dynamics studies, useful for sustainable management.

6.1 Introduction

High mountains provide test sites to study the complexities in vegetation distribution as they vary according to the slope, elevation, aspect and available climatic and edaphic conditions (Beniston and Fox 1996). It is estimated that around 23% of the Earth's forest are found in Mountains (Price et al. 2011), and that offers home to nearly 12% of the global human population (FAO 2002). The high mountains account to larger part of natural environments (Guisan et al. 1995), and thereby often behave as sensitive '*ecological indicators*' in the face of climate change (Pauli et al. 2014). Climate change could have irreversible impacts on the protective functions of mountain forests (Becker and Bugmann 1997). The vegetation of mountain regions is mainly constrained by physical components of the environment and thus, vulnerable to climatic change (Beniston 1994). The functionalities of the high mountain ecosystems are primarily determined by low temperature, and now high on climate change research agenda (Price and Barry 1997). In addition, changes in elevation alter temperature, precipitation and winds (Whiteman 2000), resulting in the mountain climate, which is different from low elevations (Barry 1992). Recent studies revealed that most of the major glaciers are retreating at an increasing pace, in the Indian Himalayan (WWF 2005; Bajracharya and Shrestha 2011). Therefore, climate change impacts on high mountain vegetation could be more prominent than on the vegetation occurring at lower elevations.

The performance of the plants in alpine regions is limited by relatively low temperatures and short growing seasons (Gugerli and Bauert 2001). Mountain vegetation (Alpine) regions are characterized by sensitive ecosystems and are vulnerable to the rise in global warming and others impacts of climate change (www.ipcc.ch, Theurillat and Guisan 2001). The alpine and subalpine flora of the Ladakh region have the greatest affinity to the flora of the Central Asiatic region (Stewart 1916; Dhar and Kacharoo 1983). The alpine pastures are an important part of the cold deserts of India, in terms of biodiversity, ecology and socio-economy of the local community (Kala and Mathur 2002; Negi 2002). Topography and climatic attributes influence community structure and ecosystem processes, consequently determining the local and regional vegetation and biodiversity distribution pattern (Champion and Seth 1968; Kachroo 1993). Further, the recent developmental intervention by the administration in terms of raising the agroforestry practices to increase the socioeconomic status of local people, may re-configure the natural vegetation pattern of the Ladakh region (Suri 2013).

Remote sensing has been the most efficient and widely used technology for mapping the vegetation type and density (Joshi et al. 2005). However, the comprehensive mapping of vegetation and land cover in hostile and inaccessible mountainous topography of the Himalayas has always been a great challenge to the researchers. In such a mountain region, earth observation-based vegetation and land cover mapping are found to be the most useful in understanding their distribution and dynamics pattern. Remote sensing has enabled mapping, studying, monitoring and managing various natural resources and thereby enabled monitoring of the environment (Lillesand and Kiefer 2000). The optical remote sensing data can be subjected to visual or digital interpretation technique for feature extraction on land and vegetation cover. Use of ratio-based indices help optimizing interpretation of certain features, thereby are used to identify specific thematic classes such as vegetation (Normalized Difference vegetation Index), snow (Normalized Difference snow Index), water body (Normalized Difference water Index). A few studies have used for rationing visible (VIS) and near-infrared (NIR) or short-wave infrared (SWIR) bands to discriminate snow and cloud features (Kyle et al. 1978).

6.1.1 Mapping of LULC

The images from satellite remote sensing platforms provide valuable information on vegetation and land cover characteristics, as they could be diligently identified and mapped using various classification techniques. In optical remote sensing, the spectral reflectance behavior of matter plays important role, and therefore linking the spectral signature of individual vegetation and land cover classes on to the image demands the interpreter's knowledge and acumen (Lambers et al. 1998). An interpreter also has to consider various resolutions of the satellite data and topography of the study area under consideration to eliminate any misidentification/classification. Further, field visit is absolute to gather training sets for use in classification and ground control points for use in classification accuracy estimate exercises. In past five decades, rapid developments in the sensor and data processing tools have led to improvement in classification and mapping accuracy.

6.1.2 Image Interpretation and Spectral Reflectance Curve

Image interpretation is an information extraction process in which a spectral class is assigned to its thematic class. An image represents the energy reflected, emitted or transmitted from an object in different parts of the spectrum, which are captured in various satellite bands. Image interpretation is essential for efficient and effective feature extraction from the data. Various interpretation keys such as tone, texture, shadow, association etc., help an interpreter in objects identification. One must also take advantage of other important characteristics of objects in order to identify and discriminate them. The elements of image interpretation are regarded of general significance, irrespective of the precise nature of the imagery and the features it portrays.

Topographical variables such as elevation, aspect and slope regulate the vegetation distribution in Mountains. The gradient analysis is suitable for the study of spatial patterns of vegetation and land cover, which in turn could explain the structural and functional variation of the vegetation in a landscape (Whittaker 1956). Often, indiscriminate and uncontrolled grazing contributes to degradation and loss of vegetation (Seth 1997). The objectives of the present study is to generate vegetation and land cover maps, using satellite remote sensing and GIS between 1975 and 2005 and to assess the changes along the elevation gradient. This was achieved in three steps: first, classification and mapping of the vegetation and land cover using satellite date for the period of 1975 and 2005, considering elevation criteria; second, three field visits were conducted to collect information for utilization in classification and accuracy assessment; third, analysis of vegetation and land cover change over three decades, with emphasis on elevation zones and agroforestry interventions.

6.2 Study Area

The Kargil district of Ladakh Union Territory, India, has been chosen as the study site that lies between 30° to 35°N latitude and 75° to 77°E longitudes (Fig. 6.1). The elevation ranges of the study site vary from 3000 to 7100 m above mean sea level, and it lies in the rain shadow of the Himalayas. Ladakh is called as the 'land of high rising passes,' and the region experiences heavy snowfall in the winter season with average snowfall of about 2–5 m. Kargil district is spread over an area of 17,303 Km^2. The region is divided into four high altitude valleys such as Suru, Drass, Indus and upper Sindh valley.

Kargil is a mountainous region situated in the north of the Himalayas, and south of the Karakoram Range. The climate of district falls under 'cold desert,' as it experiences amalgamation of both arctic and desert climate. These include wide fluctuations in diurnal and seasonal temperature, which vary from −40 °C in winter to +35 °C in summer; with extremely low precipitation with an annual 10 cm to 30 cm of snowfall (Ladakh Autonomous Hill Development Council; LAHDC 2005). The region receives the highest radiation level in the world (up to 6–7 Kwh/mm) owing to very low humidity and high elevation. These factors together explain why maximum area is nearly devoid of vegetation, except the valley floors and irrigated areas.

Most of the region remains snow-claded during the year and exhibit unique flora and fauna (Fox et al. 1991a). The area is covered with diverse vegetation cover ranging from alpine forests to alpine scrubs and pastures. These forests also support wild lives like ibex, snow leopard, musk deer, wolves, red bear and barasinghas (Fox et al.

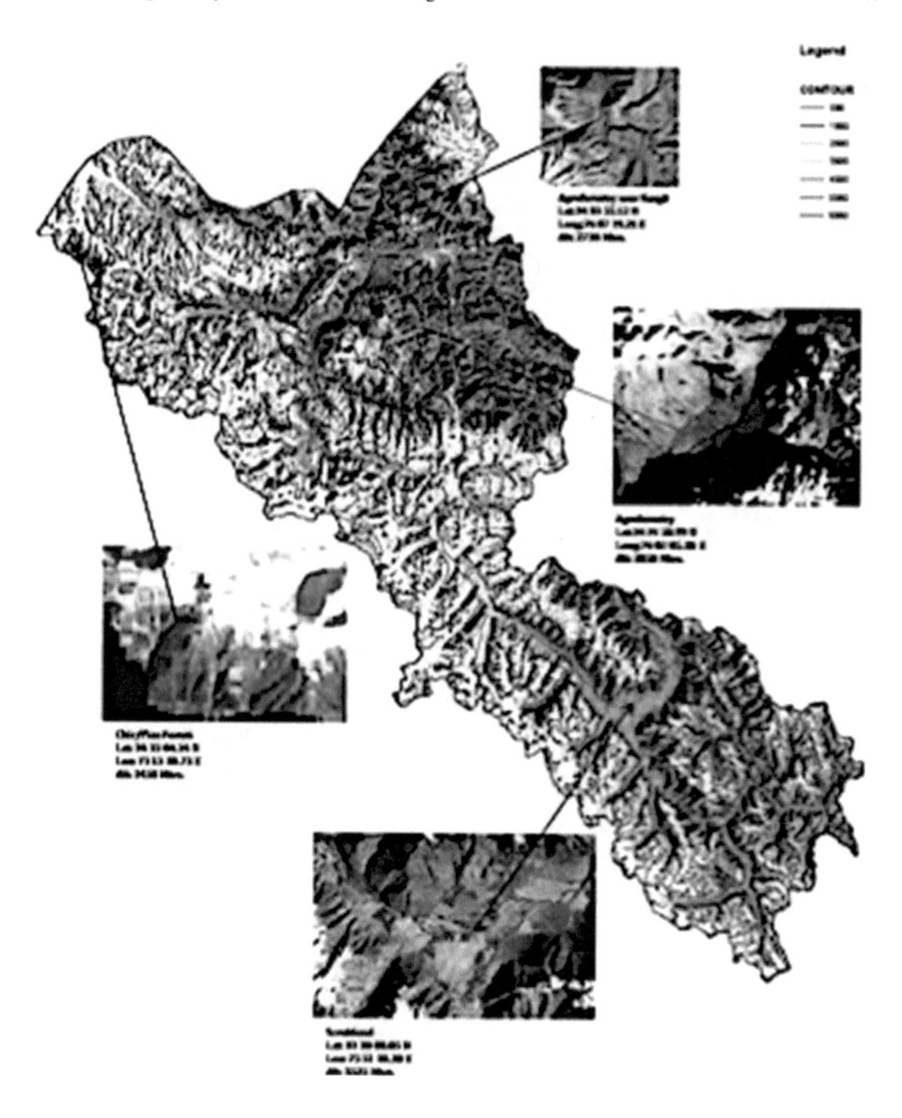

Fig. 6.1 The landsat false color composite (FCC) image of the study area showing different land cover classes

1991b). The types of people are Brogpas, Baltis, Purik, Shinas and Ladakhi (Ladakh 1999). The languages primarily spoken are Shina, Balti, Purig and Ladakhi (Fig. 6.2).

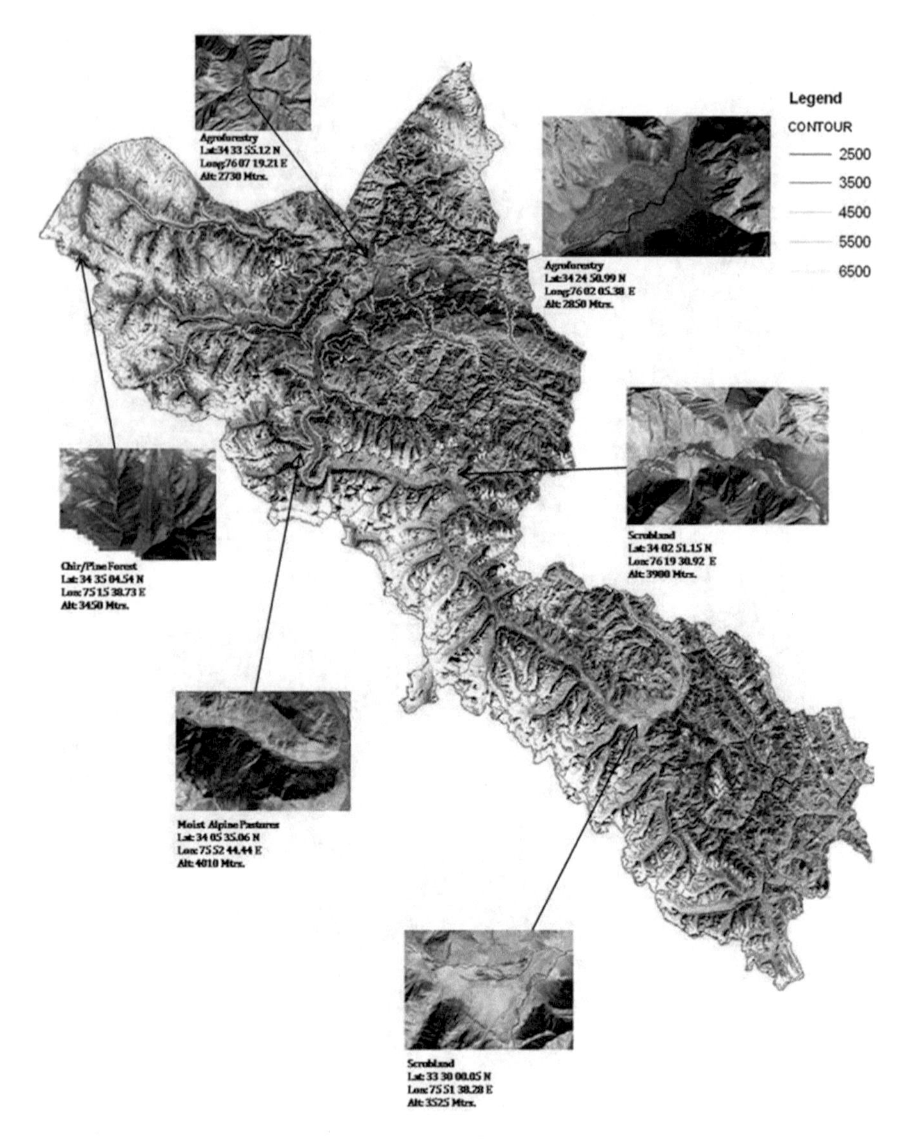

Fig. 6.2 The IRS P6 LISS III false color composite (FCC) image of the study area showing different land cover classes

6.2.1 Geological Attributes

The district of Kargil is located above the perpetual snowline extending from 3000 to 8000 m. The north-western branch of the inner Himalayas comprises of Zanskar range encompassing the territories of Ladakh, Gilgit and Baltistan. Administratively, Kargil region comprises of Zanskar and Suru valleys. The Kargil basin forms the western end of the long elongate inter-montane late tertiary basin along the Indus-suture zone. The soil is sandy, neutral to slightly alkaline, and its fertility varies widely. The soil texture is loamy clay with the pH ranging from 7.0 to 11.0.

6.2.2 Climate and People

Kargil experiences severe arctic to sub-arctic-like climatic conditions. The Kargil district experiences two prominent seasons, i.e., summer and winter, unlike other parts of Himalayas. Biogeographically, the region falls under *cold desert* category. Most of the areas are covered under thick layer of snow for almost half of a year. The snow melting usually starts during the month of April and May. Summers are warm and dry. The mean annual temperature is around 6 °C with negligible amount of rainfall every year due to the rain shadow effects of the Himalayas. At some places during winter, the temperature drops down to −40 to −50 °C. Drass, the second coldest inhabited place in the world after Siberia is located in the district.

Most of the Kargil population is inhabited by the Burig and Balti people of Tibetan origin (converted from Buddhism to Islam in the sixteenth-century) and have inter-mingled with the Dard, Mon and other Aryan people. The people of Kargil have descended from the Mongols, the Dards of central Asia and the Indo-Aryan Mon people. Some people in Kargil district might have Tibetan ancestors, though earlier Tibetan contact has left a profound influence upon the people of both Kargil and Leh.

Due to extreme climatic conditions and topography, the region offers restricted livelihood opportunities and hence is one of the least populated areas in the country, with 129 villages and two major settlements, viz. Kargil and Padum (Plate 6.1; Zanskar valley). The population of the region is primarily rural. The main economic sector of the region is irrigated agriculture and livestock rearing, which employ major workforce including agricultural laborers, herdsmen and cultivators.

6.3 Methodology

6.3.1 Classification Scheme

The first step in a vegetation type mapping exercise is to develop a simple, repetitive hierarchical classification scheme. Broadly, the vegetation distribution can be visualized in terms of biogeography that follows an equator to polar distribution pattern of vegetation type across latitude and elevation (Anon. 2008). Further, the natural vegetation must be separated from the agricultural and seasonal crops. The international vegetation classification (IVC) (formerly called the International Classification of Ecological Communities or ICEC) is based on vegetation, as it currently

Plate 6.1 Photograph showing a small village in Padumvalley

exists on the landscape. Landforms, soils and other features are not directly considered as part of the classification criteria. A vegetation classification scheme should consider the physiognomy of the area, vegetation phenology, life forms, edaphic and biotic factors as essential entities. One new class of agroforest is also introduced in to the scheme. The vegetation distribution along elevation gradient provides control in discriminating between classes and helps in defining the transition boundaries.

6.3.2 LULC Classification

The Landsat Multispectral Scanner (MSS) image of 1975 having spatial resolution of 60 m and the Indian Remote Sensing (IRS) Linear Imaging Self Scanner (LISS) III image with a spatial resolution of 23.5 m was obtained from NRSC, ISRO, Hyderabad. We performed 'image-to-image' registration using GCPs selected from ortho-rectified Landsat TM data (http://glcf.umd.edu) using a first-order polynomial transformation. The satellite data were acquired during the least cloud cover seasons to optimize discrimination of vegetation and land cover classes. We used multi-season (i.e., August–November and April–May) satellite data for classification and mapping using field-collected ground truth. The satellite datasets were georeferenced to the Universal Transverse Mercator (UTM) coordinate system and the WGS

84 datum. Top-of-Atmosphere reflectance (ToA Corrections) was performed on the false color composite images (Chavez 1996). Georeferenced satellite data was classified at 1:50,000 scale and further resampled to 30 m before proceeding for change analysis. On-screen visual interpretation technique was used for classification using interpretation based on key factors such as tone, textural, size, shape, variations, site and association etc. (Fig. 6.1). The *Shuttle Radar Topography Mission* (SRTM) 90 m DEM was used to get the variation of topography in terms of elevation, slope and aspect (http://srtm.csi.cgiar.org).

The ground truth information (GCPs) collected during the field tours was corroborated with the interpreted vegetation and land cover map. Three field tours were organized during August–September, 2009, 2011, 2015 for the purpose of reconnaissance survey, vegetation sampling and validation of different vegetation and land cover categories. The field tours included routes from Kargil to Padum located in the Zanskar valley covering places like Mulbekh, Sankoo, Rangdum and Umbala (Fig. 6.1). The GPS locations and photographs were collected from various vegetation and land cover classes for utilization in the classification and mapping exercise. The image interpretation keys provide critical reference for advanced interpretation. An image interpretation key for the study was designed prior to interpretation, which was further refined in the course of interpretation and mapping. The classification scheme utilized six LULC classes representing subalpine forests, agroforest, scrub, grassland, water, barren land, settlements and snow. Each vegetation class was checked with respect to their corresponding GCPs to judge their occurrence with respect to their locations. The classified images were further subjected to accuracy assessment and change detection.

6.4 Results and Discussion

The Kargil district accommodates unique vegetation composition owing to its characteristics environmental set up. During field visit, it was observed that the district comprises of rocky barren lands, snow-covered mountains and natural vegetation during summer season. The plants are found growing along moist river channels owing to precipitation deficiency (Fig. 6.3). The control of aspect with respect to vegetation distribution was found prominent with higher distribution of vegetation in the eastern slopes than the western slope. Champion and Seth (1968) have recorded vegetation based on elevation variation from 2000 to 6000 m. They have reported three major vegetation types in the region.

(I) Subalpine forests (14): This forest is distributed between 3000 and 4000 m elevation and represented dominantly by *Pinus roxburghii* species.

(II) Scrubland (15/C3): The shrubs are distributed in extended patches above the subalpine forests and below the alpine grasslands. The dominant species in

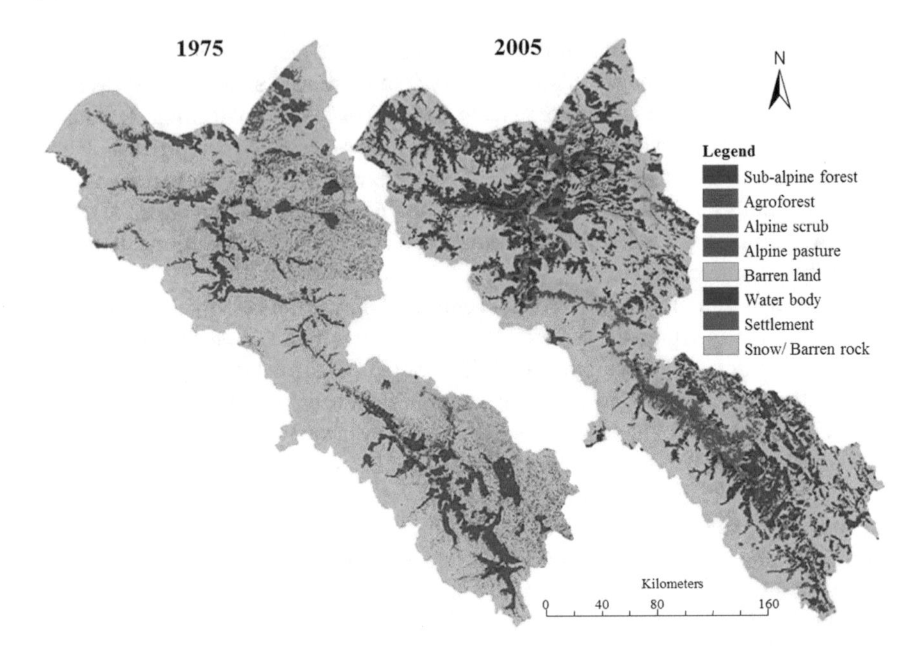

Fig. 6.3 Classified maps of the study area is showing different LULC classes (1975–2005)

this category are *Hippophae and Myricaria* species. Other species in this vegetation category are *Bromus japonicas, Hyppophae rhamnoides and Myricari germanicai.*

(III) Pastureland (16/C1): The grasslands or pastures are distributed in the moist alpine zone with dominant species composition of *Poa annua, Medicago sativa* and *Plantago lanceolata. Along with Phleum alpinum and Artemisia brevifolia* species.

(IV) Agroforest: The agroforests are practiced near the settlement areas that are close to moisture sources and with availability of flat lands. *Populas* and *Salix* are the common species occur in the fences of agricultural lands, near to the river streams.

Major crops grown in this area include grim (naked barley), wheat, peas and Oal (Alfalafa). Silvi-horticulture is also widely practiced in villages and in the low-lying areas with poplar, willow, apricot and apple as the main crop. Vegetables like cabbages, carrots, tomatoes and cauliflower have been introduced that also can thrive in the intense sunlight during the summer months. The pit is covered with plastics and allows early planting of many cool season vegetables. All agricultural activities are confined to summer, and the growing season lasts for about four to five months in a year. Production, hence, is very limited, and most of the produce is meant for self-consumption with limited agricultural surplus. The distribution of the vegetation and land cover map is presented in Fig. 6.3.

6.4.1 LULC Mapping

Of the two cloud-free data for the year 1975 and 2005, it may be observed that there were more snow cover and barren land areas during 1975 that were reduced after three decades. While more vegetation distribution can be found during 2005. For the Kargil region, it may be reported that the chance of having a good cloud-free data is higher during August to November because of minimal snow and maximal vegetation cover, and relatively clear sky with good sunshine days. This is also reflected in the relative dynamics of snow (42.43%, 16.98%), barren land (36.8%, 44.03%) and alpine pastures (18.8%, 30.02%) for the year 1975 and 2005, respectively (Table 6.1).

The alpine pastures/grasslands are the dominant vegetation types and the main component of the ecosystem. The vegetation cover of the region has a spatial extent of 6677.28 Km^2. The natural vegetation cover of the study area occupies 20.8%, 38.6% in 1975 and 2005 year of the total geographical area. A total of ten vegetation and land cover classes were delineated using remote sensing and GIS techniques such as subalpine forest (Chir pine), alpine pastures, alpine scrub, agro-forest, grassland, barren land, water body, snow and settlement (Fig. 6.3). The alpine pastures were identified as the most abundant vegetation type spread over 3248.7 km^2 (18.8%), followed by alpine scrub 256.92 km^2 (1.5%), whereas agroforest occupied only 70.4 km^2 (0.41%). The subalpine forest (Chir pine) formations occupied only 13.57 km^2 (0.08%) in 1975 that further decreased to 6.24 km^2 (0.04%) in 2005 (Table 6.1). The cold desert snow is one of the predominant classes that covered 7334.6 km^2 (42.43%) in 1975 that decreased to 2937.98 km^2 (16.8%) in 2005. The percentage distribution of vegetation and land cover classes with reference to the total geographical area for the periods 1975 and 2005 are shown in Fig. 6.4.

A decrease of 7.33 Km^2 subalpine forest area was demonstrated between 1975 (13.59 Km^2) and 2005 (6.24 Km^2), while extended agroforest area was mapped during 2005 that showed increase from 70.41 Km^2 to 190.55 Km^2 during the past three decades (Fig. 6.4). The subalpine forests were mapped between 2700 and

Table 6.1 Areal extent of LULC and change analysis 1975 to 2011 (Area in km^2)

LULC Class	1975	%	2005	%	Change	Change%
Subalpine forest	13.57	0.08	6.24	0.04	−7.33	−0.04
Agroforest	70.41	0.41	190.55	1.1	120.15	0.69
Alpine Scrubs	256.92	1.49	1285.54	7.43	1028.62	5.94
Alpine Pastures	3248.74	18.8	5194.95	30.02	1946.20	11.25
Barren land	6379.09	36.8	7619.1	44.03	1258.01	7.27
Water	0.0	0.0	58.88	0.34	58.88	0.34
Snow	7334.6	42.43	2937.98	16.98	−4396.14	−25.41
Settlement	0	0	10.08	0.06	10.08	0.06
Total	17,303.3	100	17,303.3	100		

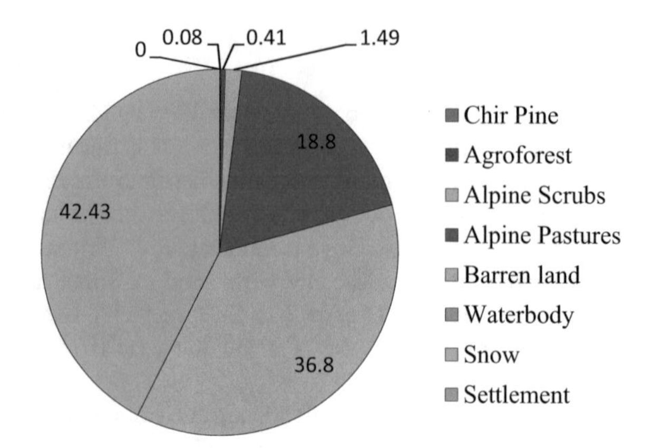

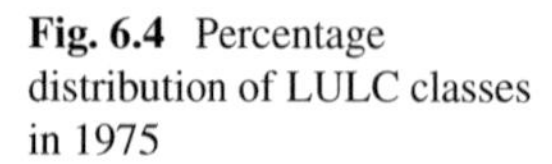
Fig. 6.4 Percentage distribution of LULC classes in 1975

3800 m elevation, along the Baramullah district boundary of Jammu and Kashmir Union Territory that is dominated with *Pinus wallichiana* and *Juniperus semiglobosa* species. Similarly, the agroforests were mapped around the settlement and water body sources at 2500–3600 m elevation. The various species found and planted here are *Salix* and *Populus* species, mixed with *Poa annua, Digitaria stewertiana, Bromus japonicus, Medicago sativa, Phleum alpinium* etc. The subalpine forest and the agroforest classes could be well discriminated with their relatively dark and bright red tone that was further confirmed using their elevation ranges of distribution.

An increase of 1038.5 Km^2 alpine scrub area was demonstrated between 1975 (1.49 Km^2) and 2005 (1285.54 Km^2), while extended alpine pasture area was mapped during 2005 that showed increase from 3248.74 Km^2 to 5194.95 Km^2 during the past three decades (Fig. 6.4). The moist alpine scrubs were mapped between 2800 and 5100 m elevation, which is dominated with *Myricaria* and *Hippophae and* formations along with composition of *Aconitum violaceum, Agrostis gigantea, Aconitum heterophyllum, Anaphalis brusus* and *Alisma graminium* species.

Similarly, the alpine pastures were mapped up to the alpine barren and snow cover areas between 2800 and 5800 m elevation. The various species found here are *Plantago lanceolata, Poa annua, Artemisia brevifolia* and *Poa pratensis.* The moist alpine scrubs could be well discriminated on satellite image with light to bright red tone, and smooth texture (Fig. 6.1). Similarly, alpine pastures reflected light red to light brown color on the false color composite that were further ascertained using the elevation range of distribution to discriminate from the moist alpine scrubs.

It was observed that the barren lands and the snow cover areas together contributed to nearly 80% of the total area in 1975, whereas in 2005, they contributed to nearly 60% area of the district. This may be attributed to more of alpine scrubs and pastures distribution in 2005 in comparison to 1975. It may so happen that availability of exact anniversary data for such comparison is near impossible for such areas, and secondly, due to global warming, the more areas in upper elevation could be suitable for holding scrub/pasture for a longer duration. The 1975 data resolution is double that of the one used for 2005, yet both the maps have shown good classification accuracy of >93%.

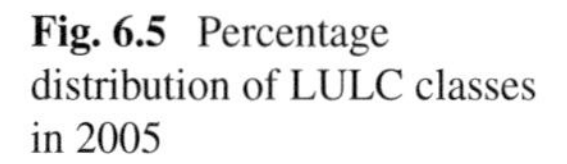

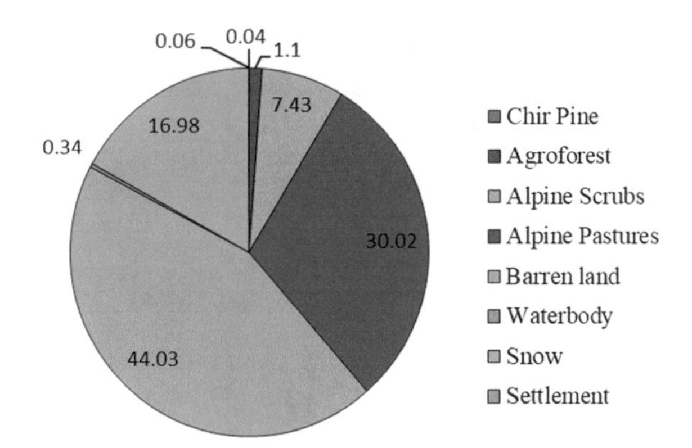

Fig. 6.5 Percentage distribution of LULC classes in 2005

Though, this might affect the minimum mappable area (MMA) of classes, but less likely to affect the comparison as post-classification comparison-based method was used in such exercise (Fig. 6.5).

The separation of classes with a sharp boundary between two adjoining classes is absolutely impossible in nature. Therefore, one can expect some misclassification at the ecotone region between two spectrally similar classes such as agroforest-scrubs and scrubs-pastures. Another noticeable observation is the mapping of water body pixels in 2005 (58.88 Km^2) in contrast to 1975, which may well be attributed to the low resolution (60 m) of satellite data used for 1975, and rise in the extent of water bodies such as alpine lakes, and more water flow through river channels owing to snow and glacier melting.

The vegetation and land cover mapping is always a challenging task due to non-availability of cloud-free data in the study area. The cloud-free satellite data for the period October 1975 and October 2005–06 used in the study provided good classification of various land use and land covers. We observed that the study area supports a unique ecosystem, the vegetation types were classified from the low to high elevation as follows: agroforests (2750–4000 m), alpine scrubs (3500–4800 m) and alpine pastures (3500–6000 m).

6.4.2 Change Detection

In LULC mapping, the post-classification-based comparison method was used that resulted in a change matrix that provided 'from-to' information. The overall result of change detection showed that settlement area increased, and the snow area shrunken during the observation period of three decades (Fig. 6.6).

The result of the vegetation and land cover change was analyzed using change detection method. The result showed that there was both positive and negative change. The change area matrix showed that land cover has significantly changed since 1975. Analysis of temperature data between 1972 and 2002 revealed the highest mean

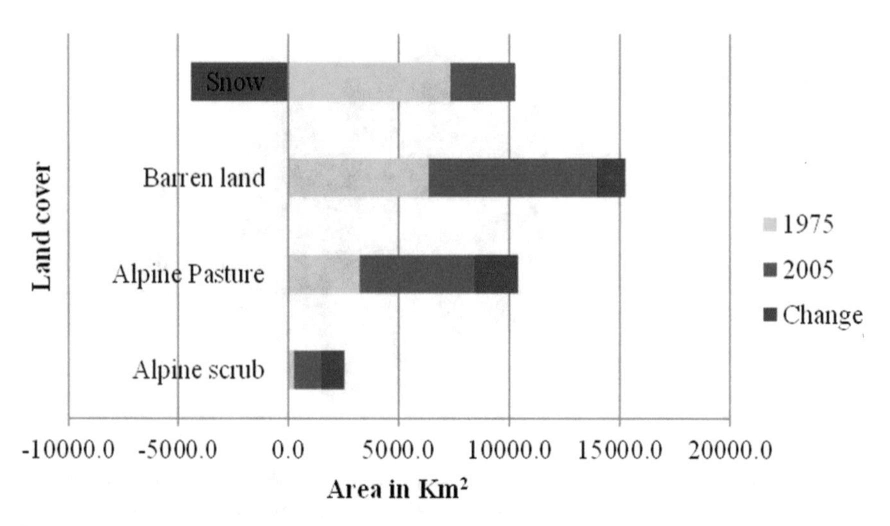

Fig. 6.6 Major LULC changes during 1975–2005

monthly temperature of about 15 °C in the month of July and August and lowest of 5 °C in the month January and February. The subalpine forest has lost about 7.33 km^2 and converted to other land uses. Agroforestry lands also showed a positive increase of 120.15 km^2 (0.7%) between 1975 and 2005 (Table 6.1). Fox et al. (1994) have reported that Ladakh has recently undergoing a substantial change from the patterns of the traditional farming method. Of the eight land cover classes, the highest negative change found in the snow cover, i.e., (4396.1 km^2) 21.4% between 1975 and 2005. The settlement area has increased by 0.06% (10.08 Km^2) in 2005. Settlement areas were greatly increased, corroborated with population and infrastructure development. The barren land has recorded a positive change between 1975 and 2005 from (6379.09 Km^2) 36.8% to 44.0% during 1975–2005 periods. This can be due to human activities, which includes grazing pressure, firewood extraction in the study area. Behera et al. (2014) reported higher plant diversity in the subalpine forests and lowest in the alpine pastures, and thereby derived a general decreasing pattern along elevation.

6.4.3 Distribution of Vegetation Types Along the Elevation Gradient

A significant increase in the vegetation cover was found during 1975–2005 with maxima in the range of 3500–4500 m (Fig. 6.7). The maximum land area was found between 4000 and 6000 m. The majority of vegetation cover is distributed in the range of 20° to 30° slope. Vegetation cover has decreased in the range of 60° to 70° slope followed by an increase in 80° to 90° slope that indicates the adaptability along

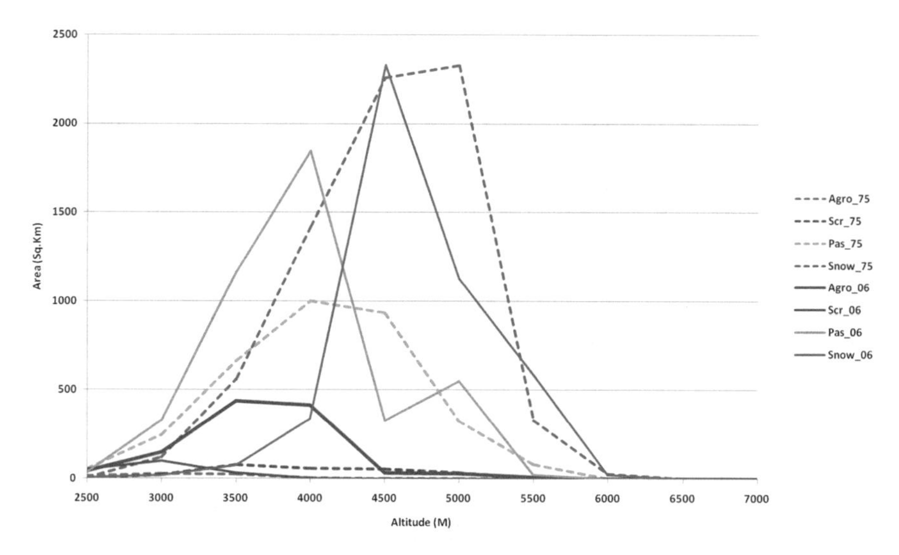

Fig. 6.7 Altitude-wise changes in different LULC classes 1975–2005

steep slopes. According to the recent results, alpine pastures distributed between 3000 and 5000 m elevation range. Alpine pastures play a significant role in the cold deserts of India, with reference to biodiversity, ecology and socio-economy of the local community (WWF 1997; Negi 2002). The highest vegetation cover was found in the range of 15° to 25° slope and the moist alpine pastures grow up to 75° slopes. Alpine pastures cover the highest vegetation fraction distributed in all aspects. The study results demonstrated increase in alpine scrub in almost all aspects. The study also quantified the maximum vegetation fraction of alpine pastures along the Eastern aspect, whereas the lowest vegetation fraction was registered along North Eastern aspect.

The multi-temporal satellite data has revealed a significant decrease in snow cover areas that resulted in an increase in non-vegetated barren lands and vegetated areas (Hall et al. 1995; Lin et al. 2012). Figure 6.4 demonstrated elevation-wise vegetation and land cover changes during 1975–2005. The overall accuracy of the vegetation and land cover maps derived for the years 1975 and 2005 was 89.7%, and 92.2%, respectively. All the kappa values were above 0.86. Anderson et al. (1976) have recommended that 85% accuracy for land cover classification using remote sensing is suitable for further studies. Kargil district provides a unique environmental and ecological setting for such a study on vegetation and land cover dynamics.

6.5 Conclusions

The present study reveals the utilization of multi-temporal earth observation datasets for detecting and monitoring changes in the vegetation and land cover distribution pattern in a Himalayan landscape. The study also indicates that for a significant conclusion to be made on 'vegetation shift' and need of a higher-resolution imagery supplemented with field observation should be carried out. The study concludes that vegetation distribution will vary owing to a different altitude, slope and aspect; therefore, physiognomic categorization holds key to the study of vegetation change/shift in tropical mountainous regions. The results showed that the Himalayan ecosystem of Kargil has been undergoing drastic changes during the study period.

During the past few decades, climate change is adversely affecting the vegetation distribution and function, including biodiversity. Therefore, management efforts will be required to prevent the future loss/damage of remaining natural ecosystems. The Himalayan ecosystem of Ladakh could be an important case to understand the impact of climate change in all its social and ecological dimensions (CBD 2011). The spatial database generated at a spatial scale 1:50,000 integrated with field data will play an important role in understanding vegetation and land cover dynamics.

Remote sensing data with enhanced spatial and spectral resolution can be used to understand the community structure even up to species level. The peculiar vegetation pattern, complex terrain and extreme weathering condition make Kargil a suitable site to understand the vegetation distribution pattern in climate change scenario. So, it can be concluded that proper conservation plan on vegetation and land cover is needed for the benefit of the region.

Acknowledgements We thank authorities of Spatial Analysis and Modelling (SAM) Laboratory, Centre for Oceans, Rivers, Atmosphere and Land Sciences (CORAL) at Indian Institute of Technology Kharagpur for providing facilities for the study.

References

Anderson JR, Hardy EE, Roach JT, Witmer RE (1976) A land use and land cover classification system for use with remote sensor data. U.S. geological survey, professional paper 964, Reston, VA

Bajracharya SR, Shrestha B (2011) The status of glaciers in the hindu kush-himalayan region. ICIMOD, Nepal (No. id: 8898)

Barry RG (1992) Mountain weather and climate. Psychology Press

Becker A, Bugmann H (eds) (1997) Predicting global change impacts on mountain hydrology and ecology: integrated catchment hydrology/altitudinal gradient studies. Workshop Report. IGBP Report 43. Stockholm, p 61

Behera MD, Matin S, Roy PS (2014) Biodiversity of Kargil cold desert in the Ladakh himalaya. In: Integrative observations and assessments. Springer, Japan, pp 253–274

Beniston M, Fox DG (eds) (1996) Impacts of climate change on mountain regions. In: Watson RT, M.C

Champion HG, Seth SK (1968) A revised survey of the forest types of India. Published by Govt. of India Press
Chavez PS (1996) Image-based atmospheric corrections—revisited and improved photogramm. Eng Rem S 62:1025–1036
Dhar U, Kachroo P (1983) Alpine flora of Kashmir himalaya
FAO (2002) The state of food insecurity in the world. Food and agriculture organization of the United Nations (FAO), Rome
Fox JL, Nurbu C, Chundawat R-S (1991a) The mountain ungulates of Ladakh, India. Biol Conserv 58:167–190
Fox JL, Nurbu C, Chundawat R-S (1991b) Status of the snow leopard Panthera uncia in northwest India. Biol Conserv 55:283–298
Fox JL, Nurbu C, Bhatt S, Chandola A (1994) Wildlife conservation and land-use changes in the Transhimalayan region of Ladakh, India. Mountain Research and Development, pp 39–60
Gugerli F, Bauert MR (2001) Growth and reproduction of Polygonum viviparum show weak responses to experimentally increased temperature at a Swiss Alpine site. Bot Helv 111(2):169–180
Guisan A, Holten JI, Spichiger R, Tessier L (eds) (1995) Potential ecological impacts of climate change in the Alps and Fennoscandian mountains. Conservatoire et Jardin botaniques, Genève, p 194
Hall DK, Riggs GA, Salomonson VV (1995) Development of methods for mapping global snow cover using moderate, resolution imaging spectroradiometer (MODIS) data. Remote Sens Environ 54:127–140
http://glcf.umd.edu/research/portal/geocover/ (Accessed on 8 Dec 2015)
http://glcfapp.umiacs.umd.edu:8080/esdi (Accessed on 8 Dec 2015)
http://srtm.csi.cgiar.org/selection/inputCoord.asp (Accessed on 8 Dec 2015)
http://www.ipcc.ch/ipccreports/tar/wg2/index.php?idp=500
https://www.google.com/earth/desktop/ (Accessed on 25 Dec 2015)
Kachroo P, Sapru BL, Dhar U (1977) Flora of Ladakh: an ecological and taxonomical appraisal. Dehra Dun: Bishen Singh Mahendra Pal Singh x, p 172. Illus, maps. Maps. Geog, 6
Kachroo P (1993) Plant diversity in Northwest Himalaya a preliminary survey. Himalayan biodiversity. Conservation strategies. pp 111–132
Kala CP, Mathur VB (2002) Patterns of plant species distribution in the trans-Himalayan region of Ladakh, India. J Veg Sci 13(6):751–754
Kyle HI, Curran RJ, Barnes WL, Escoe D (28–30 June 1978) A cloud physics radiometer, third conference on atmospheric radiation, American meteorological society, Davis, Calif., p 107
Ladakh (Jan 1999) Submitted to Leh Autonomous Hill Council by ICIMOD
LAHDC (2005) Ladhakh autonomous hill council, Leh. http://leh.nic.in.defalt/htm
Lambers H, Chapin FS III, Pons TL (1998) Plant physiological ecology. Springer-Verlag, New York
Lillesand TM, Kiefer RW (2000) Remote sensing and image interpretation. Wiley, New York, USA
Lin J, Feng X, Xiao P, Li H, Wang J, Li Y (2012) Comparison of snow indexes in estimating snow cover fraction in a mountainous area in northwestern China. IEEE Geosci Remote Sens Lett 9(4):725–729
Naqshi AR, Malla MY, Dar GH (1989) Plants of Ladakh-Nubra. J Econ Taxon Bot 13(3):539–560
Negi SS (2002) Cold deserts of India. Indus Publishing ISBN: 81-7387-127-2
Pauli H, Gottfried M, Grabherr G (2014) Effects of climate change on the alpine and nival vegetation of the Alps. J Mt Ecol 7
Price MF, Barry RG (1997) Climate change. In: Messerli B, Ives JD (eds) Mountains of the world. Parthenon, New York, pp 409–445
Price MF, Georg G, Lalisa AD, Thomas K, Daniel M, Rosalaura R (2011) Mountain forests in a changing world: realizing values, addressing challenges. Food and agriculture organization of the United Nations (FAO) with the support of the Swiss agency for development and cooperation (SDC), Rome

Seth CM (1997) Bakkerwal the guest graziers-become the owners of forests. Souvenir-A decade of service. Jammu and Kashmir Paryavaran Sanstha

Stewart RR (1916) The Flora of Ladak, Western Tibet. I. Discussion of the Flora. Bull Torrey Bot Club 43(11):571–590

Suri K (2013) Women empowerment, conflict transformation and social change in Kargil. Int J Soc Sci 2(2):119

Theurillat JP, Guisan A (2001) Potential impact of climate change on vegetation in the European Alps: a review. Clim Change 50(1–2):77–109

Whiteman CD (2000) Mountain meteorology: fundamentals and applications. Oxford University Press

Whittaker RH (1956) Vegetation of the Great Smoky mountains. Ecol Monitor Aphs 26:1–80

WWF-1997. World wide fund for nature-India. Biodiversity of Jammu and Kashmir: a profile Edited by M. Ahmedullah

WWF-2005 (2005) An overview of glaciers, glacier retreat, and subsequent impacts in Nepal, India and China. WWWF Nepal Program

Zinyowera, Moss RH (eds) Climate change 1995. Impacts, adaptations and mitigation of climate change: scientific-technical analysis. Cambridge University Press, Cambridge, pp 191–213

Chapter 7
Multi-criteria Based Land and Water Resource Development Planning Using Geospatial Technologies

N. R. Shankar Ram, Vinod K. Sharma, Khushboo Mirza, Akash Goyal, V. M. Chowdary, and C. S. Jha

Abstract Holistic planning approach for judicious use of natural resources is critical for development in the current depreciating global climate scenario. Identification of suitable areas/sites for planning land and water conservation measures is critical for long-term management and sustainable development. Information of natural resources inventoried using very high-resolution satellite images when integrated using Geographical Information Systems will provide an appropriate planning tool for sustainable management of natural resources. Particularly, planning activities that are carried out by the integration of geospatial technologies such as remote sensing and GIS help in achieving sustainable development goals. In this study, land and water resources development plans were generated for Chharba Gram Panchayat (GP) of the Dehradun district, Uttarakhand state in Northern India. Long-term surface runoff potential for different meteorological conditions was analysed spatially over the study GP. Suitable sites for land and water management practices were identified using a multi-criterion decision-based approach. Various basic and derived thematic layers that include ground water prospects, terrain characteristics and soil distribution were included in the planning. Land use land cover information generated using high-resolution satellite data at 1:2000 scale is of great help for developmental planning. Water Resource Development plan indicated that nearly 0.6% (8.43 ha) and 8% (121 ha) of the area is suitable for check dams and farm ponds, respectively. The analysis revealed that this area under single cropped areas can be converted to intensive agricultural areas while nearly 35 ha area under agricultural plantations can be converted to agro horticulture. Further, land use under sparse scrub land can be converted to agroforestry. Thus, the suggested land and water resources development plans are expected to convert the existing land use pattern into more suitable categories as per its potential without jeopardizing the environment.

N. R. Shankar Ram (✉) · V. K. Sharma · K. Mirza · A. Goyal · V. M. Chowdary
Regional Remote Sensing Centre-North, National Remote Sensing Centre, Indian Space Research Organisation, New Delhi 110049, India

C. S. Jha
National Remote Sensing Centre, Indian Space Research Organisation, Hyderabad, Telangana 500037, India
e-mail: jha_cs@nrsc.gov.in

A. Pandey et al. (eds.), *Geospatial Technologies for Land and Water Resources Management*, Water Science and Technology Library 103,
https://doi.org/10.1007/978-3-030-90479-1_7

7.1 Introduction

The steady population growth exerts unbearable pressure on land and water resources around the world. Global statistics indicate that degraded land constitutes more than 30% of the global area (Nkonya et al. 2016), while areas falling under the vulnerable categories are increasing day by day (FAO 2011). The global mean rate of agricultural production is 2.6 tons/ha. Developed nations reported a production rate of more than 4.5 tons/ha, while most of the developing countries reported a dismal figure ranging between 0.5 and 2.5 tons/ha (Jha and Peiffer 2006). In India, an estimated 44% of the total geographical area, i.e. 116 million hectares (m ha) of land suffers from the adverse effect of soil erosion and various other forms of land degradation (Mythili and Goedecke 2016). Studies such as Kalita and Sarmah (2016), reported a detachment rate of 5334 metric tons/year showing soil loss.

Active erosion resulting due to water and wind forces plays a crucial role in the loss of approximately 5.3 m tons of sub-soil per year. In addition, the remaining land has been degraded due to faulty agricultural practices, salinity, gully erosion, water logging, alkalinity, etc. It is estimated that the country also experiences nutrient losses that range between 5.37 to 8.4 m tons, reflecting food grain production losses. Thus, land resource management is crucial for the mitigation of such adversities. Land suitability evaluation classifies land into specific land utilization types based on its land potential and requirements (Akinci et al. 2013). Several studies have investigated the application of non-conventional land use planning techniques in developing countries that include remote sensing and GIS technologies (Cools et al. 2003).

Water resource development (WRD) is another crucial factor in the development of a region. Although water is abundantly available on the earth's surface, usable freshwater, accessible for utility, amounts to a meagre 1% with the bulk trapped in glaciers and snow (Pandey et al. 2011). The growing consumption and demand results in overexploitation and adversely impacts the quantity of this already minuscule resource, leading to the acute water crisis in many parts (UNFPA 2008). Such escalating problems call for strategies for addressing important issues related to land and water management in a sustainable way. Rainwater harvesting (RWH) has proven to be one of the best management strategies to replenish the dwindling surface and groundwater resources in the country (Glendenning et al. 2012). This approach has added advantages of mitigating water scarcity issues (Dalezios et al. 2018; Oweis and Hachum 2006), while simultaneously reducing the flood risks and soil erosion problems due to reduced surface runoff (Tamagnone et al. 2020). Historically, archaeological findings suggested that RWH activities have been fundamental in Indian civilizations for the past 2000 years. The indigenous water management strategy employed in RWH structures also enables effortless execution in rural settings (Gunnell et al. 2007). An integration of RWH techniques with groundwater recharge planning will certainly pave the way to address the water crisis under climate change scenarios.

Remote sensing (RS) and Geographic information system (GIS) based approaches act as valuable tools for the generation of sustainable development plans focused on natural resources management overcoming physical limitations (Ustin 2004). Developmental planning using spatial inputs generated at very high resolution (Gram Panchayat as a single unit) using RS & GIS techniques is necessary for management and judicious utilization of land, water, vegetation and other resources, in a sustainable way.

Integration of natural resources potential, socio-economic conditions and institutional constraints towards developmental planning is a complex decision-making process. Studies have successfully incorporated socio-economic factors for assessment of RWH in South Africa (Kahinda et al. 2008). Numerous GIS-based methodologies helped in the selection of appropriate RWH implementation strategies (Adham et al. 2018; De Winnaar et al. 2007; IMSD 1995; Oweis et al. 1998; Saraf and Choudhury 1998). An extensive review of methodologies and criteria used for RWH in the last three decades was discussed by Ammar et al. (2016), which provides insight in understanding various approaches. The complex processes involved in land resource development (LRD) planning also necessitates the concurrent use of decision support tools such as GIS and multi-criteria decision analysis (MCDA) or Multi-Criteria Evaluation (MCE) (Janssen and Rietveld 1990; Joerin et al. 2001; Jansen and Gregorio 2004; Malczewski 2004, 2006; Mendas and Delali 2012). MCDA is a systematic approach for practising informed planning decisions. The procedure has been broadened into the geospatial regime, where a set of possibilities are evaluated based on contradictory and incommensurate criteria. GIS facilitates thematic layer preparation, while MCE aids in the integration of these layers for generation of suitability maps based on some decision criteria. Various studies have successfully employed an integrated GIS-MCE approach for land and water resource development planning (Tienwong et al. 2009; Chowdary et al. 2013; Khoi and Murayama 2010).

Geospatial planning in rural areas on the similar lines of master plans for urban areas is crucial for economic progress that improves the quality of life. A change in hydro-meteorological parameters or occurrence of natural calamities could change the entire ecosystem. The fragile nature of natural resources has thus necessitated to carry out a comprehensive development plan for the Gram Panchayat development. It is also a well-established fact that the developmental needs are area specific that are governed by a variety of other factors related to climate, terrain characteristics, state and extent of resources, etc. Majority of studies generally focused on the development of planning strategies at a watershed scale (Gurmel et al. 1990; Farrington et al. 1999), which necessitates to work out site-specific planning at GP level. Thus, generation of thematic maps using very high-resolution data preferably at 1:4000 scale or higher scales will certainly facilitate focused and local specific development plans at GP level. Thus, this study is taken up with a specific objective of integrated Gram Panchayat (GP) spatial development planning using spatial inputs for Chharba GP, of Uttarakhand state, India. The study is aimed at optimal realization of natural resources to its potential in the light of physical, economic, social and other developmental goals. The current study focused on four major objectives namely (i) Inventory of natural resources at GP level using very high spatial resolution data and, (ii) Analysis of long-term rainfall over the GP, (iii) Assessment of surface water potential and, (iv) Generation of land and water resources development plan using multi-criteria evaluation-based evaluation approach. Geospatial layers pertaining to various natural resources such as rainfall, ground water, topography, soil characteristics were integrated.

7.2 Material and Methods

7.2.1 *Study Area*

The Chharba Gram Panchayat (GP) of the Dehradun district, Uttarakhand state in Northern India, was selected as the study region in this study. It extends from 30°23' to 30°26' N latitude and 77°46' to 77°51' E longitude and covers an area of 1567 ha (Fig. 7.1). The GP experiences a sub-temperate to temperate climate, with temperature ranging between 25 and 30 °C. Mean annual rainfall is nearly 1600 mm and is predominantly concentrated during the monsoon season. The Asan River, part of the Ganges, flows through this GP.

7.2.2 *Satellite Data Used*

Satellite data acquired from the multispectral sensors LISS-IV (Linear Imaging Self Scanner) of the Indian Remote Sensing Satellite (IRS) Resources at (5.8 m spatial resolution) and very high spatial resolution satellite data acquired using Cartosat-2S were analysed for generation of thematic maps. The Cartosat-2S (Indian Remote Sensing satellite) provides high-resolution optical data for executing applications related to resource mapping, monitoring and change detection. It provides high signal-to-noise ratio datasets consisting of multispectral (MX) and panchromatic (PAN) with a spatial resolution of 2 m and 0.65 m, respectively.

7.2.3 *Thematic Information Sources for Developmental Planning*

Value added satellite products generated under various National projects at different spatial scales namely 1:2000, 1:10000 and 1:50000 were successfully integrated for planning purposes at GP level, in this study. Details of the datasets used in this study along with data sources are shown in Table 7.1.

The datasets were standardized for integrated analysis under a GIS environment. Derived maps are generated using a combination of two or more thematic maps, or identified parameters of the different themes, as required. Subsequently, they are

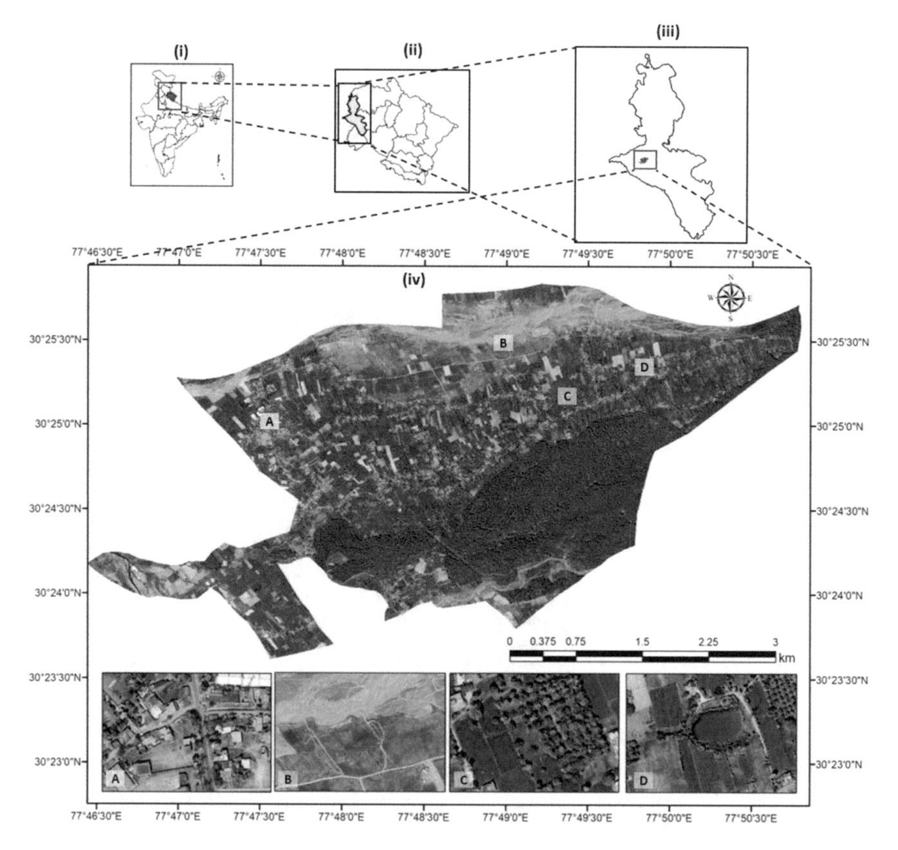

Fig. 7.1 Location map of Chharba Gram Panchayat: (i) India, (ii) Uttarakhand state, (iii) Dehradun district and (iv) GP visualization using high-resolution satellite imagery

used to generate practical and application oriented thematic datasets organised in a spatial database for querying, retrieval and analysis for generation of developmental plans.

Table 7.1 Details of thematic data used in this study for geospatial planning

Dataset/thematic map	Source	Scale/spatial resolution
Digital elevation model	Cartosat stereo data	10 m
Soil	National remote sensing center (NRSC-ISRO)	1:50000
Slope	Topographical map/IRS PAN stereo data, derived from DEM	10 m CartoDEM

(continued)

Table 7.1 (continued)

Dataset/thematic map	Source	Scale/spatial resolution
Groundwater potential	Geology, geomorphology, borewell, Lithology and yield data	1:50000
Drainage map	High-resolution satellite data	1:2000
Meteorological data	IMD	0.25° × 0.25°
Land use/land cover	Cartosat-2S (Indian remote sensing satellite) (merged product multispectral (MX) and panchromatic (PAN) with a spatial resolution of 2 m and 0.65 m, respectively	0.65 m
Derived layers		
Surface water potential	Slope, soil characteristics, land use, rainfall	
Water resource development plan	Slope, soil map, land use, drainage order, lineament, Surface runoff potential	
Land resource development plan	Soil map, land use, slope, ground water potential, surface water potential, geomorphology	

(a) *Ground Water Prospects Map*: Satellite data interpretation for likely occurrence of ground water in any region involves identifying the presence of surface features that act as an indirect indicator for groundwater availability (Ravindran and Jeyaram 1997; Das et al. 1997). Ground water potential map generated under Rajiv Gandhi Drinking Water Mission (NRSC 2012) was used for planning purposes (Fig. 7.2a). The lineaments are the surface manifestation of linear features like joints and fractures. Presence of lineaments across the alluvial zone indicates higher groundwater potential.

(b) *Digital Elevation Model (DEM)*: DEM derived from Cartosat satellite stereo data at 10 m posting interval was used in this study. DEM is the most important parameter for planning developmental activities that help in the generation of slope maps, which is an essential prerequisite for the preparation of land and water resource development plans. The elevation in the study GP ranges between 427 and 570 m (Fig. 7.2b).

(c) *Slope*: Slope map generated using CARTODEM was categorized into five classes namely (i) level to nearly level (> 0%, < = 1%), (ii) very gently sloping (>1%, < = 3%), (iii) gently sloping (>3%, < = 8%), (iv) moderately sloping (>8%, < = 15%) and (v) moderately steeply sloping (>15%, < = 30%) as per guidelines provided by IMSD (Integrated Mission for Sustainable Development) (IMSD 1995). Spatial distribution of slope classes in the study GP is presented in Fig. 7.2c.

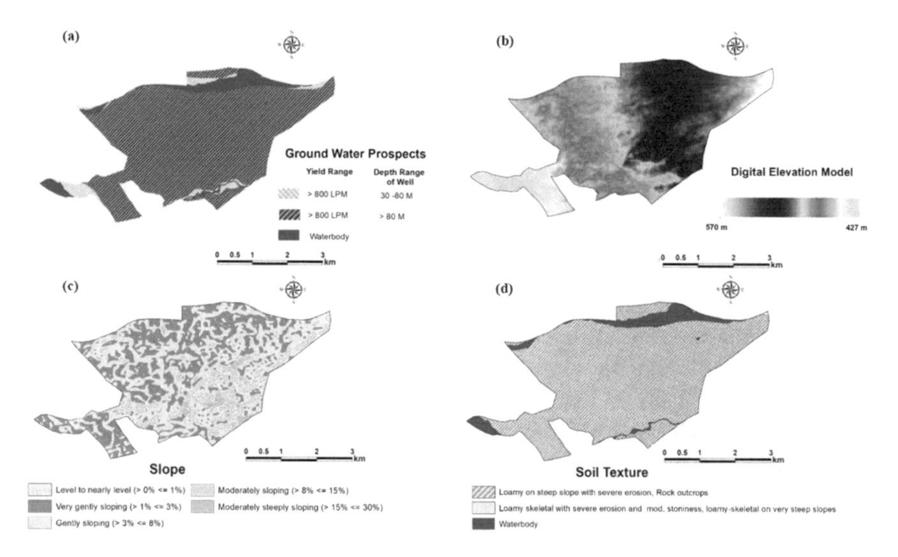

Fig. 7.2 Spatial distribution of thematic layers in Chharba GP (**a**) Groundwater prospects map (**b**) Digital elevation model (**c**)Slope map (**d**) Soil textural map

(d) *Soil Distribution*: Major soil textural class in the GP is loamy soil where the soil texture changes to loamy skeletal towards the northeast, which is prone to severe erosion. The region exhibited shallow soil depth. The soil textural map of GP is shown in Fig. 7.2d.

(e) *Meteorological data*: Indian Meteorological Department (IMD) rainfall gridded product at a daily frequency was used for evaluation of long-term rainfall analysis and also for the assessment of surface water potential in this study. IMD rainfall gridded product at $0.25^\circ \times 0.25^\circ$ spatial resolution was generated based on the daily observed rainfall records of 6329 stations spread across India (Pai et al. 2014). Long-term daily rainfall measurements for the study period (1979–2013) were analysed in the current study.

7.2.4 Evaluation of Long-Term Rainfall for Different Meteorological Conditions and Intensities

Rainfall measurements across the study region were evaluated under different meteorological conditions i.e. wet, dry and normal rainfall years. Classification was based on the long-term average (μ) of IMD rainfall measurements at an annual scale. Meteorological years that indicated rainfall deviations greater than or less than 25% from μ are classified as wet and dry years, respectively. Normal rainfall years were those measured within this 25% criteria. The number of rainfall events are classified into Light, Moderate, Rather Heavy, Heavy and Very heavy specifying rainfall intensities of <7.5 mm, 7.6 mm –35.5 mm, 35.6 mm –64.4 mm, 64.5 mm –124.4 mm and >124.5 mm, respectively.

7.2.5 Computation of long-term Surface Runoff by NRCS-CN Method

The accumulation of excess rainfall, which occurs when precipitation intensity is greater than the rate at which it is infiltrated into soil, is termed runoff. USDA Natural Resources Conservation Service (NRCS) Curve Number (CN) method (SCS 1985) is widely used globally and was used for computation of surface runoff spatially at GP scale. NRCS-CN approach computes surface runoff for individual rainfall events under assumption that potential maximum soil retention is equal to the ratio of direct runoff to available rainfall. For estimation of runoff, land use/cover, hydrological soil group and rainfall in the study area are required, where land use/cover and hydrological soil group are integrated into GIS. Subsequently, curve numbers are assigned to unique combinations of land use/cover and hydrological soil group for different antecedent moisture condition (AMC) conditions. The NRCS-CN method is based on the water balance equation of rainfall in a known interval of time Δt and the governing equations are given as below:

$$Q = \frac{(P - \lambda S)^2}{P + (1 + \lambda)S} \tag{7.1}$$

where, Q, P, λ and S represent the direct surface runoff, total precipitation, initial abstraction ratio and potential maximum retention after runoff begins, respectively. The values of λ range between 0.1 and 0.4 (Subramanya 2008). Thus, Eq. (7.1) was modified using with $\lambda = 0.3$, as follows:

$$Q = \frac{(P - 0.3S)^2}{\mathrm{P} + 0.7S} \quad \text{for} \quad P > 0.3S \tag{7.2}$$

Equation (7.2) is applicable for black soils under antecedent moisture condition of type I (AMC I) while it is applicable for all other soils having AMC I, II and III conditions. In this study, Eq. (7.2) was used to compute direct surface runoff in the study area. Further, the runoff resulting from each rainfall event was summed up to get annual runoff. Runoff potential was estimated using the following equation:

$$\text{Runoff potential}(\%) = \text{Runoff coefficient} \times 100 \tag{7.3}$$

where, Runoff coefficient = Annual runoff/Annual rainfall.

7.2.6 GIS Approach for Land and Water Resource Development Planning

Generation of thematic maps such as land use land cover, water bodies, drainage network, slope classes using satellite data and their integration in GIS were integrated to generate developmental plans for optimum management of land and water resources at GP level. Area specific activities are generally those regions, where a certain type of LRD or WRD activity is recommended for implementation. A GIS-based multi-criteria decision analysis was carried out using the thematic maps to delineate suitable areas for adoption of developmental activity.

7.2.6.1 Water Resources Development Plan Generation

Water harvesting measures such as check dams, Nala-bunds, percolation tanks and farm ponds form the most prevalent categories of RWH techniques (Oweis et al. 2012) across globe. Suitable zones for location of RWH and recharge structures are identified by integrating thematic layers viz., drainage order map with buffer, land use land cover, slope, soil and runoff potential maps in a GIS environment. Guidelines for planning area specific activities and selection of sites were widely reported in the literature (IMSD 1995; Chowdary et al. 2009; Chowdhury et al. 2010; Shankar and Mohan 2005) and are presented in Table 7.2. Planning of measures focused on groundwater recharge depends mostly on soil textural and hydraulic properties and hence suitable locations for percolation tanks should be identified in highly permeable soils (Sankar and Mohan 2005).

7.2.6.2 Land Resources Development Plan Generation

Recognizing the paramount importance of land resources for meeting the rapidly growing demand for more food production, priority should be towards land productivity improvement, while conserving soil resources and optimum land utilization. Land use land cover information derived from very high spatial resolution satellite images help planners and decision makers to develop short to long-term sustainable land resources development plan. A decision model includes logical integration of thematic maps based on the identified conditional operators that help to evaluate the suitability of a particular land use category in the study GP. Towards this, essential prerequisites such as land use land cover, soil, slope and groundwater potential maps were integrated into the GIS using boolean conditions.

Ground water potential map generated by integrating geomorphological, hydrogeological and land use data with geophysical investigations at 1:50000 scale was used as one of the important parameters in this study. Subsequently, study GP was divided into Composite Mapping Units (CMU) by integrating different thematic layers viz., land use land cover, soil, slope and groundwater potential maps that are essential

Table 7.2 Guidelines for site selection for planning water resource development activities

WRD activity	Slope in %	Land use	Soil permeability	Drainage	Favourable settings
Check dams	<10%	River/Drainage stream (near farm lands)	Moderate to high	Up to 3rd order	Suitable zones may be nearly 25 ha in area. Irrigation wells may be located at downstream sites of the proposed structure. Regions experiencing high water table fluctuations preferred
Farm pond	0–3%	Arable areas	Low	1st order/sheet wash area	Regions containing Lineament/fracture may be avoided
Underground barrier	0–3%	Stream bed	High	4th–7th order	Sandy and gravel bed. Wide stream bed or perennial streams. Across the streams having sufficient linear stretch Suitable over areas with stream bed thickness > 5 m
Percolation tank	0–3%	Drainage course, Open land/waste land	High	2nd and 3rd order	Lineaments/fracture zones preferable. Areas near perennial streams preferred. Stream bed may be used if sufficient catchment is available

Source Chowdary et al. (2009)

for generation of land resources development plan. CMU is a three-dimensional homogeneous landscape unit characterized by a unique combination of land, water and vegetation with distinct boundaries. This coupled with surface water potential, helps in the generation of alternate land resource development plans. Methodology adopted from the GIS-based land use planning project initiated in India titled 'Integrated Mission for Sustainable Development', which generates, analyses and integrates natural resource thematic data at 1:50000 scale, in conjunction with satellite imagery forms the major basis for the present study (IMSD 1995). The guiding factors described for land use plan generation have been presented in tabular form in Table 7.3. Further, the information on land capability classes and recommended land treatment management practices (Pretall and Polius 1981) also served as guiding tools.

Table 7.3 Decision rules for land resource development plan generation

GWP code	SL code	SD code	Existing land use	Proposed land use
4	1	1	Single cropped	Intensive agriculture
			Wasteland	Agro horticulture
		2	Single cropped	Intensive agriculture
			Wasteland	Horticulture
	2	1	Single cropped	Intensive agriculture
			Wasteland	Horticulture
		2	Single cropped	Intensive agriculture
			Wasteland	Agro horticulture
	3,4	1,2	Fallow	Agro horticulture
			Wasteland	Agro horticulture
	5,6,7	1,2	Fallow	Social forestry
			Wasteland	Social forestry
3	1	1	Single cropped	Intensive agriculture
			Wasteland	Agro horticulture
		2	Fallow	Agro horticulture
			Wasteland	Agro horticulture
	2	1,2	Fallow	Agro horticulture
			Wasteland	Social forestry
	3	1	Fallow	Agro horticulture with bunding
			Wasteland	Social forestry
		2	Fallow	Social forestry
			Wasteland	Social forestry
	4,5,6,7	1,2	Fallow	Social forestry
			Wasteland	Social forestry
2	1	1,2	Fallow	Agro horticulture
			Wasteland	Social forestry
	2	1	Fallow	Agro horticulture
			Wasteland	Social forestry
		2	Fallow	Fodder crop
			Wasteland	Social forestry
	3,4,5,6,7	1,2	Fallow	Fodder crop
			Wasteland	Social forestry
1	1	1	Fallow	Agro horticulture
			Wasteland	Agro horticulture
		2	Fallow	Fodder crop
			Wasteland	Social forestry

(continued)

Table 7.3 (continued)

GWP code	SL code	SD code	Existing land use	Proposed land use
	2	1,2	Fallow	Fodder crop
			Wasteland	Social forestry
1	3	1,2	Fallow	Fodder crop
			Wasteland	Social forestry
	4,5,6,7	1,2	Fallow	Social forestry
			Wasteland	Social forestry

Ground water potential code (GWP Code) 4 corresponds to good, 3 corresponds to moderate, 2 corresponds to moderate to poor and 1corresponds to poor; Slope code (SL Code) 1 corresponds to 0–1%; 2 corresponds to 1–3%; 3 corresponds to 3–5%; 4 corresponds to 5–10%; 5,6,7 corresponds greater than 10%; Soil depth code (SD Code) 1 corresponds to shallow; 2 corresponds to deep

7.3 Result and Discussion

An inventory of thematic layers namely land use land cover, geology, geomorphology, water bodies, soil, slope was made and subsequently integrated for the generation of derived layers namely surface water potential map, water resource development and land resource development plans maps.

7.3.1 Land Use/Land Cover

Land Use/Land Cover (LULC) refers to classification or labelling of human activities and natural elements on the landscape within a definite time period of analysis of relevant source data. Various land use and land cover features viz., urban or built-up land, agricultural land, forest land constitute a typical LULC map. Knowledge of the existing LULC classes and their spatial distribution forms the basis for any developmental planning. LULC mapping of the Chharba GP was carried out at 1:2000 scale using very high-resolution Cartosat-2S satellite data at sub-meter spatial resolution (Fig. 7.3). The LULC statistics of the study GP are given in Table 7.4.

7.3.2 Long-Term Rainfall Analysis

Long-term rainfall data of the GP extracted from IMD rainfall grid was analysed for the period 1979–2003. Annual rainfall in the study GP ranges between 970 mm in the year 1987 to a maximum value of 2281 mm in the year 2011, indicating high temporal variability (Fig. 7.4).

Mean annual rainfall of the study area computed based on gridded rainfall data corresponding to a study period of 35 years (1979–2013) is nearly 1640 mm. Number

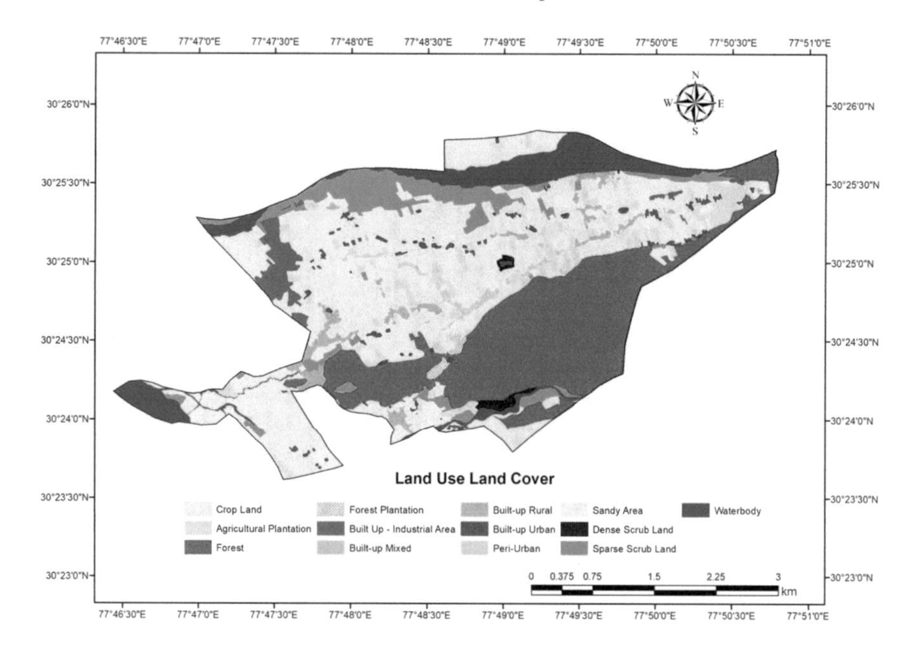

Fig. 7.3 Spatial distribution of LULC classes in Chharba GP

Table 7.4 Land use/land cover area statistics in the Chharba GP, Uttrakhand

S.No.	Land use/land cover Class	Area (ha)	(%) area
1	Agricultural area	629	43.6
2	Agricultural plantations	72	5.0
3	Built-up mixed	0.65	0.0
4	Built-up rural	47	3.3
5	Built-up urban	20	1.4
6	Built-up—industrial area	32	2.2
7	Peri-urban	19	1.3
8	Forest	355	24.6
9	Forest plantations	5.4	0.4
10	Sparse scrub land	153	10.6
11	Dense scrub land	7	0.5
12	Sandy area	15	1.0
13	Waterbody	86	5.9

of rainy days in the study GP was also computed for different years that represent wet, dry and normal years. Mean number of rainy days are observed to be 110, 124 and 98, respectively for dry, normal and wet years (Table 7.5). Further rainfall events categorized into different classes namely Light, Moderate, Rather Heavy, Heavy and

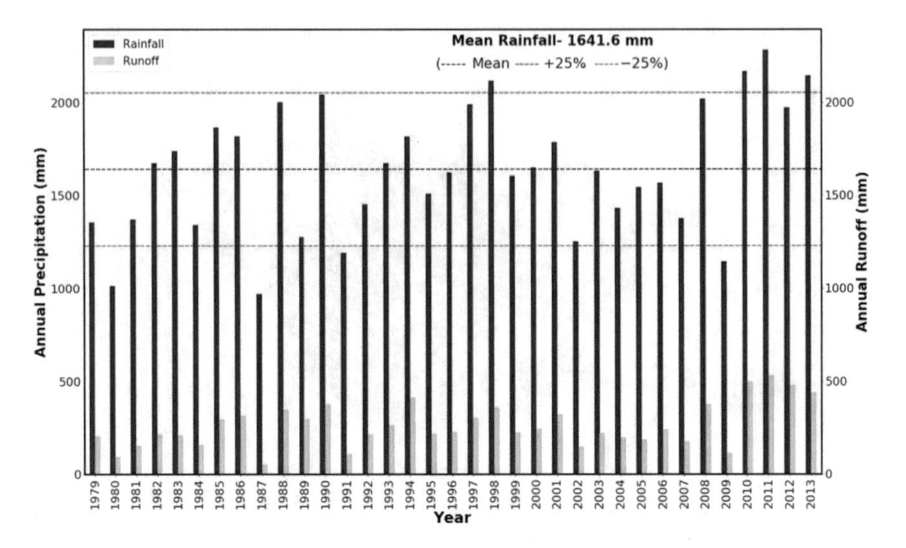

Fig. 7.4 Temporal distribution of annual rainfall and computed surface runoff for the period 1979–2013

Very Heavy rainfall events were computed for the entire study period and observed to be 82, 52, 8, 2 and 1, respectively for the entire study period (Fig. 7.5).

7.3.3 *Surface Water Potential Assessment*

The runoff resulting from the highest rainfall event (162 mm) during the span of 35 years' recurrence interval period (1979–2013) was estimated. Spatial distribution of runoff corresponding to highest rainfall event in the study GP indicated that a major portion of the study area yielded runoff of <50 mm, 50–100 mm and 100–150 mm during AMC I, AMC II and AMC III conditions, respectively (Fig. 7.6a–c). Further, the spatial distribution of runoff in the study GP for three meteorological conditions that signify dry year (1991), normal year (1988) and wet year (2010) were also computed (Fig. 7.6d–f), respectively. Annual runoff computations indicated that a major portion of study GP contributed <100 mm, 300–400 mm and 500–800 mm of runoff during representative dry, normal and wet meteorological years, respectively.

It is observed that the GP experiences high surface runoff during the monsoon season, which needs to be conserved for meeting the water demand during the lean season. A comprehensive assessment of hydrological parameters at GP level that include rainfall intensity, runoff variations provide key guidelines for planning suitable soil and water conservation measures. The approximate runoff potential and future scope for water harvesting for the GP was worked out. Thus, surface runoff assessment spatially at annual scale and also for highest rainfall event under various

Table 7.5 Rainfall analysis for dry, wet and normal conditions (1979–2003)

Statistics	Meteorological condition								
	Dry years			Wet years			Normal years		
	Rainfall (mm)	Runoff (mm)	Rainy days	Rainfall (mm)	Runoff (mm)	Rainy days	Rainfall (mm)	Runoff (mm)	Rainy days
Mean	1080	93	110	2177	457	124	1263	218	98
±Standard Dev.	105	28	10	72	75	9	737	140	53

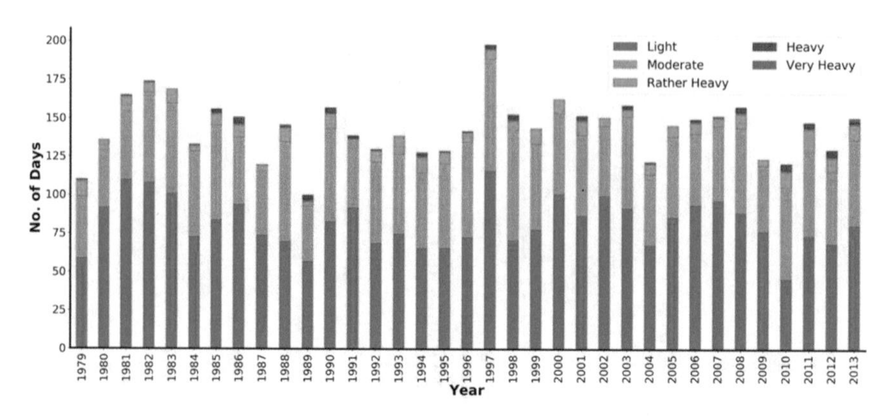

Fig. 7.5 Temporal variation of number of rainfall events during the study period (1979–2013)

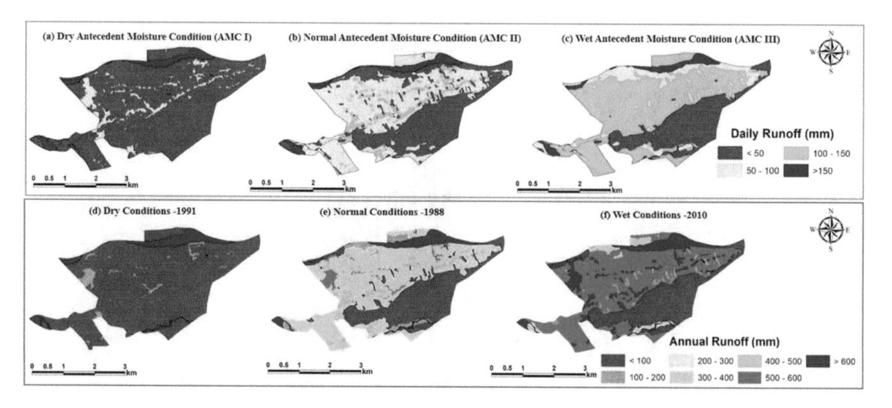

Fig. 7.6 Spatial distribution of surface runoff (**a**) Antecedent moisture condition (AMC I) for highest rainfall event, (**b**) Antecedent moisture condition (AMC II)) for highest rainfall event, (**c**) Antecedent moisture condition (AMC III)) for highest rainfall event, (**d**) Dry meteorological conditions (Year 1991), (**e**) Normal meteorological conditions (Year 1988), (**f**) Wet meteorological conditions (Year 2010)

meteorological and AMC conditions not only helps in the identification of suitable recharge zones but is also useful in the design of soil conservation structures.

7.3.4 Water Resources Development Plan Over the Region

Favourable conditions suitable for identification of suitable sites for adoption of soil and water conservation measures were integrated using boolean conditions as given in Table 7.2. The WRD plan for the study GP is spatially represented in Fig. 7.7. The plan indicated that nearly 0.6% (8.43 ha) and 8% (121 ha) of area is suitable for check

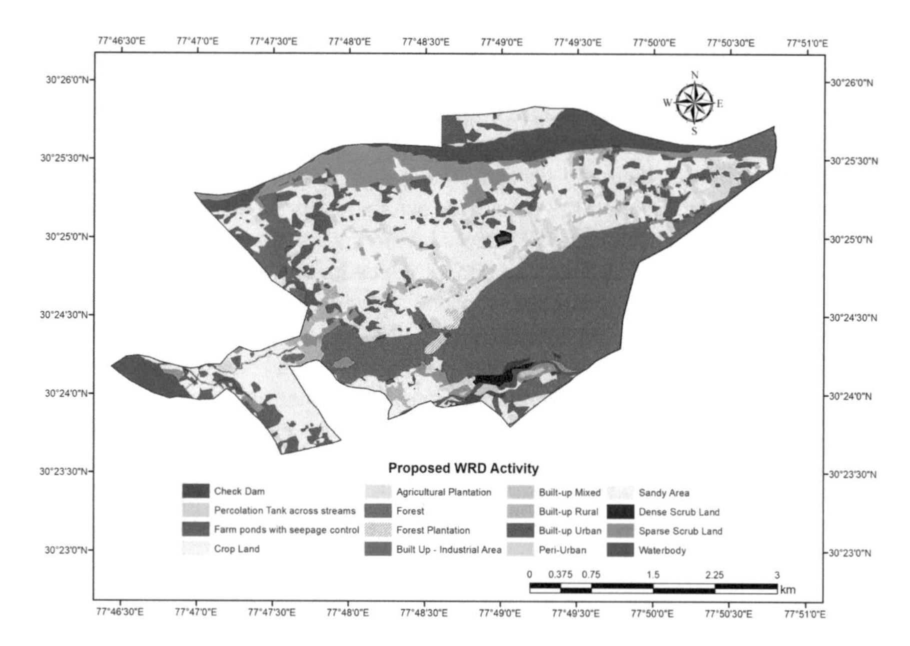

Fig. 7.7 Water resource development plan for the Chharba GP

dams and farm ponds, respectively. Rainwater harvesting measures such as check dams reduce runoff velocity while retaining the water behind the check dam, thereby allowing it to percolate into the aquifer. This not only minimizes soil erosion but also replenishes groundwater reserves at downstream sites. Implementation of farm ponds attempts to mitigate the temporal discontinuity between rainfall availability and irrigation crop demand. Thus, adoption of such rainwater harvesting measures to aid in the recycling of water helps in the intensification of cropping systems in terms of double and triple cropping systems. This will further enhance the scope for conversion of single cropped areas to agro-horticulture systems.

7.3.5 Land Resources Development Plan Over the Region

Geospatial inputs that include groundwater prospects, slope, existing LULC and soil conditions were integrated to generate the LRD plan for the study GP as per the guidelines given in Table 7.3. LULC cover generated using very high-resolution data for the study GP is shown as Fig. 7.3. Cropped area and area under agricultural plantations are nearly 629 ha and 72 ha, respectively (Table 7.4). The study GP indicated good ground water prospects (Fig. 7.2a) and a major portion of the study area is covered under the slope categories below 8%. Overall, the slope and soil conditions are also favourable for transformation of single cropped areas into intensive agriculture. Land resource development plan for the study GP is spatially represented in Fig. 7.8. The analysis revealed that this area under single cropped areas can be converted to intensive agricultural areas while nearly 35 ha under agricultural plantations can be converted to agro horticulture. Further, the land use category of sparse scrubland can be converted to agroforestry.

Implementation of land and water resource development plans at GP level not only helps to meet the long-term objectives but also helps to mobilize the financial resources and government policy to achieve hierarchical goals. Further, involvement of local people is quite necessary as part of education, awareness and consensus. Implementation, monitoring and maintenance of the plans and evaluation of implemented schemes for their end benefits should also be part of the strategic planning activities. The geo-referenced cadastral map at 1:4000 scale, when overlaid with land and water resources plans may directly be correlated with ground information for effective implementation. Continuous monitoring of the available water resources is necessary and should also address the environmental degradation of the GP. The programme should also be updated as per the priorities and government policies. The methodology is generic in nature and can be adapted for other GPs' as well.

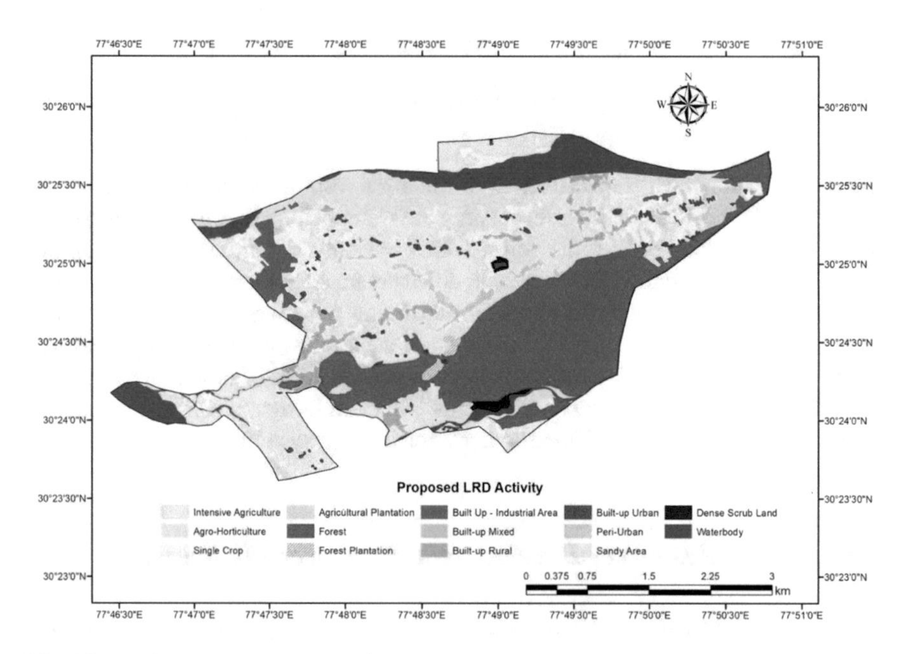

Fig. 7.8 Land resource development plan for the Chharba GP

7.4 Conclusions

The Chharba GP, in Dehradun district of Uttrakhand state was considered in this study. The study GP has its specific issues, potentials and developmental needs, defined and not limited by a set of geophysical, climatic and socio-economic factors. Inventory of natural resources in the study GP using very high-resolution satellite image enabled accurate mapping of natural resources, which forms the basis for the generation of land and water resource development plans with higher accuracy. Multiple thematic data pertaining to land and water resources, topography, climate and soil characteristics were integrated using MCE analysis in the GIS platform so as to identify suitable zones for adoption of RWH measures. Long-term rainfall data from IMD grid was analysed for different meteorological conditions and was used to evaluate the surface runoff spatially. The high annual rainfall with good number of rainy days, along with the sufficient surface water potential improves the scope for planning rainwater harvesting. The relatively large vegetation cover and adoption of intensive agriculture in the GP will help in holistic development of the region. Thus, the suggested land and water resources development plans are expected to convert the existing land use practices into more sustainable categories, that will meet future requirements without jeopardizing the environment. Conjunctive use of surface and ground water resources along with the implementation of a policy to earmark a percentage of land for agro- horticulture system plays a crucial role for sustainable land use planning. The framework for both land and water resource development planning are generic in nature and has the potential to be readily implemented across other agro-climatic zones to address resource scarcity and challenges arising due to climate change.

Acknowledgements Authors express sincere thanks to the National Remote Sensing Centre (NRSC) for the necessary resources and data for carrying out this study.

References

Adham A, Sayl KN, Abed R, Abdeladhim MA, Wesseling JG, Riksen M, Fleskens L, Karim U, Ritsema CJ (2018) A GIS-based approach for identifying potential sites for harvesting rainwater in the Western Desert of Iraq. Int Soil Water Conserv Res 6(4):297–304

Akinci H, Özalp AY, Turgut B (2013) Agricultural land use suitability analysis using GIS and AHP technique. Comput Electron Agric 97:71–82

Ammar A, Riksen M, Ouessar M, Ritsema C (2016) Identification of suitable sites for rainwater harvesting structures in arid and semi-arid regions: a review. Int Soil Water Conserv Res 4(2):108–120

Chowdary VM, Ramakrishnan D, Srivastava YK, Chandran V, Jeyaram A (2009) Integrated water resource development plan for sustainable management of Mayurakshi watershed, India using remote sensing and GIS. Water Resour Manage 23(8):1581–1602

Chowdhury A, Jha MK, Chowdary VM (2010) Delineation of groundwater recharge zones and identification of artificial recharge sites in West Medinipur district, West Bengal, using RS, GIS and MCDM techniques. Environ Earth Sci 59(6):1209–1222

Chowdary VM, Chakraborthy D, Jeyaram A, Murthy YK, Sharma JR, Dadhwal VK (2013) Multi-criteria decision making approach for watershed prioritization using analytic hierarchy process technique and GIS. Water Resour Manage 27(10):3555–3571

Cools N, De Pauw E, Deckers J (2003) Towards an integration of conventional land evaluation methods and farmers' soil suitability assessment: a case study in northwestern Syria. Agr Ecosyst Environ 95(1):327–342

Das S, Behera SC, Kar A, Narendra P, Guha S (1997) Hydrogeomorphological mapping in ground water exploration using remotely sensed data—a case study in keonjhar district, orissa. J Indian Soc Remote Sens 25(4):247–259

Dalezios NR, Angelakis AN, Eslamian S (2018) Water scarcity management: part 1: methodological framework. Int J Global Environ Issues 17(1):1–40

De Winnaar G, Jewitt GPW, Horan M (2007) A GIS-based approach for identifying potential runoff harvesting sites in the Thukela river basin, South Africa. Phys Chem Earth, Parts a/b/c 32(15–18):1058–1067

FAO (2011) The state of the world's land and water resources for food and agriculture: managing systems at risk. Earthscan

Farrington J, Turton C, James AJ (1999) Participatory watershed development. Challenges for the twenty-first century

Glendenning CJ, Van Ogtrop FF, Mishra AK, Vervoort RW (2012) Balancing watershed and local scale impacts of rain water harvesting in India—a review. Agric Water Manag 107:1–13

Gurmel S, Venkataramanan C, Sastry G, Joshi BP (1990) Manual of soil and water conservation practices. *Manual of soil and water conservation practices*

Gunnell Y, Anupama K, Sultan B (2007) Response of the South Indian runoff-harvesting civilization to northeast monsoon rainfall variability during the last 2000 years: instrumental records and indirect evidence. The Holocene 17(2):207–215

IMSD (1995) Integrated mission for sustainable development technical guidelines. National remote sensing agency, Department of space, Govt. of India

Janssen R, Rietveld P (1990) Multicriteria analysis and geographical information systems: an application to agricultural land use in the Netherlands. In: Geographical information systems for urban and regional planning. Springer, Dordrecht, pp 129–139

Jansen LJ, Di Gregorio A (2004) Obtaining land-use information from a remotely sensed land cover map: results from a case study in Lebanon. Int J Appl Earth Obs Geoinf 5(2):141–157

Jha MK, Peiffer S (2006) Applications of remote sensing and GIS technologies in groundwater hydrology: past, present and future. Bayreuth, BayCEER, p 201

Joerin F, Thériault M, Musy A (2001) Using GIS and outranking multicriteria analysis for land-use suitability assessment. Int J Geogr Inf Sci 15(2):153–174

Kahinda JM, Lillie ESB, Taigbenu AE, Taute M, Boroto RJ (2008) Developing suitability maps for rainwater harvesting in South Africa. Phys Chem Earth, Parts a/b/c 33(8–13):788–799

Kalita N, Sarmah R (2016) Soil loss sensitivity in the Belsiri river basin using Universal soil loss equation in GIS. Int J Curr Res 8(3):28831–28838

Khoi DD, Murayama Y (2010) Delineation of suitable cropland areas using a GIS based multi-criteria evaluation approach in the Tam Dao national park region, Vietnam. Sustainability 2(7):2024–2043

Malczewski J (2004) GIS-based land-use suitability analysis: a critical overview. Prog Plan 62(1):3–65

Malczewski J (2006) GIS-based multicriteria decision analysis: a survey of the literature. Int J Geogr Inf Sci 20(7):703–726

Mendas A, Delali A (2012) Integration of multicriteria decision analysis in GIS to develop land suitability for agriculture: application to durum wheat cultivation in the region of mleta in Algeria. Comput Electron Agric 83:117–126

Mythili G, Goedecke J (2016) Economics of land degradation in India. In: Economics of land degradation and improvement–a global assessment for sustainable development. Springer, Cham, pp 431–469

Nkonya E, Mirzabaev A, von Braun J (2016) Economics of land degradation and improvement: an introduction and overview. In: Economics of land degradation and improvement—a global assessment for sustainable development. Springer, Cham, pp 1–14

NRSC (2012) Project report on" Groundwater prospects mapping under Rajiv Gandhi drinking water mission. Study commissioned by ministry of rural development, GOI

Oweis T, Hachum A (2006) From water use efficiency to water productivity: issues of research and development. AARINENA water use efficiency network, p 14

Oweis TY, Prinz D, Hachum AY (2012) Rainwater harvesting for agriculture in the dry areas. CRC press

Oweis T, Oberle A, Prinz D (1998) Determination of potential sites and methods for water harvesting in central Syria. Advan GeoEcol 31:83–88

Pandey A, Behra S, Pandey RP, Singh RP (2011) Application of GIS for watershed prioritization and management—a case study. Inter J EnvSci Dev Monit 2(1):25–42

Pai DS, Rajeevan M, Sreejith OP, Mukhopadhyay B, Satbha NS (2014) Development of a new high spatial resolution (0.25 × 0.25) long period (1901–2010) daily gridded rainfall data set over India and its comparison with existing data sets over the region. Mausam 65(1):1–18

Pretall O, Polius J (1981) Land resources in St. Lucia: land caspability classification and crop allocation. In: St. Lucia development atlas, department of regional development, organization of American states, Washington DC, USA, p 29

Ravindran KV, Jeyaram A (1997) Groundwater prospects of Shahbad tehsil, Baran district, eastern Rajasthan: a remote sensing approach. J Indian Soc Remote Sens 25(4):239–246

Saraf AK, Choudhury PR (1998) Integrated remote sensing and GIS for groundwater exploration and identification of artificial recharge sites. Int J Remote Sens 19(10):1825–1841

SCS (1985) National engineering handbook, section 4: hydrology. US soil conservation service, USDA, Washington, DC

Shankar MR, Mohan G (2005) A GIS based hydrogeomorphic approach for identification of site-specific artificial-recharge techniques in the Deccan volcanic province. J Earth Syst Sci 114(5):505–514

Subramanya K (2008) Engineering hydrology. 7 West Patal Nagar. New Delhi 110(008)

Tamagnone P, Comino E, Rosso M (2020) Rainwater harvesting techniques as an adaptation strategy for flood mitigation. J Hydrol 586:124880

Tienwong K, Dasananda S, Navanugraha C (2009) Integration of land evaluation and the analytical hierarchical process method for energy crops in Kanchanaburi, Thailand. Sci Asia 35(2):170–177

UNFPA (United Nations Population Fund) (2008) State of the world population, culture, gender and human rights. UNFPA, 108

Ustin SL (ed) (2004) Manual of remote sensing, remote sensing for natural resource management and environmental monitoring, vol. 4. Wiley

Chapter 8
Delineation of Waterlogged Areas Using Geospatial Technologies and Google Earth Engine Cloud Platform

Neeti Neeti, Ayushi Pandey, and V. M. Chowdary

Abstract The plethora of remotely sensed datasets, specifically the Sentinel-2 data, provides an opportunity for assessment of waterlogging areas. Surface waterlogging is one of the main hazards in north Bihar, specifically in Gandak and Kosi command areas, which necessitates to map and monitor the waterlogging situation accurately at a fine scale. Present study is carried out with an objective of mapping the permanent and seasonal waterlogging areas in a flood-prone Vaishali district, Bihar state using Sentinel-2 multi-spectral data available at 10 m spatial resolution for 2020. The permanent and seasonal waterlogged area mapping were carried out using two approaches, namely spectral index (NDWI) and Otsu method based on the wetness of KT transformed images. The result suggests that both techniques could identify similar spatial patterns of waterlogged areas, while differed on area statistics. The total area under permanent and seasonal waterlogging using NDWI was less compared to the Otsu-KT approach. The higher estimate of Otsu-KT approach could be related to saturated areas, where pixels with higher moisture content during the binary thresholding approach may get classified. Although Otsu algorithm proved to be effective for delineation of water bodies and river network, its efficacy for delineation of waterlogged areas with saturated areas needs to be studied extensively.

N. Neeti (✉)
TERI School of Advanced Studies, New Delhi, India
e-mail: neeti@terisas.ac.in

A. Pandey
Graphic Era (Deemed To Be University), Dehradun, India

V. M. Chowdary
Regional Remote Sensing Center-North, NRSC, New Delhi, India

A. Pandey et al. (eds.), *Geospatial Technologies for Land and Water Resources Management*, Water Science and Technology Library 103,
https://doi.org/10.1007/978-3-030-90479-1_8

8.1 Introduction

Surface water logging is the condition of stagnant water in the depression that arises due to the accumulation of surface runoff (Lohani et al. 1999; Chatterjee et al. 2003). Waterlogging could occur due to surface flooding or a rise in the groundwater table. Generally, waterlogging occurs in shallow depressions situated in proximity to water bodies such as rivers and lakes. Waterlogging can also occur in agricultural areas due to canal irrigation or drainage impedance due to an impervious layer of clay due to improper alignment of the canal. The impact of waterlogging can be seen in terms of crop damage, salinization due to increased amounts of salts such as sodium chloride and sodium sulfate leading to excess salt than what is required for normal crop growth.

In India, approximately, 11.6 million ha of the net sown area is under waterlogged condition (Planning Commission, 2011), out of which more than 20% area is reported from the eastern part of India, i.e., Bihar and West Bengal, part of Odisha, and Uttar Pradesh states thereby making area unproductive. These states constitute a significant part of the Ganga Meghna-Brahmputra (GMB) basin and are home to 500 million of the world's poorest people (Chowdhury et al. 2018). Thus, the water logging condition directly leads to issues related to food security.

Waterlogging is one of the major disasters in north Bihar, specifically in the Gandak and Kosi command areas (Chowdary et al. 2008). Some of the reasons for waterlogging in these areas are silting in canals (Dutta et al. 2004), seepage from water conveyance systems (Brahmabhatt et al. 2000), among others. Being highly populated with the very low socio-economic status of the region, and its coverage over a larger area, waterlogging hazard in this region is considered to be a major disaster. Thus, periodic mapping of waterlogging conditions and subsequent planning for adoption of measures for its mitigation is essential for this region.

Over the years, state-of-the-art remote sensing and GIS technologies have made possible the accurate detection of surface waterlogging conditions. Initial studies considered extraction of waterlogged areas through visual interpretation techniques using Landsat or IRS LISS III dataset (Chowdary et al. 2008). Xu and Ma (2021) conducted a Coarse-to-fine waterlogging probability assessment using remote sensing images and social media data for Wuhan, China. However, the possibility of detection of waterlogged conditions at a very fine scale has vastly improved with the availability of a fine Spatio-temporal resolution Earth Observation dataset. The number of spectral indices has been developed over time using different multispectral bands of remotely sensed data. One such important index is the Normalized Difference Water Index (NDWI) that uses the higher reflectance of water in green and complete absorption in near-infrared spectral bands to identify water features. NDWI is largely affected by building signals in the image, although it could efficiently suppress signals associated with vegetation and soil (McFeeters 1996). Han-Qiu (2005) introduced Modified Normalized Difference Water Index (MNDWI) for urban water extraction with higher accuracy, using a shortwave infrared band. Jingqi et al. (2011) constructed an automatic water extraction model based on Kauth Thomas transformation. Recently, Li et al. (2021) proposed another index based on K-T transformation for water extraction. In this approach, the threshold for the wetness component of K-T transformation was identified using the Otsu approach, while removing shadow and cloud using the support vector machine-based approach.

In recent times, Sentinel-2 images from the Copernicus program of the European Commission and the European Space are widely used for water body extraction. The fine spatial and temporal resolution of these images relative to Landsat-8 images, made them acceptable globally. Delineation of the river from Sentinel-2 is more accurate than Landsat-8 (Yang and Chen 2017). The Sentinel-2 images are having visible and near-infrared spectral bands at 10 m resolution, while the shortwave infrared band is available at 20 m. Hence, water body extraction based on the shortwave infrared band is less accurate up to a certain extent (Li et al. 2021). However, NDWI generated at 10-m resolution using green and near-infrared band is also known for misclassification errors (Fisher et al. 2016). Though band sharpening could reduce misclassification (Du et al. 2016), it is a tedious, time-consuming process. Studies on the relative performance of remote sensing-based approaches for identifying permanent and seasonal waterlogged areas are limited. Google Earth Engine (GEE) is an open-source tool and allows to consult and work simultaneously with multiple datasets (Fattore et al. 2021). GEE is a powerful high-performance computing tool, and it is accessed and controlled through a web-based accessible application programming interface (API) (Fattore et al. 2021). Hence, this study is carried out with the specific objective to map permanent and seasonal waterlogged areas in the Vaishali district of Bihar using the open-source GEE platform. The study envisaged spectral index and dynamic thresholding-based approach Otsu for mapping of waterlogged areas in this region.

8.2 Materials and Methods

8.2.1 Study Area

The study area, Vaishali district of Bihar state, India, is geographically situated between 83° 19.83′ to 88° 17.67′ E longitude and 24° 20.17′ to 27° 31.25′ N latitude (Fig. 8.1). The study district, Vaishali is located in the northern part of the state, a major flood-prone area, and experiences severe seasonal waterlogging problems. Kosi, Gandak, and Burhi Gandak river basins constitute major sub-basins of the Upper Ganga basin within North Bihar region. Vaishali district constitutes a part of the Ganga river basin with two sub-basins, namely Gandak and Burhi–Gandak, where major portion of the study area falls under Burhi–Gandak basin.

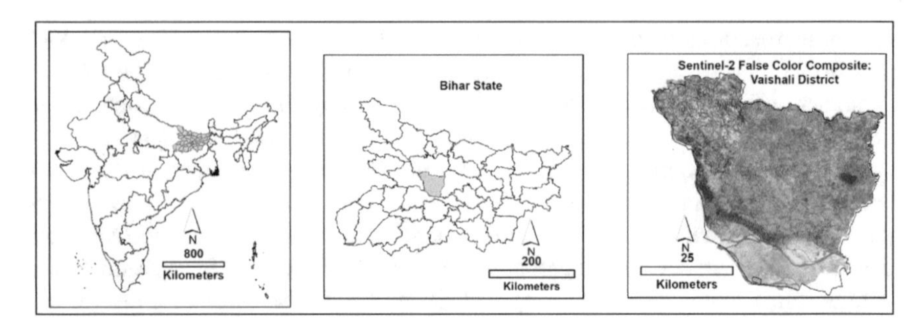

Fig. 8.1 Location map of study area

8.2.2 Data Used

Multi-temporal Sentinel-2 data corresponding to study area for the year 2020 was used in the analysis. The data of year 2020 was selected for extraction of the seasonal and permanent waterlogging area as they correspond to wet year. Total annual rainfall from Climate Hazards Group Infrared Precipitation with Stations (CHIRPS) rainfall gridded product at a spatial resolution of 0.05° (~5 km) during the year 2020 is observed to be nearly 1500 mm. Sentinel-2 consisting of two satellites, namely 2A and 2B, was launched in 2015 and 2017, respectively, and have a revisit time of 5 days. Sentinel-2 carries a multi-spectrometer with 13 spectral bands spanning across visible, near-infrared, and shortwave infrared parts of the electromagnetic spectrum. These images are available freely from Sentinel Scientific data hub (https://scihub.copernicus.eu) and are the "COPERNICUS/S2" data available at the Google Earth Engine platform were analyzed.

8.2.3 Methodology

The Sentinel-2 images from April and May for 2020 were analyzed for dry season (pre-monsoon) waterlogged areas, while data available during November and December months were considered for investigating wet season (post-monsoon) waterlogged areas. The entire analysis was carried out on Google Earth Engine (GEE) platform. GEE is a cloud platform made available by Google for online analysis of global-scale geospatial data (Gorelick et al. 2017). The platform stores freely available Earth Observation datasets and provide efficient computing power for online data processing. The GEE platform's major advantage is that it works on cloud mode, where the analysis does not require downloading the images and processing them in a local computer that requires more time and physical space in the local system.

8.2.4 Preprocessing

The "COPERNICUS/S2" data was used to produce seasonal cloud-free composites for dry and wet seasons. The Quality Assessment (QA) band and median reducer in GEE were used to improve data quality. QA values help in the indication of pixels that might be cloud contaminated. In addition, GEE can be instructed to pick the median pixel value in the stack to reject values that are too bright (e.g., due to cloud) or too dark (e.g., shadows) and picks the median pixel values in each band over time. The study area did not have a shadow problem, and most of the cloud-contaminated images were not considered during the selection of the post-monsoon season dataset. Using the QA band and median reducer, cloud contaminations were further reduced in the composite images.

8.2.5 K-T Transformation of Sentinel-2 Images

K-T transformation, proposed by Kauth and Thomas, is an orthogonal transformation technique also known as Tasseled cap transformation similar to Principal Component Analysis. K-T transformation rotates the coordinate space and points to the direction of the plant growth process and the soil. The three components of K-T transformation are brightness, greenness, and wetness. The brightness component represents the total radiation level. The greenness component represents vegetation growth, and wetness represents the humidity of ground objects and is commonly used for water information extraction. There are two sets of conversion coefficients of the three K-T transformation components available. Nedkov (2017) and Shi and Xu (2019) deduced the conversion coefficients of the three components for Sentinel-2 using Gram Schmidt Orthogonalization (GSO) and Procrustes Analysis (PCP) method, respectively. The coefficients for KT transformation using PCP for the three components are given in Table 8.1 (Li et al. 2021).

The wetness component derived using the PCP method is more successful in discriminating water bodies from background objects than the GSO approach (Li et al. 2021). Therefore, in this study, GSO approach was adopted for K-T transformation of multi-temporal Sentinel-2 images for water body detection.

8.2.6 Otsu-Based Approach for Identification of Dynamic Threshold for Waterbody Extraction

Otsu is a dynamic threshold determination approach to determine image threshold by maximizing class variance (Otsu 1979). This approach is not affected by differences in image contrast (Xie et al. 2016), where the brightness value of an image is divided into two parts according to brightness level. The principle behind this approach is to

Table 8.1 K-T transformation coefficients of Sentinel-2 images using PCP approach

Spectral bands	PCP		
	Brightness	Greenness	Wetness
B1-Coastal	0.2381	−0.2266	0.1825
B2-Blue	0.2569	−0.2818	0.1763
B3-Green	0.2934	−0.302	0.1615
B4-Red	0.302	−0.4283	0.0486
B5-RE-1	0.3099	−0.2959	0.017
B6-RE-2	0.374	0.1602	0.0223
B7-RE-3	0.418	0.3127	0.0219
B8-NIR-1	0.358	0.3138	−0.0755
B8A-NIR-2	0.3834	0.4261	−0.091
B9-WV	0.0103	0.1454	−0.1369
B10-Cirrus	0.002	−0.0017	0.0003
B11-SWIR-1	0.0896	−0.1341	−0.771
B12-SWIR-2	0.078	−0.2538	−0.5293

Source Li et al. (2021)

minimize differences in brightness value within each part and maximize differences between the two parts. The process of threshold computations using the Otsu method can be given in the following steps:

$$\sigma^2 = P_{nw}(G_{nw} - G)^2 + P_w(G_w - G)^2 \tag{8.1}$$

$$M = P_{nw}G_{nw} + P_wG_w \tag{8.2}$$

$$P_{nw} + P_w = 1 \tag{8.3}$$

$$t = \text{ArgMax}\{P_{nw}(G_{nw} - G)^2 + P_w(G_w - G)^2\} \tag{8.4}$$

where σ^2 is the class variance between non-water and water. P_{nw} and P_w are the probability of a pixel that belongs to the non-water and water classes, respectively. G_{nw} and G_w are average brightness levels of non-water and water pixels, respectively. G is the average brightness level of the image pixel.

8.2.7 *Spectral Water Index Method*

The principle of the spectral water index method is based on the differential spectral characteristics of water bodies in different multi-spectral image bands. Generally, a spectral index is constructed by considering the contrast between high absorption and high reflectance of water in different bands. One of the commonly used spectral index is the Normalized Difference Water Index (NDWI) that uses green and near-infrared bands of the multi-spectral image and is computed as follows:

$$NDWI = \frac{\rho_{\text{green}} - \rho_{\text{NIR}}}{\rho_{\text{green}} + \rho_{\text{NIR}}}$$

where, ρ_{green} and ρ_{NIR} are the reflectance in green and Near-Infrared bands, and its value ranges between -1 and $+1$. Pure water has maximum absorption in the NIR band, and higher reflectance in the green band, and therefore such pixel in an image will always have positive NDWI. The water information is extracted from NDWI by threshold-based segmentation. Generally, 0 is taken as the threshold for differentiating between water and non-water pixels. In the case of Sentinel-2, both green and NIR bands are available at 10 m resolution, and therefore, NDWI was used for analysis.

8.3 Assessment of Permanent and Seasonal Waterlogged Areas

The waterlogged area was assessed for the two seasons using two approaches: (1) Otsu-based approach on wetness component of K-T transformation; (2) NDWI approach. Waterbodies in pre-monsoon and post-monsoon months were extracted using multi-spectral Sentinel-2 images using the two approaches. In former approach (1), all the pixels for which wetness value (W) was greater than the Otsu-based threshold (T) were considered as water. All the permanent water bodies were masked out using a published land use/land cover map for the study area (NRSC). Permanently, waterlogged areas were identified if a pixel was waterlogged in pre-monsoon months. However, if a pixel was waterlogged only in the post-monsoon season, it was considered seasonally waterlogged.

8.4 Results and Discussion

The surface waterlogged areas in the study area were delineated using pre-and post-monsoon Sentinel-2 imageries for 2020. The areas experiencing surface waterlogging during both pre-and post-monsoon periods are considered as permanently waterlogged, whereas the post-monsoon surface waterlogged areas after eliminating permanently waterlogged areas are classified as seasonally waterlogged areas.

The permanent and seasonal waterlogged area mapping were carried out using two approaches, namely spectral index (NDWI) and Otsu method based on the wetness of KT transformed images.

The spatial pattern of pre-monsoon and post-monsoon waterlogged areas for the year 2020 is given in Fig. 8.2. Most of the waterlogged areas can be seen in the eastern and southern parts of the district.

The waterlogged area in 2020 is given in Table 8.2. Nearly, 1500 ha was identified as permanent waterlogged when NDWI approach was used. Otsu-KT approach identified nearly 1000 ha more permanent waterlogged area relative to NDWI approach. As described earlier, the higher estimate due to the Otsu-KT approach could be related to wetness associated with moisture content. Therefore, the pixels with higher moisture content during the binary thresholding approach may get classified as water, while it may not be accurate. Total seasonally waterlogged area in 2020 was estimated as 12% of the district area using NDWI and 18.3% of the district area using the

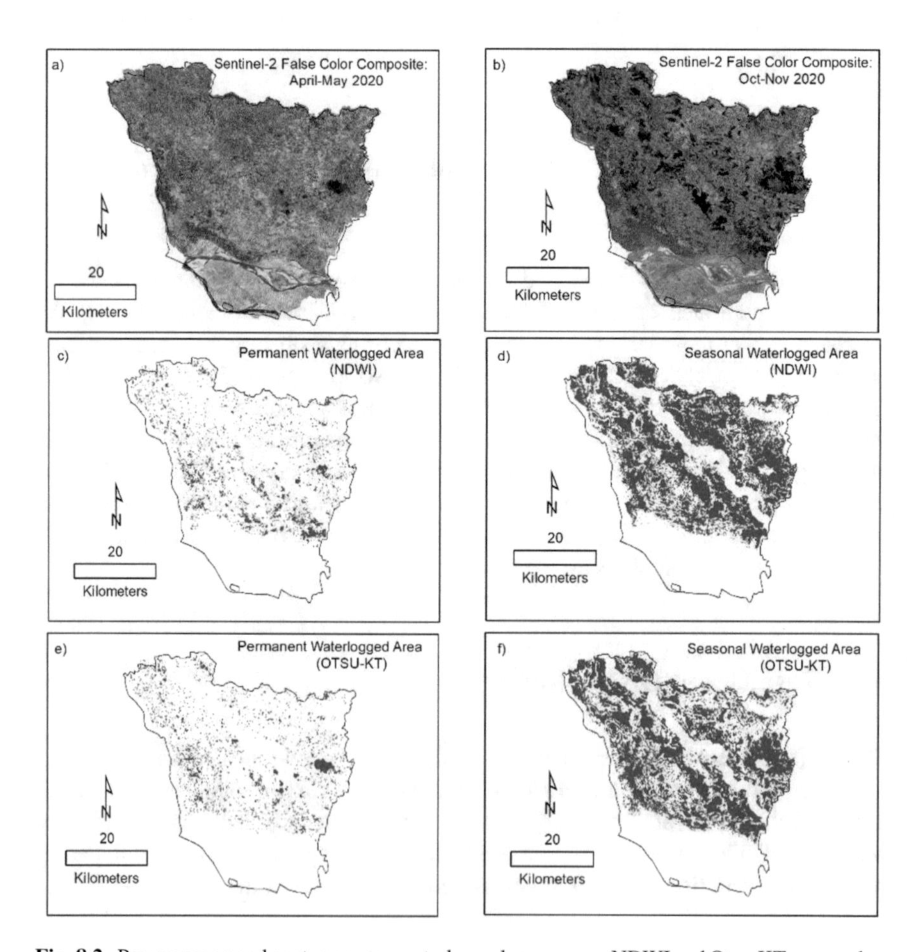

Fig. 8.2 Pre-monsoon and post-monsoon waterlogged areas as per NDWI and Otsu-KT approaches

Table 8.2 Permanent and seasonal waterlogged area in 2020

Approach	Permanent waterlogged area in hectare (% of the study area)	Seasonal waterlogged area in hectare (% of the study area)
NDWI	1495.83 (0.74%)	24,045.36 (11.95%)
OTSU-KT	2610.14 (1.29%)	36,845. 65 (18.31%)

Otsu-KT approach. Zhou et al. (2020) also reported that Otsu approach can be effective for extraction of water surface using Sentinel-1 data. However, this approach is a unsupervised method needs to be evaluated with respect to supervised learning algorithms. Further, accuracy of Otsu approach depends on the identification of wet ness component of K-T transformation and as well as aquatic vegetation. Serious commission errors by NDWI approach were reported by Li et al. (2021) in case of river extraction. Further Li et al. (2021) observed significant improvement in the accuracy when wetness component of the image was obtained by K-T transformation, where threshold T is computed by Otsu algorithm. Although Otsu algorithm proved to be effective for delineation of water bodies and river network, its efficacy for delineation of waterlogged areas with saturated areas needs to be studied extensively. The prevailing waterlogging situation, would affect the soil's salinity, thereby affecting the agricultural practices and cropping patterns in the area.

8.5 Conclusions

This study attempts to map and monitor the waterlogging situation in a flood-prone Vaishali district in Bihar. The study was carried out using a fine resolution Sentinel-2 dataset using a spectral index-based NDWI approach and the Otsu-KT approach. To overcome the drawbacks of data storage and computation facilities, the utilization of Google Earth Engine (GEE) is beneficial for prompt implementation of various algorithms and analyzes. Although Otsu algorithm proved to be effective for delineation of water bodies and river network, its efficacy for delineation of waterlogged areas with saturated areas needs to be studied extensively. The study reveals that the waterlogging situation would ultimately affect the soil salinity and thus the crop productivity. Therefore, spatio-temporal assessment of waterlogged areas is necessary to improve and revise the management and the preparedness strategies.

References

Brahmabhatt VS, Dalwadi GB, Chhabra SB, Ray SS, Dadhwal VK (2000) Land use/land cover change mapping in Mahi canal command area, Gujarat, using multi-temporal satellite data. J Indian Soc Rem Sens 28(4):221–232

Chatterjee C, Kumar R, Mani P (2003) Delineation of surface waterlogged areas in parts of Bihar using IRS-1C LISS-III data. J Indian Soc Rem Sens 31(1):57–65

Chowdhury SR, Nayak AK, Brahmanand PS, Mohanty RK, Chakraborty S, Kumar A, Ambast SK (2018) Delineation of waterlogged areas using spatial techniques for suitable crop management in Eastern India. ICAR Res Bull 79

Chowdary VM, Chandran RV, Neeti N, Bothale RV, Srivastava YK, Ingle P, Singh R et al (2008) Assessment of surface and sub-surface waterlogged areas in irrigation command areas of Bihar state using remote sensing and GIS. Agricult Water Manag 95(7):754–766

Du Y, Zhang Y, Ling F, Wang Q, Li W, Li X (2016) Water bodies' mapping from Sentinel-2 imagery with modified normalized difference water index at 10-m spatial resolution produced by sharpening the SWIR band. Rem Sens 8(4):354. https://doi.org/10.3390/rs8040354

Dutta D, Sharma JR, Bothale RV, Bothale V (2004) Assessment of waterlogging and salt affected soils in the command areas of all major and medium irrigation project in India. Technical document. Regional Remote Sensing Service Centre, Jodhpur, India, 1–74

Fattore C, Abate N, Faridani F, Masini N, Lasaponara R (2021) Google earth engine as multi-sensor open-source tool for supporting the preservation of archaeological areas: the case study of flood and fire mapping in Metaponto, Italy. Sensors 21:1791

Fisher A, Flood N, Danaher T (2016) Comparing landsat water index methods for automated water classification in eastern Australia. Rem Sens Environ 175:167–182. https://doi.org/10.1016/j.rse.2015.12.055

Gorelick N, Hancher M, Dixon M, Ilyushchenko S, Thau D, Moore R (2017) Google Earth Engine: planetary-scale geospatial analysis for everyone. Rem Sens Environ 202:18–27

Han-Qiu X (2005) A study on information extraction of water body with the modified normalized difference water index (MNDWI). J Rem Sens 5:589–595

Jingqi Z, Wei G, Ping S (2011) Automatic water bodies extraction model based on KT transformation. Sci Soil Wat Conserv 3:88–92

Li J, Peng B, Wei Y, Ye H (2021) Accurate extraction of surface water in complex environment based on Google Earth Engine and Sentinel-2. Plos One 16(6):e0253209

Lohani AK, Jaiswal RK, Jha R (1999) Waterlogged area mapping of Mokama group of Tals using remote sensing and GIS. J Inst Eng 80:133–137

McFeeters SK (1996) The use of the normalized difference water index (NDWI) in the delineation of open water features. Int J Remote Sens 17(7):1425–1432. https://doi.org/10.1080/01431169608948714

Nedkov R (2017) Orthogonal transformation of segmented images from the satellite Sentinel-2. Comptes Rendus De L'academie Bulgare Des Sciences 70(5):687–692

Otsu N (1979) A threshold selection method from gray-level histograms. IEEE Trans Syst Man Cybern 9(1):62–66

Shi T, Xu H (2019) Derivation of tasseled cap transformation coefficients for Sentinel-2 MSI at-sensor reflectance data. IEEE J Sel Top Appl Earth Observ Rem Sens 12(10):4038–4048. https://doi.org/10.1109/JSTARS.2019.2938388

Wang J, Ding J, Yu D, Ma X, Zhang Z, Ge X, Teng D, Li X, Liang J, Lizaga I et al (2019) Capability of Sentinel-2 MSI data for monitoring and mapping of soil salinity in dry and wet seasons in the Ebinur Lake region, Xinjiang, China. Geoderma 353:172–187

Xie H, Luo X, Xu X, Pan H, Tong X (2016) Evaluation of Landsat 8 OLI imagery for unsupervised inland water extraction. Int J Remote Sens 37(8):1826–1844. https://doi.org/10.1080/01431161.2016.1168948

Xu L, Ma A (2021) Coarse-to-fine waterlogging probability assessment based on remote sensing image and social media data. Geospat Inf Sci 24(2):279–301

Yang X, Chen L (2017) Evaluation of automated urban surface water extraction from Sentinel-2A imagery using different water indices. J Appl Remote Sens 11(2):026016. https://doi.org/10.1117/1.JRS.11.026016

Zhou S, Kan P, Silbergagel J, Jin J (2020) Application of image segmentation in surface water extraction of fresh water lakes using Radar data. Int. J. Geo-Inf. 9(424):1–16. https://doi.org/10.3390/ijgi9070424

Chapter 9
Utility of Satellite-Based Open Access Data in Estimating Land and Water Productivity for a Canal Command

P. K. Mishra, Subhrasita Behera, P. K. Singh, and Rohit Sambare

Abstract Increasing competition for land and water resources is expected in future due to rising demands for food and bioenergy production, biodiversity conservation, and changing production conditions due to climate change. Growing competition for water in many sectors reduces its availability for irrigation. Thus, efficient approaches are required for effective management of water in every sector, particularly in agriculture. To achieve efficient and effective water use, it requires increasing crop water productivity (WP) and crop yield through improved crop varieties. Only high water productivity values carry little importance if they are not associated with increased or acceptable yields. Such association of high (or moderate) water productivity values with high (or moderate) yields has considerable implications on water's effective use. Land productivity and water productivity increment are the most efficient solution for meeting increasing food demand and climate variation. Water consumption (evapotranspiration (ET)) is an influencing factor for productivity estimation across a command area. Thus, in the present analysis, ET was estimated first, followed by land and water productivity assessment for the Hirakud canal command situated in the Mahanadi basin. Water accounting plus (WA+) Framework, jointly developed by the UNESCO-IHE, IWMI and FAO, has been utilized to assess the total water consumptions, agricultural water consumptions (using green water and blue water), and estimation of land and water productivity for a period of 12 years, i.e., 2003–2014. WA+ is a python based framework designed to provide explicit spatial information on water depletion and the net withdrawal process of a region using globally available open access data. The results showed that land productivity and

P. K. Mishra (✉) · P. K. Singh
Water Resources Systems Division, National Institute of Hydrology, Roorkee, Uttarakhand 247667, India

S. Behera
Water Resources and Environmental Engineering, Indian Institute of Science, Bengaluru, Karnataka 560012, India

R. Sambare
Research Management and Outreach Division, National Institute of Hydrology, Roorkee, Uttarakhand 247667, India

A. Pandey et al. (eds.), *Geospatial Technologies for Land and Water Resources Management*, Water Science and Technology Library 103,
https://doi.org/10.1007/978-3-030-90479-1_9

water productivity varies from 1973 to 2219 (kg/ha) and 0.41 to 0.55 (kg/m^3) for irrigated cereals (paddy), respectively, across the Hirakud canal command area.

9.1 Introduction

9.1.1 Land and Water Productivity?

Productivity is a ratio between a unit of output and a unit of input. Here, water productivity is used exclusively to denote the product's amount or value over the volume or value of water depleted or diverted. The value of the product might be expressed in different terms (biomass, grain, money). For example, the so-called "crop per drop" approach focuses on the amount of product per unit of water. However, water productivity, defined as kilogram per drop (kg/m^3), is a useful concept when comparing the productivity in different parts of the same system or river basin and comparing the productivity of water in agriculture with other possible water uses. Again land productivity also indicates the amount of production per unit of land (kg/ha). The purpose of defining these terms is to measure these resources' existing performance (Molden et al. 2004, 2010).

9.1.2 The Need for Land and Water Productivity Assessment

Increasing competition for land and water resources is expected in the coming future due to rising demands for food and bioenergy production, biodiversity conservation, and changing production conditions due to climate change (Allen et al. 1983). Growing competition for water in many sectors reduces its availability for irrigation. Thus, efficient approaches are required for effective management of water in every sector. To achieve efficient and effective water use, it requires increasing crop water productivity (WP) and drought tolerance by genetic improvement and physiological regulation. But only high-water productivity values carry little or no interest if they are not associated with increased or acceptable yields. Such association of high (or moderate) productivity values with high (or moderate) yields has significant implications on water's effective use. Land productivity and water productivity increment is the most efficient solution for meeting increasing food demand and climate variation. Water consumption (evapotranspiration (ET)) is an influencing factor for productivity estimation across a basin. Thus, ET estimation needs to be calculated first before productivity assessment.

9.1.3 *What is Water Accounting Plus (WA+)?*

Water accounting is a systematic acquisition, analysis and communication of information relating to stocks, flows and fluxes of water (from sources to sinks) in natural, disturbed or heavily engineered environments.

Water accounting frameworks	
System of Environmental Economic Accounting for Water (SEEAW)	United Nations Statistics Division (UNSD) in collaboration with the London Group on Environmental Accounting
Aquastat	FAO
Water Accounting Standard of Australia	Water Accounting Standards Board, Bureau of meteorology, Govt. of Australia
Water Accounting Plus	UNESCO-IHE, IWMI and FAO

While the former three are mostly flow accounting frameworks, WA+ is a comprehensive depletion accounting framework for assessing, reporting, communicating, and analyzing water resources status based on open access datasets and with standard terminology.

For communicating water resources related information and services obtained from consumptive use in a geographical domain to users, water accounting (WA) is the best process WA is a tool to take a sound water management decision. The IWMI WA framework was originally designed for irrigation schemes within a basin but was later used for basin analysis. For instance, water depletion at the irrigation service scale represents only crop evapotranspiration, while at basin scale, it also includes municipalities, industries, fisheries, forestry, dedicated wetlands, and all other uses. As a result, parts of the critical information in a basin context is not covered in the original IWMI WA framework.

Thus, WA+ is a modified and upgraded version of water accounting which IWMI has developed (Karimi et al. 2013a, b) based on original initiatives taken by the Delft University of Technology (Bastiaanssen and Ali 2003). Water accounting plus (WA+) is a framework designed to provide explicit spatial information on water depletion and the net withdrawal process from river basins. It provides the link between water balance, land use, water use, and management options to modify it by grouping land use classes with common management characteristics. WA+ tracks water depletions rather than withdrawals, and it goes past flow and runoff accounting. Water is depleted by ET, flows to sinks, incorporates into products, or becomes non-recoverable due to quality degradation. Water depletions are divided into beneficial and non-beneficial according to the type of use and intended purpose. WA+ has eight fact sheets. Sheet 2 and Sheet 3 represent the evapotranspiration and agricultural services (land productivity and water productivity) assessment, respectively, Wardlow and Egbert (2008), Zhao et al. (2019). WA+ framework has been developed using python codes and available in https://github.com/wateraccounting.

9.2 Materials and Methods

9.2.1 *Study Area*

The geographic location of the study area extends from 83° 04′ E 21° 34′ N to 83° 33′ E 20° 54′. It consists of Left Bank Command (LBC) area and Right Bank Command (RBC) area. The total geographic area of LBC is 39,5000 ha and for RBC is 198,000 ha. The river stretch is around 80 km across the command area. Sambalpur is the largest city situated within the command area, followed by the Bargarh. Hirakud command area spread across the four districts of the Odisha state, i.e., Sambalpur, Bargarh, Sonepur and Bolangir. Four canals originate from the Hirakud dam, i.e., Bargarh canal, Hirakud distributary, Sambalpur distributary and Sason canal. The longest canal is Bargarh main canal originates near Burla, which is approximately 107 km. There are three distributaries, i.e., Godbhaga, Attabira, and Bargarh distributaries diverting from this main canal. This canal irrigates about 133,540 ha of land from Sambalpur, Bargarh and Sonepur districts. This network of the main canal and distributaries serves mainly irrigation and power generation in the RBC area. Their waters also used for the Municipal water supply and industries.

On the LBC area, there is Sason main canal which is around 86 km in length, starts from the dam and flows across the command area. This canal flows in the LBC area of the Sambalpur district. The culturable Command Area (CCA) of this canal is 24,487 ha. This canal system also consists of several distributaries and network of minors for mainly irrigation purposes. In the LBC area, Sambalpur distributary starts from the Hirakud dam having discharge capacity of 3.40 cumec and the flows toward the Sambalpur city. This distributary irrigates around 4100 ha of land surrounding the Sambalpur town. The entire Hirakud dam project provides irrigation to 155,635 ha of land in Kharif season and 108,385 ha of land in the Rabi season.

Additionally, the water released after hydroelectric power generation irrigates the 436,000 ha of land in the Mahanadi delta region. Studies on Hirakud canal commands are limited (Asokan and Dutta 2008). The location map of the Hirakud command is shown in Fig. 9.1.

9.2.2 *Data Used*

Different types of data are the root inputs for doing effective water management planning. But the staggered and poor availability of in-situ data is the primary cause of ineffective water management of a basin. Generally, the interpretation and communication are inadequate of water-related data to researchers, planners, policymakers. Application of various satellite data products leads to time, cost-effective and nearly accurate watershed evaluation. Data sets used for this evaluation are freely available and accessible.

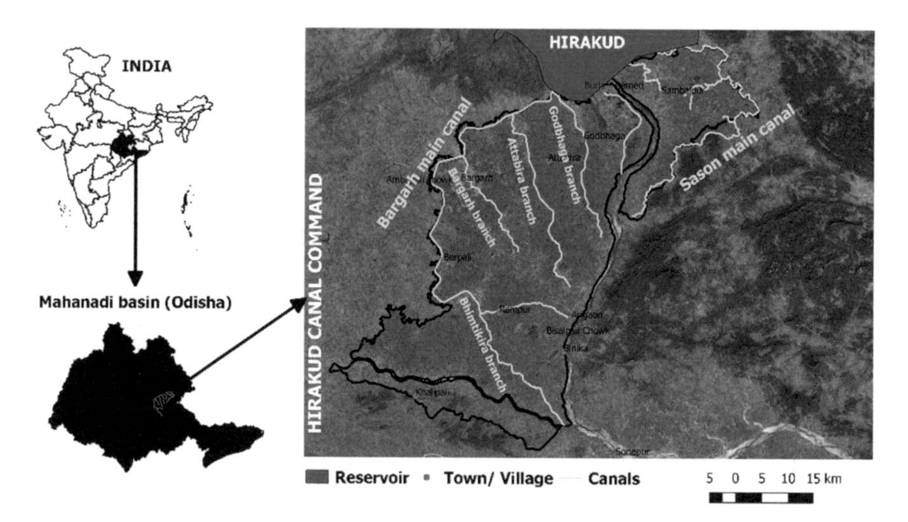

Fig. 9.1 Location map of the study area

9.2.2.1 Land Use and Land Cover

Land use and land cover (LULC) is the main parameter, which affects the hydrological cycle and the services and benefits for society and the environment. Thus, spatially distributed information on LULC is necessary for running WA+. Based on remotely sensed data and different algorithms, several regional and global land cover databases have been generated (Arino et al. 2010). There are different data products like Globcover, global map of irrigation area (GMIA), moderate resolution imaging spectroradiometer (MODIS), international water management institute (IWMI) crop maps, monthly irrigated and rainfed crop area (MIRCA), world database on the protected area (WDPA), world population data were used to prepare WA+ based LULC, see Table 9.1. In terms of water management, the LULC classes have been classified into four major clusters: protected land use (PLU), utilized land use (ULU), modified land use (MLU), managed water use (MWU).

9.2.2.2 Precipitation

Gross precipitation is a primary input for WA+. Climate hazards group infrared precipitation with station data (CHIRPS), tropical rainfall measurement mission (TRMM) remote sensing rainfall products are available for this accounting procedure (Mitra et al. 2009).

Table 9.1 Details of land use land cover data product

S. No.	Data product name	Period of availability	Spatial resolution	File format	Remarks
1	Globcover	Dec'04–June'06; Jan–Dec' 2009	300 m	tiff	Based on ENVISAT MERIS, 22 style classes
2	GMIA (Global Map of Irrigated Areas)	Since 01–10–2013	10 km	shp	gives information about % irrigated areas or hectare per pixel
3	MIRCA (Monthly Irrigated and Rainfed Crop Areas)	1998–2002	10 km	shp	% of each cell monthly covered by each of 26 nos. Irrigated or rainfed crops
4	WDPA (World Database on Protected Areas by United Nations Environmental Program)	2010–2016	–	shp	29 descriptors, referred to as data attributes; global spatial dataset on terrestrial and marine protected areas
5	JRC (Joint Research Center Data Catalog)	2000–2016	1 km	GeoTIFF	Flood hazard maps- Total 13 datasets available with different return period frequency
6	The international water management institute (Global Reservoir and Dam Database)	2000–2010	230 m	tiff	Contains 6862 reservoirs/dams of a capacity >0.1 km^3
7	MODIS (Moderate Resolution Imaging Spectroradiometer)-MCD12Q1	2001–2013	500 m (0.05 deg)	tiff	17 classes
8	World POP	2010; 2015; 2020	100 m (0.000833 deg)		People per hectare' (pph) datasets

9.2.2.3 Evapotranspiration (ET)

Over the past decades, there are various methods and algorithms to calculate actual evapotranspiration (ET). ET ensemble product is an accurate product which was created by linear averaging of seven individual ET products: (1) Modis Global Terrestrial Evapotranspiration Algorithm (MOD16), (2) Atmosphere-Land Exchange Inverse Model (MOD16), (3) Global Land Evaporation Amsterdam Model (GLEAM), (4) Operational Simplified Surface Energy Balance (SSEBop), (5) CSIRO MODIS Reflectance-based Evapotranspiration (CMRSET), and (6) Surface Energy Balance System (SEBS), (7) ETmonitor and subsequently downscaled to 0.0025° using the MODIS-based, normalized difference vegetation index (NDVI) data. The period of analysis was from 2003–01–01 to 2014–12–31. Potential evapotranspiration (ET Ref) is also available from the climate forecast system reanalysis (CFSR) dataset.

9.2.2.4 Leaf Area Index (LAI), Net Dry Matter (Biomass Production) (NDM)

Leaf area index (LAI), and net dry matter (NDM), which give information on leaf area per unit ground surface area and mass of carbon per unit area, also were used for this sheet preparation to determine vegetation coverage and amount of production from agriculture, respectively. NDM was created from gross primary production (GPP) and net primary production (NPP), which are MODIS datasets. The details of precipitation, evapotranspiration, and meteorological data product are presented in Table 9.2.

9.2.3 Methodology

9.2.3.1 WA+ Based Land Use and Land Cover (WALU)

In terms of water management, the LULC classes have been classified into four major clusters: protected land use (PLU), utilized land use (ULU), modified land use (MLU), managed water use (MWU), as shown in Fig. 9.2. Protected land use represents areas set aside for no/minimal disturbance by humans. It includes natural ecosystems or biomes earmarked for conservation and coastal protection. Examples are national parks, coastal dunes, game reserves, glaciers. The group "utilized land use" represents a land use that provides a range of ecosystem services and which has had little interference by man. However, people often use such land for their services, like food production or fuelwood and nomads on natural pastures. Examples include grassland or savanna (for grazing or wood) and forest land (for timber). The group "modified land use" refers to land that is significantly modified by human activity for the sake of food, feed, fiber, (bio-)fuels, and fish production. It also

Table 9.2 Details of precipitation, evapotranspiration, and meteorological data product

S. No.	Data product name	Period of availability	Spatial resolution	Temporal resolution	File type	Remarks
A	*Rainfall*					
1	Tropical Rainfall Measuring Mission (TRMM)	1998 to April 2015	27 km (0.25 deg)	3-hourly, daily, monthly	tiff	NASA & JAXA mission
2	Climate Hazards Group Infrared Precipitation with Station data (CHIRPS)	1981 onwards	5 km	daily, monthly	tiff	Funded by USGS and USAID
B	*Evapotranspiration*					
1	MOD 16	2000 onwards	1 km	8 daily, monthly, annual	tiff	MODIS Product
2	Global Land Evaporation Amsterdam Model (GLEAM)	2003–2012	27 km (0.25 deg)	daily, monthly	tiff	
3	ETensemble (ETensV1.0)	January 2003 until December 2014	250 m (0.0025 deg)	monthly	tiff	Global
C	*Meteorological*					
1	Global Land Data Assimilation System (GLDAS)	1948 onwards	110 km (1 deg)	3-hourly, monthly	tiff	Measures Meteorological parameters such as atmospheric pressure, radiation, temp, etc
2	Climate Forecast System Reanalysis (CFSR)	1979–2009	0.5 deg	hourly	tiff	Provides the output at atmospheric, oceanic and land surface (10 data products)
3	MOD 15-17 Vegetation	2000 onwards	1 km	8 daily, monthly	tiff	NDVI, LAI, FPAR, GPP, NPP

Fig. 9.2 Representation of 4 major LULC classes in WALU

includes improved road networks to connect growing populations, dump sites, and increasing space for leisure and socio-economic growth in the most general terms. Water diversions and withdrawals do not occur in the "modified land use" group, but by modifying vegetation density, hydrological processes such as ET, drainage, percolation, and recharge are affected. Changes in ET in the "modified land use" class can have a significant impact on groundwater levels, streamflow, and downstream water availability. Rainfed cropping systems, deforestation, creation of plantation forests, the establishment of lanes and parks, home gardens and wind shelters typically fall in the "modified land use" class. The group "managed water use" represents the land use classes in which the natural water cycle is manipulated by physical infrastructure; water is intentionally retained, withdrawn, pumped, diverted and spilled by pumping stations, valves, pipes, dams, weirs, gates, canals, sluices, culverts and drains for specific objectives. Examples are drinking water supply schemes, irrigation systems, hydropower storage, maintaining water levels for navigation, flood storage in wetlands, etc. Managed water use includes domestic water use in urban areas and villages, irrigated agriculture, expanding industries for economic development and golf courses.

9.2.3.2 ET Separation

By knowing the quantity of ET from different land use, the next attempt is to estimate the benefits obtained from various water use. It means every land use land cover needs a value to describe both aspects of ET, i.e., beneficial and non-beneficial fraction. In general, evaporation (E) is counted as non-beneficial, as a considerable quantity of E generates from wet soils (fallow or partially filled land). Like evaporation from water bodies, which are meant for fishing, leisure, water sports are considered beneficial. In this manner, each LULC class assigned a percentage of beneficial and non-beneficial use of water based on its intended purpose. Again the beneficial evapotranspiration (ET) is expressed in agriculture, environment, economic, energy, and leisure. In addition to that independent assessment of transpiration (T), interception (I) and evaporation (E) from soil and water helps to prepare effective water management policies and strategies.

9.2.3.3 ET Mapping (Sheet 2)

Sheet 2 (ET sheet) informs details about beneficial and non-beneficial ET as per prevailing LULC and total water consumption. The major inputs for the generation of Water Accounting Sheet 2 are satellite-based ET, LAI, Net primary production (NPP), Gross primary production (GPP), daily precipitation and a LULC map. ET sheet also shows the separation of ET into evaporation (E), transpiration (T), interception (I) for each LULC class. To calculate both aspects of ET, each LULC class has been assigned a percentage of beneficial and non-beneficial uses based on their intended purpose.

9.2.3.4 Land and Water Productivity Mapping (Sheet 3)

To find productivity separately from the rainfall and irrigation water, water consumption (ET) from rainfed and irrigated water was calculated as ET green and ET blue (Fig. 9.3). The assessment of WP and land productivity also involves the application of green water (rainfed) and blue water (irrigated) concepts using the Budyko framework (Falkenmark and Rockström 2006). Unlike other water accounting frameworks, WA+ framework recognizes the essential difference between Blue and Green water. Spatial maps of land and water productivity provide the areas (progressive farmers) performing well in a large basin. This rich information enables water managers to understand the local level interventions by progressive and less progressive farmers. The calculation of yield and water productivity is done as given below.

$$\text{Fresh yield} = \frac{\text{Harvest index} \times \text{Biomass dry weight}}{1 - \text{Moisture content}} \quad (\text{kg/ha}) \tag{9.1}$$

$$\text{Water productivity} = \frac{\text{Yield}}{\text{Water consumption}} \quad \left(\text{kg/m}^3\right) \tag{9.2}$$

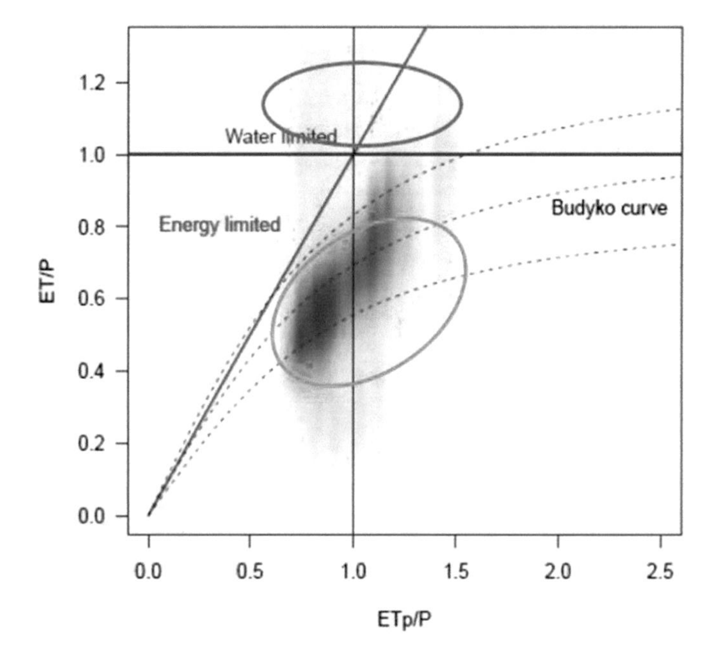

Fig. 9.3 Budyko curve (Budyko 1974)

9.3 Results and Discussion

9.3.1 *Computations of Average Water Yield Based on CHIRPS Rainfall and ET Ensemble*

Initially, the CHIRPS rainfall data and ET ensemble data for duration 2003–2014 were analyzed for average water yield in the canal command. The seasonal variability of rainfall indicates the four monsoon months (JJAS) during which the area receives significant rainfall sustaining the Kharif crops (Fig. 9.4). The winter rainfall is insignificant. The rainfall analysis indicated the year 2008 was a wet year followed by 2010 and 2007 as the dry and average years.

9.3.2 *WA+ Based Land Use and Land Cover (WALU)*

The WALU based LULC was generated, as shown in Fig. 9.5. The command comprises 79.6% of the area under crops, mostly paddy. 15.7% of the built-up areas and other lands, followed by shrublands (2.7%) and water bodies (2%).

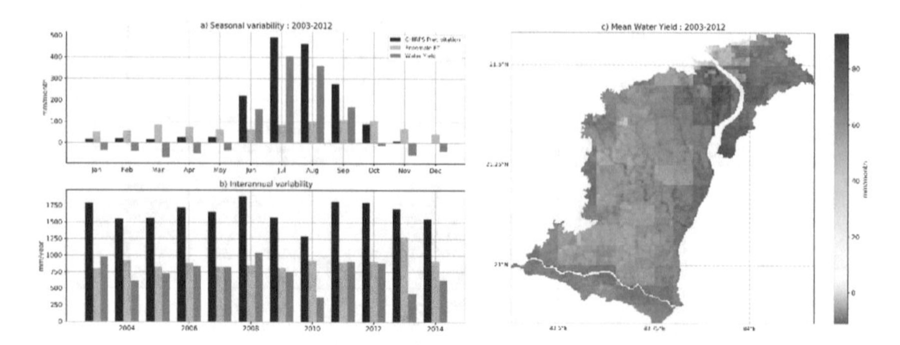

Fig. 9.4 Seasonal and Interannual variability of CHIRPS rainfall data across Hirakud canal command

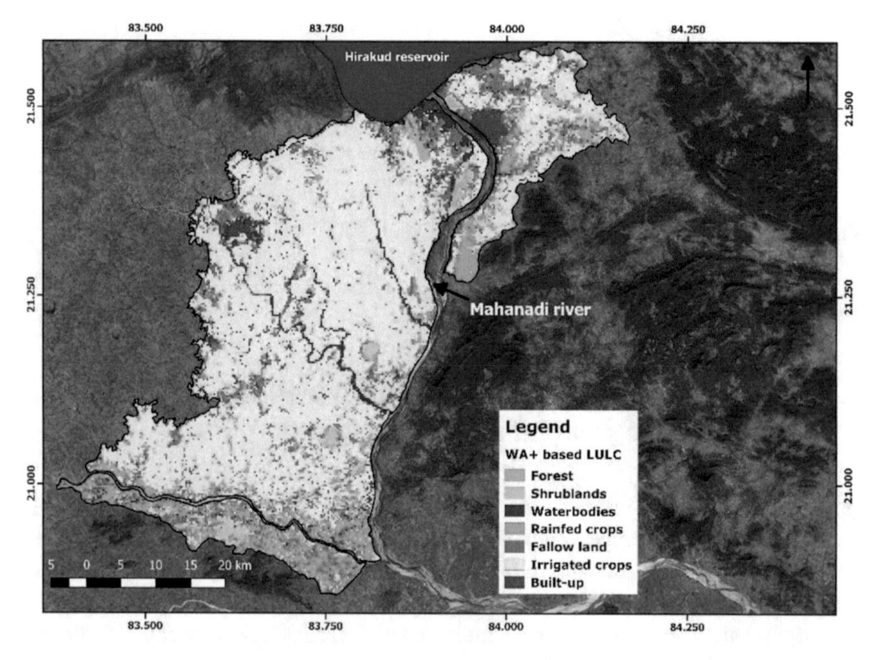

Fig. 9.5 WA+ based land use and land cover (WALU)

9.3.3 ET Separation

The WA+ framework provides details on the total consumptive losses due to evaporation, transpiration, and interception. This information is vital to plan and manage the beneficial and non-beneficial water consumptions in a basin. It can be seen from Table 9.3 that the water use in the canal command is mainly managed (managed water use/

Table 9.3 Consumptive uses for different WALU categories (Units in km^3/year)

Year	Total ET	Non-manageable/protected land use		Manageable/utilized land use		Managed			
						Modified land use		Managed water use	
		ET	T	ET	T	ET	T	ET	T
2003–04	23.9	0.0	0.0	2.0	1.1	6.8	4.1	15.1	9.1
2004–05	21.5	0.0	0.0	1.8	1.0	6.2	3.8	13.6	8.4
2005–06	22.1	0.0	0.0	1.8	1.0	6.3	3.8	13.9	8.3
2006–07	21.4	0.0	0.0	1.8	1.0	6.0	3.7	13.7	8.6
2007–08	21.8	0.0	0.0	1.8	1.1	6.3	4.0	13.7	8.7
2008–09	20.9	0.0	0.0	1.7	0.8	5.8	3.2	13.4	7.1
2009–10	22.7	0.0	0.0	1.8	0.7	6.3	2.8	14.5	6.1
2010–11	23.7	0.0	0.0	2.0	1.1	6.7	4.3	15.0	9.2
2011–12	23.5	0.0	0.0	1.9	1.0	6.5	3.9	15.0	9.0
2012–13	30.1	0.0	0.0	2.5	1.3	8.6	4.9	19.0	11.5
2013–14	26.3	0.0	0.0	2.2	1.3	7.7	5.1	16.4	10.8
Ann avg	23.45	0.00	0.00	1.94	1.04	6.65	3.96	14.85	8.80

irrigated-63.3%; modified land use/rainfed-28.3%). The transpiration percentage, otherwise treated as the beneficial use of water consumption for agriculture, is 58.8% of the total ET consumption (23.45 km^3/year).

9.3.4 Computations of Total Water Consumptions, Land Productivity, and Water Productivity

Before computing the land and water productivity (WP) of the basin, Sheet 2 (Total Water Consumptions) was estimated. Sheet 2 shows that the total water consumptions in the basin for the wet year was 23.45 km^3/year from different LULCs. The average non-beneficial consumption in the Hirakud canal command is 11.5 km^3/year, whereas the beneficial consumptions are 12.0 km^3/year. This indicates an enormous scope to adopt water conservation practices in the basin to minimize non-beneficial consumptions. Non-beneficial consumptions of water are the water consumed for purposes other than the intended ones. E.g., in an irrigated agriculture, the sole aim of water consumptions (intended purpose) is to augment the crop yield only. Any losses in terms of soil evaporation, although unavoidable, is the non-beneficial consumption. Finally, the agricultural water consumptions and land productivity and water productivity (WP) of the different crops in the canal command were estimated using green water and blue water concepts. Figure 9.6 (Sheet 3) shows the estimates of land productivity (kg/ha/year) and WP (kg/m^3) separately for rainfed (green water) and irrigated (blue water) cereals (mainly rice) during a dry rainfall year, i.e., 2010–11. It can be observed from Fig. 9.6 that the command has land productivity of 1979 kg/ha/year and 2095 kg/ha/year, respectively, for rainfed and irrigated cereals. Overall, the land productivity is found to vary from 1778 to 2524 kg/ha/year and 1158 to 2254 kg/ha/year during 2003–04 to 2013–14, respectively for rainfed and irrigated cereals. Similar results were also reported in the Water Productivity Mapping of Major Indian Crops report of NABARD, and ICRIER (Sharma et al. 2018). The land productivity of the irrigated rice is 2200 kg/ha/year for Odisha.

The spatial variability of land productivity is also shown in Fig. 9.6a for the year 2010–11. It can be observed that a few patches in the basin have the land productivity of the order of 3249 kg/ha/year. The Sheet 3 (Figure also shows the estimates of WP (kg/m^3) during the dry year, i.e., 2010–11 for rainfed (green water) and irrigated (blue water) cereals. The command has a WP of 0.42 kg/m^3 and 0.53 kg/m^3, respectively, for rainfed and irrigated cereals. Overall, the WP is found to vary from 0.39 to 0.55 kg/m^3 and 0.28 to 0.57 kg/m^3 during 2003–04 to 2013–14, respectively, for rainfed and irrigated cereals. As per the Water Productivity Mapping of Major Indian Crops report of NABARD and ICRIER (Sharma et al., 2018), the rice's WP has been reported as 0.35 kg/m^3 in the state of Odisha.

It can be seen from Fig. 9.6b that the WP is higher in the periphery or tail end of main and branch canals. The higher values of WP may be attributed to several reasons, including improved and advanced irrigation systems and conservation practices. But, one point is vivid that these reasons being situated at the tail-end, the irrigation water reaching the fields are probably less in comparison to the upper reaches of distributaries. This is perhaps increasing the overall productivity in terms of higher yield with less application of water. These spatial maps help us identify the best performing farmers, thereby their practices that can be replicated to other parts of the command to increase land productivity and WP.

9.4 Conclusions

Nowadays, the utility of satellite-based open access data in estimating and planning water resources, especially in canal command, is gaining importance because of data scarcity and high cost associated. This study has applied WA+ Framework to the Hirakud canal command to estimate the spatial and temporal variability of land productivity and water productivity. Apart from this, assessments were also made for the agricultural water consumptions (using green water and blue water), separation of ET into evaporation, transpiration, interception, and estimation of beneficial and non-beneficial ET based on the WA+ based land use land cover (WALU) for a period

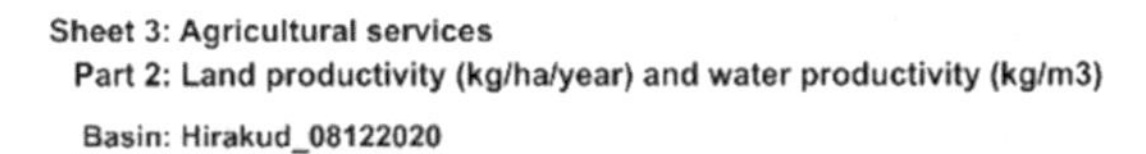

Crop

	Cereals	Non-cereals: Root / tuber crops	Non-cereals: Leguminous crops	Non-cereals: Sugar crops	Non-cereals: Merged	Fruit & vegetables: Vegetables & melons	Fruit & vegetables: Fruits & nuts	Fruit & vegetables: Merged	Oil-seeds	Feed crops	Beverage crops	Other crops		
Land productivity	1979	-	-	-	-	15785	-	-	-	-	-	-	Yield	rainfed
	1264	-	-	-	-	-	-	-	-	-	-	-	Yield from rainfall	irrigated
	831	-	-	-	-	-	-	-	-	-	-	-	Incremental yield	irrigated
	2095	-	-	-	-	-	-	-	-	-	-	-	Total yield	irrigated
Water productivity	0.42	-	-	-	-	3.43	-	-	-	-	-	-	WP	rainfed
	0.95	-	-	-	-	-	-	-	-	-	-	-	WP from rainfall	irrigated
	0.31	-	-	-	-	-	-	-	-	-	-	-	Incremental WP	irrigated
	0.53	-	-	-	-	-	-	-	-	-	-	-	Total WP	irrigated

Non-crop

	Livestock: Meat	Livestock: Milk	Fish (Aquaculture)	Timber		
Land productivity	-	-	-	-	Yield	rainfed
	-	-	-	-	Yield from rainfall	irrigated
	-	-	-	-	Incremental yield	irrigated
	-	-	-	-	Total yield	irrigated
Water productivity	-	-	-	-	WP	rainfed
	-	-	-	-	WP from rainfall	irrigated
	-	-	-	-	Incremental WP	irrigated
	-	-	-	-	Total WP	irrigated

Fig. 9.6 Sheet 3 (Part II) showing land and water productivity of cereals in a dry year. **a** Spatial variability of land productivity for irrigated rice in Hirakud canal command. **b** Spatial variability of water productivity for irrigated rice in Hirakud canal command

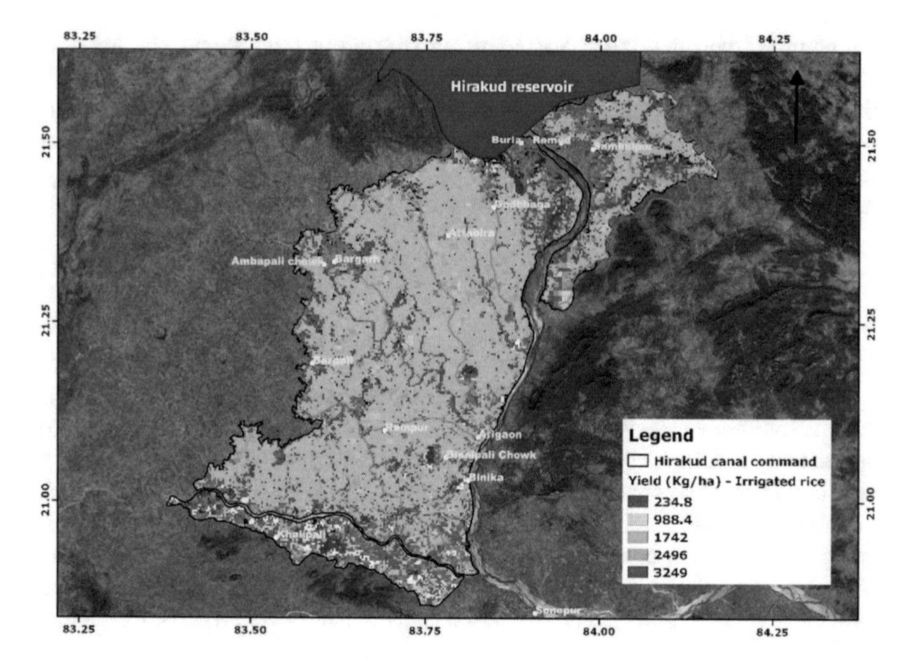

Fig. 9.6 (continued)

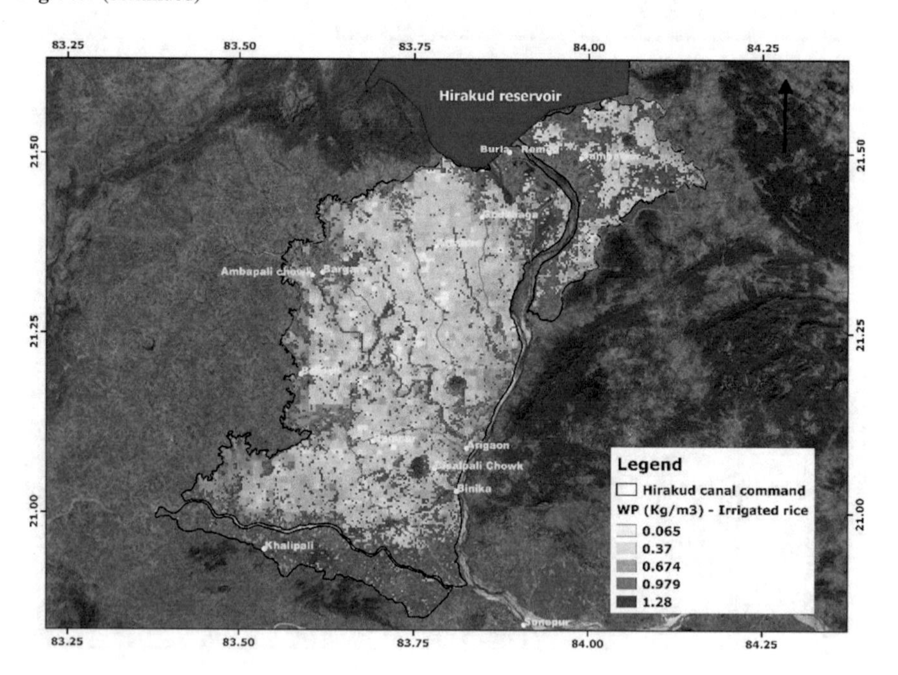

Fig. 9.6 (continued)

of 12 years, i.e., 2003–2014. In the Hirakud canal command basin, although the beneficial water consumption is higher than the non-beneficial water consumptions, we still have ample scope to minimize the water use in increasing the irrigated rice yield. It has also been seen that the water productivity of tail-end reaches of the canal is higher than the upper reaches implying that other regions can also use less water for the existing yield. Overall, the land productivity is found to vary from 1778 to 2524 kg/ha/year and 1158 to 2254 kg/ha/year during 2003–04 to 2013–14, respectively, for rainfed and irrigated cereals.

References

Allen RG, Brockway CE, Wright JL (1983) Weather station siting and consumptive use estimates. J Water Resour Plan Manag 109(2):134–146

Arino O, Perez JR, Kalogirou V, Defourny P, Achard F (2010) Globcover 2009. In: ESA living planet symposium, vol (1), pp 1–3

Asokan SM, Dutta D (2008) Analysis of water resources in the Mahanadi River Basin, India under projected climate conditions. Hydrol Process 22(18):3589–3603

Bastiaanssen WGM, Ali S (2003) A new crop yield forecasting model based on satellite measurements applied across the Indus Basin, Pakistan. Agr Ecosyst Environ 94(3):321–340

Budyko M (1974) Climate and life. Academic Press, San Diego, CA, USA, pp 72–191

Falkenmark M, Rockström J (2006) The new blue and green water paradigm: breaking new ground for water resources planning and management. J Water Resour Plan Manag 132(3):129–132

Karimi P, Bastiaanssen WGM, Molden D, Cheema MJM (2013) Basin-wide water accounting based on remote sensing data: An application for the Indus Basin. Hydrol Earth Syst Sci 17(7):2473–2486

Karimi P, Bastiaanssen WGM, Molden D (2013) Water Accounting Plus (WA+)—A water accounting procedure for complex river basins based on satellite measurements. Hydrol Earth Syst Sci 17(7):2459–2472

Mitra AK, Bohra AK, Rajeevan MN, Krishnamurti T N (2009) Daily indian precipitation analysis formed from a merge of rain-gauge data with the TRMM TMPA satellite-derived rainfall estimates. J Meteorol Soc Jpn 87A:265–279

Molden D, Oweis T, Steduto P, Bindraban P, Hanjra MA, Kijne J (2010) Improving agricultural water productivity: between optimism and caution. Agric Water Manag 97(4):528–535

Molden D, Murray-Rust H, Makin L (2004) Water productivity in agriculture: limits and opportunities for improvement. Choice Rev (Online)

Sharma BR, Gulati A, Mohan G, Manchanda S, Ray I, Amarasinghe U (2018) Water productivity mapping of major Indian crops. A report by NABARD and ICRIER © pp 1–182

Wardlow BD, Egbert SL (2008) Large-area crop mapping using time-series MODIS 250 m NDVI data: an assessment for the U.S. Central Great Plains. Remote Sens Environ 112(3):1096–1116

Zhao J, Chen X, Zhang J, Zhao H, Song Y (2019) Higher temporal evapotranspiration estimation with improved SEBS model from geostationary meteorological satellite data. Sci Rep 9(1):1–15

Chapter 10
Performance Evaluation of a Minor of Upper Ganga Canal System Using Geospatial Technology and Secondary Data

Randhir Jha, Ashish Pandey, and Srinivasulu Ale

Abstract Food and fiber demand is increasing due to population growth, which compels us to optimize the irrigation system's performance to get more yield of food and fiber from the available resource. The high yielding varieties of crop requires a timely and adequate supply of water. A well-performing, well-managed irrigation system is a prerequisite to ensure a timely and proper water supply. The development of technology makes us ease in assessing the performance of the irrigation system. The availability of geospatial data in conjunction with the other ground data helps assess the performance minutely in spatial and temporal approaches. Performance evaluation of a canal irrigation system can be carried out by evaluating its actual water dynamics, water use, and productivity. Gadarjudda and Lakhnauta minor canal systems of the Upper Ganga Canal (UGC) are old systems and still performing in the right way with general maintenance works. The minor systems are decade old, and ungauged water management is carried out effectively. The physical condition of systems is not good; however, the canal carries the designed discharge. The results show that depleted fraction of Gadarjudda and Lakhnauta minor is nearly 1 to 1.30 kg/m^3 for paddy productivity (actual water consumed), followed by Wheat and Sugarcane productivity by 1.61 kg/m^3 and 6.08 kg/m^3, respectively. The spatial maps of the indicator generated in the GIS environment give the information of the irrigation system's performance over command areas and help for evaluation. Thus, the study reveals the excellent performance of the UGC system.

R. Jha (✉) · A. Pandey
Department of Water Resources Development and Management, IIT Roorkee, Roorkee, India

S. Ale
Texas A&M University, College Station, TX 77843, USA

A. Pandey et al. (eds.), *Geospatial Technologies for Land and Water Resources Management*, Water Science and Technology Library 103,
https://doi.org/10.1007/978-3-030-90479-1_10

10.1 Introduction

Achieving a balance between the increasing demand for food and its supply is a challenge for the whole world. The world is paying attention to address these problems by gaining optimum outcomes from available water resources. However, due to climate change, the uneven distribution of rainfall and reduction in the number of available water resources is inducing challenges for the agriculture sector (Kumar et al. 2005). In regions with insufficient rainfall, an efficient irrigation method can only meet the crop water requirement.

Irrigation plays a major role in achieving sustainable and efficient agriculture production (Oweis and Hachum 2003). Irrigated farming is one of the prime sectors that distress from the scarcity of freshwater and its application loss. On the contrary, improper irrigation practices and management may cause waterlogging and salinity in the agricultural area. Thus, optimal use of available freshwater for irrigation plays a vital role in conserving land and water and maximizing the agriculture yield with the available resources (Oweis and Hachum 2003). Thus, monitoring irrigation systems is essential to achieve the best performance with the available inputs (Bos et al. 1993).

The crop water productivity is an important parameter to demonstrate the real consequences of scenarios to evaluate an irrigation system performance and can be expressed in physical or monetary units (Ahmad et al. 2009). By general irrigation practice, a nearly 30% increase in agricultural production is only due to timely water (avoiding other agriculture inputs). Water productivity is the proportion between the unit yield to the unit of the contribution of water. The term irrigation water productivity is utilized to indicate horticulture/agriculture production over the volume of water exhausted or depleted (Zwart and Bastiaanssen 2004).

The ability of Remote Sensing (RS) and Geographic Information systems (GIS) to analyze and visualize spatial and temporal data has made them useful tools for the performance evaluation of irrigation systems (Bastiaanssen 1998). Many studies on RS and GIS techniques have been conducted in recent years, which contributes significant information for evaluating irrigation system performance (Ray et al. 2002; Pandey and Mogerekar 2021). The generation of irrigation network inventory maps, water productivity, crop water use conditions using field and secondary data, and high-resolution satellite imageries are useful in evaluating the irrigation system performance by grading the scale of indicators.

The development of any irrigation system needs a large number of investments; its output performance would help in the recovery of investment and increase the income of beneficiaries through agro-based farming. Therefore, in this study, the performance evaluation of a minor canal of the Upper Ganga Canal (UGC) system has been carried out with the objectives to (1) Examine the health of the irrigation system (2) Determine crop parameters using remote sensing and GIS. (3) Evaluation of the irrigation system by analyzing performance indicators.

10.2 Study Area

The Gadarjudda and Lakhnauta minors of Upper Ganga Canal (UGC), old decadal systems and still performing in the right way with general maintenance works, are selected as command areas in the study. The Gadarjudda minor having 0.28 m^3/s head discharge is approximately 5.4 km long and fully lined, in the command area of 526 ha. The Lakhnauta minor having a head discharge of 0.31 m^3/s is approximately 4.8 km long, in the command area of 715 ha. The study area is situated about 10 km southwest of Roorkee city in Uttarakhand, India. The location map of the study areas is given in Fig. 10.1.

Britishers constructed the Upper Ganga canal with a head discharge of 170 m^3/s during 1842 and 1854. The Upper Ganga channel has since been expanded segment steadily for the present head release of 295 m^3/s. The canal network comprises a 438 km long main canal and around 6437 km long distribution channels. It irrigates about 9 lakh-ha of rich agricultural land in Uttar Pradesh and Uttarakhand states. This channel has been the wellspring of horticultural success a quite bit in these States. The Upper Ganga canal system's selected minors draw their supply from the Sadhauli distributary of the Deoband branch.

The system command area receives good rainfall in the monsoon and very little in the winter. The climate of the command area is humid sub-tropical, as per the Köppen-Geiger climate classification system. The average annual rainfall and average annual temperatures are 1170 mm and 23.7 °C, respectively, and paddy, sugarcane, and wheat are the main crops grown. Intercropping of mustard with wheat is also practiced in the study area. Other crops grown are berseem and oats as fodder crops. The salient features of Gadarjudda Minor and Lakhnauta minor are given in Table 10.1.

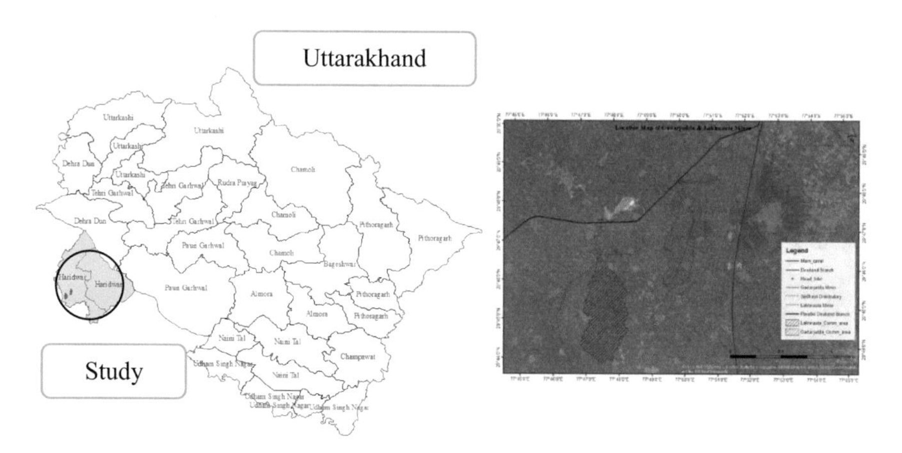

Fig. 10.1 Location map of the study area

Table 10.1 The salient features of Gadarjudda Minor and Lakhnauta Minor

Features	Details of Gadarjudda Minor	Details of Lakhnauta Minor
Location	Latitude: 29° 47′ 23.53″ N Longitude: 77° 47′ 24.42″ E	Latitude: 29° 47′ 24″ N Longitude: 77° 46′ 30″ E
Discharge (m^3/s)	0.28	0.31
Total canal length (km)	5.4	4.8
Culturable command area (ha)	526	715
Outlets	18 pipe outlets	22 pipe outlets

10.3 Materials and Methods

The various field data, secondary data, meteorological data, and satellite data are obtained from different sources. The methodology flowchart is given in Fig. 10.2. The sources and type of data acquisition are given in Table 10.2.

The methodology adopted for performance evaluation of the irrigation systems is given below:

i. Study of the physical condition of the system,
ii. Study of the system Canal network, discharge, and efficiency,
iii. Study of the crop, cropped area, and yield,
iv. Analysis of meteorological data,
v. Monitoring of the soil type, moisture content, and infiltration capacity,
vi. Land use/cover information from Landsat 8 Imagery.

The crop water use and productivity parameters of performance indicators are determined by the analysis and observation of the canal network and its condition, potential evapotranspiration (ET_o) and actual evapotranspiration (ET_c), total and useful or effective rainfall, irrigation water supply, and information regarding the production of the crop. The systematic methodology adopted for the analysis is described below:

10.3.1 Canal Network and Its Physical Health

Initially, we collected the canal irrigation system's information regarding the physical condition of infrastructure, service delivery, and fulfilment of its initially designed purpose. It includes making inventory over the physical condition, design discharge, the system's efficiency, number of structures within the system, their purpose from irrigation office, farmer interaction, and self-inspection of the system from head to tail.

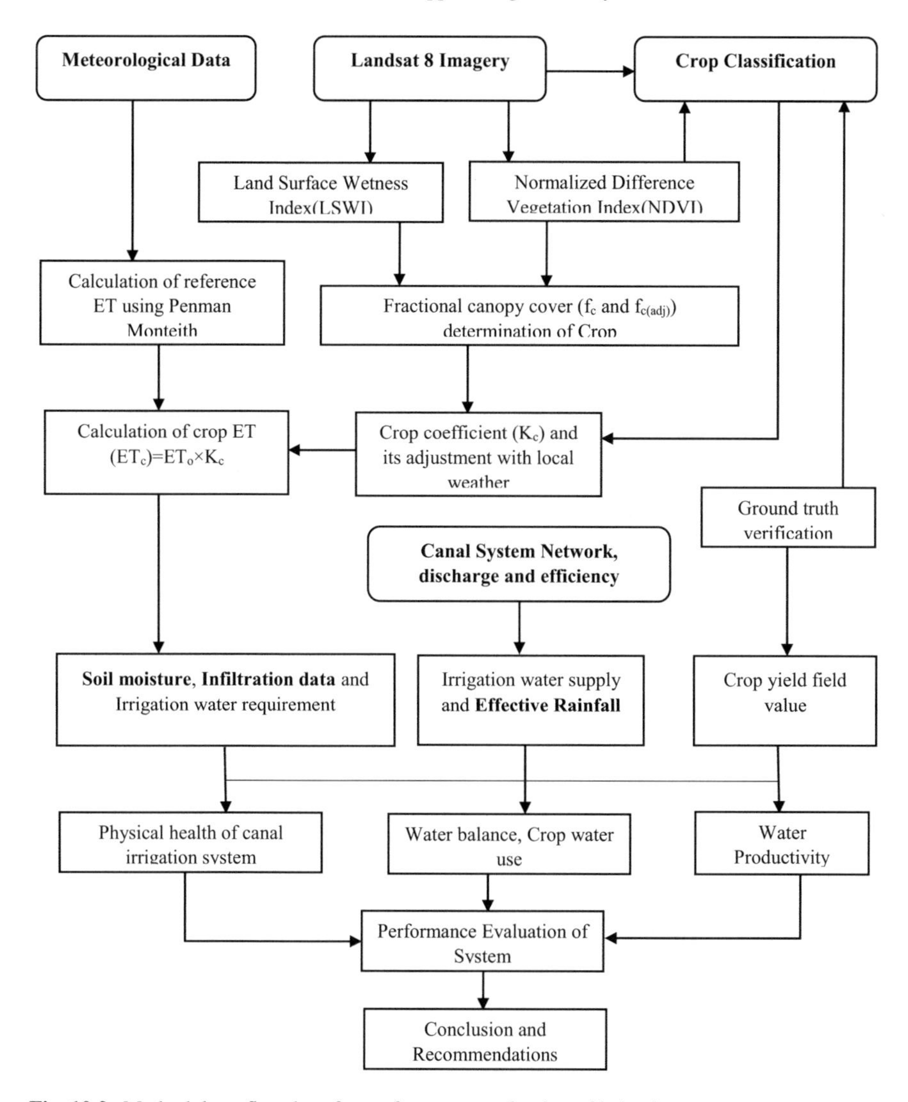

Fig. 10.2 Methodology flowchart for performance evaluation of irrigation system

10.3.2 Estimation of Reference Evapotranspiration ET_o

Ascertaining of the irrigation water demand is essential to determine the crop water requirement spatially. The procedure requires estimating evapotranspiration and reference evapotranspiration, which can be confirmed from a substantial number of accessible strategies and equations.

Table 10.2 Data/information and source of acquisition

Data/information	Type of data	Source
Crop type in each season and cropping area	Cropping details	State Irrigation Office—Muzaffarnagar, Manglor and Farmer Interview
Climate data (Temperature, rainfall, solar radiation, sunshine hours, etc.)	Monthly values	National Institute of Hydrology (NIH)—Roorkee
Canal network, irrigation supply, efficiency of the irrigation system, soil moisture, infiltration rate	Flow rates, field data	Command area map, Field measurements, state irrigation Office, and reference reports
Red, infrared, and thermal bands of satellite data	Imagery	www.glovis.usgs.gov
NDVI, LSWI, and maps from satellite images	Processed imagery	www.glovis.usgs.gov

10.3.2.1 FAO Penman–Monteith Equation

FAO prescribes the Penman–Monteith equation for assurance of reference evapotranspiration and is generally acknowledged universally. The condition requires information for a substantial number of climate parameters. The FAO Penman–Monteith equation for ascertaining ET_o as, given FAO 56 by Allen et al. (1998) is given below:

$$\mathrm{ET}_o = \frac{0.408\Delta(R_n - G) + \gamma \frac{900}{T+273} u_2 (e_s - e_a)}{\Delta + \gamma(1 + 0.34u_2)}$$

where, ET_o is grass evapotranspiration (mm day^{-1}), R_n is net radiation at the crop surface (mm day^{-1}), G is soil heat flux density (MJ/m^2/day), T is mean daily air temperature at 2 m height (°C), u_2 is the wind speed at 2 m height (m/s), Δ is slope vapor pressure–temperature curve (kPa/°C), e_s is saturated vapor pressure (kPa), e_a is actual vapor pressure (kPa), and γ is psychrometric constant (kPa/°C).

10.3.3 Estimation of Crop Coefficient (K_c) and Local Weather Adjustment

The crop coefficient is associated with the crop type, characteristics, sowing or planting date, development rate and stage, growing period, and weather conditions. The frequency of irrigation/rainfall, duration of irrigation/rainfall, the magnitude of the event, and evapotranspiration of crop, the crop coefficient is interpolated from FAO 56 crop coefficient curve (Allen et al. 1998). Crop coefficients for the mid-development stage, $K_{c(\mathrm{mid})}$ and late season $K_{c(\mathrm{end})}$ are taken from FAO-56 (Allen et al. 1998). The crop development occurs in four stages i.e., starting stage, crop development stage, mid-stage, and late-stage. The monthly crop coefficient (K_c) was assessed using the tenets and rules given in the irrigation water system and drainage paper FAO-56 (Allen et al. 1998) for Paddy and Wheat crops. For the specific environment, change where slightest relative humidity (RH_{min}) contrasts from 45% or where wind speed measured at 2 m stature (u_2) is larger or smaller than 2.0 m s^{-1}, the K_c values at mid and end of the season are adjusted and the equation applied for adjustment in K_c is given below:

$$K_{cb} = K_{cb(\text{tab})} + [0.04(u_2 - 2) - 0.004\{\text{RH}_{\min} - 45\}]\left(\frac{h}{3}\right)^{0.3}$$

where,

K_{cb} is adjusted basal crop coefficient for mid and late seasons growth stage, $K_{cb(\text{tab})}$ is the standard tabulated value for $K_{c(\text{mid})}$ and $K_{c(\text{end})}$ (if greater than or equal to 0.45), u_2 is mean daily wind speed at 2 m height during the mid or late season growth stages (m/s) for 1 m/s $\leq u_2 \leq$ 6 m/s, and RH_{min} is mean value of wind speed at 2 m height during mid or late season growth stages % for 20% $\leq RH_{min} \leq$ 80%, and h is mean plant height during mid or late season growth stages for 20% $\leq RH_{min} \leq$ 80%.

10.3.4 Soil Water Availability

Soil water availability is the water accessibility for the plant held in the soil. Distinctive soil has a diverse limit of water holding. The measure of water held by the soil against gravitational drive is termed field capacity. The crop takes the water held by the soil in the root zone. As water take-up proceeds with the water content at the root zone diminishes, the remaining water in the soil is held with large force making it hard to separate for the plant. A point where the plant can't extract remaining water from the soil is called a wilting point. The available total soil moisture in the root zone is the moisture available between the soil water content at field capacity and wilting point. The condition for the aggregate available water (Allen et al. 2005) is given by:

$$\text{TAW} = 1000(\theta_{\text{FC}} - \theta_{\text{WP}}) \times Z_r \quad \text{and}$$
$$\text{TAW} = p \times \text{RAW}$$

where,

TAW is total available soil water in the root zone (mm), Z_r is the rooting depth (m), θ_{WP} is the water content at wilting point (m^3 m^{-3}), θ_{FC} is the water content at field capacity (m^3 m^{-3}), RAW is the readily available soil moisture in the root zone (mm), and p is a normal portion of total available soil water (TAW) that can be exhausted from the root zone before soil water stress happens.

10.3.5 Irrigation Requirement

Irrigation water requirement incorporates the crop (ET_c), keeping up a great salt adjust inside the root zone. With the assessed estimation of ET_c, the irrigation water requirement of a crop can be figured; (Doorenbos and Kassam 1979) adopted the following relationship (using identical units of depth, i.e., mm/cm/inches):

$$I = \mathrm{ET}_c - P + \mathrm{RO} + \mathrm{DP} + L + D_r\left(\theta_f - \theta_i\right)$$

where,

I is irrigation water requirement, ET_c is evapotranspiration of the crop, P is effective precipitation, RO is runoff from the crop field from irrigation and/or rainfall, DP is deep percolation in the field from irrigation and/or rainfall, L is leaching requirement, θ_i is root zone soil moisture content before irrigation, θ_f is root zone soil moisture content at field capacity, and D_r is a depth of the root system.

The calculation steps for the water balance study are summarized in the flowchart presented in Fig. 10.3.

10.3.6 Crop Yield

The information over crop yield is obtained from farmer discussion, Irrigation Office, and Revenue Office.

10.3.7 The Indicators of Performance

The performance indicators are used to evaluate water use, yield, and impacts of intervention in irrigation systems to know performance and changes in a system's performance over time. It also helps to observe the performance of the system in different areas and at different levels. Performance indicators provide a general notation about the relative performance of the irrigation system. This will allow an initial screening of systems, which perform well in different environments (Molden et al. 1998). This can be categorized as follows (Table 10.3).

10.4 Results and Discussion

Gadarjudda minor system is lined, while Lakhnauta minor is an earthen canal system. The most common problems in both minors are the silting and growth of weeds. These problems need the attention of the beneficiaries as well as the service provider for

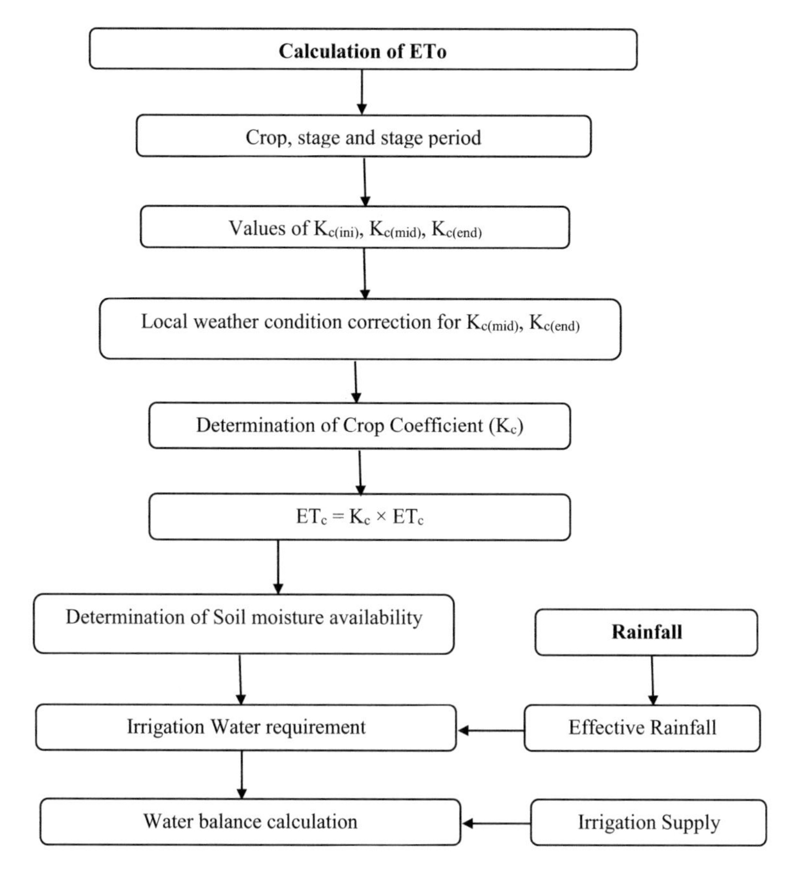

Fig. 10.3 Methodology used for water balance study

Table 10.3 Performance Indicators used to evaluate water use and yield (Bos et al. 2005)

(i) Water use indicators	*Formula*
Relative water supply	$\frac{\text{Water Supply(Irrigation+ Rainfall)}}{\text{Crop demand}}$
Relative irrigation supply	$\frac{\text{Irrigation Supply}}{\text{Irrigation Demand}}$
Depleted fraction	$\frac{ET_c}{\text{Total Rainfall +Irrigation Supply}}$
Crop water deficit	$ET_0 - ET_c$
(ii) Production indicators	*Formula*
Crop yield over total water supply	$\frac{\text{Yield}}{\text{Total water Supply}}$
Crop yield over water consumed	$\frac{\text{Yield}}{ET_a}$

routine annual maintenance. The cracks formed inclined canal systems are another problem of water leakage from the canal. Since the systems are century old, they need rehabilitation to reshape and modernize structures for optimum operation. The canal water reaches each field of command area by rotation. There has been a high intensity of cropping and mixed cropping in these command areas. The absence of the water-logged regions shows the excellent performance of systems in water supply and distribution.

Some social and physical phenomena cause maintenance issues, which are observed in the canal system. Plates 10.1, 10.2, 10.3 and 10.4 show damages to the canal and their causes.

i. Farmers are putting an obstruction inflow to get more water than one's share causing side cutting of the earthen canal bank (Plate 10.1).
ii. Cutting of canal banks to draw more water illegally than one's share causes damage to the canal bank (Plate 10.2).

Plate 10.1 Bank cutting due to obstruction in the flow

Plate 10.2 Illegal withdrawal of water

iii. Direct use of the canal for animal drinking and bathing purpose without proper provision and flow causes erosion in the canal's bank (Plate 10.3).
iv. Cracks inclining of the canal cause leakage of water and waterlogging in adjoining fields (Plate 10.4).

Water accounting is a fundamental exercise of irrigation water management and the operation of the canal infrastructure. In both the minor system, roasters at the distributary level are based on water availability during the crop period. In Kharif season, the canal's roaster is 2 weeks ON & 2 weeks OFF; and for Rabi 1 week ON & 3 weeks OFF. Sometimes without any prior notice, the roaster may differ depending upon water availability in the canal system.

The optimal use of irrigation water can be achieved by water balance study either by implementing rescheduling in supply or by adopting other crops, which suits

Plate 10.3 Cattle bathing in the canal and thereby causing bank erosion

Plate 10.4 Cracks in the lining of the canal

Table 10.4 Performance evaluation of Gadarjudda Minor

Crop name	Command area	Relative water supply	Relative irrigation supply	Depleted fraction	Crop water deficit	Productivity	
						Irrigation supply	Water consumed
	(ha)	RWS	RIS	DF	(mm)	(kg/m^3)	(kg/m^3)
Year: 2013–14							
Paddy	72	2.57	3.35	0.36	6.60	1.34	1.25
Wheat	170	1.25	1.13	0.75	0.70	2.67	1.49
Sugarcane	275	1.19	1.04	0.74	106.90	8.93	5.69
Year: 2014–15							
Paddy	74	2.18	4.3	0.43	2.10	1.21	1.17
Wheat	169	1.22	1.24	0.77	0.80	2.3	1.6
Sugarcane	273	1.27	1.44	0.73	36.20	9.19	6.08
Year: 2015–16							
Paddy	85	1.91	4.19	0.49	6.50	1.36	1.22
Wheat	161	0.73	0.72	1.07	52.40	1.81	1.45
Sugarcane	247	1.2	1.16	0.67	178.70	7.68	6.02
Year: 2016							
Paddy	84	3.13	14.2	0.30	1.20	1.53	1.3

the supply schedule. It is also necessary to make a roaster as per the weather forecast because irrigation supply should need to increase/decrease whenever there was less/more rainfall in the command area. The findings of performance indicators and their significance in evaluating performance are provided in Tables 10.4 and 10.5.

10.4.1 Development of Indicators Map Using GIS

The spatial maps of the indicator generated in the GIS environment give the information of the irrigation system's performance over command areas and help for evaluation. The maps of irrigation requirement, crop water deficit mapping, the productivity of years 2015 and 2016 were developed in GIS and are presented in Figs. 10.4, 10.5 and 10.6.

10.4.1.1 Irrigation Requirement Mapping

For the planner, irrigation requirement mapping is essential for the spatial distribution of irrigation supply for crops in different seasons. The irrigation water requirement map of wheat and sugarcane crops of 2015–16 and Paddy crop for the year 2016

Table 10.5 Performance evaluation of Lakhnauta Minor

Crop name	Command area	Relative water supply	Relative irrigation supply	Depleted fraction	Crop water deficit	Productivity	
						Irrigation supply	Water consumed
	(ha)	RWS	RIS	DF	(mm)	(kg/m³)	(kg/m³)
Year: 2013–14							
Paddy	133	2.35	2.51	0.40	1.90	1.8	1.24
Wheat	183	1.31	1.26	0.72	0.70	2.39	1.49
Sugarcane	372	1.08	0.79	0.75	193.70	10.83	5.63
Year: 2014–15							
Paddy	133	1.95	3.24	0.48	2.10	1.55	1.13
Wheat	182	1.15	1.11	0.81	3.20	2.57	1.61
Sugarcane	355	1.22	1.28	0.73	81.60	9.86	6.03
Year: 2015–16							
Paddy	156	1.67	3	0.57	1.20	1.9	1.21
Wheat	170	0.59	0.55	1.18	79.90	2	1.41
Sugarcane	321	1.12	0.89	0.65	282.40	8.78	5.86
Year: 2016							
Paddy	156	3.04	13.68	0.31	1.20	1.53	1.3

of Gadarjudda and Lakhnauta minor is presented in Fig. 10.4 for paddy, wheat, and sugarcane.

10.4.1.2 Crop Water Deficit Mapping

The crop water deficit map helps the planner overcome the deficit of water where it is in deficit. It also helps in the equity of distribution of irrigation water over the command areas. Crop water deficit (CWD) mapping also allows finding the amount of water deficit of respective crops. The CWD is the indicator to determine the irrigation system's performance based on the deficit of water to the crop. If the deficit is more, the performance of the irrigation system is poor and vice-versa. The CWD map of Gadarjudda and Lakhnauta minor is presented in Fig. 10.5.

10.4.1.3 Productivity Mapping

The physical crop water productivity is the ratio of crop yield (ton/ha) to the amount of water used (m^3/ha). Productivity mapping helps to find the actual water consumed/delivered to crop per unit of its yield. It can be generally represented in the unit of kg/m^3. The productivity map represents the spatial distribution of crop yield per unit of water consumed. GIS also helps to find the crop water productivity

Fig. 10.4 Irrigation water requirement

using remote sensing data. In this study yield values obtained from farmer interaction, irrigation office is taken as yield of the crops and actual water use/supply comes from the simulation of system condition in various software used for the calculation of water productivity of crops.

In the Gadarjudda command, paddy productivity is found to be 1.30 kg/m^3, wheat 1.45 kg/m^3, and sugarcane 6.02 kg/m^3. These values are good in terms of their range of productivities recommended by FAO. Similarly, in the Lakhnauta command, paddy productivity is found to be 1.30 kg/m^3 in 2016. The productivity of wheat and sugarcane was 1.41 kg/m^3, and 5.86 kg/m3, respectively, for 2015–16. These values are also good in terms of their range of productivities suggested by FAO. Comparing these two minor systems, Gadarjudda performs in a good manner than Lakhnauta in terms of productivity. The spatial distribution of paddy productivity, wheat, and sugarcane in both the minors is shown in Fig. 10.6.

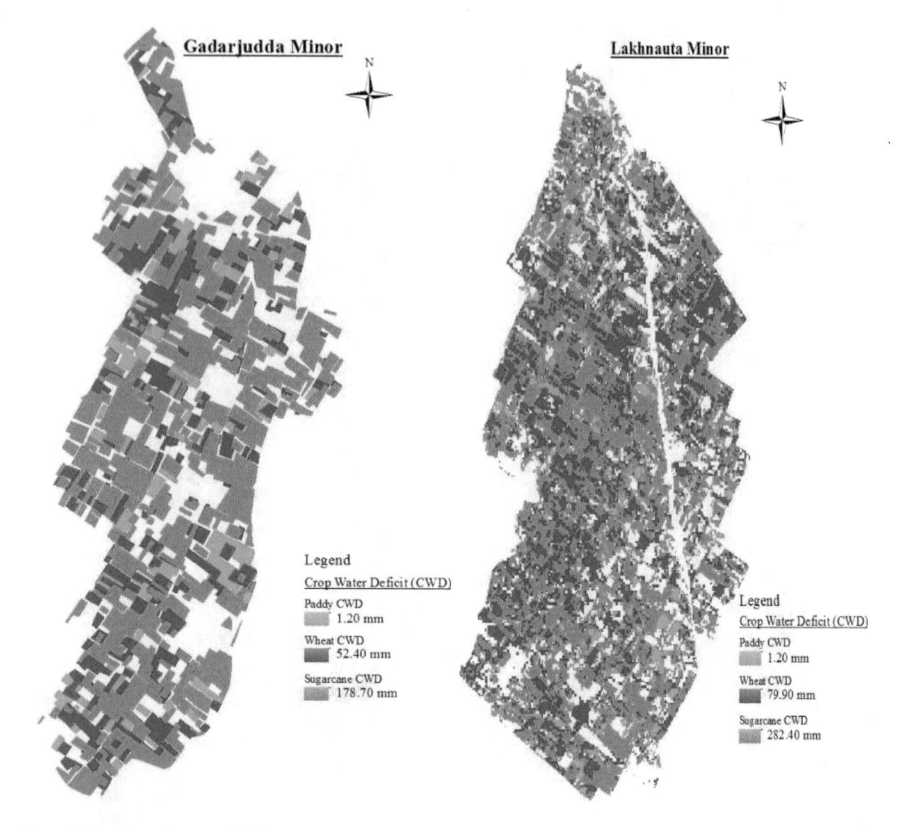

Fig. 10.5 Crop water deficit

10.5 Conclusions

Performance evaluation of the Gadarjudda and Lakhnauta irrigation systems' was carried out based on the physical condition of water supply and distribution, water use, and productivity indicators derived using satellite, meteorological, and ground survey secondary data. The time-to-time inspection of the canal system, including discharge measurement in the field, the system's efficiency obtained from reference study, and other various information, helps compare the system's present performance with the designed one and evaluate the physical condition of the system. The values of actual evapotranspiration and crop coefficient were obtained from field data, meteorological data, and FAO guidelines. The simulation of crop water demand and supply data in field conditions using various software gave various performance indicators for evaluating Gadarjudda and Lakhnauta minor irrigation systems. Following conclusions are drawn from the present study:

i. Gadarjudda and Lakhnauta systems are well managed except for some exceptions. These systems are very old, so they need rehabilitation work for their

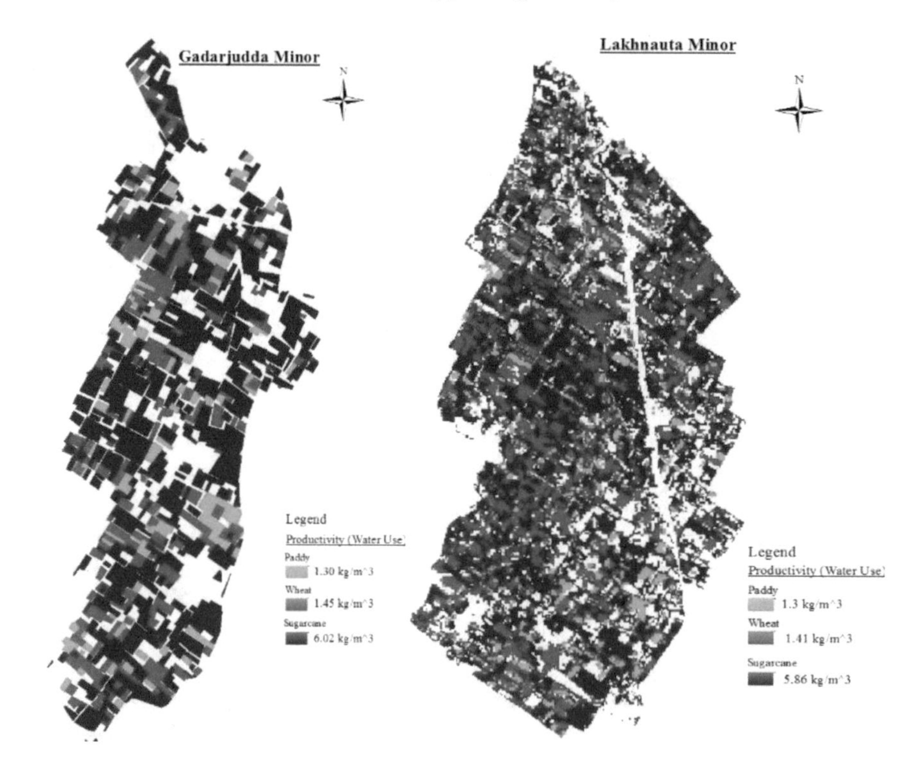

Fig. 10.6 Water productivity (actual water use)

optimization, i.e., to increase efficiency, easy operation, etc. Agriculture in the command is good, giving the high intensity of cropping and better yield.

ii. The irrigation system's good physical condition plays a vital role in supplying timely and quantitatively water supply to crop to increase the system performance. The high intensity of cropping trend within both minor command areas shows the system's performance despite theft and illegal water withdrawal. The annual maintenance made systems carry water at designed capacity even after decades of their construction.
iii. The relative water supply (RWS) and relative irrigation supply (RIS) values show higher values mean better system performance with less efficiency. The RIS is less than 1 for wheat and sugarcane, indicating a gap between crop water demand and supply. However, the RWS of more than 1 shows the uneven distribution of rainfall during the crop period.
iv. Depleted fraction (DF) indicates the condition of actual water use from the supplied quantity. DF value 1 means no deficit of water in a crop, less than 1 means more water supply, and more than 1 means water stress condition in the crop. This is also shown by the crop water deficit values, which are higher or lower.

v. The RWS and RIS values are greater than one for abundant water situations, while the scarce situation, low RWS, and RIS values are observed. The low depleted fraction in the water abundant rice field shows water loss, which could be due to untimely irrigation. The higher value of crop water deficit shows the gap of supply that remains to fulfill.

vi. As per the FAO, paddy productivity varies from 0.6 to 1.6 kg/m^3, and the productivity of wheat varies from 1.0 to 2.2 kg/m^3. The productivity of paddy in 2016 is 1.30 kg/m^3, and wheat in 2014–15 (1.6 kg/m^3) shows the best productivity during study years in Gadarjudda minorsss. Almost the same result was observed for Lakhnauta minor. While comparing both minor systems, the productivity of these crops is more in Gadarjudda minor.

References

Ahmad MUD, Turral H, Nazeer A (2009) Diagnosing irrigation performance and water productivity through satellite remote sensing and secondary data in a large irrigation system of Pakistan. Agric Water Manag 96(4):551–564

Allen RG, Pereira LS, Raes D, Smith M (1998) Crop Evapotranspiration-Guidelines for computing crop water requirements-FAO Irrigation and drainage paper 56. FAO, Rome 300(9):D05109

Allen RG, Pereira LS, Smith M, Raes D, Wright JL (2005) FAO-56 dual crop coefficient method for estimating evaporation from soil and application extensions. J Irrig Drain Eng 131(1):2–13

Bastiaanssen WG (1998) Remote sensing in water resources management: the state of the art. International Water Management Institute, pp 1–118

Bos MG, Murray-Rust DH, Merrey DJ, Johnson HG, Snellen WB (1993) Methodologies for assessing performance of irrigation and drainage management. Irrig Drain Syst 7(4):231–261

Bos MG, Burton MA, Molden DJ (2005) Irrigation and drainage performance assessment: practical guidelines. Cabi Publishing, pp 26–46

Doorenbos J, Kassam AH (1979) Yield response to water. Irrigat Drainage Paper 33:257

Kumar R, Singh RD, Sharma KD (2005) Water resources of India. Curr Sci 89(5):794–811

Molden DJ, Sakthivadivel R, Perry CJ, De Fraiture C (1998) Indicators for comparing performance of irrigated agricultural systems. In: IWMI, vol 20, pp 1–26

Oweis TY, Hachum AY (2003) 11 Improving water productivity in the dry areas of West Asia and North Africa. Water Prod Agric: Limits Opportunities Improvement 1:179

Pandey A, Mogarekar N (2021) Development of a spatial decision system for irrigation management. J Ind Soc Remote Sens. https://doi.org/10.1007/s12524-020-01305-

Ray SS, Dadhwal VK, Navalgund RR (2002) Performance evaluation of an irrigation command area using remote sensing: a case study of Mahi command, Gujarat, India. Agric Water Manage 56(2):81–91

Zwart SJ, Bastiaanssen WG (2004) Review of measured crop water productivity values for irrigated wheat, Paddy, cotton and maize. Agric Water Manag 69(2):115–133

Chapter 11
Role of Geospatial Technology for Enhancement of Field Water Use Efficiency

Debasis Senapati and Ashish Pandey

Abstract Enhancement of water use efficiency in the agricultural field is essential for the sustainable management of water resources. This tedious task can only be possible by either increasing output (crop yield) or by decreasing input (irrigation) as per the water resources engineering perspective. Crop yield is restricted to various factors apart from irrigation viz. nutrient supplement, crop variety, property of soil, soil health, etc., which is challenging to manage. Irrigation can be managed by adopting various agricultural water management techniques (e.g. alternative wet and dry, spate irrigation, deficit irrigation, precision irrigation, etc.) at field or plot scale. However, agricultural water management techniques are challenging to adopt for a regional scale due to the lack of real-time soil moisture and evapotranspiration data. This difficulty can be overcome with the help of Geospatial Technology. In this chapter, a detailed role of remote sensing and GIS to enhance field water use efficiency is explained.

11.1 Introduction

Water demand is increasing in every sector, leading to a severe water scarcity problem in the world. However, a glimpse of water scarcity has been observed in India's twenty-one major cities in various scenarios as per the Composite Water Management Index (CWMI) report released by National Institution for Transforming India (NITI) Aayog in 2018. Groundwater is depleting at a rate of 17.7 $\pm$ 4.5 Gigatonnes/year in Northwest India (Rodell et al. 2009) due to a significant rise in irrigation practices. Enhancement of irrigation water use efficiency or water productivity in the agriculture sector is a governing major for sustainable management

D. Senapati (✉) · A. Pandey
Department of Water Resources Development and Management, Indian Institute of Technology Roorkee, Roorkee 247667, India
e-mail: dsenapati@wr.iitr.ac.in

A. Pandey
e-mail: ashish.pandey@wr.iitr.ac.in

A. Pandey et al. (eds.), *Geospatial Technologies for Land and Water Resources Management*, Water Science and Technology Library 103,
https://doi.org/10.1007/978-3-030-90479-1_11

of water resources in developing countries. Most of India's overall water use efficiency is 38%, whereas 50–60% in developed countries as per Ministry of Jal Shakti in 2019 (http://nwm.gov.in/sites/default/files/1.%20National-water-mission-%20%20%20water-use-efficiency.pdf). Therefore, it is very important to take necessary steps toward enhancement of irrigation efficiency. Enhancement of water use efficiency at the field scale requires accurate assessment with an effective management strategy (Patel and Rajput 2013; Gao et al. 2019; Wang et al. 2020). However, remote sensing and GIS play a vital role in assessing water use efficiency at a regional and global scale.

Irrigation must be supplied as per the net irrigation water requirement to improve water use efficiency without sacrificing yield (Jensen et al. 1990). Irrigation scheduling advisory forecasting in location-specific might helpful for farmers to manage irrigation in the field. Calculation of irrigation water requirement and irrigation scheduling forecasting requires adequate information of available soil moisture and evapotranspiration (ET), which is dynamic and exhaustive to measure. In field and plot, accurate ET can be determined with the help of modern instruments like large-aperture scintillometer, sap flow and eddy covariance flux tower. Irrigation scheduling of an agricultural field using ET obtained by using the modern instruments could save 50% of irrigation water compared to ET from the Penman–Monteith equation (Ragab et al. 2017). However, remote sensing data plays a significant role in the determination of two major parameters, i.e., ET and soil moisture, for a larger scale. In this chapter, lucid methods of determination of actual evapotranspiration (AET) and soil moisture using remote sensing are reviewed, and some future suggestion for enhancement of water use efficiency is discussed.

11.2 Estimation of Actual Evapotranspiration

Empirical equations are used to determine ET, among all FAO-56 (Allen et al. 1998) is widely used for its applicability. FAO-56 is a standard method and used for validation of the new ET method. A detailed evolution of the AET estimation methodology using remote sensing is summarized in Table 11.1. Surface energy balance and surface temperature (Ts)-vegetation space index (VI) method are two widely used algorithms for the determination of AET.

11.2.1 Surface Energy Balance Models

The application of remote sensing for the determination of ET was started by Choudhury (1989); Moran and Jackson (1991); Kustas and Norman (1996), Bastiaanssen et al.(1998). As per Jackson et al.(1997), these classical methods are suitable for small scale due to the requirement of reliable field data for calibration and lack of Spatio-temporal scalability.

Table 11.1 Evolution of application of remote sensing for determination of AET

Author (year)	Study area	Methodology	Major findings
Bastiaanssen et al. (1998)	Spain, Niger, and China	**One source SEB method: SEBAL** He formulated and successfully validated SEBAL algorithm with four different approaches viz. turbulent surface fluxes measurement, airborne turbulent flux measurements, soil moisture profiles measurements, hydrological models	• Spatial variation of hydro-meteorological parameters is estimated empirically with limited accessible parameters, i.e., short wave atmospheric transmittance, vegetation height, and surface temperature • No numerical simulation models • Calculate the fluxes independently from land cover and can handle thermal infrared images
Roerink et al. (2000)	Italy	**One source SEB method: S-SEBI** The atmospheric conditions over the area can be considered constant, and the area reflects sufficient variations in surface hydrological conditions	• S-SEBI calculated heat fluxes are systematically higher than the measured values • The large discrepancies between the net radiation measurements and the measured energy balance closure values
Su (2002)	USA	**One source SEB method: SEBS** SEBS is developed for the determination of evaporative fraction and atmospheric turbulent fluxes using satellite data. The reliabilities of SEBS are assessed from four experimental data sets	• The mean error is expected to be around 20% relative to the mean sensible heat flux (H), when the input geometrical and physical variables are reliable at point scale, whereas at 50% at a regional scale
Norman et al. (2003)	USA, El Reno, Oklahoma	**Two source SEB method: disALEXI** Disaggregated atmosphere land exchange inverse model (DisALEXI) is a two-step approach (1) Surface brightness-temperature-change is used to estimate average surface flux (2) Disintegrate the surface flux to determine vegetative index and surface temperature (3) Estimates ET by combining low and high-resolution RS data with a 10^1–10^2 m scale without any local observation	• Algorithm was tested with ground-based eddy covariance flux measurements collected in a rangeland landscape • The root mean square deviation between model estimates and measurements was comparable with the observational accuracy typically associated with micro-meteorological flux measurement techniques for all flux components

(continued)

Table 11.1 (continued)

Author (year)	Study area	Methodology	Major findings
Duchemin et al. (2006)	Central Morocco	Established a relationship between LAI, AET (eddy flux tower derived), ET_o (Penman–Monteith equation), and NDVI. Space and time variability of crop phenology and water requirement were estimated using the established relationship between plant transpiration coefficients (K_{cb}), LAI and NDVI with Landsat-7 data	• The relationship between LAI ~ NDVI and LAI ~ Kcbwas found exponential, whereas the NDVI ~ Kcb was linear
Allen et al. (2007)	Global	**One source SEB method: METRIC** It is an image processing model for the determination of ET. It uses the SEBAL surface energy balance process. It is calibrated internally using ground observation of ET	• METRIC is highly reliant on the selection of appropriate cold and hot pixels. It can apply to complex landscape regions
Er-Raki et al. (2007, 2010)	Center of Morocco	To test three methods based on the FAO-56 "dual" crop coefficient (K_{cb}) and a fraction of soil surface covered by vegetation (f_c) approach viz. no calibration, local calibration, and NDVI-calibration to estimate AET. The AET is compared with the Eddy correlation system. This study was extended in 2010 for the generation of the spatial distribution of seasonal crop ET using SPOT and Landsat images	The local calibrated method is best, and NDVI-calibrated method is acceptable. The crop coefficient (Kcb)-NDVI relationship employed for estimation of crop water requirement
Gonzalez-Dugo and Mateos (2008)	Spain	Evaluated convenient ways to derive crop coefficients using multispectral vegetation indices obtained by remote sensing data such as NDVI and SAVI	There is no relationship between ET and yield in sugarbeet and cotton fields despite the homogeneous irrigation scheme in Southern Spain
Miralles et al. (2011)	Global	A GLEAM (Global Land surface Evaporation: the Amsterdam Methodology) was developed using Priestley and Taylor (PT), interception model, soil water module, and stress module for estimation of daily evaporation on a global scale	Validation of evaporation against one year of eddy covariance measurements from 43 FLUXNET stations shows daily and monthly R values of 0.83 and 0.90, respectively
Liaqat et al. (2015)	Indus-basin	Assessed Spatio-temporal distribution of AET using SEBS and validated with an advection-aridity method using nine meteorological sites for 2009–2010	SEBS model underestimates during Kharif season and overestimate during rabi

(continued)

Table 11.1 (continued)

Author (year)	Study area	Methodology	Major findings
Cruz-Blanco et al. (2014)	Southern Spain	Developed a tool MA + LSE for estimation of reference ET	This tool is useful where intensive meteorological data are not available with simply solar radiation and air temperature
Senay (2018)	Southern California	Used SSEBop ET model and Landsat image to study water used dynamics over 1984–2014 and validated with gridded flux data and water balance method	R^2 up to 0.88, root mean square error as low as 14 mm/month
Park et al. (2017)	East Asia	Satellite-based crop coefficient and ET are estimated by developing a relation between (NDVI, LAI, and soil moisture) and (NDVI and LAI). The satellite-derived results are compared with 4 flux tower-derived crop coefficients and ET	NDVI, LAI, and soil moisture-based K_c show good results compared with (NDVI and LAI) derived K_c based on statistical analysis

Bastiaanssen et al. (1998) formulated and successfully validated the Surface Energy Balance Algorithm for Land (SEBAL) method to estimate ET for meso and macro scale, which is still used. SEBAL estimates most of the hydro-meteorological parameters empirically, requires surface temperature and vegetation height.

The simplified Surface Energy Balance Index (S-SEBI) algorithm was developed by Roerink et al. (2000), which estimates latent heat of evaporation instantaneously. This approach is different from classical surface energy balance models because it does not need to compute sensible heat. In S-SEBI, the relation between observed land surface temperature and surface reflectance is used. The relationship is confined with two boundaries, i.e., dry radiation controlled condition and evaporation controlled wet condition. S-SEBI model would be applied successfully in those areas where atmospheric conditions are reasonably constant across satellite images, and the image must contain wet and dry areas.

The surface Energy Balance System (SEBS) was developed by Su (2002). The tools of SEBS can easily utilize spectral reflectance and radiance for the determination of land surface parameters. SEBS consists of a model for determining roughness length for heat transfer and evaporative fraction for some cases. SEBS facilitates the determination of sensible heat and latent heat of evaporation from satellite data.

11.2.2 Surface Temperature (Ts)-Vegetation Space Index (VI) Methods

Evaporative fraction is one of the parameters through which ET can be estimated. Sensible heat flux and latent heat of evaporation data are not required in this method which is challenging to obtain. Triangular and trapezoidal are the two major types Ts-VI methods available in the literature. The triangular method has been widely used (Carlson et al. 1995; Jiang and Islam 1999, 2001, 2003). Determination of dry and wet edges is the distinctive limitation of the triangular method, where the empirical regression approaches used for the determination of edges (Garcia et al. 2014). The surface energy balance principle addresses the issue, but a clear-sky condition is required for uniform atmosphere forcing. Time-Domain Triangular Method (TDTM) is proposed by Minacapilli et al. (2016) to overcome the disadvantages of the space domain triangular method (SDTM). Zhu et al. (2017) proposed a universal triangle method and found better results than TDTM and SDTM.

Effect of water stresses on canopy transpiration is not considered in the triangular method, and a trapezoidal method was developed (Moran et al. 1994). Long and Singh (2012) developed a two-source trapezoid model by considering soil wetness isolines and VI/Ts space. A hybrid dual-source trapezoidal method was developed, combining trapezoidal and triangular on-patch approaches in an equal aerodynamic resistance framework (Yang and Shang 2013).

11.2.3 Estimation of Soil Moisture

Electromagnetic reflectance and land surface temperature are the main parameters used to estimate soil moisture content from optical and thermal remote sensing data (Elzeiny and Effat 2017; Amato et al. 2015). However, an active remote sensor such as radar used many factors viz. backscatter signal, surface roughness, vegetation cover, and frequencies for determination of soil moisture, depending on the backscattering capability and method of polarization various active sensors viz. Sentinel-1 Synthetic Aperture Radar (SAR) systems, Advanced Synthetic Aperture Radar (ASAR), Advance scatterometer (ASCAT) are used for the determination of soil moisture. Near-surface soil moisture up to 0–5 cm using passive and active microwave data viz. Passive and Active L-band Sensor (PALS), Electronically Scanned Thinned Array Radiometer (ESTAR), Polarimetric Scanning Radiometer (PSR), Advanced Microwave Scanning Radiometer (AMSRE), Soil Moisture and Ocean Salinity (SMOS), AQUARIUS and Soil Moisture Active and Passive (SMAP). Soil moisture estimation with the help of remote sensing can further utilized for the determination of soil properties (Mohanty 2013). Summary of application of different remote sensing for retrieval of soil moisture is tabulated in Table 11.2.

Table 11.2 Application of different types of remote sensing for retrieval of soil moisture

Optical remote sensing	Zhang et al. (2013)	Visible and Shortwave infrared Drought Index (VSDI) was developed for monitoring soil and vegetation moisture using optical remote sensing
	Lesaignoux et al. (2013)	A semi-empirical soil model was developed for the determination of soil moisture from the spectral signature
	Sadeghi et al. (2015)	Kubelka–Munk two flux radiative transfer theory-based physical model was derived for retrieval of soil moisture using the optical remote sensing data
Thermal remote sensing	Verstraeten et al. (2006)	Incorporated maximum and minimum thermal inertia in soil moisture saturation index (SMSI) to determine surface soil moisture from thermal remote sensing data (METEOSAT imagery), Markov type filter converts surface soil moisture to profile value
	Sugathan et al. (2014)	Studied the relationship between surface albedo and surface soil moisture and found that surface albedo is falling exponentially with an increase in soil moisture
Microwave remote sensing	Crapolicchio and Lecomte (2004)	First time studied the application of scatterometer data (ASCAT) for the determination of soil moisture
	Entekhabi et al. (2010)	Studied the concept and applicability of Soil Moisture Active and Passive (SMAP) mission
	Wagner et al. (2013)	Compared three satellite soil moisture data sets (SMOS, AMSR-E, and ASCAT) and ECMWF forecast soil moisture data to in-situ measured soil moisture and found that SMOS is best
	Chan et al. (2016)	Described the validation of SMAP products and assess the accuracy of satellite products

Accurate crop water requirement estimation is required to enhance water use efficiency, which requires soil moisture not only at the soil surface but also in the different profiles up to the root zone. *I*- and *P*-band signal of opportunity of active and passive microwave remote sensing can be helpful for the determination of root zone soil moisture (Yueh et al. 2019; Garrison et al. 2018). Moreover, physical models like the soil vegetation atmosphere transfer model combining with assimilated thermal infrared (TIR) and synthetic aperture radar (SAR) data can able to determine root zone soil moisture and water stress (Lei et al. 2020).

11.3 Irrigation Scheduling Forecast

Irrigation scheduling is an important management strategy for managing irrigation water, improving field water use efficiency. Irrigation scheduling is nothing but an optimized way of timing and amount of irrigation. Farmers of India mostly do not aware of the importance of water in a natural water-rich region. Advance short-term (same day) and long-term (6 days before) weather forecasts should incorporate during the determination of irrigation scheduling. Irrigation scheduling can be determined either based on available soil moisture or ET. There are many studies (Marino et al. 1993; Cao et al. 2019) used weather forecast for the development of real-time irrigation schedule.

Wang and Cai (2009) used both short and long-term weather forecast to schedule irrigation in the Havana Lowlands region, Illinois. The soil water atmosphere plant (SWAP) model is used to simulate yield for 2002–2006. Irrigation scheduling is determined based on available soil moisture information. The result shows that by incorporating real-time or short-term weather forecast in irrigation scheduling, the yield of maize can improve to 16%, and it can further increase to 21% if long-term weather forecast is used in irrigation scheduling.

Cai et al. (2011) described a modeling framework for real-time decision support for irrigation scheduling. A stochastic optimization program based on soil moisture and crop water stress is used for irrigation scheduling. The weather forecast results in three different years viz. dry, normal, and wet are observed. More economic gains are observed in normal and wet years as compared with dry years.

Belaqziz et al. (2013) developed the Irrigation Priority Index (IPI) based on the normalized indicator for evaluating the actual irrigation schedule in Morocco. The normalized indicator takes into account the water stress coefficient and time of irrigation. IPI value ranges from -1 to 1 and decreases with irrigation priority. The result shows a clear relationship between IPI and grain yield, which signifies the developed IPI from remote sensing is an optimized way of irrigation scheduling.

Cruz-Blanco et al. (2014) developed a new tool (MA + LSE) by combing remote sensing and weather forecast for estimation of ET. Irrigation scheduling and yield have efficiently been estimated using the new tool. Average seasonal irrigation volume and yield were underestimated 2.6% and 2.2%, respectively, by the MA + LSE method as compared to the PM-FAO56 method.

Lorite et al. (2015) evaluated weather forecasts for estimation of ET and developed irrigation scheduling. The optimized irrigation scheduling was done with the help of the Aquacrop model. In this study, simulated crop yield and irrigation scheduling are compared using weather data from automatic weather stations and forecasted weather data.

11.4 Conclusion

Geospatial technology is an effective tool for the development of field water use efficiency in a large area. It is also proving its applicability in the remote area where field data is inaccessible. In this short review, an approach to enhance field water use efficiency with the help of optimizing irrigation scheduling forecast based on remote sensing ET and available soil moisture is disused. The following are future recommendations from the short review.

1. Although there are various algorithms and tools for determining AET from remote sensing, the accuracy of the estimation should be improved.
2. Determination of irrigation scheduling forecast at field scale using near prediction of climatic variables can further scope of future research.
3. Application of irrigation scheduling forecast based on remote sensing AET and soil moisture should be taken place in most water-rich regions of India.
4. Determination of accurate soil moisture up to root zone depth from remote sensing data can further scope future research.

References

Allen RG, Tasumi M, Trezza R (2007) Satellite-based energy balance for mapping evapotranspiration with internalized calibration (METRIC)—Model. J Irrig Drain Eng 133(4):380–394

Allen RG, Raes D, Smith M (1998) crop evapotranspiration: guidelines for computing crop requirements. FAO Irridation and Drainage Paper No. 56. FAO, Rome

Amato F, Havel J, Gad AA, El-Zeiny AM (2015) Remotely sensed soil data analysis using artificial neural networks: a case study of El-Fayoum depression, Egypt. ISPRS Int. J. Geo-Inf. 4(2):677–696

Bastiaanssen WG, Menenti M, Feddes RA, Holtslag AAM (1998) A remote sensing surface energy balance algorithm for land (SEBAL). 1. Formulation. Journal of Hydrology 212:198–212

Belaqziz S, Khabba S, Er-Raki S, Jarlan L, Le Page M, Kharrou MH, Adnani E, Chehbouni A (2013) A new irrigation priority index based on remote sensing data for assessing the networks irrigation scheduling. Agric Water Manag 119:1–9

Cai X, Hejazi MI, Wang D (2011) Value of probabilistic weather forecasts: Assessment by real-time optimization of irrigation scheduling. J Water Resour Plan Manag 137(5):391–403

Cao J, Tan J, Cui Y, Luo Y (2019) Irrigation scheduling of paddy rice using short-term weather forecast data. Agric Water Manag 213:714–723

Carlson TN, Gillies RR, Schmugge TJ (1995) An interpretation of methodologies for indirect measurement of soil water content. Agric for Meteorol 77(3–4):191–205

Chan SK, Bindlish R, O'Neill PE, Njoku E, Jackson T, Colliander A, Chen F, Burgin M, Dunbar S, Piepmeier J, Yueh S, Entekhabi D, Cosh MH, Caldwell T, Walker J, Wu X, Berg A, Rowlandson T, Pacheco A, McNairn H, Thibeault M, Martinez-Fernandez J, Gonzalez-Zamora A, Seyfried M, Bosch D, Starks P, Goodrich D, Prueger J, Palecki M, Small EE, Zreda M, Calvet JC, Crow WT, Kerr Y (2016) Assessment of the SMAP passive soil moisture product. IEEE Trans Geosci Remote Sens 54(8):4994–5007

Choudhury BJ (1989) Estimating evaporation and carbon assimilation using infrared temperature data. In: Asrar G (ed) Vistas in modeling, in theory and applications of optical remote sensing. Wiley, New York, pp 628–690

Crapolicchio R, Lecomte P (2004) The ERS-2 scatterometer mission: events and long-loop instrument and data performances assessment. In: Proceedings of the ENVISAT & ERS symposium, pp 6–10

Cruz-Blanco M, Lorite IJ, Santos C (2014) An innovative remote sensing based reference evapotranspiration method to support irrigation water management under semi-arid conditions. Agric Water Manag 131:135–145

Duchemin B, Hadria R, Erraki S, Boulet G, Maisongrande P, Chehbouni A, Simonneaux V (2006) Monitoring wheat phenology and irrigation in Central Morocco: on the use of relationships between evapotranspiration, crops coefficients, leaf area index and remotely-sensed vegetation indices. Agric Water Manag 79(1):1–27

El-Zeiny AM, Effat HA (2017) Environmental monitoring of spatiotemporal change in land use/land cover and its impact on land surface temperature in El-Fayoum governorate, Egypt. Remote Sens Appl Soc Environ 8:266–277

Entekhabi D, Njoku EG, O'Neill PE, Kellogg KH, Crow WT, Edelstein WN, Entin JK, Goodman SD, Jackson TJ, Johnson J, Kimball J, Piepmeier JR, Koster RD, Martin N, McDonald KC, Moghaddam M, Moran S, Reichle R, Shi JC, Spencer MW, Thurman SW, Tsang L, van Zyl J (2010) The soil moisture active passive (SMAP) mission. Proc IEEE 98(5):704–716

Er-Raki S, Chehbouni A, Boulet G, Williams DG (2010) Using the dual approach of FAO-56 for partitioning ET into soil and plant components for olive orchards in a semi-arid region. Agric Water Manag 97(11):1769–1778

Er-Raki S, Chehbouni A, Guemouria N, Duchemin B, Ezzahar J, Hadria R (2007) Combining FAO-56 model and ground-based remote sensing to estimate water consumptions of wheat crops in a semi-arid region. Agric Water Manag 87(1):41–54

Gao H, Yan C, Liu Q, Li Z, Yang X, Qi R (2019) Exploring optimal soil mulching to enhance yield and water use efficiency in maize cropping in China: a meta-analysis. Agric Water Manag 225:105741

Garcia M, Fernandez N, Villagarcia L, Domingo F, Puigdefabregas J, Sandholt I (2014) Accuracy of the temperature-vegetation dryness index using MODIS under water-limited vs. energy-limited evapotranspiration conditions. Remote Sens Environ 149:100–117

Garrison JL, Piepmeier JR, Shah R (2018) Signals of opportunity: enabling new science outside of protected bands. In: 2018 International conference on electromagnetics in advanced applications (ICEAA). IEEE Sept 2018, pp 501–504

González-Dugo MP, Mateos L (2008) Spectral vegetation indices for benchmarking water productivity of irrigated cotton and sugarbeet crops. Agric Water Manag 95(1):48–58

Hess T (1996) A microcomputer scheduling program for supplementary irrigation. Comput Electron Agric 15(3):233–243

Jackson TJ, Schmugge J, Engman ET (1997) Remote sensing applications to hydrology: soil moisture. Hydrol Sci J 41(4):517–530

Jensen ME, Burman RD, Allen RG (1990) Evapotranspiration and irrigation water requirements, vol 1. FAO, Rome, Italy, pp 54–60

Jiang L, Islam S (1999) A methodology for estimation of surface evapotranspiration over large areas using remote sensing observations. Geophys Res Lett 26(17):2773–2776

Jiang L, Islam S (2001) Estimation of surface evaporation map over southern Great Plains using remote sensing data. Water Resour Res 37(2):329–340

Jiang L, Islam S (2003) An intercomparison of regional latent heat flux estimation using remote sensing data. Int J Remote Sens 24(11):2221–2236
Kustas WP, Norman JM (1996) Use of remote sensing for evapotranspiration monitoring over land surfaces. Hydrol Sci J 41(4):495–516
Lei F, Crow WT, Kustas WP, Dong J, Yang Y, Knipper KR, Anderson MC, Gao F, Notarnicola C, Greifeneder F, McKee LM (2020) Data assimilation of high-resolution thermal and radar remote sensing retrievals for soil moisture monitoring in a drip-irrigated vineyard. Remote Sens Environ 239:111622
Lesaignoux A, Fabre S, Briottet X (2013) Influence of soil moisture content on spectral reflectance of bare soils in the 0.4–14 μm domain. Int J Remote Sens 34(7):2268–2285
Liaqat UW, Choi M, Awan UK (2015) Spatio-temporal distribution of actual evapotranspiration in the Indus Basin Irrigation System. Hydrol Process 29(11):2613–2627
Long D, Singh VP (2012) A modified surface energy balance algorithm for land (M-SEBAL) based on a trapezoidal framework. Water Resour Res 48(2)
Lorite IJ, Ramírez-Cuesta JM, Cruz-Blanco M, Santos C (2015) Using weather forecast data for irrigation scheduling under semi-arid conditions. Irrig SCi 33(6):411–427
Marino MA, Tracy JC, Taghavi SA (1993) Forecasting of reference crop evapotranspiration. Agric Water Manag 24(3):163–187
Minacapilli M, Consoli S, Vanella D, Ciraolo G, Motisi A (2016) A time domain triangle method approach to estimate actual evapotranspiration: application in a Mediterranean region using MODIS and MSG-SEVIRI products. Remote Sens Environ 174:10–23
Miralles DG, Holmes TRH, De Jeu RAM, Gash JH, Meesters AGCA, Dolman AJ (2011) Global land-surface evaporation estimated from satellite-based observations. Hydrol Earth Syst Sci 15(2):453–469
Mohanty BP (2013) Soil hydraulic property estimation using remote sensing: a review. Vadose Zone J 12(4):1–9
Moran MS, Jackson RD (1991) Assessing the spatial distribution of evapotranspiration using remotely sensed inputs. J Environ Qual 20(4):725–737
Moran MS, Clarke TR, Kustas WP, Weltz M, Amer SA (1994) Evaluation of hydrologic parameters in a semiarid rangeland using remotely sensed spectral data. Water Resour Res 30(5):1287–1297
Aayog N (2018). Composite water management index, a tool for water management, June 2018. https://www.niti.gov.in/writereaddata/files/document_publication/2018-05-18-Water-index-Report_vS6B.pdf.
Norman JM, Anderson MC, Kustas WP, French AN, Mecikalski JOHN, Torn R, Tanner BCW et al (2003) Remote sensing of surface energy fluxes at 101-m pixel resolutions. Water Resour Res 39(8)
Park J, Baik J, Choi M (2017) Satellite-based crop coefficient and evapotranspiration using surface soil moisture and vegetation indices in Northeast Asia. CATENA 156:305–314
Patel N, Rajput TBS (2013) Effect of deficit irrigation on crop growth, yield and quality of onion in subsurface drip irrigation. Int J Plant Prod 7(3):417–436
Ragab R, Evans JG, Battilani A, Solimando D (2017) Towards accurate estimation of crop water requirement without the crop coefficient Kc: New approach using modern technologies. Irrig Drain 66(4):469–477
Rodell M, Velicogna I, Famiglietti JS (2009) Satellite-based estimates of groundwater depletion in India. Nature 460(7258):999–1002
Roerink GJ, Su Z, Menenti M (2000) S-SEBI: A simple remote sensing algorithm to estimate the surface energy balance. Phys Chem Earth Part B 25(2):147–157
Sadeghi M, Jones SB, Philpot WD (2015) A linear physically-based model for remote sensing of soil moisture using short wave infrared bands. Remote Sens Environ 164:66–76
Senay GB (2018) Satellite psychrometric formulation of the operational simplified surface energy balance (SSEBop) model for quantifying and mapping evapotranspiration. Appl Eng Agric 34(3):555–566

Su Z (2002) The Surface Energy Balance System (SEBS) for estimation of turbulent heat fluxes. Hydrol Earth Syst Sci 6(1):85–100

Sugathan N, Biju V, Renuka G (2014) Influence of soil moisture content on surface albedo and soil thermal parameters at a tropical station. J Earth Syst Sci 123(5):1115–1128

Verstraeten WW, Veroustraete F, van der Sande CJ, Grootaers I, Feyen J (2006) Soil moisture retrieval using thermal inertia, determined with visible and thermal spaceborne data, validated for European forests. Remote Sens Environ 101(3):299–314

Wagner W, Hahn S, Kidd R, Melzer T, Bartalis Z, Hasenauer S, Figa-Saldana J, De Rosnay P, Jann A, Schneider S, Komma J (2013) The ASCAT soil moisture product: a review of its specifications, validation results, and emerging applications. Meteorologische Zeitschrift

Wang D, Cai X (2009) Irrigation scheduling—role of weather forecasting and farmers' behavior. J Water Resour Plan Manag 135(5):364–372

Wang Y, Guo T, Qi L, Zeng H, Liang Y, Wei S, Gao F, Wang L, Zhang R, Jia Z (2020) Meta-analysis of ridge-furrow cultivation effects on maize production and water use efficiency. Agric Water Manag 234:106144

Yang Y, Shang S (2013) A hybrid dual-source scheme and trapezoid framework–based evapotranspiration model (HTEM) using satellite images: algorithm and model test. J Geophys Res Atmos 118(5):2284–2300

Yueh S, Shah R, Xu X, Elder K, Starr B (2019) Experimental demonstration of soil moisture remote sensing using P-band satellite signals of opportunity. IEEE Geosci Remote Sens Lett 17(2):207–211

Zhang D, Zhou G (2016) Estimation of soil moisture from optical and thermal remote sensing: a review. Sensors 16(8):1308

Zhang K, Kimball JS, Running SW (2016) A review of remote sensing based actual evapotranspiration estimation. Wiley Interdiscip Rev Water 3(6):834–853

Zhang N, Hong Y, Qin Q, Liu L (2013) VSDI: a visible and shortwave infrared drought index for monitoring soil and vegetation moisture based on optical remote sensing. Int J Remote Sens 34(13):4585–4609

Zhu W, Jia S, Lv A (2017) A universal Ts-VI triangle method for the continuous retrieval of evaporative fraction from MODIS products. J Geophys Res Atmos 122(19):10–206

Chapter 12
Estimating Evapotranspiration in Relation to Land-Use Change Using Satellite Remote Sensing

Dheeraj K. Gupta, Nitesh Patidar, Mukunda Dev Behera, Sudhindra Nath Panda, and V. M. Chowdary

Abstract Evapotranspiration (ET) estimation at river basin scale with respect to various land use and land cover (LULC) provides useful conservation prescriptions. With the advancement of satellite remote sensing, ET estimation has gained tremendous attention. Satellite remote sensing-based methods can map spatially distributed ET over different land use and thus are helpful for inaccessible areas. In this chapter, a commonly used surface energy balance-based modified Priestley–Taylor algorithm was demonstrated to estimate LULC-wise ET in two eastern river basins of India, named Brahmani and Baitarani. The potential impact of cloud cover on the performance of the ET estimation was also assessed. The results showed that the forest accounted for the highest ET followed by water body/moist riverbed in both the river basins. The ET estimates were found reasonable during non-monsoon season; however, during monsoon season, an underestimation was observed due to cloud cover, revealing that a denser time-stack of satellite images is required for an accurate estimation of ET during monsoon season. The assessment of the effects of cloud cover on ET estimates revealed that the method used in the study require cloud-free satellite images for accurate estimates of ET. With the increased availability of data

D. K. Gupta
Tropical Forest Research Institute, Jabalpur, India

N. Patidar
National Institute of Hydrology, Roorkee, India

M. D. Behera (✉)
Centre for Oceans, Rivers, Atmosphere and Land Sciences (CORAL), Indian Institute of Technology Kharagpur, Kharagpur, West Bengal 721302, India
e-mail: mdbehera@coral.iitkgp.ac.in

S. N. Panda
Agricultural and Food Engineering Department, Indian Institute of Technology Kharagpur, Kharagpur, West Bengal 721302, India

V. M. Chowdary
Regional Remote Sensing Centre (North), NRSC-ISRO, New Delhi, India

A. Pandey et al. (eds.), *Geospatial Technologies for Land and Water Resources Management*, Water Science and Technology Library 103,
https://doi.org/10.1007/978-3-030-90479-1_12

from different satellites from recent launches, a dense time-stack of data can be generated by fusing multisensor datasets. Such fusion may improve the accuracy of ET estimates considerably with better information about the spatio-temporal variability.

12.1 Introduction

Evapotranspiration (ET) is one of the vital components of the water cycle. The actual ET is a key parameter in the water balance study, describing the processes within the soil–water–atmosphere–plant environment. ET variations, both in space and time, provide useful information to water resources managers and researchers. Unfortunately, ET estimation under actual field conditions is still a challenging task for scientists and water managers due to the complexities associated with it that has led to the development of various methods during the past decades (Chen et al. 2019; Zhang et al. 2016).

The methods for estimating ET can generally be grouped into four categories, i.e., the hydrological methods (water balance), direct measurement (lysimeters), micrometeorological (energy balance), and empirical or combination methods (Thornthwaite) (Mauder et al. 2020; Rahman and Zhang 2019). Most of these methods can only provide point estimates of ET that are not sufficient for system-level water management. Distributed physically based hydrological models can compute ET patterns but require enormous datasets, which are often unavailable. As the direct measurement of actual ET is difficult and it only provides point estimates, the remote sensing-based ET estimation seems to be a feasible alternative. It has been considered as the most promising tool for ET estimation from local to global scales (Calera et al. 2017; Chen and Liu 2020). Various remote sensing-based ET estimation methods have been developed during the past decades (Huang et al. 2019; Jose et al. 2020; Madugundu et al. 2017). The main advantage of such approaches is that the large areas can be covered without extensive field monitoring networks.

With the continuous efforts by many researchers, ET has been estimated on scales from the regional (Ambast et al. 2002; Matsushima 2007) to the global level (Dong and Dai 2017; Raoufi and Beighley 2017; Raoufi 2019; Trabucco and Zomer 2018; Senay et al. 2020; Wei et al. 2017; Zhang et al. 2019) using remote sensing data. In past two decades, ET algorithms have been improved to handle heterogeneity in surface conditions and to enhance spatial and temporal resolution of the estimates (Javadian et al. 2019; Numata et al. 2017; Losgedaragh and Rahimzadegan 2018).

The algorithms for ET estimation can be divided into three groups: (i) models based on Penman–Monteith (P-M) equation, (ii) models based on surface energy balance method, and (iii) models based on Priestley–Taylor (P–T) equation. In the P-M-based methods, evaporation fraction (EF) is expressed as a ratio of actual evapotranspiration (ET) to the available energy (sum of ET and sensible heat flux) which is derived from satellite data. The "VI-Ts" (vegetation index-surface temperature) diagram is used to estimate EF of bare soil. In surface energy balance methods, all variables can be obtained from remote sensing data, except aerodynamic resistance.

Ambast et al. (2002) presented a remote sensing-based simplified operational procedure to estimate sensible heat flux incorporating the local meteorological conditions. The model utilized the surface reflectance in visible, infrared, and thermal bands to generate surface albedo, T_s and leaf area index to determine regional ET. The P–T equation can be considered as the simplified version of the Penman equation. It includes only five variables, which are net radiation, soil heat flux, two variables related to air temperature, and the dimensionless P–T coefficient. Among these variables, the most difficult issue is to determine the P–T coefficient.

The methodology followed in this study is based on Priestley–Taylor (P–T) equation in which the P–T coefficient was obtained using the relationship between remotely sensed vegetation index (VI) and surface temperature (T_s-VI diagram). Advantages of T_{s-}VI triangle over the residual method of surface energy balance for ET estimation are: (i) a direct calculation of evaporation fraction (EF), defined as the ratio of latent heat flux to surface available energy, can be obtained; (ii) estimations of net radiation (R_n) and EF are independent from each other; (iii) no ground-based near surface measurements are needed; and (iv) atmospheric correction are not essential (Zhang and Zhong 2017).

12.2 Need of ET Modeling at River Basin Scale in Relation to LULC

Most of the water resources management strategies and policies are formulated at river basin scale, and hence, there is a need to quantify the spatial variation of various components of the hydrological cycle in a river basin. Planning at the basin scale involves larger area with considerable variations in both climate and LULC. Measurements of ET are rarely available and are unlikely to be sufficient to describe either the spatial variation or the influence of LULC on the ET regime. In the absence of measurements, an alternative solution is to use mathematical models to estimate the variations in ET with (i) meteorological data describing variations in the climate and (ii) LULC data describing variations in land, vegetation, and water. Increasing competition for water increases the importance of the RBs as the appropriate unit of analysis to address the challenges faced for water resources management. Modeling at this scale can provide essential information to policymakers in their resource allocation decisions. LULC change has a direct effect on hydrology through its link with the ET regime. In a large river basin (RB), there may be considerable variation in both climate and land use across the region. The present study aims (i) to estimate ET at basin scale and (ii) to analyze the pattern of spatio-temporal variation of ET in relation to various LULC.

12.3 Brahmani and Baitarani River Basins

The Brahmani and Baitarani RBs together extend over an area of 51,822 km^2 (Fig. 12.1). The independent drainage areas of Brahmani and Baitarani RBs are 39,033 km^2 and 12,879 km^2, respectively. The basins are bounded on the north by the Chhotnagpur plateau, on the west and south by the ridge separating it from the Mahanadi basin and on the east by the Bay of Bengal. The basin lies in the states of Odisha and Jharkhand. The Brahmani river rises near Nagri village in Ranchi district of Jharkhand and has a total length of 799 km, whereas the Baitarani river rises in the hill ranges of Keonjhar district of Odisha and has a length of about 355 km. Both river systems fall into the Bay of Bengal forming a common delta. The maximum elevation reaches up to 1181 m as revealed by the SRTM-DEM (Fig. 12.1).

In these basins, about 80% of the annual normal rainfall occurs during south-west monsoon (June to September). The annual normal rainfall varies from 1250 to 1750 mm over the Brahmani river basin and from 1250 to 1500 mm over the Baitarani river basin. The average annual rainfall of Brahmani and Baitarani RBs are 1305 and 1488 mm, respectively (India Meteorological Department).

The mean maximum temperature varies from 35 °C to 42.5 °C over the catchment areas of rivers Brahmani and Baitarani in the month of May. The mean minimum temperature varies from 15 °C to 10 °C over Brahmani catchment and from 15 °C to 12.5 °C over Baitarani catchment in the month of January. The extreme temperatures in summer (max.) and winter (min.) recorded in May and January are of the order of 47.5 °C and 50 °C, respectively.

The humidity is about 80–90% during monsoon season and 40–70% during non-monsoon season, the higher humidity being over the coastal region. The dominant soil types in the basins are red and yellow soils, red sandy and loamy soils, mixed red and black soils, and coastal alluvium.

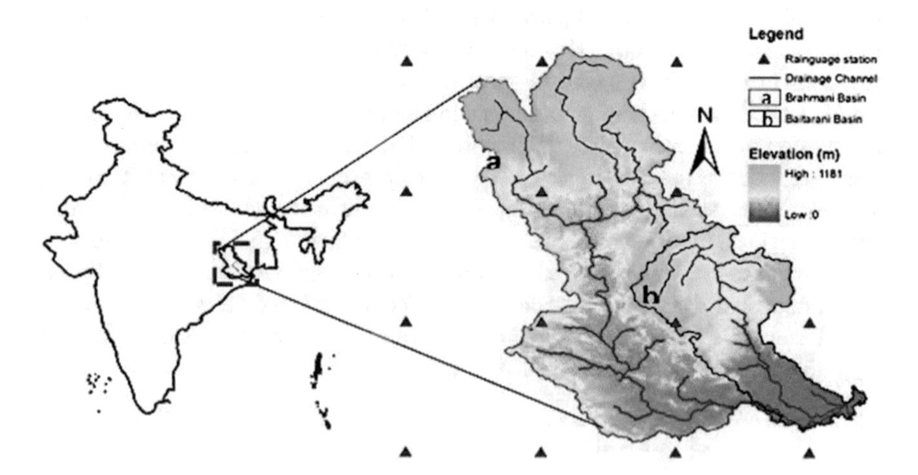

Fig. 12.1 Location map. **a** Brahmani and **b** Baitarani river basins with elevation

12.4 Data Acquisition and Analysis

Land surface temperature, albedo, and emissivity were derived from composite MODIS Terra land products. The downward solar (longwave and shortwave) radiations were obtained from the NASA Langley Research Center POWER project funded through the NASA earth science directorate applied science program (NASA climatology resource for agro-climatology daily averaged data). The downward solar radiation was taken at 1-degree intervals and further interpolated to derive continuous surface. MODIS data used in this study was downloaded from the NASA data center having good temporal resolution of 8 days and acceptable spatial resolution of 1000 m.

Shortwave white sky albedo (WSA) data was retrieved from MODIS product (MCD43A3), and land surface temperature and emissivity were retrieved from 1-km resolution brightness temperature in MODIS band 31 and 32 (MOD11A1, onboard Terra satellite). Descriptions of these products can be found from the MODIS WIST Web site. The images were re-projected from the Sinusoidal (SIN) projection to the Universal Transverse Mercator (UTM with Zone45) and resampled to 1 km resolution. LULC map derived from LISS-III imagery was gathered from Indian Space Research Organization-Geosphere Biosphere Program (ISRO-GBP) sponsored project on "Land use and land cover change modeling using climatic and socio-economic variables." The spatial resolution of LULC map was downscaled from 62.5 m to 1 km for utilization in ET estimation. Rainfall data of a decade (1996 to 2005) was received from IMD. The RB boundaries and physiognomy were delineated from SRTM-derived DEM at 90 m resolution using ArcGIS package. Descriptions of SRTM DEM can be found from the *consortium for spatial information* (CGIAR-CSI).

12.5 Modified Priestley–Taylor Algorithm: Theory and Application

The flowchart depicting the estimation procedure of ET is shown in Fig. 12.2. Daily evapotranspiration (ET_{24}) in mm/day is determined from total daily available energy as

$$ET = \frac{LE}{\lambda \rho_w} \times 86{,}400 \times 10^3 \quad (12.1)$$

where λ is the latent heat of vaporization (KJ/kg); ρ_w is the density of water (kg/m^3);

LE is the instantaneous latent heat flux, residual term of the energy budget calculated as:

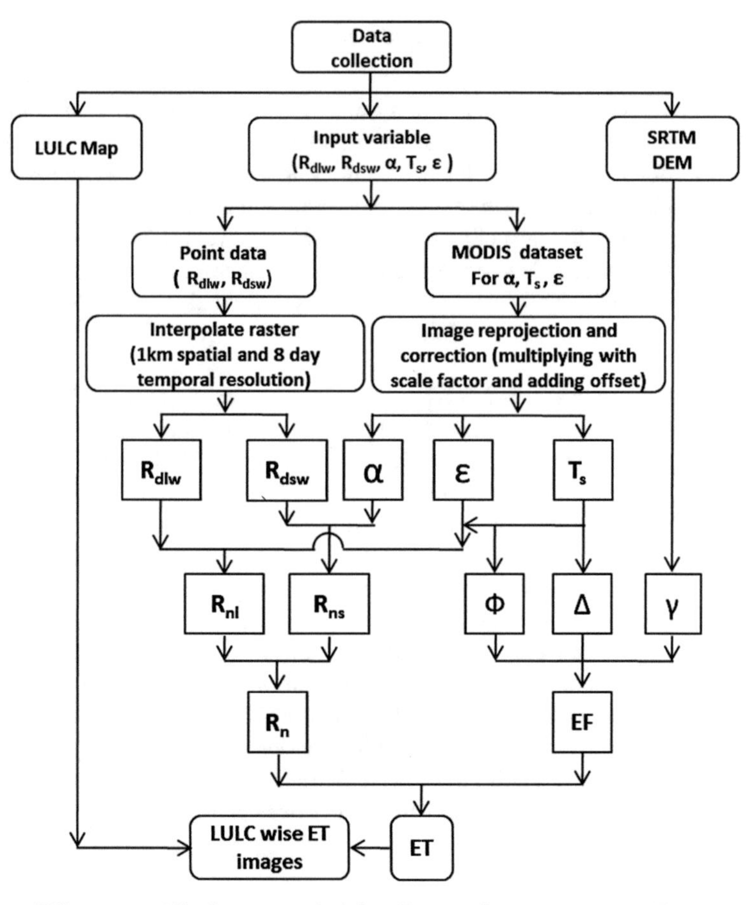

(Where, α = Albedo, ε = emissivity, Ts = surface temperature, R_{dlw} = downward long wave radiation, R_{dsw} = downward shortwave radiation, R_{ns} = net shortwave radiation, R_{nl} = net long wave radiation, R_n = net radiation, Φ = combined-effect parameter that accounts for aerodynamic resistance, Δ = saturation vapour pressure curve Vs temperature, γ = psychrometric constant, EF = evapotranspiration fraction)

Fig. 12.2 Flowchart depicting methodology for estimation of ET

Net available energy

$$R_n - G = \text{LE} + H \tag{12.2}$$

$$\text{LE} = R_n - H - G \tag{12.3}$$

where R_n is the surface net radiation (Wm^{-2}); G is the soil heat flux (Wm^{-2}); and H is sensible heat flux (Wm^{-2}) calculated as

$$H = \rho C_p \frac{T_s - T_a}{R_a} \tag{12.4}$$

where ρ is air density, C_p is the specific heat capacity of air, T_s, T_a are the aerodynamic surface and air temperatures, and R_a is the aerodynamic resistance. LE estimated by Priestley–Taylor type models as

$$\mathrm{LE} = \phi\left((R_n - G)\frac{\Delta}{\Delta + \gamma}\right) \tag{12.5}$$

where Δ is the slope of saturated vapor pressure versus air temperature (kPa C^{-1}); γ is the psychrometric constant (kPa C^{-1}); ϕ is the combined effect parameter that accounts for aerodynamic resistance (dimensionless). This equation is applicable for a range of surface conditions where $R_n - G$ is the driving force for ET. It should be noted that all quantities involved in the right-hand side of Eq. (12.5) can be derived from remotely sensed data alone.

Evaporative fraction (EF), defined as the ratio between evaporation and available energy, can be directly estimated from Eq. (12.5) as:

$$\mathrm{EF} = \phi\frac{\Delta}{\Delta + \gamma} \tag{12.6}$$

By reshuffling Eqs. (12.5) and (12.6)

$$\mathrm{LE} = \mathrm{EF} \times (R_n - G) \tag{12.7}$$

Although parameter ϕ in Eq. (12.5) looks apparently the same as α in P–T equation, there is a distinct difference in the physical meaning between these two parameters. In P–T equation, α is generally interpreted as the ratio of actual evaporation to the equilibrium evaporation and a series of paper has demonstrated this parameter with a good approximate to be 1.26.

P–T equation is generally applicable for wet surfaces, whereas Eq. (12.5) holds true for a wide range of surface evaporative conditions with ϕ varying from 0 to $(\Delta + \gamma)/\Delta$ (since $0 \leq \mathrm{EF} \leq 1$, and hence $0 \leq \phi \leq \frac{\Delta+\gamma}{\Delta}$) when significant advection and convection are absent. A zero value for ϕ corresponds to pixels with no ET, i.e., $\mathrm{LE} = 0$ and $H = R_n - G$, while a value of $(\Delta + \gamma)/\Delta$ corresponds to pixels with maximum ET, i.e., $\mathrm{LE} = R_n - G$ and $H = 0$ (Eq. 12.3).

$$\phi = \phi_{\max}\left(\frac{T_{\max} - T_s}{T_{\max} - T_{\min}}\right) \tag{12.8}$$

where $\phi_{\max} =$

$$\phi = 1.26\left(\frac{T_{\max} - T_s}{T_{\max} - T_{\min}}\right) \tag{12.9}$$

Net radiation (R_n)

The net radiation, R_n, is the difference between incoming and outgoing radiation of both shorter and longer wavelengths. It is the balance between the energy absorbed, reflected, and emitted by the earth's surface or the difference between the incoming net shortwave (R_{ns}) and the net outgoing longwave (R_{nl}) radiation. R_n is normally positive during the daytime and negative during the nighttime. The total daily value for R_n is almost always positive over a period of 24 h, except in extreme conditions at high latitudes.

$$R_n = \begin{cases} R_{ns} + R_{nl} \\ \left(R_s^{\downarrow} - R_s^{\uparrow}\right) + \left(R_l^{\downarrow} - R_l^{\uparrow}\right) \\ (1 - \alpha)R_s^{\downarrow} + R_l^{\downarrow} - \sigma\varepsilon_s T_s^4 \end{cases} \tag{12.10}$$

where T_s is the land surface temperature (°K), respectively; α is the surface albedo; $R_s^{\downarrow}$ is the downward shortwave radiation (Wm^{-2}); $R_l^{\downarrow}$ is the downward longwave radiation (Wm^{-2}); σ is the Stefan–Bolzmann constant (5.67×10^{-8} Wm^{-2} K^{-4}); ε_s is the surface emissivity of the surface. The parameters α, ε_s, and T_s can be derived from remote sensing data of visible and thermal infrared spectral ranges.

Albedo (α) and net short/long wave radiations (R_{ns}/R_{nl})

Albedo is defined as the fraction of solar energy (shortwave radiation) reflected from the earth back into space. It is a measure of the reflectivity of the earth's surface and depends upon many factors, including the color and roughness of the terrain, angle of incidence or slope of the ground surface, the extent of forest or agriculture cover, and the amount of cloud and snow cover. Clouds, ice, and snow reflect a greater proportion of radiation than do land and ocean surfaces.

The net shortwave radiation, R_{ns}, is the fraction of the shortwave radiation $R_s^{\downarrow}$ that is not reflected from the surface. Its value is $(1 - \alpha)\, R_s^{\downarrow}$. The difference between outgoing and incoming long wave radiation is called the net long wave radiation, R_{nl}. As the outgoing long wave radiation is almost always greater than the incoming long wave radiation, R_{nl} represents an energy loss.

Soil heat flux (G)

The soil heat flux, G, is the energy that is utilized in heating the soil. G is positive when the soil is warming and negative when the soil is cooling. Although the soil heat flux is small compared to R_n and may often be ignored, the amount of energy gained or lost by the soil in this process should theoretically be subtracted or added to R_n during ET estimation. For 1 day or longer timescale, G can be ignored and net available energy ($R_n - G$) reduces to net radiation (R_n). At daily timescales, $G_{24} = 0$; LE_{24} (mm/day) can be computed as:

$$\mathrm{LE}_{24} = \mathrm{EF}_{24} \times R_{n24} \tag{12.11}$$

Slope of Vapor Pressure Curve (Δ)

The slope of saturated vapor pressure versus air temperature

$$\Delta = \frac{4098e^{\circ}(T)}{(T_a + 237.3)^2} = \frac{2504\exp\left(\frac{17.27T_a}{T_a+237.3}\right)}{(T_a + 237.3)^2} \tag{12.12}$$

where T_a is the air temperature (°C); and $e°(T)$ = saturation vapor pressure at T_a °C (kPa). For the sake of simplicity, the above equation can be written in °K (°K = °C + 273.15)

$$\Delta = \frac{4098e^{\circ}(T)}{(T_a - 35.85)^2} = \frac{2504}{(T_a - 35.85)^2}\exp\left(\frac{17.27(T_a - 273.15)}{T_a - 35.85}\right) \tag{12.13}$$

The sensitivity of $\Delta/(\Delta + \gamma)$ on the variation of temperature is very small, and air temperature (T_a) required in Eq. (12.12) to calculate Δ so as to calculate $\Delta/(\Delta + \gamma)$ in Eq. (12.6). This can be obtained either by a linear regression between T_s and $T_s - T_a$ or by using mean surface temperature or mean water surface temperature as a surrogate. In this work, taking into account the small sensitivity of $\Delta/(\Delta + \gamma)$ and the correlation of T_s with air temperature, remotely sensed T_s was used to estimate the parameter Δ instead of the use of air temperature.

Psychrometric Constant (γ)

The psychrometric constant (γ) relates the partial pressure of water in air to the air temperature. This lets one interpolate actual vapor pressure from paired dry and wet thermometer bulb temperature readings (Allen et al. 1998).

$$\gamma = \frac{C_P P}{m_w \lambda} \times 10^{-3} \tag{12.14}$$

where γ is the psychrometric constant (kPa °C^{-1}); C_p is the specific heat of moist air (1.013 kJ kg^{-1} °C^{-1}); P is the atmospheric pressure (kPa); m_w is the ratio molecular weight of water vapor/dry air (0.622); and λ is the latent heat of vaporization (MJ kg^{-1}).

Atmospheric Pressure (P)

Air pressure above sea level calculated as.

$$P = 101.3\left[\frac{293 - 0.0065Z}{293}\right]^{5.26} \tag{12.15}$$

where z is the elevation (m). Elevation has been retrieved from SRTM DEM data.

Surface emissivity (ε_s)

Surface emissivity (ε_s) is the ratio of energy radiated by a particular material to energy radiated by a black body at the same temperature, which can be obtained from MODIS band 31 and 32 as follows

$$\varepsilon_s = 0.237 + 1.778\epsilon_{31} - 1.807\epsilon_{31}\epsilon_{32} - 1.037\epsilon_{32} + 1.774 \quad (12.16)$$

where ε_{31} is the surface emissivity in MODIS band-31; and ε_{32} is the surface emissivity in MODIS band-32.

All the above parameters and variables were framed in the "Model Maker" using ERDAS imagine image processing software in accordance with the methodology to estimate the ET (Fig. 12.2).

Spatio-temporal cover fraction (STCF)

The spatio-temporal cover fraction (STCF) represents the area under space *vs* time graph (Fig. 12.4a). It depicts the areal and temporal extent for clear sky days. Therefore, the sum of STCF and percent cloud cover would be equal to unity. For example, the area under the gray color in Fig. 12.4a shows 67.7% areal and temporal extent meaning clear sky condition, while area under black color is 32.3% showing the presence of cloud in data. In a year, 46 MODIS images are available at 8-day temporal resolution. For one time period, if it covers the whole area (100% space, i.e., no cloud cover), the value of STCF would be 100% (1 × 100%), thus STCF and percent cloud cover are complementary to each other if one increases, other will decrease and vice versa. At annual scale, the mean value of STCF of 46 images has to be calculated.

12.6 Annual Estimates of ET in Brahmani and Baitarani Basins

Brahmani RB is mainly occupied by forest and cropland, whereas Baitarani RB is dominated by cropland as revealed from LULC maps. All five LULC classes, i.e., forests (deciduous forest, evergreen forest, and mixed forest and plantation), water bodies/moist riverbed (dam, lakes, rivers etc.), croplands (different types of crops, irrigated and rainfed/dryland); barren land (urban areas, fallow land, and wasteland) and shrub lands have shown significant variation in ET.

The annual ET showed significant spatial variations, ranging from less than 18 mm in barren areas to 1182 mm in water bodies/moist riverbed such as reservoirs or rivers, with the annual mean ET of 325 mm for the whole study area with spatial and temporal cover of 67.7% (Figs. 12.3, 12.4 and 12.5a). The results indicated high ET for forest; water bodies/moist riverbed and low ET for urban area. The mean ET values for Brahmani and Baitarani RBs were estimated 329 mm and 322 mm with spatio-temporal coverage of 68% and 67%, respectively (Figs. 12.4 and 12.5a).

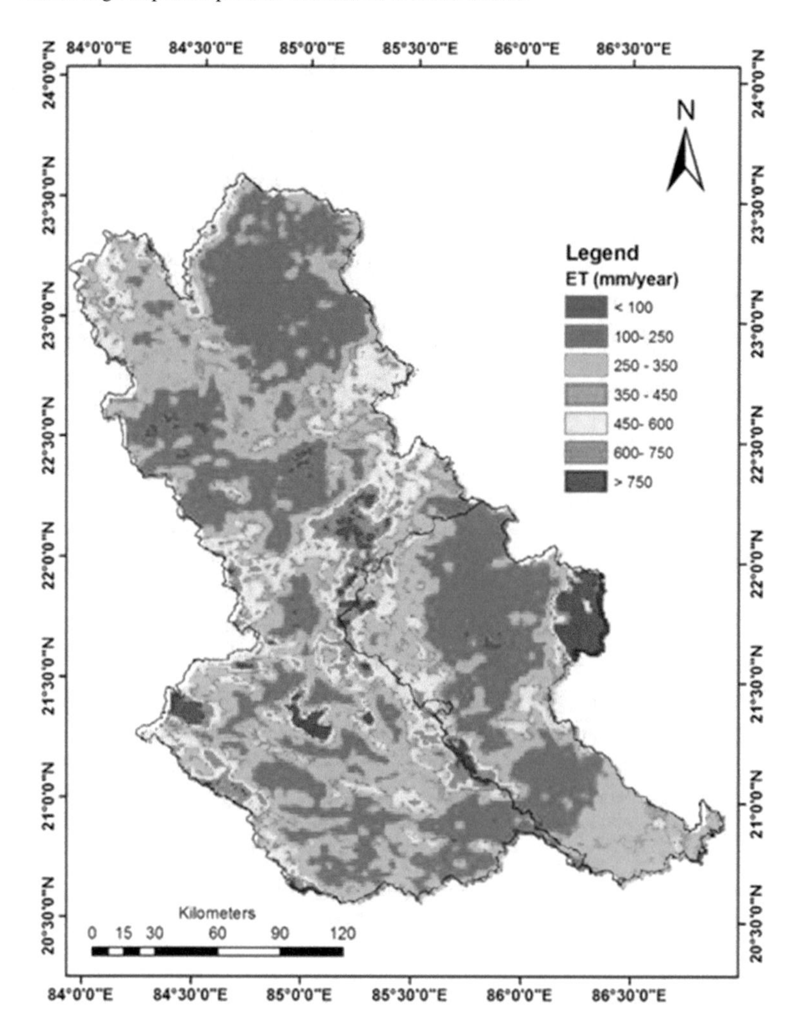

Fig. 12.3 Annual evapotranspiration estimation for Brahmani and Baitarani river basins for the year 2004

Seasonal and annual ET can vary over different types of LULC such as cropland, rivers, lakes, and barren lands at the field scale.

The results showed that mean annual ET loss has exceeded 300 mm for water bodies, forests, and scrubland. However, the annual ET values for water body/moist riverbed (347 mm) were lower than those for forest (431 mm). The forests cover 37.5% area. ET from shrub lands (covering 7% area) is estimated as 323 mm, while crop ET is estimated as 263 mm from an area of 47% of the basin area (Fig. 12.5a). The average annual ET estimated for LULC, i.e., water body/moist river bed, forest, scrubland, cropland, and barren land are 347, 431, 323, 263, and 226 mm, respectively (Fig. 12.5a). The area proportions of these classes are shown in Table 12.1. The

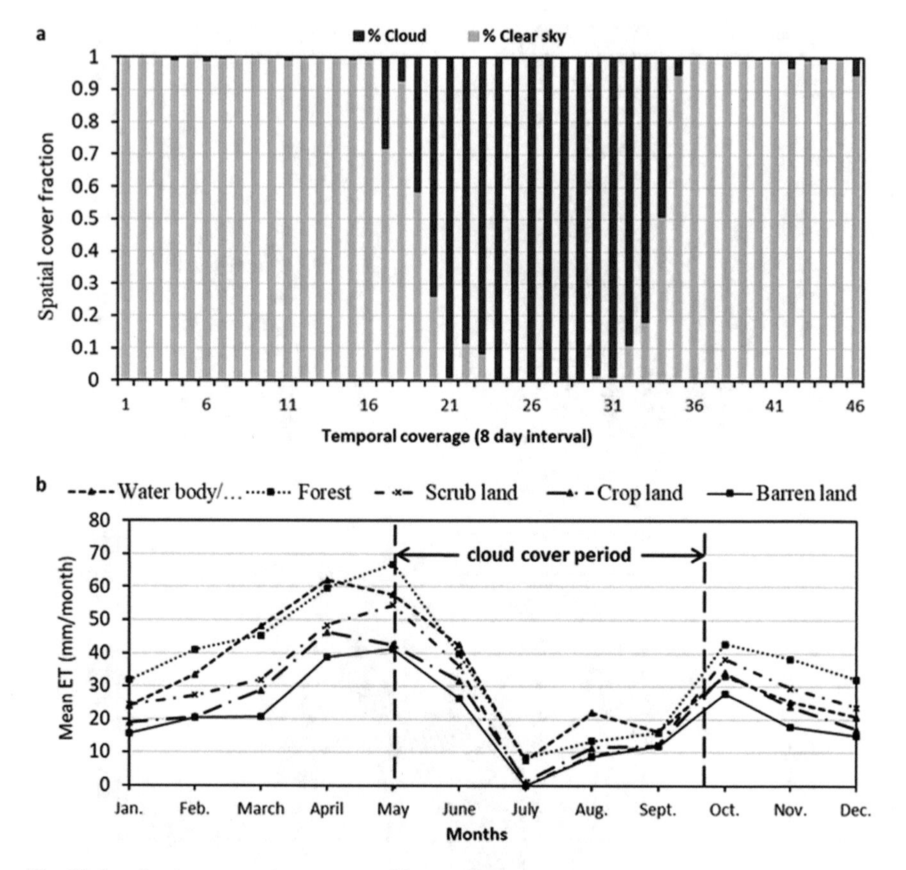

Fig. 12.4 **a** Spatio-temporal coverage and **b** monthly average ET variation with land use and land cover (the underestimation due to cloud cover is highlighted)

estimation indicates higher ET for forest, water bodies/moist river bed, and cropland, while for barren land ET estimate was low. This indicates that the ET is controlled by LULC with varied water availability. Underestimation in ET was observed during monsoon due to the cloud contamination in data.

12.7 Spatial Variation of ET in Relation to LULC

The ET pattern of LULC in Brahmani and Baitarani RBs almost follows the usual trend except water body/moist river bed (347 mm) compared to forest (431 mm). However, this unexpected response of ET for water body/moist river bed can be attributed to the difference in receipt of incoming net solar radiation, evaporation fraction, and spatio-temporal cover fraction. ET primarily depends upon two parameters, i.e., net solar radiation, evaporation fraction (Eq. 12.11), and available water.

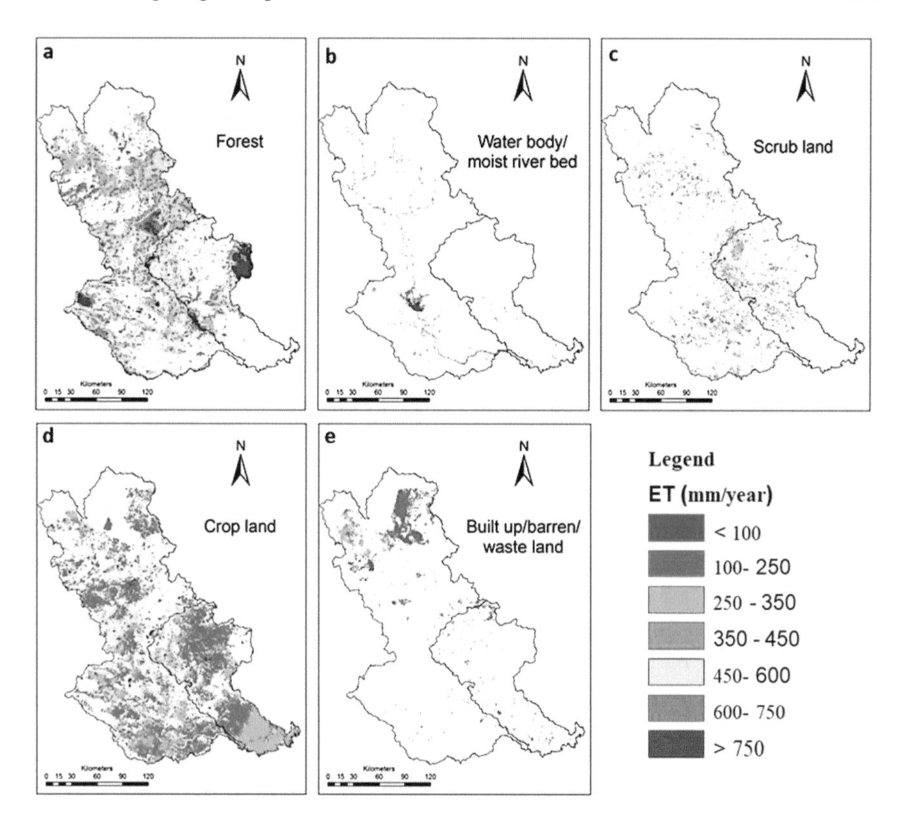

Fig. 12.5 LULC-wise annual ET variation **a** forest; **b** water body/moist river bed; **c** cropland; **d** scrubland; **e** built up/barren/wasteland

Table 12.1 LULC-wise area statistics for Brahmani and Baitarani river basins and the whole area

LULC	Brahmani river basin		Baitarani river basin		Total area	
	Area (km^2)	% Area	Area (km^2)	% Area	Area (km^2)	% Area
Forest	13,725	41.7	3272	25.3	16,745	36.9
Water body/moist river bed	1020	3.1	102	0.8	1113	2.5
Scrubland	1993	6.1	1191	9.2	3135	6.9
Cropland	13,363	40.6	7977	61.6	21,206	46.7
Built up/barren/wasteland	2790	8.5	415	3.2	3175	7
Total	32,891	100	12,957	100	45,374	100

Here, the water body/moist river bed receives less net radiation (77.4 W/m^2) compared to that of forest (84.7 W/m^2; Fig. 12.5b). This may be explained by the fact that net radiation is a function of incoming shortwave and long wave radiation along with spatially varying parameters, i.e., surface albedo, T_s, and surface emissivity

(Eq. 12.10). There is no significant spatial variation in the incoming shortwave and longwave radiation due to non-dependency on ground features, whereas surface albedo, surface temperature, and surface emissivity significantly varied with LULC resulting into the different value of R_n for each LULC classes. Further, the average temperature for water body/moist river bed (303.45 °K) differs significantly from forest (301.16 °K). Higher value of T_s significantly lowers the value of ϕ (as ϕ is inversely proportional to T_s, Eq. 12.9), thereby resulting in lower EF value (EF is directly proportional to ϕ, Eq. 12.6). Decreasing ET means decrease of escaping energy flux from surface as latent heat (LE), and it induces the increase of local temperature, since the net available energy ($R_n - G$) term is equal to the sum of latent (LE) and sensible heat (H) flux for a pixel and period of time (say 1 day, Eq. 1.2). Thus, for a surface the available energy would be utilized in two forms, either in latent heat or sensible heat flux as per the availability of water. For instance, if net available energy is utilized in the form of latent heat, then ET value would be higher, while sensible heat would be lower as their sum to remain constant. For a dry pixel, there would be no scope for LE (ET = 0), H would be equal to $R_n - G$. In other way, if LE is less, then H will increase thus raising T_s, (as H is directly proportional to T_s). Mathematically, it can be rewritten as $LE_{water\ body}$ (347 mm) < LE_{forest} (431 mm; Fig. 12.6a) so $H_{water\ body} > H_{forest}$ results in $Ts_{water\ body}$ (303.45°K) > Ts_{forest} (301.16 °K). That is why forests are felt to be cooler at the cost of ET loss. Thus, when water is spread on a warm surface, water vapors seem to be rising up by absorbing its heat and leaving it cooled.

Moreover, a low value of spatio-temporal cover for water body/moist river bed (67.5%) and to that of forest (70%) (Fig. 12.5e) making estimated $ET_{water\ body}$ < ET_{forest}. Therefore, the cumulative effects of aforementioned parameters resulted into exhibiting this unlikely response of higher ET loss for forests compared to that of water body/moist river bed (Fig. 12.6a). For rest of the classes, the ET response was as usual ET_{forest} (431) > $ET_{scrubland}$ (323 mm) > $ET_{cropland}$ (263 mm) > $ET_{barrenland}$ (226 mm), which can be described as above in terms of their corresponding R_n and EF.

Water body/moist riverbed

A notable difference in mean ET values for water body/moist river bed located in Brahmani (350 mm) and Baitarani (307 mm) is observed (Fig. 12.5b). Brahmani is showing 14% more ET to that of Baitarani without any significant difference in net radiation (77.3 and 77 Wm^{-2}, respectively in Brahmani and Baitarani RBs). The difference in ET can be attributed to the difference in evaporation fraction and spatio-temporal cover fraction. The evaporation fraction for water body/moist river bed present in Brahmani and Baitarani RBs differs significantly (0.43 and 0.38, respectively) owing to a difference of 0.48 °K in temperature. Higher temperature of water body/moist river bed present in Baitarani RB (303.9 °K) to that of Brahmani RB (303.42 °K) may be attributed to the difference in water depth and thereby change in the water temperature. Since evaporation fraction is directly proportional to ϕ (Eq. 12.6), ϕ is inversely proportional to T_s (Eq. 12.8), and EF is inversely proportional to T_s. Higher the value of T_s, lower is the evaporation fraction. Hence,

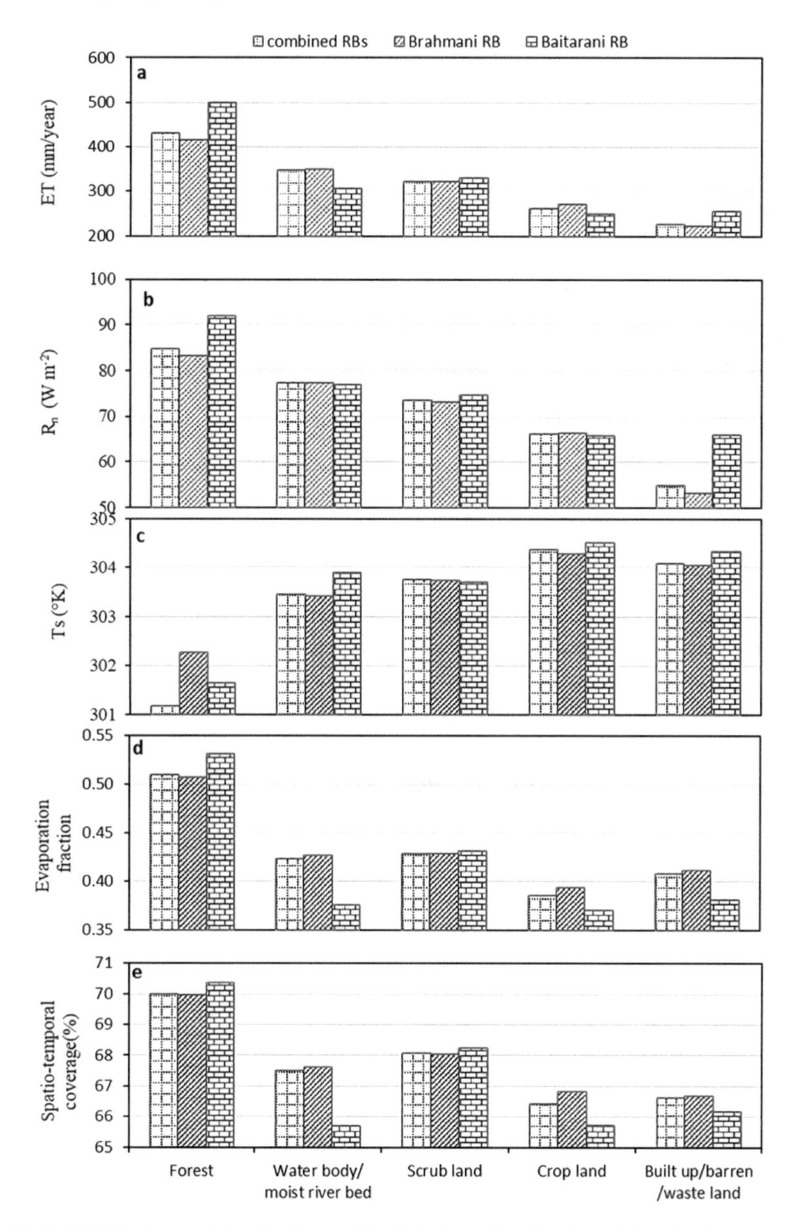

Fig. 12.6 LULC-wise variation in the combined Brahmani and Baitarani river basins. **a** Annual ET; **b** net radiation; **c** surface temperature; **d** evaporation fraction; **e** spatio-temporal fraction

Baitarani RB with scattered and shallower water body/moist riverbed showed higher temperature resulting in lower evaporation fraction than Brahmani RB with 1.9% larger and permanent (Rengali dam) water body/moist riverbed.

Forest

The average annual ET from the forests present in Baitarani RB (500 mm) is 20.2% higher than that of Brahmani RB (416 mm) which might be explained by the above facts explained above (Fig. 12.5a). The forest in Baitarani RB occupies 25% area showing higher ET than that of Brahmani RB having 42% area (Table 12.1) which can be better explained by the difference in incoming net solar radiation received by them. As forest in Baitarani RB received 91.9 Wm^{-2} net solar radiation compared to that of Brahmani RB (83.2 Wm^{-2}) causing more ET (Fig. 12.5a), the forest present in Baitarani RB is denser compared to that of Brahmani RB, with dense natural forests and combination of mixed forest and plantation, respectively. Thus, mean temperature in forest of Baitarani RB is lower by 0.62 °K to that of Brahmani RB. This temperature difference has affected ϕ, EF and finally ET.

Scrubland

The average ET for scrubland present in Brahmani RB (321 mm) and Baitarani RB (329 mm) does not differ significantly (2.4%, Fig. 12.5c). EF values showed almost equal value (0.43) while R_n also showed 73.2 and 74.74 Wm^{-2} for Brahmani and Baitarani RBs, respectively. Here, the ratio of ET (0.97) for these two basins is almost in proportion of the ratio of their corresponding R_n (0.97) with 68.05 and 68.23% of spatio-temporal cover fraction. Thus, for scrubland in both the RBs, all the parameters, i.e., net radiation and evaporative fraction showed almost similar trends.

Cropland

Cropland is present in Brahmani and Baitarani RBs exhibiting 271 and 250 mm mean annual ET with 66.8 and 65.7% of spatio-temporal cover, respectively (Fig. 12.5d). Net radiations received by them are 66.4 and 65.7 Wm^{-2}, while EF values are 0.39 and 0.37, respectively. The ratio of ET for Brahmani and Baitarani river basin is 1.08 which is in proportion to multiplication of their corresponding R_n and EF (as explained above).

Barren land

This class accommodates barren land, settlement, wasteland and fallow land together. The mean annual ET exhibited by Brahmani and Baitarani RBs is 222 and 255 mm, respectively, with a little difference of spatio-temporal cover of 66.7% and 66.2%, respectively (Fig. 12.5e). Net radiation received by barren lands is comparatively less for Brahmani RB (53.24 Wm^{-2}) from Baitarani RB (66 Wm^{-2}; Fig. 12.6b). Interestingly, the EF of this class is higher for Brahmani RB (0.41) than Baitarani RB (0.38; Fig. 12.6d). Brahmani to Baitarani RB ratios of R_n, EF, and ET are 0.806, 1.08, and 0.87, respectively. Hence, the complementary nature of R_n (0.806) and EF (1.08) will compensate together in the result of ET (0.87) for both the basins.

12.8 Temporal Variation of ET in Relation to LULC

Figure 12.7 showed the monthly ET variations of LULC in the RBs. In general, all LULC types showed similar seasonal trends in ET throughout the year. The values of ET started to rise rapidly in April, attained peak in May, and then declined to the lowest during rainy season due to cloud cover in MODIS (Terra) derived albedo, surface temperature, and emissivity.

The ET loss from the barren land was low as the area was devoid of vegetation. During the winter, ET rates from all LULC are low, owing to lower incoming radiation. Forest transpires at higher rate than other vegetation, as they maintain canopy cover throughout the year and thus supersedes scrublands and croplands (Dunn and Mackay, 1995). In the two RBs, the cropping season is primarily concentrated during monsoon (kharif) period of 3–4 months (mid-June to mid-October) contributing to standing crop. For the remaining 8–9 month's period, cropland remains barren and the ET loss can be compared with barren lands.

This could be mainly due to the fact that the ET values for the rainfed croplands during non-growing seasons are lower than those at the other two vegetated lands. The

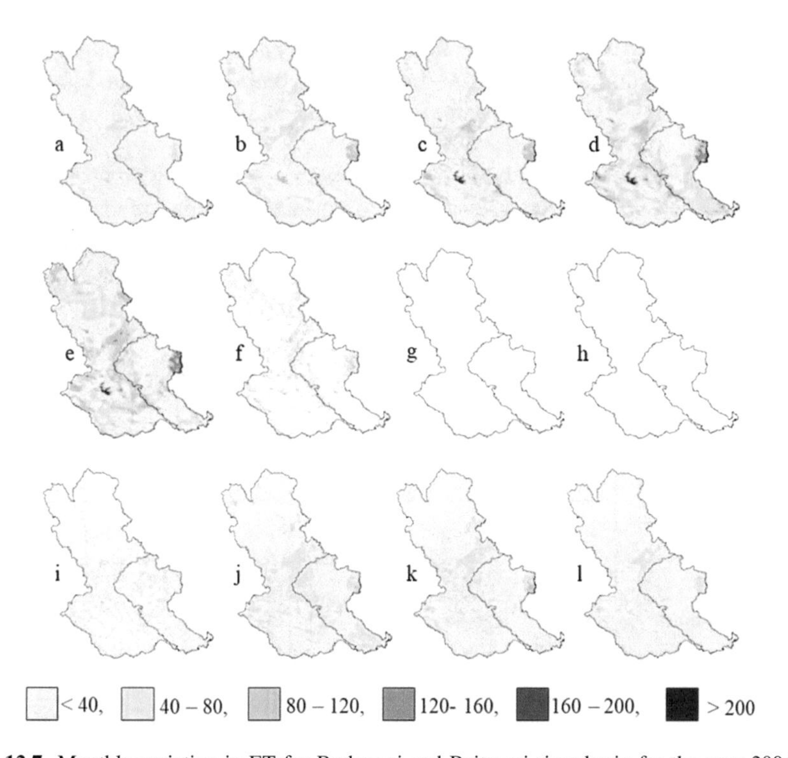

Fig. 12.7 Monthly variation in ET for Brahmani and Baitarani river basin for the year 2004; **a** January; **b** February; **c** March; **d** April; **e** May; **f** June; **g** July; **h** August; **i** September; **j** October; **k** November; **l** December

barren lands generally have the lower ET values because of the lower soil moisture availability. Thus, the energy transformation is mainly limited in the form of sensible heat exchange.

Figure 12.7 revealed a general increasing trend from the month of January to May. However, the effect of ET loss from vegetation especially forest surface is more prominent in the month of January. As the forests in these RBs are deciduous in nature, the leaf fall generally starts from early February and the new leaves are received by mid of April. This effect of seasonal vegetation phenology is prominent in the monthly temporal ET images (Fig. 12.6). The effect of cloud cover, which is a major limitation of optical remote sensing data such as MODIS, is prominent in monsoon season and thereby leading to underestimation in ET for June to September.

12.9 Projected ET

Projected ET value depicts what would be the scenario of ET with 100% spatio-temporal cover fraction. It is ratio of the estimated ET value to its spatio-temporal cover fraction. The projected ET by taking spatio-temporal cover fraction into account for each LULC classes, i.e., forests-595 and 711 mm, water body/moist river bed 517 and 467 mm, scrubland-472 and 482 mm, cropland-405 and 380 mm, and built up/barren land-333 and 385 mm for Brahmani and Baitarani RBs, respectively (Table 12.2).

Annual ET estimated for both the RBs together was 325 with 67.7% spatio-temporal cover. Now taking projection into consideration the annual ET can be 480 mm for the whole Brahmani and Baitarani RB (Table 12.3). Projected annual ET for Brahmani and Baitarani RBs by taking spatio-temporal cover fraction and area

Table 12.2 LULC-wise projected ET for Brahmani and Baitarani river basins

LULC	Brahmani basin			Baitarani basin		
	Annual ET (mm/year)	STCF (%)	Projected ET (mm/year)	Annual ET (mm/year)	STCF (%)	Projected ET (mm/year)
	a	b	$c = a/b \times 100$	p	q	$r = p/q \times 100$
Forest	416	70.0	595	500	70.4	711
Water body/moist river bed	350	67.6	517	307	65.7	467
Scrubland	321	68.0	472	329	68.2	482
Cropland	271	66.8	405	250	65.7	380
Built up/barren/wasteland	222	66.7	333	255	66.2	385

STCF spatio-temporal cover fraction

Table 12.3 River basin-wise estimated and projected annual ET

S. No.	Basin	STCF (%)	Annual ET (mm)	
			Estimated	Projected
		a	*b*	$c = b/a \times 100$
1	Brahmani	68	329	484
2	Baitarani	67	322	481
3	Combined	67.7	325	480

fraction of each LULC class into consideration are 484 and 481 mm, respectively (Table 12.3).

12.10 Mean Runoff Coefficient (C)

The proportion of total rainfall that becomes runoff during a storm event represents the runoff coefficient or ratio between peak or quick flow and rainfall volume on an event basis. The water resulting from infiltration and percolation of rainfall stored in root zone is used in ET throughout the year, and rest of water converted into runoff. In the absence of peak or quick flow, mean runoff flow was estimated by subtracting projected ET from annual rainfall. Runoff coefficient varies with LULC, aspect and antecedent soil moisture conditions. Estimated runoff coefficient provides a rough idea of how much water resulting into runoff. For instance, the annual rainfall of Brahmani RB is 1305 mm and projected ET 484 mm would result into 821 mm runoff and the ratio of runoff to rainfall, i.e., runoff coefficient 0.63. Based on average annual rainfall and projected ET, the runoff coefficient for Brahmani and Baitarani basins were estimated to be 0.63 and 0.68, respectively (Table 12.4). Higher value of runoff coefficient for the Baitarani basin is due to its high rainfall and higher slope (drop of 900 m in 360 km) in comparison to Brahmani RB (drop of 600 m in 799 km).

Table 12.4 River basin-wise runoff coefficient

	Annual rainfall (mm)	Annual projected ET (mm)	Annual runoff (mm)	Runoff coefficient
River basin	*a*	*b*	$c = a - b$	c/a
Brahmani	1305	484	821	0.63
Baitarani	1488	481	1007	0.68

12.11 Discussion and Conclusions

This study presents the estimates of remotely sensed ET and examines the spatio-temporal variations of ET in terms of five types of LULC in Brahmani and Baitarani RBs in India. Different types of LULC reflect different ET. Overall, forests have the highest ET. The water body/moist river bed presents a higher rate of ET than scrublands, while agricultural lands and the barren lands have the lowest ET.

Seasonal estimation of water-use quantified (ET) for different types of LULC could help in developing managerial strategies to improve water use management. The study on quantification of spatial and temporal variability of ET will be useful in various disciplines, including hydrologic budgeting, water resource planning, irrigated agriculture, and ecological system risk management.

However, the effect of scale and spatial resolution is a limitation to the boundary classes leading to some amount of under- and overestimation. The underestimation of ET for water body/moist river bed and overestimation for forest class could be due to the working scale of 1 km. On downscaling, the water body/moist river bed pixels have accommodated lots of non-water bodies/moist river bed in Baitarani RB owing to smaller dimension of the water channels in contrast to that of Brahmani RB. There is effect of seasonality that influences different LULC categories such as river in dry season with dry sand bed and in wet season with water-filled condition. A river channels/bed consists of boulders and cobbles in mountainous region, where gravel, sand, and silts in plain region will contribute very less ET value for seasonal rivers, while in non-monsoon season, the flow width shrinks and becomes limited compared to river width, mainly sand remains visible. Thus, a long and seasonal river may contribute less ET value, thereby affecting overall lower ET value for that class, thus resulting in underestimation of ET for water body/moist river bed class and overestimation for forest class. Further, Brahmani basin accommodates a large reservoir (i.e., Rengali dam), where Baitarani basin accommodates a small dam (i.e., Hadagarh) with a smaller reservoir. The method used in this study was based on P–T equation in which P–T coefficient was obtained using the relationship between remotely sensed surface temperature and vegetation index (T_s-VI diagram). Such a determination of the P–T coefficient may increase uncertainties and errors in ET estimations because the determination of an ideal T_s-VI diagram is largely dependent on the heterogeneity of land surface. Also, cloud cover is another limitation of optical satellite image that could result in underestimation of ET during monsoon season. The reason behind choosing this method is the ease of data accessibility as selecting a model for ET estimation depends not only on how important the potential controls are for the system of study, but also on what data is available to run the model (Fisher et al. 2010). Therefore, the overall errors in ET can be traced back to EF and R_n separately (Tang et al. 2010).

ET as an indicator of the rate of change of the global water cycle represents a substantial portion of the water budget for the ecosystem (Jiang et al. 2009). For example, in comparison to 80% of the annual normal rainfall occurring during the four months of monsoon season (June to September), the estimated ET seems to be underestimated due to cloud cover which is a limitation of optical remote sensing. ET for non-monsoon season (eight months) with remaining 20% rainfall is estimated in range 322 mm and 329 mm. The projected ET (varies from 481 to 484 mm) is underestimated and would result in overestimation of runoff for the Brahmani and Baitarani RBs. Depending on water availability, climate regimes, and landscape conditions, ET can represent a substantial portion of the regional water budget in the study area.

The average annual ET for whole study area was estimated to be 325 mm with spatio-temporal cover of 67.7%, while 329 mm and 322 mm with spatio-temporal cover of 67 and 68% for Brahmani and Baitarani RB, respectively. Among all LULC classes forest showed highest ET value of 431 mm compared to 347 mm of water body/moist river bed for whole area owing to more net radiation received more EF and STCF for forest class compared to that of water body/moist river bed. Mean monthly ET varied between 32 and 104 mm with maximum in the month of May and minimum in November excluding the monsoon months (June to September). Between two RBs, highest ET was estimated in the forest class (500 mm) situated in Baitarani basin as compared to that of Brahmani basin (416 mm) owing to significant difference in net radiation (91.9 and 83.2 W/m^2) received by them with 0.53 and 0.51 EF value and spatio-temporal cover of 70.4 and 70%, respectively. Application of the ET estimation algorithm was limited to clear sky days. As observed in month of June, July, August, and September, desired accuracy could not be obtained due to cloud cover in monsoon season.

All the LULC classes showed increasing trend of ET from January to before onset of monsoon in June and found to be constant after the month of October in the year 2004. Among all LULCs, forest showed highest ET followed by water body/moist river bed, scrubland, cropland, and lowest by barren land. The spatio-temporal distribution of ET in different land cover/land use types shows how water is being used. This information is particularly crucial for water resources planning and management because it shows when and where water is being used as intended.

Acknowledgements The land use and land cover data received from ISRO-GBP project is thankfully acknowledged. Various input data received from NASA are also duly acknowledged. The study has benefitted from discussion among researchers of Spatial Analysis and Modelling (SAM) Lab, CORAL, Indian Institute of Technology Kharagpur, are also thankfully acknowledged.

References

Allen RG, Pereira LS, Raes D, Smith M (1998) Crop evapotranspiration, guidelines for computing crop water requirements. Irrigation and Drainage Paper No, 56: FAO, Rome, Italy, 300 p

Ambast SK, Keshari AK, Gosain AK (2002) An operational model for estimating regional evapotranspiration through surface energy partitioning (RESEP). Int J Remote Sens 23(22):4917–4930

Calera A, Campos I, Osann A, D'Urso G, Menenti M (2017) Remote sensing for crop water management: from ET modelling to services for the end users. Sensors 17(5):1104

Chen JM, Liu J (2020) Evolution of evapotranspiration models using thermal and shortwave remote sensing data. Rem Sens Environ 237:111594

Chen X, Su Z, Ma Y, Middleton EM (2019) Optimization of a remote sensing energy balance method over different canopy applied at global scale. Agric Forest Meteorol 279:107633

Dong B, Dai A (2017) The uncertainties and causes of the recent changes in global evapotranspiration from 1982 to 2010. Clim Dyn 49(1):279–296

Fisher JB, Whittaker RJ, Malhi Y (2010) ET come home: potential evapotranspiration in geographical ecology. Glob Ecol Biogeogr 20(1):1–18

Huang J, Gómez-Dans JL, Huang H, Ma H, Wu Q, Lewis PE, Xie X (2019) Assimilation of remote sensing into crop growth models: Current status and perspectives. Agric Forest Meteorol 276:107609

Javadian M, Behrangi A, Gholizadeh M, Tajrishy M (2019) METRIC and WaPOR estimates of evapotranspiration over the Lake Urmia Basin: comparative analysis and composite assessment. Water 11(8):1647

Jiang L, Islam S, Guo W, Julta AS, Senarath SUS, Ramsay BH, Eltahir E (2009) A satellite-based daily actual evapotranspiration estimation algorithm over South Florida. Global Planet Change 67:62–77

José JV, Oliveira NPRD, Silva TJDAD, Bonfim-Silva EM, Costa JDO, Fenner W, Coelho RD (2020) Quantification of cotton water consumption by remote sensing. Geocarto Int 35(16):1800–1813

Losgedaragh SZ, Rahimzadegan M (2018) Evaluation of SEBS, SEBAL, and METRIC models in estimation of the evaporation from the freshwater lakes (Case study: Amirkabir dam, Iran). J Hydrol 561:523–531

Madugundu R, Al-Gaadi KA, Tola E, Hassaballa AA, Patil VC (2017) Performance of the METRIC model in estimating evapotranspiration fluxes over an irrigated field in Saudi Arabia using Landsat-8 images. Hydrol Earth Syst Sci 21(12):6135–6151

Matsushima D (2007) Estimating regional distribution of surface heat fluxes by combining satellite data and a heat budget model over the Kherlenriver basin, Mongolia. J Hydrol 333:86–99

Mauder M, Foken T, Cuxart J (2020) Surface-energy-balance closure over land: a review. Bound-Layer Meteorol 177:395–426

Numata I, Khand K, Kjaersgaard J, Cochrane MA, Silva SS (2017) Evaluation of Landsat-based METRIC modeling to provide high-spatial resolution evapotranspiration estimates for Amazonian forests. Remote Sens 9(1):46

Rahman MM, Zhang W (2019) Review on estimation methods of the Earth's surface energy balance components from ground and satellite measurements. J Earth Syst Sci 128(4):1–22

Raoufi R, Beighley E (2017) Estimating daily global evapotranspiration using penman–monteith equation and remotely sensed land surface temperature. Remote Sens 9(11):1138

Raoufi R (2019) Estimating daily global evapotranspiration. Doctoral dissertation, Northeastern University

Senay GB, Kagone S, Velpuri NM (2020) Operational global actual evapotranspiration: development, evaluation, and dissemination. Sensors 20(7):1915

Tang R, Li Z, Tang B (2010) An application of the T-s-VI triangle method with enhanced edges determination for evapotranspiration estimation from MODIS data in and and semi-arid regions: implementation and validation. Remote Sens Environ 114(3):540–551

Trabucco A, Zomer RJ (2018) Global aridity index and potential evapotranspiration (ET0) climate database v2. CGIAR Consort Spat Inf 10

Wei Z, Yoshimura K, Wang L, Miralles DG, Jasechko S, Lee X (2017) Revisiting the contribution of transpiration to global terrestrial evapotranspiration. Geophys Res Lett 44(6):2792–2801

Zhan S, Song C, Wang J, Sheng Y, Quan J (2019) A global assessment of terrestrial evapotranspiration increase due to surface water area change. Earth's Future 7(3):266–282

Zhang J, Zhong Y, Gu C (2017) Energy balance method for modelling of soft tissue deformation. Comput Aided Des 93:15–25

Zhang Y, Peña-Arancibia JL, McVicar TR, Chiew FH, Vaze J, Liu C, Pan M (2016) Multi-decadal trends in global terrestrial evapotranspiration and its components. Sci Rep 6(1):1–12

Chapter 13
Application of Remote Sensing and GIS in Crop Yield Forecasting and Water Productivity

Kapil Bhoutika, Dhananjay Paswan Das, Arvind Kumar, and Ashish Pandey

Abstract Sugarcane is one of India's most important cash crops and one of the major crops of Uttarakhand state. Accurate crop yield forecasting is essential for making appropriate government policies. Statistical regression method using meteorological parameters is one of the most widely used crop yield forecasting methods. With the help of statistical regression, it is possible to forecast the sugarcane yield a few months before the harvest. But there is no direct cause–effect relationship between meteorological parameters and crop yield, so uses of other independent parameters can increase the crop yield accuracy. Evapotranspiration is one of the most crucial independent parameters, which can be easily estimated using remote sensing. The benefit of remote sensing over other fields and empirical methods for evapotranspiration is the easy availability of data over a large area as data availability becomes critical in other methods. Crop water efficiency can be easily found by crop water productivity. The developed Sugarcane yield actual evapotranspiration (AET) model using regression techniques for the F2 stage and both with and without AET model for F3 stage except 2019–20 in Haridwar district and the developed sugarcane yield model with and without AET using regression techniques for the F2 and F3 stage in Dehradun district showed a good relationship between predicted and observed values of yield which is below 5% deviation. From the study of crop water productivity, we can easily mark the areas with low water productivity and used different planning to increase the water efficiency to fulfill the need of people in reducing water availability.

13.1 Introduction

Sugarcane (Saccharum officinarum L.) is one of India's most important economic beneficial crops, which plays a vital role in the country's agriculture and industrial development (Natarajan et al. 2016). Global production of sugarcane in 2018

K. Bhoutika (✉) · D. P. Das · A. Kumar · A. Pandey
Department of Water Resources Development and Management, Indian Institute of Technology Roorkee, Roorkee 24667, India

A. Pandey et al. (eds.), *Geospatial Technologies for Land and Water Resources Management*, Water Science and Technology Library 103,
https://doi.org/10.1007/978-3-030-90479-1_13

was 1.90 billion tons over an area of 26.26 million hectares which are continuously increasing over the years (FAO 2018; Naseri et al. 2021). India produces 18.17% (341.20 million tons) of the world's total production, whereas bazil holds the second position, i.e., 39.38% (739.27 million tons). The production of sugarcane in Uttarakhand is 71.42 lakh tons (Department of Agriculture and Farmers Welfare Report 2017–2018). Sugarcane has a large cropping season, so it undergoes all seasons, i.e., summer, rainy, and winter. The major weather components which control the crop yield are rainfall, temperature, and humidity (Bhatla et al. 2018). Ministry of agriculture in 2006 estimated the sugarcane yield as 283.4 million, which is further revised to 355.5 million tons due to which there is a ban on sugarcane exports at the time of high international sugarcane market rate for which farmers suffered a huge loss (Suresh and Krishna Priya 2009), so it is necessary to forecast accurate crop yield.

Crop yield is defined as the crop produced per unit area, basically taken in kg/Ha. The crop yield forecast is an important parameter to make the policy and to import and export the crop for food security (Verma et al. 2021). Crop yield forecasting needs historical yield data and weather parameters in which the relationship is made by different models (Jayakumar et al. 2016). Previous studies show that very less notable work is done on sugarcane in India as compared to any other cereal crops. For crop yield forecasting, the regression model is a simple yet effective technique to make the relationship between weather parameters and actual crop yield (Wisiol 1987).

The regression method is quite simple and powerful, which is mostly used in crop yield forecasting, but in most cases, the accuracy from these simple regression models is not satisfactory. So for better accuracy, we need to use the statistical regression model with other different independent parameters other than the basic meteorological parameters. Different remote sensing parameters can be used for advancements in crop yield forecasting. Many worldwide researchers have stated that various remote sensing-based products can be used for crop yield forecasting (Mulianga et al. 2013; Morel et al. 2014). For the Indian condition, Rao et al. 2002 used NDVI for sugarcane yield forecasting. Recent agricultural studies focused on MODIS data (Doraiswamy et al. 2004; Potgieter et al. 2011; Mkhabela et al. 2011; Kouadio et al. 2012; Vintrou et al. 2012; Johnson 2014; Mosleh and Hassan 2014; Whitcraft et al. 2014), because of its high temporal resolution and free availability (Potgieter et al. 2011).

Evapotranspiration (ET) is the exchange of water and energy between the atmosphere, land surface, and soil by the processes of transpiration (from plants) and evaporation (Liu et al. 2019; Gunawardhana et al. 2021). It can be used as an independent parameter in the regression models. Nowadays, there are a number of remote sensing-based evapotranspiration product which is freely available and can be used to estimate spatially distributed region-scale evapotranspiration (Mu et al. 2011).

As we know that agriculture accounts for the largest share (85%) of global freshwater consumption. The availability of freshwater is continuously decreasing with an increase in industrialization and urbanization (Seckler et al. 1998; Toung and Bhuiyan 1994; Brar et al. 2012). So it is need of the hour to know the crop water productivity of different crops. Crop water productivity is the ratio of actual crop yield and actual evapotranspiration. Recent development in land and water management improves water use efficiency (Zwart and Bastiaanssen 2004). When water is limited, it is essential to know the crop water use efficiency (Brauman et al. 2013). The study of crop yield with crop water productivity is beneficial to know the relationship between these two components (Rockstorm et al. 2007; Monfreda et al. 2008; Speelman et al. 2008).

Keeping in mind the above points, the objectives of the study are

- To develop the sugarcane yield forecast model using the statistical regression method.
- To compare the accuracy of the sugarcane yield forecasting model with and without actual evapotranspiration as independent parameters.
- To study the changes of crop water productivity of sugarcane in the study area.

13.2 Materials and Methods

13.2.1 Study Area

The study area is shown in Fig. 13.1, which lies between 77° 20′ E—79° 00′E Longitudes and 29° 30′ N—30° 20′ N Latitudes. It covers Haridwar and Dehradun districts of Uttarakhand. The rainfall in the Haridwar and Dehradun districts ranges from 1500–2000 mm. The temperature is varying from below 5 to above 40 °C. The Tarai area's soils are deep, well-drained, a small amount of alkaline, and a mixture of coarse and fine loamy soils. At places, the problems of wetness, overflow, and erosion are also observed. Tarai soils are one of the most productive soils in the country (Pareek et al. 2019). Sugarcane is one of the dominant crops in the districts of Haridwar and Dehradun, which is the study area.

13.2.2 Data Collection and Preprocessing

13.2.2.1 Weather Data

Historical weather data (rainfall, maximum and minimum temperatures, maximum and minimum relative humidity) at a daily scale was collected from AMFU Roorkee for Haridwar district, whereas the gridded data (rainfall, maximum and minimum temperatures at a spatial resolution of 0.25 × 0.25 and 1 × 1 degree, respectively) for Dehradun district data were downloaded from Indian Meteorological Department (IMD) Web site for the period of 2001–20.

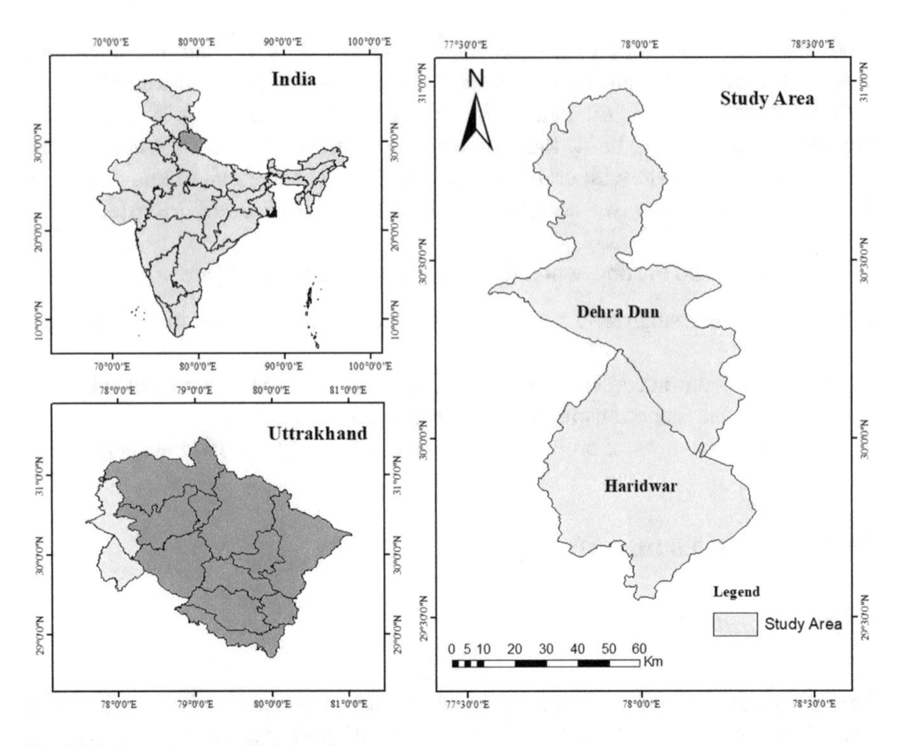

Fig. 13.1 Location map of the study area

13.2.2.2 Crop Yield Data

District-level sugarcane yield data for Haridwar and Dehradun districts were obtained from the Directorate of Agriculture, Uttarakhand, for the period of 2001–19. Figure 13.2 shows the officially reported crop yield statistics of Haridwar and Dehradun districts for the past 19 years. It represents the trend and changes in sugarcane productivity over the years. It is quite clear from Fig. 13.2 that sugarcane productivity shows an increasing trend for both the district, which may be probably due to recent scientific advancements, proper agriculture management, and crop pattern improvement.

13.2.2.3 Land Use Land Cover (LULC) Map

LULC map of the Haridwar and Dehradun districts was taken from the https://livingatlas.arcgis.com/landcover/ Web site for the year 2020. This map was developed by Impact Observatory for Esri. © 2021 Esri. This dataset is produced by the National Geographic Society in partnership with Google and the World Resources Institute for the Dynamic World Project using ESA Sentinel-2 imagery at 10 m spatial resolution.

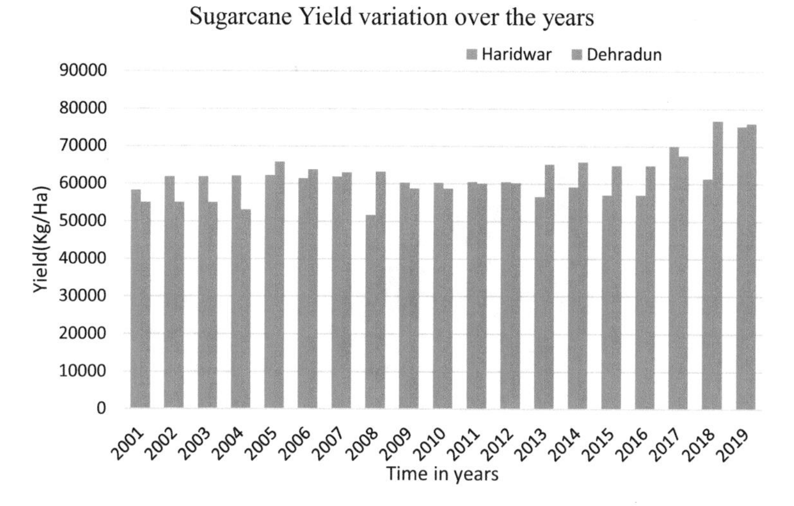

Fig. 13.2 Sugarcane productivity of Haridwar and Dehradun districts (2001–19)

There is 40.75% and 5.78% of the total geographical area which is an agricultural area in Haridwar and Dehradun districts, respectively, which is shown in Fig. 13.3.

13.2.2.4 Evapotranspiration Dataset

MODIS MOD16A2 V6 8 day interval at 500 m resolution evapotranspiration data is downloaded for the period of 2001–20 from https://lpdaac.usgs.gov/. The MOD16 evapotranspiration data is based on the Penmen–Monteith equation (Monteith 1965) where the inputs such as land cover, vegetation property, and albedo are taken from MODIS remotely sensed products and other daily inputs taken from meteorological reanalysis. These images are firstly preprocessed in the QGIS platform and then 0.1 is multiplied to MOD16 data in order to get the actual evapotranspiration in mm/8 day.

13.2.3 Methodology

13.2.3.1 Statistical Regression Models

Statistical regression techniques are one of the widely used crop yield forecasting techniques in which regression equations are made between crop yield and one or more meteorological variables. This is a straightforward method, and also the requirement of data is less as compared to other methods. The limitation of this method lies in that the long-range of historical yield and weather data are needed for accurate crop yield forecasting. Care is needed to give priority to agronomic

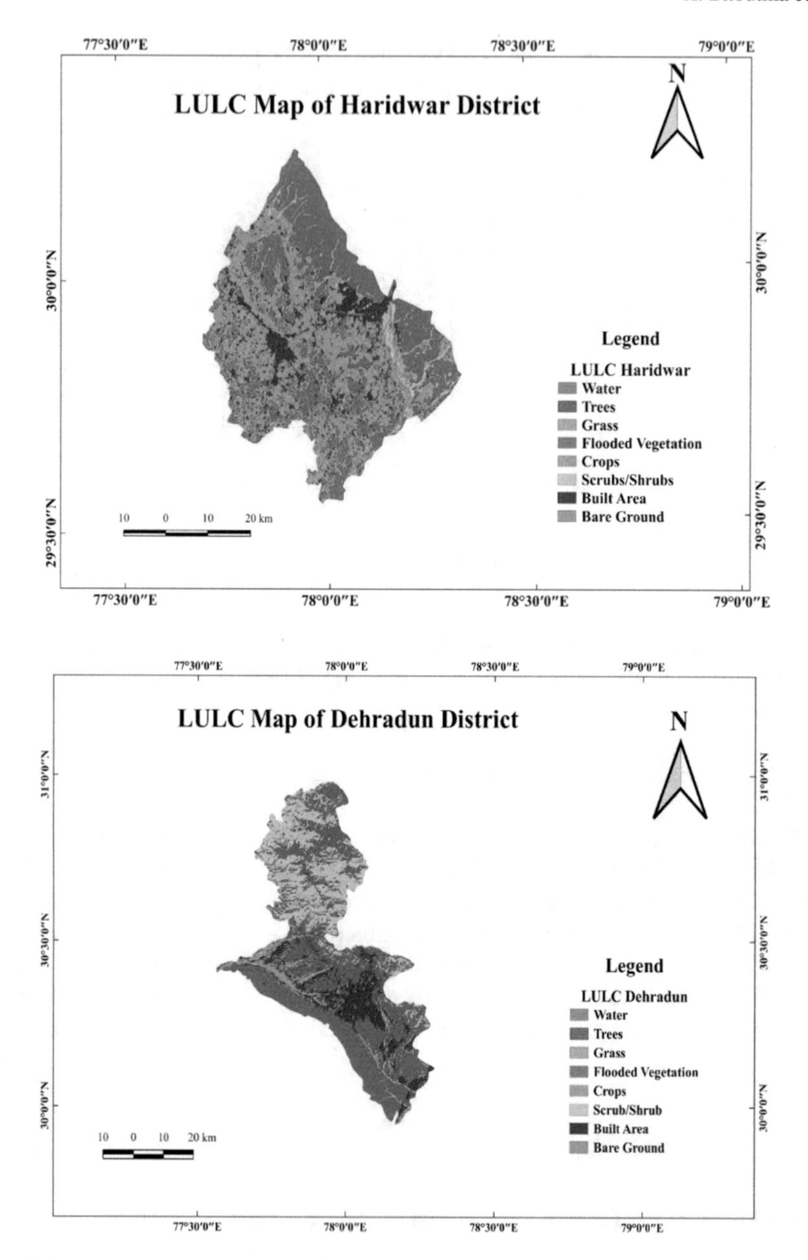

Fig. 13.3 Land use and land cover (LULC) map of Haridwar and Dehradun districts

significance as compared to statistical significance; otherwise, they will provide unrealistic forecasted values.

In statistical regression models, different independent parameters are used other than the meteorological parameters to increase the accuracy of the model, which are generally used in statistical regression models. In this study, actual evapotranspiration was used as independent parameters to improve the accuracy of the crop model for sugarcane.

This model using weighted and unweighted averages for stepwise regression analysis, which are going in the model one by one. This method requires at least 15 years of actual yield data and weekly meteorological data during the cropping period of the individual years (Verma et al. 2021). In this study, 20 years' historical yield data and the weekly meteorological data during the crop period of the individual years are used, in which two years are used for validation of the model. The method involves few steps laid down in a flowchart.

Figure 13.4 presents a flowchart showing yield forecasting of sugarcane using a simple regression model. The regression analysis used weekly weather data for the

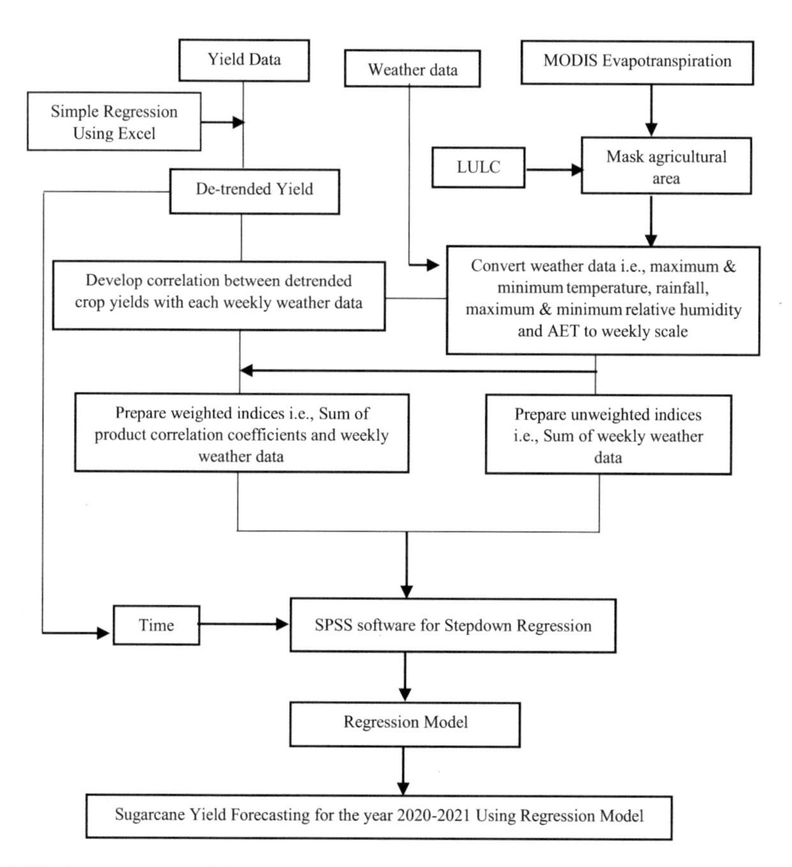

Fig. 13.4 Sugarcane crop yield forecasting using simple regression model

historical period (2001–20) followed by detrended yield data for the same period. Evapotranspiration for the respective crop period was evaluated from the MODIS dataset, whereas humidity (max, min) values of Haridwar district were used due to the unavailability of data at Dehradun district. After correcting crop yield data, weighted and unweighted indices were obtained and fed into SPSS software so as to develop the regression equations. Eventually, the indices of the current years were substituted in the developed equations so as to forecast crop yield.

13.3 Crop water Productivity

Crop water productivity of sugarcane is defined as the ratio of actual crop yield and actual crop evapotranspiration. Generally, the unit of water productivity is in kg/m^3. High crop water productivity shows high crop water efficiency. It is calculated by the formula (Zwart and Bastiaanssen 2004), which is described below-

$$\text{Crop Water productivity} = \frac{\text{Actual Crop Yield}}{\text{Actual Crop Evapotranspiration}}$$

13.4 Results and Discussions

13.4.1 Sugarcane Statistical Forecast Model

The sugarcane crop yield forecasting was divided into two stages: F2 and F3. For sugarcane, the F2 stage was taken from the first week of February to the second week of July, whereas the F3 stage was from the first week of February to the third week of December. The crop yield forecasting was done by SPSS Software using stepwise regression both with and without AET as an independent parameter which is given in Table 13.1. The correlation between various meteorological parameters and yield was computed, and its significance was tested using a t-test at F2 and F3 stages for the Haridwar district. Various statistical parameters are given in Table 13.2.

Table 13.1 Sugarcane forecast model summary for F3 stage

Crop	Districts	AET	R	R square	Adjusted R square	Std. error
Sugarcane	Haridwar	Without	0.819^b	0.670	0.629	3166.837
		With	0.819^b	0.670	0.629	3166.837
	Dehardun	Without	0.972^d	0.945	0.929	1717.232
		With	0.964^d	0.930	0.910	1939.315

Table 13.2 t-test and regression coefficients for Haridwar district

Stage	AET	Model	Unstandardized coefficients		t	sig
			B	SE		
F2	Without	Constant	4597.942	15788.850	0.291	0.775
		Z120	0.988	0.297	3.330	0.004
		Z11	447.624	161.389	2.774	0.014
	With	Constant	44476.682	2688.295	16.545	0.000
		Z361	0.679	0.108	6.287	0.000
		Z240	0.621	0.102	6.119	0.000
		Z30	−6.075	2.787	−2.180	0.048
F3	Without	Constant	26187.088	8930.439	2.932	0.010
		Z451	0.708	0.224	3.162	0.006
		Z41	47.182	15.531	3.038	0.008
	With	Constant	26187.088	8930.439	2.932	0.010
		Z451	0.708	0.224	3.162	0.006
		Z41	47.182	15.531	3.038	0.008

Similarly, the correlation between various meteorological and yield was computed, and its significance is given in Table 13.3 for the Dehradun district.

From Table 13.4, it is quite clear that in the F2 stage, with AET model was performing better than the without AET model based upon different statistical indicators (R^2 and standard error), whereas for the F3 stage, the performance of both the models is almost similar, which is given in Table 13.1.

It was observed that in both with and without AET cases, the percentage deviation is varying within $\pm$ 5%, and some cases lie within $\pm$ 10%, which was under the permissible limit (Fig. 13.5). Overall, considering all the statistical parameters, i.e., coefficient of determination, standard error, and percentage deviation regression model, with AET model is performing comparatively better than without the AET model, which is given in Tables 13.1 and 13.4.

Using the collected data, regression models were developed and validated for sugarcane for Haridwar and Dehradun districts. Several regression models with and without AET considered an independent variable for the Haridwar and Dehradun districts are generated using SPSS, and the best model is shown below.

For F2 stage, with AET for Haridwar ($R^2 = 0.906$ and Std. error = 1268.29) and Dehradun ($R^2 = 0.903$ and Std. error = 2203.86) performed well as compared to without AET models for Haridwar ($R^2 = 0.675$ and Std. error = 3145.10) and same for Dehradun ($R^2 = 0.903$ and Std. error = 2203.86) district.

For F3 stage, with AET for Haridwar ($R^2 = 0.728$ and Std. error = 2734.45) and Dehradun ($R^2 = 0.930$ and Std. error = 1939.31) performed well as compared to without AET models for Haridwar ($R^2 = 0.670$ and Std. error = 3166.83) slightly less for Dehradun ($R^2 = 0.945$ and Std. error = 1717.23) district.

Table 13.3 t-test and regression coefficients for Dehradun district

Stage	AET	Model	Unstandardized coefficients		t	sig
			B	SE		
F2	Without	Constant	64178.445	3012.418	21.305	0.000
		Time	632.483	106.457	5.941	0.000
		Z31	46.800	9.794	4.779	0.000
		Z151	1.930	0.530	3.639	0.002
	With	Constant	64178.445	3012.418	21.305	0.000
		Time	632.483	106.457	5.941	0.000
		Z31	46.800	9.794	4.779	0.000
		Z151	1.930	0.530	3.639	0.002
F3	Without	Constant	58400.269	6753.432	8.647	0.000
		Time	912.031	84.320	10.816	0.000
		Z351	0.443	0.072	6.176	0.000
		Z41	179.757	38.648	4.651	0.000
		Z40	−39.766	9.587	−4.148	0.001
	With	Constant	57058.580	6563.376	8.693	0.000
		Time	988.133	93.490	10.569	0.000
		Z61	157.411	25.884	6.081	0.000
		Z51	19.633	6.260	3.136	0.007
		Z60	8.403	3.148	2.670	0.018

Table 13.4 Sugarcane forecast model summary for F2 stage

Crop	Districts	AET	R	R square	Adjusted R square	Std. error
Sugarcane	Haridwar	Without	0.821[b]	0.675	0.634	3145.100
		With	0.952[c]	0.906	0.885	1268.292
	Dehradun	Without	0.950[c]	0.903	0.883	2203.867
		With	0.950[c]	0.903	0.883	2203.867

From this, it was evident that the F2 stage model for sugarcane was better performing as compared to the F3 stage. AET model was better than without AET model, except in the F3 stage without the AET model, which was slightly better (Tables 13.5 and 13.6).

The validation model shows that, therefore, Haridwar district was good with AET on 2018–2019, and on 2019–2020, AET model is slightly underperformed. In Dehradun district, the percentage deviation was very less for both AET and without the AET model. Without AET model is slightly better in Dehradun. Overall performance of the AET model is acceptable (−3.27% to −0.33%) except in Haridwar district's sugarcane yield forecasting on 2019–2020 (−17.94% and −10.3852%) (Table 13.7).

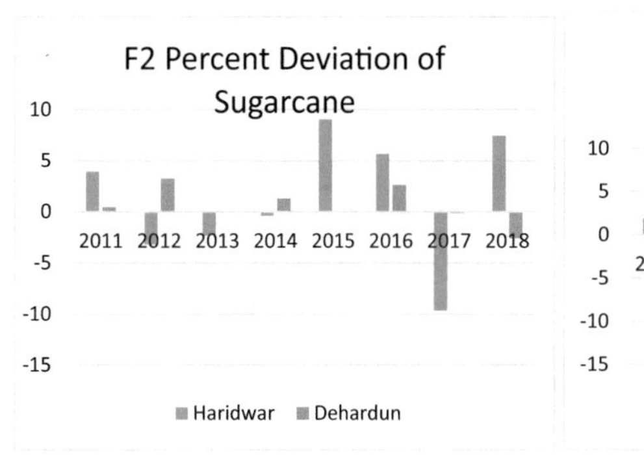

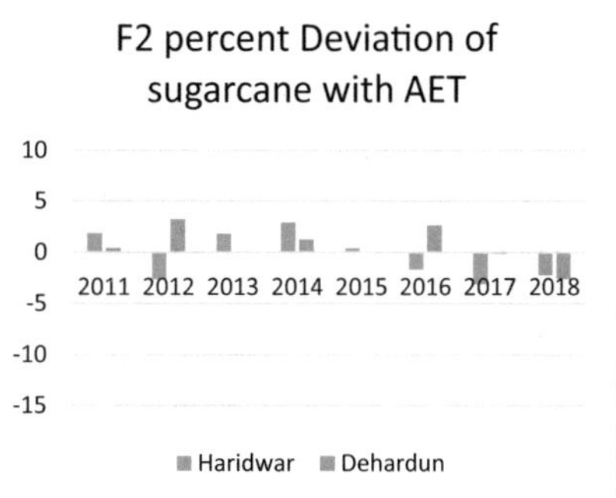

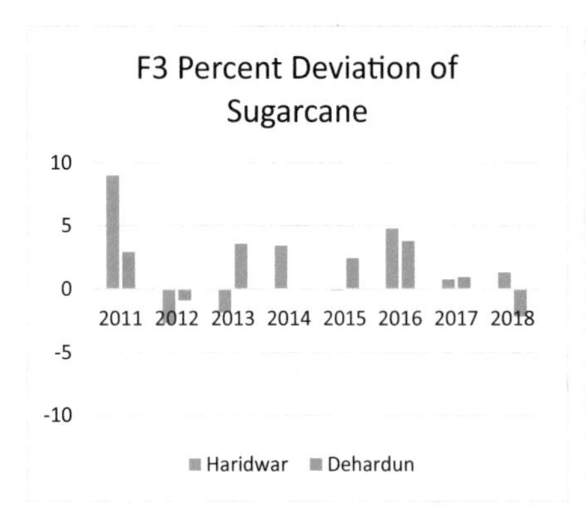

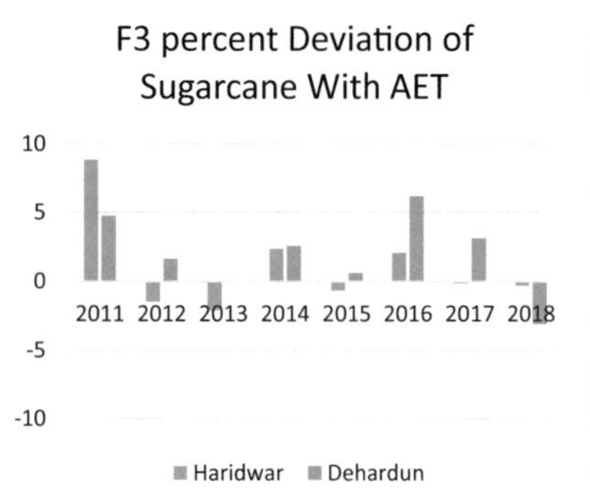

Fig. 13.5 Sugarcane forecasted yield deviation from actual yield for F2 and F3 stages

Table 13.5 Validation of statistical regression model of sugarcane yield forecasting

Districts	Year	Actual yield (kg/ha)	F2 stage		F3 stage	
			Without AET	With AET	Without AET	With AET
Haridwar	2018–19	70,000	73,182.44	66,579.14	62,213.78	61,193.5
	2019–20	68,100	62,956.23	61,786.25	67,479.93	67,479.93
Dehradun	2018–19	67,500	74,810.87	74,810.87	75,138.35	74,422.23
	2019–20	76,800	75,234.6	75,234.6	73,492.93	73,514.54

Table 13.6 Deviation percentage in the validation of statistical regression model of sugarcane yield

Districts	Year	F2 stage		F3 stage	
		Without AET	With AET	Without AET	With AET
Haridwar	2018–19	7.4632	−2.23328	1.325383	−0.33633
	2019–20	−16.3928	−17.9466	−10.3852	−10.3852
Dehradun	2018–19	−2.59001	−2.59001	−2.16361	−3.09606
	2019–20	−1.0071	−1.0071	−3.29878	−3.27034

Table 13.7 Forecasted yield for sugarcane for the year 2020–21

Districts	Yield (Kg/ha) from regression model without AET		Yield (Kg/ha) from regression model with AET	
	F2	F3	F2	F3
Haridwar	61,788.82	68,821.9	81,990.18	85,751.89
Dehradun	65,301.82	71,094.0	65,301.82	74,008.8

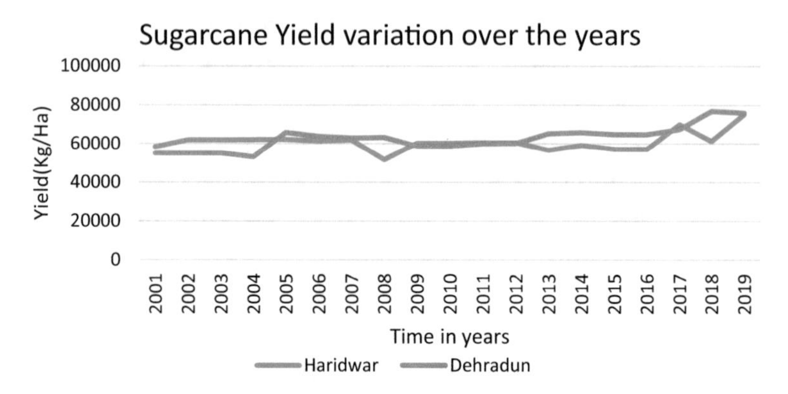

Fig. 13.6 Yield trend of sugarcane in Haridwar and Dehradun districts (2001–2019)

13.4.2 Crop Water Productivity

From the crop water productivity, we can see that crop water productivity is increased in the past several years in the study area. Dehradun's crop water productivity for sugarcane crops is slightly better than Haridwar district. The continuous decrease in water has had a significant impact on the agricultural sector. India's water use efficiency is significantly less for the agricultural sector, so we need to increase crop water productivity, especially for sugarcane which is a water-intensive crop. Though from this, we can see that the crop water productivity of Sugarcane for Haridwar and Dehradun district is increasing, we have to increase it more. With the use of advanced methods and equipment, the yield of the study area is increasing, which also increases the water requirement. It is need of the hour to improve the crop water productivity of the crop. By using remote sensing, we can easily get the data for large areas for our study, which can be used for this type of study (Figs. 13.6 and 13.7).

13.5 Conclusion

Remote sensing uses in crop yield forecasting at the regional scale are continuously increasing. This study shows that remote sensing-based MODIS evapotranspiration

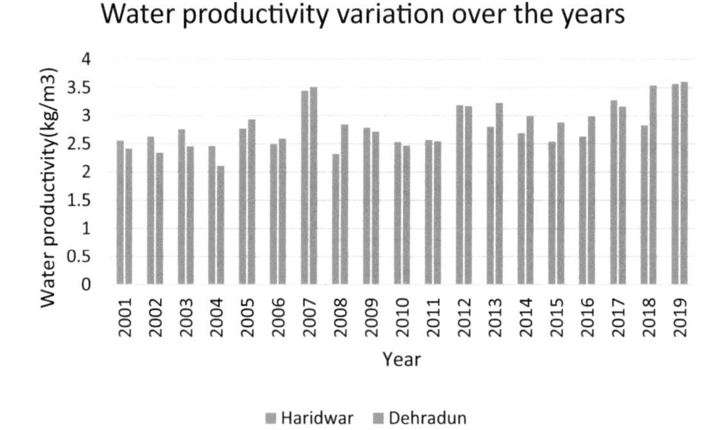

Fig. 13.7 Crop water productivity variation of sugarcane in Haridwar and Dehradun districts

data can be utilized for sugarcane yield forecasting. From this, we conclude that step-wise regression with AET as an independent parameter can be successfully applied for sugarcane yield forecasting, and in F2 stage, it gives good result for Haridwar and Dehradun districts. For the F3 stage, other remote sensing parameters can be used for improving the forecasting accuracy. The model discussed in this paper reasonably reduces the error in yield forecasting, giving a high R^2 value and less standard error. This model developed using remote sensing-derived evapotranspiration gives promising result which can be used for yield forecasting of other crops like wheat and rice. The use of remote sensing-based evapotranspiration is used for finding the crop water productivity of sugarcane. The variation of sugarcane water productivity from 2.2 kg/m^3 to 3.5 kg/m^3 shows an enormous scope for an increase in water productivity. This paper successfully shows the integration of remote sensing data in crop yield forecasting and calculation of crop water productivity.

Acknowledgements The authors wish to acknowledge the Ministry of Earth Science (MoES) for providing the funds and to the Directorate of Agriculture, Uttarakhand, for providing historical crop yield data. We are also grateful to AMFU Roorkee, IIT Roorkee, and Indian Meteorological Department (IMD) for providing weather data.

Appendix 1

Weather Indices

Weather variable	Weather indices	
	Unweighted weather indices	Weighted weather indices
Tmax	Z10	Z11
Tmin	Z20	Z21
Rain	Z30	Z31
RHmax	Z40	Z41
RHmin	Z50	Z51
AET	Z60	Z61
Tmax_Tmin	Z120	Z121
Tmax_Rain	Z130	Z131
Tmax_RHmax	Z140	Z141
Tmax_RHmin	Z150	Z151
Tmax_AET	Z160	Z161
Tmin_Rain	Z230	Z231
Tmin_RHmax	Z240	Z241
Tmin_Rhmin	Z250	Z251
Tmin_AET	Z260	Z261
Rain_RHmax	Z340	Z341
Rain_RHmin	Z350	Z351
Rain_AET	Z360	Z361
RHmax_Rhmin	Z450	Z451
RHmax_AET	Z460	Z461
RHmin_AET	Z560	Z561

References

Bhatla R, Dani B, Tripathi A (2018) Impact of climate on sugarcane yield over Gorakhpur District, UP using statistical model. Vayu Mandal 44(1):11–22

Brar SK, Mahal SS, Brar AS, Vashist KK, Sharma N, Buttar GS (2012) Transplanting time and seedling age affect water productivity, rice yield and quality in north-west India. Agric Water Manag 115:217–222

Brauman KA, Siebert S, Foley JA (2013) Improvements in crop water productivity increase water sustainability and food security—a global analysis. Environ Res Lett 8(2):024030

Department of Agriculture and Farmers Welfare Report, 2017–2018: https://agricoop.nic.in/

Doraiswamy PC, Hatfield JL, Jackson TJ, Akhmedov B, Prueger J, Stern A (2004) Crop condition and yield simulations using Landsat and MODIS. Remote Sens Environ 92:548–559

FAO (2018) Food and Agricultural Organization Statistical Yearbook 2. http://www.fao.org
Gunawardhana M, Silvester E, Jones OA, Grover S (2021) Evapotranspiration and biogeochemical regulation in a mountain peatland: insights from eddy covariance and ionic balance measurements. J Hydrol Reg Stud 36:100851
Jayakumar M, Rajavel M, Surendran U (2016) Climate-based statistical regression models for crop yield forecasting of coffee in humid tropical Kerala, India. Int J Biometeorol 60(12):1943–1952
Johnson DM (2014) An assessment of pre- and within-season remotely sensed variables for forecasting corn and soybean yields in the United States. Remote Sens Environ 141:116–128
Kouadio L, Duveiller G, Djaby B, El Jarroudi M, Defourny P, Tychon B (2012) Estimating regional wheat yield from the shape of decreasing curves of green area index temporal profiles retrieved from MODIS data. Int J Appl Earth Obs Geoinf 18:111–118
Liu YJ, Chen J, Pan T (2019) Analysis of changes in reference evapotranspiration, pan evaporation, and actual evapotranspiration and their influencing factors in the North China Plain during 1998–2005. Earth Space Sci 6(8):1366–1377
Mkhabela MS, Bullock P, Raj S, Wang S, Yang Y (2011) Crop yield forecasting on the Canadian Prairies using MODIS NDVI data. Agric for Meteorol 151:385–393
Monfreda C, Ramankutty N, Foley JA (2008) Farming the planet: 2. Geographic distribution of crop areas, yields, physiological types, and net primary production in the year 2000. Global Biogeochem Cycles 22(1)
Monteith JL (1965) Evaporation and environment. In: Symposia of the society for experimental biology, vol 19. Cambridge University Press (CUP), Cambridge, pp 205–234
Morel J, Todoroff P, Bégué A, Bury A, Martiné JF, Petit M (2014) Toward a satellite-based system of sugarcane yield estimation and forecasting in smallholder farming conditions: a case study on Reunion Island. Remote Sens 6(7):6620–6635
Mosleh M, Hassan Q (2014) Development of a remote sensing-based “Boro” rice mapping system. Remote Sens 6:1938–1953
Mu Q, Zhao M, Running SW (2011) Improvements to a MODIS global terrestrial evapotranspiration algorithm. Remote Sens Environ 115(8):1781–1800
Mulianga B, Bégué A, Simoes M, Todoroff P (2013) Forecasting regional sugarcane yield based on time integral and spatial aggregation of MODIS NDVI. Remote Sens 5(5):2184–2199
Naseri H, Parashkoohi MG, Ranjbar I, Zamani DM (2021) Energy-economic and life cycle assessment of sugarcane production in different tillage systems. Energy 217:119252
Natarajan R, Subramanian J, Papageorgiou EI (2016) Hybrid learning of fuzzy cognitive maps for sugarcane yield classification. Comput Electron Agric 127:147–157
Pareek N, Raverkar KP, Bhatt MK, Kaushik S, Chandra S, Singh G, Joshi HC (2019) Soil nutrient status of Bhabhar and hill areas of Uttarakhand. ENVIS Bull Himalayan Ecol 27
Potgieter A, Apan A, Hammer G, Dunn P (2011) Estimating winter crop area across seasons andregions using time-sequential MODIS imagery. Int J Remote Sens 32:4281–4310
Rao PK, Rao VV, Venkataratnam L (2002) Remote sensing: A technology for assessment of sugarcane crop acreage and yield. Sugar Tech, 4(3):97–101
Rockström J, Lannerstad M, Falkenmark M (2007) Assessing the water challenge of a new green revolution in developing countries. Proc Natl Acad Sci 104(15):6253–6260
Seckler D, Amarasinghe U, Molden D, De Silva R, Barker R (1998) World water demand and supply, 1990 to 2025: scenarios and issues. Res Rep19. Int Water Manag Inst Colombo, Sri Lanka
Speelman S, D’Haese M, Buysse J, D’Haese L (2008) A measure for the efficiency of water use and its determinants, a case study of small-scale irrigation schemes in North-West Province, South Africa. Agric Syst 98(1):31–39
Suresh KK, Krishna Priya SR (2009) A study on pre-harvest forecast of sugarcane yield using climatic variables. Stat Appl 7&8 (1&2)(New Series):1–8
Toung TP, Bhuiyan SI (1994) Innovations towards improving water-use efficiency in Rice. In: Paper presented at the World Bank’s 1994 Water Resource Seminar, Landsdowne, VA, USA, 13–15 Dec 1994

Verma AK, Garg PK, Prasad KH, Dadhwal VK, Dubey SK, Kumar A (2021) Sugarcane yield forecasting model based on weather parameters. Sugar Tech 23(1):158–166
Vintrou E, Desbrosse A, Bégué A, Traoré S, Baron C, Seen DL (2012) Crop area mapping in West Africa using landscape stratification of MODIS time series and comparison with existing global land products. Int J Appl Earth Obs Geoinf 14:83–93
Whitcraft AK, Becker-Reshef I, Justice CO (2014) Agricultural growing season calendars derived from MODIS surface reflectance. Int J Dig Earth. https://doi.org/10.1080/17538947.2014.894147
Wisiol K (1987) Choosing a basis for yield forecasts and estimates. In: Wisiol K, Hesketh JD (eds) Plant growth modelling for resource management, vol 1. CRC Press, Boca Raton, pp 75–103
Zwart SJ, Bastiaanssen WG (2004) Review of measured crop water productivity values for irrigated wheat, Rice, cotton and maize. Agric Water Manag 69(2):115–133

Chapter 14
Performance Evaluation of SM2RAIN-ASCAT Rainfall Product Over an Agricultural Watershed of India

Deen Dayal, Gagandeep Singh, Ashish Pandey, and Praveen Kumar Gupta

Abstract Precipitation is an essential climatic variable for any hydrological study. For hydrological studies, obtaining the observed precipitation, whether real-time or historical, has been quite challenging. In this regard, satellite-based precipitation estimates play an essential role in enhancing the present hydrologic prediction capability. They are available at a high spatiotemporal resolution with global coverage. In the present study, a satellite soil moisture (retrieved from scatterometer data)-based rainfall product (SM2RAIN-ASCAT) was evaluated for its performance to the gauge-based India Meteorological Department (IMD) gridded dataset over the Betwa river basin. The evaluation of satellite-based daily rainfall was carried out based on qualitative (based on contingency table) and quantitative indicators for 2007 to 2019. In general, the SM2RAIN-ASCAT rainfall product is excellent in detecting the daily rainfall events over the Betwa basin, although some of the events are falsely detected. A fair to the good agreement has been observed between satellite rainfall against IMD rainfall product ($d = 0.51$ to 0.72). The mean absolute error (MAE) in satellite rainfall was found to be in the range of 2.05–3.63 mm/day. The SM2RAIN-ASCAT rainfall product is good for low rainfall events. However, high rainfall events are always underestimated. Thus, the analysis indicates that the SM2RAIN-ASCAT rainfall product can be utilized in hydrological studies of ungauged regions.

14.1 Introduction

Water governs life on earth. Precipitation is a natural process responsible for the presence of fresh water on the earth's surface. It is one of the most critical components which governs the hydrology of any area. The Global Climate Observing System

D. Dayal (✉) · G. Singh · A. Pandey
Department of Water Resources Development and Management, Indian Institute of Technology Roorkee, Roorkee, Uttarakhand 247 667, India
e-mail: ddayal@wr.iitr.ac.in

P. K. Gupta
Space Applications Centre, ISRO, Ahmedabad 380015, India

A. Pandey et al. (eds.), *Geospatial Technologies for Land and Water Resources Management*, Water Science and Technology Library 103,
https://doi.org/10.1007/978-3-030-90479-1_14

(GCOS) ranks it at the top in the list of essential climate variables (Maggioni and Massari 2018). The importance of precipitation can be realized by the fact that it is an imperative dataset that finds application in hydrology, climatology, meteorology, ecology, and environmental assessment (Himanshu et al. 2018a, b, 2019; Anjum et al. 2019; Sahoo et al. 2021; Swain et al. 2021). Conventionally, the most reliable source of rainfall data is obtained from the direct measurement using rain gauges. Still, many a time, this data is not very useful to study the spatial distribution of rainfall due to the sparse density of the gauging stations in a particular area.

A significant limitation of gauge-based observations is a discontinuity in the recorded observations, which often renders the data unproductive (Thakur et al. 2021). The satellite-based precipitation products (SPPs) address the concerns of sparse and missing data. These products can provide continuous time series of data at fine spatial resolution (Sun et al. 2018) even at inaccessible places. However, these datasets are prone to various uncertainties attributed to recording time errors, sensor calibration errors, and errors incorporated by the algorithms (Porcù et al. 2014; Gebremichael et al. 2005). The accuracy of the SPPs is affected by these uncertainties. Still, considering the advantages discussed above, these products can be employed for various applications after incorporating necessary corrections and conducting accurate data validation using the ground-based rainfall measurements.

Microwave remote sensing (active and passive) and infrared sensors play a significant role in developing satellite-based precipitation products. There are numerous SPPs available in the open-source domain viz Tropical Rainfall Measuring Mission (TRMM), TRMM Multi-satellite Precipitation Analysis (TMPA), Global Satellite Mapping of Precipitation (GSMaP), Climate Prediction Center (CPC) Morphing (CMORPH), Precipitation Estimation from Remotely Sensed Information using Artificial Neural Networks (PERSIANN), Tropical Application of Meteorology Using Satellite Data (TAMSAT), Integrated Multi-Satellite Retrievals for Global Precipitation Measurement (GPM) (IMERG) that have been evaluated and used for various applications (Joshi et al. 2013; Nair et al. 2009; Prakash et al. 2014,2016; Himanshu et al. 2017a, b; Santos et al. 2021; Guptha et al. 2021; Dayal et al. 2021).

In this study, one of the recently developed remotely sensed rainfall products, SM2RAIN-ASCAT, is evaluated over the Betwa river basin in India. This product estimates the rainfall using the soil moisture to rain algorithm (Brocca et al. 2019a). The performance of this dataset has been assessed in comparison with the ground-based gridded dataset of the India Meteorological Department using various statistical indices, namely probability of detection (POD), false alarm ratio (FAR), critical success index (CSI), Pearson's correlation coefficient (R), agreement index (d), mean absolute error (MAE), percent bias (PBias), Kling-Gupta efficiency (KGE), and RMSE to standard deviation ratio (RSR).

14.2 Material and Methods

14.2.1 Study Area

Betwa river basin is positioned in the central part of India. Basin's total area is about 43,930 km^2, of which 68.9% is located in Madhya Pradesh (under ten districts of the state) and 31.1% is under five districts of southern Uttar Pradesh (Palmate et al. 2017). It extends from 22°52′ N to 26°03′ N latitude and 77°6′ E to 80°14′ E longitude as shown in Fig. 14.1.

The river's full length from its origin to its confluence with the Yamuna River is about 590 km, out of which 232 km lies in Madhya Pradesh and the rest 358 km in Uttar Pradesh (Dayal et al. 2018). Most of the basin area (about 69%) is under cultivation (Palmate et al. 2017). It has land covers of flat wheat-growing agriculture to steep forest hilly areas with varying vegetation and topography in a complex pattern. The climate of the Betwa basin is moderate, mostly dry except in monsoon season (Chaube et al. 2011). The average annual rainfall over the basin is 953 mm, out of which nearly 80% occurs in the monsoon season. The average annual evaporation losses are about 1830 mm (Pandey et al. 2021). The daily mean temperature ranges from a minimum of 8.1 °C to a maximum of 42.3 °C.

14.2.2 Ground-Based Rainfall Dataset

Ground rainfall data used in this study has been obtained from India Meteorological Department (IMD), Pune. The dataset for 2007 to 2019 has been downloaded from IMD, Pune Web site (https://www.imdpune.gov.in/Clim_Pred_LRF_New/Grided_Data_Download.html).

The gridded dataset has been prepared at a spatial resolution of 0.25° using the daily rainfall records from 6995 rain gauge stations (with varying availability periods) spread over the Indian main land (Pai et al. 2014). However, the station density varied from year to year, and on average, about 2600 stations per year were available to prepare the gridded data. The gridded data has been prepared using the inverse distance weighted interpolation (IDW) scheme (proposed by Shepard 1968 and locally modified by Rajeevan et al. 2006), and a maximum of four nearest neighbor stations, within a radius of 1.5 around the grid point, have been utilized.

Betwa river basin covers 90 grids of IMD dataset, and the spatial distribution of average annual rainfall over the basin is shown in Fig. 14.2. The average annual rainfall over the basin from 2007 to 2019 varies from the lowest 536 mm to the highest 1303 mm, with a mean value of 953 mm.

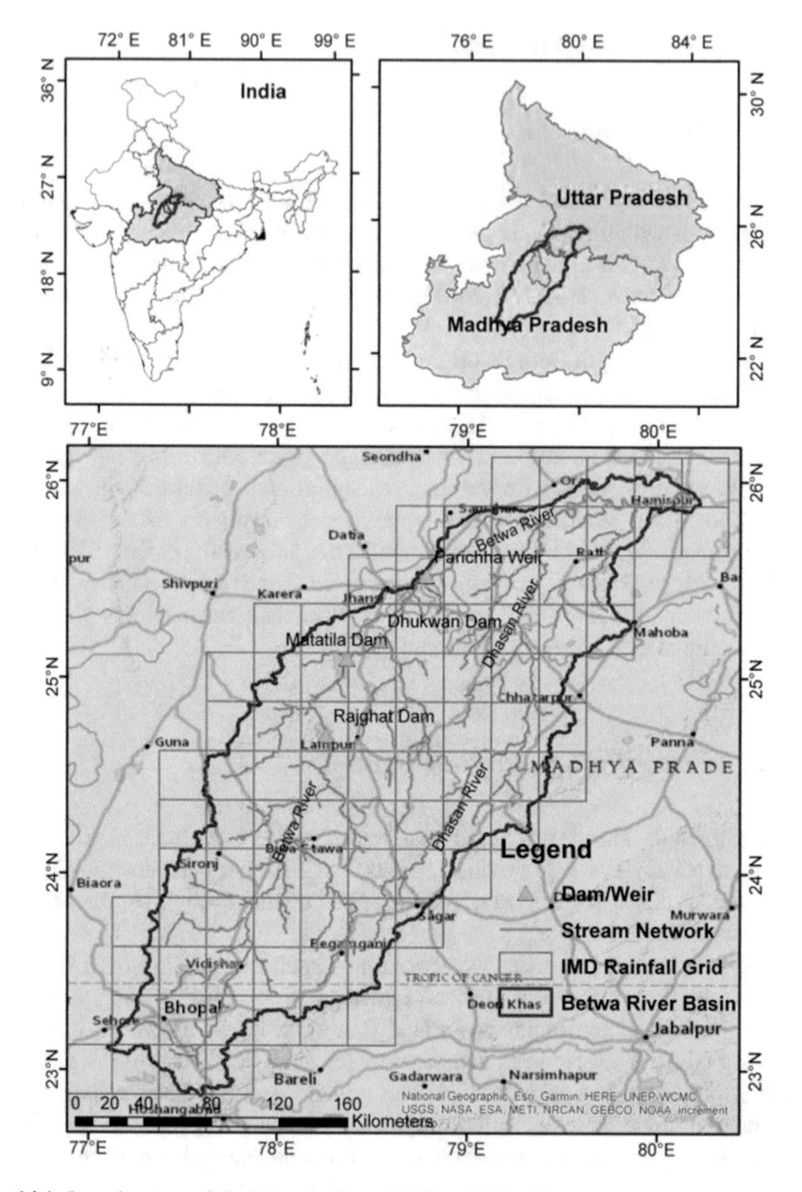

Fig. 14.1 Location map of the Betwa basin and IMD rainfall grids

14.2.3 SM2RAIN-ASCAT Data

The SM2RAIN-ASCAT rainfall dataset has been developed using a bottom-up approach (M2RAIN or Soil Moisture to Rain) applied to satellite soil moisture observations (Brocca et al. 2013a, 2019a). For the development of this dataset, soil moisture

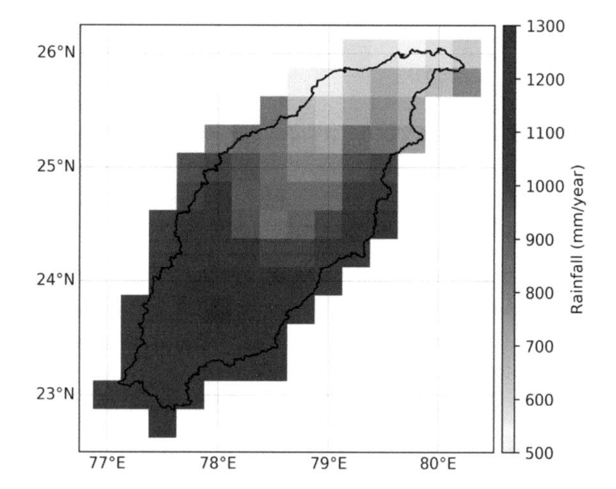

Fig. 14.2 Average annual rainfall over Betwa river basin

estimates from advanced sccatterometer (ASCAT) onboard MetOp A/B/C satellites (launched in 2006, 2012, and 2018, respectively) have been utilized (Wagner et al. 2013). For the application of the SM2RAIN algorithm, the ASCAT surface soil moisture product has been linearly interpolated in time to get the regular time series and filtered with an exponential filter (Wagner et al. 1999; Brocca et al. 2013b) or Daubechies-based wavelet filter (Massari et al. 2017). The SM2RAIN algorithm is based on the soil water balance equation (Eq. 14.1) and estimates the amount of water entering the soil by using in situ or satellite-based soil moisture observations (Brocca et al. 2013a, 2014,2015).

$$nZ\frac{\mathrm{d}S(t)}{\mathrm{d}t} = p(t) - g(t) - sr(t) - e(t) \tag{14.1}$$

where n, Z, S, t, p, g, sr, and e are soil porosity, soil layer depth (mm), relative soil moisture, time, rainfall (mm/day), sub-surface runoff (mm/day), surface runoff (mm/day), and actual evapotranspiration (mm/day), respectively. The climatological correction of rainfall estimate for reference rainfall has been carried out using the cumulative density function (CDF) matching approach (Brocca et al. 2011). Betwa river basin is covered by 461 grid points of SM2RAIN-ASCAT version-1.3 product, and data for the period of 2007 to 2019 has been utilized in this study. The SM2RAIN-ASCAT v1.3 dataset is available on a daily time resolution at a grid size of 0.1° × 0.1° (Brocca et al. 2019b), and it can be freely downloaded from: https://zenodo.org/record/3972958#.YBqeu_szbeQ.

14.2.4 Methodology

For the evaluation of purpose, the original grid size of 0.1° × 0.1° of SM2RAIN-ASCAT data has been resampled at 0.25° × 0.25° to align with the IMD grid. The performance of the daily precipitation estimates was evaluated based on several qualitative and quantitative statistical metrics. The statistical indices which have been used in this study are presented in Table 14.1.

where H is number of hit events, M is number of miss events, F is number of false precipitation events, P_{obs} is observed precipitation, $\overline{P_{obs}}$ is mean observed precipitation, P_{sat} is satellite precipitation, $\overline{P_{sat}}$ is mean satellite precipitation, n is the total number of events, α is a measure of the flow variability error, β is a bias term, RMSE is the root mean squared error, and σ is the standard deviation.

14.3 Result and Discussion

To carry out quantitative performances, the SM2RAIN-ASCAT rainfall product was evaluated against the observed rainfall product using contingency table indices (Schaefer 1990) and the continuous statistical indices. A brief description of the contingency table indices and the quantitative statistical indices are presented in

Table 14.1 Statistical indices with formulas and their respective range

Statistical indices	Formula	Theoretical range	Ideal value
Probability of detection (POD)	$\text{POD} = \frac{H}{H+M}$	0 to 1	1
False alarm ratio (FAR)	$\text{FAR} = \frac{F}{H+F}$	0 to 1	0
Critical success index (CSI)	$\text{CSI} = \frac{H}{H+M+F}$	0 to 1	1
Pearson's correlation coefficient (R)	$R = \frac{\sum(P_{obs}-\overline{P_{obs}})\times(P_{sat}-\overline{P_{sat}})}{\sqrt{\sum(P_{obs}-\overline{P_{obs}})^2}\times\sqrt{\sum(P_{sat}-\overline{P_{sat}})^2}}$	-1 to 1	-1 or 1
Agreement index (d)	$d = 1 - \frac{\sum(P_{obs}-P_{sat})^2}{\sum(\lvert P_{sat}-\overline{P_{obs}}\rvert+\lvert P_{obs}-\overline{P_{obs}}\rvert)^2}$	-∞ to 1	1
Mean absolute error (MAE)	$\text{MAE} = \frac{\sum\lvert P_{obs}-P_{sat}\rvert}{n}$	0 to ∞	0
Percent bias (PBias)	$\text{PBias} = \frac{\sum(P_{obs}-P_{sat})}{P_{obs}} \times 100$	-∞ to ∞	0
Kling-Gupta efficiency (KGE)	$\text{KGE} = 1 - \sqrt{(R-1)^2+(\alpha-1)^2+(\beta-1)^2}$	-∞ to 1	1
RMSE to standard deviation ratio (RSR)	$\text{RSR} = \frac{\text{RMSE}}{\sigma(P_{sat})}$	0 to ∞	0

Table 14.1. Before the applicability of satellite-based precipitation estimates for any hydrological application, their performance evaluation must be carried out against the observed rainfall dataset. In this study, a comprehensive assessment of the SM2RAIN-ASCAT rainfall product has been carried out with the IMD dataset for 2007–2019. The evaluation has been carried out on the basis of a contingency table and quantitative indices. The gridded as well basin-averaged rainfall on a daily time scale has been considered for the evaluation against the IMD rainfall product.

14.3.1 Evaluation of Gridded SM2RAIN-ASCAT Rainfall

Detection capabilities of SM2RAIN-ASCAT for rain events were evaluated using contingency table indices. The evaluation indices result for assessing daily gridded SM2RAIN-ASCAT rainfall against the gridded IMD rainfall dataset are presented in Table 14.2. The statistical parameters, viz., POD, FAR, CSI, *R*, *d*, MAE, Pbias, KGE, and RSR were estimated for daily SM2RAIN-ASCAT rainfall against observed rainfall (IMD gridded product) over each grid cell within the Betwa river basin. The spatial variation of POD, FAR, and CSI has been presented in Fig. 14.3. From Fig. 14.3, it can be observed that SM2RAIN-ASCAT performed well in capturing the rainfall events over the Betwa river basin. The number of events that SM2RAIN-ASCAT falsely detected is less as compared to the truly detected rainfall events. The accuracy in capturing the amount of daily rainfall by SM2RAIN-ASCAT has been assessed by *R*, *d*, MAE, Pbias, KGE, and RSR and is summarized in Table 14.2. The spatial variation of the accuracy in daily rainfall measured over the Betwa river basin is presented in Fig. 14.4. The agreement between gridded SM2RAIN-ASCAT daily rainfall and observed one over Betwa basin is satisfactory to fair as Pearson's

Table 14.2 Summary of performance evaluation of SM2RAIN-ASCAT rainfall product

Performance statistic	Evaluation of daily rainfall at grid level			Evaluation of basin-averaged daily rainfall
	Minimum	Maximum	Median	
POD	0.89	0.96	0.92	0.95
FAR	0.48	0.75	0.54	0.35
CSI	0.25	0.5	0.44	0.62
R	0.36	0.64	0.53	0.71
d	0.51	0.72	0.62	0.78
MAE	2.05	3.63	2.8	1.92
PBias	−19.48	63.98	−0.81	0.56
KGE	0.08	0.42	0.31	0.53
RSR	0.77	0.97	0.85	0.71

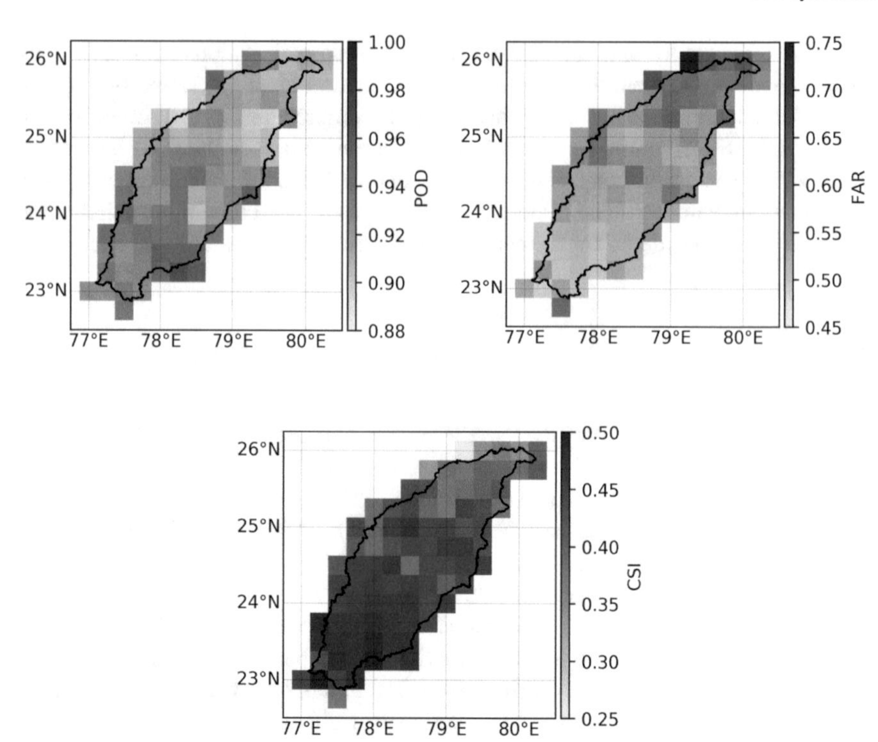

Fig. 14.3 Performance of SM2RAIN-ASCAT for detecting rainfall events

correlation (R), and agreement index (d) varies from 0.36 to 0.64 and 0.51 to 0.72, respectively, with median values of 0.53 and 0.62.

The spatial variation of POD depicts higher values in the southwestern part of the basin, while the majority grids of the northern part of the basin represent comparatively lower values. Therefore, it can be inferred that SM2RAIN-ASCAT performs better in the southern part of the basin as compared to the northern part; however, since the minimum POD value in the basin is 0.89, it indicates that overall, the range of POD is indicating a good rainfall detection capability of satellite data product. Spatial variation of FAR in the basin shows that the entire area below 24 deg N latitude produces low false alarms. In comparison, the area above 25 deg N latitude has high false alarms. This pattern indicates that the satellite product performs better in the lower part of the basin. In the Betwa basin, the critical success index values range between 0.25 and 0.5, but the majority of the basin shows values close to 0.5. Ideally, CSI values should be 1, but with a maximum of 0.5 in the region, it can be inferred that the product is unable to capture all the rainfall events.

The spatial variation of the agreement index over the study area depicts the values ranging from 0.5 to 0.75. The majority of the grids possess a decent level of agreement, i.e., greater than 0.6. Ideally, $d = 1$ indicates a perfect match, and 0 indicates no agreement at all. Error estimation in satellite rainfall is carried out using MAE

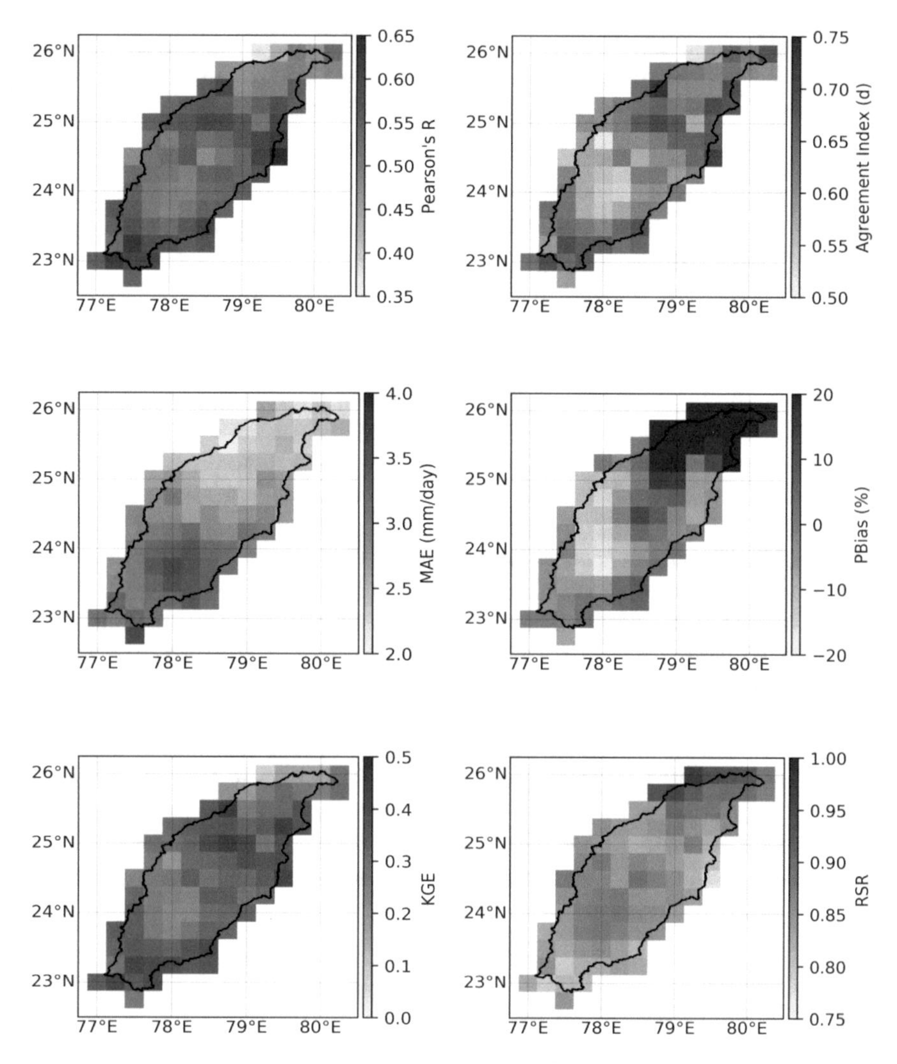

Fig. 14.4 Performance of SM2RAIN-ASCAT rainfall based on quantitative statistical indices

(expressed in mm/day). The spatial variation of both MAE indicates that the error is more in the southern part of the basin, while the northern part reflects a smaller error. This observation is not in line with the qualitative assessments, which can be attributed to two major reasons: higher rainfall variation in the southern part of the basin as observed in the RMSE to standard deviation ratio plot. The other reason is the high average annual rainfall in this part of the basin as shown in Fig. 14.2, which consequently introduces error in rainfall detection from the satellite data observations. The spatial variation in the percentage bias over the basin shows that there is a positive bias in the northern part of the basin while the southern part shows negative bias. Finally, the spatial variability of the Pearson's R and KGE shows overall satisfactory performance of the SM2RAIN-ASCAT for detecting rainfall events.

14.3.2 Evaluation of Basin-Averaged SM2RAIN-ASCAT Rainfall

In this study, an evaluation of the basin-averaged SM2RAIN-ASCAT daily rainfall was also conducted to observe how the satellite product is performing compared to the IMD dataset in the study area. The POD indicates a very high probability (95%) of rainfall detection. Furthermore, FAR of 0.35 signifies a decent score for evaluating the performance of the product. CSI of 0.62 indicates that the product is not able to capture all the rainfall events. R and d values were found to be 0.71 and 0.78, respectively, indicating a good correlation and level of agreement. The error representing statistical indices is concerned which MAE shows a value of 1.92 mm/day, highlighting the degree of mismatch in the satellite data product. Percent bias of 0.56 in the average daily rainfall product indicates that the product overestimates the rainfall in the basin. The scatterplot (Fig. 14.5) between the SM2RAIN-ASCAT rain and IMD shows that the dataset performs very well in capturing the rainfall of up to 25 mm/day, but the higher intensity of rainfall was underestimated.

This is also evident from the time series plot shown in Fig. 14.6, wherein the satellite product matches the shape of the peaks displayed in IMD data, but the magnitude of the peaks is underestimated. Therefore, these results indicate that the basin average daily SM2RAIN-ASCAT rainfall dataset has a satisfactory capability for capturing low-intensity rainfall for the spatial representation over the study area.

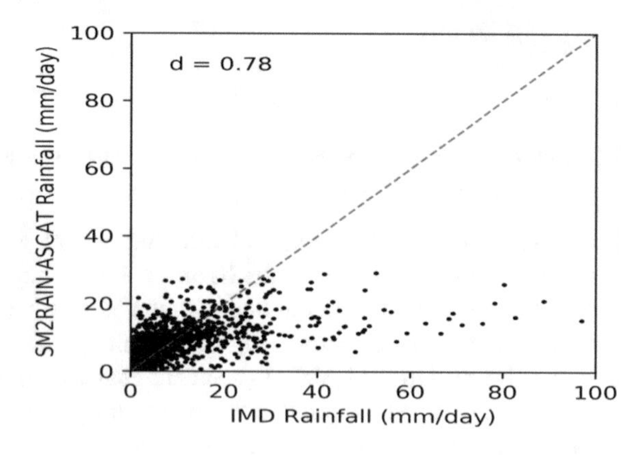

Fig. 14.5 Scatterplot between basin-averaged SM2RAIN-ASCAT rainfall and IMD rainfall

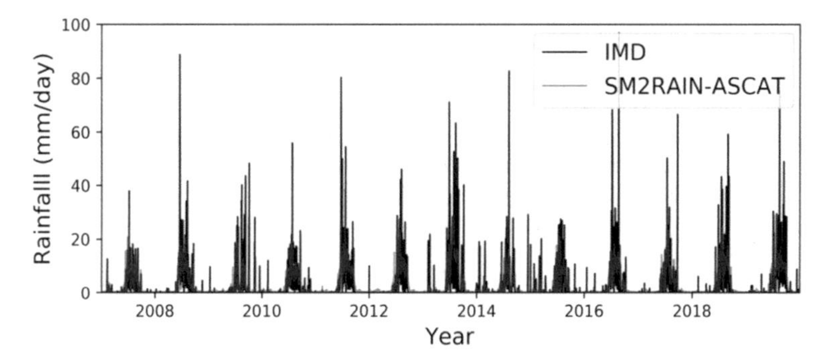

Fig. 14.6 Time series plot of basin-averaged SM2RAIN-ASCAT rainfall and IMD rainfall

14.4 Summary and Conclusion

In this study, a satellite soil moisture-based rainfall product (SM2Rain-ASCAT) has been evaluated with respect to the observed rainfall data. The evaluation was carried out at grid level rainfall as well as basin-averaged rainfall using the IMD rainfall product. The satellite rainfall evaluation has been carried out on the basis of some qualitative as well as quantitative performance indices. The following points have been observed from the present study:

- SM2RAIN-ASCAT rainfall product is excellent in detecting the daily rainfall events over the Betwa basin as POD was found to be in the range of 0.89–0.96.
- The satellite rainfall product performed better in southwestern part of the Betwa basin for detecting the daily rainfall events.
- The results also indicate that there is a possibility of detecting false rainfall events (when there is no rainfall in the observed data) by the satellite rainfall product (FAR value is in the range of 0.48–0.75).
- There is a fair agreement between the daily satellite rainfall product and the observed one, as the value of the agreement index is found to be in the range of 0.51–0.72 over the Betwa basin.
- The mean absolute error in the satellite rainfall product is found to be in the range of 2.05–3.63 mm/day.
- The percent bias indicated that the satellite rainfall product underestimates the rainfall as the value of PBias over the Betwa basin is in the range of −19.48–63.98, with a median of −0.81.
- The satellite rainfall product is very good in detecting low rainfall events, but the high rainfall events are highly underestimated.

Acknowledgements The authors are thankful to the Department of WRD&M, IIT Roorkee and Space Applications Centre (SAC), ISRO, Ahmedabad, for providing the laboratory facilities and financial support to conduct this study. We extend our sincere gratitude to IMD Pune, India, and Hydrology group of IRPI, CNR, Perugia, Italy, for providing the open-access rainfall datasets.

References

Anjum MN, Ahmad I, Ding Y, Shangguan D, Zaman M, Ijaz MW, Sarwar K, Han H, Yang, M (2019) Assessment of IMERG-V06 precipitation product over different hydro-climatic regimes in the Tianshan Mountains, North-Western China. Remote Sens 11(19):2314

Brocca L, Melone F, Moramarco T, Wagner W, Albergel C (2013a) Scaling and filtering approaches for the use of satellite soil moisture observations. Remote Sens Energy Fluxes Soil Moisture Content 411:426

Brocca L, Moramarco T, Melone F, Wagner W (2013b) A new method for rainfall estimation through soil moisture observations. Geophys Res Lett 40(5):853–858

Brocca L, Hasenauer S, Lacava T, Melone F, Moramarco T, Wagner W, Dorigo W, Matgen P, Martínez-Fernández J, Llorens P, Bittelli M (2011) Soil moisture estimation through ASCAT and AMSR-E sensors: An intercomparison and validation study across Europe. Remote Sens Environ 115(12):3390–3408

Brocca L, Ciabatta L, Massari C, Moramarco T, Hahn S, Hasenauer S, Kidd R, Dorigo W, Wagner W, Levizzani V (2014) Soil as a natural rain gauge: estimating global rainfall from satellite soil moisture data. J Geophys Res Atmos 119(9):5128–5141

Brocca L, Filippucci P, Hahn S, Ciabatta L, Massari C, Camici S, Schüller L, Bojkov B, Wagner W (2019a) SM2RAIN–ASCAT (2007–2018): global daily satellite rainfall data from ASCAT soil moisture observations. Earth Syst Sci Data 11(4):1583–1601

Brocca L, Filippucci P, Hahn S, Ciabatta L, Massari C, Camici S, Schüller L, Bojkov B, Wagner W (2019b) SM2RAIN-ASCAT (2007-June 2020): global daily satellite rainfall from ASCAT soil moisture (Version 1.3). Zenodo. https://doi.org/10.5281/zenodo.3972958

Brocca L, Massari C, Ciabatta L, Moramarco T, Penna D, Zuecco G, Pianezzola L, Borga M, Matgen P, Martínez-Fernández J.(2015) Rainfall estimation from in situ soil moisture observations at several sites in Europe: an evaluation of the SM2RAIN algorithm. J Hydrol Hydromechanics 63(3):201–209

Chaube UC, Suryavanshi S, Nurzaman L, Pandey A (2011) Synthesis of flow series of tributaries in Upper Betwa basin. Int J Environ Sci 1(7):1459

Dayal D, Pandey A, Himanshu SK, Palmate SS (2018) Long term historic changes of precipitation and aridity index over an Indian River Basin. World environmental and water resources congress 2018: Groundwater, sustainability, and hydro-climate/climate change. American Society of Civil Engineers, Reston, VA, pp 262–272

Dayal D, Gupta PK, Pandey A (2021) Streamflow estimation using satellite-retrieved water fluxes and machine learning technique over monsoon-dominated catchments of India. Hydrol Sci J 66(4):656–671

Gebremichael M, Krajewski WF, Morrissey ML, Huffman GJ, Adler RF (2005) A detailed evaluation of GPCP 1 daily rainfall estimates over the Mississippi River Basin. J Appl Meteorol Climatol 44(5):665–681

Guptha GC, Swain S, Al-Ansari N, Taloor AK, Dayal D (2021) Evaluation of an urban drainage system and its resilience using remote sensing and GIS. Remote Sensing Applications: Society and Environment 23:100601. https://doi.org/10.1016/j.rsase.2021.100601

Himanshu SK, Pandey A, Shrestha P (2017a) Application of SWAT in an Indian river basin for modeling runoff, sediment and water balance. Environ Earth Sci 76(1). https://doi.org/10.1007/s12665-016-6316-8

Himanshu SK, Pandey A, Yadav B (2017b) Assessing the applicability of TMPA-3B42V7 precipitation dataset in wavelet-support vector machine approach for suspended sediment load prediction. J Hydrol 550:103–117

Himanshu SK, Pandey A, Dayal D (2018a) Evaluation of satellite-based precipitation estimates over an agricultural watershed of India. World environmental and water resources congress 2018: watershed management, irrigation and drainage, and water resources planning and management. American Society of Civil Engineers, Reston, VA, pp 308–320

Himanshu SK, Pandey A, Patil A (2018b) Hydrologic Evaluation of the TMPA-3B42V7 Precipitation Data Set over an Agricultural Watershed Using the SWAT Model. J Hydrol Eng 23(4):05018003. https://doi.org/10.1061/(ASCE)HE.1943-5584.0001629

Himanshu SK, Pandey A, Yadav B, Gupta A (2019) Evaluation of best management practices for sediment and nutrient loss control using SWAT model. Soil and Tillage Res 192:42–58. https://doi.org/10.1016/j.still.2019.04.016

Joshi MK, Rai A, Pandey AC (2013) Validation of TMPA and GPCP 1DD against the ground truth rain-gauge data for Indian region. Int J Climatol 33(12):2633–2648

Maggioni V, Massari C (2018) On the performance of satellite precipitation products in riverine flood modeling: a review. J Hydrol 558:214–224

Massari C, Su CH, Brocca L, Sang YF, Ciabatta L, Ryu D, Wagner W (2017) Near real time de-noising of satellite-based soil moisture retrievals: an intercomparison among three different techniques. Remote Sens Environ 198:17–29

Nair S, Srinivasan G, Nemani R (2009) Evaluation of multi-satellite TRMM derived rainfall estimates over a western state of India. J Meteorol Soc Japan. Ser. II, 87(6):927–939

Pai DS, Sridhar L, Rajeevan M, Sreejith OP, Satbhai NS, Mukhopadhyay B (2014) Development of a new high spatial resolution (0.25× 0.25) long period (1901–2010) daily gridded rainfall data set over India and its comparison with existing data sets over the region. Mausam 65(1):1–18

Palmate SS, Ashish P, Mishra SK (2017) Modelling spatiotemporal land dynamics for a transboundary river basin using integrated Cellular Automata and Markov Chain approach. Appl Geogr 82:11–23

PandeyA, Dayal D, Palmate SS, Mishra SK, Himanshu SK, Pandey RP (2021) Long-term historic changes in temperature and potential evapotranspiration over betwa river Basin. In: Climate impacts on water resources in India. Springer, Cham, pp 267–286

Porcù F, Milani L, Petracca M (2014) On the uncertainties in validating satellite instantaneous rainfall estimates with raingauge operational network. Atmos Res 144:73–81

Prakash S, Sathiyamoorthy V, Mahesh C, Gairola RM (2014) An evaluation of high-resolution multisatellite rainfall products over the Indian monsoon region. Int J Remote Sens 35(9):3018–3035

Prakash S, Mitra AK, Rajagopal EN, Pai DS (2016) Assessment of TRMM-based TMPA-3B42 and GSMaP precipitation products over India for the peak southwest monsoon season. Int J Climatol 36(4):1614–1631

Rajeevan M, Bhate J, Kale JD, Lal B (2006) High resolution daily gridded rainfall data for the Indian region: Analysis of break and active monsoon spells. Curr Sci 296–306

Sahoo S, Swain S, Goswami A, Sharma R, Pateriya B (2021) Assessment of trends and multi-decadal changes in groundwater level in parts of the Malwa region, Punjab, India. Groundwater for Sustain Dev 14:100644. https://doi.org/10.1016/j.gsd.2021.100644

Santos CAG, Neto RMB, do Nascimento TVM, da Silva RM, Mishra M, Frade TG (2021) Geospatial drought severity analysis based on PERSIANN-CDR-estimated rainfall data for Odisha state in India (1983–2018). Sci Total Environ 750:141258

Schaefer JT (1990) The critical success index as an indicator of warning skill. Weather Forecast 5(4):570–575

Shepard D (1968) A two-dimensional interpolation function for irregularly-spaced data. In: Proceedings of the 1968 23rd ACM national conference, pp 517–524

Sun Q, Miao C, Duan Q, Ashouri H, Sorooshian S, Hsu KL (2018) A review of global precipitation data sets: data sources, estimation, and intercomparisons. Rev Geophys 56(1):79–107

Swain S, Mishra SK, Pandey A (2021) A detailed assessment of meteorological drought characteristics using simplified rainfall index over Narmada River Basin, India. Environ Earth Sci 80(6):1–15

Thakur PK, Garg V, Kalura P, Agrawal B, Sharma V, Mohapatra M, Kalia M, Aggarwal SP, Calmant S, Ghosh S, Dhote PR, Sharma R, Chauhan P (2021) Water level status of Indian reservoirs: a synoptic view from altimeter observations. Advances in Space Research 68(2):619–640. https://doi.org/10.1016/j.asr.2020.06.015

Wagner W, Lemoine G, Rott H (1999) A method for estimating soil moisture from ERS scatterometer and soil data. Remote Sens Environ 70(2):191–207

Wagner W, Hahn S, Kidd R, Melzer T, Bartalis Z, Hasenauer S, Figa-Saldaña J, De Rosnay P, Jann A, Schneider S, Rubel F (2013) The ASCAT soil moisture product: a review of its specifications, validation results, and emerging applications. Meteorol Z 22(1):5–33

Chapter 15
Curve Numbers Computation Using Observed Rainfall-Runoff Data and RS and GIS-Based NRCS-CN Method for Direct Surface Runoff Estimation in Tilaiya Catchment

Ravindra Kumar Verma, Ashish Pandey, and Surendra Kumar Mishra

Abstract Natural Resource Conservation Services-Curve Number (NRCS-CN) method is commonly used in hydrologic and geological sciences for surface runoff estimation in ungauged watersheds for various hydrological applications because it requires very few parameters in comparison with other hydrological methods. Curve Number (CN) is one of the most sensitive parameters of the method because it accounts for most of the watershed runoff producing response characteristics. Therefore, this study aims to compute different CN*s* to estimate surface runoff in the Tilaiya catchment of the Barakar river basin, Jharkhand, in eastern India. For this, firstly, two approaches were applied to compute CN*s* values for different antecedent soil moisture conditions: (i) from observed rainfall-runoff (*P*–*Q*) events data (CN_{PQ}), (ii) GIS-based NRCS-CN method (CN_{RS}) using RS inputs in the observed period 2001–2010, and secondly, surface runoff was estimated by applying the computed CN*s* values at daily and event scale. As a result of calculating the datasets of CN*s* values for the years, 2001–2010 shows a large variability of CN. On the other hand, the GIS-based NRCS-CN method of estimating CN*s* presented in this study shows the characteristics of the catchment and actual rainfall. Estimated runoff using the CN_{PQ} exhibited better results than fixed values applied in the GIS-based NRCS-CN method and could be used for planning, developing, and managing the water resource in the Tilaiya catchment.

R. K. Verma (✉) · A. Pandey · S. K. Mishra
Department of Water Resources Development and Management, Indian Institute of Technology, Roorkee, Uttarakhand 247667, India
e-mail: ravindra.wt@sric.iitr.ac.in

A. Pandey et al. (eds.), *Geospatial Technologies for Land and Water Resources Management*, Water Science and Technology Library 103,
https://doi.org/10.1007/978-3-030-90479-1_15

15.1 Introduction

An accurate surface runoff or rainfall excess estimation in ungauged watersheds is an essential hydrologic variable that play a significant role in hydrologic and geological sciences such as designing of small hydraulic structures, land use planning, agriculture use efficiency, water balance calculation models, and watershed management (Nagarajan and Poongothai 2012; Verma 2018). Numerous simple and complex methods have been developed so far to understand runoff process and its flow characteristics in a watershed, but most of them are costly, time-consuming, and difficult to apply because of a lack of adequate data. Therefore, simple methods for estimating runoff depth from watersheds are mainly imperative and often feasible in hydrologic engineering, hydrological modeling, and many hydrologic applications. The most popular method for meeting this need is the Natural Resource Conservation Services-Curve Number (NRCS-CN) method [earlier known as Soil Conservation Services (SCS), USDA (1986)], because it requires a minimum number of parameters. It was pioneered and developed by a US-based agency, namely Natural Resource Conservation Service (USDA-NRCS), in 1954.

The NRCS-CN method is often used for runoff depth or excess runoff estimation from a depth of precipitation in given soil textural classification, land use/land cover, and watershed wetness. The method combines the watershed parameters and climatic factors in one entity called the CN, which varies with cumulative rainfall, land use, soil cover, and initial soil moisture content (Hawkins 1978; Chow et al. 1988). It is used worldwide because of its simplicity, convenience, accounting major runoff producing watershed characteristics such as soil type, land use/treatment, hydrological condition, and antecedent moisture conditions (AMC) (Ponce and Hawkins 1996; Mishra and Singh 2003; Sahu et al. 2007; Verma et al. 2018).

For its simplicity and agency method, and applicability, the method has gained much attention among researchers, hydrologists, and policymakers since its inception. Later on, it has been adopted into many complex models, viz AGNPS, SWAT, CREAMS, ANSWERS, QUALHYMO, and other environmental areas, viz sediment yield estimation, soil moisture accounting, water quality management, environmental flow estimation, irrigation, long-term simulation, rainwater harvesting, urban hydrology, hydrograph simulation, partitioning of heavy metals, the impact of forests fire on runoff responses, and effects of land uses changes on hydrologic responses (Sahu et al. 2007; Verma et al. 2020, 2021).

Over two decades, the applicability of NRCS-CN has also expanded beyond its original scope, and it is now also used for large area and inaccessible areas by integrating RS/GIS techniques. RS technique provides a source of input data, e.g., digital LU/LC and hydrological soil group, basin morphology to estimate runoff response. However, GIS is used to interpret and display spatially distributes CN and even estimates runoff profile at each pixel level. Therefore, many researchers or hydrologists have preferred the GIS-based SCS-CN method over the traditional technique for estimating runoff characteristics such as depth or volume for various uses, viz artificial recharge, flood estimation, urban hydrology with better accuracy, and results in a short time (Ragan and Jackson 1980; Vojtek and Vojteková 2016; Thakur et al. 2020; Moniruzzaman et al. 2021).

Recently, Anbazhagan et al. (2005) estimated total runoff volume using GIS-based SCS-CN approach in different sub-watersheds and prioritized sub-watershed for artificial recharge in the Ayyar river basin based on available runoff volume. Vojtek and Vojteková (2016) also used a GIS-based NRCS-CN method to estimate high runoff volume in Vyčoma catchment of Western Slovakia and assessed risk for potential flood occurrence in built-up areas. Melesse and Shin and Moniruzzaman et al. (2020) stated that the land use/land cover is an important input of the NRCS-CN model for understanding watershed runoff responses. They estimated decadal land use changes and their impact on runoff responses in S-65A sub-basin of the Kissimmee river basin, South Florida, and Dhaka city catchment, Bangladesh.

Many studies (Tiwari et al. 1991; Pandey and Sahu 2002; Tripathi et al. 2002; Jasrotia et. al. 2002; Nayak and Jaiswal 2003; Pandey et al. 2003; Anbazhagan et al. 2005; Mohan Zade et al. 2005; Patil et al. 2008; Gupta and Panigrahy 2008; Gajbhiye and Mishra 2012; Nagarajan and Poongothai 2012; Deshmukh et al. 2013; Gajbhiye et al. 2017; Verma et al. 2018) effectively utilized GIS-based NRCS-CN approach to estimate surface runoff for Indian watersheds. For example, Pandey and Sahu (2002) used NRCS-CN approach to estimate runoff for Karso watershed in the Barakar catchment and found minimum and maximum deviations between the observed and predicted runoff which were 3.27% and 28.33%, respectively. Verma et al. (2018) and Gajbhiye and Mishra (2012) estimated unadjusted CN and slope-adjusted CN values using RS and GIS application in Kalu watershed, Maharastra, and Kanhaiya watershed, respectively. They found that computed runoff using adjusted CN values was more appropriate than unadjusted CN values.

The literature shows that RS and GIS tools for surface runoff estimation have received increasing attention in the past three decades because it includes spatiotemporal and geomorphologic variations, which facilitate hydrologists to deal with large scale, complex, and spatially distributed hydrological processes. In other words, it is also applicable for a large area after dividing it into sub-watersheds. Originally, the NRSC-CN method's applicability was limited to 250 km^2 catchment. Therefore, the objectives of this study are: (i) to compute appropriate CN*s* for Tilaiya catchment using *P*-*Q* data and GIS-based NRCS method and their comparison, (ii) to estimate the surface runoff by accounting the spatial variation of rainfall within the catchment, and (iii) to develop relationships between rainfall and runoff.

15.2 Materials

15.2.1 Study Area

The present study was conducted in the Tilaiya reservoir catchment, which is the upper most part of the Barakar river basin. It has a total drainage area of approximately

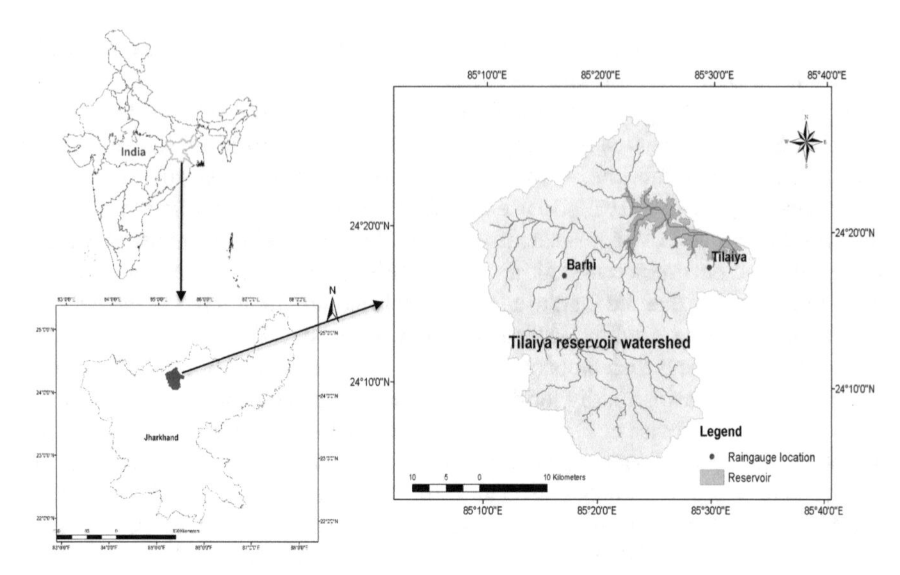

Fig. 15.1 Location of the study area and its drainage patterns

960 km^2, and geographical boundary lies in between 24° 05′ N, 85° 09′ E and 24° 28′ N, 86° 23′ E (Fig. 15.1). The catchment has a multipurpose dam constructed by the Damodar Valley Corporation (DVC) in 1953 for moderate flood and irrigation in the lower region of its catchment. The dam is situated in the uppermost reach of the Barakar river, about 64.4 km below its source.

The Barakar river originates near the Padma in the Hazaribag district of Jharkhand and runs 200 km parallel to the Damodar river and bends to the east to meet the Damodar river. It flows west to east direction and meets the river Damodar river in the Burdwan district of West Bengal. The study area's geography is hilly, rugged topography, and unstable slopes and deposits of mineral resources. However, the climate is characterized by moderate winters and hot, humid summers. Normal temperature varies between 40 and 42 °C in the summers (May and June), whereas 13–26 °C in the cold months (December and January). Rainfall is the main source of runoff. The annual rainfall varies from 671 to 1445 mm (mean 1072 mm) for 2001–2010, which is unevenly distributed across the catchment.

About 75–80% of annual rainfall occurred in the monsoon season (mid-June to October), and a significant. A significant amount of runoff occurred in the study area and distributed unevenly in space and time. A very low sedimentation rate of about 0.229 ha m/100 km^2/year and a trapping efficiency varying between 95 and 100% were observed in the Tilaiya reservoir, which indicates less erosion in the catchment (CWC report 1997).

15.2.2 Watershed Database

Remote Sensing and Collateral Data

Required databases for this study were collected from various sources. A brief description of each database is given as follows:

- Daily rainfall data (2001–2010) from Barhi and Tilaiya rain gauge stations located in the study area were collected from the DVC, Maithon, Dhanbad (Fig. 15.1).
- Daily discharge data of the same period at the inflow point of Tilaiya dam were also collected from the DVC, Maithon. The DVC monitors both rainfall stations and discharges gauge sites.
- Remote sensing data: Landsat ETM + imageries (30 m resolution) were downloaded from Earth Explorer Web site (http://earthexplorer.usgs.gov/) of a period December 09 to December 12, 2009. Furthermore, mosaic all images for preparing LU/LC map. The images' path and row are (140, 43), (141, 43).

15.3 Method

15.3.1 NRCS-CN Method

The equation for runoff estimation given by NRCS-CN method is

$$Q = \frac{(P - I_a)^2}{P - I_a + S} \text{ for } P \geq I_a, \text{ 0 otherwise} \tag{15.1}$$

$$Q = 0 \tag{15.2}$$

Parameters S is mapped to the CN as

$$S = 25.4\left(\frac{1000}{\text{CN}} - 10\right) \tag{15.3}$$

where P = rainfall (mm), Q = rainfall excess or direct runoff (mm), S and CN = dimensionless variables, which is varying from 0 to 100.

For Indian watersheds, an initial loss is 30% of S is suggested; it can be expressed as

$$I_a = 0.3S \tag{15.4}$$

Using Eq. (15.4), the Eq. (15.1) can be recast as,

$$Q = \frac{(P - 0.3S)^2}{(P + 0.7S)} \quad (15.5)$$

The above conceptual method indicates that higher is S, lower will be CN, and vice versa. It infers that the runoff potential increases with an increase in CN and decreases with a decrease in CN. S can be computed by Eq. (15.6), which is suggested by Mishra and Singh.

$$S = \frac{P}{\lambda} + \frac{(1 - \lambda)Q - \sqrt{(1 - \lambda)^2 Q^2 + 4\lambda P Q}}{2\lambda^2} \quad (15.6)$$

15.3.2 Computation of CN

Many researchers have proved that CN is the most sensitive parameter in the NRCS-CN method. On the other hand, CN is also varying with the soil type, land use, hydrological condition, AMC, and hydrologic soil group (HSG). Therefore, accurate determination of this parameter is very important considering watershed zes characterize, viz land use, soil type, AMC, and HSG. AMC is expressed by three classes (I (dry), II (normal), and III (wet)), according to rainfall limits for dormant and growing seasons. CN values for AMC classes I, II, and III correspond to 90, 50, and 10% cumulative probability of exceedance of surface runoff for a given rainfall, respectively (Hjelmfelt et al. 1982). In other words, the higher the antecedent moisture or rainfall amount, the higher is CN and, therefore, the high runoff potential of the watershed and vice versa. In the present study, the following two approaches are adopted for CN estimation.

15.3.2.1 CN Computation Using Observed P-Q Data

A seven steps procedure is used for estimating CN from *P-Q* event data, which is outlined here as: (i) separate base flow from daily discharge data using free available software, namely Web-based hydrograph analysis tool (WHAT) (https://engineering.purdue.edu/mapserve/WHAT/) to estimate direct surface runoff, (ii) generate events from the daily rainfall and obtained corresponding direct surface runoff data, (iii) compute values of runoff and corresponding rainfall for each event, (iv) use Mishra and Singh formula (Eq. 15.6) to estimate S, (v) estimate CN for each selected event using Eq. 15.3, (vi) repeat above steps (ii) to (v) for each year, and (vii) compute annual mean CN as representative CN. It was found that 2008 observed eight high events, whereas only two high events were observed in 2009. The duration for the selected events varies from 4 to 13 days, as given in Table 15.1.

Table 15.1 43 generated events

S. No.	Event name	Event duration (day/month/year)	S. No.	Event name	Event duration (day/month/year)
1	2001_Evt1	15.06.2001–21.06.2001	23	2005_Evt4	04.08.2005–08.08.2005
2	2001_Evt2	27.06.2001–01.07.2001	24	2006_Evt1	02.06.2006–09.06.2006
3	2001_Evt3	13.07.2001–21.07.2001	25	2006_Evt2	08.07.2006–14.07.2006
4	2001_Evt4	17.08.2001–22.08.2001	26	2007_Evt1	02.07.2007–08.07.2007
5	2001_Evt5	23.08.2001–04.09.2001	27	2007_Evt2	14.07.2007–19.07.2007
6	2001_Evt6	10.09.2001–16.09.2001	28	2007_Evt3	10.08.2007–17.08.2007
7	2001_Evt7	29.09.2001–06.10.2001	29	2007_Evt4	06.09.2007–12.09.2007
8	2002_Evt1	30.06.2002–06.07.2002	30	2007_Evt5	22.09.2007–27.09.2007
9	2002_Evt2	31.07.2002–06.08.2002	31	2008_Evt1	15.06.2008–21.06.2008
10	2002_Evt3	10.08.2002–17.08.2002	32	2008_Evt2	28.06.2008–04.07.2008
11	2002_Evt4	22.09.2002–30.09.2002	33	2008_Evt3	06.07.2008–11.07.2008
12	2003_Evt1	26.06.2003–01.07.2003	34	2008_Evt4	14.07.2008–17.07.2008
13	2003_Evt2	22.07.2003–03.08.2003	35	2008_Evt5	06.08.2008–12.08.2008
14	2003_Evt3	09.08.2003–14.08.2003	36	2008_Evt6	15.08.2008–18.08.2008
15	2003_Evt4	17.08.2003–21.08.2003	37	2008_Evt7	20.08.2008–23.08.2008
16	2004_Evt1	17.06.2004–26.06.2004	38	2008_Evt8	22.09.2008–26.09.2008
17	2004_Evt2	10.09.2004–13.09.2004	39	2009_Evt1	28.06.2009–02.07.2009
18	2004_Evt3	15.09.2004–20.09.2004	40	2009_Evt2	13.08.2009–22.08.2009
19	2004_Evt4	24.09.2004–28.09.2004	41	2010_Evt1	17.06.2010–21.06.2010
20	2005_Evt1	12.07.2005–20.07.2005	42	2010_Evt2	26.07.2010–29.07.2010
21	2005_Evt2	15.08.2005–19.08.2005	43	2010_Evt3	18.08.2010–26.08.2010
22	2005_Evt3	20.09.2005–25.09.2005			

15.3.2.2 CN Computation Using RS and GIS-Based NRCS-CN Method

To estimate CN*s* from the NRCS-CN method coupled with RS and GIS, the following procedure was adopted (Fig. 15.2). For this, the LULC map of 2009 has been derived from LANDSAT ETM + imageries (Fig. 15.3a).

Figure 15.3a shows that the study area is mostly dominated by built-up and fellow land, covering 36% and 31%, respectively. Shrubs and agriculture area are also present in the watershed in a scattered way. A small amount of mining area is also identified of about 19 km^2 (0.02%) (Fig. 15.3a), whereas the soil map shows soil textures of the study area, which composed of loamy (84%) and loamy skeletal (16%).

The thematic map of HSG was also generated using the NBSS and LUP classification and vector to raster conversion tools of GIS, because it is essential to understand infiltration capacity and transmission of surface water in the aquifer of the study area

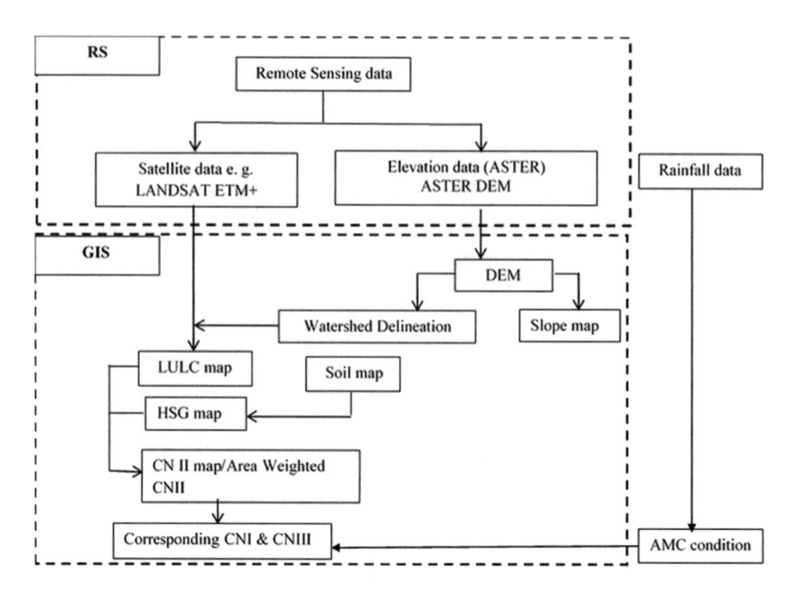

Fig. 15.2 Procedure for runoff estimation RS and GIS-based NRCS-CN method

based on soil cover and subsoil conditions for predicting runoff depth. The criteria adopted for such classification is illustrated in Table 15.2 (Chow et al. 1988).

In general, several types of soils and land use combinations are found in a watershed. For this type of watershed, computation of an area-weighted curve number (CN_w) is commonly used for identifying CN for the entire watershed using Eq. 15.7 (Pandey and Sahu 2002; Nagarajan and Poongothai 2012; Verma et al. 2018).

$$CN_{WII} = \frac{CN_1 A_1 + CN_2 A_2 + \ldots.. + CN_n A_n}{A} \quad (15.7)$$

where CN_w = weighted curve number, CN_1 = curve number of first sub-watershed, A_1 = area of first watershed having CN_1, CN_n = curve number of nth watersheds, A_n = area for the n_{th} watershed, and A = the total area of the watershed.

To compute the $CN_{WII(RS)}$ for AMCII condition of the whole study area, an overlay map was prepared by overlaying HSG and LULC using GIS software. After overlay, a map with new polygons representing the merged soil hydrologic group and land use was generated. The next step was to assign the appropriate CN value for each polygon of the map with the use of the SCS-CN table (Chow et al. 1988), and based on the map of HSG and map of land use classes, the resulting CN*s* were obtained. Finally, $CN_{wII(RS)}$ value for AMCII was computed using the zonal statistics tool of GIS software for the Tilaiya catchment (Table 15.2).

Later computed CN*s* were also adjusted based on the season and AMC. In this study, AMC is estimated based on daily rainfall data. AMC II (normal) is assigned

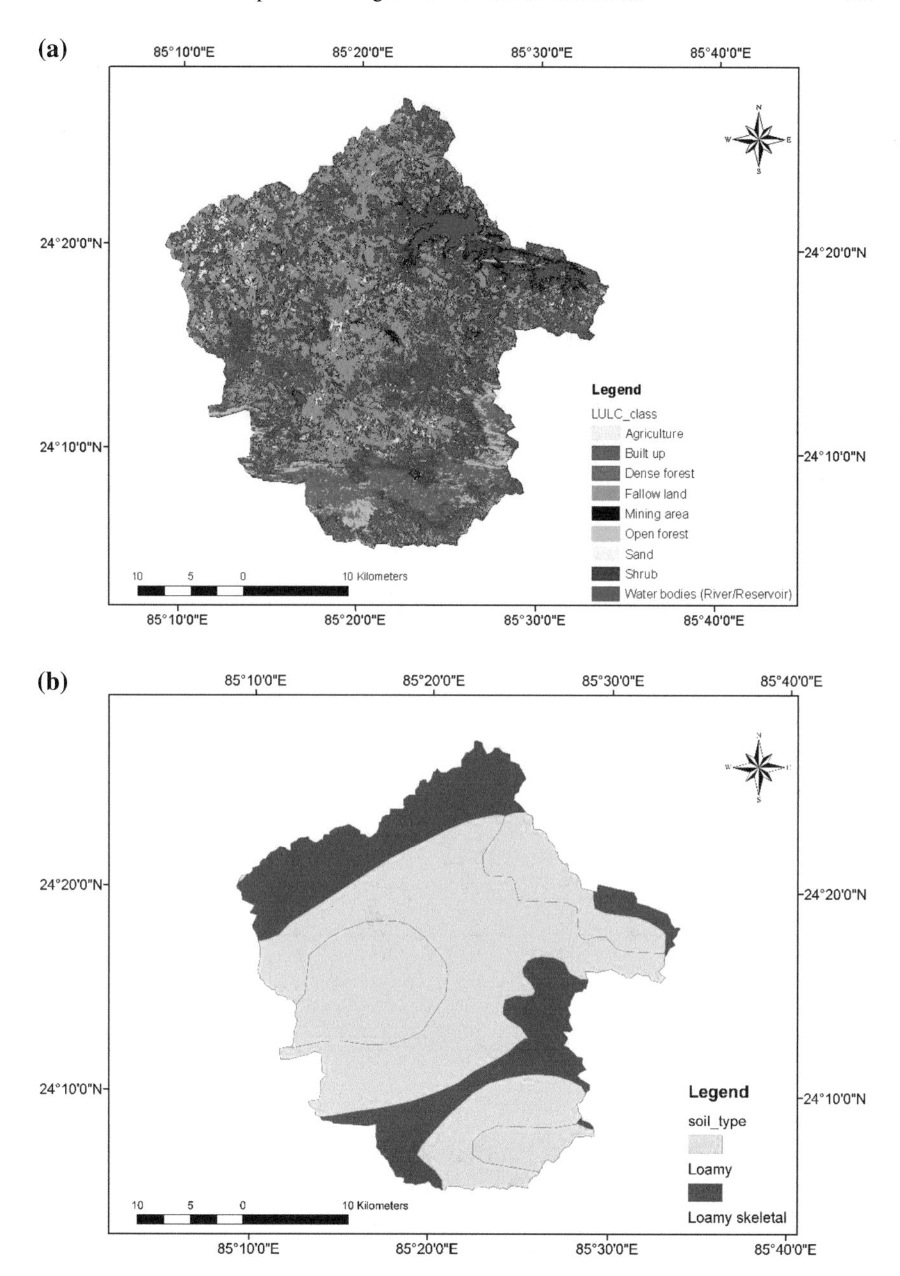

Fig. 15.3 **a** LULC use map of Tilaiya watershed. **b** Soil map of Tilaiya watershed

Table 15.2 LULC and their assigned CN values and computed CN_w for different AMC for the Indian conditions

S. No.	Land use	CN values according to HSG				Area (km^2)	CN_w
		HSG					
		A	B	C	D		
		Soil textural class					
		Sand, gravels	Shallow loess, sandy loam, loamy skeletal	Clay loams, shallow sandy loam	Soils having slow infiltration rates, soils		
1	Dense forest	26	40	58	61		$CN_{wI} = 52.44$ $CN_{wII} = 71.5$ $CN_{wIII} = 85.36$
2	Open forest	28	44	60	64		
3	Shrubs	33	47	64	61		
4	Agriculture (vegetable lands, crop fields, and other cultivated lands)	67	77	83	87		
5	Built up	77	86	94	98		
6	Fellows land, (barren land, open space)	77	86	91	93		
7	Sandy area	96	96	96	96		
8	Mining area	71	86	91	93		
9	Water bodies (river, pond and reservoir)	100	100	100	100		

for rainfall whose value is between 36 and 53 mm, whereas AMC I (dry) and AMC III (wet) for whose values less than 36 mm and greater than 56 mm, respectively.

Later, CN_{wII} was also adjusted for another two AMC: AMC I and AMC III using CN conversion formulae given by Mishra et al. (2008).

For AMC (I) and AMC (III), the CN_{wII} was adjusted using Eqs. (15.8) and (15.9), respectively.

$$CN_{w1} = \frac{CN_{wII}}{2.2754 - 0.012754 CN_{w2}} \tag{15.8}$$

$$CN_{wIII} = \frac{CN_{wII}}{0.430 - 0.0057CN_{w2}} \quad (15.9)$$

15.3.3 *Surface Runoff Calculation*

For assessing the study area's runoff potential, the first AMC condition was defined for each day based on five-day antecedent rainfall. The CN for AMC II and its corresponding CN*s* values were also calculated for each event. Using computed CN values, *S* was also calculated using Eq. 15.6. Finally, the runoff value was calculated using Eq. (15.5), validated for the Indian condition.

15.4 Results and Discussion

15.4.1 *CN* Estimation

Before estimating surface runoff in Tilaiya catchment at inflow site near Tilaiya reservoir, first CN*s* were estimated using the above-mentioned two approaches: (i) from *P*-*Q* data (CN_{PQ}) and (ii) RS and GIS-based NRCS-CN procedure (PQ_{RS}). Using the first approach, CN was computed for each 43 rainfall events (Table 15.3). Different CN*s* values ranged from 27.01 to 80.49 observed in the study area. Whereas annual mean CNs as representative annual CN ranged from 33.64 to 80.21, they were significantly different. It is to be noted here that temporal CN*s* variation is observed because 43 rainfall events characterize such as magnitude and spatiotemporal variation. This approach demonstrates advantages over the method in which a single CN value is often taken for the study area (traditional approach) for computing surface runoff. Results also support findings of the McCuen (2002) study that CN*s* are different for different events.

Using the second approach, $CN_{II(RS)}$ map was created for the study area (Fig. 15.4), which depicts spatial distribution of $CN_{II\ (RS)}$. It varies from 40 to 100. Like the previous approach, it is to be noted here that spatial CN*s* variation is found because major CN affecting factors such as daily spatiotemporal characterizes of rainfall such as magnitude and different watershed characterizes such as soil type, soil depth, AMC, land use, and HSG were considered. To compute the surface runoff of the study area, first, CN_{wII} value for AMC II was computed using the zonal statistics tool of GIS software as representative CN, and its value is observed 71.5. Later, its corresponding CN_{wI} and CN_{wIII} were also computed using Eqs. 15.8 and 15.9, respectively.

Table 15.3 CNs of the 43 rainfall events and annual mean CNs

S. No.	Event name	P (mm)	Q (mm)	CN	Mean CN	S. No.	Event name	P (mm)	Q (mm)	CN	Mean CN
1	2001_Evt1	74.14	7.903	66.59	53.07	23	2005_Evt4	81.3	9.26	33.60	33.64
2	2001_Evt2	191.04	16.50	41.82	53.07	24	2006_Evt1	223.55	66.14	80.44	33.64
3	2001_Evt3	134.43	14.30	52.36	53.07	25	2006_Evt2	123.55	11.05	33.64	33.64
4	2001_Evt4	54.31	2.45	67.97	53.07	26	2007_Evt1	74.0	18.84	75.78	74.57
5	2001_Evt5	182.67	40.18	53.07	53.07	27	2007_Evt2	110.0	41.52	74.57	74.57
6	2001_Evt6	95.14	2.36	51.98	53.07	28	2007_Evt3	96.1	24.69	70.80	74.57
7	2001_Evt7	96.74	8.00	58.33	53.07	29	2007_Evt4	65.3	15.75	77.35	74.57
8	2002_Evt1	316.8	64.54	38.76	52.83	30	2007_Evt5	137.0	33.17	62.01	74.57
9	2002_Evt2	118	14.47	56.95	52.83	31	2008_Evt1	302.13	27.16	71.07	80.21
10	2002_Evt3	143.6	47.23	66.39	52.83	32	2008_Evt2	282.23	25.38	63.60	80.21
11	2002_Evt4	182.85	28.43	48.70	52.83	33	2008_Evt3	376.27	33.83	81.94	80.21
12	2003_Evt1	98.45	9.96	59.56	47.33	34	2008_Evt4	135.31	12.17	87.06	80.21
13	2003_Evt2	222.7	19.99	38.45	47.33	35	2008_Evt5	644.28	57.93	64.50	80.21
14	2003_Evt3	144.7	6.56	44.36	47.33	36	2008_Evt6	233.19	20.97	82.26	80.21
15	2003_Evt4	124.35	8.03	50.31	47.33	37	2008_Evt7	455.74	40.97	79.93	80.21
16	2004_Evt1	208.6	38.75	47.70	47.33	38	2008_Evt8	113.76	10.23	80.49	80.21
17	2004_Evt2	77.4	8.01	65.19	57.20	39	2009_Evt1	74.50	1.569	57.42	39.78
18	2004_Evt3	219.4	54.03	50.77	57.20	40	2009_Evt2	179.43	9.16	39.78	39.78
19	2004_Evt4	118.5	25.54	63.64	57.20	41	2010_Evt1	57.24	2.54	27.01	39.78
20	2005_Evt1	156	7.10	41.36	57.20	42	2010_Evt2	65.27	1.91	65.43	56.22
21	2005_Evt2	44.4	2.96	44.88	44.88	43	2010_Evt3	127.73	5.33	47.00	56.22
22	2005_Evt3	125.5	8.72	64.57	44.88						

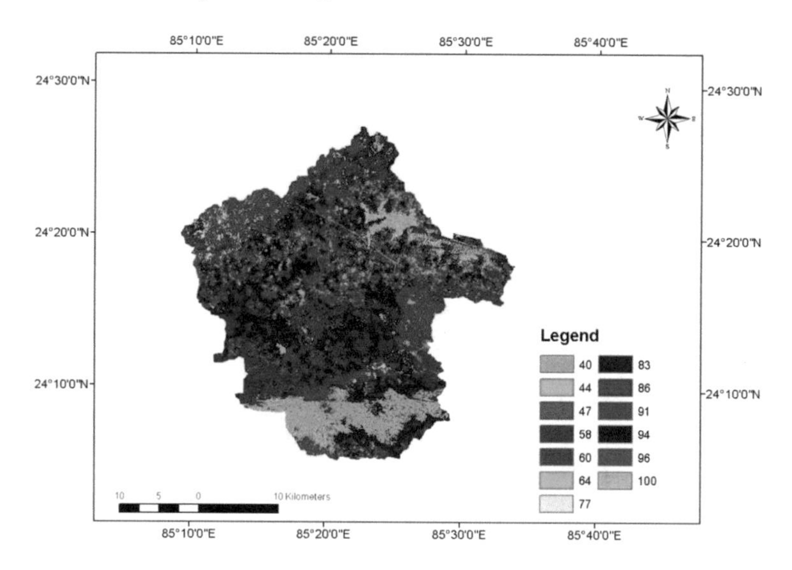

Fig. 15.4 Spatial distribution of $CN_{II(RS)}$ in Tilaiya watershed

15.4.2 Surface Runoff Estimation

However, in the second approach, CNw (CN_{WI} for AMCI, CN_{WII} for AMCII, and CN_{WIII} for AMCIII) and daily rainfall values were used to compute daily surface runoff in the study area. The parameter CN_w used in this approach was derived from remote sensing data. Using the CN_w value, *S* was computed. Furthermore, computed *S* values were used in Eq. (15.5), and daily runoff values were computed. Table 15.4 shows the daily and monthly computed runoff of the study area.

15.4.3 Validation of Used Approaches

To validate the computed surface runoff of the whole Tilaiya catchment using both approaches, it is compared with the observed runoff for the 43 rainfall events. The agreement between $Q_{computed}$ and $Q_{observed}$ is depicted in Fig. 15.5 with a correlation coefficient (R^2). The value of R^2, 0.61 for this case, indicates a satisfactory relationship between observed and computed runoff.

Figure 15.6 shows a comparison between observed and computed runoff by both approaches. Runoff estimated using CN*w* shows some difference from observed value in comparison with runoff estimate from the CN derived from *P*-*Q* data approach. This may be reason of high percentage of loamy soil (*C* type of HSG) predominantly covered throughout the study area, which is mainly comprised of built up (CN = 94) and fellow land (CN = 91). It is also evident that study area is also covered high runoff producing land use such as sandy area (CN = 91) and

Table 15.4 Sample of daily and monthly rainfall-runoff computation using GIS-based NRCS-CN

Date	Mean *P* (mm)	P_5 (cm)	AMC	CN_w	S	I_a	Daily *Q* (mm)	Monthly *Q* (mm)
1-Jul-01	3.28	0.328	III	85.36	43.87	13.16	0	
2-Jul-01	3.35	0.335	III	85.36	43.87	13.16	0	
3-Jul-01	11.03	1.103	III	85.36	43.87	13.16	0	
4-Jul-01	5.55	0.56	III	85.36	43.87	13.16	0	
5-Jul-01	1.26	0.13	III	85.36	43.87	13.16	0	
6-Jul-01	6.09	0.609	I	52.44	230.46	69.14	0	
7-Jul-01	0.4	0.04	I	52.44	230.46	69.14	0	
8-Jul-01	4.01	0.40	I	52.44	230.46	69.14	0	
9-Jul-01	3.4	0.34	I	52.44	230.46	69.14	0	
10-Jul-01	33.75	0.375	II	71.5	101.24	30.37	0.11	
11-Jul-01	12.83	1.28	III	85.36	43.87	13.16	0	
12-Jul-01	6.68	0.67	III	85.36	43.87	13.16	0	
13-Jul-01	3.04	0.30	III	85.36	43.87	13.16	0	
14-Jul-01	6.30	0.63	III	85.36	43.87	13.16	0	
15-Jul-01	15.80	1.58	III	85.36	43.87	13.16	0.15	
16-Jul-01	40.94	4.09	III	85.36	43.87	13.16	10.77	
17-Jul-01	28.841	2.88	III	85.36	43.87	13.16	4.13	
18-Jul-01	7.59	0.76	III	85.36	43.87	13.16	0	
19-Jul-01	0.69	0.07	III	85.36	43.87	13.16	0	
20-Jul-01	8.26	0.83	III	85.36	43.87	13.16	0	
21-Jul-01	22.98	0.23	III	85.36	43.87	13.16	1.79	
22-Jul-01	4.57	0.46	III	85.36	43.87	13.16	0	
23-Jul-01	7.40	0.74	II	71.5	101.24	30.37	0	
24-Jul-01	1.72	0.17	II	71.5	101.24	30.37	0	
25-Jul-01	5.40	0.54	II	71.5	101.24	30.37	0	
26-Jul-01	17.65	0.18	III	85.36	43.87	13.16	1.79	
27-Jul-01	3.26	0.33	II	71.5	101.24	30.37	0	
28-Jul-01	13.20	0.13	II	71.5	101.24	30.37	0	
29-Jul-01	14.22	0.14	III	85.36	43.87	13.16	0	
30-Jul-01	12.80	1.28	III	85.36	43.87	13.16	0	
31-Jul-01	7.33	0.73	III	85.36	43.87	13.16	0	
								18.74
1-Aug-01	1.06	0.11	II	71.5	101.24	30.37	0	
2-Aug-01	6.88	0.69	III	85.36	43.87	13.16	0	
3-Aug-01	5.97	0.60	II	71.5	101.24	30.37	0	

(continued)

Table 15.4 (continued)

Date	Mean *P* (mm)	P_5 (cm)	AMC	CN_w	S	I_a	Daily *Q* (mm)	Monthly *Q* (mm)
4-Aug-01	13.85	0.14	II	71.5	101.24	30.37	0	
5-Aug-01	1.04	0.10	II	71.5	101.24	30.37	0	
6-Aug-01	1.88	0.19	I	52.44	230.46	69.14	0	
7-Aug-01	0.72	0.07	I	52.44	230.46	69.14	0	
8-Aug-01	0.21	0.02	1	52.44	230.46	69.14	0	
9-Aug-01	7.61	0.76	1	52.44	230.46	69.14	0	
10-Aug-01	16.74	1.674	1	52.44	230.46	69.14	0	
11-Aug-01	1.11	0.1	1	52.44	230.46	69.14	0	
12-Aug-01	0.15	0.02	1	52.44	230.46	69.14	0	
13-Aug-01	0.012	0.001	1	52.44	230.46	69.14	0	
14-Aug-01	5.40	0.54	1	52.44	230.46	69.14	0	
15-Aug-01	27.18	2.72	2	71.5	101.24	30.37	0	
16-Aug-01	3.67	0.37	2	71.5	101.24	30.37	0	
17-Aug-01	3.33	0.33	2	71.5	101.24	30.37	0	
18-Aug-01	19.50	0.20	3	85.36	43.87	13.16	0.8	
19-Aug-01	30.15	3.02	3	85.36	43.87	13.16	4.7	
20-Aug-01	1.09	0.11	3	85.36	43.87	13.16	0	
21-Aug-01	0.224	0.22	3	85.36	43.87	13.16	0	
22-Aug-01	0	0	3	85.36	43.87	13.16	0	
23-Aug-01	2.26	0.23	3	85.36	43.87	13.16	0	
24-Aug-01	6.42	0.64	2	71.5	101.24	30.73	0	
25-Aug-01	22.74	0.23	1	52.44	230.46	69.14	0	
26-Aug-01	21.57	0.22	3	85.36	43.87	13.16	1.35	
27-Aug-01	2.89	0.29	3	85.36	43.87	13.16	0	
28-Aug-01	30.83	0.31	3	85.36	43.87	13.16	5.07	
29-Aug-01	22.50	0.25	3	85.36	43.87	13.16	1.64	
30-Aug-01	22.39	0.23	3	85.36	43.87	13.16	1.60	
31-Aug-01	21.07	2.11	3	85.36	43.87	13.16	1.19	
								16.35

mining area (CN = 91). These results also support many researchers' recommendation that CN*s* values computed from measured *P*-*Q* data is always better than CN*s* selected from the NEH tables.

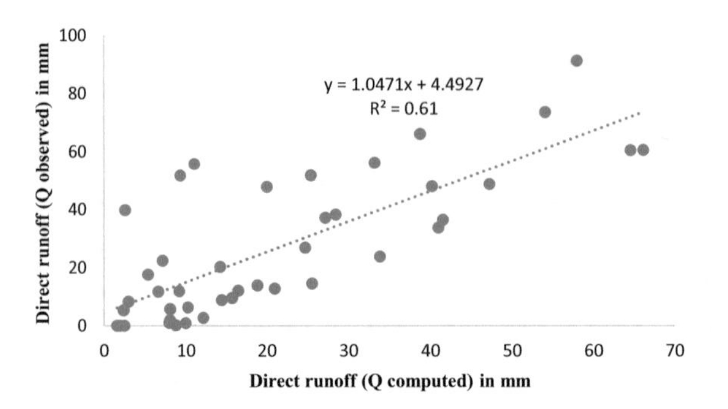

Fig. 15.5 Comparison of computed and observed runoff for a period (2001–2010)

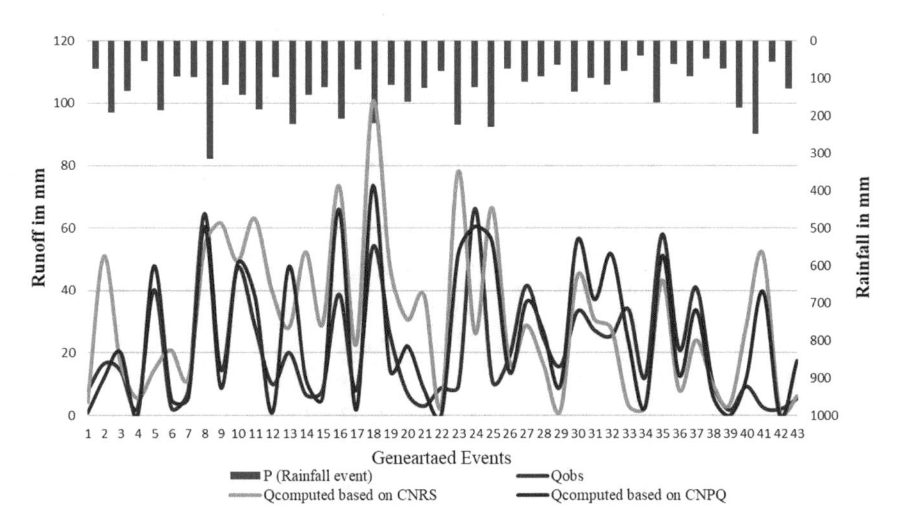

Fig. 15.6 Rainfall-runoff graph (observed versus computed) of the study area

15.5 Conclusions

The following conclusions can be derived from this study:

- In this study, various CN*s* were computed for three different AMC (I, II, and III) using two approaches: (i) from *P-Q* data and (ii) RS and GIS-based NRCS-CN method for estimating surface runoff in Tilaiya catchment.
- Using 43 derived multi-day events in the observed period 2001–2010 shows variation in CN*s* in the studied catchment, and its computed values vary from 27.01 to 80.49.
- RS data and GIS-coupled NRCS-CN approach facilitate estimation of spatially distributed CN within a watershed, and CN_w demonstrates advantages over the traditional method in which single CN value is often taken for a watershed.

- Spatially distributed CN varies from 40 to 100 with $CN_w = 71.5$ for AMC II condition, which shows remarkable differences in CN estimation as well as surface runoff computation.
- Runoff estimated values using both approaches show some difference; however, estimated runoff by using measured *P-Q* data can be used in water resource management effectively. Furthermore, research is needed to study of the effect of slope on CN and other empirical methods of CN estimation.

References

Anbazhagan S, Ramasamy SM, Das Gupta S (2005) Remote sensing and GIS for artificial recharge study, runoff estimation and planning in Ayyar basin, Tamil Nadu, India. Environ Geol 48:158–170. https://doi.org/10.1007/s00254-005-1284-4

Chow VT, Maidment DR, Mays LW (1988) Applied hydrology. Mc-Graw-Hill, New York

CWC (1997). Capacity survey of Tilaiya reservoir. A Report by Central Water Commission

Deshmukh SD, Chaube UC, Hailu AE, Gudeta DA, Kassa MK (2013) Estimation and comparison of curve numbers based on dynamic land use land cover changes, observed rainfall-runoff data and land slope. J Hydrol 49:89–101

Gajbhiye S, Mishra SK (2012) Application of NRSC-SCS curve number model in runoff estimation using RS & GIS. In: Proceedings of IEEE conference. http://ieeexplore.ieee.org/

Gajbhiye Meshram S, Sharma SK, Tignath S (2017) Application of remote sensing and geographical information system for generation of runoff curve number. Appl Water Sci 7:1773–1779. https://doi.org/10.1007/s13201-015-0350-7

Gupta PK, Panigrahy S (2008) Predicting the spatio-temporal variation of run-off generation in India using remotely sensed input and soil conservation service curve number model. Curr Sci 95(11):1580–1587

Hawkins RH (1978) Runoff curve number with varying site moisture. J Irrig Drainage Eng ASCE 104(4):389–398

Hjelmfelt AT Jr, Kramer KA, Burwell RE (1982) Curve numbers as random variables. In: Proceedings international symposium on rainfall-runoff modelling. Water Resources Publication, pp 365–373

Jasrotia AS, Dhiman SD, Aggarwal SP (2002) Rainfall-runoff and soil erosion modeling using remote sensing and GIS technique—a case study of Tons watershed. J Indian Soc Remote Sens 30(3):167–179

McCuen RH (2002) Approach to confidence interval estimation for curve numbers. J Hydrol Eng 7(1):43–48. https://doi.org/10.1061/(ASCE)1084-0699(2002)7:1(43)

Mishra SK, Singh VP (2003) Soil conservation service curve number (SCS-CN) methodology. Kluwer Academic Publishers, Dordrecht, The Netherlands. ISBN 1-4020-1132-6

Mishra SK, Pandey RP, Jain MK, Singh VP (2008) A rain duration and modified AMC dependent SCS-CN procedure for long duration rainfall- runoff events. Water Resou Manag 22:861–876

Moniruzzaman M, Thakur P K, Kumar P, Ashraful Alam M, Garg V, Rousta I, Olafsson H (2020) Decadal urban land use/land cover changes and Its impact on surface runoff potential for the Dhaka city and surroundings using remote sensing. Remote Sens 13:83. https://doi.org/10.3390/rs13010083

Moniruzzaman M, Thakur PK, Kumar P, Ashraful Alam M, Garg V, Rousta I, Olafsson H (2021) Decadal urban land use/land cover changes and its impact on surface runoff potential for the Dhaka City and surroundings using remote sensing. Remote Sens 13:83. https://doi.org/10.3390/rs13010083

Nagarajan N, Poongothai S (2012) Spatial mapping of runoff from a watershed using SCS-CN method with remote sensing and GIS. J Hydrol Eng 17(11):1268–1277

Nayak TR, Jaiswal RK (2003) Rainfall-runoff modeling using satellite data and GIS for Bebas river in Madhya Pradesh. J Inst Eng India 84:47–50

Pandey A, Sahu AK (2002) Generation of curve number using remote sensing and geographic information system. Water resources, Map This e-mail address is being protected from spambots. You need JavaScript enabled to view it India Conference. http://mapindia2002@gisdevelopment.net/

Pandey A, Dabral PP, Chowdary VM, Mal BC (2003) Estimation of run-off for agricultural watershed using SCS curve number and geographical information system. http://www.gisdevelopment.net

Patil JP, Sarangi A, Singh OP, Singh AK, Ahmad T (2008) Development of a GIS interface for estimation of runoff from watersheds. Water Resour Manage 22:1221–1239

Ponce VM, Hawkins RH (1996) Runoff curve number: has it reached maturity? J Hydrol Eng ASCE 1(1):11–19

Ragan RM, Jackson TM (1980) Runoff synthesis using Landsat and SCS model. J Hydraul Eng ASCE 106(5):667–678

Sahu RK, Mishra SK, Eldho TI, Jain MK (2007) An advanced soil moisture accounting procedure for SCS curve number method. J Hydrol Process ASCE 21:2827–2881

Thakur PK, Garg V, Kalura P, Agrawal B, Sharma V, Mohapatra M, Kalia M, Aggarwal SP, Calmant S, Ghosh S, Dhote PR (2020) Water level status of Indian reservoirs: a synoptic view from altimeter observations. Adv Space Res. https://doi.org/10.1016/j.asr.2020.06.015

Tiwari KN, Kumar P, Sibastian M, Paul K (1991) Hydrological modeling for runoff determination, remote sensing technique. J Water Resour Plan Manag 7(3):178–184

Tripathi MP, Panda RK, Pradhan S, Sudhakar S (2002) Runoff modelling of a small watershed using satellite data and GIS. J Indian Soc Remote Sens 30(1):39–52

USDA, SCS (1986) Urban hydrology for small watersheds. Technical Release No. 55, 2nd edn. Washington DC

Verma S, Singh A, Mishra SK, Singh PK, Verma RK (2018) Efficacy of slope-adjusted curve number models with varying abstraction coefficient for runoff estimation. Int J Hydrol Technol 8(4):317–338

Verma S, Singh PK, Mishra SK, Jain SK, Berndtsson R, Singh A, Verma RK (2018) Simplified SMA-inspired 1-parameter SCS-CN model for runoff estimation. Arab J Geosci 11:1–19

Verma S, Mishra SK, Verma RK (2020) Improved runoff curve numbers for large watersheds of the United States. Hydrol Sci J. https://doi.org/10.1080/02626667.2020.1832676

Verma RK, Verma S, Mishra SK, Pandey A (2021) SCS-CN based improved models for direct surface runoff estimation from large rainfall events. Water Resour Manage 35:2149–2175. https://doi.org/10.1007/s11269-021-02831-5

Vojtek M, Vojteková J (2016) GIS-based approach to estimate surface runoff in small catchments: a case study. Quaestiones Geographicae 35(3), Bogucki Wydawnic two Naukowe, Poznań, 97–116, 12

Zade M, Ray SS, Dutta S, Panigrahy S (2005) Analysis of runoff pattern for all major basins of India derived using remote sensing data. Curr Sci 88(8):1301–1305

Chapter 16
Assessment of Hydrologic Flux from the Haldi Catchment into Hooghly Estuary, India

S. K. Pandey, Chiranjivi Jayaram, A. N. V. Satyanarayana, V. M. Chowdary, and C. S. Jha

Abstract Monitoring the impact of hydrologic flux from river basin toward the coastal environment is useful for assessment of terrestrial impacts on coastal ecosystem. In this study, hydrologic flux from the Haldi catchment into Hooghly estuarine system was simulated for the period 1989–1994 using semi-distributed watershed model, the Soil and Water Assessment Tool (SWAT) integrated with ArcGIS, i.e., ArcSWAT (2009.93.5). Sensitivity and auto-calibration analysis was also carried out to investigate optimal model parameters, while global sensitivity was carried out using SWAT-CUP Sequential Uncertainty Fitting Algorithm (SUFI2) module. Prior to calibration, observed stream flow was partitioned into two major components namely surface runoff and base flow using an automatic recursive digital filter technique. Base flow ranges between 13 and 29% of the observed stream flow during the period 1990–94. A detailed uncertainty analysis was carried out to assess the performance of SWAT model over the study area. Higher surface runoff is inferred from the enhanced calibrated parameter CN2, whereas reduction in the soil evaporation compensation factor (ESCO) was observed. This signifies the increased extraction of water from the lower soil moisture profile that results in meeting the requirements of evaporative demand. Thus, the output of the model is substantially affected by soil water capacity (SOL_AWC), where lower SOL_AWC implies reduction of soil

S. K. Pandey (✉)
Indian Institute of Technology-Bhubaneswar, Bhubaneswar, Odisha, India

C. Jayaram
Regional Remote Sensing Centre—East, NRSC, ISRO, Kolkata, India

A. N. V. Satyanarayana
Indian Institute of Technology-Kharagpur, Kharagpur, West Bengal, India
e-mail: anvsatya@coral.iitkgp.ernet.in

V. M. Chowdary
Regional Remote Sensing Centre—North, NRSC, ISRO, New Delhi, Delhi, India

C. S. Jha
National Remote Sensing Centre, Hyderabad, India

A. Pandey et al. (eds.), *Geospatial Technologies for Land and Water Resources Management*, Water Science and Technology Library 103,
https://doi.org/10.1007/978-3-030-90479-1_16

retention capacity associated with increased surface runoff. Model performance indicator parameters such as R^2 and NSE show that the model performed satisfactory and good respectively at daily and monthly time scales.

16.1 Introduction

Monitoring the impact of hydrologic flux from river basins toward the coastal environment is useful for assessment of terrestrial impacts on coastal ecosystem. Assessment of flux into estuary from upper catchment can be simulated through models that characterize the dynamics of the physical, hydrological, and bio-geochemical processes. Conceptually, such models describe water fluxes and associated sediment and nutrient fluxes from the land surface to the outlet of the catchment. A large number of input parameters are required for modeling the hydrological processes that make the structure of modeling framework as complex. The temporal and spatial resolutions of the hydrological and biochemical processes play a vital role in defining the complexity of a given watershed simulation model. More specifically, the method to convert the excessive rainfall to runoff is generally analyzed using lumped models that assume the spatial homogeneity of the inputs, thereby introducing the errors while simulating rainfall-runoff process. Natural Resources Conservation Service-Curve Number (NRCS-CN) method is one of the most popular tools for computing runoff depth (Greene and Cruise 1995; USDA 1972) as inadequate data is a major limitation in hydrology for the quantitative description of the hydrologic processes.

Geospatial technologies enable the management and optimization of the inputs for the hydrological models that are spatially distributed. The spatiotemporally distributed natural resources data and physical terrain parameters acquired from the remote sensing (RS) platforms and subsequent analysis using Geographic Information Systems (GIS) could provide parameterization to hydrologic models. Efforts on modeling runoff from agricultural watersheds and its impacts on coastal environment are required from environmental perspective.

Hydrological and water quality models not only provide valuable insights on hydrological processes but also helps in the assessment of anthropogenic and climate effects on these processes. Several hydrological and soil erosion based studies were carried out using simple model such as USLE (Pandey et al. 2007) to others of more sophisticated in nature viz CREAMS (Knisel 1980), ANSWERS (Beasley and Huggins 1982), AGNPS (Young et al. 1987, 1989), SWRRB-WQ (Arnold et al. 1990), WEPP (Nearing et al. 1989), and SWAT model (Arnold et al. 1998) for understanding hydrological and erosion processes thus helping the conservation planner. Several researchers envisaged distributed parameter models such as SWAT, AGNPS, and WEPP, which has capability for assessing hydrologic and sediment flux at spatial and temporal scale. The distributed modeling could be able to simulate the heterogeneous rainfall distribution and the corresponding catchment properties that provide a better runoff hydrograph simulation. Studies pertaining to simulation of hydrologic processes were carried out using SWAT model in the watersheds of varying spatial scales, i.e., varied from field to river basin scales all over the globe (Abbaspour et al. 2007; Mango et al. 2011; Easton et al. 2010; Grunwald and Norton 2000; Gebremicael et al. 2013; Griensven et al. 2013; Brighenti et al. 2019). Arnold et al. (2012) reported various uses of SWAT and its applicability at various watershed scales. Although, SWAT model's stability and its usefulness was well documented, but calibration and uncertainty analysis is a major challenge (Abbaspour et al. 2018). Distributed parameter model, SWAT model, was selected in the present study for modeling hydrologic flux.

Haldi catchment that is located in the lower Ganga basin, West Bengal, India, is selected as a case study area. The study area, i.e., Haldi catchment, discharges substantial amount of hydrologic flux into Hooghly estuarine system influencing the ecosystem of the region. Hooghly estuary is the first deltaic offshoot of the Ganges–Brahmaputra–Meghna delta that flows through the marshy coastal regions. The present study is taken up with the specific objectives (i) to simulate runoff using a semi-distributed physical based watershed model, Soil and Water Assessment Tool (SWAT), (ii) to carry out sensitivity and auto-calibration analysis and to investigate optimal model parameters; and (iii) to conduct uncertainty analysis to evaluate the model performance in simulating the stream flow over the study region.

16.2 Study Region

The Haldi catchment is a part of the Damodar sub-basin of lower Ganges, located on the southern bank of the Damodar River (Fig. 16.1). Geographically the study area is located between 21.45° N–23.30° N latitude and 85.45° E–88.15° E longitude. Total catchment area is around 10,200 km^2 and spreads in Puruliya, Bankura,

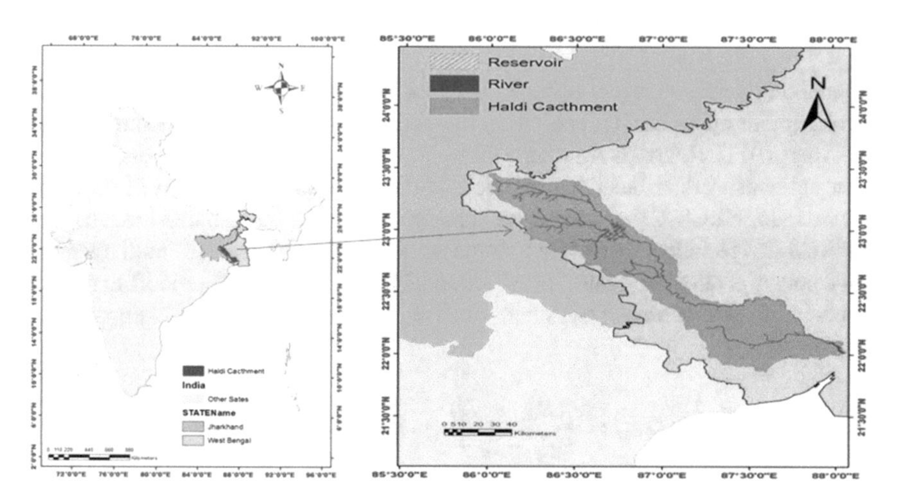

Fig. 16.1 Location map of study area

Pashchim Medinipur, and Purba Medinapur districts of West Bengal as well as Purbi Singhbhum district of Jharkhand. Haldi river is contributed by the confluence of three rivers namely Kansai, Kumari, and Tongo, which is called Kansabati river and joined by Kalaghai river in the downstream. The study area lies in tropical region and experiences all three seasons with minimum and maximum temperatures range between 10 and 40 °C during winter and summer, respectively. The study region has an average annual rainfall of ~1200 mm.

16.3 Model Description

SWAT model is a comprehensive, semi-distributed, conceptual, physically based continuous time, and watershed model, which operates on daily or sub-daily time steps capable of predicting the impact of soil and water conservation practices on sediment, agricultural, and chemical yields in large watersheds land use land cover (Arnold et al. 1998). Simulation of the hydrological processes could be grouped land phase and routing phase. Water, sediment, nutrient, and pesticide loading to the main channel of every sub-basin is controlled by land phase, while routing phase decides movement of water, sediment, etc., through the watershed channel network to watershed outlet. The computational components of SWAT are: hydrology, weather, sedimentation, soil temperature, crop growth, agricultural management, and reservoir routing. Most important operational component of SWAT model is hydrologic response unit (HRU), which is identified by integrating land use land cover (LULC), soil and land slope classes for the case study area.

16.3.1 Hydrology Component: Surface Runoff

Simulation of surface runoff using SWAT model incorporates two methods (i) SCS curve number approach (USDA 1972) (ii) Green–Ampt infiltration method (Green and Ampt 1911). Runoff is simulated for every hydrologic response unit and directed to the outlet using distributed hydrologic modeling. Due to unavailability of sub-daily rainfall data, NRCS-CN approach was used in this study. Fundamental assumption of NRCS-CN technique is that for a single storm, the ratio of actual rainfall retention to the potential maximum retention (S) is equal to the ratio of direct runoff to rainfall minus initial abstraction. Assessment of surface runoff using SCS-CN approach is given as follows:

$$\begin{aligned} Q_{\mathrm{surf}} &= \frac{(R_{\mathrm{day}} - I_a)^2}{R_{\mathrm{day}} - I_a + S'} \quad \text{when } R_{\mathrm{day}} > I_a \\ &= 0 \quad \text{when } R_{\mathrm{day}} \le I_a \end{aligned} \tag{16.1}$$

where *Qsurf* is accumulated runoff on rainfall excess, R_{day} rainfall depth at the day, *Ia* is initial abstraction, and *S* is retention parameter all in mm. S can be represented as function of curve number (CN) that is obtained from the following expression:

$$S = 25.4 \times \left(\frac{1000}{CN} - 10\right) \tag{16.2}$$

Curve number is a relative measure of retention of water by a given Soil–Vegetation–Land Use complex, and its values range between 0 and 100. CN is a function of soil permeability, antecedent soil water condition, and land use/land cover. Hence, antecedent moisture condition plays an important role, while selecting SCS-CN for a particular event. SCS defined three antecedent moisture conditions (AMC), where AMC I, II, and III refers to I-dry (wilting point), II-Average moisture, and III-wet (field capacity), respectively. Curve numbers for AMC I and AMC III conditions were calculated using CN_2 corresponding to AMC II condition and are given as follows:

$$CN_1 = CN_2 - \frac{20 \times (100 - CN_2)}{\left(100 - CN_2 + e^{[2.533 - 0.0636 \times (100 - CN_2)]}\right)} \tag{16.3}$$

$$CN_3 = CN_2 . e^{[0.636 \times (100 - CN_2)]} \tag{16.4}$$

where CN1, CN2, and CN3 are curve numbers for AMC I, AMC II, and AMC III moisture conditions, respectively.

16.4 Spatial Data Inputs for SWAT Model

16.4.1 Meteorological Data

SWAT model uses important meteorological parameters such as precipitation, air temperature, solar radiation, relative humidity, and wind speed for computing hydrologic flux that could be directly read from the file by the model. It could also derive the values using mean monthly data analyzed for a number of years. A weather generator model is incorporated that parallelly computes the precipitation for the day, followed the maximum and minimum temperature, solar radiation, and relative humidity based on the presence or absence of rain for the day and then wind speed. Indian Meteorological Department (IMD) rainfall gridded product at 0.25° × 0.25° was used in this study. Very high resolution daily gridded (0.25° × 0.25°) rainfall data for the Indian region was generated using the data collected from 6000 stations spread across the country (NCC 2008). In this study, daily rainfall values corresponding to four grid locations that lies within the catchment were selected.

16.4.2 Digital Elevation Model

Shuttle Radar Topography Mission (SRTM) Digital Elevation Model (DEM) at 90 m grid resolution (Jarvis et al. 2008) was used to generate topographic information of the catchment. Depressionless DEM is a prerequisite not only for delineating watersheds into sub-watersheds but also important for hydrological modeling. Hence, depressionless DEM was generated that was used for of watershed boundary delineation and stream network generation. Further, slope length, slope gradient, channel width, and length were estimated. Elevation and slope range between 0–647 m and 0–5%, respectively, in the study region, while nearly 50% of the total area is flat terrain (Fig. 16.2).

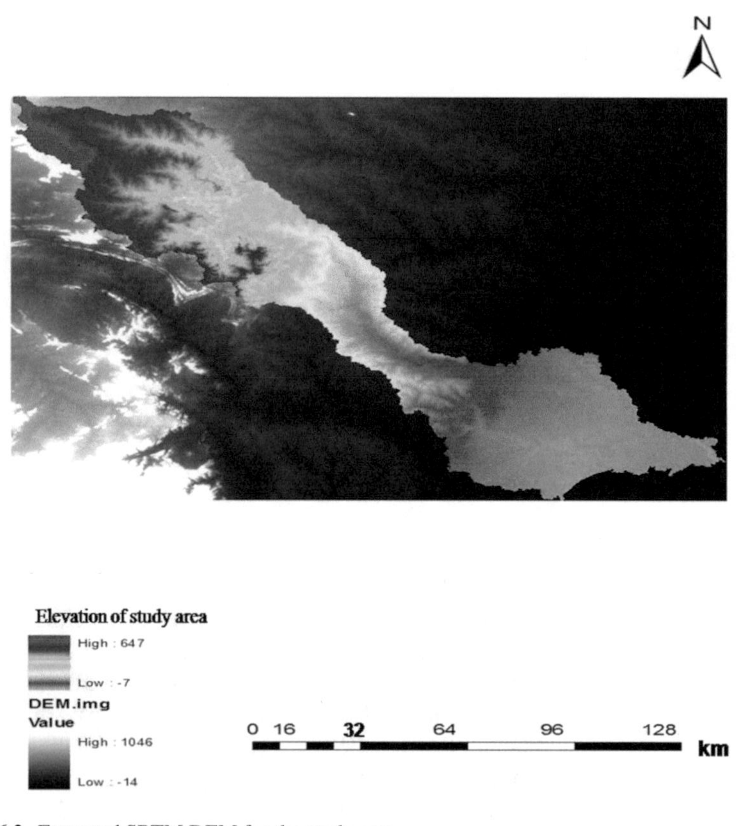

Fig. 16.2 Extracted SRTM DEM for the study area

16.4.3 Land Use/Land Cover

Land use/land cover map of the study area generated as a part of the National Land Use /Land Cover mapping project by National Remote Sensing Centre, India, for the year 2005 at 1:50,000 scale which was used in this study (Fig. 16.3). Further, major land use/land cover classes of the study area were reclassified into land use classes as

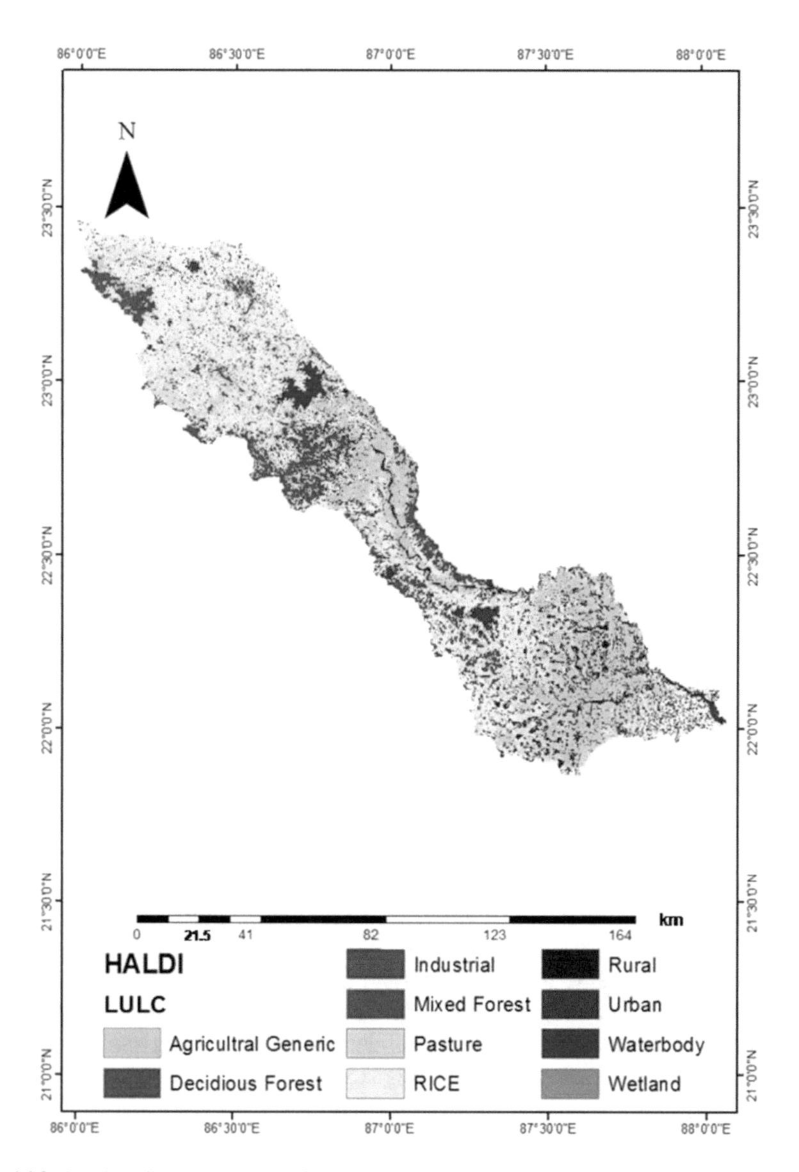

Fig. 16.3 Land use/land cover map of the study area

per SWAT model requirements. Area under paddy cultivation nearly occupies more than 50% of the study area during monsoon months. Nearly 24% of the study area is under other agricultural crops followed by 8% of forest area.

16.4.4 Soil Map

Soil map of the study area at 1:250,000 scale that is generated by the National Bureau of Soil Survey and Land Use Planning (NBSS and LUP) was used in this study (Fig. 16.4). Soil parameters such as hydrological conductivity, bulk density, and available water capacity corresponding to NBSS and LUP were obtained from the literature.

16.4.5 Observed Stream Discharge

Daily stream discharge data for the study area was obtained from Central Water Commission, Govt. of India, for the period 1991–1995. Location of stream gauging stations is shown in Fig. 16.5.

16.4.6 Watershed Delineation

Watershed delineator tool in ArcSWAT was used for generation of stream network from the DEM of the study watershed followed by watershed delineation. Eight-pour point algorithm was used to delineate the watershed that comprises the removal of peak, filling of sink, and the generation of gridded flow direction and accumulation from the DEM (Jenson and Domingue 1998). Depressionless DEM is used to calculate the flow direction from each pixel to one of the eight surrounding pixels (Fairfiled and Leymarie 1991). Subsequently, a flow accumulation grid is generated where each output grid cell indicates a cell value that identifies the number of grid cells contributing runoff to a particular grid cell. Pixel values that are greater than the defined threshold value as suggested by Mark (1983) are used to extract the drainage network. Sub-watershed can be automatically defined by giving outlets at the confluence of streams and also at user-defined pour point locations where water flows out of watershed area (Bhatt and Ahmed 2014). Sub-watersheds are defined as the incremental drainage area of each outlet, and in the present study, study area was delineated into 15 sub-watersheds (Fig. 16.5).

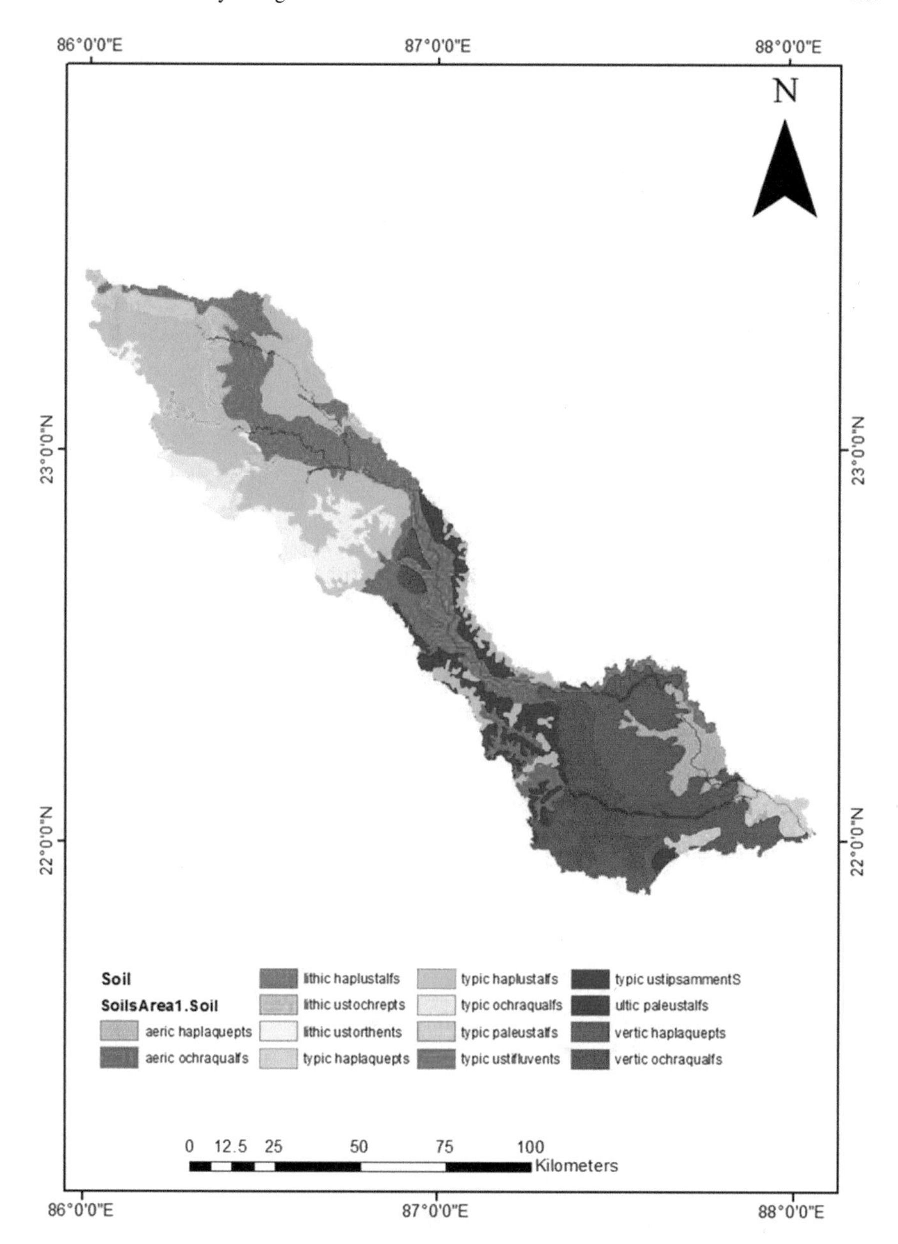

Fig. 16.4 Spatial distribution of soil textural classes

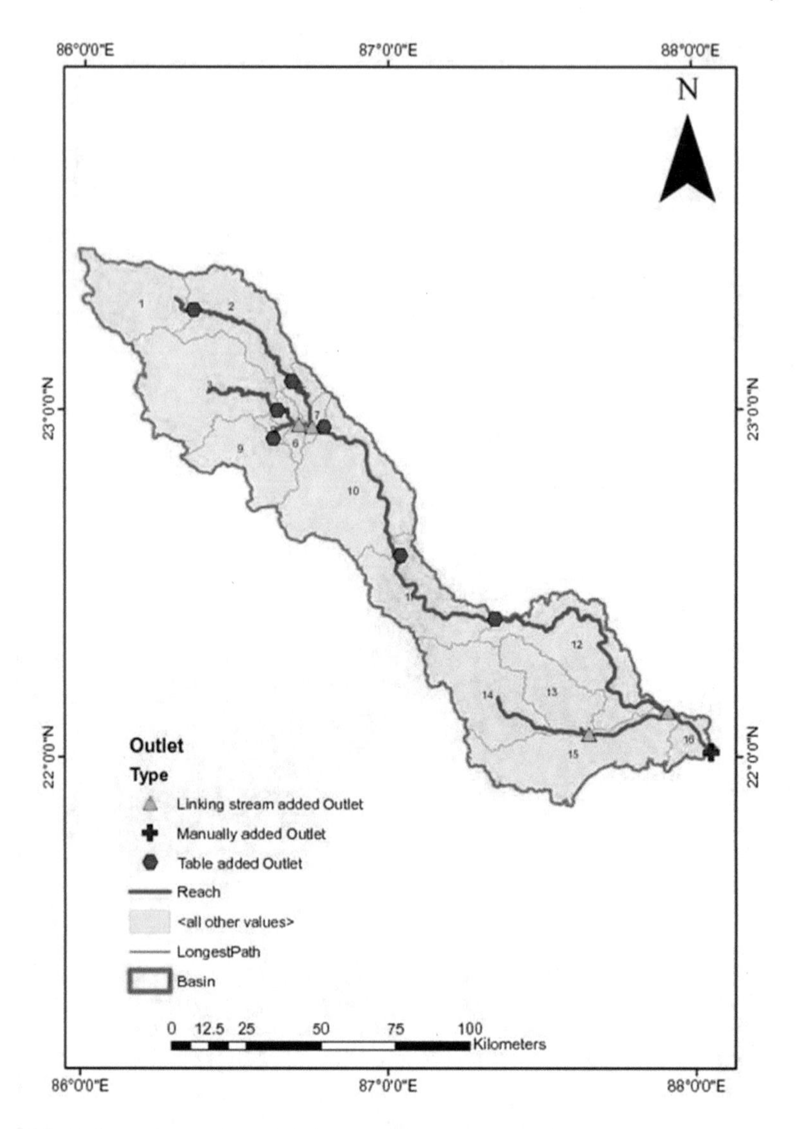

Fig. 16.5 Delineated watershed of study area with stream network

16.4.7 Generation of Hydrologic Response Unit (HRU)

HRU indicates unique combination of land use/land cover, soil, and topography that has hydrologically homogeneous properties and can be demarcated based on the percentage threshold value defined by user. In this study, 10%, 10%, and 5% threshold values are used for land use, soil, and slope, respectively. Grid cells having the similar soil and land use category within each sub-basin were grouped together and converted into polygons representing HRUs. Assigning threshold values to the land use and soil themes control the number of HRUs in each sub-basin. Once after the HRUs were created, the model allows the user to input the precipitation and the temperature data. Runoff is simulated for individual hydrologic response unit and directed toward the outlet, using distributed hydrologic modeling.

16.4.8 Model Inputs

The SWAT model allows changes in a large number of parameters such as soil data (.sol), weather generator data (.wgn). sub-basin general data (.sub), HRU general data(.hru), main channel data (.rte), groundwater data(.gw), water use data(.wus), management data(.mgt), soil chemical data(.chm), pond data(.pnd), and stream water quality data(.swq) for each sub-watershed. Further, selection of a particular method for runoff estimation and evapotranspiration can be made.

16.5 Simulation of Daily and Monthly Hydrologic Fluxes Using SWAT Model

Initially, model was simulated considering rainfall and temperature data for two years, i.e., 1989–90, 90–91 as warm up period, apart from the weather generator data file as specified by the SWAT manual. Manning's roughness coefficient of 0.014 and heat unit value of 1800 were adopted for all land use/cover classes within the watershed as per SWAT user manual. In the current work, runoff is estimated by following NRCS-CN method rather than the Green–Ampt infiltration method due to non-availability of rainfall data on a sub-daily scale. For evapotranspiration (ET) estimation, three methods, namely Hargreaves, Penmann–Monteith, and Priestly–Taylor methods, are presently available in the model. Kannan et al. (2006) reported that the CN method together with the Hargreaves ET estimation provided a better estimate of runoff compared to other combinations. Hence, Hargreaves ET method was considered in this study, while Muskingum method was used for stream flow routing.

16.5.1 Partitioning of Observed Stream Flow into Surface Runoff and Base Flow Components

Prior to calibration analysis, observed stream flow needs to be partitioned into two major components: surface runoff and base flow. Most of the analytical methods that were developed to delineate base flow from total stream flow (McCuen 1989) are based on the physical meaning, but still separation techniques are found to be subjective (Arnold et al. 1995). Manual separation of the stream flow hydrograph into surface and ground water flows is difficult, and often such estimates were not replicated by researchers (White and Sloto 1990). Hence, in this study, an automated base flow separation program (Arnold et al. 1995) was used following the automatic recursive digital filter method (Nathan and McMahon 1990). Lyne and Hollick (1979) originally used this particular recursive filter in signal analysis and processing. Although the technique has no true physical basis, it is objective and reproducible. Differentiating the surface runoff (high frequency signals) from base flow (low frequency signals) is similar to the separation of high frequency signals in signal analysis and processing (Arnold et al. 1995). The basic filter equation is given as follows:

$$q_t = \beta q_{t-1} + ((1 + \beta)/2) * (Q_t - Q_{t-1}) \tag{16.5}$$

where q_t is filtered surface runoff (quick response) at time t, Q_t is original stream flow, and β is filter parameter. Subsequently, base flow, b_t, can be calculated using the equation:

$$b_t = Q_t - q_t \tag{16.6}$$

A three time pass of the filter over the stream flow is made in forward, backward, and forward directions resulting three values of base flow. Arnold et al. (1995) suggested that the original base flow lies in between pass 1 and pass 3 and accordingly base flow resulting from pass 2 was chosen in the present work.

16.5.2 Global Sensitivity Analysis

Sensitivity analysis of SWAT parameters needs to be carried out prior to the calibration of the model to identify and rank the parameters that could impact the model outputs like streamflow and the sediment yield (Saltelli et al. 2000). Global sensitivity was carried out using SWAT-CUP Sequential Uncertainty Fitting Algorithm (SUFI2) module. In this analysis, multiple regression system, which regress with the Latin hypercube generated parameters against the objective function values. Previous studies by Easton et al. (2010), Betrie et al. (2011), Gebremicael et al. (2013), and lower and upper bound values of the SWAT model database were considered to obtain the initial parameters. P-values and t-test are used to identify the relative significance of each parameter on the output based on the linear approximations. Larger absolute value of t value is directly proportional to the sensitivity of that parameter while value of P close to zero has more significance.

16.5.3 Model Calibration Using SWAT-CUP

SWAT stream flow was evaluated using 20 various statistical methods by Coffey et al. (2004). SUFI2, GLUE, ParaSol, and MCMC are widely used approaches for uncertainty analysis and calibration. Sequential Uncertainty Fitting Algorithm (SUFI2) developed by Abbaspour et al. (2007) was applied in this study due to its ease of application with various sources of uncertainty that arise during the measurement of the variables, theoretical model, and parameter uncertainty. SUFI2 incorporates the Generalized Likelihood Uncertainty Estimation (GLUE) procedure (Beven and Binlley 1992) that include the gradient approach elements (Kool and Parker 1988) and were adapted to facilitate the wider search (Talebizadeh et al. 2010). This approach is employed on a set of parameters than the individual values to explicitly account for the interaction of the parameters. Further, SUFI2 requires only the least number of simulation experiments to arrive at better calibration and the uncertainty prediction (Yang et al. 2008).

16.5.4 Performance Evaluation of Arc SWAT Model for Stream Flow Simulation

Calibration and validation of ArcSWAT model was carried out in terms of statistical indicators P-factor, Nash–Sutcliffe coefficient (Nash and Sutcliffe 1970), and coefficient of determination (R^2) at daily and monthly time scales. These statistical indicators are defined as mathematical measures of how well a model simulation fits the available observation. NSE shows the confidence of 1:1, between the observed versus simulated values. The RMSE and R^2 indicates the association between the observed and model data (Alansi et al. 2009). The value of NSE ranges from $-\infty$ to 1 where 1.0 lies for perfect fit. The model is considered to be good provided the NSE $\geq$ 0:70 and satisfactory if 0:35 $\leq$ NSE $\leq$ 0:70 (Van Liew and Garbrecht 2003; Demissie et al. 2013). Apart from these two indicators, P-factor and R-factor introduced by Abbaspour et al. (2007a) were also used for testing the uncertainty analysis performance.

P-factor: P-factor is an indicator of the percentage of observations covered by the 95PPU. This is computed at 2.5 and 97.5% levels of the cumulative distribution that is a result of the Latin hypercube sampling after omitting the 5% of poor simulations.

R-factor: The mean distance d from the upper to the lower 95PPU along with the goodness of fit was obtained from uncertainty measures that are deduced from the percentage of measured data, which is in the 95PPU band:

$$R - factor = \frac{d_X}{\sigma_X}$$

R-factor should be less than 1.

When the 100% of the observations fall under the 95 PPU category with a d value that is ~0, then d is expressed as:

$$d_x = \frac{1}{k}\left(\sum_{l=1}^{k} (X_U - X_L)_l\right)$$

where k is the number of observed data points. Detailed procedure on SWAT-CUP SUF12 procedure is presented in Abbaspour et al. (2007a).

16.6 Application of ArcSWAT Model for Case Study Area

16.6.1 ArcSWAT Model Calibration and Validation

Study region was delineated into 16 sub-basins using SRTM DEM data (Fig. 16.5). The study area was divided into 211 HRUs, which are basic simulation units for SWAT model. Further, base flow was separated from the stream discharge for calibration and validation of the simulated model discharge.

16.6.2 Partitioning Base Flow Component

In this study, base flow was partitioned based on the discharge flow stream for the period 1991–1994 (Fig. 16.6). Base flow was quantified from the observed stream discharge for different filters and presented in Table 16.1. Arnold et al. (1995) suggested that the original base flow lies in between pass 1 and pass 3 and accordingly base flow resulting from pass 2 was chosen in this study. Base flow from filter 1 and filter 2 ranges between 31 to 47% and 13 to 29% of the observed stream flow, respectively, during the period 1990–94, which indicated higher range for filter 1.

16.6.3 Sensitivity Analysis

Sensitivity analysis helps in identifying the most sensitive parameters for stream flow simulation. The higher the relative sensitivity value of a parameter, the more sensitive it is toward model output. All the selected parameters are related to runoff and groundwater flow processes since the stream flow simulation by ArcSWAT model is influenced by both runoff and groundwater flow process parameters. Table 16.2 shows the sensitive parameters and their optimized values of the SWAT model to simulate flow

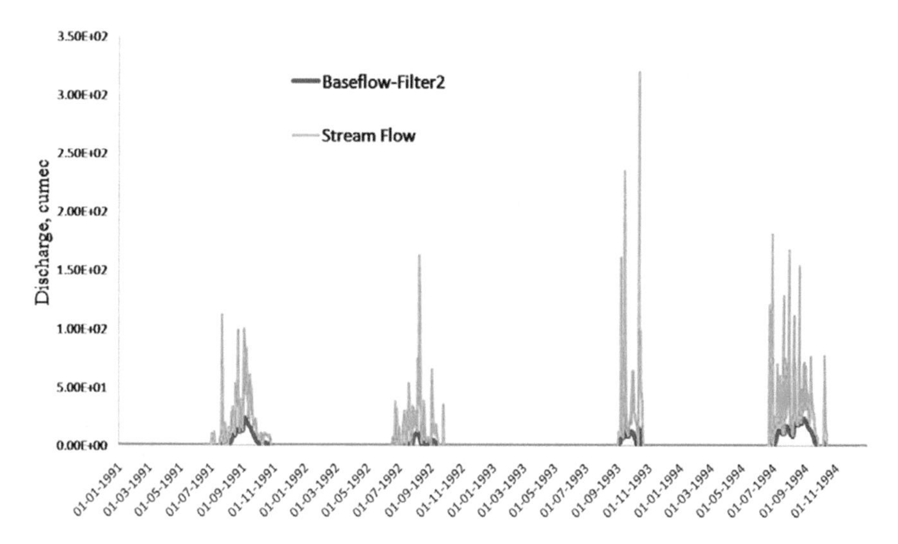

Fig. 16.6 Partitioning of base flow component from the stream flow

Table 16.1 Quantification of base flow component from observed stream flow (%)

Year	Base flow component (%)		
	Filter 1	Filter 2	Filter 3
1990–91	47	29	20
1991–92	31	13.4	6
1992–93	36.5	16.2	9
1993–94	45	29	21.6

The reduction of the soil retention capacity (SOL_AWC) and the subsurface runoff lag (SURLAG) and the other parameters like SCS curve number (CN2), base flow alpha factor (ALPHA_BF), soil evaporation compensation factor (ESCO), threshold water depth (GWQMN), deep aquifer percolation fraction (RCHRG_DP), and available capacity of water is lower for the flow predictions in the study area. From the table, it is seen that CN2 is the most sensitive parameter as absolute value of t-statistics is observed to be 46.67. According to Seibert and McDonnel (2010), such change indicates significant changes of the catchment response behavior. The change observed in the surface runoff parameters (e.g., CN2, ESCO, and SOL_AWC) is relatively high. Higher surface runoff is attributed to enhanced calibrated parameter CN2, whereas the decline in the soil evaporation compensation factor (ESCO) indicates the increased water extraction from the lower soil that could reach the demand of evaporation associated with the decline in the soil water. Thus, the output is affected by the available soil water capacity (SOL_AWC) significantly. The reduced retention capacity of the soil and subsequent increase in the surface runoff is inferred by the lower SOL_AWC value. Initially, physically plausible wide ranges were assigned

Table 16.2 Global sensitivity analysis statistics

Parameters name	*t*-state	*p*-value
r_CN2.mgt	−46.67	0.0
v_SOL_AWC (1).sol	−1.87	0.06
r_SLSUBBSN.hru	−1.48	0.13
v_REVAPMN.gw	−1.23	0.21
r_HRU_SLP.hru	−0.75	0.45
v_CH_N2.rte	−0.72	0.46
a_SOL_Z (1).sol	−0.57	0.56
v_EPCO.hru	0.038	0.96
v_GW_DELAY.gw	0.039	0.96
r_OV_N.hru	0.16	0.87
v_GWQMN.gw	0.44	0.65
v_ CH_K2.rte	0.64	0.51
v_SURLAG.bsn	0.65	0.51
v_ALPHA_BF.gw	0.73	0.45
v_ESCO.hru	0.82	0.40
v_SOL_ALB (1).sol	0.88	0.37
v_GW_REVAP.gw	1.044	0.29
r_SOL_K (1).sol	2.20	0.02

to the sensitive parameters. After performing recursive iteration, ranges were subsequently reduced in order to meet the objective function and final parameter ranges are presented in Table 16.3.

16.6.4 Calibration and Validation of the Model

Distributed watershed modeling systems often suffer from large uncertainties. Hence, uncertainties in the final model calibrated parameters were accounted by carrying out initial simulations for the period (1989–1990) as warm up period in order to account for aquifer recharge, which were not counted in the model efficiency. Hence, uncertainty band was reduced successively by increasing the number of iterations in SWAT-CUP. Further, SWAT model was calibrated for daily and monthly observed values of discharge data during calibration period (1991–1992). Subsequently, model was validated for the period 1993–1994 for both daily and monthly time scale. Flow hydrographs during model calibration period at daily and monthly time scales were shown in Figs. 16.7a–c and 16.8, respectively. These figures show the ability of the model to capture the flow pattern with some exceptional sudden fluctuations within the 95 PPU band, where final parameter values achieved after calibration of the model were used. The daily runoff hydrographs both for low and high flows were captured

Table 16.3 Parameter range during calibration

S. No.	Parameters name	Fitted value	Min.	Max.
1	r_CN2.mgt	−0.356	−0.500	0.400
2	v_SOL_AWC(1).sol	0.341	0.285	0.389
3	r_SOL_K(1).sol	−0.035	−0.047	0.268
4	v_SOL_ALB(1).sol	0.194	0.187	0.214
5	a_SOL_Z(1).sol	8.614	6.887	8.645
6	v_GWQMN.gw	3553.230	3546.136	3638.864
7	v_GW_REVAP.gw	0.144	0.099	0.147
8	v_REVAPMN.gw	181.208	160.078	185.278
9	v_ALPHA_BF.gw	0.849	0.724	0.906
10	v_GW_DELAY.gw	271.167	268.970	315.220
11	v_ESCO.hru	0.707	0.684	0.816
12	v_EPCO.hru	0.652	0.630	0.945
13	r_SLSUBBSN.hru	0.155	0.153	0.249
14	r_OV_N.hru	−0.022	−0.223	0.059
15	r_HRU_SLP.hru	0.184	0.086	0.260
16	v_CH_N2.rte	0.077	0.067	0.212
17	v_CH_K2.rte	139.522	99.790	145.538
18	v_SURLAG.bsn	20.204	20.109	24.0

by the model. Also, the model performances at monthly and daily scale during the calibration period were good and satisfactory, respectively, as per NSE (Table 16.4). The performance of the model was satisfactory and agreed with previous studies.

Observed and simulated stream discharges during the model validation period at daily and monthly time scales were shown in the Figs. 16.9 and 16.10, respectively. Base flow components were accounted in the stream flows that were observed during both calibration and validation. Parameters that establish the performance of the model, such as R^2 and NSE, show that the model performed satisfactorily and good, respectively, at daily and monthly time scales. Particularly, R^2 values show better agreement during calibration period compared to validation period during both the time scales. Model is said to be calibrated for monthly and daily simulation at least for the simulation which is said to be best by the model in the total 1000 run of the model with given range of parameters. But if we consider *P*-factor and *R*-factor, it is quite evident that these results are true for only a part of the total number of observations (e.g., 0.10 for daily simulation), while *R*-factor is below 0.5 that can be taken as satisfactory because it is less distant from observational value. Especially for daily simulation, one peculiarity was observed about the nature of the NSE and R^2. The NSE and R^2 are more sensitive toward the higher values, so larger values in a dataset are overestimated which leads to an overestimation of model performance. Therefore, model performance seems better statistically during calibration period and

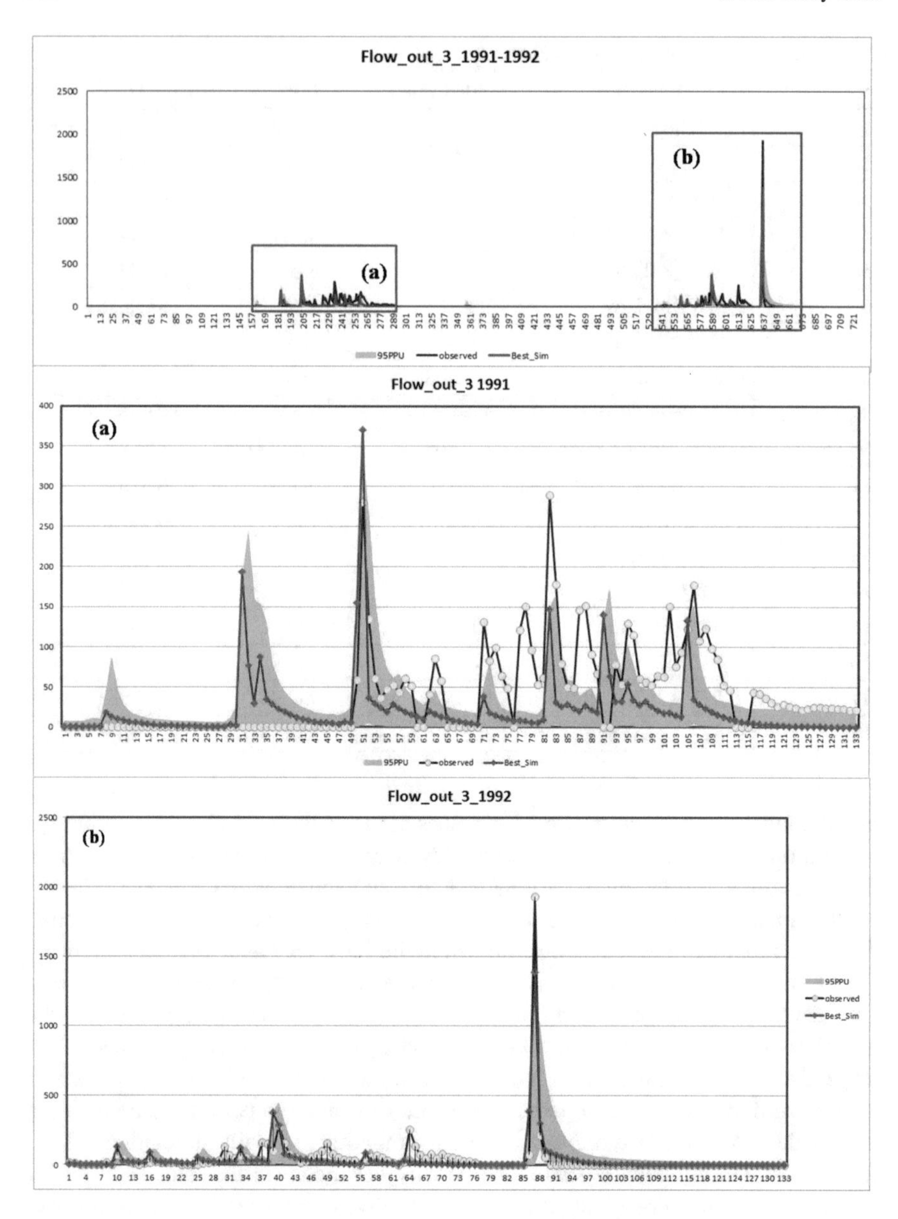

Fig. 16.7 Calibration of model **a** 1990–91, **b** 1991–92 with daily discharge data

to an acceptable value during validation period. However, model is well calibrated and validated for the monthly data series as far as R^2 and NSE are considered for the evaluation of model performance statistically. *P* and *R* factors indicated a similar trend while a greater number of observational data points are bracketed into the uncertainty band of thickness less than 0.5. However, the model is inadequate to

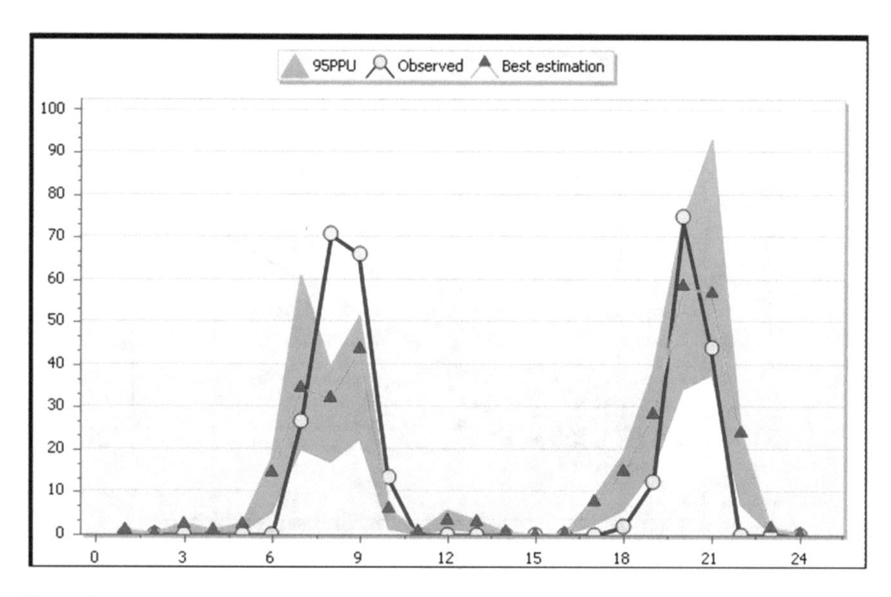

Fig. 16.8 Calibration of the model at monthly time scale

Table 16.4 Performance statistics for simulating stream flows during calibration and validation periods

Statistical parameter	Time step-daily		Time step-monthly	
	Calibration	Validation	Calibration	Validation
P-factor	0.10	0.06	0.25	0.14
R-factor	0.38	0.21	0.49	0.50
R^2	0.78	0.69	0.82	0.79
NSE	0.76	0.61	0.81	0.75

simulate detailed, single-event flood routing. Analysis shows that a poor correlation for daily scale in comparison with monthly time scale. In this study, contrary to several studies, stream flow was portioned into surface runoff and base flow for calibration and validation using automated base flow separation program (Arnold et al. 1995) using the automatic recursive digital filter technique following Nathan and McMahon (1990). It has been reported that SWAT daily simulation flow in general is not as good as monthly simulations (Peterson and Hamlet 1998; King et al. 1999; Spruill et al. 2000; Jain et al. 2010). The monthly total tends to smooth the data which increase the NSE. Hydrologic flux and sediment transport from the land to ocean via riverine system is a key link in the coastal ecosystem and is significantly influenced by the land use land cover (Wilk and Hughes 2002).

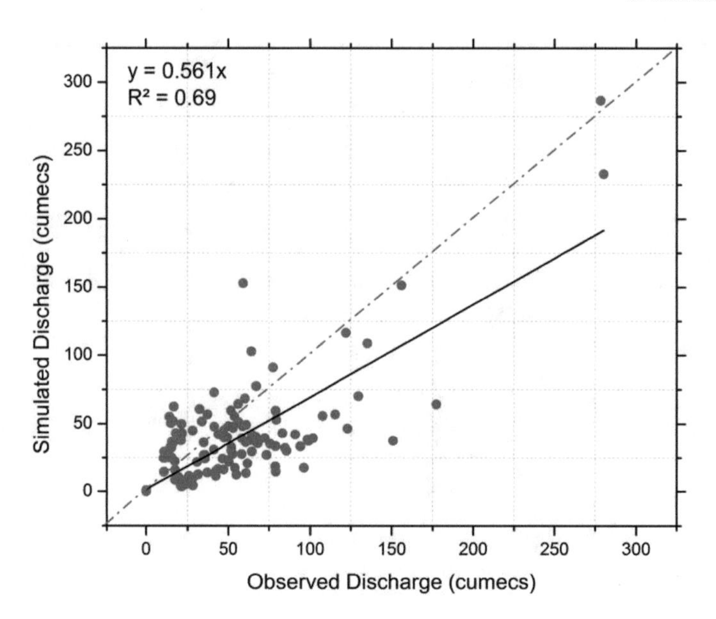

Fig. 16.9 Validation of the model with daily discharge data

16.7 Conclusions

Land cover and vegetation types in catchments also play a pivotal role in determining the runoff fluxes into the channel. Major source of discharge in the river is monsoonal rainfall. In this study, hydrologic flux from the Haldi catchment area to the Hooghly estuarine system during monsoon period of 1989–1994 was simulated using SWAT model (viz ArcSWAT 2009.93.5). The model simulated fluxes are validated with the available in situ observations. A detailed uncertainty analysis to assess the performance of SWAT model for simulation of runoff was carried out. Calibration of ArcSWAT model for Haldi catchment indicated that the model simulates hydrological characteristics of the study area satisfactorily. Flow component was partitioned from the observed stream discharge flow for the period 1991–1994 (Fig. 16.6). Base flow was quantified from the observed stream discharge using different filters, and base flow resulting from pass 2 was chosen in this study. Base flow from filter 1 and filter 2 ranges between 31 to 47% and 13 to 29% of the observed stream flow, respectively, during the period 1990–94, which indicated higher range for filter 1. By the mean of different statistical performance evaluators, it is observed that monthly mean hydrology is better represented than daily simulations. Uncertainty analysis has shown that the model is more sensitive to the sudden peak magnitudes of the rainfall simulations during the study period.

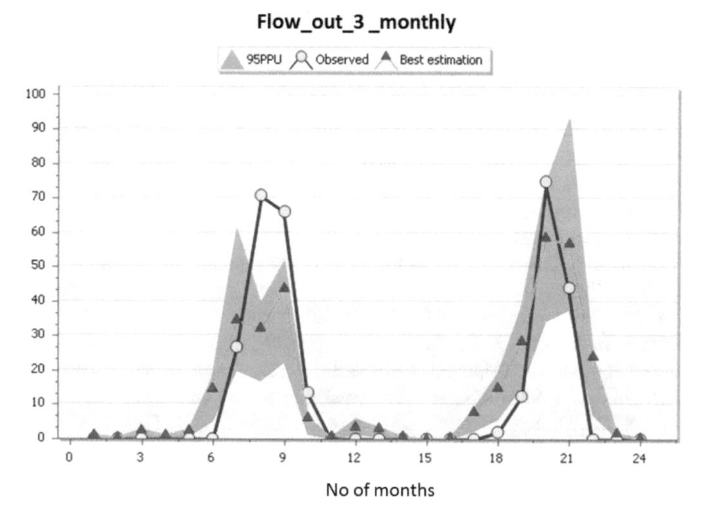

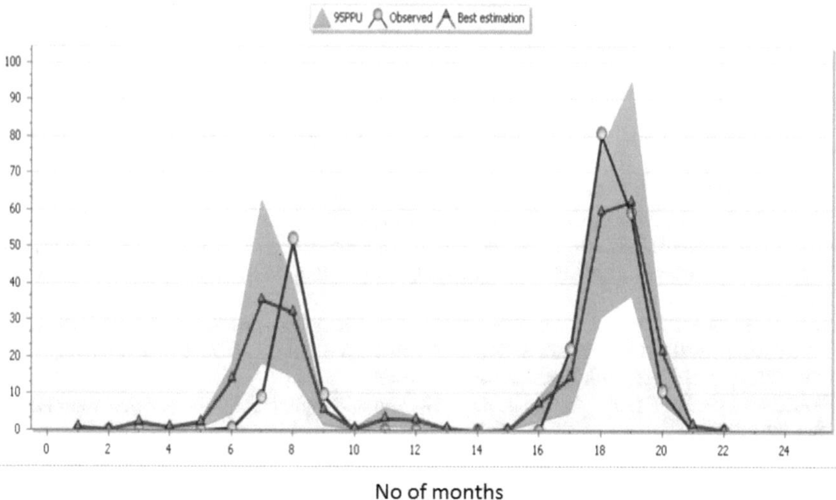

Fig. 16.10 Validation of the model with monthly discharge data

References

Abbaspour K, Yang J, Maximov I, Siber R, Bogner K, Mieleitner J, Zobrist J, Srinivasan R (2007) Modelling hydrology and water quality in the pre-alpine/alpine Thur watershed using SWAT. J Hydrol 333:413–430

Abbaspour KC, Vaghefi SA, Srinivasan R (2018) A guideline for successful calibration and uncertainty analysis for soil and water assessment: a review of papers from the 2016 international SWAT conference. Water 10(6):1–18

Abbaspour KC, Vejdani M, Haghighat S, Yang J (2007a) SWAT-CUP Calibration and Uncertainty Programs for SWAT. In: Oxley L, Kulasiri D (eds) MODSIM 2007 international congress on modelling and simulation, modelling and simulation society of Australia and New Zealand, pp 1596–1602

Alansi AW, Amin MSM, AbdulHalim G, Shafri HZM, Thamer AM, Waleed ARM, Aimrun W, Ezrin MH (2009) The effect of development and land use change on rainfall-10 runoff and runoff-sediment relationships under humid tropical condition. Eur J Sci Res 31(1):88–105

Arnold JG, Williams JR, Nicks AD, Sammons NB (1990) SWRRB: a basin scale simulation model for soil and water resource management. Texas A & M Univ. press, College station, TX

Arnold JG, Allen PM, Muttiah R, Bernhardt G (1995) Automated baseflow separation and Re and recession analysis techniques. Groundwater 33:1010–1018

Arnold JG, Moriasi DN, Gassman PW, Abbaspour KC, White MJ, Srinivasan R, Santhi C, van Harmel RD, Van Griensven A, Van Liew MW, Kannan N, Jha MK (2012) SWAT: model use, calibration, and validation. Trans ASABE 55(4):1491–1508

Arnold JG, Srinivasan R, Muttiah RS, Williams JR (1998) Large area hydrologic modeling and assessment. Part I: model development. J Am Water Resour Assoc 34(1): 73–89

Beasley DB, Hugins LF (1982) ANSWERS: Areal non-point source watershed environmental response simulation. In: User's manual. USEPA Rep. 905/9–82–001. USEPA, Chicago, IL

Betrie G, Mohamed Y, van Griensven A, Srinivasan R, Mynett A (2011) Sediment management modelling in Blue Nile Basin using SWAT model. Hydrol Earth Syst Sci 7:5497–5524

Beven K, Binley A (1992) The future of distributed models-model calibration and uncertainty prediction. Hydrol Process 6(3):279–298

Bhatt S, Ahmed SA (2014) Morphometric analysis to determine floods in the Upper Krishna basin using Cartosat DEM. Geocarto Int 29(8):878–894. https://doi.org/10.1080/10106049.2013.868042

Brighenti TM, Bonumá NB, Srinivasan R, Chaffe PLB (2019) Simulating sub-daily hydrological process with SWAT: a review. Hydrol Sci J 64(12):1415–1423. https://doi.org/10.1080/02626667.2019.1642477

Coffey ME, Workman SR, Taraba JL, Fogle AW (2004) Statistical procedures for evaluating daily and monthly hydrologic model predictions. Trans ASAE 47(1):59–68

Demissie TA, Saathoff F, Sileshi Y, Gebissa A (2013) Climate change impacts on the streamflow and simulated sediment flux to Gilgel Gibe 1 hydropower reservoir—Ethiopia. Eur Int J Sci Technol 2(2):63–77

Easton Z, Fuka D, White E, Collick A, Asharge B, McCartney M, Awulachew S, Ahmed A, Steenhuis T (2010) A multibasin SWAT model analysis of runoff and sedimentation in the Blue Nile Ethiopia. Hydrol Earth Syst Sci 14:1827–1841

Fairfiled J, Leymarie P (1991) Drainage network from grid digital elevation models. Water Resour Res 30:1681–1692

Gebremicael TG, Mohamed YA, Betrie GD, van der Zaag P, Teferi E (2013) Trend analysis of runoff and sediment fluxes in the Upper Blue Nile basin: a combined analysis of statistical tests, physically-based models and landuse maps. J Hydrol 482:57–68

Green WH, Ampt GA (1911) Studies on soil physics. J Agric Sci 4(1):1–24

Greene RG, Cruise JF (1995) Development of a geographic information system for urban watershed analysis. Photogramm Eng Remote Sens 62(7):863–870

Griensven AI, Popescu MRA, Ndomba PM, Beevers L, Betrie GD (2013) Comparison of sediment transport computations using hydrodynamic versus hydrologic models in the Simiyu River in Tanzania. Phys Chem Earth 61–62:12–21

Grunwald S, Norton LD (2000) Calibration and validation of non-point source pollution model. Agric Water Manage 45:17–39

Jain SK, Tyagi J, Singh V (2010) Simulation of runoff and sediment yield for a Himalayan watershed using SWAT model. J Water Resour Protect 2:267–281

Jarvis A, Reuter HI, Nelson A, Guevara E (2008) Hole filled SRTM for the globe version 4, available from the CGIAR-CSI SRTM 90 m Database. http://srtm.csi.cgiar.org

Jenson S, Domingue J (1998) Extracting topographic structure from digital elevation data for geographic information system analysis. Photogramm Eng Remote Sens 54:1593–1600

Kannan N, White SM, Worrall F, Whelan MJ (2006) Pesticide modelling for a small catchment using SWAT-2000. J Environ Sci Health B 41(7):1049–1070. https://doi.org/10.1080/03601230600850804

King KW, Arnold JG, Bingner RL (1999) Comparison of Green-Ampt and curve number methods on Goodwin creek watershed using SWAT. Trans ASAE 42(4):919–925

Knisel W (ed) (1980) CREAMS: a field scale model for chemicals, runoff and erosion from agricultural management systems. Conserv Res Rep 20. USDA-ARS, Washington, DC

Kool JB, Parker JC (1988) Analysis of the inverse problem for transient unsaturated flow. Water Resour Res 24(6):817–830

Lyne V, Hollick M (1979) Stochastic time variable rainfall runoff modeling. In: Hydrology and water resources symposium, Berth, Australia, 1979, Proceedings. National Committee on Hydrology and Water Resources of the Institution of Engineers, Australia, pp 89–92

Mango LM, Melesse AM, McClain ME, Gann D, Setegn SG (2011) Land use and climate change impacts on the hydrology of the upper Mara River Basin, Kenya: results of a modeling study to support better resource management. Hydrol Earth Syst Sci 15:2245–2258. https://doi.org/10.5194/hess-15-2245-2011

Mark DM (1983) Relations between field-surveyed channel networks and map-based geomorphometric measures, Inez, Kentucky. Ann Assoc Am Geogr 73:358–372. https://doi.org/10.1111/j.1467-8306.1983.tb01422.x

McCuen RH (1989) Hydrologic analysis and design. Prentice Hall, pp 355–360

Nash J, Sutcliffe J (1970) River flow forecasting through conceptual models part I—a discussion of principles. J Hydrol 10:282–290

Nathan RJ, McMahon TA (1990) Evaluation of automated techniques for baseflow and recession analysis. Water Resour Res 26(7):1465–1473

NCC (2008) A high resolution daily gridded rainfall dataset (1971–2005) for mesoscale meteorological studies. National Climate centre Research Report No. 9, pp 12

Nearing MA, Foster GR, Lane LJ, Finkner SC (1989) A process-based soil erosion model for USDA water erosion prediction project technology. Trans ASAE 32(5):1587–1593

Pandey A, Chowdary VM, Mal BC (2007) Identification of critical erosion prone areas in the small agricultural watershed using USLE. GIS Rem Sens Water Resour Manage 21:729–746

Peterson JR, Hamlet JM (1998) Hydrologic calibration of the SWAT model in a watershed containing Fragipan soils. J Am Water Resour Assoc 34(3):531–544

Saltelli A, Scott EM, Chan K, Marian S (2000) Sensitivity analysis. Chichester, Wiley & Sons

Seibert J, McDonnell JJ (2010) Land-cover impacts on streamflow: a change detection modeling approach that incorporates parameter uncertainty. Hydrol Sci J 55(3):316–332

Spruill CA, Workman SR, Taraba JL (2000) Simulation of daily and monthly stream discharge from small watersheds using the SWAT model. Trans ASAE 43(6):1431–1439

Talebizadeh M, Morid S, Ayyoubzadeh SA, Ghasemzadeh M (2010) Uncertainty analysis in sediment load modeling using ANN and SWAT model. Water Resour Manage 2010(24):1747–1761

USDA (1972) USDA soil conservation service. In: National engineering handbook

Van Liew MW, Garbrecht J (2003) Hydrologic simulation of the Little Washita River experimental watershed using SWAT. J Am Water Resour Assoc 39(2):413–426

White K, Sloto PA (1990) Baseflow frequency characteristics of selected pennsylvania streams. U.S. Geological Survey Water Resources Investigation Report 90-4160, 66 pp

Wilk J, Hughes DA (2002) Simulating the impacts of land use and climate change on water resource availability for a large south Indian catchment. Hydrol Sci J 47:19–30

Yang J, Reichert P, Abbaspour KC, Xia J, Yang H (2008) Comparing uncertainty analysis techniques for a SWAT application to the Chaohe Basin in China. J Hydrol 358(1–2):1–23

Young RA, Onstad CA, Bosch DD, Anderson WP (1987) AGNPS: agricultural non-point source pollution model. A watershed analysis tool. Conserv Res Rep 35. USDA-ARS, Washington, DC

Young RA, Onstad CA, Bosch DD, Anderson WP (1989) AGNPS: agricultural non-point source pollution model for evaluating agricultural watersheds. J Soil Water Conserv 44:168-–173

Chapter 17
Covariation Between LULC Change and Hydrological Balance in River Basin Scale

Nitesh Patidar, Pulakesh Das, Poonam Tripathi, and Mukunda Dev Behera

Abstract Land use and land cover (LULC) change has one of the key modes of human modifications and has raised several queries related to its impact on hydrology and climate. Since the LULC plays a vital role in partitioning energy and water fluxes at the land surface into different components, the changes in LULC can impact the water and energy cycles to a significant level. However, at a basin scale, the severity of such consequences depends on the scale, type, and heterogeneity of LULC changes. An assessment of the impacts of LULC change on hydrology in three major river basins—Brahmaputra, Ganga, and Mahanadi of India is performed in this study using the variable infiltration capacity (VIC) model, a physically-based distributed hydrological model. The assessment reveals an important compensating effect in the hydrologic changes resulted from LULC transformations. The negative consequences (e.g., increased surface runoff due to urbanization) at one place are compensated by opposite positive changes (e.g., decreased surface runoff due to forest plantation) at other places at the basin scale. Such compensation leads to insignificant hydrologic changes at the basin scale; however, the consequences are significant at the local and sub-basin scales.

N. Patidar
National Institute of Hydrology, Roorkee, India

P. Das
Sustainable Landscapes and Restoration, World Resources Institute India, Delhi, India

P. Tripathi
International Centre for Integrated Mountain Development, Kathmandu, Nepal

M. D. Behera (✉)
Centre for Oceans, Rivers, Atmosphere and Land Sciences (CORAL), Indian Institute of Technology Kharagpur, Kharagpur, India
e-mail: mdbehera@coral.iitkgp.ac.in

A. Pandey et al. (eds.), *Geospatial Technologies for Land and Water Resources Management*, Water Science and Technology Library 103,
https://doi.org/10.1007/978-3-030-90479-1_17

17.1 Introduction

The land use land cover (LULC) is a boundary layer between the earth's subsurface and the atmosphere, which governs the partitioning of energy and water fluxes at the land surface. With the increased demand for land and urban infrastructures to cater increasing population, the natural LULC has been transformed into managed lands such as agriculture and builtup (Li et al. 2015; Patidar and Keshari 2020; Tsarouchi et al. 2014). Such modifications have led to significant physiological variations due to altered vegetation type, coverage, and density (Jacobson 2011; John et al. 2016). Conversion of LULC types results in altered canopy coverage, rooting depth, albedo, surface roughness, and biomass. Moreover, the LULC changes modify soil compactness, porosity, infiltration capacity, hydraulic conductivity, and organic contents of the topsoil layer (Ghimire et al. 2014; Liu et al. 2002). Consequently, the partitioning of incoming solar radiation into sensible and latent heat fluxes, precipitation to runoff and infiltration, wind speed, soil moisture, and heat storage capacity is changed (Mishra et al. 2010; Patidar and Behera 2019). These direct and indirect impacts of LULC change can negatively affect the hydrology and climate at local and regional scales (Bosmans et al. 2017; Sterling et al. 2013). A few severe alterations in hydrology and climate induced by LULC changes observed at different parts of the world include (1) increased surface runoff and flood frequency due to deforestation and urbanization (Schilling et al. 2010; Yang et al. 2010), (2) increased evapotranspiration (ET) due to agriculture expansion (Jaksa and Sridhar 2015; Lei et al. 2015), (3) changes in baseflow contribution to rivers due to deforestation (Das et al. 2017; Peña-Arancibia et al. 2019), (4) decreased groundwater recharge due to increased impervious surfaces associated with urbanization (Minnig et al. 2018), and (5) decreased diurnal temperature due to urbanization (Kalnay and Cai 2003). Although, there are mixed opinions on the consequences of LULC changes such as on the baseflow, temperature, and precipitation.

Since the LULC plays a vital role in exchanging water between the surface and the subsurface through infiltration and root water uptake, the hydrologic impacts' assessment became important during the past few decades. Rapid urbanization and agriculture expansions made the LULC change an essential consideration in water resources assessment and allocation, flood and drought forecasting, future planning for sustainable water management, etc. LULC changes can affect water availability by altering streamflow and groundwater (Dawes et al. 2012; Schilling et al. 2010). The impervious surfaces can increase the surface runoff and decrease groundwater recharge by blocking the infiltration, leading to short-term increased streamflow during the rainy season, and decreased streamflow and groundwater during the dry season. Such changes also lead to increased flood magnitude and frequency (Bloschl et al. 2007). Similarly, deforestation results in increased surface runoff due to decreased canopy interception and ET. The consequences of LULC changes are serious when the hydrologic responses of the transformed LULC are considerably different from that of the previous LULC distribution. Generally, the conversion of forest to settlement is the most sensitive to affect hydrology; although, the magnitude of such impacts varies due to other factors such as topography and climate (Patidar and Behera 2019). The precipitation characteristics may alter the sensitivity of the LULC changes to its effects on hydrology, where the impacts are more prominent in the areas with uniform temporal distribution (Wagner et al. 2013). It implies that hydrologic changes could be different even with similar LULC transformations in different basins.

The Indian River basins mostly receive precipitation through the southwest monsoon. The spatio-temporal distribution of precipitation in India is highly erratic, which induces variable hydrologic responses in the different river basins. Moreover, due to high heterogeneity, the multidirectional changes are evident in large Indian River basins. Therefore, assessing the sensitivity and impacts of LULC change to hydrology under Indian climatic conditions observed during the past few decades is necessary for designing effective policies for sustainable water management. The key objectives of this study were to investigate the sensitivity of various LULC transformations and assess their impacts on the surface runoff and ET in three major Indian River basins as Ganga, Brahmaputra and Mahanadi, and east-flowing river basins.

17.2 Study Area and Data Used

The study area includes three major river basins–Ganga, Brahmaputra, and Mahanadi–and other associated eastern flowing river basins such as Brahmani-Baitarani, Subarnarekha, Nagavali, and Vamsadhara of India (Fig. 17.1). The assessment was done only for the Indian region of the basins having a total geographic area of 10, 26, 938 km^2, covering approximately 31% of the total geographic area of India. The Ganga–Brahmaputra basin extends from Himalaya in the north to the Bay of Bengal in the south covering diverse topography, LULC, and climate. The elevation ranges of the study area vary between 0 and ~7322 m (above m.s.l). The climate condition varies from per-humid in the southeast to arid in the east (Raju et al. 2013). The climate shows significant variations in space and time, with a daily average temperature varying between −5 °C and 27 °C. The precipitation mostly occurs during mid-June to September, and the total annual ranges from 543 mm (in the west) to more than 2300 mm (in the northeast). The natural vegetation varies with latitude from deciduous to evergreen forests.

The Mahanadi and east-flowing river basins, hereafter referred to as MEB, are bounded by Ganga in the north, the Bay of Bengal in the east with a complex drainage system near the coast. The average annual precipitation in MEB is ~1360 mm, with an average temperature ranging from 4 °C (winter) to 42 °C (summer). The monsoonal precipitation accounts for ~86% of total annual precipitation in MEB.

Various remote sensing and in-situ datasets were used in this study. The gridded precipitation and temperature data generated by the India Metrological Department (IMD) were used. These datasets are daily gridded precipitation (0.25°) and temperature (1°) derived from an extensive network of observation sites (Pai et al. 2014). The altitude and related information, such as slope and river networks, were extracted

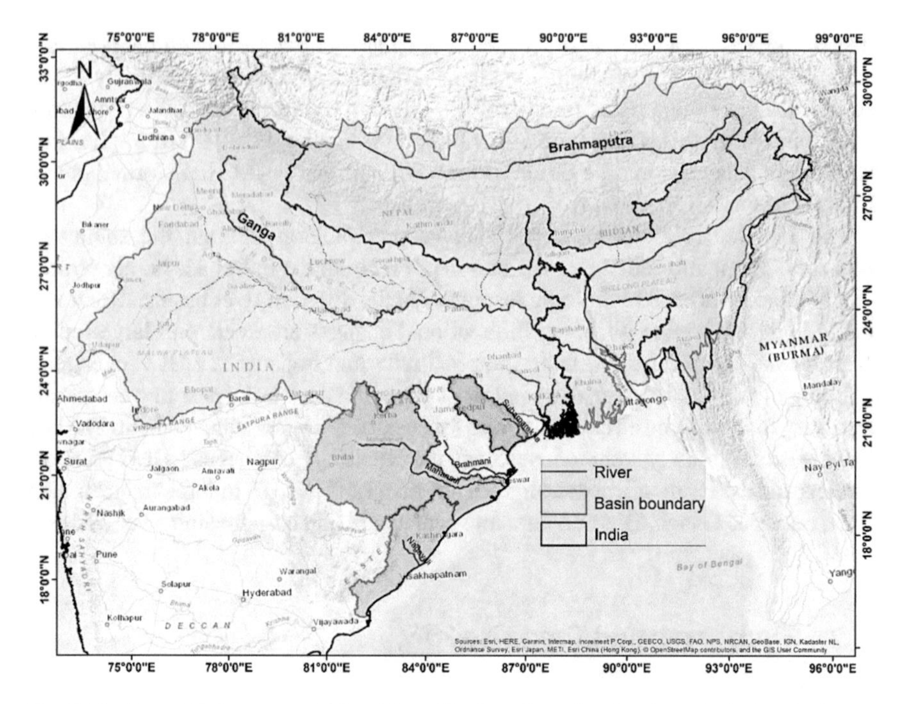

Fig. 17.1 Ganga, Brahmaputra, Mahanadi, and basins of east flowing rivers including Brahmani-baitarani, Subarnarekha, Nagavali, and Vamsadhara. The study area is the Indian part of these river basins

from the shuttle radar topographic mission (SRTM) digital elevation model (DEM). The soil maps of the Food and Agriculture Organization (FAO) and the National Bureau of Soil Survey and Land Use Planning (NBSS and LUP), India were used. The vegetation parameters, such as leaf area index (LAI) and albedo, were extracted using moderate resolution imaging spectroradiometer (MODIS) LAI and albedo data product. The stream discharge data of the Central Water Commission (CWC) were used for model calibration and validation in the Mahanadi River basin.

The LULC maps derived by Behera et al. (2014) for Ganga, Roy et al. (2015) for the Brahmaputra, and Das et al. (2020) for MEB were used. The LULC maps for Ganga were available for the years 1975 and 2010, and for Brahmaputra and MEB for the years 1985 and 2005.

17.3 Methodology

The hydrologic impacts of LULC change can be assessed either using long-term in-situ observations or hydrological modeling. Although the former method provides a more realistic scenario, it is rarely used due to the unavailability of long-term observations and the influence of other factors like climate variability. The observed changes in hydrologic fluxes need to be precise to get convincing information about the changes (Ashagrie et al. 2006). On the other hand, hydrological modeling provides a systematic way of estimating the impacts of LULC changes. This method is commonly used to simulate the differential effects of climate conditions and LULC changes on hydrologic parameters; wherein, the accuracy is subjected to the precision of inputs, parameters, and efficacy of the model.

Here, the variable infiltration capacity (VIC) hydrological model is used. Simulations were performed taking the LULC maps of two time periods keeping the other variables and parameters unchanged. The LULC map of 1975 for Ganga and 1985 for Brahmaputra and MEB was used in the first simulation, while the LULC map of 2010 for Ganga and 2005 for Brahmaputra and MEB was used in the second simulation. The simulated ET and runoff were analyzed for quantitative and comparative assessment. This method assures that the change in ET and runoff is isolated from human interventions as water diversions, storages, and climate variability. The following sub-sections elaborate on the working principle of the VIC model and its implementation in the study area.

17.3.1 VIC Model Description

The VIC model is developed by the University of Washington and is based on the concept of variable infiltration curve developed by Ren-jun et al. (1980). The model has evolved substantially since its development in 1990's and has been tested in various river basins throughout the world (Gao et al. 2009; Ren-jun et al. 1980). VIC is a grid-based model in which the simulations for each grid are performed independently, and a routing model is coupled externally to derive the streamflow (Lohmann et al. 1998b). VIC is capable of handling the sub-grid heterogeneity by considering the proportions of LULC within the grid cells, which make it suitable for assessing the impacts of LULC change. This model also includes multiple soil layers for a better simulation of soil moisture and water infiltration. The other inputs, such as meteorological and topographic variables, are also provided at a grid scale. The ET is simulated using the Penman–Monteith equation. Thus, the ET is sensitive to any change in vegetation parameters such as LAI, albedo, root depth, and stomatal resistance. The parameters for respective LULCs within the grids are read from the look-up table. ET over a grid cell is simulated as the sum of the evaporation from the canopy layer of the nth vegetation tile, transpiration from the nth vegetation tile, and evaporation from the bare soil weighted by the respective surface cover fractions. The canopy evaporation from the nth vegetation tile is estimated as a function of potential ET and other resistances weighted by the vegetation fraction. A few additional factors on canopy resistance and soil moisture are added to estimate transpiration from the respective vegetation tile. While the runoff is estimated using the ARNO model (Todini 1996).

17.3.2 *Assessment of the Hydrologic Impacts Using VIC Model*

The model was run to simulate ET and runoff under two LULC scenarios, and change was detected by subtracting the fluxes. The overall schematic of the methodology is shown in Fig. 17.2. A two-layer model was developed, wherein the soil parameters were defined using the soil texture information from NBSS and LUP (for the top layer) and FAO (for the bottom layer) soil texture maps. The soil hydraulic properties, such as saturated hydraulic conductivity, field capacity, and wilting point, for each soil texture were defined using the index provided by Cosby et al. (1984). Vegetation parameters are based on the Global Land Data Assimilation System (GLDAS) as described by Rodell et al. (2004). The LAI value, which is the most sensitive vegetation parameter to alter water balance in the VIC, is extracted from the moderate resolution imaging spectroradiometer (MODIS) LAI data product [MCD15A3] (Patidar and Behera 2019). The vegetation parameters are provided on a monthly scale irrespective of the soil type. The changes in LULC are addressed by changing the LULC fractions under each scenario while keeping the same parameter table (Fig. 17.2).

The simulations in each grid are performed independently in VIC and induce no flux exchange from/to the neighboring grids. The grids are interconnected only during the post-processing, where the VIC-simulated runoff is estimated at the basin outlet using the routing module (Lohmann et al. 1998a). The simulations can be done without considering the entire basin, although the entire catchment is required when estimating streamflow. Therefore, the VIC was suitable for this study as only

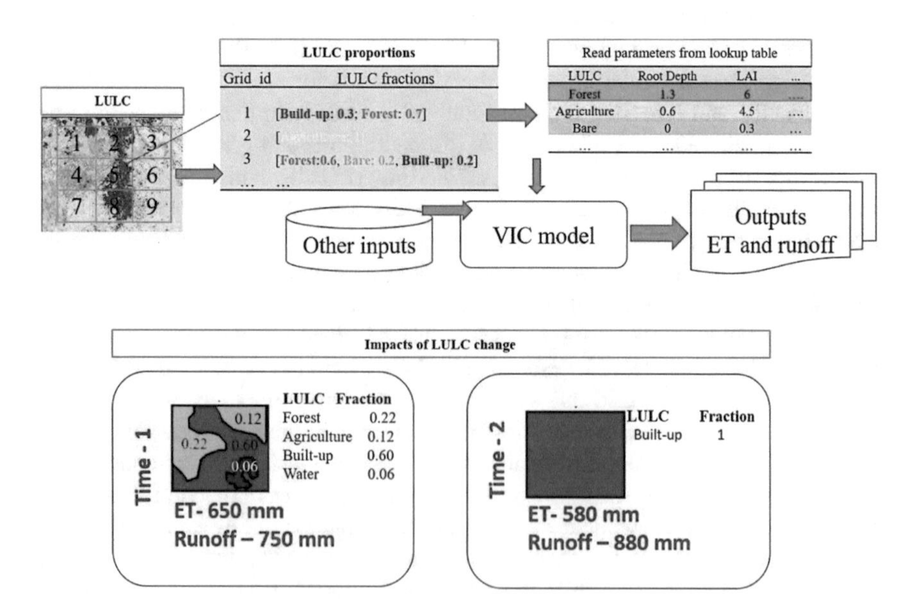

Fig. 17.2 Schematic of methodology for assessing hydrologic impacts of LULC change

the Indian part of Ganga and Brahmaputra River basins were considered due to data constraints. The model was calibrated and validated in the Mahanadi River basin using the streamflow data accessed from the Central Water Commission (CWC) at Basantpur, Sundergarh (from 2001 to 2010), and Rhondia (from 1971 to 1979) gauging sites. The Nash–Sutcliffe efficiency (NSE) was estimated as 0.56 and 0.65 at Basantpur and Sundergarh, respectively, during calibration. During the validation, the NSE is observed to be 0.76 at Rhondia, which shows an acceptable performance.

17.4 LULC Transformations

Agriculture is the most dominated LULC in all the river basins, except the Brahmaputra, dominated by forest covers. Agriculture occupies 73% of Ganga, 55% of MEB (Mahanadi and eastern flowing river basins), and 16% of Brahmaputra River basin, as per the recent assessment done by Behera et al. (2014, 2018). The Brahmaputra basin is dominated by forest, which occupies 62% of the total geographic area of the basin. The Ganga and MEB also accommodate considerable forest areas. Ganga has 20% and MEB has 35% of the area under forest. A generalized LULC map is provided in Fig. 17.3. A detailed discussion on the LULC of the study area can be found in Behera et al. (2014, 2018).

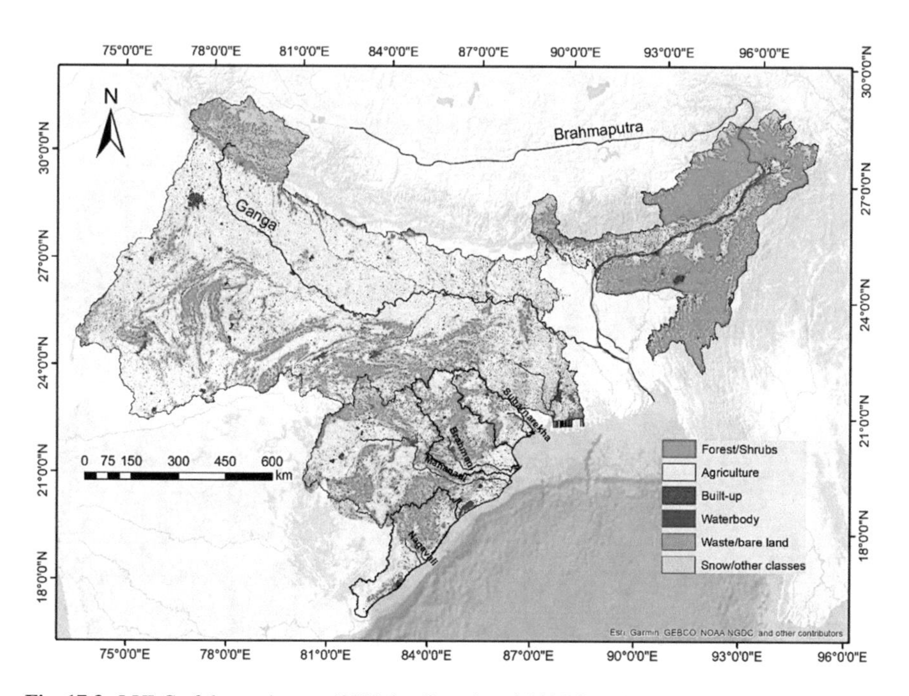

Fig. 17.3 LULC of the study area (2010 for Ganga) and 2005 for other river basins

The key transformations during the past, from 1975 to 2010 in Ganga and 1985 to 2005 in Brahmaputra and MEB, include deforestation and the expansion of agricultural and urban areas. Figure 17.4 highlights the location and extent of these changes. Although the builtup has a small coverage with respect to the total geographic area of the basins, it shows a considerable percentage increase in almost all basins. In Ganga, an increase of 45% built-up area is observed while the MEB and Brahmaputra show an increase of 16% and 6%, respectively. Deforestation is another important change which occurred in all the basins. A net change of 4% in the forest in Ganga, 1.5% in MEB, and 1.6% in the Brahmaputra is observed. A common trend has been to expand the agricultural lands at the cost of the nearby wastelands and forest/shrublands. The agriculture and waste lands also lost a considerable proportion to built-up lands. All changes are multidirectional at basin scale, except the built-up expansion. There has been a decrease in green areas due to deforestation in one place, while an increase at other places due to forest regeneration, afforestation, and plantations is observed. Such complementary changes lead to a small net change when analyzed at the basin scale. Therefore, the changes are more visible at a local scale, for instance, there are many patches of deforestation as shown in Fig. 17.4; however, the net change in the forest is less due to afforestation and plantations at other places.

The change map also indicates that deforestation is the most common change in all the basins. The other changes are relatively more evident in the Ganga basin as compared to other basins. These changes are mostly driven by the need for land for

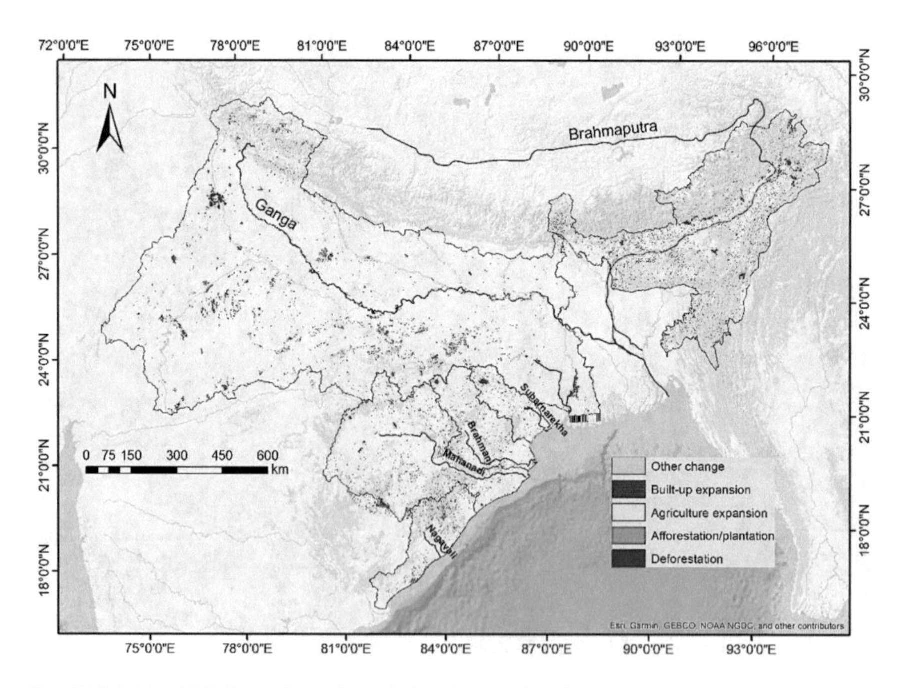

Fig. 17.4 Major LULC transformations during the past decades

agriculture and urban infrastructures. The Ganga has the most fertile lands among all basins with suitable topography, climate and availability of surface water, and groundwater resources for irrigation. The suitability of land for agriculture could be an important driving factor behind the expansions of agriculture patches at the cost of nearby natural vegetation or waste lands. An analysis was done by Behera et al. (2014) to investigate the change in agriculture patches of different sizes between 1975 and 2010, and it was found that the mean patch size of agriculture has been increased from 0.76 to 0.82 for small patches (<50 km^2). Similarly, the urban expansion has also been the highest in the Ganga basin than the other two basins. Being the most populated basin, accommodating 40% of India's population, the Ganga basin has the highest human settlements. Expansions at the periphery of such existing settlements with an increased density led to a significant increase in built-up areas (Patidar and Keshari 2020).

17.5 Potential Impacts

Quantification of the potential impacts of LULC change is important for making effective policies for sustainable water management. An experiment was performed using the VIC model to investigate the effect of various LULC transitions on ET and runoff under Indian climatic conditions. The simulations are performed by changing the LULC in the model grid literately for 10 LULC classes using constant climatic conditions at 23°15′1″ N and 83°14′59″ E. Using the daily simulations from 1975 to 2005, the annual average ET and runoff were calculated and compared for each LULC class.

The change matrix of the impacts is shown in Fig. 17.5. The conversion from vegetation to non-vegetation shows the highest sensitivity to affect the ET and runoff. A conversion from evergreen broad-leaved forest (EBF) to builtup (BU) can reduce the annual ET up to 377 mm, which is the highest among all conversions. The conversion from other forest classes, such as evergreen needle-leaved (ENF), mangroves

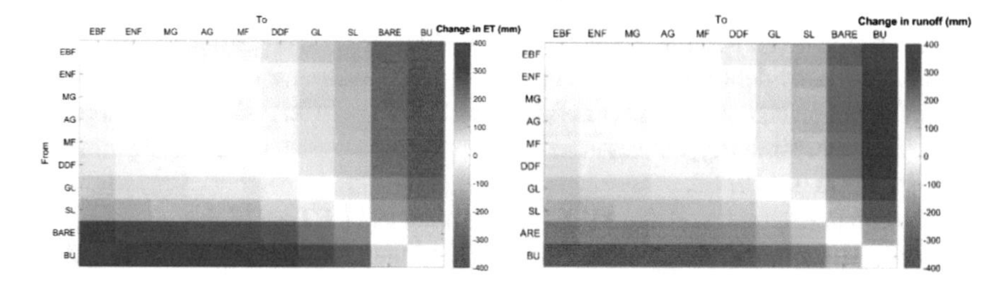

Fig. 17.5 Change matrix showing the hydrologic responses of various LULC transitions. EBF: evergreen broad-leaved forest, ENF: evergreen needle-leaved forest, MG: mangroves, AG: agriculture, MF: mixed forest, DDF: deciduous forest, GL: grassland, SL: scrubland, BARE: bare ground, BU: builtup

(MG), mixed forest (MG), and deciduous forest (DDF), to bare/built up also leads to a significant change in ET as highlighted by red color in the matrix (Fig. 17.5). Conversion from one vegetation class to the other vegetation class is relatively less sensitive. The impact on runoff is just opposite to that on the ET. The highest impact is observed due to evergreen broad-leaved forest (EBF) conversion to builtup (BU), which can increase the annual runoff up to 377 mm. Such alterations could impact the streamflow and flood magnitude to a considerable extent. Similar to ET, the transitions from vegetation to non-vegetation, or vice-versa, have the largest impact on runoff.

Theoretically, the magnitude of impacts depends on multiple factors including soil hydraulic properties, meteorological conditions, and topography. In this experiment, the other factors were kept unchanged to understand the LULC transition-wise impacts. However, it is obvious to have slight variations in the magnitude of the impacts when some of the aforementioned factors are changed; although, the order of sensitivity of impacts is expected to be the same. It is also known that precipitation is the most sensitive to affect runoff (Bloschl et al. 2007). Similarly, the soil moisture availability in the root zone is among the key governing factors for ET (Jaksa and Sridhar 2015). Therefore, under real conditions, when the influence of all these factors is combined, the impacts of small LULC change might not be very significant.

17.6 Changes in ET and Runoff

The ET and runoff simulated using the LULC maps of two time periods were compared to study the effect of LULC change (Figs. 17.6 and 17.7). In the Ganga basin, annual ET has a maximum increase of 74 mm and a maximum decrease of 123 mm. An opposite impact was observed on the runoff with a maximum increase of 122 mm to a decrease of 73 mm. In the Brahmaputra basin, the range of change in ET was observed between −99 and 7 mm (where negative indicates a decrease). The MEB has the lowest impact on ET, which varies between −4 to 5 mm. An opposite impact on runoff is observed. The runoff change varies between −73 to 122 mm in Ganga, −21 to 13 mm in the Brahmaputra, and between ±2 mm in MEB. Although the impacts are evidently visible at a grid scale, the impacts are minimal at the basin scale. The net change in ET and runoff in Ganga is found to be insignificant (approximately 1 mm). Similarly, the net impacts are within 1.5 mm in the Brahmaputra and within 0.5 mm in MEB.

It is seen in Brahmaputra and MEB that the impacts on ET are higher as compared to runoff. However, in Ganga, the magnitude of the changes in ET and runoff is comparable. This indicates the role of precipitation characteristics in the impact assessment of LULC change. The differential impacts in the Ganga and Brahmaputra/MEB could be attributed to a higher number of precipitation events in Brahmaputra and MEB, which needs to be studied further. The more number of rainy days

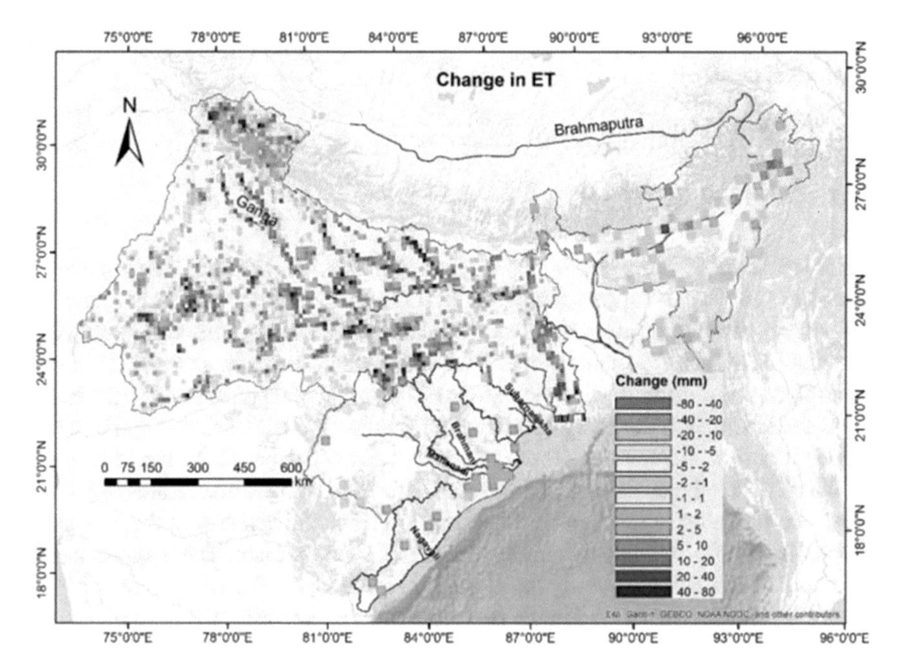

Fig. 17.6 Changes in ET induced by LULC change

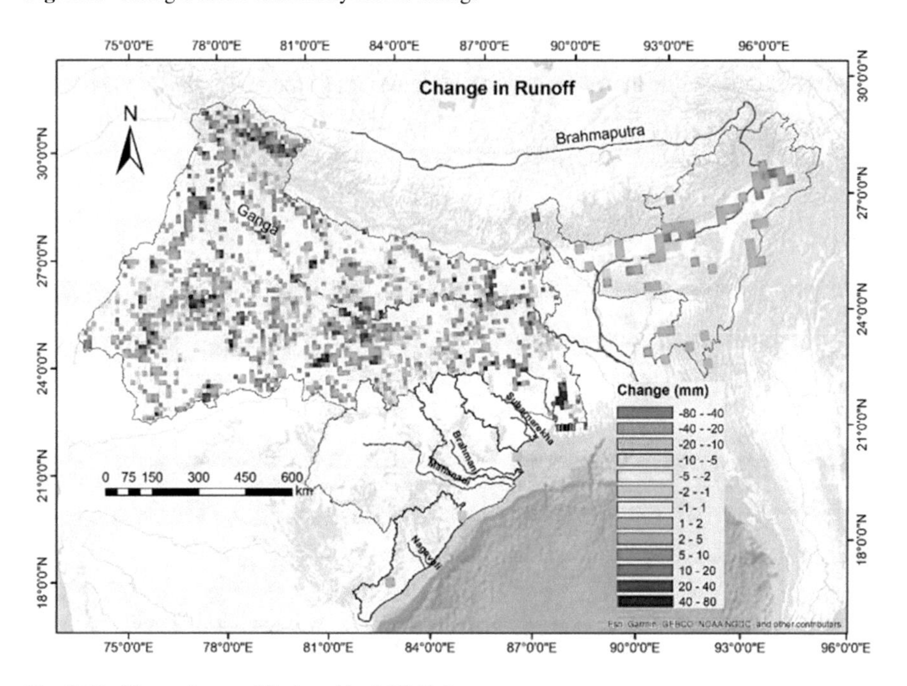

Fig. 17.7 Change in runoff induced by LULC change

in a year can provide more soil moisture to plants leading to a higher ET from the same vegetation.

17.7 The Compensation Effect

The assessment of the impacts in Indian River basins showed insignificant impacts on ET and runoff despite various changes in LULC, which highlighted a very important mechanism of compensating negative impacts by positive impacts. Due to multidirectional LULC transformations in all the basins, the resulting increase in ET/runoff at one place was nullified by the decrease in ET/runoff at other places within the basin. The decrease in ET induced by deforestation was mainly compensated by increased ET resulted from plantations. The conversion from waste/bare land led to increase ET, which also reduced the overall impact of deforestation. The histograms of ET and runoff are plotted to understand this mechanism (Fig. 17.8), showing the change in ET/runoff on abscissas and the number of grids on the ordinate. The magnitude and the extent of changes are clearly highlighted in these plots. The bars in the left indicate a decrease (negative) and in the right indicates an increase. A perfect symmetry of the histogram indicates all increase is compensated by the decrease, which is true for Ganga and MEB. On the other hand, the Brahmaputra shows slight skewness toward negative for ET and toward positive for runoff, indicating a net decrease in ET and increase in runoff. It can also be observed that most of the changes are of small magnitude while the large magnitudes have small frequencies in all the basins,

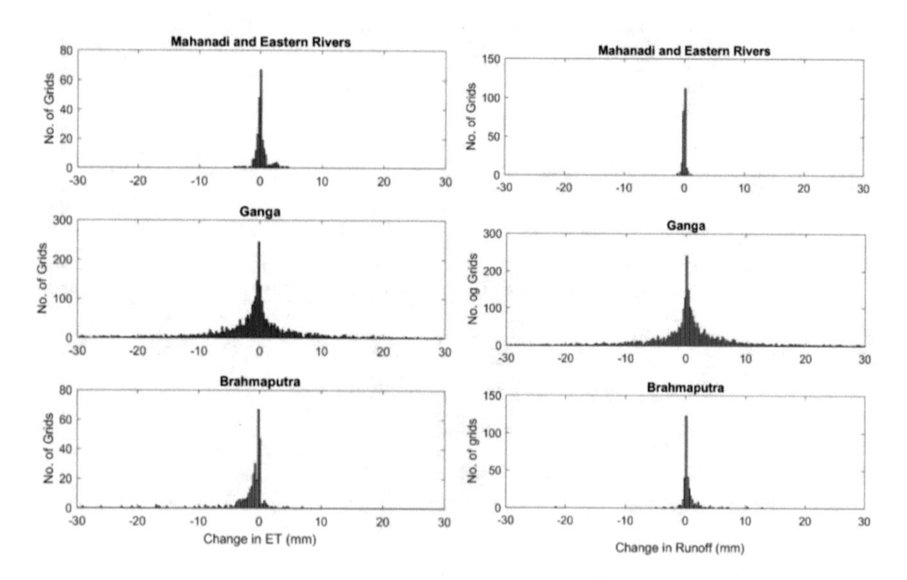

Fig. 17.8 Histograms of change in ET and runoff

as indicated by the near-zero peaks. This is another reason behind the small net impacts.

In all the basins, the decrease in the ET induced by the conversion of forest/agriculture to wasteland/builtup/scrubland was mostly compensated by the increase in ET due to increased plantation/agriculture areas at the basin scale. However, the impacts are more prominent at a local scale, for instance, at the sub-basin level, the changes in annual ET range from a decrease of 8.5 mm to an increase of 9.2 mm, which shows that the small sub-basins are more sensitive to LULC changes due to less compensation effect (Patidar and Behera 2019). A similar phenomenon was observed in Meuse River basin in Western Europe (Ashagrie et al. 2006). Similarly, an assessment of the impacts of LULC change in an agriculture catchment, named Mula and Mutha, upstream of Pune, India, using soil and water assessment tool (SWAT), also revealed that the impacts are too small at the basin level due to compensation effects, even after substantial LULC changes (Wagner et al. 2013).

17.8 Conclusions

The knowledge about the LULC induced changes in the hydrological regime has become important in the changing world due to increasing concern about water availability and hydrological disasters. In this context, the information about the shifts in the hydrological regime of major Indian River basins–Ganga–Brahmaputra and Mahanadi–could be an important input to the upcoming water management policies.

The assessment of the LULC induced changes in ET and runoff in these river basins reveals a very important compensating effect occurring due to multidirectional LULC changes. The negative consequences are nullified or at least reduced by the positive changes in ET and runoff at the basin scale. Fortunately, despite various LULC alterations, this compensating effect led to insignificant changes at basin scale in almost all river basins. The net change in ET and runoff is within 1.5 mm/year. However, the changes are significant at local scales and range up to 123 mm/year. This indicates that a bottom-up approach would be more effective for managing natural landscapes and dependent systems. This would require more detailed investigations of the changes in LULC and hydrology at watershed and sub-basin levels.

Acknowledgements The first author would like to thank the National Institute of Hydrology Roorkee, India, for providing infrastructure support while writing this chapter. Spatial Analysis and Modeling (SAM) Laboratory, CORAL, Indian Institute of Technology Kharagpur is also thankfully acknowledged for providing infrastructure support for hydrological modeling used in this study.

References

Ashagrie AG, De Laat PJM, De Wit MJM, Tu M, Uhlenbrook S (2006) Detecting the influence of land use changes on floods in the Meuse River Basin—the predictive power of a ninety-year rainfall-runoff relation. Hydrol Earth Syst Sci 3(2):529–559. https://doi.org/10.5194/hessd-3-529-2006

Behera MD, Patidar N, Chitale VS, Behera N, Gupta D, Matin S, Tare V, Sen DJ, Panda SN (2014) Increase in agricultural patch contiguity over the past three decades in Ganga River Basin, India. Curr Sci 107(3):502–511

Behera MD, Tripathi P, Das P, Srivastava SK, Roy PS, Joshi C, Behera PR, Deka J, Kumar P, Khan ML, Tripathi OP, Dash T, Krishnamurthy YVN (2018) Remote sensing based deforestation analysis in Mahanadi and Brahmaputra river basin in India since 1985. J Environ Manage 206:1192–1203. https://doi.org/10.1016/j.jenvman.2017.10.015

Bloschl G, Ardoin-Bardin S, Bonell M, Dorninger M, Goodrich D, Gutknecht D, Matamoros D, Merz B, Shand P, Jan S (2007) At what scales do climate variability and land cover change impact on flooding and low flows? Hydrol Process 21:1241–1247. https://doi.org/10.1002/hyp

Bosmans JHC, Van Beek LPH, Sutanudjaja EH, Bierkens MFP (2017) Hydrological impacts of global land cover change and human water use. Hydrol Earth Syst Sci 21(11):5603–5626. https://doi.org/10.5194/hess-21-5603-2017

Cosby BJ, Hornberger GM, Clapp RB, Ginn TR (1984) A statistical exploration of the relationships of soil moisture characteristics to the physical properties of soils. Water Resour Res 20(6):682–690. https://doi.org/10.1029/WR020i006p00682

Das P, Behera MD, Pal S, Chowdary VM, Behera PR, Singh TP (2020) Studying land use dynamics using decadal satellite images and Dyna-CLUE model in the Mahanadi River basin, India. Environ Monit Assess 191(3):804. https://doi.org/10.1007/s10661-019-7698-3

Das P, Behera MD, Patidar N, Sahoo B, Tripathi P, Behera PR, Srivastava SK, Roy PS, Thakur P, Agrawal SP (2017) Changes in evapotranspiration, runoff and baseflow with LULC change in Eastern Indian River basins during 1985–2005 using variable infiltration capacity approach. In: 38th Asian conference on remote sensing—space applications: touching human lives, ACRS 2017

Dawes W, Ali R, Varma S, Emelyanova I, Hodgson G, McFarlane D (2012) Modelling the effects of climate and land cover change on groundwater recharge in south-west Western Australia. Hydrol Earth Syst Sci 16(8):2709–2722. https://doi.org/10.5194/hess-16-2709-2012

Gao H, Tang Q, Shi X, Zhu C, Bohn T, Su F, Sheffield J, Pan M, Lettenmaier D, Wood EF(2009) Water Budget record from variable infiltration capacity (VIC) model algorithm theoretical basis document. Department of Civil and Environmental Engineering, University of Washington, Seattle, WA

Ghimire CP, Bruijnzeel LA, Lubczynski MW, Bonell M (2014) Negative trade-off between changes in vegetation water use and infiltration recovery after reforesting degraded pasture land in the Nepalese Lesser Himalaya. Hydrol Earth Syst Sci 18(12):4933–4949. https://doi.org/10.5194/hess-18-4933-2014

Jacobson CR (2011) Identification and quantification of the hydrological impacts of imperviousness in urban catchments: a review. J Environ Manage 92(6):1438–1448. https://doi.org/10.1016/j.jenvman.2011.01.018

Jaksa WT, Sridhar V (2015) Effect of irrigation in simulating long-term evapotranspiration climatology in a human-dominated river basin system. Agric For Meteorol 200:109–118. https://doi.org/10.1016/j.agrformet.2014.09.008

John R, Chen J, Kim Y, Ou-yang ZT, Xiao J, Park H, Shao C, Zhang Y, Amarjargal A, Batkhshig O, Qi J (2016) Differentiating anthropogenic modification and precipitation-driven change on vegetation productivity on the Mongolian Plateau. Landsc Ecol 31(3):547–566. https://doi.org/10.1007/s10980-015-0261-x

Kalnay E, Cai M (2003) Impact of urbanization and land-use change on climate. Nature 423(6939):528–531. https://doi.org/10.1038/nature01675

Lei H, Yang D, Yang H, Yuan Z, Lv H (2015) Simulated impacts of irrigation on evapotranspiration in a strongly exploited region: a case study of the Haihe River basin, China. Hydrol Process 29(12):2704–2719. https://doi.org/10.1002/hyp.10402

Li X, Gong P, Liang L (2015) A 30-year (1984–2013) record of annual urban dynamics of Beijing City derived from Landsat data. Remote Sens Environ 166:78–90. https://doi.org/10.1016/j.rse.2015.06.007

Liu SL, Fu BJ, Lü YH, Chen LD (2002) Effects of reforestation and deforestation on soil properties in humid mountainous areas: a case study in Wolong Nature Reserve, Sichuan province, China. Soil Use Manag 18(4):376–380. https://doi.org/10.1079/sum2002148

Lohmann D, Raschke E, Nijssen B, Lettenmaier DP (1998) Regional scale hydrology: I. Formulation of the VIC-2L model coupled to a routing model. Hydrol Sci J 43(1):131–141. https://doi.org/10.1080/02626669809492107

Lohmann D, Raschke E, Nijssen B, Lettenmaier DP (1998) Regional scale hydrology: II. Application of the VIC-2L model to the Weser River, Germany. Hydrol Sci J 43(1):143–158. https://doi.org/10.1080/02626669809492108

Minnig M, Moeck C, Radny D, Schirmer M (2018) Impact of urbanization on groundwater recharge rates in Dübendorf, Switzerland. J Hydrol 563:1135–1146. https://doi.org/10.1016/j.jhydrol.2017.09.058

Mishra V, Cherkauer KA, Niyogi D, Lei M, Pijanowski BC, Ray DK, Bowling C, Yang G (2010) A regional scale assessment of land use land cover and climatic changes on water and energy cycle in the upper Midwest United States. Int J Climatol

Pai DS, Sridhar L, Rajeevan M, Sreejith OP, Satbhai NS, Mukhopadhyay B (2014) (1901–2010) daily gridded rainfall data set over India and its comparison with existing data sets over the region. Mausam 1(January):1–18

Patidar N, Behera MD (2019) How significantly do land use and land cover (LULC) changes influence the water balance of a River Basin? A study in Ganga River Basin, India. Proc Natl Acad Sci India Sect A Phys Sci 89(2). https://doi.org/10.1007/s40010-017-0426-x

Patidar N, Keshari AK (2020) A rule-based spectral unmixing algorithm for extracting annual time series of sub-pixel impervious surface fraction. Int J Remote Sens 41(10):3970–3992. https://doi.org/10.1080/01431161.2019.1711243

Peña-Arancibia JL, Bruijnzeel LA, Mulligan M, van Dijk AIJM (2019) Forests as 'sponges' and 'pumps': assessing the impact of deforestation on dry-season flows across the tropics. J Hydrol 574(February):946–963. https://doi.org/10.1016/j.jhydrol.2019.04.064

Raju BM, Rao KV, Venkateswarlu B, Rao AV, Rao CR, Rao VU, Rao BB, Kumar NR, Dhakar R, Swapna N, Latha P(2013) Revisiting climatic classification in India: a district-level analysis. Curr Sci 105(4):492–495

Ren-jun Z, Yi-lin Z, Le-run F, Xin-ren LIU, Quan Z (1980) The Xinanjiang model. Hydrological forecasting. Prévisions Hydrologiques 129:351–356

Rodell M, Houser PR, Jambor U, Gottschalck J, Mitchell K, Meng C-J, Arsenault K, Cosgrove B, Radakovich J, Bosilovich M, Entin JK, Walker JP, Lohmann D, Toll D (2004) The global land data assimilation system. Bull Am Meteor Soc 85(3):381–394. https://doi.org/10.1175/BAMS-85-3-381

Roy PS, Roy A, Joshi PK, Kale MP, Srivastava VK, Srivastava SK, Dwevidi RS, Joshi C, Behera MD, Meiyappan P, Sharma Y, Jain AK, Singh JS, Palchowdhuri Y, Ramachandran RM, Pinjarla B, Chakravarthi V, Babu N, Gowsalya MS, Kushwaha D (2015) Development of decadal (1985–1995–2005) land use and land cover database for India. Remote Sensing 7(3):2401–2430. https://doi.org/10.3390/rs70302401

Schilling KE, Chan K, Liu H, Zhang Y (2010) Quantifying the effect of land use land cover change on increasing discharge in the Upper Mississippi River. J Hydrol 387(3–4):343–345. https://doi.org/10.1016/j.jhydrol.2010.04.019

Sterling SM, Ducharne A, Polcher J (2013) The impact of global land-cover change on the terrestrial water cycle. Nat Clim Chang 3(4):385–390. https://doi.org/10.1038/nclimate1690

Todini E (1996) The ARNO rainfall—runoff model. J Hydrol 175(1–4):339–382. https://doi.org/10.1016/S0022-1694(96)80016-3
Tsarouchi GM, Mijic A, Moulds S, Buytaert W (2014) Historical and future land-cover changes in the Upper Ganges basin of India. Int J Remote Sens 35(9):3150–3176. https://doi.org/10.1080/01431161.2014.903352
Wagner PD, Kumar S, Schneider K (2013) An assessment of land use change impacts on the water resources of the Mula and Mutha Rivers catchment upstream of Pune, India. Hydrol Earth Syst Sci 17:2233–2246. https://doi.org/10.5194/hess-17-2233-2013
Yang G, Bowling LC, Cherkauer KA, Pijanowski BC, Niyogi D (2010) Hydroclimatic response of watersheds to urban intensity: an observational and modeling-based analysis for the White River Basin, Indiana. J Hydrometeorol 11:122–138. https://doi.org/10.1175/2009JHM1143.1

Chapter 18
Reservoir Monitoring Using Satellite Altimetry: A Case Study Over Mayurakshi Reservoir

A. Sai Krishnaveni, Chiranjivi Jayaram, V. M. Chowdary, and C. S. Jha

Abstract Satellite altimetry-based studies on monitoring the inland water bodies are limited in the Indian context due to small and medium size of these water bodies. The present study demonstrates the utilization of altimeters in monitoring reservoirs with a surface area <100 km^2 and Altimeter data from SARAL/AltiKa and RA-2 radar altimeter of ENVISAT having similar ground track were used in this study for the period 2002–2010 and 2013–2014, respectively. This study was carried out for both monsoon and non-monsoon seasons to assess the seasonal dependence between in-situ and satellite data. Reservoir water levels obtained from ENVISAT and SARAL were compared with in-situ gauge data. The root mean square error (RMSE) between the in-situ and ENVISAT data is 0.34 with R^2 of 0.97 during the monsoon season compared to the non-monsoon values of 0.70 and 0.88, RMSE and R^2, respectively) Further, it was observed that the 40 Hz data of SARAL/AltiKa is in good agreement (RMSE of 0.44 m and $R^2 = 0.95$) with in-situ data than ENVISAT 20 Hz data. Further, elevation-capacity rating curves were generated using both ENVISAT and SARAL/AltiKa observations. Overall, a 35-day time resolution of ENVISAT and SARAL/AltiKa data are insufficient for reservoir monitoring at a daily scale. However, this study has showcased the application of altimeter data in estimating the capacity of smaller water bodies.

A. Sai Krishnaveni (✉) · C. S. Jha
National Atmospheric Research Laboratory (NARL), Gadanki 517112, India
e-mail: saikrishnaveni_a@narl.gov.in

C. Jayaram
Regional Remote Sensing Centre—East, NRSC/ISRO, New Town, Kolkata 700156, India

V. M. Chowdary
Regional Remote Sensing Centre—North, NRSC/ISRO, Sadiq Nagar, New Delhi, India

A. Sai Krishnaveni
National Remote Sensing Centre, ISRO, Balanagar, Hyderabad 500037, India

A. Pandey et al. (eds.), *Geospatial Technologies for Land and Water Resources Management*, Water Science and Technology Library 103,
https://doi.org/10.1007/978-3-030-90479-1_18

18.1 Introduction

Reservoir water storage monitoring is an integral part of the catchment scale water resources management. The local datum referenced in-situ reservoir water level measurements are made at a daily scale with an accuracy of ± 10 cm. However, they are often inconsistent given the logistical and downtime due to the irregular maintenance (Salami and Nnadi 2012). Further, ground observations are sometimes unavailable due to lack of proper information, transmission network, and sometimes due to administrative constraints. Reduction in the hydro-meteorological observational network and the availability of consistent and accurate ground based observations from inaccessible regions would often lead to impracticalities in the monitoring of water resource projects (Ghosh et al. 2015).

Thus, monitoring of inland water bodies requires a supplementary solution that is economical and should be able to provide an alternative to the traditional measurements. Further, the solution should provide accurate and long term, all weather data that could be implemented with minimum human intervention (Salami and Nnadi 2012). Satellite remote sensing can overcome this problem with the availability of global data in open source domain that could substantiate and complement the in-situ data pertaining to precipitation temperature, reference evapotranspiration topography soil moisture and total water storage (Tapley et al. 2004; Milzow et al. 2011).

Spaceborne techniques provide multi-temporal observations of interferometric radar (Alsdorf et al. 2001); synthetic aperture radars, microwave radiometers that facilitate the measurement of water extent together with the altimeter data records (Alsdorf 2003; Bates et al. 2006; Alsdorf et al. 2007; Smith and Pavelsky 2008). Satellite altimetry is a useful tool to measure water level variability in rivers (Dubey et al. 2015; Frappart et al. 2015) and wetlands despite having limitations due to spatial resolution, related to both along-track path length and coarse across-track spacing. Further, data acquisition through altimeters not only gets affected due to its type but also due to the satellite orbit, atmospheric propagation and instrumental errors (Alsdorf et al. 2007; Tang et al. 2009). Though the satellite altimetry was envisaged to measure the ocean topography and sea surface heights, it has provided consistent data on water levels over large lakes and reservoirs (Crétaux et al. 2017, Bogning et al. 2018; Sai Krishnaveni et al. 2016).

Although researchers in the past primarily used altimetric geophysical data records (GDR) for ocean studies, with the availability of data over rivers and lake surfaces, attempts were made to derive the water level by retracking of the returned radar waveforms effectively. Brooks (1982) for the first time employed SEASAT altimetry data to determine water elevations over Canadian inland water bodies. Similarly, bathymetry studies were carried out on the Tanganyika lake, Caspian Sea and the great lakes using SEASAT data (Rapley et al. 1987; Olliver 1987; Au et al. 1989). Berry et al. (2005) had deduced the retrieval of water levels from rivers and smaller water bodies by retracking the individual echoes. GEOSAT altimeter data was used to derive the time series observations of Grand Lac in Cambodia to substantiate the temporally sparse in-situ measurements (Mason et al. 1990) Similarly, GEOSAT data was used to obtain the water levels from large lakes with an area greater than 100 km^2 by Birkett (1994). Applicability of altimeter data for lake monitoring that has diameter greater than 25 km was established by Morris and Gill (1994a, b). Further, Birkett (2000) had applied altimeter data for studying water level dynamics across Lake Chad that spreads around 2000 km^2 of permanent water body and nearly 10,000 km^2 area under inundation during the wet season. It was opined that the radar altimetry collected under good conditions could be used to forecast the lake level fluctuations. Cai and Ji (2009) observed that the water levels in Poyang lake (~4078 km^2) were derived using altimetry data at higher accuracy and augmented the capability to get water levels from the ungauged areas. Thus, the altimetry technique was well explored for monitoring of rivers, lakes and wetlands by several researchers across the globe (Morris and Gill 1994a, b; Birkett 1995; Koblinsky et al. 1993; Coe and Birkett 2004; Leon et al. 2006; Frappart et al., 2006).

The altimetry derived water elevations over Kainji reservoir, which is one of the largest reservoirs in Africa (~1270 km^2) indicated that RMSE varied between 0.50 and 0.83 m during both dry and wet seasons (Salami and Nnadi 2012). Kouraev et al. (2004) and Zakharova et al. (2006) generated rating curves using altimetry-based water levels and in-situ river flux data. This has enabled the discharge estimates to be made in the absence of in-situ stage measurements. Michailovsky (2012) derived river stage measurements over rivers that have a width of 80 m using retracked ENVISAT altimetry data across Zambezi river basin, South Africa. Further, as opined by Vignudelli et al. (2019), and Bogning et al. (2018), the availability of in-situ gauge measurement of river discharge and river levels is reducing with time, thereby hindering the sustained management of the reservoirs. Satellite based altimeters could bridge this gap with a reliable accuracy as demonstrated across many regions (Birkinshaw et al. 2014; Papa et al. 2008; Chowdary et al. 2017).

Replacement of in-situ reservoir bathymetry data with radar altimeter data provided alternate means for conventional survey in the reservoir capacity estimation studies (Abileah et al. 2011). Thus, this study was taken up for capacity estimation of Mayurakshi Reservoir remotely using ENVISAT and SARAL/AltiKa altimeter data. Most of the previous studies were carried out using satellite altimeters over large inland water bodies. However, there are limited studies in the Indian context, where most of the water bodies are medium to small in size. Current study demonstrates the application of spaceborne radar altimeters to monitor small and medium reservoirs with a surface area <100 km^2. Intra-seasonal variation of derived altimetric levels was also analyzed for the case study reservoir. Further, rating curves that directly relate the altimetry derived water elevations with in-situ reservoir capacity were also generated for the case study reservoir.

18.2 Study Area

The case study reservoir (Mayurakshi Reservoir: 24° 6.6′ N latitude and 87° 18.9′ E longitude) is located in Dumka (Santhal Pargana) district of Jharkhand state,

India (Fig. 18.1) over the Mayurakshi river, with catchment area of ~1860 km^2. The reservoir water spread area is ~68 km^2. The live and dead storages of the reservoir at full reservoir level (FRL) of 121.31 m and dead storage level of 106.38 m above mean sea level (msl) are 547.59 mm^3 and 67.65 mm^3, respectively. Mean elevation at reservoir location is nearly 122.40 m amsl. The region is prevailed by tropical climate with three seasons: (i) summer (March–June), (ii) rainy season

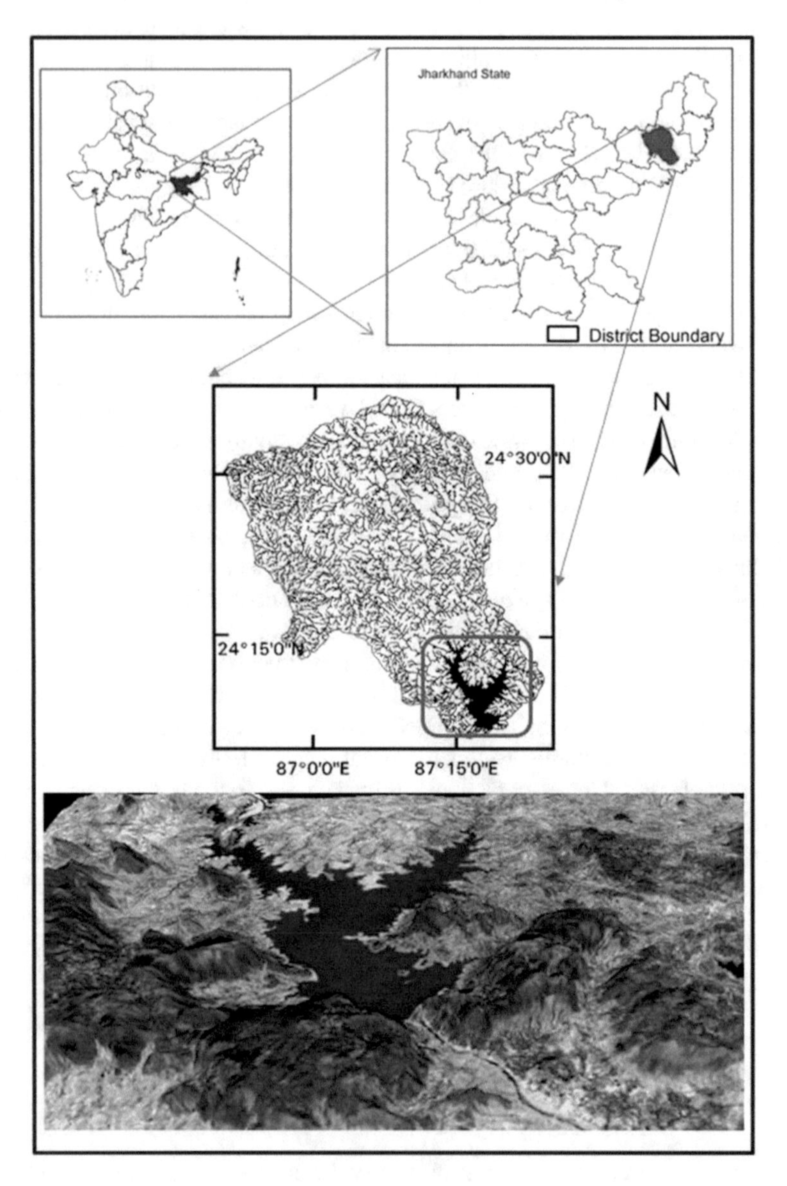

Fig. 18.1 Geographic location of the Mayurakshi Reservoir

(July–October), and (iii) winter season (November–February). Mean annual rainfall in the reservoir catchment is ~1400 mm. Drainage pattern in the reservoir catchment is dendritic in nature with high drainage density and drainage frequency. This reflects most of the homogeneity of lithology. Soil degradation in the region due to reduced forest cover and gully erosion has resulted in reservoir sedimentation.

18.3 Methodology

18.3.1 Working Principle of the Radar Altimeters

Radar altimeters are designed to monitor ocean surfaces topography and sea ice. These sensors are nadir-viewing radars with a narrow field of view that are capable of receiving the return pulse from the ocean/water surfaces that are directly below the satellite track (Rapley 1990). The range between the satellite and target water surface is obtained from the travel time between the emitted pulse and its echo. Thus, it is surmised that the returned power of the radar pulse is sampled in delay time across a range "window" that is organized into different range "bins." Pulse echo at the altimeter is also termed as 'waveform' that signifies the spread of the power in the range window. Altimeter tracks and captures the leading edge of the waveform within the range window with the varying surface topography. However, the waveform tracking cannot resolve the echo for the inland/coastal waters owing to the heterogenous reflecting surface. To overcome these errors, the individual waveforms need to be retracked to obtain better estimates of the water levels for inland water bodies (Birkett 1995, 2000). Hence, the retrieval of water heights from altimetry is an averaging process of the waveforms over the given footprint that is arrived at after removing the range from the altitude estimate as shown in Fig. 18.2. Further details on the working principle of satellite altimeters are discussed in Birkett (1995, 2000), SARAL/Altika Hand book (2013).

18.3.2 Altimetry Data Used

Altimetric height measurements for Mayurakshi Reservoir were retrieved for the time periods 2002–2010 and 2013–2014 using ENVISAT and SARAL/AltiKa, respectively as both of these satellites follow a similar ground track at a repeat period of 35 days (Fig. 18.3). ENVISAT 18 Hz and SARAL/AltiKa 40 Hz waveform data available at 170 m and 360 m footprint, respectively were used for estimation of mean surface heights. Every single altimeter measurement corresponding to fixed geographical coordinates indicates the mean of 100 waveforms. Thereafter, precise orbit information, calculated range and the height above reference ellipsoid were used for computation of surface heights.

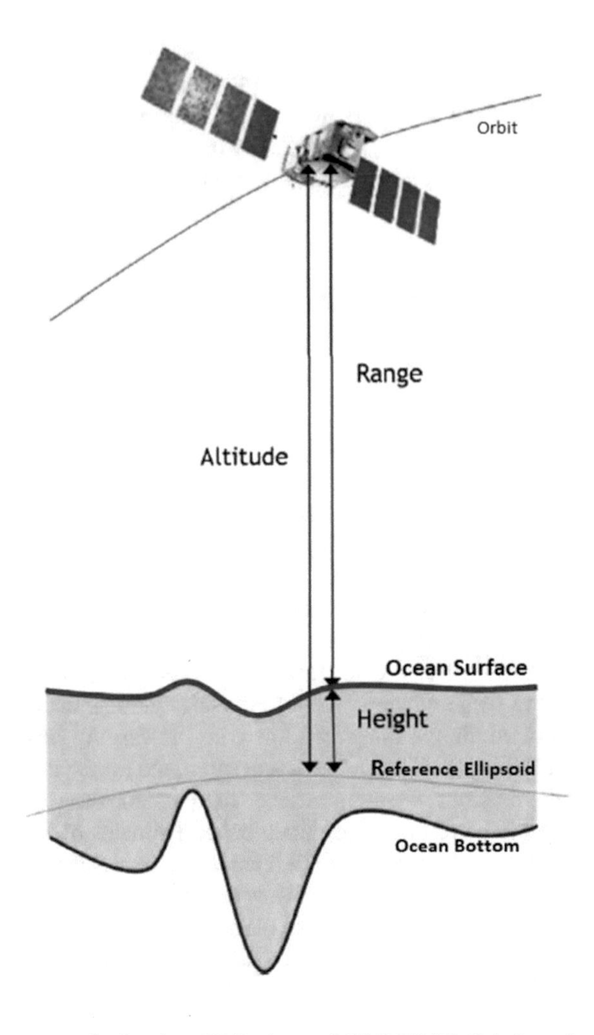

Fig. 18.2 Altimetry heights, altitude and range with reference to ocean bottom and reference. *Source* SARAL/Altika Hand book (2013)

Necessary corrections were made in the GDR data of ENVISAT RA-2 and SARAL/AltiKa for obtaining altimetric water surface heights in this study. Given the orbit cycle of the altimeter missions (±1 km), a time series of water surface height could be derived for a particular region for further applications. Hence, the data that is likely to be affected due to interference of coastline or island was omitted. Cloud free MODIS/IRS P6/AWiFS imagery that is near the RA 2 and SARAL/AltiKa acquisition dates were used to verify that the altimeter points were confined to the water surface.

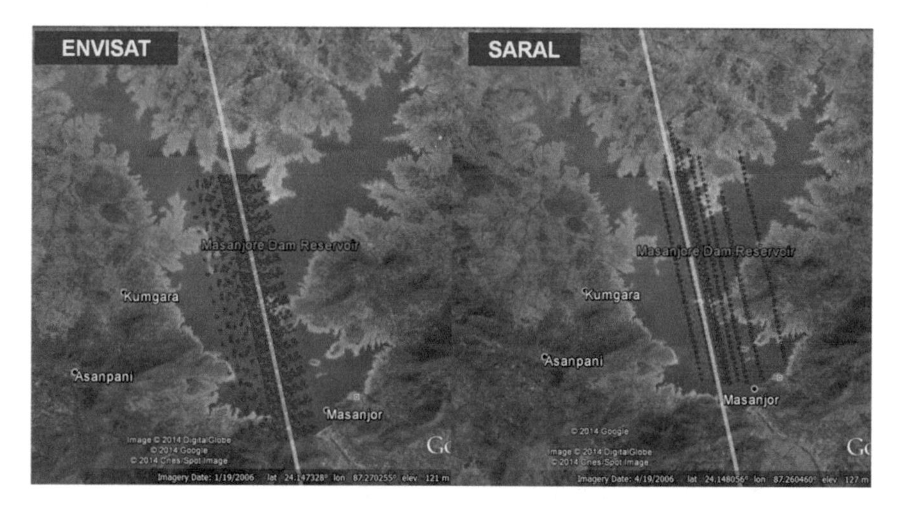

Fig. 18.3 Footprint of ENVISAT and SARAL/AltiKa satellites over the Mayurakshi Reservoir

18.3.3 Extraction of Reservoir Elevation Levels

The waveform echoes that emit from the surface of the reservoir are complex and different from the single peak of the echo emitting from the ocean surface. Therefore, these multi-peak waveforms need to be retracked for river/wetland studies to obtain better results. Particularly retracking is necessary for small water bodies and water bodies that were located in complex terrain (Berry et al. 2005). Studies by Frappart et al. (2006), Cheng et al. (2010), Silva et al. (2011) reported that the ICE-1 algorithm derived altimetric ranges matched well with the in-situ observations. Further, it was stated that the average of over 100 pulses of 18 Hz frequency instead of the 1 Hz frequency remove several ranges derived from radar echoes that are affected by non-aquatic reflectors. Hence, ICE-1 retracking algorithm was applied on the 18 Hz of ENVISAT and 40 Hz of SARAL/AltiKa data in this study. Further, along-track altimetry profiles were processed for deriving water elevation levels of the reservoir. Initially, the data for the crossing of satellite track and the reservoir was selected and the outliers were manually edited. Following Cheng et al. (2010), the mean of all altimetry measurements was computed for every satellite crossover, while excluding outliers.

For processing altimetry data, the water surface height from the altimeter (S_h) was derived by following the expression Cai and Ji (2009):

$$S_h = S_a - (R_m + I_d + W_{\text{tmod}} + D_t + P_t + E_t) - G \quad (18.1)$$

where S_h is the water surface height, S_a is the altitude of the satellite, R_m is the range measured using 1 Hz data and I_d is the DORIS ionospheric correction, W_{tm} is wet tropospheric model, D_t the dry troposphere, P_t the polar tide, E_t the solid earth tide, and *G* the reference geoid height obtained from EGM2008. Details of altimetry corrections are discussed in Birkett (1998). Inter-annual and intra-seasonal variations were analyzed from 2002 to 2010 pertaining to absolute and relative water level changes and the results are evaluated with in-situ data as well. For studying the seasonal variations of altimetric levels, the datasets were categorized into monsoon and non-monsoon. Further, changes in relative heights were deduced using the data acquired at periodic intervals as per the revisit capability of the altimeter suggested by Birkett (1994).

18.3.4 Computation of Reservoir Capacity

Reservoir capacity is the critical variable for hydrology and therefore, the altimetric height needs to be converted to reservoir capacity as established by the previous studies (Manavalan et al. 1993; Jain and Goel 2002; Peng et al. 2006; UNESCO 2010). Abileah et al. (2011) and Chowdary et al. (2017) had further extended these methods by replacing the in-situ reservoir levels with altimeter data to estimate reservoir capacity that could substitute the traditional method of ground surveying. In this scenario, elevation–capacity curves obtained for the reservoir are noteworthy. A similar approach was adopted for the conversion of altimetry data to estimate reservoir capacities, with the availability of a rating curve of the crossover between satellite track and the reservoir. Water levels are usually recorded daily and capacity is obtained through the rating curve. Multiple regression of discharge measurements obtained from the hydraulic variables that were estimated from satellite altimeter data is used to derive the discharge (Michailovsky et al. 2012). The systematic biases that are caused by the differences in the reference datum and the instrument errors are removed from both the gauge and altimeter data before computing the absolute difference between the altimeter measurements and the gauge data Subsequently, the whole set of altimetric observations are divided into two datasets, where one dataset was used to establish a relationship between altimetry data and reservoir capacity, while the other dataset was used for validation.

18.4 Results and Discussion

18.4.1 Time Series Analysis

18.4.1.1 Inter-Annual Variability

In order to apply the satellite altimeter data for reservoir management, it is imperative that the data need to be validated with the in-situ data. In this connection, the reservoir water levels extracted from ENVISAT and SARAL/AltiKa were evaluated with season-wise gauge level observations during the periods 2002–2010 (Fig. 18.4a, b) and 2013–2014, respectively (Fig. 18.5a, b). From Fig. 18.4a, a minor yet consistent bias between the altimetry and in-situ data was observed, which was due to the difference in some lake waveforms that are dissimilar to the oceanic waveforms. The difference between the ENVISAT altimeter record and the gauge data in terms of their absolute average and root mean square error, were compared respectively. The difference in the datum that was used for the gauge and the altimeter data had resulted towards the scatter in the time series. There exist some errors, despite the correction of altimeter derived water level to the same datum using EGM 2008. Further research is necessary focusing on the conversion of the reference ellipsoid or geoid based water levels to the datum used by the in-situ gauge data. Thus, variations between the relative water levels were computed (Fig. 18.4b). It was observed that the variability in the amplitude is analogous during the majority study duration except for low stage that could be attributed to the seasonal dependency of a few waveforms. Variations among time series data were observed to be 0.02–1.97 m with an RMSE of 1.27 m, standard deviation of 0.52 m and R^2 is 0.93.

Relative height changes (altimeter and gauge) are comparable to each other to a greater extent as observed by Salami and Nnadi (2012). Here, the high RMSE can be

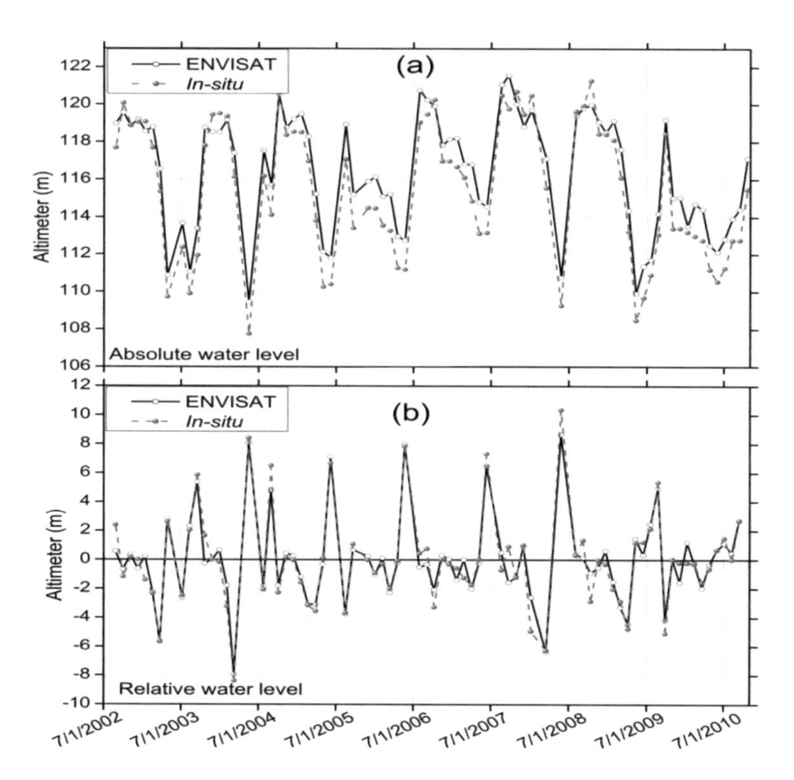

Fig. 18.4 Inter-annual variations of study reservoir **a** absolute and **b** relative water levels between ENVISAT and Gauge observations

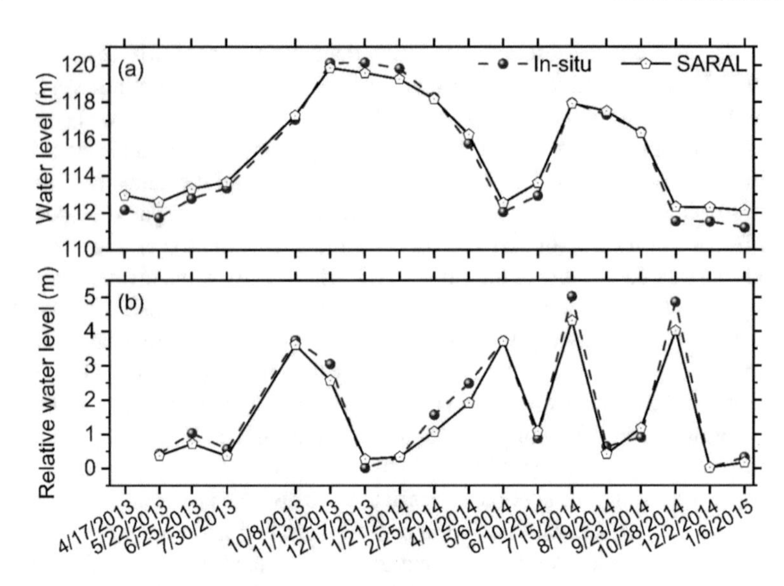

Fig. 18.5 Evaluation of inter-annual variations of reservoir: **a** absolute and **b** relative water levels between SARAL/AltiKa and Gauge observations

attributed to the surface area of the reservoir and the complex topography, where the satellite observes only a narrow expanse of water and the same was reported by Birkett (1998). Further, from the size of the Mayurakshi Reservoir and its geographical location, it is inferred that complex terrain conditions could be encountered within the satellite footprint that would contaminate the satellite retrievals. To account for the seasonal variability in the water surface area of the reservoir, seasonal analysis was carried out.

Thus, it could be stated that the water level measurements obtained from the radar altimeters would serve as the virtual gauge that provides discrete measurements at the repeat cycle corresponding to the satellite ground track (35-day in the case of ENVISAT). These values represent the average height within the altimeter footprint, which can be a few hundred meters to a few kilometres wide, depending on the surface roughness. Although, several earlier studies highlighted the utility of altimetry data for monitoring larger lakes and water bodies more than a hundred sq. kms, but this study indicated that retracked altimetry can be of great help for the extraction of reservoir elevations and reservoir capacity. Datasets that exceeded the maximum reservoir level during monsoon season were avoided, which is not possible. In order to omit the large errors, a threshold of 2 m is proposed that resulted in an RMSE of 0.76 m for constructing rating curve using ENVISAT data.

18.4.1.2 Seasonal Variability

To study the seasonal variation of extracted reservoir altimetric levels from ENVISAT, the altimetry data was divided into two seasons namely monsoon (June–November) and non-monsoon season (December–May). During non-monsoon season, the comparison of ENVISAT with gauge height variations was shown in Fig. 18.6a. Accuracy for this period ranges from 0.08 to 1.96 m with R^2 of 0.88, RMSE of 0.7 m and a standard deviation of 0.47 m. ENVISAT and gauge height variations for the monsoon season were compared (Fig. 18.6b). Analysis of monsoon season showed an RMSE of 0.36 m and the standard deviation is of 0.22 m with high correlation coefficient (0.96). The inter-comparison analysis has revealed that the temporal data for the period, 1992–2014 have varying accuracy in terms of RMSE ranging from 0.4 to 06 m Inter-comparison between relative water levels obtained from ENVISAT altimeter data and in-situ data for different seasons were presented (Fig. 18.7a–c), which indicated good R^2 of 0.96 during monsoon season. The largest error reflects the period of low water level. Overall, the RMSE is <15% of the maximum observed change, which is a small proportion of range of stages. Also, it is evident that there exists seasonal and inter-annual variability in RMSE.

In general, it is observed that the altimetric measurements during monsoon season are in good agreement with gauge measurements compared to non-monsoon period (Fig. 18.6a, b). During the periods of higher water level, the longer sequence of narrow peak waveforms gets reduced into two to three broad peak echoes at minima periods. Significant variability was observed in the waveform shape during lower water elevation that could be attributed to contamination of altimeter measurements in the non-monsoon season due to the contribution of dry land within the altimeter footprint (Fig. 18.6a). Summary statistical parameters corresponding to intra-seasonal analysis was given in Table 18.1. The better results during monsoon season (Fig. 18.6b) are due to the comparatively full stage of the reservoir together with the calm inflow within the reservoir. Further, during the monsoon season, backscatter effects from the vegetation are also relatively less that might have improved the validation results

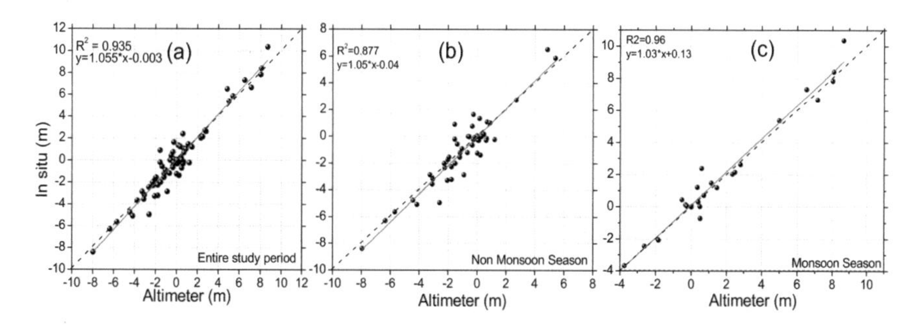

Fig. 18.6 Inter-comparison between relative water levels obtained from ENVISAT Altimeter and In-situ data for **a** entire study period, **b** non-monsoon period and **c** monsoon period

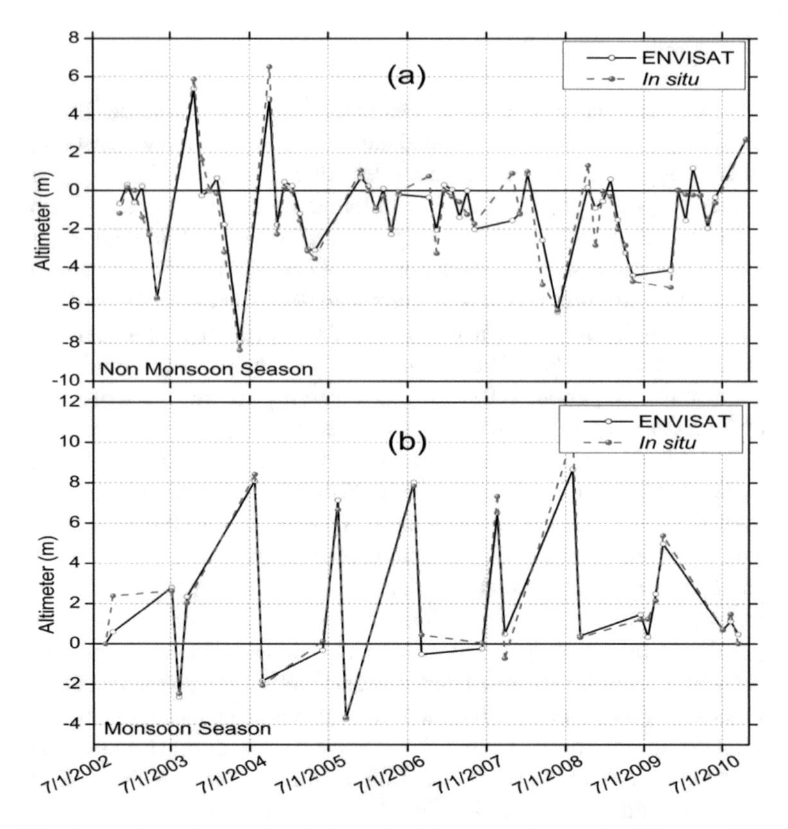

Fig. 18.7 Comparison of seasonal variations of relative reservoir water levels between ENVISAT and Gauge during **a** non-monsoon and **b** monsoon season

Table 18.1 Summary of the statistical parameters

	Non-monsoon	Monsoon
Accuracy range (m)	0.018–1.96	0.016–0.883
Standard deviation	0.477	0.223
RMSE	0.701	0.363
R^2	0.879	0.965

(Peng et al. 2006). Thus, the number of radar echoes that helps to improve accuracy depends on the size of the target water body.

18.4.2 Capacity

Water resource assessment studies require not only the stage that is derived from the altimeter data but also the corresponding capacity. In-situ gauge observations and the corresponding capacity for the study period were collected from the reservoir authorities. These data sets were used to derive the elevation-capacity rating curve and used to calibrate and validate the altimeter products. Rating curve is evaluated by applying the appropriate fit with the rest of the gauge height data. Figure 18.8a shows the relationship between measured capacity and ENVISAT derived elevation for the study area during the study period. The regression equation is obtained between the elevation and capacity. This equation is then used to obtain the reservoir capacity based on the ENVISAT measurements. The capacity derived based on altimetry observations using the regression equation compares very well with the observed capacity as shown in Fig. 18.8b. The same equation is also used to derive the reservoir capacity based on the SARAL/AltiKa measurements and the results are observed to be matching well with an R^2 of 0.93 as shown in Fig. 18.8c. Similarly, the variability of capacity was accurately estimated with a good R^2 value for datasets and are within ±20% of the observed capacity. From this analysis, it is inferred that the accurate estimation of capacity is achieved due to the minimal changes in the rating curve with time. An RMSE of 0.44 m represents an error of 7.2% in capacity in the wet season. Although 35-day, repeativity of ENVISAT and SARAL/AltiKa data are insufficient for reservoir monitoring at a daily scale, this study demonstrates the potential of altimeters for estimating the capacity of ungauged smaller water bodies or reservoirs.

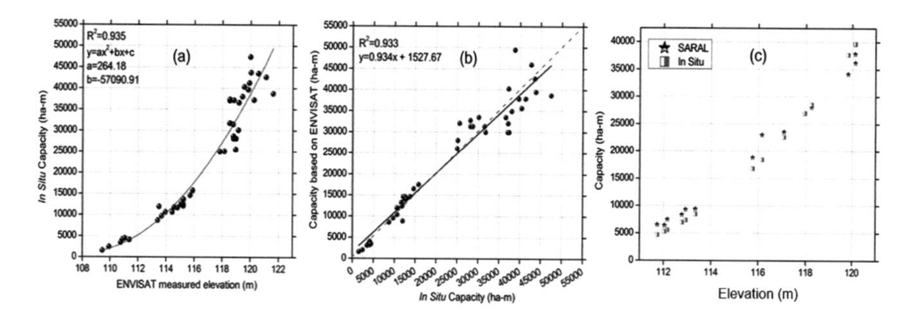

Fig. 18.8 Inter-comparison between the capacity estimated from in-situ observations and **a** ENVISAT elevation, **b** In-Situ capacity and ENVISAT estimated capacity and **c** In-situ elevation and SARAL/AltiKa estimated capacity

18.5 Conclusions

Present study establishes that the potential of radar altimeters in providing long-term data on the relative heights over smaller reservoirs particularly during monsoon season. It was also observed that the altimeter could resolve both the seasonal and inter-annual variability. However, the applicability of altimetry in reservoir monitoring at a daily scale is often limited due to a reduction in the accuracy during non-monsoon period when the reservoir water level is low. Particularly during non-monsoon season, the altimeter level variations from the Mayurakshi Reservoir are indicative of the combined floodplain and water body rather than the reservoir alone. Therefore, it could be stated that altimeter cannot accurately measure low water levels and also every chance of an increase in the error during dry period due to the emergence of plain land. However, altimetric data during non-monsoon season are still admissible but must be used selectively. Coarse spatial resolution of the altimeter data would render it invalid in monitoring the frequent changes in water surface area during dry seasons. Application of altimeter data from multiple missions could overcome some of the current limitations. The high frequency retracked waveform SARAL/AltiKa data could solve these major problems, however further studies are required on waveform retracking to improve the results further.

Acknowledgements Authors sincerely appreciate reservoir authorities for extending help during execution of this study. Thanks, are also due to RRSC-East and NRSC authorities for providing the necessary facilities for conducting the study.

References

Abileah B, Vignudelli S, Scozzari A (2011) A completely remote sensing approach to monitoring reservoir water volume. In: 15th International water technology conference, IWTC 15. Alexander, Egypt

Alsdorf DE (2003) Water storage of the central Amazon floodplain measured with GIS and remote sensing imagery. Ann Assoc Am Geogr 93(1):55–66

Alsdorf DE, Smith LC, Melack JM (2001) Amazon foodplain water level changes measured with interferometric SIR-C radar. IEEE Trans Geosci Remote Sens 39(2):423–431

Alsdorf DE, Rodrıguez E, Lettenmaier DP (2007) Measuring surface water from space. Rev Geophys **45**:RG2002. https://doi.org/10.1029/2006RG000197

Au AY, Brown RD, Welker JE (1989) Analysis of altimetry over inland seas. NASA Tech Memo, 100729

Bates PD, Wilson MD, Horritt MS, Mason DC, Holden N, Currie A (2006) Reach scale floodplain inundation dynamics observed using airborne synthetic aperture radar imagery: data analysis and modelling. J Hydrol 328:306–318

Berry PAM, Garlick JD, Freeman JA, Mathers EL (2005) Global in—land water monitoring from multi-mission altimetry. Geophys Res Lett 32:L16401. https://doi.org/10.1029/2005GL022814

Birkett CM (1994) Radar altimetry: a new concept in monitoring lake level changes. Eos, Trans Am Geophys Union 75(24):273. https://doi.org/10.1029/94EO00944

Birkett CM (1998) Contribution of the TOPEX NASA Radar Altimeter to the global monitoring of large rivers and wetlands. Water Resour Res 34(5):1223–1239. https://doi.org/10.1029/98WR00124

Birkett CM (2000) Synergistic remote sensing of Lake Chad: variability of basin inundation. Remote Sens Environ 72:218–236

Birkett CM (1995) The contribution of TOPEX/POSEIDON to the global monitoring of climatically sensitive lakes. J Geophys Res 100(95)

Birkinshaw SJ, Moore P, Kilsby CG, O'Donnell GM, Hardy AJ, Berry PAM (2014) Daily discharge estimated at ungauged river sites using remote sensing. Hydrol Processes 28:1043–1054

Bogning S, Frappart F, Blarel F, Nino F, Mahe G, Bricquet J-P, Seyler F, Onguéné R, Etame J, Paiz M-C, Braun J-J (2018) Monitoring water levels and discharges using radar altimetry in an ungauged river basin: the case of the Ogooué. Remote Sens 10(2):350. https://doi.org/10.3390/rs10020350

Brooks RL (1982) Lake elevations from satellite radar altimetry from a validation area in Canada, report. Geosci. Res. Corp., Salisbury, Maryland, U.S.A., 1982

Cai X, Ji W (2009) Wetland hydrologic application of satellite altimetry—a case study in the Poyang Lake watershed. Progr Nat Sci 19:1781–1787

Cheng KC, Kuo CY, Tseng HZ, Yi Y, Shum CK (2010) Lakesurface heigh calibration of Jason-1 and Jason-2 over the great lakes. MarGeodesy 33:186–203

Chowdary VM, Sai krishnaveni A, Suresh Babu AV, Sharma RK, Rao VV, Nagaraja R, Dadhwal VK (2017) Reservoir capacity estimation using SARAL/AltiKa altimetry data coupled with Resourcesat P6-AWiFS and RISAT 1 microwave data. Geocarto Int 32(9):1034–1047. https://doi.org/10.1080/10106049.2016.1188165

Coe MT, Birkett CM (2004) Calculation of river discharge and prediction of lake height from Satellite radar altimetry: example for the Lake Chad basin. Water Resour Res 40:W10205

Cretaux J-F, Nielson K, Frappart F, Papa F, Calmant S, Benveniste J (2017) Hydrological applications of satellite altimetry: rivers, lakes, man-made reservoirs, inundated areas. In: Satellite Altimetry over Oceans and Land surfaces, Stammer D, Cazenave A (eds) Earth observation of global changes; Boca Raon, FL, USA, CRC Press, pp. 459–504

Dubey AK, Gupta P, Dutta S, Singh RP (2015) Water level retrieval using SARAL/AltiKa observations in the braided Brahmaputra River, Eastern India. Marine Geodesy 1008156. https://doi.org/10.1080/01490419.2015

Frappart F, Calmant S, Cauhope M, Seyler F, Cazenave A (2006) Validation of ENVISAT RA-2 derived water levels over the Amazon Basin. Remote Sens Environ 100(2):252–264

Frappart F, Papa F, Marieu V, Malbeteau V, Jordy F, Calmant S, Durand F, Bala S (2015) Preliminary assessement of SARAL/AltiKa observations over the Ganges-Brahmaputra and Irrawaddy Rivers. Mar Geodesy. https://doi.org/10.1080/01490419.2014.990591

Ghosh S, Thakur PK, Garg V, Nandy S, Aggarwarl S, Saha SK, Sharma R, Bhattacharya S (2015) SARAL / AltiKa waveform analysis to monitor inland water levels: a case study of Maithon Reservoir, Jharkhand, India. *M*arine Geodesy.: https://doi.org/10.1080/01490419.2015.1039680

Jain SK, Goel MK (2002) Assessing the vulnerability to soil erosion of the Ukai dam catchment using remote sensing and GIS. Hydrol Sci J 47(1):31–40

Koblinsky CJ, Clarke RT, Brenner AC, Frey H (1993) Measurement of river level variations with satellite altimetry. Water Resour Res 29(6):1839–1848

Kouraev AV, Zakharova EA, Samain O, Mognard NM, Cazennave A (2004) Ob' River discharge from Topex-Poseidon satellite altimetry (1992–2002). Remote Sens Environ 93:238–245

Leon JG, Calmant S, Seyler F, Bonnet MP, Cauhopé M, Frappart F (2006) Rating curves and average water depth at the Upper Negro river from satellite altimetry and model leddis charges. J Hydrol 328(3–4):481–496

Manavalan P, Sathyanath P, Rajegowda GL (1993) Digital image analysis techniques to estimate the water-spread for capacity evaluation of reservoirs. PE&RS. 59:1389–1395

Mason IM, Harris A, Birkett CM, Cudlip WC, Rapley CG (1990) Remote sensing of lakes for the proxy monitoring of climatic change. In: Remote sensing and climate change, Proceedings of the

16th annual conference of the remote sensing society, University College Swansea, pp 314–324, Remote Sens Soc., Nottingham, England

Michailovsky CI, McEnnis S, Berry PAM, Smith R, Bauer Gottwein P (2012) River monitoring from satellite radar altimetry in the Zambezi river basin. Hydrol Earth Syst Sci Discuss 9:3203–3235

Milzow C, Krogh PE, Bauer-Gottwein P (2011) Combining satellite radar altimetry, SAR surface soil moisture and GRACE total storage changes for hydrological model calibration in a large poorly gauged catchment. Hydrol Earth Syst Sci 15:1729–1743

Morris CS, Gill SK (1994a) Variation of Great Lakes water levels derived from Geosat altimetry. Water Resour Res 30(4):1009–1017

Morris CS, Gill SK (1994b) Evaluation of the TOPEX/POSEIDON altimeter system over the Great Lakes. J Gephys Res 99(C12):24527–24539

Olliver JG (1987) An analysis of results from SEASAT altimetry over land and lakes, paper presented at IAG Symposium, IUGG XIX General Assembly, Int. Assoc. of Geod., Vancouver

Papa F, Durand F, Rossow WB, Rahman A, Bala SK (2008) Satellite altimeter-derived monthly discharge of the Ganga-Brahmaputra River and its seasonal to interannual variations from 1993 to 2008. J Geophys Res (oceans) 115:C12013

Peng D, Guo S, Liu P, Liu T (2006) Reservoir storage curve estimation based on remote sensing data. J Hydrol Eng 11(2):165–172

Rapley CG (1990) Satellite radar altimeters. In: Vaughan RA (ed) Microwave remote sensing for oceanographic and marine weather-forecast models, pp 45–63. Kluwer Acad., Norwell, Mass

Rapley CG, Guzkowska MAJ, Cudlip W, Mason IM (1987) An exploratory study of inland water and land altimetry using seasat data. Rep. 6483/85/NL/BI, Eur. Space Agency, Paris

Sai Krishnaveni A, Chowdary VM, Dutta, D, Sharma JR., Dadhwal VK (2016) ARAL/AltiKa altimetry data for monitoring of inland water body: a case study of Mayurakshi Reservoir, India. J Indian Soc Remote Sens 44:797–802. https://doi.org/10.1007/s12524-015-0535-4

Salami YD, Nnadi FN (2012) Seasonal and interannual validation of satellite measured reservoir levels at the Kainji dam. Int J Water Resour Environ Eng 4(4):105–113

SARAL/Altika Products Hand book (2013) CNES: SALP-MU-M-OP-15984-CN, ISRO, ISS 2.4 9, Dec 2013. Available at https://www.aviso.altimetry.fr/fileadmin/documents/data/tools/SARAL_Altika_products_handbook.pdf.

Silva JS, Calmant S, Seyler F, Rotunno Filho OC, Cochonneau G, Mansur WJ (2011) Water levels in the Amazon basin derived from the ERS 2 and ENVISAT radar altimetry missions. Remote Sens Environ 114:2160–2181

Smith LC, Pavelsky TM (2008) Estimation of river discharge, propagation speed, and hydraulic geometry from space: Lena River Siberia. Water Resour Res 44:W03427. https://doi.org/10.1029/2007WR006133

Tang QH et al (2009) Remote sensing: hydrology. Prog Phys Geogr 33(4):490–509. https://doi.org/10.1177/0309133309346650

Tapley BD, Bettadpur S, Watkins M, Reigber C (2004) The gravity recovery and climate experiment: mission overview and early results. Geophys Res Lett 31:L09607

UNESCO (2010) Application of satellite remote sensing to support water resources management in Africa: results from the TIGER Initiative. Technical Documents in Hydrology, no 85, UNESCO, Paris, p 153

Vignudelli S, Scozzari A, Abileah R, Gomez-Enri J, Benveniste J, Cippollini P (2019) Water surface elevation in coastal and inland waters using satellite radar altimetry. In: Maggioni V, Massari C (eds) Extreme Hydroclimatic events and multivariate hazards in a changing environment: a remote sensing approach. Elsevier, pp 87–127. ISBN: 9780128148990

Zakharova EA, Kouraev AV, Cazenave A, Seyler F (2006) Amazon river discharge estimated from TOPEX / Poseidon altimetry. Comptes Rendus Geosci 338:188–196

Chapter 19
Geospatial Technologies for Assessment of Reservoir Sedimentation

Rajashree Vinod Bothale, V. M. Chowdary, R. Vinu Chandran, Gaurav Kumar, and J. R. Sharma

Abstract Soil erosion in the catchment area of the reservoir catchment leads to sedimentation problem in the reservoir thereby affecting its both live and dead storage capacities that reduces the designed life span and planned economic benefits. Conventional techniques for assessment of reservoir sedimentation not only involve lot of manpower but also time intensive and costly to implement. Remote sensing techniques by virtue of its synoptic coverage and multidate observations are reported to be quite useful for computation of reservoir live capacity. These surveys are fast, economical and reliable. The present study was taken up to update the stage—area—capacity curves (estimating loss in the live storage capacity) for 30 reservoirs spread across India, where delineated waterspread area corresponding to satellite pass forms the important basis. The difference between the present satellite measured waterspread area and that of a previous survey (obtained through hydrographic survey) is the areal extent of silting at these levels. Integrating the area over different levels gives an estimate of volume of silting observed by satellite between the maximum and minimum reservoir level.

R. V. Bothale
National Remote Sensing Centre, Hyderabad, India

V. M. Chowdary (✉)
Regional Remote Sensing Centre-North, NRSC, New Delhi, Delhi, India

R. Vinu Chandran
National Remote Sensing Centre, Hyderabad, India

Former Scientist, Regional Remote Sensing Centre-East, NRSC, Kolkata, India

J. R. Sharma
Former Scientist, National Remote Sensing Centre, Hyderabad, India

G. Kumar
Regional Remote Sensing Centre-West, NRSC, Jodhpur, India

A. Pandey et al. (eds.), *Geospatial Technologies for Land and Water Resources Management*, Water Science and Technology Library 103,
https://doi.org/10.1007/978-3-030-90479-1_19

19.1 Introduction

Gradual reduction of reservoir capacities due to silting is great concern, and efforts should be made to study its impact on all the water resources development projects globally. Hence, accurate assessment of the sedimentation rate is necessary not only for studying its effect on the reservoir life but also required for planning reservoir operations optimally. Particularly, impact of sedimentation on multipurpose reservoirs is significant as silting affects both dead storage and live storage capacities, which in turn induce long and short-range impacts on reservoir operations. Particularly, sedimentation adversely affects planning for long-term utilization of reservoir capacity for irrigation, power generation, drinking water supply and flood moderation. Sedimentation in the reservoir also affects the river regime, which increases the back water levels in head reaches of reservoir and formation of island deltas. Some reservoirs in the world have been silted up so fast that have become useless. Yasuka reservoir in Japan has lost 85% capacity in less than 13 years (CWC 2002). Several reservoirs spread in India are experiencing capacity losses at the rate of 0.2–1% per annum. Nearly 40,000 minor tanks in Karnataka have lost more than 50% of their capacities. Therefore, it is necessary that the surveys be conducted in all the existing reservoirs for ascertaining siltation rate and their useful life. Thus, periodical surveys will enable selection of appropriate measures for controlling sedimentation along with efficient management and operation of reservoirs thereby deriving maximum benefits for the society.

Knowledge on the characteristics of reservoir sedimentation that include amount, distribution configuration and configuration of reservoir deposits is necessary to understand the sedimentation mechanism. Velocity of water that enters into the reservoir is inversely proportional to channel cross-sectional area. Owing to different densities of clear stored reservoir water and muddy inflows into reservoir, heavy turbid water flows along the channel bottom towards the dam under the influence of gravity (Fig. 19.1) (Varshney 1997). It is believed that sediment deposition takes place at the bottom of reservoir affecting dead storage, but contrary to earlier belief that sediment is deposited throughout the reservoir thereby reducing the incremental capacity at all elevations. Several factors such as sediment loading, size distribution, stream discharge fluctuations, reservoir shape, stream valley slope, land cover at the reservoir head, reservoir size and its location, outlets control the sediment movement and deposition pattern in the reservoir.

Recent past, much attention was paid to issues such as reservoir sedimentation that affects live storage capacity of reservoirs due to incoming sediments (Garde 1995; Varshney 1997; Morris and Fan 1997; Goel et al. 2002; Jain and Goel 2002). Narayan and Babu (1983) reported that soil erosion rate in India is nearly 0.16 t/km^2/year, where 10% of sediments contributes to reservoir sedimentation problems and 29% is deposited into the sea. CWC (2002) reported that nearly 20% of the live storage capacity of major and medium reservoirs in India was lost due to sedimentation, which means a loss of irrigation potential of about 4 M ha. Several reservoirs in India are experiencing capacity losses at the rate of 0.2–1.5% per annum (CWC

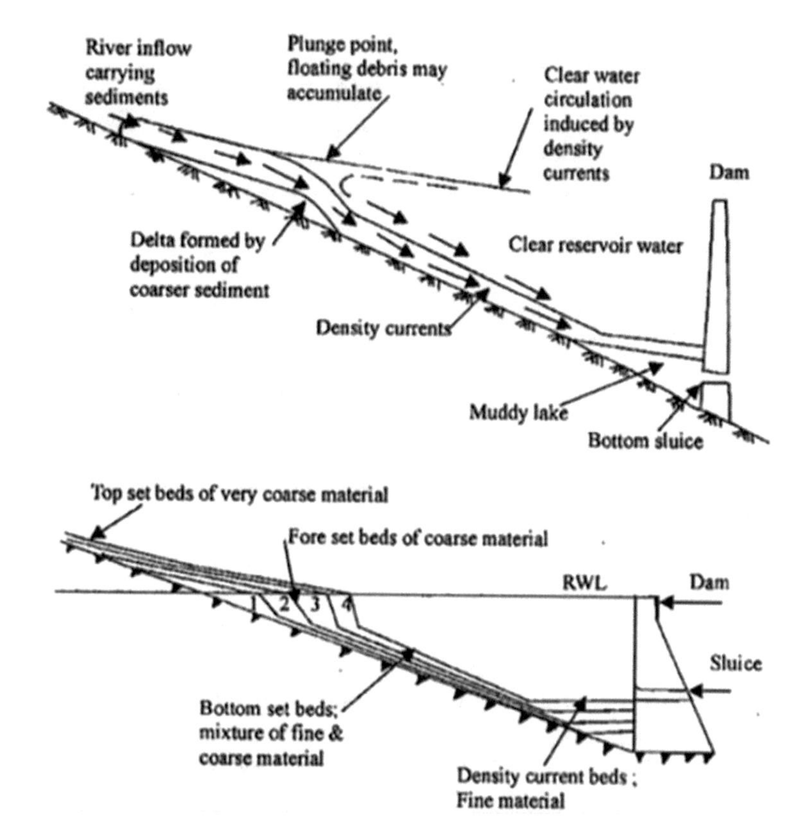

Fig. 19.1 Conceptual sketch of density currents and sediments in a reservoir (*Source* Varshney 1997)

2001, Goel et al. 2002). Shangle (1991) analysed sedimentation surveys based on the 43 reservoirs, which revealed that sedimentation rates range between 0.34 and 27.85 ha m/100 km^2/year, 0.15 and 10.65 ha m/100 km^2/year and 1.0 and 2.3 ha m/100 km^2/year for major, medium and minor reservoirs, respectively. Analysis of capacity surveys also shows a wide variation in sedimentation rates that may be attributed to various factors like physiography, climate, hydrometeorology, etc. Region-wise average rate of siltation in India is given in Table 19.1.

19.2 Sedimentation Surveys

Sedimentation surveys can be carried out either through conventional ways or by using remote sensing data.

Table 19.1 Region-wise average rate of siltation

Region	No. of reservoirs	Average rate of siltation (000m^3/km^2/year)
Himalayan region (Indus, Ganges and Brahmaputra basin)	31	1.22
Indo-Gangetic plains	15	0.95
East flowing rivers up to Godavari (Excluding Ganges)	5	0.76
Deccan Peninsular east flowing rivers including Godavari and south Indian rivers	114	2.27
West flowing rivers up to Narmada	53	1.12
Narmada and Tapi basins	10	2.84
West flowing rivers beyond Tapi and south Indian rivers	35	3.07

(*Source* CWC 2020)

19.2.1 Conventional Surveys

Conventional surveys for reservoir capacity estimation conducted so far are time intensive and often require up to three years to complete the survey of a major reservoir. These surveys involve the conventional equipment such as theodolite, plane table, sextant, range finders, sounding rods, echo-sounders and slow moving boat. However, field surveys are difficult to carry out in the dense forests due to presence of wild life and heavy rainfall during monsoon season. This problem is further compounded due to difficult intervisibility land steep foreshores. In general, these surveys are supposed to be conducted at the normal frequency of five years. However, in practice, it is found that the interval between two successive surveys varies between 2 and 15 years. Modern survey techniques such as hydrographic data acquisition systems (HYDAC) were also used for capacity surveys. This system uses ultrasonic waves for depth measurement. A short sound pulse is emitted by a transducer in the form of a beam vertically towards the bottom. Part of sound energy is reflected from the bed and returns as an echo to the same transducer, which operates both as transmitter and receiver. This eliminates the possibility of angular errors in shallow depths. Depth is computed from the time recorded between emission of pulse and return of its echo. With the help of graphic recorder, a continuous curve presenting a true picture of the bottom is obtained. The range of the stem is up to 1500 m (33 kHz) and 250 m (210 kHz), and its accuracy is ±5 cm. The system is also equipped with transducers, which can detect and allow the measurement of low-density sediment deposits. Use of global positioning system (GPS) has also become common in HYDAC system. Simultaneous use of bathymeter and differential GPS (DGPS) gives x, y, z coordinates. Hydrographic surveys based on DGPS allow faster data acquisition with better accuracy than any previous hydrographic techniques. Thus, DGPS-based surveys eliminate the need to establish line of sight between base stations to boat, while achieving centimetre accuracy data. Central water commission carried out surveys of eight reservoirs, viz. Matalia (UP), Konar and Tilaiya (Bihar), Idukki and Kakki (Kerala), Balimela (Odisha), Linganamakki (Karnataka) and Jayakwadi (Maharashtra) using HYDAC system.

19.2.2 Remote Sensing Survey

Remote sensing data by virtue of its synoptic viewing capability and large area coverage at periodic time periods is of great help for reservoir capacity assessment. It also helps in the acquisition of timely, reliable and comprehensive data on various natural resources. Particularly, its capability to map and monitor inaccessible areas helps to acquire data on reservoir waterspread, which is an important parameter for reservoir capacity surveys. Employing geospatial technologies for reservoir capacity surveys has gained importance and popularity in the recent past globally, particularly with the launch of satellite missions such as Landsat, SPOT, and IRS (Goel et al. 2002). Mohanty et al. (1986) reported that the area capacity curve derived from the multidate Landsat imagery for Hirakud reservoir, Orissa, was nearly coincided with the curve obtained through conventional means. Suvit (1988) correlated the Landsat-MSS-derived surface areas of Ubolratana reservoir in Thailand with the corresponding water levels for computing reservoir capacity. Managond et al. (1985) analysed Landsat-MSS data for estimating the waterspread areas of Malaprabhas reservoir as on different dates and compared with the field surveyed figures. Rao et al. (1988) used multidate Landsat-MS imageries for evaluating capacity of Sriramsagar reservoir. Chandrasekhar et al. (2000) calculated capacity loss for Tawa reservoir as 14.2% in 20 years since its impoundment in 1975. Remote sensing-based approach was envisaged by many researchers in the past for reservoir capacity assessment (Managond et al. 1985, Manavalan et al. 1993, Goel and Jain 1996, 1998, Gupta 1999, Jain et al. 2002, Rajashree et al. 2003, 2004). Peng et al. (2006) derived new storage curve for Fengman reservoir, China, using Landsat imagery and reported that the remote sensing-based storage curve estimation is reasonable and has relatively high precision.

Contrary to remote sensing and conventional approaches, Jothiprakash and Garg (2009) used artificial neural network (ANN) model for assessment of sediment volume retained in the Gobindasagar reservoir on the Satluj river, India. In this approach, annual rainfall annual inflows and reservoir capacities were used as inputs. Several studies used *gauge* water levels and optical or microwave remote sensing data derived waterspread areas for computing reservoir capacities (Bates et al. 2006; Alsdorf et al. 2007; Smith et al. 2008). Recently, satellite altimetry provides an alternative means to monitor water-level variations over large water-bodies from space. Monitoring of rivers and large water bodies was demonstrated through SARAL/AltiKa and Envisat altimetry missions (Arsen et al. 2015; Dubey et al. 2015; Ghosh et al. 2015). Further, waterspread areas derived using SAR images or multispectral images were used in conjunction with altimetry-based water levels for reservoir capacity estimation of large lakes (Frappart et al. 2015; Ding and Li 2011; Duan and Bastiaanssen 2013; Chowdary et al. 2017). Overall, these studies highlighted that remote sensing-based surveys are less tedious and yielded good accuracy relative to traditional approaches. The present study was carried out with a specific objective of updating the stage—area—capacity curves (estimating loss in the live storage capacity) for 30 reservoirs spread across India due to sedimentation through satellite remote sensing.

19.3 Reservoirs Studied

Analysis of sedimentation was carried out for total 30 reservoirs distributed all over Indian river basins located in different geographical regions (Fig. 19.2a, b).

List of reservoirs analysed in this study along with basic details was presented in Table 19.2.

19.4 Satellite Data Used

Satellite data (IRS-1D LISS III and IRS P6 LISS III) at 23.5 spatial resolution data was selected based on their acquisition dates and cloud free data availability. Details of the satellite data used in this study are given in Table 19.3.

19.5 Methodology

Reservoir sedimentation surveys involve mapping of waterspread areas using satellite data for different water elevations between minimum draw down level (MDDL) and full reservoir level (FRL). In general, decrease of reservoir waterspread area at any elevation due to sediment deposition forms the basis for remote sensing-based surveys. Hence, delineation of waterspread areas corresponding to satellite passes on different dates using remote sensing is prerequisite for remote sensing analysis. Out of all land features, discrimination of land and water features on the image is considered to be easy due to high spectral variability between land and water features in near-infrared (NIR) band. This property enables to delineate waterspread area at different reservoir levels and hereby helps to revise elevation–capacity curve using this approach. Difference in the revised and previously computed reservoir capacity values (previous survey) can be attributed to the siltation problem in the reservoir. Remote sensing-based reservoir capacity assessment approach includes suitable satellite data selection, time series database creation, waterspread area delineation, capacity and capacity loss computations.

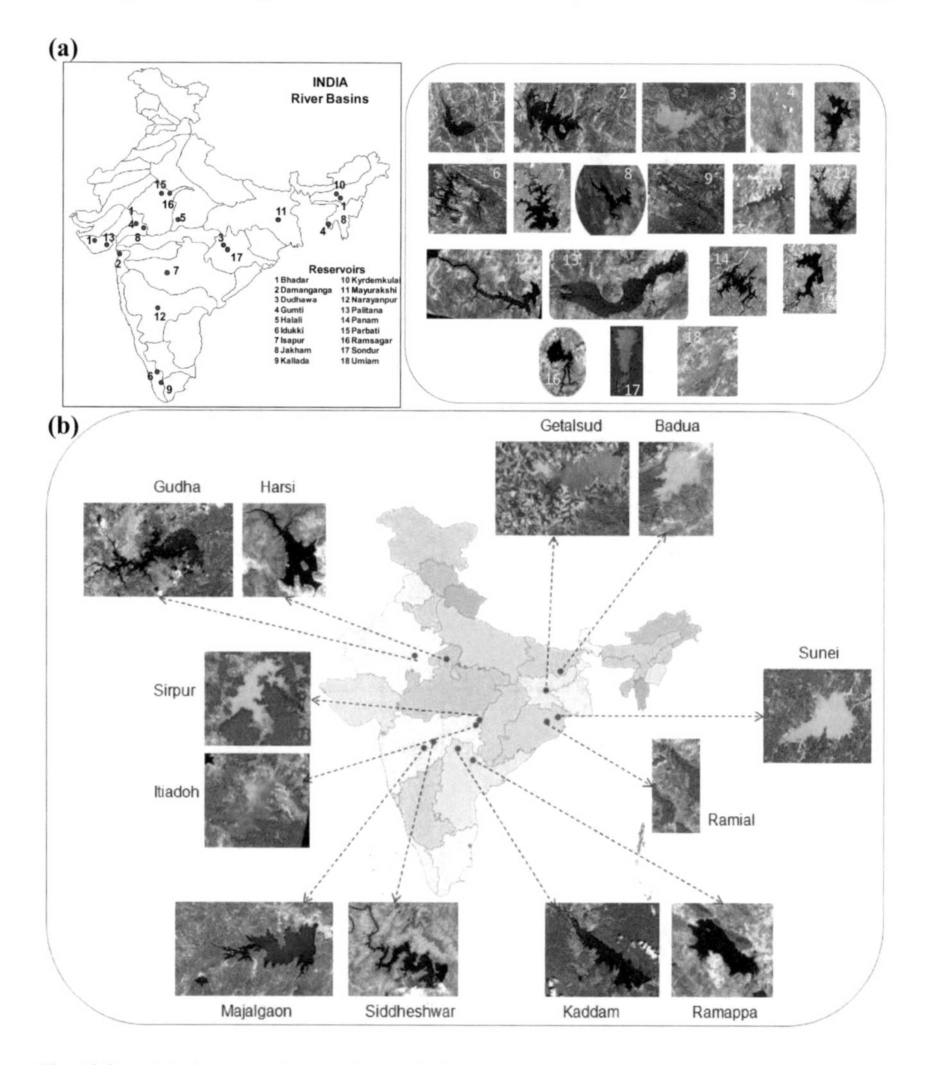

Fig. 19.2 **a**, **b** Index map of reservoirs studied

19.5.1 Criteria for Satellite Data Selection

Selection of suitable data is imperative for accurate reservoir capacity estimation. Before data procurement, first it is necessary to study the reservoir operation schedule from the reservoir authorities. The dates when the reservoir was at or near MDDL and FRL must be recorded. Satellite passes coinciding with MDDL and FRL (if available) are the best choice for analysis. If overpass does not occur on the same day then nearby dates can be chosen. All the available dates between MDDL and FRL spanning over either one operation year or two operation years can be taken up for

Table 19.2 Basic details of reservoirs studied

S. No.	Reservoir	State	River on which constructed	Region	Location	Year of impounding	Designed live capacity (Mm^3)	Catchment area (km^2)
1	Badua	Bihar	Badua	Ganges	25°30′ N/86°30′ E	1965	122.731	480.5
2	Bhadar	Gujarat	Bhadar	Dra. to Arabian sea	21°50′ N/70°46′ E	1964	223.703	2435.00
3	Damanganga	Gujarat	Damanganga	Dra. to Arabian sea	20°10′ N/73^{0}05’ E	1983	502.00	1813.00
4	Dudhawa	Chattisgarh	Mahanadi	Drainage to Bay of Bengal	20°18′ N/81°46′ E	1964	284.130	625.27
5	Getalsud	Jharkhand	Subaranrekha	Drainage to Bay of Bengal	23°27′ N/85°33′ E	1971	230.661	725
6	Gudha	Rajasthan	Chambal	Ganges	25°29′ N/75°28′ E	1956	93.587	744.96
7	Gumti	Tripura	Gumti	Brahmaputra	23°25′ N/91°49′ E	1984	312.9	338.00
8	Halali	Madhya Pradesh	Halali	Ganges	23°30′ N/77°30′ E	1977	226.940	699.00
9	Harsi	Madhya Pradesh	Parbati	Ganges	25°58′ N/77°58′ E	1935	192.66	880
10	Idukki	Kerala	Periyar	Dra. to Arabian sea	09°50′ N/76°58′ E	1974	1461.81	649.31
11	Isapur	Maharashtra	Penganga	Drainage to Bay of Bengal	19°43′ N/77°27′ E	1983	928.262	4650.00
12	Itiadoh	Maharashtra	Garvi	Drainage to Bay of Bengal	20°47′ N/80°10′ E	1971	317.874	704.48

(continued)

Table 19.2 (continued)

S. No.	Reservoir	State	River on which constructed	Region	Location	Year of impounding	Designed live capacity (Mm^3)	Catchment area (km^2)
13	Jakham	Rajasthan	Jakham	Dra. to Arabian sea	24°10′ N/74°35′ E	1986	132.28	1010.00
14	Kaddam	Telangana	Kaddam	Drainage To Bay of Bengal	19°07′ N/78°47′ E	1959	116.609	2590
15	Kallada	Kerala	Kallada	Dra. to Arabian sea	08°49 N/09°17′	1985	487.92	549.00
16	Kyrdemkulai	Meghalaya	Umtru	Brahmaputra	25°44′ N/91°48′ E	1983	3.824	150.00
17	Majalgaon	Maharashtra	Sindphana	Drainage to Bay of Bengal	19°15′ N/76° 23′ E	1990	312	3840
18	Mayurakshi	Jharkhand	Mayurakshi	Ganges	24°06′ N/87°17′ E	1955	547.59	1860.00
19	Narayanpur	Karnataka	Krishna	Drainage to Bay of Bengal	16°10′ N/76°21′ E	1982	867.889	46,850.00
20	Palitana	Gujarat	Shetrunji	Dra. to Arabian sea	21°28′ N/71°52′ E	1959	374.832	4317.00
21	Panam	Gujarat	Panam	Dra. to Arabian sea	23°03′ N/73°42′ E	1977	689.567	2312.0
22	Parbati	Rajasthan	Parbati	Ganges	26°37′ N/77°26′ E	1963	102.893	786.00
23	Ramappa	Telangana	Godavari	Drainage To Bay of Bengal	18°15′ N/79°56′31″ E	1919	79.918	183.81
24	Ramial	Odisha	Ramial	Drainage To Bay of Bengal	21°06′ N/85° 35′	1985	75.835	328

(continued)

Table 19.2 (continued)

S. No.	Reservoir	State	River on which constructed	Region	Location	Year of impounding	Designed live capacity (Mm^3)	Catchment area (km^2)
25	Ramsagar	Rajasthan	Bamni	Ganges	26°35′ N/77°35′ E	1905	29.39	176.01
26	Siddheshwar	Maharashtra	Purna	Drainage To Bay of Bengal	19° 35′ N/76° 57 E	1963	80.822	440
27	Sirpur	Maharashtra	Bagh	Drainage To Bay of Bengal	20°03′ N/80° 27′	1971	192.52	432.53
28	Sondur	Chattisgarh	Sondur	Drainage To Bay of Bengal	20°14′ N/82°06′ E	1988	179.611	518.00
29	Sunei	Odisha	Sono	Drainage To Bay of Bengal	21°28′ N/87° 28′ E	1991	61.6	227
30	Umiam	Meghalaya	Umiam	Brahmaputra	25°37′ N/91°51′ E	1965	131.70	2214.44

Table 19.3 Details of satellite pass used in the study

Reservoir	Path/row	Satellite data acquisition dates
Badua	106/54	07 October 07, 13 September 07, 24 November 07, 28 February 08, 05 March 07, 03 June 08, 14 February 06
Bhadar	91/57	13 October 02, 7 November 02, 2 December 02, 15 February 03, 26 May 03, 16 May 02, 10 June 02
Damanganga	94/58	19 October 01, 03 December 03, 04 October 02, 27 January 02, 18 March 02, 03 March 03, 17 May 03
Dudhawa	102/58	15 October 03, 20 October 01, 3 January 02, 19 March 02, 19 March 02, 8 May 02, 30 October 02, 4 March 03
Getalsud	105/55	07 October 06, 31 October 06, 24 November 06, 18 December 06, 17 April 07, 18 March 08, 29 May 08
Gudha	95/54	11 September 06, 22 November 06, 30 September 07, 09 January 07, 02 February 07, 02 June 07
Gumti	112/55	27 July 03, 14 December 02, 02 February 03, 24 March 03, 13 May 03, 08 April 02
Halali	97/55	05 October 03, 24 November 03, 25 September 02, 14 November 02, 27 Feb 01, 03 April 02, 02 June 03
Harsi	97/53	15 October 06, 08 November 06, 02 December 06, 03 November 07, 10 October 07, 17 June 06, 12 June 07
Idukki	100/67	15 December 01, 28 February 02, 15 March 01, 25 December 02, 10 March 03, 29 April 03
Isapur	98/58	01 November 01, 22 September 02, 31 December 02, 19 February 03, 5 May 03, 8 June 04
Itiadoh	100/57	19 August 06, 06 October 06, 17 December 06, 27 February 07, 23 March 07, 10 May 07
Jakham	94/55	19 October 01, 13 November 01, 04 October 02, 21 February 02, 18 March 02, 01 June 02
Kadam	100/59	06 October 06, 25 October 07, 18 November 07, 05 January 08, 17 March 08
Kallada	100/67	09 December 03, 25 December 02, 26 September 03, 10 March 03, 04 April 03, 29 April 03, 14 May 02
Kyrdemkulai	111/53	13 October 03, 08 September 02, 17 March 02, 16 January 01, 22 November 02, 11 January 03
Majalgaon	97/59	15 October 06, 02 December 06, 19 January 07, 08 March 07, 25 April 07, 19 May 07
Mayurakshi	107/55	13 January 02, 09 November 02, 17 February 03, 04 March 02, 08 April 03, 28 May 03
Narayanpur	98/61	6 December 02, 10 April 03, 11 November 02, 16 March 03, 6 March 02, 31 March 02, 20 May 02
Palitana	92 57	10 October 02, 04 November 02, 18 January 03, 09 March 03, 14 November 03, 28 April 03, 23 May 03
Panam	94/56	14 October 03, 28 December 03, 01 April 04, 04 October 02, 22 April 03, 11 June 03

(continued)

Table 19.3 (continued)

Reservoir	Path/row	Satellite data acquisition dates
Parbati	97/53	05 October 03, 25 September 02, 14 November 02, 22 February 03, 13 April 03, 02 June 03
Ramappa	101/60	11October 06, 28 November 06, 08 February 07, 04 March 07, 28 March 07, 08 June 07, 15 April 08, 09 May 08
Ramial	106/57	12 October 06, 05 November 06, 24 November 07, 16 January 07, 05 March 07, 29 March 07, 22 April 07, 03 June 08
Ramsagar	97/53	05 October 03, 25 September 02, 14 November 02, 22 February 03, 13 April 03, 02 June 03
Siddheshwar	98/59	20 October 06, 07 December 06, 31 December 06, 30 April 07, 06 April 07
Sirpur	101/57	30 October 07, 10 January 08, 22 December 06, 08 February 07, 28 March 07, 15 May 07
Sondur	103/58	12 October 03, 22 September 01, 6 December 01, 10 April 02, 5 May 02, 20 April 03, 9 June 03
Sunei	106/57	12 October 06, 24 November 07, 18 December 07, 16 January 07, 04 February 08, 05 March 07, 29 March 07, 22 April 07, 16 April 08
Umiam	111/53	08 September 02, 13 October 03, 26 January 02, 20 February 02, 17 March 02, 31 May 02

analysis. More the number of data, better will be the accuracy of sediment estimation between different water levels. The height interval between consecutive water levels should be around 1–3 m. Hence, optimum dates for acquisition of satellite data were selected based on the information about the minimum and maximum reservoir levels attained during last three years that was collected from reservoir authorities and water management directorate of CWC.

19.5.2 Extraction of Reservoir Spread Area

Researchers envisaged image thresholding, spectral indices, supervised and unsupervised classification-based approaches to maximum extent for monitoring water bodies (Nagarajan et al. 1993; Sharma et al. 1996; Goel et al. 2002; Manju et al. 2005). Modelling approaches based on multiple spectral bands were used to resolve boundary issues between land and water features (Goel et al. 2002). Threshold values can be found out for single band and ratioed image, and subsequently water pixels can be classified based on logical criteria that satisfy in both images. In general, both the supervised and unsupervised classification techniques can be used for demarcation of waterspread area. Conventional maximum likelihood classifier is not always appropriate for surface water mapping. Though the spectral signatures of water are distinguishable from other land features, it may fail to identify water pixels clearly at the land and water interface. Further, it depends on the expert's interpretation capability. Deep water bodies can be easily identifiable, while shallow water bodies can be falsely classified as soil. Contrary to this, likely chance of saturated soil being misclassified as water is high particularly along the reservoir periphery. Hence, spectral indices based on green and near IR bands can be effectively used for delineation of boundary water pixels. Ratioing of two spectral bands and thresholding water (having ratio higher than threshold) can help in enhancing a particular feature or class from the satellite data. Hence, in this study, thresholding technique is used for extraction of reservoir area.

In general, thresholds for single band or spectral index are identified through visual interpretation by experts. However, identification of thresholds for each satellite data pass is dynamic and needs to be done for each dataset independently. Further, aquatic vegetation may pose problems for clear demarcation of land and water boundary. Spectral reflectance of both shallow and deep waterbodies in the NIR region is minimum compared to red and green spectra bands, which signifies that the DN values for water features are significantly less than other land cover pixels. Normalized Difference Vegetation Index (NDVI) (Tucker 1979) and Normalized Difference Water Index (NDWI) (McFeeters 1996) are some of the spectral indices, which enhance vegetation and water features, respectively. Spectral indices NDVI and NDWI can be computed as given below:

$$\mathrm{NDVI} = (\mathrm{NIR} - \mathrm{R})/\mathrm{NIR} + \mathrm{R}) \quad (19.1)$$

where NIR and R are radiance values in near-infrared band in red band, respectively.

Normalized difference water index, NDWI, can be given as:

$$\mathrm{NDWI} = (\mathrm{G} - \mathrm{NIR})/(\mathrm{G} + \mathrm{NIR}) \quad (19.2)$$

where 'G' is radiance value in green band.

Density slicing technique is used for delineation of waterspread areas, where image pixel grey values are categorized based on expert's identified class intervals. Minimum and maximum pixel values were identified through histogram analysis, where water pixels that occupy lower histogram ranges in NDVI image and higher histogram ranges in NDWI image. A pixel is classified as a water pixel if it meets the threshold condition in both NDVI and NDWI ratioed images, else rejected from water class.

19.5.3 Refinement of Extracted Waterspread Area

Removal of isolated water pixels from the classified image is necessary as the area is continuous within the contour. Further, water pixels located at the mouth of the reservoir that do not account to reservoir spread area also need to be eliminated (Goel et al. 2002). In this study, manual editing process was adopted to refine the reservoir waterspread image. Further, the main river and numerous small rivulets that join the reservoir from multiple directions at the reservoir tail end are slightly at higher elevations and needs to be removed for better accuracy (Goel et al. 2002). Thus, waterspread area was quietly demarcated for study reservoirs by manually editing the few extended channels around the reservoir periphery.

19.6 Assessment of Revised Live Storage Capacity of Reservoir

Waterspread areas for each individual optical image for all the study reservoirs were identified. The waterspread area corresponding to each satellite data pass was computed by multiplying the number of water pixels by the pixel area (23 × 23). In general, reservoir capacity or storage volume in the period of satellite survey can be estimated through either trapezoidal, prismoidal or cone formula. In this study, modified trapezoidal formula was adopted for reservoir capacity assessment in this study and is given as below:

$$V = \sum h/3 * (A_1 + A_2 + \mathrm{SQRT}(A_1 * A_2)) \tag{19.3}$$

where V is the reservoir capacity between two successive elevations h_1 and h_2, h is the elevation difference $(h_1 - h_2)$, A_1 and A_2 are reservoir waterspread areas at elevation h_1 and h_2, respectively. Reservoir capacity assessment using this approach necessitates to compute areal spread at periodic intervals. Hence, waterspread areas at regular intervals were assessed using a best-fit second-order polynomial equation that was built based on area and elevation values.

The original elevation capacity tables before the impoundment of respective dams spread across India were obtained from the project authorities. Initially, original reservoir elevations and capacities corresponding to satellite pass date were used to construct best-fit second-order polynomial equation and subsequently capacities were computed for intermediate elevations. Difference between the original and revised volumes in each zone is attributed to capacity loss due to sedimentation. It was further assumed that the original and revised reservoir capacities at the lowest observation level are same. Subsequent aggregation of reservoir capacities at consecutive levels between the lowest and the highest observation levels yield cumulative revised reservoir capacity.

19.7 Effect of Erroneous Waterspread Area Estimation on Reservoir Capacity

Remote sensing-based reservoir sedimentation estimate is highly dependent on important factors such as delineated waterspread area accuracy, water elevation and the original elevation area capacity table. Thus, precise revision of elevation area capacity curves is quite possible through exact waterspread delineation. Thus, analysis was carried out to evaluate the change in the reservoir capacity as a result of underestimation or overestimation of reservoir spread area by $\pm 10\%$ through remote sensing.

19.8 Customized Software for Reservoir Capacity Estimation

Periodical assessment of reservoir capacity assessment using remote sensing necessitates the development of customized package that can help the user to carry out capacity estimation studies in a timely and reliable way. The analysis was carried out using software 'KSHAMTA', a PacKage for JalaSHay CApacity EstiMation and sTorage Loss Analysis. Environment for visualization of Images (ENVI) software along with its development tool Interactive Data Language (IDL) were used for development of semi-automated package 'KSHAMTA' for reservoir capacity estimation. KSHAMTA (Fig. 19.3) is a tailor-made application software package for estimation of the reservoir capacity and storage loss analysis. It provides end-to-end facility for multidate processing of satellite images starting from loading of satellite data to the area *Vs* capacity curve generation.

19.9 Results and Discussion

19.9.1 Extraction of Waterspread Area (WSA)

Waterspread area (WSA) was computed for each satellite date of pass for all the reservoirs. However, WSA needs be computed at regular intervals for capacity estimations, thereby WSA was computed using best-fit second-order polynomial equation using water levels for each satellite data pass and its corresponding WSA. Minimum and maximum WSA at FRL were observed for Kyrdemkulai reservoir as 0.787 M m^2 and Narayanpur reservoir as 123.333 M m^2. Different reservoir types were considered for analysis. Table 19.4 shows the categorization of reservoirs as per WSA. Analysis for Halali reservoir was carried out beyond FRL due to availability of satellite data. Kallada reservoir was analysed from penstock level, which is the actual operating level for Kallada.

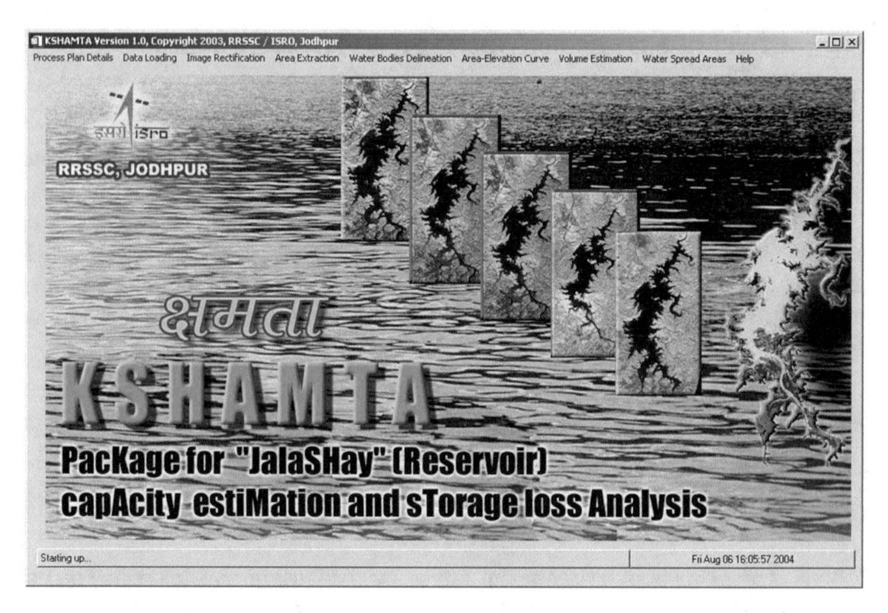

Fig. 19.3 Main screen of KSHAMTA software

Table 19.4 Categorization of reservoirs as per waterspread area

Reservoirs with WSA <1 sqkm	Reservoirs with WSA between 1 and 10 km^2	Reservoirs with WSA between 10 and 50 km^2	Reservoirs with WSA between 50 and 100 km^2	Reservoirs with WSA between 100 and 150 km^2
2	4	15	8	1

19.9.2 Live Storage Capacity Loss Due to Sedimentation

Cumulative reservoir capacity was compared with the capacity either from original elevation–area–capacity curve generated at the time of construction of reservoir or with the capacities computed from previous surveys. Change in the capacity gives overall loss due to sedimentation. Capacity loss in live capacity for all 30 reservoirs spread across India was carried out (Table 19.5). Sedimentation rates were also computed for these reservoirs between live storage level and MDDL (Table 19.5). Rate of silt deposited over the years was computed using catchment area of reservoirs.

Temporal distribution of Mayurakshi reservoir spread, Jharkhand and Sirpur reservoir spread in Maharashtra for different satellite data passes is shown in Figs. 19.4 and 19.5, respectively, as an example. The reservoir waterspread contours extracted from the satellite data along with revised elevation–area and elevation–capacity curves for both Mayurakshi and Sirpur reservoirs are shown as Figs. 19.4 and 19.5, respectively.

Remote sensing-based survey for Mayurakshi reservoir conducted during 2002–2003 revealed that the storage capacity has gone down from 547.6 in 1955 to 485.4 M

Table 19.5 Live capacity loss and rate of silt deposition

S. No.	Reservoir	Original capacity (Mcum)	Calculated capacity using RS (Mcum)	Capacity loss in live storage (%)	Number of years in operation	Rate of silt deposited (mm/year)
1	Badua	122.731	102.112	16.8	43	1.00
2	Bhadar	225.06	191.74	14.28	38	0.345
3	Damanganga	502	476.133	5.15	19	0.271
4	Dudhawa	284.13	284.806	*	39	*
5	Getalsud	230.661	200.775	12.96	36	1.15
6	Gudha	93.587	68.392	26.92	49	0.69
7	Gumti	312.9	249.07	20.4	19	9.94
8	Halali	226.94	188.583	16.902	27	2.032
9	Harsi	192.66	171.016	11.23	72	0.34
10	Idukki	1461.8	1464.385	*	29	*
11	Isapur	928.26	899.629	3.08	20	0.308
12	Itiadoh	317.874	318.292	*	35	*
13	Jakham	132.28	175.45	*	17	*
14	Kaddam	116.609	97.544	16.35	49	0.15
15	Kallada	423.95	376.705	11.14	18	4.78
16	Kyrdemkulai	3.824	3.414	10.72	19	0.144
17	Majalgaon	312	287.813	7.75	17	0.37
18	Mayurakshi	547.59	485.41	11.35	48	0.69
19	Narayanpur	867.89	842.254	2.954	21	0.025
20	Palitana	374.83	324.307	13.48	44	0.265
21	Panam	689.57	660.993	4.14	26	0.475
22	Parbati	102.89	86.405	16.02	40	0.524
23	Ramappa	79.918	74.850	6.342	88	0.31
24	Ramial	75.835	71.372	5.89	22	0.62
25	Ramsagar	29.397	24.663	16.1	98	0.274
26	Siddheshwar	80.822	79.131	2.09	44	0.09
27	Sirpur	192.52	159.165	17.325	35	2.20
28	Sondur	179.61	134.788	24.95	15	5.768
29	Sunei	61.6	59.498	3.412	16	0.58
30	Umiam	131.7	130.124	1.19	37	0.02

*Calculated capacity is more than designed capacity

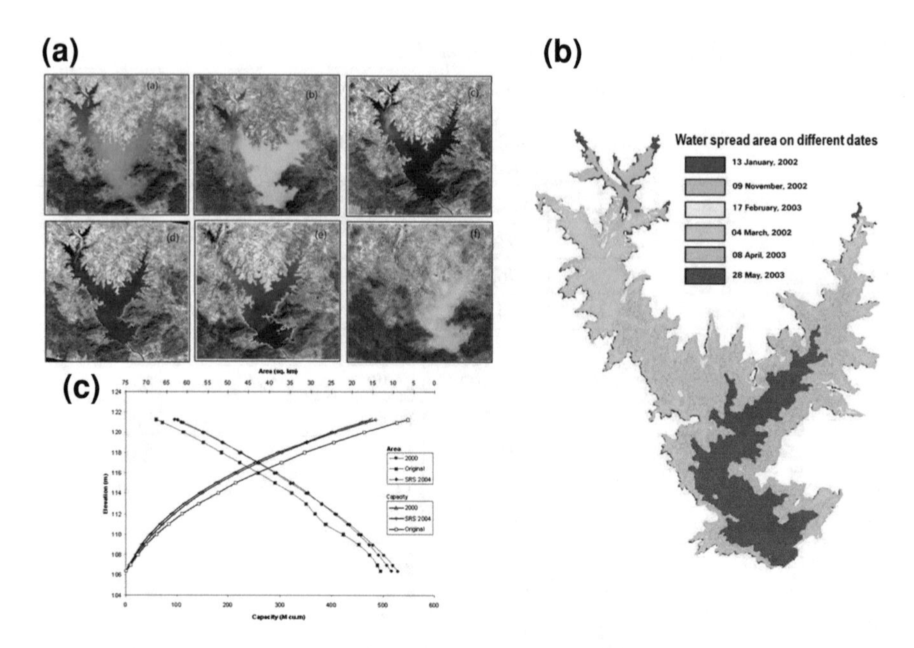

Fig. 19.4 Mayurakshi reservoir, Jharkhand state **a** false colour composites acquired on different dates **b** waterspread areas on different dates **c** elevation–area–capacity curves

m^3 in 2002–2003 (Table 19.5). The hydrographic survey also proved that the storage capacity has reduced from 547.6 M m^3 in 1955 to 474.8 M m^3 in 1999–2000, which substantiated the suitability of remote sensing-based approach. Overall remote sensing-based approach revealed that the loss of live storage capacity of reservoir is around 11.4% within a span of 47 years. While hydrographic survey indicated nearly 13.3% of total live capacity loss during the period 1955–2000, remote sensing-based survey (2002–2003) revealed that the rate of siltation is 1.3 M m^3/year while the rate of silting is found to be 1.6 M m^3/year as per conventional survey conducted during the year 2000. Sirpur reservoir that was constructed across Bagh river in Maharashtra state indicated capacity loss in live storage of nearly 17.325% (Table 19.5).

Original and computed area capacity values for reservoirs and percent change in the capacity are shown in Figs. 19.6 and 19.7, respectively.

Umiam reservoir in Maghalaya state has minimum capacity loss of 1.19% in the span of 37 years (Table 19.5). Gudha reservoir has maximum capacity loss of 26.92% in a life span of 49 years whereas Sondur reservoir has capacity loss of 24.95% in life span of 15 years only, which is quite alarming (Fig. 19.7). Large capacity reduction is observed in Halali, Gumti and Sirpur reservoirs also (Fig. 19.7). Catchment characteristics play an important role and are responsible for erosion processes in the catchment and subsequent capacity loss in the live capacity of reservoirs. Capacity loss was observed to be low for Narayanpur as this reservoir is in succession to other reservoir, thereby getting less silt. Most of the reservoirs followed normal decreasing trend of capacity between different time periods (Fig. 19.7). Reservoirs in Gujarat

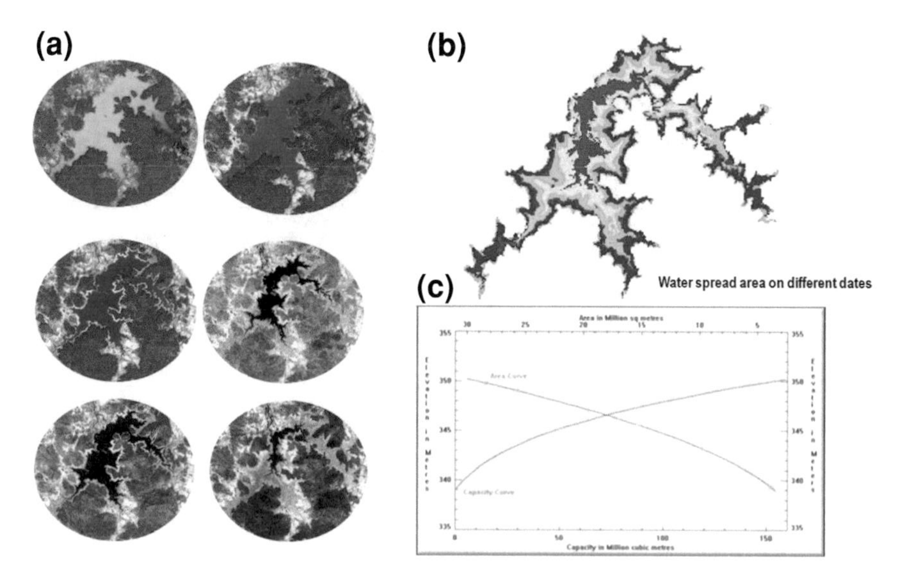

Fig. 19.5 Sirpur reservoir, Maharashtra state **a** false colour composites acquired on different dates **b** waterspread areas on different dates **c** elevation–area–capacity curves

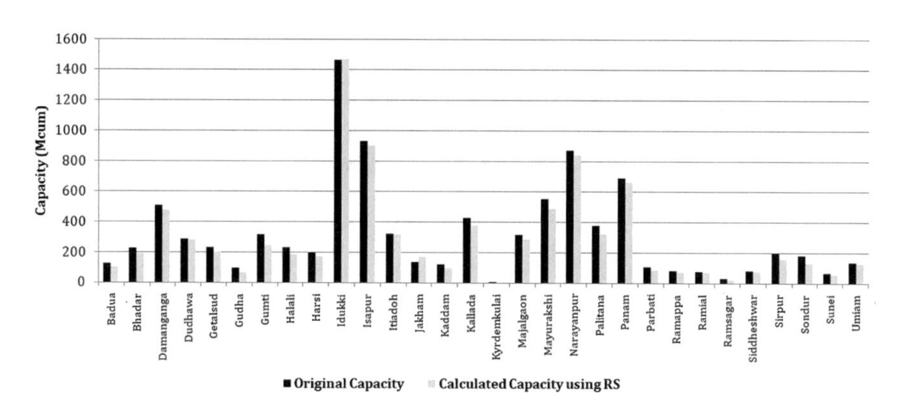

Fig. 19.6 Original capacity and calculated capacity using RS

that were analysed by one agency show more capacity loss when compared to the capacity obtained using remote sensing. Jakham reservoir shows absolute reverse trend, where present capacity is much higher than the original capacity itself. To cross check such results, original capacity was computed using original area values and the capacity formula. Most of the discrepancies could be explained with this approach.

Rate of sedimentation was compared with number of years of reservoir operation (Fig. 19.8). Rate of sedimentation is more for reservoirs, which are relatively young with less operating years. This may be attributed to the fact that during the initial

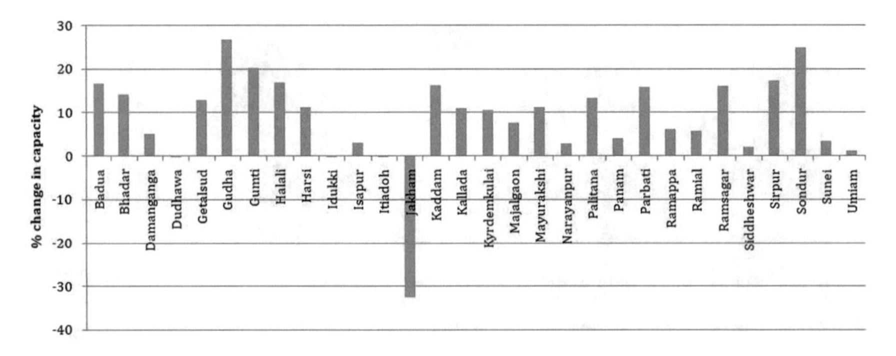

Fig. 19.7 Percent change in the capacity of reservoirs

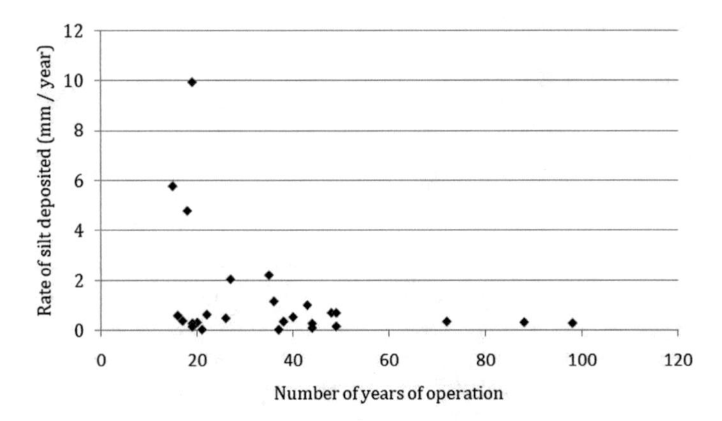

Fig. 19.8 Rate of sedimentation versus life of reservoir

stages of reservoir, sedimentation from catchment is generally more. As the reservoir life increases, stability of the regime changes with time and the rate of sedimentation gradually reduces.

Remote sensing-based method is promising as satellite-derived reservoir spreads are more realistic, whereas conventional hydrographic surveys suffer from certain inherent limitations. Particularly, these surveys face difficulties in fixing survey ranges, movement of boat along survey lines, dynamic and complex nature of silt deposits at reservoir bottom which effects on reflectance of sound waves during echo sounding. It was reported by Rajeev et al. (1992) that remote sensing storages are relatively underestimated. Higher rate of siltation than the assumed value at planning stage can be attributed to severe soil erosion in the reservoir catchments due to inadequate forest cover, denudation of forest, uncontrolled grazing and negligence of available pasture land. Conservation measures, viz. construction of retention terraces, bench terraces, gull control works and silt detention dams were adopted in some of the catchment areas, still the catchments are experiencing soil degradation.

Table 19.6 Sensitivity analysis of waterspread area and reservoir capacity for Mayurakshi reservoir

RS-based reservoir capacity survey (M m^3)	Error in overestimation of waterspread area (%)	Live capacity w.r.t overestimation of waterspread (M m^3)	Increase in the capacity (%)	Error in underestimation of waterspread area (%)	Live capacity w.r.t underestimation of waterspread (M m^3)	Decrease in the capacity (%)
485.4	2.5	497	2.4	2.5	467	3.8
	5.0	512	5.5	5.0	457	6.0
	7.5	527	8.6	7.5	452	6.8
	10.0	542	11.8	10.0	442	9

The remote sensing-based approach highlights the need to adopt appropriate measures for controlling soil erosion in the reservoir catchment areas thereby increasing the life of reservoir. Further, effect of over- or underestimation of reservoir spread area on reservoir capacity was also studied for Mayurakshi reservoir (Table 19.6). It may be observed that the change in the reservoir live capacity is less than 4% of our actual estimate, when the error in waterspread area estimation is not more than ±2.5%, while changes in the reservoir spread area beyond 2.5% has brought noticeable changes in reservoir capacity (Table 19.6). It signifies that any possible error associated with border pixels while delineation of waterspread will significantly affect the reservoir capacity estimation, which can be overcome by using higher spatial resolution satellite. Further, remote sensing images can be chosen at closer time intervals so that the revised waterspread area can be computed for smaller elevation intervals, which may lead to increased accuracy in sedimentation assessment.

Although remote sensing-based approach is promising, it is limited to assessment of reservoir live capacity only. However, this limitation can be insignificant as reservoir levels rarely fall below the MDDL during normal years. Further, sediment deposition in the live storage zone is of great significance considering the reservoir operations between MDDL and FRL. Remote sensing approach coupled with hydrographic survey can be envisaged for reservoir capacity assessment both in live and dead storage zones.

19.10 Conclusions

Multispectral satellite data available at higher spatial and temporal resolutions promises fast and reliable estimation of waterspread areas required for capacity estimation of reservoirs as remote sensing images can be obtained at frequent interval for better representation of ground profile. This study signifies that any possible error associated with border pixels while delineation of waterspread will significantly affect the reservoir capacity estimation, which can be overcome by using higher spatial resolution satellite for waterspread area delineation. One of the major limitations with remote sensing technique is that it provides accurate estimates for fan-shaped reservoir where there is a considerable change in the waterspread area with small change in the water level. For U- or V-shaped reservoirs where steep banks exist, there may not be remarkable change in waterspread area with fluctuation in water level. In such conditions, capacity estimation may not provide accurate results. Remote sensing-based survey being applicable between MDDL and FRL, remote sensing surveys can be carried out in conjunction with hydrographic surveys at shorter and longer intervals, respectively, for assessment of sedimentation for the entire reservoir zone.

Acknowledgements Authors sincerely acknowledge Central Water Commission, Ministry of Water Resources, New Delhi, to entrust us the current study.

References

Alsdorf DE, Rodrıguez E, Lettenmaier DP (2007) Measuring surface water from space. Rev Geophys 45:RG2002. https://doi.org/10.1029/2006RG000197

Arsen A, Cretaux JF, Abarca del Rio R (2015) Use of SARAL/AltiKa over mountainous lakes, Intercomparison with Envisat Mission. Marine Geodesy. https://doi.org/10.1080/01490419.2014.1002590

Bates PD, Wilson MD, Horritt MS, Mason DC, Holden N, Currie A (2006) Reach scale floodplain inundation dynamics observed using airborne synthetic aperture radar imagery: data analysis and modelling. J Hydrol 328:306–318

Chandrsekhar K, Jeyaseelan AT, Jain J, Chakraborthi AK (2000) Sedimentation study of Tawa reservoir through remote sensing. In: Proceedings 10th national symposium on hydrology with focal theme on urban hydrology held during July 18–19, 2000. New Delhi, pp 523–532

Chowdary VM, Sai Krishnaveni A, Suresh Babu AV, Sharma RK, Rao VV, Nagaraja R, Dadhwal VK (2017) Reservoir capacity estimation using SARAL/AltiKa altimetry data coupled with Resources at P6-AWiFS and RISAT microwave data. Geocarto Int 32(9):1034–1047. https://doi.org/10.1080/10106049.2016.1188165

CWC (Central Water Commission) (2020) Compendium on sedimentation of reservoirs in India, Watershed & Reservoir Sedimentation Directorate, Central Water Commission, New Delhi

Ding X, Li X (2011) Monitoring of the water area variations of lake Dongting in China with ENVISAT ASAR images. Int J Appl Earth Obs 13:894–901

Duan Z, Bastiaanssen WGM (2013) Estimating water volume variations in lakes and reservoirs from four operational satellite altimetry databases and satellite imagery data. Remote Sens Environ 134:403–416

Dubey AK, Gupta P, Dutta S, Singh RP (2015) Water level retrieval using SARAL/AltiKa observations in the Braided Brahmaputra river, Eastern India. Marine Geodesy. https://doi.org/10.1080/01490419.2015.1008156

Frappart F, Papa F, Marieu V, Malbeteau Y, Jordy F, Calmant S, Durand F, Bala S (2015) Preliminary assessment of SARAL/AltiKa observations over the Ganges-Brahmaputra and Irrawady rivers. Mar Geodesy. https://doi.org/10.1080/01490419.2014.990591

Ghosh S, Kumar Thakur P, Garg V, Nandy S, Aggarwal S, Saha SK, Bhattacharyya S (2015) SARAL/AltiKa waveform analysis to monitor inland water levels: a case study of Maithon reservoir, Jharkhand India. Mar Geodesy 38(sup1):597–613. https://doi.org/10.1080/01490419.2015.1039680

Garde RJ (1995) Reservoir sedimentation, state of art report of Indian National Committee on Hydrology (INCOH), No. INCOH/SAR-6/95. National Institute of Hydrology, Roorkee
Goel MK, Jain SK (1996) Evaluation of reservoir sedimentation using multi-temporal IRS–1A LISS–II data. Asian Pacific Remote Sens GIS J 8(2):39–43
Goel MK, Jain SK (1998) Reservoir sedimentation study for Ukai dam using satellite data. Report no. UM-1/97–98', National Institute of Hydrology, Roorkee, India
Goel MK, Jain SK, Agarwal PK (2002) Assessment of sediment deposition rate in Bargi reservoir using digital image processing. Hydrol Sci 47(S):S81–S92
Gupta SC (1999) State paper on reservoir sedimentation assessment using remote sensing techniques. In: Proceedings national workshop on reservoir sedimentation assessment using remote sensing data, National Institute of Hydrology, Roorkee, India
Jain SK, Goel MK (2002) Assessing the vulnerability to soil erosion of the Ukai dam catchment using remote sensing and GIS. Hydrol Sci J 47(1):31–40
Jain SK, Singh P, Seth SM (2002) Assessment of sedimentation in Bhakra reservoir in the western Himalayan region using remotely sensed data. Hydrol Sci J 47(2):203–212
Jothiprakash, Garg V (2009) Reservoir sedimentation estimation using artificial neural network. J Hydrol Eng 14(9):1035–1040
Managond MK, Alasingrachar MA, Srinivas MG (1985) Storage analysis of Malaprabha reservoir using remotely sensed data. In: Proceedings 19th international symposium on remote sensing of environment. Ann Arbor, Michigan, pp 749–756
Manavalan P, Sathyanath P, Rajegowda GL (1993) Digital image analysis techniques to estimate the water-spread for capacity evaluation of reservoirs. PE&RS 59:1389–1395
Manju G, Chowdary VM, Srivastava YK, Selvamani S, Jeyaram A, Adiga S (2005) Mapping and characterization of inland wetlands using remote sensing and GIS. J Indian Society Remote Sens 33(1):45–54
McFeeters SK (1996) The use of normalised difference water index (NDWI) in the elineation of open water features. Int J Remote Sens 17(7):1425–1432
Mohanty RB, Mohapatra G, Mishra D, Mohapatra SS (1986) Application of remote sensing to sedimentation studies in Hirakud reservoir. In: Joint report of Orissa remote sensing applications centre and Hirakud research station
Morris GI, Fan J (1997) Reservoir handbook-design and management of dams, reservoirs and watersheds of sustainable use. Tata McGraw-Hill, New York, USA
Nagarajan R, Marathe T, Collins WG (1993) Identification of flood prone regions of Rapti river using temporal remotely sensed data. Int J Remote Sens 14(7):1297–1303
Narayan DVV, Babu R (1983) Estimation of soil erosion in India. J Irrig Drain Eng 109(4):419–431
Peng D, Guo S, Liu P, Liu T (2006) Reservoir storage curve estimation based on remote sensing data. J Hydrol Eng 11(2):165–172
Rajashree VB, Manavalan P, Sharma JR, Adiga S (2003) Reservoir capacity surveys through remote sensing. Technical report, ISRO-NNRMS-TR-104–2003, p 52
Rajashree VB, Sharma JR et al (2004) Sedimentation analysis through satellite remote sensing, for different reservoirs, RCJ/PR/2004/1-19
Rajeev S, Sathwara PK, Purohit, Mehra OS (1992) Reservoir storage capacity estimation using satellite data—a case study in the Ukai reservoir. Scientific note Space Applications Centre (ISRO), Ahmedabad. SAC/RSA/RSAG-LRD/SN/03/92
Rao TH, Rao CR, Vishwanathan R (1988) Capacity evaluation of Sriramasagar reservoir by using remote sensing techniques. In: Proceedings of 54th R&D session, CBIP, Ranchi, vol 1 (Civil)
Shangle AK (1991) Reservoir sedimentation status in India. Jalvigyan Sameeksha 5:63–70
Sharma PK, Chopra R, Verma VK, Thomas A (1996) Flood management using remote sensing technology: the Punjab (India) experience. Int J Remote Sens 17(17):3511–3521
Smith LC, Pavelsky TM (2008) Estimation of river discharge, propagation speed, and hydraulic geometry from space: Lena River Siberia. Water Resour Res 44:W03427. https://doi.org/10.1029/2007WR006133

Suvit V (1988) The reservoir capacity of Ubolratana Dam between 173 and 180 meters above mean sea level. Asian-Pacific Remote Sens J 9(1):1–6

Tucker CJ (1979) Red and photographic infrared linear combination for monitoring vegetation. Rem Sens Environ 8:127–150

Varshney RS (1997) Impact of siltation on the useful life of large reservoirs. State of art report of INCOH, No. INCOH/SAR-11/97, National Institute of Hydrology, Roorkee

Chapter 20
Management Strategies for Critical Erosion-Prone Areas of Small Agricultural Watershed Based on Sediment and Nutrient Yield

M. K. Sarkar, R. K. Panda, Ayushi Pandey, and V. M. Chowdary

Abstract Identification of the critical areas is essential for the effective implementation of watershed management programs. In the present study, monitoring of runoff, sediment, and nutrient yield (NO_3–N and soluble P) was carried out from a small agricultural watershed to identify the critical areas on the basis of sediment and nutrient yield. Land use/land cover information was generated from Indian Remote Sensing Satellite data (IRS-1D-LISS-III). Universal Soil Loss Equation (USLE) was used to estimate soil loss from the study area, while remote sensing and GIS techniques were used for parameterization of USLE model. Delivery ratio was computed for the watershed to compare with the observed sediment yield. Soil samples were collected and tested from a selected portion of the wasteland for measuring soil fertility status (NPK), soil reaction (pH), organic carbon content, etc. The sediment delivery ratio in the study watershed ranges between 0.44 and 0.66. All the sub-watersheds fall within "slight" soil erosion class (0–5 ton ha^{-1} $year^{-1}$). Sub-watershed I yielded more nutrient followed by Sub-watershed II and Sub-watershed III. The wastelands of the study area have a moderate amount of organic carbon (within 0.5–0.75%) and low available nutrient (NPK), initially, farmers can go for leguminous crop cultivation for increasing the nitrogen status of the field, and also, sabaigrass-based intercropping systems can be adopted in small plots of the wasteland.

M. K. Sarkar
Indian Institute of Technology Kharagpur, Kharagpur, West Bengal, India

R. K. Panda
Indian Institute of Technology Bhubaneswar, Bhubaneswar, Odisha, India

A. Pandey
Graphic Era (Deemed to be University), Deharadun, India

V. M. Chowdary (✉)
Regional Remote Sensing Centre-North, National Remote Sensing Centre, New Delhi, Delhi, India

A. Pandey et al. (eds.), *Geospatial Technologies for Land and Water Resources Management*, Water Science and Technology Library 103,
https://doi.org/10.1007/978-3-030-90479-1_20

20.1 Introduction

Land and water are the two most vital natural resources of the world (Assis et al. 2021), and these resources must be conserved and sustainably managed for environmental protection and ecological balance. Prime soil resources of the world are finite, non-renewable, and prone to degradation through misuse and mismanagement. Due to population pressure, the availability of arable land and clean water is decreasing day by day. The situation in India is more alarming, with a population density of about seven times that of the world's average. Further, about 10–40% of the total rainfall is unutilized annually, and 4.8 million tons of nutrients are lost through soil erosion. In West Bengal, out of 8.875 m ha of total geographical area, 4.677 m ha is affected by soil erosion and degradation problems (Suresh 1997). Rao (1981) reported that annual soil loss in West Bengal is as high as 16.5 ton/ha from bare soil and 4.6 ton/ha under dense crops. Therefore, the problem of land degradation due to soil erosion is very serious and with increasing population pressure, exploitation of natural resources. In India, satellite-based remote sensing inputs have played important role in the management of natural resources. Geospatial information can be well-integrated with the conventional database for modeling of runoff and sediment loss to take up appropriate soil and water conservation measures (Pandey et al. 2007, 2021). Srinivasan et al. (2021) used weighted index overlay technique to estimate the soil erosion probability zones in arid part of South Deccan Plateau, India.

Pandey et al. (2021) investigated the effect of different multi-source digital elevation models (DEMs) on soil erosion modeling by employing the revised universal soil loss equation (RUSLE) and geospatial technologies. Fook et al. (1992) envisaged geospatial technique and the USLE for generation of soil erosion susceptibility map of the Upper Klang Valley that encompasses 456 km^2 situated in Kuala Lumpur, the Federal Territory, and parts of the districts of Gombak and Ulu Langat. Das et al. (2021a) prioritized Gomti River basin for soil and water resources conservation using RS and GIS techniques. Opena (1992) generated a soil erosion susceptibility map of the Tamlung River catchment using a geographic information system to test the Comprehensive Resource Inventory and Evaluation System (CRIES) applicability in determining soil loss using the USLE. Aerial photography was also envisaged for deriving USLE parameters (Stephens et al. 1985). Negash et al. (2021) estimated the amount of soil loss per year using soil data, satellite image, ASTER DEM, and rainfall data. Endalamaw et al. (2021) used RUSLE model with ArcGIS and multi-criteria evaluation technique to identify high erosion-risk areas and estimated the annual soil loss rate in Gilegel beles watershed, to prioritize high erosion-risk areas for conservation planning. Bouamrane et al. (2021) identified the soil erosion hazard map for the Mellah watershed, Northeastern Algeria, employing RUSLE, AHP, and frequency ratio.

Kumar et al. (2020) carried out a comparative study to estimate the spatial pattern of annual soil erosion rate and sediment yield using remote sensing and GIS techniques. Mongkolsawat et al. (1994) used the USLE and GIS to establish spatial information of soil erosion in Huai Sua Ten watersheds, Northeast Thailand. Rao et al. (1994) undertook a study to develop a watershed prioritization scheme for conservation planning based on the sediment yield potential estimates of different sub-watersheds of the Saluli watershed in the Western Ghats of India using the Universal Soil Loss Equation. Das et al. (2021b) used Revised Universal Soil Loss Equation (RUSLE) and Modified Morgan–Morgan–Finney (MMF) model to estimate soil loss and found MMF model estimates 7.74% less erosion than the USLE model. Ogawa et al. (1997) evaluated land degradation/erosion conditions in the Pothwar tract, Northern Pakistan, using remote sensing, GIS techniques, and the USLE model. Rinos et al. (1999) identified priority areas based on soil erosion intensity in the Bata River basin using the USLE model. Geospatial technologies were widely used for estimation of soil erosion spatially (Fistikoglu and Harmancioglu 2002; Angima et al. 2003; Cohen et al. 2005; Pandey et al. 2007; Jhariya et al. 2020). Panda et al (2021) prioritized the sub-watershed of Upper Subarnarekha catchment using the simulated sediment yield using spatially distributed slope, soil texture, and land use.

Wolfe et al. (1995) proposed that best management practices can be structural or management-oriented, where structural practices include terracing, streambank stabilization, and animal waste systems and management practices include crop rotation, tillage practices. Mitchell et al. (1993) integrated agricultural non-point source (AGNPS) model with the geographic resources analysis support system (GRASS) and GIS to develop a decision support tool to assist with the management of runoff and erosion from agricultural watersheds. Mass et al. (1985) stated that critical areas of non-point source pollution could be defined both from the land resource and the water quality perspectives. From the land resources perspective, critical areas are those land areas where the soil erosion rate exceeds the soil loss tolerance value. From the water quality perspective, critical areas are areas where the greatest improvement can be achieved with the least capital investment in the best management practices. Elnashar et al (2021) developed a Revised Universal Soil Loss Equation framework fully implemented in the Google earth engine cloud platform (RUSLE-GEE) for soil erosion assessment and found that approximately 27% of the Blue Nile required soil protection measures as these areas fall under moderate to high risk of erosion class. Bekele and Gemi (2021) conducted a study in Dijo watersheds, Ethiopia, using geographic information system and Universal Soil Loss Equation to identify soil erosion hotspot areas for proper planning. Tim et al. (1992) used an average value of 9.0 ton ha^{-1} $year^{-1}$ as soil tolerance for identifying the critical areas from the land resource perspectives. Important driving factors such as steep slopes, climate (e.g., long dry periods followed by heavy rainfall) coupled with improper land use land cover patterns may lead to soil erosion (Renschler and Harbor 1999). This further may lead to soil fertility and organic matter loss. Thus, this study was carried out with specific objectives of (i) assessment of soil erosion in the study watershed using USLE model, while remote sensing and GIS techniques are used for parameterization, (ii) monitoring of nutrient losses through surface runoff from the study watershed, and (iii) recommendation of management strategy for the erosion-prone area in the watershed.

20.2 Study Area Description

A small agricultural watershed located in Eastern India was used to quantitatively assess non-point source pollution of water. The whole watershed drains water through a single well-defined outlet that falls in the boundary of Kapgari village. Hence, the watershed was named as Kapgari watershed. The watershed is located in Jamboni block, Midnapore district of West Bengal state, India, and bounded by geographical co-ordinates 86° 50′ and 86° 55′ E longitude and 22° 30′ and 22° 35′ N latitude (Fig. 20.1). Based on the drainage network and topography, the watershed was subdivided into three sub-watersheds. The region's mean annual rainfall is 1250 mm, and nearly 80% of the annual rainfall is received from June to September. The daily mean temperature ranges from 20 to 38 °C, while the daily mean relative humidity

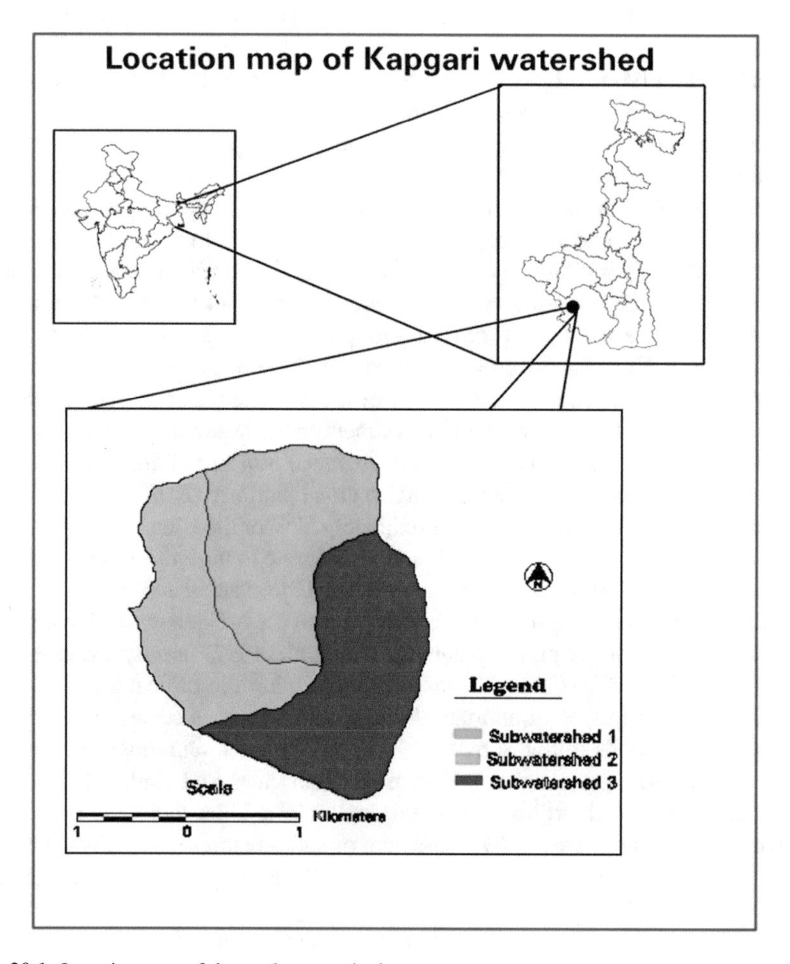

Fig. 20.1 Location map of the study watershed

varies between 40 and 88%. The climate is sub-humid subtropical. The topography of the watershed consists of uplands, medium lands, lowlands, and forestland. The dominant crop in the study area is paddy, which is generally cultivated during the monsoon period. This area was chosen for the study as agriculture is a major land use in the area and is experiencing nitrate contamination problems for some time. The farmers of this region have been cultivating high-yield varieties of paddy intensively with high doses of nitrogenous fertilizer. The predominant form of fertilizer used by the farmers in the watershed is urea, single super phosphate, and muriate of potash. In winter, some oilseed crops such as mustard and groundnut, pulses, wheat, maize and some vegetables like potato and brinjal are grown with high doses of fertilizers. Because of cattle dung, phosphorus pollution is also a major concern in the study area. The predominant soil type is sandy loam. However, in some areas, sandy, loamy sand, loam, and clay loam soils are also found.

20.3 Data Used

20.3.1 Land Use Land Cover Data

Linear image self-scanner (IRS-1D-LISS-III) data (path-107, row-56) corresponding to November 4th, 2003, ground features with different spectral reflectance properties gets distinctly separated in multispectral mode for generating the land use land cover map. It is the process of sorting pixels into a finite number of classes or categories of data based on their reflectance values in different spectral bands. The maximum likelihood supervised classification technique was used in this study.

20.3.2 Digital Elevation Model

Digital elevation model (DEM) was generated using 20 m elevation contours available from the topographic map at 1:50,000 scale and spot heights.

20.3.3 Soil Texture and Soil Nutrient Analysis

Runoff and soil erosion processes depend on the surface soil texture. Undisturbed soil samples from a depth range 0–15 cm were collected at different watershed locations (Mouza (village)-wise) and were cleaned from pebbles, roots, and other foreign materials. Chemical analysis was carried out for textural classification of soil into sand, silt, and clay at the Department of Agriculture and Food Engineering, IIT, Kharagpur. Textural analysis of soil samples collected from different study area locations was used to update the National Bureau of Soil Survey and Land Use Planning NBSS and LUP soil map at 1:250,000 scale.

20.3.4 Runoff, Sediment Sampling, and Nutrient Loss Analysis

Rainfall data of monsoon season for the year 2003 was collected from automatic rain gage station located in Kapgari watershed, Jamboni block, West Midnapore, West Bengal. Continuous discharge from the sub-watersheds was carried out using a stage-level recorder, and the current meter was used to measure the water velocity passing through the culverts. Bottle samples were used to collect the water samples for each storm (rainfall event) from the channel bottom. Analyses of runoff and sediment samples were carried out periodically to determine the pollutants (NO_3–N and P) and sediment loading in the runoff water. Water samples collected from the outlets of sub-watersheds were passed through filter paper (Whatman No. 1), and subsequently, suspended sediments retained on the filter paper were dried, and weight was taken. Thus, the amount of sediments obtained from a certain volume of water was measured, and finally, it was converted into tons/ha of sediment yield from the watershed. The NO_3–N and phosphorus concentration in the water samples were also determined by the ion-chromatography apparatus.

20.4 Methodology

20.4.1 Assessment of Sediment Yield Using Universal Soil Loss Equation (USLE)

The erosion prediction model, Universal Soil Loss Equation (USLE), was widely used across the globe for the estimation of soil losses from a specified land in a specified cropping and management system. This model predicts the losses from sheet and rill erosion under specified conditions. It computes the soil loss for a given site, as a product of six major factors, whose most likely values at a particular location can be expressed numerically (Wischmeier and Smith 1978) as given below:

$$A = R \times K \times L \times S \times C \times P \quad (20.1)$$

where *A* is the computed soil loss per unit area, expressed in the units selected for *K* and the period selected for *R*. In practice, these are usually so selected that they compute *A* in metric tons per ha per year, but other units can be selected, R is the rainfall erosivity factor, *K* is the soil erodibility factor, *L* is the slope length factor, *S* is the slope steepness factor, C is the crop management factor, and P is the conservation practice factor. Schematic representation of methodology for assessment of soil erosion is shown in Fig. 20.2. In this study, soil loss was assessed for 24 individual rainfall events spread over the year 2003. Sediment yield (SY) from the study watershed was assessed by using the equation as given below:

$$SY = SDR \times R \times K \times L \times S \times C \times P \quad (20.2)$$

SDR is sediment delivery ratio, and its assessment from the study watershed is necessary for computing sediment yield from the watershed. SDR is defined as the proportion of the detached soil material from the source area reaching the sink area through either overland or channel flow. SDR is computed by dividing the observed sediment yield by the estimated soil loss. Soil loss was estimated for twelve rainfall

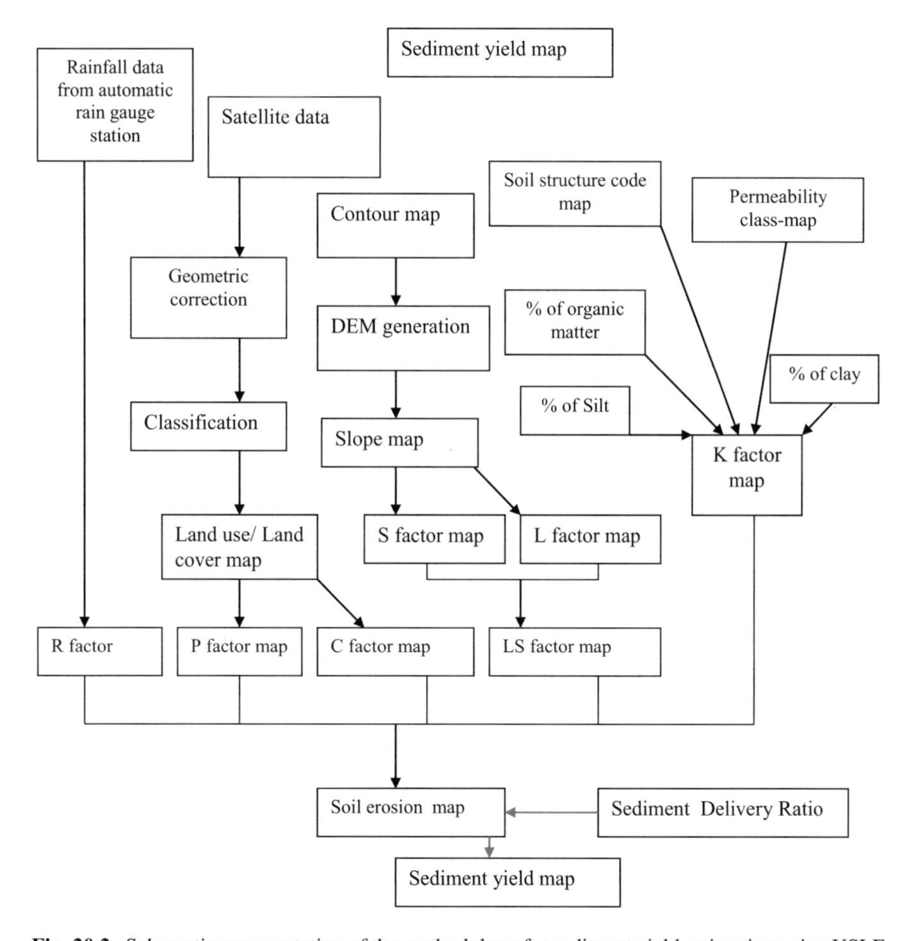

Fig. 20.2 Schematic representation of the methodology for sediment yield estimation using USLE model

events, and sediment delivery ratio was found out for the individual event. The mean sediment delivery ratio for the study area was computed based on 12 rainfall events and used for the assessment of sediment yield for the remaining 12 rainfall events.

20.4.2 USLE Model Parametrization

(a) **Rainfall erosivity factor (*R*)**

The rainfall erosivity factor (R) is computed as the product of kinetic energy (KE) of the storm event (Wischmeier and Smith 1978) and the 30 min maximum rainfall intensity (I_{30}) during the storm. The EI_{30} can be expressed as

$$\mathrm{EI}_{30} = \mathrm{KE} \times I_{30} \tag{20.3}$$

$$\mathrm{KE} = 0.119 + 0.0873 \log_{10} I \quad \text{for } I < 76 \tag{20.4}$$

$$\mathrm{KE} = 0.283 \quad \text{for } I > 76 \tag{20.5}$$

where KE is kinetic energy in megajoules/ha-mm and EI_{30} is the measure of rainfall erosivity factor and is summed up to obtain values at daily, weekly, monthly, seasonally, and yearly scales, and I is rainfall intensity in mm/h.

(b) **Soil erodibility factor (*K*)**

The soil erodibility factor (K) is the rate of soil loss per unit of R or EI for a specified soil as measured on a unit plot, 22.1 m length of uniform 9% slope continuously in cleaned tilled fallow land. These standard plot sizes were adopted for runoff plot studies. Runoff plots were earlier made to be of 1/100 acre with dimensions 6 ft. wide and 72.6 (nearly 22.1 m) long. Thus, the standard conditions assumed have no special significance but a historical accident. Therefore, K has units of mass per area per erosivity unit. In SI system, one set of units (metric ton ha hour/ha megajoule mm) can be abbreviated as (t ha h/ha MJ mm). In this study, K was computed using the following equation (Foster et al 1981):

$$\mathrm{K} = 2.8 \times 10^{-7} \times M^{1.14}(12 - a) + 4.3 \times 10^{-3}(b - 2) + 3.3 \times 10^{-3}(c - 3) \tag{20.6}$$

where M is particle size parameter (% silt + % very fine sand) (100—% clay), a is % organic matter, b is soil structure code (very fine granular, 1; fine granular, 2; medium or coarse granular, 3; blocky, platy, or massive, 4), and c is profile permeability class (rapid, 1; moderate to rapid, 2; moderate 3; slow to moderate, 4; slow, 5; very slow, 6).

(c) **The slope length factor (*L*) and slope steepness factor (*S*)**

Although, the effects of slope length and gradient are represented in the USLE as *L* and *S* factors, respectively. However, evaluated as a single topographic factor (*LS*), the *LS* factor can be computed (Wischemeir and Smith 1978) as given below:

$$LS = (L/22)^m * \left(0.065 + 0.045 * s + 0.0065 * s^2\right) \quad (20.7)$$

where m is 0.5 if slope $\geq$5%; 0.4 if slope $\leq$5% and >3%; 0.3 if slope $\leq$3% and >1%; 0.2 if slope <1%, s is slope (%), and L is slope length (m).

(d) **The crop management factor (*C*)**

Crop management factor represents the ratio of soil loss under a given crop to that from bare soil, which includes the effects of cover, crop sequence, tillage practices residue management.

(e) **Conservation practice factor (*P*)**

The support practice factor is the ratio of soil loss with a support practice like contouring, strip cropping, or terracing to that with straight row farming up and down the slope.

20.4.3 Watershed Management Strategies

A major portion of the study area is either fallow or under wastelands. Hence, optimum utilization of wastelands or degraded lands is necessary to meet the growing human needs of food, fodder, fiber, and fuelwood, etc. Thus, the fertility status of soil (available NPK), soil reaction (pH), organic carbon content, etc. were analyzed. Soil physical properties like bulk density were also tested. The main aim was to suggest the farmers some specific crops for cultivation based on soil test data.

20.5 Results and Discussion

20.5.1 Computation of USLE Parameters

Sediment yield was assessed using the USLE model corresponding to 24 rainfall events in the year 2003. USLE factors such as rainfall erosivity factor (*R* factor), soil erodibility factor (*K* factor), slope and slope length factor (*LS* factor), crop management factor (*C* factor), and conservation practice factor (*P* factor) were spatially parametrized using remote sensing and GIS techniques. USLE parameters are a function of logically related geographic features, where these attributes were either

Table 20.1 Rainfall erosivity factor (*R*) for selected rainfall events

Date	Rainfall (mm)	EI30 (MJ mm/ha h)	Date	Rainfall (mm)	EI30 (MJ mm/ha h)
Jun 18	35	224.26	Aug 22	20.75	160.034
Jun 19	18.5	87.2134	Aug 26	24.5	334.915
Jun 20	58	426.7129	Sep 1	19	158.2709
Jul 2	16.7	77.14179	Sep 6	16.5	79.0548
Jul 10	33.2	496.2128	Sep 8	27	188.97
Jul 13	28.25	214.95	Sep 23	28.75	127.834
Jul 22	44	680.883	Sep 25	17.5	60.4721
Jul 23	49.25	500.8	Oct 2	18	101.5186
Jul 25	30	178.3781	Oct 4	23	244.8
Jul 29	32	302	Oct 6	22	142.83
Jul 31	44	395.2	Oct 7	63	385.02
Aug 16	33	345.6	Oct 9	28	173.88

collected from existing information or extracted from the classification of LISS-III-IRS-1D satellite data. Further, each USLE factor attribute was digitally encoded using GIS.

(a) **Rainfall erosivity factor (*R*)**

The rainfall erosivity factor (*R*) was computed as the product of the kinetic energy of the storm and the 30 min maximum intensity occurring during the storm. Rainfall data was collected from the automatic rain gage station situated at the study area. EI_{30} for each rainfall event was computed based on the rain gage chart (Table 20.1).

(b) **Soil erodibility factor (*K*) map**

The soil erodibility factor (*K*) defines the inherent susceptibility of the soil to erosion. The *K* values of different soils vary due to its significant differences among soil properties such as texture, structure, permeability, and organic matter content. The saturated hydraulic conductivity of the soil samples collected from the study area was also measured (Mouza-wise). Soil properties of different Mouzas located in the study watershed are given in Table 20.2. Permeability classes were observed from the results using the standard table. Different soil property maps such as organic matter, percentage of clay, silt, and soil structure code, and permeability classes were generated based on the ground truth information collected from the study area (Figs. 20.3, 20.4, 20.5, 20.6 and 20.7). The *K* factor layer was spatially computed by integrating different governing factors as Foster et al. (1981) proposed. Spatially distributed *K* map in the study watershed is given in Fig. 20.8.

(c) **Slope length and slope steepness (*LS*) factor map**

Slope and its corresponding slope length were computed from digital elevation model (DEM). In this study, the slope length was fixed as 23 m, since soil loss was estimated

Table 20.2 Soil properties of Kapgari watershed (Mouza-wise)

Mouza name	Soil texture	% of sand	% of silt	% of clay	% of organic matter	Saturated hydraulic conductivity (mm/h)
Chondsor	Loamy sand	80.8	11.6	7.6	0.843	13.45
Beliaguri	Sandy loam	79.2	15.4	5.4	0.948	12.26
Beragari	Loam	48.2	38.8	13	1.165	2.74
Gagansuli	Sandy loam	70.6	24.6	4.8	1.167	11.53
Kukurmuri	Loam	48.6	38.2	13.2	1.2	3.36
Kapri	Loamy sand	81.4	14	4.6	0.862	13.68
Parusuli	Sandy loam	70	24.6	4.4	1.08	11.32
Govindpur	Sandy loam	76	13.7	10.3	1.034	9.46
Bagda	Sandy loam	76.4	13.2	10.4	0.924	10.95
Latiabad	Clay loam	41.6	37.2	21.2	1.189	1.31
Gumuria	Clay loam	41.5	38	20.5	1.293	1.28
Jorsa	Sandy loam	75.6	14.6	9.8	0.99	9.83
Chalakchuan	Sandy	81	14.8	4.2	0.914	14.2
Bagheri	Loam	47.8	36.4	15.8	1.12	3.58

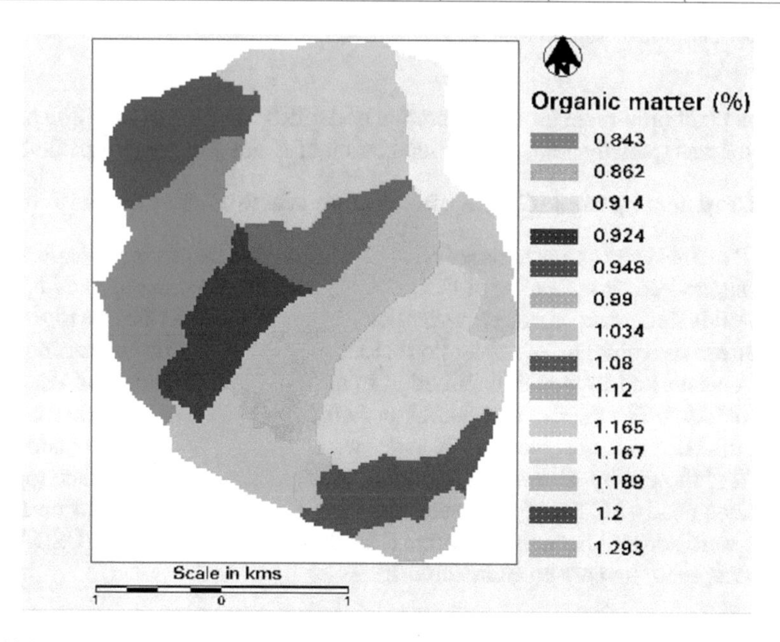

Fig. 20.3 Organic matter map of Kapgari watershed

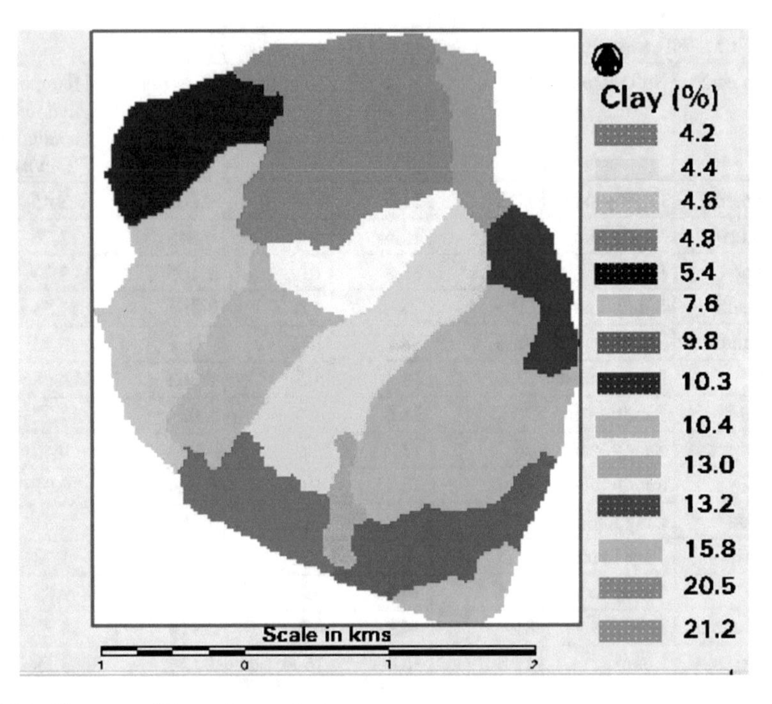

Fig. 20.4 Clay map of Kapgari watershed

for each pixel (minimum ground resolution of the IRS-1D-LISS-III satellite image). *LS* factor was spatially computed by integration of *L* and *S* factors (Fig. 20.9).

(d) **Crop management (*C*) and conservation practice factor (*P*) map**

C and *P* factors were assigned based on the LULC classes that were generated based on the supervised classification of FCC of LISS-III-IRS-1D images. The impact of land cover is significant on surface water flow, where LULC describes the top layer of land surface on which surface water flows. LULC map is categorized into nine major LULC classes, and its spatial distribution in the watershed is shown as Fig. 20.10 and Table 20.3. The relative distribution of LULC classes in the watershed is given in Fig. 20.11. It was observed that Kapgari watershed is dominated by paddy field (36.04%) followed by fallow land (31.36%). Wasteland and degraded land together constitutes nearly 16.2% of the watershed. The *C* and *P* values based on LULC classes were adopted from the literature (Rao 1981; Singh et al. 1981; USDA-SCS Handbook 1969) and are given in Table 20.4.

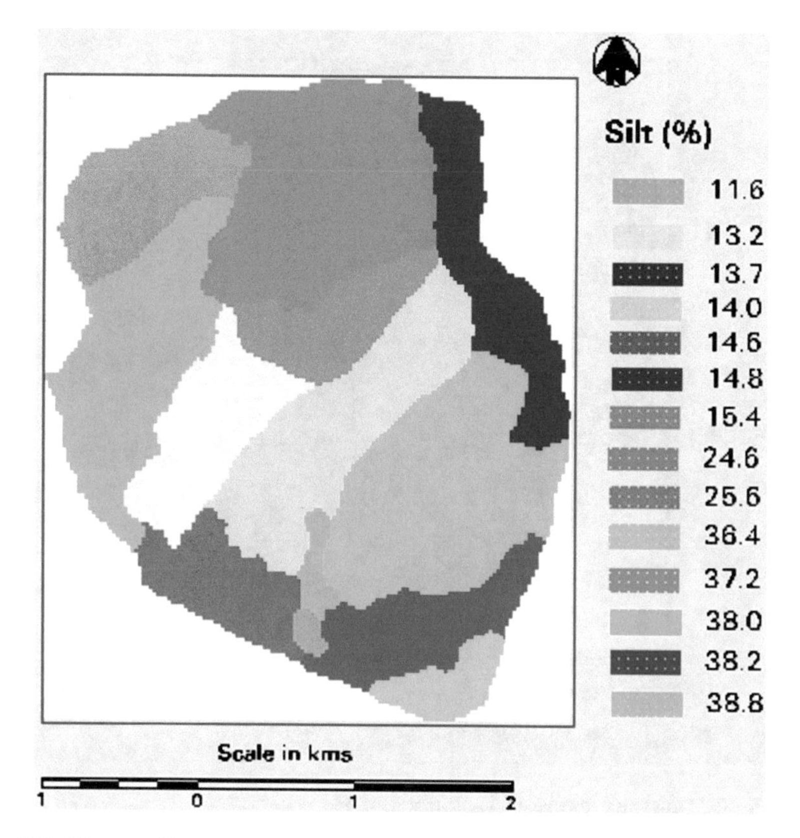

Fig. 20.5 Silt map of Kapgari watershed

20.5.2 Assessment of Sediment Yield

Sediment loss from the watershed was monitored during the monsoon season (June–October) of the year 2003. Soil loss was assessed for all 24 rainfall events using USLE model in the year 2003 (Table 20.5). Spatially distributed sediment yield for an individual event corresponding to 58 mm rainfall that occurred on June 28th was shown as Fig. 20.12. The scattered diagram between observed and the estimated sediment yield values (Fig. 20.13) shows that the simulated sediment yield data points are above the 1:1 line for all the rainfall events indicating over prediction. It may be because USLE does not take the deposition of sediment into account. Hence, the sediment delivery ratio was computed for twelve rainfall events. The average sediment delivery ratio was computed, and it was multiplied with the estimated soil losses corresponding to the remaining rainfall events and compared with the observed sediment yield. The mean sediment delivery ratio in the watershed was computed to be 0.63 (Table 20.6).

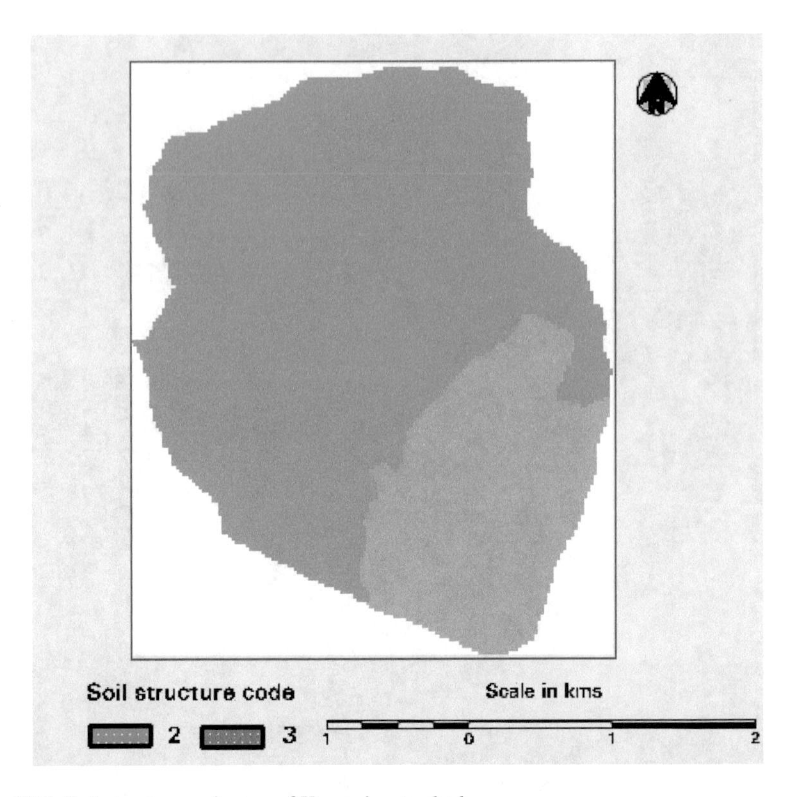

Fig. 20.6 Soil structure code map of Kapgari watershed

Sub-watershed III yielded the highest sediment followed by Sub-watershed II and Sub-watershed I, which is due to the undulating topography. A major portion of this sub-watershed consist of barren land, which might have contributed to soil loss. The contribution of sediment yield from Sub-watershed II is more compared to Sub-watershed I because this sub-watershed constitutes degraded forestland as well as settlements. However, the lowest contribution of sediment from Sub-watershed I could be attributed to the presence of bunded paddy fields, in which sediments get deposited.

Further, sediment yield was computed for the remaining 12 rainfall events after accounting SDR. RMSE of 0.003 indicated a good agreement between estimated and observed sediment yields after accounting SDR for sediment yield assessment (Table 20.7). CRM value was also very low (−0.07), which shows a little tendency of over prediction. The scatter diagram between observed and estimated sediment yield for the remaining rainfall events shows that all the simulated sediment yield data points are found to be close to the 1:1 regression line (Fig. 20.14). A similar trend was observed for the sub-watersheds as shown in Figs. 20.15, 20.16, 20.17 and 20.18. Sub-watershed III has bare land with undulating topography, which is susceptible to erosion. Sub-watershed II also contributes to sediment yield significantly, as it is

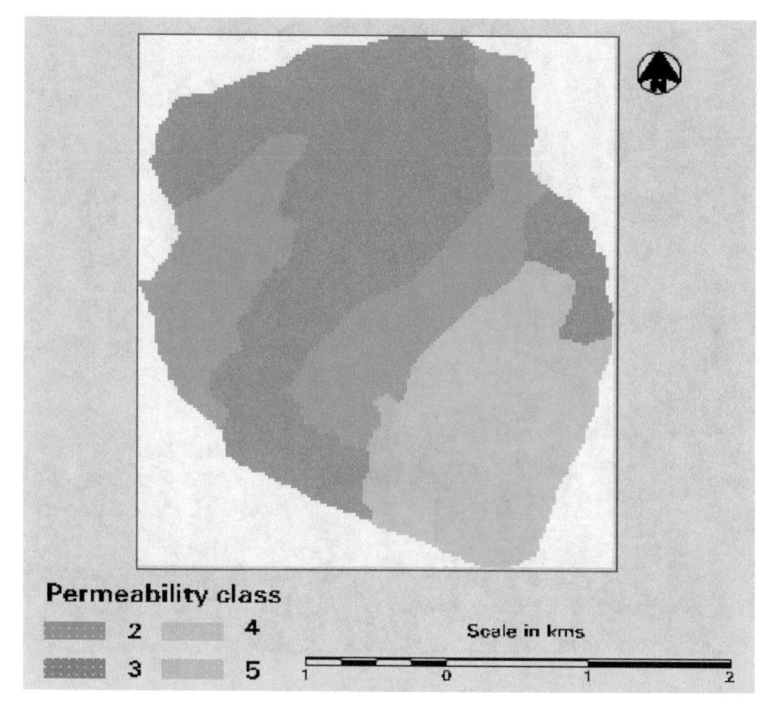

Fig. 20.7 Permeability class map of Kapgari watershed

predominantly occupied by open forest and degraded forest. Conservation practice is also observed to be poor in this watershed. Sediment yield is observed to be low in Sub-watershed I, as this watershed is predominated by paddy fields. This can be attributed to the fact that sediment deposition takes place within the bunded paddy fields. However, it was also inferred that all the sub-watersheds fall within "slight" soil erosion class (0–5 ton ha^{-1} $year^{-1}$), and hence, none of the sub-watershed was found to be critical from a sediment yield point of view. Proper management needs to be taken up for the wastelands where topsoil is washed away due to erosion.

20.5.3 Analysis of Nutrient Losses Through Surface Runoff

Nutrient loss from the watershed was monitored during the monsoon season (June–October) of 2003. The water samples were taken from the outlets of sub-watersheds and also from the main outlet. NO_3–N and phosphorus concentrations in the water samples were determined using the ion-chromatography systems. Temporal variation of nutrient loss from the study watershed and the sub-watersheds is shown in Figs. 20.19, 20.20 and 20.21. From these figures, it was seen that there is some initial concentration of NO_3–N and soluble phosphorus within the first few days, which

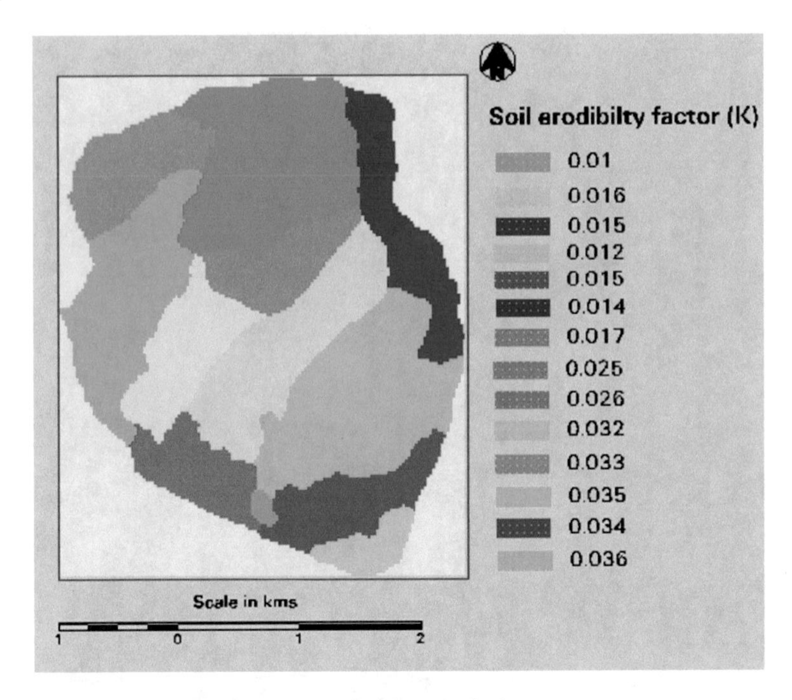

Fig. 20.8 Soil erodibility factor (K) map of study watershed

may be due to the presence of residual nitrogen and phosphorus as a result of fertilizer application in the vegetable crop during pre-monsoon season. Cow dung is also contributed to phosphorus. Nutrient concentration in the runoff water got increased gradually and reached a peak value with fertilizer application. But, with time, the concentration of the nutrient in the runoff got reduced. SSP and urea are commonly used fertilizers where SSP is applied fully and urea in 50% as basal dose. Rest dose of urea was applied 20–30 days after sowing. Therefore, the concentration of nitrate–nitrogen again increased gradually. However, the time to attain the peak value of nutrient concentration was not the same for all the sub-watersheds, which may be due to variations in the fertilizer application dates by different landowners. It was seen that Sub-watershed I contributed more nutrient loss because the area is predominated with paddy cultivation, and a higher amount of fertilizers are applied. Sub-watershed II also contributed significant soluble phosphorus; sewage water drained from the settlements and cow dung might have contributed to high phosphorous concentration. So from a nutrient loss point of view, Sub-watershed I is critical, followed by Sub-watershed II. The peak of nutrient loss (nitrate–nitrogen and soluble phosphorous) was attained during monsoon season, where monitoring was done immediately after fertilizer application.

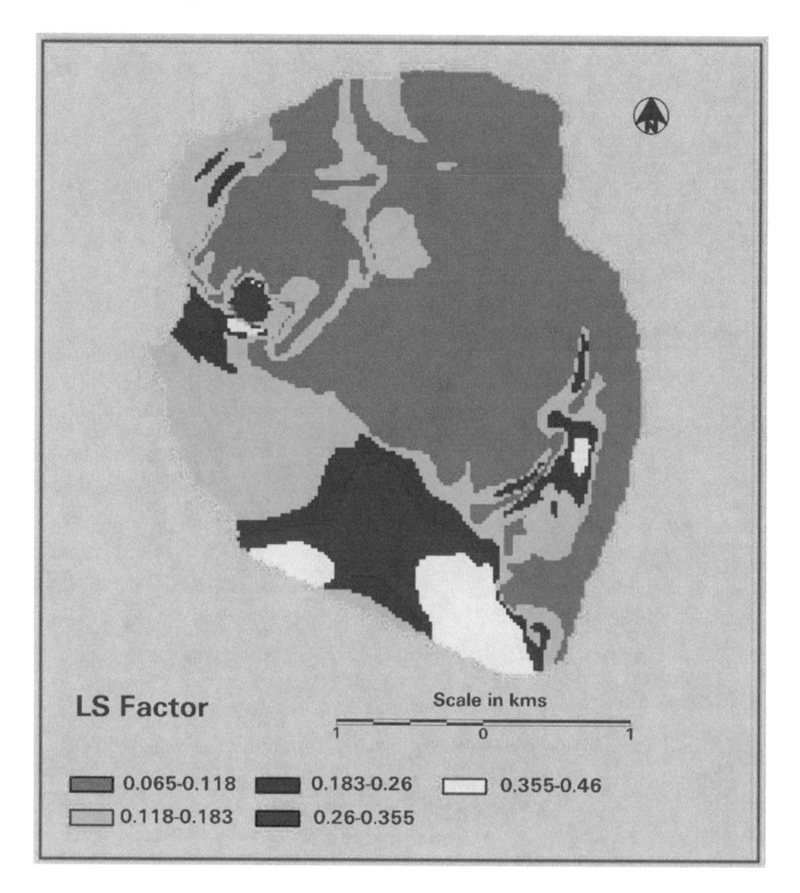

Fig. 20.9 Spatial distribution of *LS* factor map

20.5.4 Watershed Management Strategies

A major portion of the study watershed is under the wasteland category. Keeping the above facts, soil samples were collected for assessment of nutrient status (NPK), soil reaction (pH), organic matter content, etc. in the study watershed. The main aim was to implement management strategies for properly utilizing resources, maintaining soil fertility, improving land productivity, and accruing maximum financial benefits. The soil at the experimental site is lateritic, and mainly, the soil texture is of sandy loam type; however, at some places, loam and sandy loam-type soil were also seen (Table 20.8). The selected area of the wasteland is constituted mainly of coarse-textured particles and is susceptible to erosion. The soil physical property like bulk density was also tested. pH results show that the soil is slightly acidic in reaction (pH 5.5–6.5) (Table 20.8).

Soils in the study watershed were categorized based on organic carbon and NPK status (Table 20.9). Nutrient status in terms of NPK and organic carbon was given

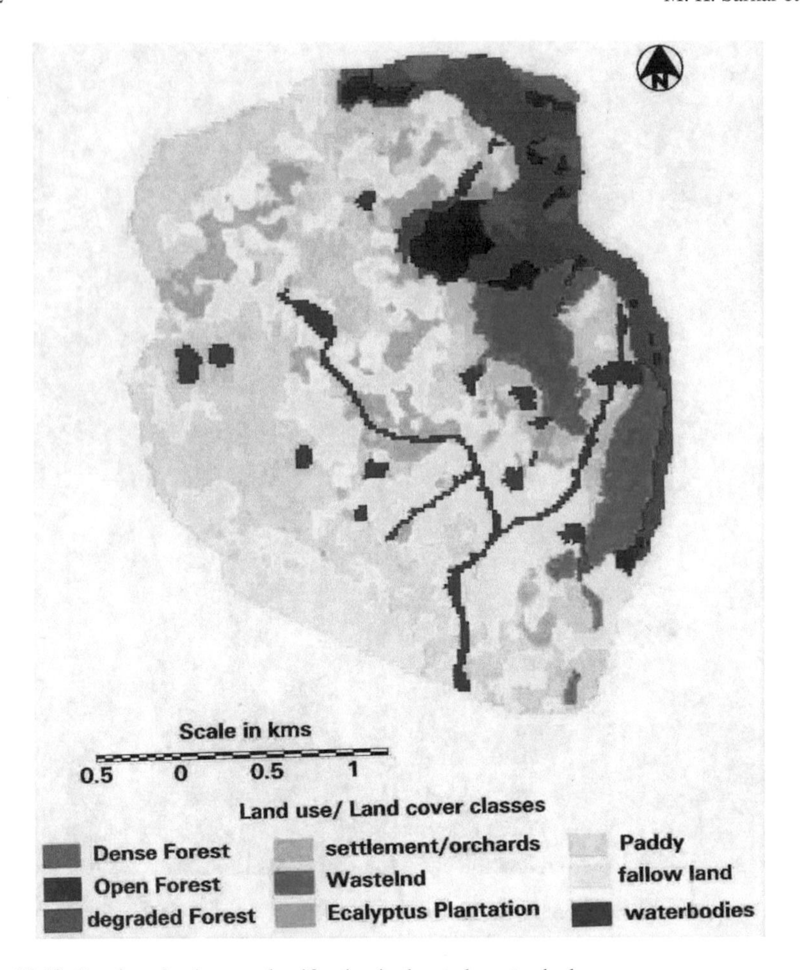

Fig. 20.10 Land use land cover classification in the study watershed

Table 20.3 Land use/land cover classification statistic of Kapgari watershed

Land use/Land cover classes	Area (ha)	% area
Dense forest	14.55	1.49
Open forest	29.52	3.03
Degraded forest	84.79	8.71
Village/orchards	66.97	6.88
Water bodies	44.22	4.54
Wasteland	72.68	7.48
Plantation	4.60	0.47
Fallow land	305.34	31.36
Paddy	350.83	36.04
Total	973.5	100

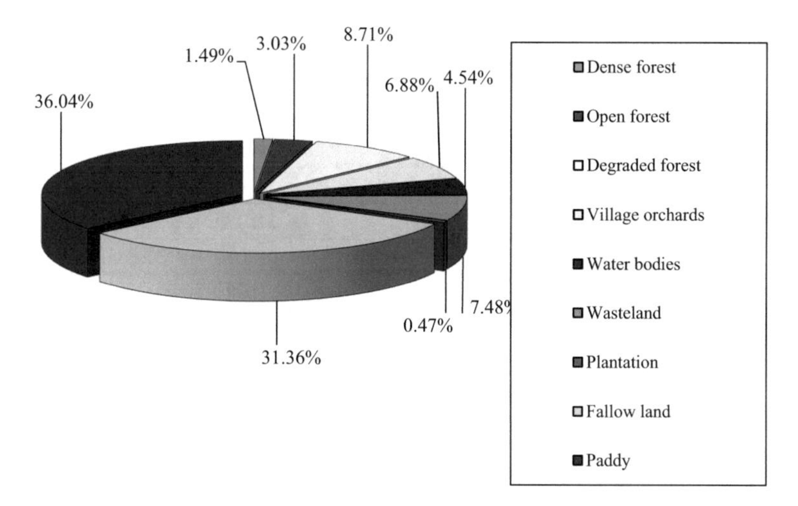

Fig. 20.11 Land use/land cover classification statistics of Kapgari watershed

Table 20.4 *C* and *P* factors for different land use/land cover classes

Land use land cover	*C* value	*P* value
Dense forest	0.01	0.8
Open forest	0.02	0.85
Degraded forest	0.13	0.9
Water bodies	0	0
Paddy	0.28	0.03
Wasteland	0.14	1
Plantation	0.13	0.8
Village/orchards	0.5	0.1
Fallow	0.28	0.03
River bed	1	1

in Tables 20.10 and 20.11. From the soil test results, it is seen that the experimental sites have organic carbon of medium range (within 0.5–0.75%). However, available nutrient (NPK) is very low (less than 280 kg/ha, 45 kg/ha, and 150 kg/ha for available N, P_2O_5, and K_20, respectively). Since the land was forestland long before, the organic carbon is in the medium range due to the decomposition of forest leaves. But due to the absence of any fertilizer application or cultivation, available NPK content is very low. To make the land fertile and cultivable, proper management and fertilizer application is necessary. The soil is slightly acidic in reaction containing low N, P, and K levels. Such land can support selected species of shrubs and trees besides some annual crop plants, with due soil conservation measures and agronomic management. Therefore, the application of lime may bring changes in land use. Further, nitrogen

Table 20.5 Observed versus estimated sediment yield for the main watershed

Date	Rainfall (mm)	EI30 (MJ mm/ha h)	Sediment yield (t/ha)	
			Estimated	Observed
Jun 18	35	224.26	0.036	0.023
Jun 19	18.5	87.2134	0.014	0.009
Jun 20	58	426.7129	0.069	0.047
Jul 2	16.7	77.14179	0.012	0.006
Jul 10	33.2	496.2128	0.08	0.055
Jul 13	28.25	214.95	0.034	0.021
Jul 22	44	680.883	0.12	0.069
Jul 23	49.25	500.8	0.081	0.059
Jul 25	30	178.3781	0.03	0.018
Jul 29	32	302	0.049	0.031
Jul 31	44	395.2	0.064	0.043
Aug 16	33	345.6	0.056	0.032
Aug 22	20.75	160.034	0.026	0.02
Aug 26	24.5	334.915	0.054	0.031
Sep 1	19	158.2709	0.026	0.018
Sep 6	16.5	79.0548	0.013	0.009
Sep 8	27	188.97	0.03	0.012
Sep 23	28.75	127.834	0.021	0.009
Sep 25	17.5	60.4721	0.01	0.005
Oct 2	18	101.5186	0.016	0.008
Oct 4	23	244.8	0.039	0.02
Oct 6	22	142.83	0.023	0.016
Oct 7	63	385.02	0.062	0.04
Oct 9	28	173.88	0.028	0.016
Statistical parameter				
Root-mean-square error (RMSE)			0.018	
Coefficient of residual mass (CRM)			−0.6	

efficiency can be improved either by improving crop uptake or by changing the timing and method of application (Mabler and Bailey 1994).

With the present scenario, the productivity of crops grown under the above condition is stagnant. Adopting appropriate agronomic practices with agro-ecologically suitable crops and cropping systems may be the only alternative to traditional rainfed agriculture for proper utilization of resources, maintaining soil fertility, improving land productivity, and accruing maximum benefits. For resource-poor farmers, these practices are appropriate in as much as they guarantee a harvest and are borne of traditional wisdom.

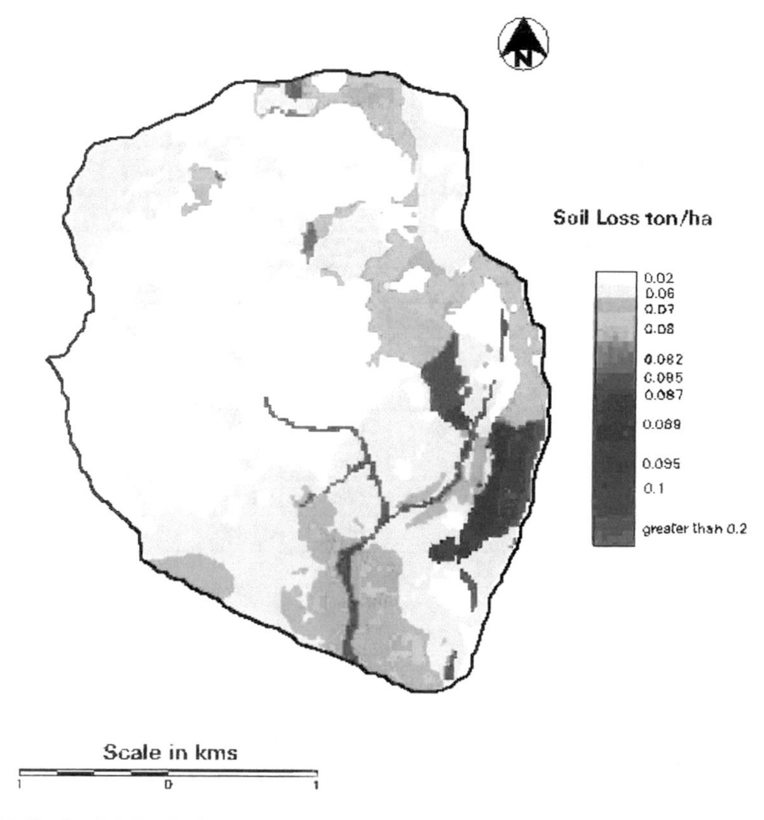

Fig. 20.12 Spatial distribution of soil loss in the study area on 06-20-2003 for the rainfall event 58 mm

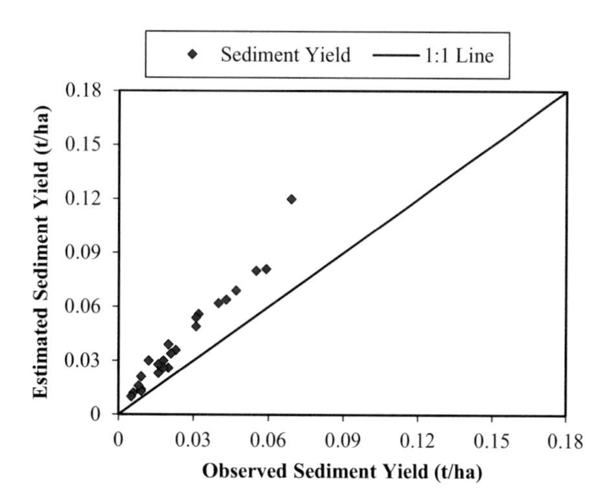

Fig. 20.13 Observed versus estimated sediment yield for the main watershed

Table 20.6 Computation of sediment delivery ratio

Date	Rainfall (mm)	EI30 (MJ mm/ha h)	Sediment yield (t/ha)		Sediment delivery ratio
			Estimated	Observed	
Jun 18	35	224.26	0.036	0.023	0.638
Jun 19	18.5	87.2134	0.014	0.009	0.642
Jun 20	58	426.7129	0.069	0.047	0.681
Jul 2	16.7	77.14179	0.012	0.006	0.5
Jul 10	33.2	496.2128	0.08	0.055	0.687
Jul 13	28.25	214.95	0.034	0.021	0.617
Jul 22	44	680.883	0.12	0.069	0.575
Jul 23	49.25	500.8	0.081	0.059	0.728
Jul 25	30	178.3781	0.03	0.018	0.6
Jul 29	32	302	0.049	0.031	0.632
Jul 31	44	395.2	0.064	0.043	0.671
Aug 16	33	345.6	0.056	0.032	0.571
Mean sediment delivery ratio					0.63

Table 20.7 Estimated versus observed sediment yield for the remaining rainfall events after accounting sediment delivery ratio

Date	Rainfall (mm)	Sediment yield (t/ha)	
		Estimated	Observed
Aug 22	20.75	0.0163527	0.02
Aug 26	24.5	0.0339633	0.031
Sep 1	19	0.0163527	0.018
Sep 6	16.5	0.00817635	0.009
Sep 8	27	0.0188685	0.012
Sep 23	28.75	0.01320795	0.009
Sep 25	17.5	0.0062895	0.005
Oct 2	18	0.0100632	0.008
Oct 4	23	0.02452905	0.02
Oct 6	22	0.01446585	0.016
Oct 7	63	0.0389949	0.04
Oct 9	28	0.0176106	0.016
Statistical parameters			
Root-mean-square error (RMSE)			0.003
Coefficient of determination (CRM)			−0.07

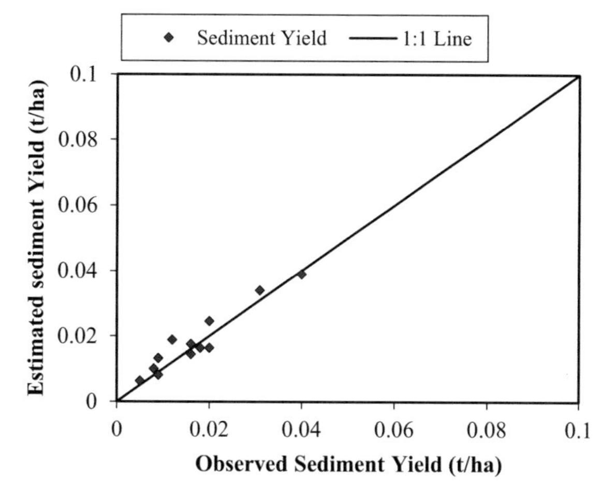

Fig. 20.14 Observed versus estimated sediment yield for the main watershed (after sediment delivery ratio)

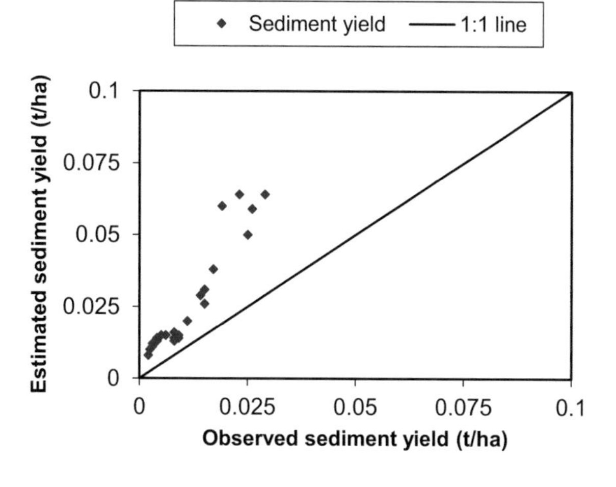

Fig. 20.15 Observed versus estimated sediment yield for the sub-watershed I

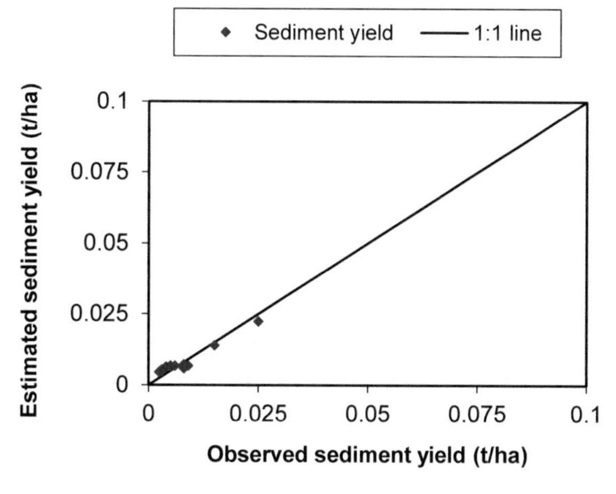

Fig. 20.16 Observed versus estimated sediment yield for the sub-watershed I (after sediment delivery ratio)

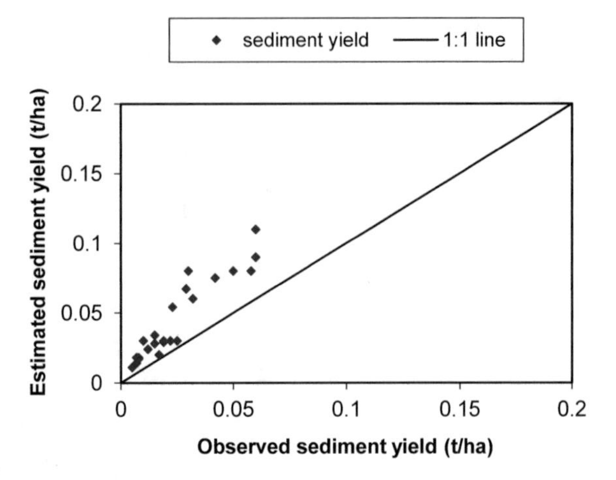

Fig. 20.17 Observed versus estimated sediment yield for the Sub-watershed II

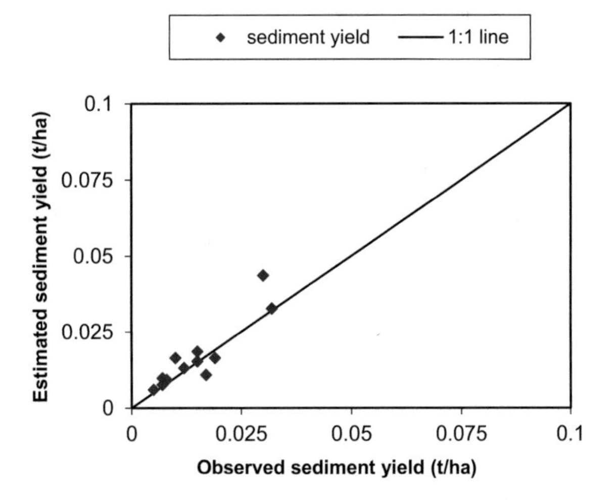

Fig. 20.18 Observed versus estimated sediment yield for the Sub-watershed II (considering sediment delivery ratio)

The land can be effectively utilized for planting various types of trees, which provide multi-purpose uses such as fuelwood, leaves, timber. Agroforestry systems can be adopted in this area. *Acacia auriculoformis* (Akashmani), *Eucalyptus spp.*, etc. can be grown in such agroclimatic zone. The foliage of some of the species can be pruned and used as green leaf manure, e.g., *Leucaena leucocephala, Sesbania grandiflora.* Sabaigrass (***Eulaliopsis binata***) is a perennial grass that can grow in hot and dry climates with much stress. The thin and long leaves are harvested and used for paper industries, making ropes, and other handicrafts items for several uses. The plants are generally one meter to one and a half meters long. It produces a huge number of thin, long, and green leaves. Reduction of erosion with no till and reduced tilling systems in soyabean, corn, grain sorghum, and cotton was reported by

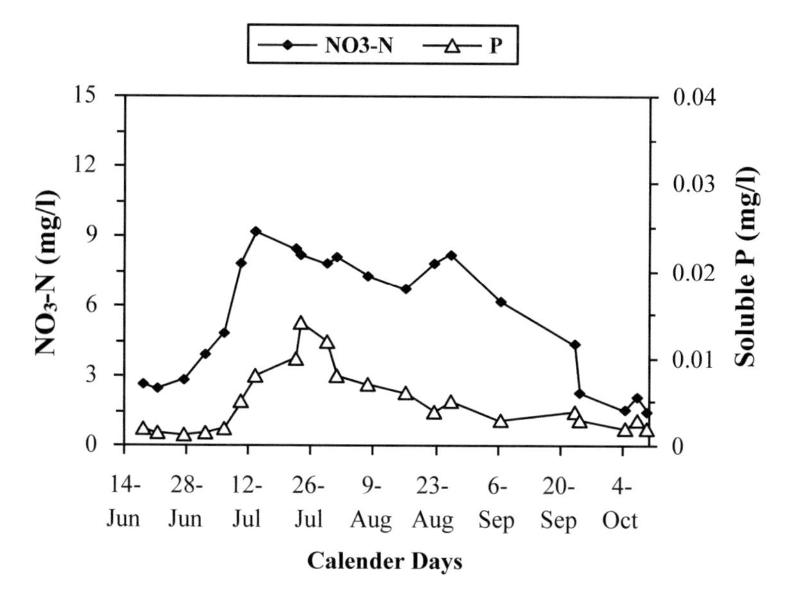

Fig. 20.19 Observed nutrient loss for the main watershed

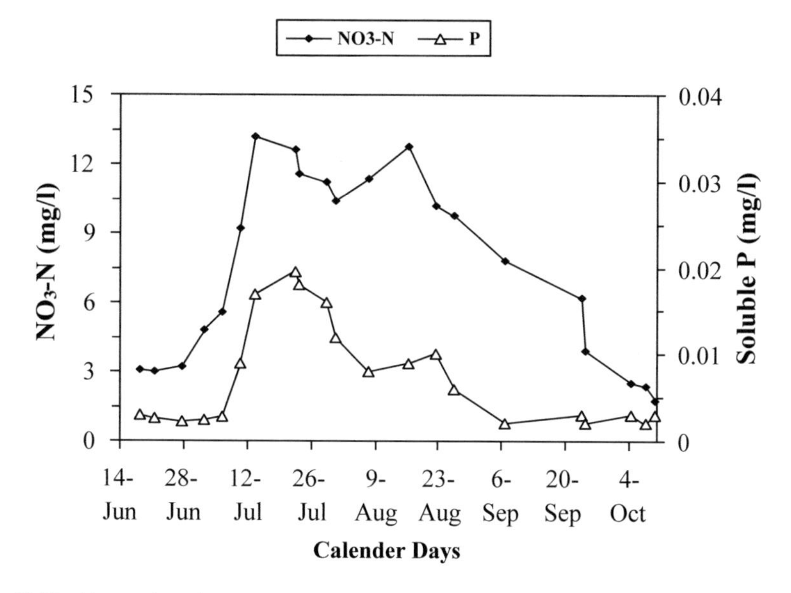

Fig. 20.20 Observed nutrient loss for the Sub-watershed I

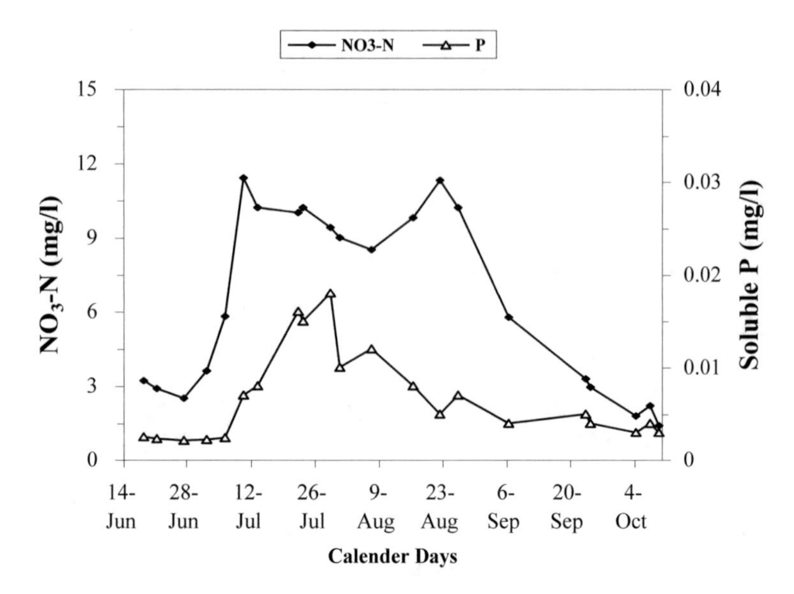

Fig. 20.21 Observed nutrient loss for the Sub-watershed II

Table 20.8 Soil test value of the wasteland (Kapgari watershed)

Serial no	Soil texture	Bulk density (g/cc)	Soil pH	Category
Site 1	Sandy loam	1.53	5.89	Slightly acid
Site 2	Sandy loam	1.56	5.76	Slightly acid
Site 3	Sandy loam	1.49	5.84	Slightly acid
Site 4	Sandy loam	1.51	5.92	Slightly acid
Site 5	Loam	1.59	6.1	Slightly acid
Site 6	Loamy sand	1.61	5.50	Slightly acid
Site 7	Loamy sand	1.63	5.53	Slightly acid
Site 8	Sandy loam	1.51	5.61	Slightly acid
Site 9	Sandy loam	1.52	5.58	Slightly acid
Site 10	Sandy loam	1.49	5.78	Slightly acid

Table 20.9 Soil category concerning organic carbon, P_2O_5, K_2O, and N

Categories	Organic carbon (%)	Available N (kg ha^{-1})	Available P_2O_5 (kg ha^{-1})	Available K_2O (kg ha^{-1})
High	>0.75	>450	>90	>340
Medium	0.5–0.75	280–450	45–90	150–340
Low	<0.5	<280	<45	<150

Table 20.10 Nutrient status of the soil (organic carbon and available nitrogen)

S. No.	Organic carbon (%)	Category	Available N (Kg/ha)	Category
Site 1	0.52	Medium	180.5	Low
Site 2	0.56	Medium	193.6	Low
Site 3	0.546	Medium	187.2	Low
Site 4	0.58	Medium	196.4	Low
Site 5	0.62	Medium	207.3	Low
Site 6	0.5	Medium	176.4	Low
Site 7	0.51	Medium	178.8	Low
Site 8	0.536	Medium	188.9	Low
Site 9	0.568	Medium	195.4	Low
Site 10	0.64	Medium	208.3	Low

Table 20.11 Nutrient status of the soil. (Available P_2O_5 and available K_2O)

S. No.	Available P_2O_5 (Kg/ha)	Category	Available K (Kg/ha)	Category
Site 1	10.3	Low	110.2	Low
Site 2	11.5	Low	80.5	Low
Site 3	12.3	Low	90.2	Low
Site 4	14.6	Low	93.4	Low
Site 5	13.7	Low	82.6	Low
Site 6	11.3	Low	100.4	Low
Site 7	15.8	Low	88.7	Low
Site 8	14.9	Low	85.4	Low
Site 9	13.7	Low	93.6	Low
Site 10	15.2	Low	87.8	Low

McGregor et al. (1999). Langdale et al. (1979) concluded that a shift from tilled to no tilled crop systems significantly reduced erosion and sediment transported nutrients.

Besides grass, sabai has many other advantages when cultivated. It helps protect soil from erosion, changes the physical condition of soil, and increases the soil's water-holding capacity. This grass bears a huge number of fibrous roots, which develop soil-binding effect, and thereby reduces erosion. For this reason, sabaigrass may be selected as one of the important erosion-controlling crop for the soil erosion-prone red lateritic/murrum lands. To have efficient utilization of the natural resources and accrue maximum benefit, several seasonal crops like black gram (Phaseolus *mungo*), groundnut *(Arachis hypogea*), niger (Giuzota *abyssinicia Cass*) are compatible with sabaigrass for intercropping during *Kharif* season in the selected area of Kapgari watershed. On the other hand, a few perennial fruit plants like drumstick (*Moringa oleifera*), guava (*Psidium guajava*), sapota (*Achras zapota L.*), aonla or amalaki (*Emblica officinalis*) can be planted in the same field of sabaigrass for creating permanent sources of income and making the venture sustainable farming systems. Such sabaigrass-based cropping systems in rainfed farming can provide pulses, fruits, and vegetables besides a suitable raw material for paper industries and for making of rope and grass-based handicrafts high-value utility items that may promote employment and income in rural areas.

References

Anima SD, Stott DE, O'Neill MK, Ong CK, Weesies GA (2003) Soil erosion prediction using RUSLE for central Kenyan highland conditions. Agr Ecosyst Environ 97(1–3):295–308

Assis KGO, da Silva YJAB, Lopes JWB, Medeiros JC, Teixeira MPR, Rimá FB, Singh VP (2021) Soil loss and sediment yield in a perennial catchment in southwest Piauí, Brazil. Environ Monit Assess 193(1):1–11

Bekele B, Gemi Y (2021) Soil erosion risk and sediment yield assessment with universal soil loss equation and GIS: in Dijo watershed, Rift valley Basin of Ethiopia. Model Earth Syst Environ 7(1):273–291

Bouamrane A, Bouamrane A, Abida H (2021) Water erosion hazard distribution under a semi-arid climate condition: case of Mellah Watershed, North-eastern Algeria. Geoderma 403:115381

Cohen MJ, Shepherd KD, Walsh MG (2005) Empirical formulation of the universal soil loss equation for erosion risk assessment in a tropical watershed. Geoderma 124:235–252

Das B, Singh S, Jain SK, Thakur PK (2021a) Prioritization of sub-basins of Gomti river for soil and water conservation through morphometric and LULC analysis using remote sensing and GIS. J Indian Soc Remote Sens 1–20

Das S, Deb P, Bora PK, Katre P (2021b) Comparison of RUSLE and MMF soil loss models and evaluation of catchment scale best management practices for a mountainous watershed in India. Sustainability 13(1):232

Endalamaw NT, Moges MA, Kebede YS, Alehegn BM, Sinshaw BG (2021) Potential soil loss estimation for conservation planning, upper Blue Nile Basin, Ethiopia. Environ Chall 100224

Elnashar A, Zeng H, Wu B, Fenta AA, Nabil M, Duerler R (2021) Soil erosion assessment in the Blue Nile Basin driven by a novel RUSLE-GEE framework. Sci Tot Environ 148466

Fistikoglu O, Harmancioglu NB (2002) Integration of GIS with USLE in assessment of soil erosion. Water Resour Manag 16:447–467

Fook KL, Bolhassan J, Mahmood NN (1992) Soil erosion mapping using remote sensing and GIS techniques for land-use planners. Asian-Pac Remote Sens J 5(1):105–115

Foster GR, McCool DK, Renard KG, Moldenhauer WC (1981) Conversion of universal soil loss equation to SI metric units. J Soil Water Cons 36:355–359

Jhariya DC, Kumar T, Pandey HK (2020) Watershed prioritization based on soil and water hazard model using remote sensing, geographical information system and multi-criteria decision analysis approach. Geocarto Int 35(2):188–208

Kumar T, Jhariya DC, Pandey HK (2020) Comparative study of different models for soil erosion and sediment yield in Pairi watershed, Chhattisgarh, India. Geocarto Int 35(11):1245–1266

Langdale GW, Leonard RA, Fleming WG, Jackson WA (1979) Nitrogen and chloride movement in small upland piedmont watersheds: II. Nitrogen and chloride transport in runoff. J Environ Qual 8:57–63

Mabler RL, Bailey FG (1994) Nutrient management in Idaho. Supplement. J Soil Water Cons 49:89–92

Mass RP, Smolen MD, Still DA (1985) Selecting critical areas for non point source pollution control. J Soil Water Cons 40(I):60–71

McGregor KC, Cullum RF, Mutchler CK (1999) Long-term management effects on runoff, erosion and crop production. Trans ASAE 42(I):99–105

Mitchell JK, Engel BA, Srinivasan R, Wang SSY (1993) Validation of AGNPS for small watersheds using integrated AGNPS/ GIS system. Water Resour Bull 29(5):833–841
Mongkolsawat C, Thirangoon P, Sriwongsa S (1994) Soil erosion mapping with universal soil loss equation and GIS. In: Proceedings of the 15th Asian conference on remote sensing, AARS, 17–23 Nov 1994, Bangalore, India
Negash DA, Moisa MB, Merga BB, Sedeta F, Gemeda DO (2021) Soil erosion risk assessment for prioritization of sub-watershed: the case of Chogo Watershed, Horo Guduru Wollega, Ethiopia. Environ Earth Sci 80(17):1–11
Ogawa S, Saito G, Mino N, Uchida S, Khan NM, Shafiq M (1997) Estimation of soil erosion using USLE and LANDSAT TM in Pakistan. In: Proceedings of the 18th Asian conference on remote sensing, AARS, 20–24 Oct 1997, Malaysia
Opena FT (1992) The application of Geographic Information Systems to soil erosion susceptibility mapping. Asian-Pac Remote Sens J 5(1):117–122
Panda C, Das DM, Raul SK, Sahoo BC (2021) Sediment yield prediction and prioritization of sub-watersheds in the Upper Subarnarekha basin (India) using SWAT. Arab J Geosci 14(9):1–19
Pandey A, Gautam AK, Chowdary VM, Jha CS, Cerdà A (2021) Uncertainty assessment in soil erosion modelling using RUSLE, multisource and multiresolution DEMs. J Indian Soc Remote Sens 49(7):1689–1707
Pandey A, Chowdary VM, Mal BC (2007) Identification of critical erosion prone areas in the small agricultural watershed using USLE, GIS, and remote sensing. Water Resour Manage 21(4):729–746
Rao VV, Chakroborti AK, Vaz N, Sarma U (1994) Watershed prioritization based on sediment yield modelling and IRS-1A LISS data. Asian-Pacific Remote Sens J 6(2):101–105
Rao YP (1981). Evaluation of cropping management factor in USLE under natural condition of Kharagpur, India. In: Proceedings of Southeast Asian regional symposium on problems of soil erosion and sedimentation. AIT, Bangkok, pp 241–254
Renschler CS, Harbor J (2002) Soil erosion assessment tools from point to regional scales the role of geomorphologists in land management research and implementation. Geomorphology 47(2):189–209
Rinos MH, Aggarwal SP, De Silva RP (1999) Application of remote sensing and GIS on soil erosion assessment at Bata River Basin, India. Proceedings of the 19th Asian Conference on Remote Sensing, AARS, 22–25 Nov 1999, Hong Kong, China
Singh G, Babu R, Chandra S (1981) Soil loss prediction research in India. Indian Council of Agricultural Research, Bulletin of Central Soil and Water Conservation Research and Training Institute, Dehradun, India, no T-12/D-9
Srinivasan R, Karthika KS, Suputhra SA, Chandrakala M, Hegde R (2021) Mapping of soil erosion and probability zones using remote sensing and GIS in Arid part of South Deccan Plateau, India. J Indian Soc Remote Sens 1–17
Stephens PR, MacMillan JK, Daigle JL, Cihlar J (1985) Estimating universal soil loss equation factor values with aerial photography. J Soil Water Cons 40(3):293–296
Suresh R (1997) Soil and water conservation engineering. Standard publisher Distributor, Delhi, pp 2–3
Tim US, Moostaghimi S, Shanholtz VO (1992) Identification of critical non point pollution source areas using geographic information systems and water quality modeling. Water Resour Bull 25(5):877–887
USDA Soil Conservation Service (1969) Hydrology. In: SCS national engineering handbook. U.S. Department of Agriculture, Washington, D.C., Sect. 4
Wischmeier WH, Smith DD (1978) Predicting rainfall erosion losses, a guide to conservation planning. Agricultural handbook 537. U. S. Department of Agriculture, Washington, D. C.
Wolfe ML, Batchelor WD, Dillaha TA, Mosthaghimi S, Heatwle CD (1995) A farm-scale water quality planning system for evaluating based management practices. In: Proceedings of the international symposium on water quality modelling. ASAE. Orlando

Chapter 21
Hydrological Change Detection Mapping and Monitoring of Ramganga Reservoir, Pauri Gharwal, Uttarakhand, Using Geospatial Technique

Manish Rawat, Ashish Pandey, Basant Yadav, Praveen Kumar Gupta, and J. G. Patel

Abstract Wetlands serve an essential role in conserving the ecological balance of both biotic and abiotic lives in both inland and coastal environments. Hence, understanding their existence and the spatial extent of change in the wetland ecosystem is vital and monitored using remote sensing technology. A study was performed for the Ramganga reservoir, located in Uttarakhand, India, one of the important wetlands in Uttarakhand that comes under Ramsar sites with many bird species and is rich in biodiversity. The present study modeled the spatiotemporal changes of the reservoir using the multi-temporal Landsat TM (1992, 2000, and 2010) and Landsat OLI-TIRS (2020) imageries. While performing spatiotemporal analysis, the applicability of various satellite-derived indexes such as the Normalized Difference Water Index (NDWI), Modified NDWI (MNDWI), Normalized Difference Vegetation Index (NDVI), and Normalized Difference Turbidity Index (NDTI) has been employed for retrieval of wetland elements including the wetland extent boundary, water-spread area, aquatic vegetation, and turbidity level of the reservoir during pre- and post-monsoon seasons by using remote sensing and hierarchical decision tree algorithm. In the post-monsoon season, the minimum and maximum water-spread areas were 59.98 km^2 in 1992 and 74.30 km^2 in 2010, respectively. Similarly, in the pre-monsoon season, the reservoir had a minimum, and the maximum water-spread area was 28.697 km^2 in 2010 and 58.536 km^2 in 2020, respectively. The total aquatic vegetation of wetland was increased from 3.41 to 4.14 km^2 in the years 1992 and 2020 during post-monsoon. Throughout the study period, the medium turbidity level in the reservoir was observed to be less than 1 km^2. The result indicates that these different satellite-derived spectral indices are less time-consuming and provide more accurate mapping and wetlands monitoring. This requires constant monitoring of the

M. Rawat (✉) · A. Pandey · B. Yadav
Department of Water Resources Development and Management, Indian Institute of Technology Roorkee, Roorkee, Uttarakhand, India

P. K. Gupta · J. G. Patel
Space Applications Centre, Indian Space Research Organisation, Ahmedabad, India

A. Pandey et al. (eds.), *Geospatial Technologies for Land and Water Resources Management*, Water Science and Technology Library 103,
https://doi.org/10.1007/978-3-030-90479-1_21

structural components of wetlands and urgently focuses on wetland conservation, rehabilitation, and management.

21.1 Introduction

Wetlands are one of the most important natural habitats, serving various functions and providing diverse ecosystem goods and services essential to biodiversity, nutritional balance, and human well-being (Wang et al. 2012). Wetlands are defined as areas of land where water is covered temporarily or permanently. Most natural waterbodies such as lakes, rivers, mangroves, coastal lagoons, man-made coral reefs and man-made waters bodies, such as reservoirs, farming pools, canals, pools, comprise the wetland ecosystems in India (Ramsar Convention, 2012). Wetlands are perceived as one of the most prolific but challenging ecosystems. With ballooning population and increasing pressure on arable lands, approximately, 50% of the world's wetlands are being transformed into agricultural, industrial, and housing lands during the last hundred years (Kingsford 2011). In India, wetlands are distributed from the Himalayas range to the Deccan Plateau in various terrestrial regions. They account for 4.7% of the overall landscape extent in the country (Bassi et al. 2014). Natural habitats are crucial for the survival of fish and birds and valuable resources for sustaining biological diversity, flood control management, groundwater recharge, and the improvement of water quality. Considering the significant role of wetlands in people's livelihoods and ecosystem protection, information for their safety and management is imperative.

However, the past few years have seen significant land use and land cover area changes due to anthropogenic activities such as urbanization and agricultural practices. It also resulted in subsequent degradation and shrinkage of wetlands, aquatic vegetation growth, deterioration of water quality, and sedimentation have become a cumulative strain on wetlands as a result of a decline in biological diversity, reduced the population of migratory birds, the productivity of fish and other species, and the conservation of biodiversity. Wetlands are also harmed by drought, salinization, eutrophication, and pollution (Alvarez-Cobelas et al. 2008).

Long-term changes in wetland characteristics better understand wetland patterns and dynamics, allowing management organizations to plan restoration or conservation programs. In 2011, the Space Application Centre (SAC) prepared the National Wetland Atlas, which specified the extent of wetlands was 15.26 million ha or 4.64% of the total geographical area. The decline of wetland areas is mainly attributed to anthropogenic activities (Sivakumar and Ghosh 2016; Mondal et al. 2018). Wetlands are characterized as ecotones between aquatic and terrestrial systems due to their hydrological condition and function. Their significance in regulating climate patterns and mitigating the effects of climate change is widely acknowledged (Du et al. 2019). Thus, a detailed inventory and mapping of wetland systems is beneficial for understanding the spatial distribution of various wetlands and their interaction with other landscape units.

Remote sensing technology has proven to be effective measures for inventorying, monitoring, and managing wetland environments in the past few decades and also provides the most viable option when little or no prior information about the wetland is available. This technique is most effective for wetlands studies involving monitoring water-spread area, aquatic vegetation, turbidity level, and nutritional status of wetland ecosystems. The various methods are used for extracting wetlands spatiotemporal change information from remote sensing imagery: single-band methods and multi-band methods. One of the most accurate methods for wetland delineation is based on the indices-based classification for enhancement and extraction of structural components of wetland elements.

A semiautomatic method has been used to extract the decipher information of wetlands in spatial, spectral, and temporal patterns. In this research, four spectral indices (NDVI, NDWI, MNDWI, NDTI) have been generated for pre- and post-monsoon satellite imageries. Combining these indices enhances the extent of water spread, the status of aquatic vegetation, and the water turbidity level in the wet and dry seasons (post-monsoon and pre-monsoon periods), which can be easily interpreted and manually interpreted digitized. The status of aquatic vegetation and turbidity level is an important factor for analyzing the eutrophication of wetlands and important aspects of wetland management and sustainable development (Manju et al. 2005; Chopra and Jakhar 2016; Ray et al. 2012). The turbidity of water is an effective measure of the wetland health conditions. The spatial distribution of the suspended particles is a substantial source of uncertainty in turbidity estimation. Therefore, a wetland inventory is imperative.

The Modified Normalized Difference Water Index (MNDWI) effectively improves the surface water characteristics, which reduce and suppress built-up land, vegetation, and soil disturbance. However, the waterbodies estimation by NDWI is frequently mixed with the terrestrial disturbance that causes inaccuracy in the extraction of water-spread area. As a result, the MNDWI is the best suitable for retrieving extending water areas from the water availability region and reducing built-up land noise over the NDWI. In general, wetland boundary was defined to include the water-spread area and adjacent wetland areas with marshy land that can be visually interpreted with the assistance of the MNDWI in the GIS environment. Huising (2002) states that mapping and monitoring wetland boundaries and their location are imperative. The NDVI is calculated as a normalized relationship between the near-infrared and the red spectrum (Ahmed 2016). Several studies have utilized NDVI to assess the crop-coverage area (Shao et al. 2016). The vegetational index is a key measurement metric for environmental resource management that gives information about the vegetative-coverage area in each pixel on a satellite image (Bhandari et al. 2012).

The objective of the present study is to analyze the decadal changes in the Ramganga reservoir, Uttarakhand. This study includes perceiving spatiotemporal variations in the water-coverage area, aquatic vegetation, and the water turbidity level of Ramganga reservoir in the pre- and post-monsoon seasons by integrating four indices (MNDWI, NDWI NDVI, and NDTI) using Landsats 5 and 8 satellite imageries over the last twenty-eight years data from 1992 to 2020. This knowledge of wetland spatial variability and changes occurring with periods is essential to the wetlands' sustainable conservation and rehabilitation planning.

21.2 Data and Methodology

21.2.1 Study Area and Dataset

This study was carried out for the Ramganga reservoir wetland. It is situated in the Corbett National Park range near Ramnagar city in the Pauri Garhwal district of Uttarakhand, with a latitude of 29° 35′ 29.45′ N and longitude of 78° 42′ 27.78′ E (Fig. 21.1). The study area is situated on the foothills of the Himalayas in the Tarai region, at an elevation of 347 m above mean sea level (m.s.l.).

The temporal Landsat 5 TM and Landsat 8 OLI imageries of the pre-monsoon season (April/May) and post-monsoon season (October/November) were used in the present study to analyze and classify wetland structural components of the Ramganga reservoir. The available cloud-free satellite imageries of Ramganga reservoir have been acquired in 1992, 2000, 2010, and 2020 shown in Table 21.1. These Landsat time series imageries were used to assess the spatiotemporal changes in water-spread area, aquatic vegetation, and the water turbidity level in the study area. The performances of various bi-spectral indices, including MNDWI, NDWI, NDVI, and NDTI, were

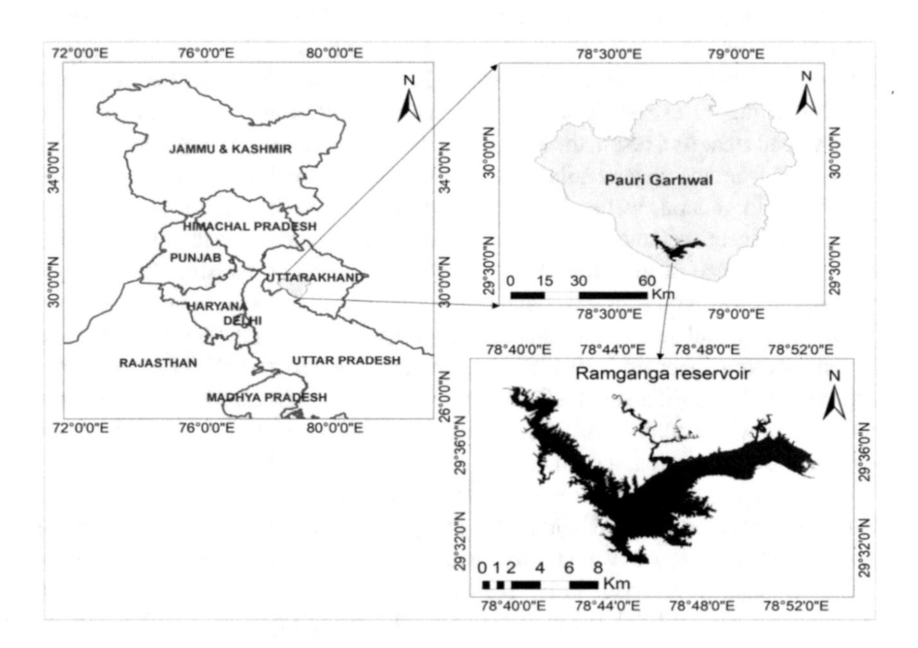

Fig. 21.1 Location map of Ramganga reservoir wetland, Uttarakhand, India

Table 21.1 Landsat image acquisition dates for the study period

Satellite ID	Sensor ID	Path/row	Acquisition date (pre-monsoon season)	Acquisition date (post-monsoon season)	Spatial resolution
Landsat 5	TM	146/39	March 16, 1992	Nov 11, 1992	30
Landsat 7	TM	146/39	May 1, 2000	Nov 25, 2000	30
Landsat 5	TM	146/39	April 3, 2010	Oct 28, 2010	30
Landsat 8	OLI	146/39	April 11, 2020	Oct 7, 2020	30

analyzed to extract Ramganga reservoir wetland elements from the Landsat data series. The methodology used in this study is shown in Fig. 21.4.

Figure 21.2 depicts a false color composite (FCC) map of Landsat images of the Ramganga reservoir catchment and its surrounding area that were acquired in the pre- and post-monsoon seasons of 1992, 2000, 2010, and 2020, respectively.

A false color composite (FCC) map of Landsat imageries of Ramganga reservoir of the study dated 1992, 2000, 2010, and 2020, respectively, is as shown in Fig. 21.3. These Landsat imageries have been employed for analyzing the spatiotemporal changes in wetland boundary. All images were acquired in pre- and post-monsoon seasons to provide a perspective of wetland manifestation in remote sensing data and a synoptic perspective of the area (Fig. 21.4).

21.2.2 *Indices-Based Classification Using Decision Tree Algorithm*

In this study, a decision tree algorithm has been developed to classify water-spread area, vegetation, and water turbidity level using different bi-spectral indexes, including NDWI, NDVI, and NDTI, Fig. 21.5. The decision tree classification (DTC) algorithm is a supervised machine learning algorithm. It can be used for both a classification problem as well as for a regression problem. The decision tree approach is a non-parametric classifier and an instance of the machine learning algorithm. This is a tree-like structure consisting of root nodes, internal nodes, and leaves, which may have one or more feature input combinations. A new input vector moves down through successive branches from the root node until it is positioned in a specific class.

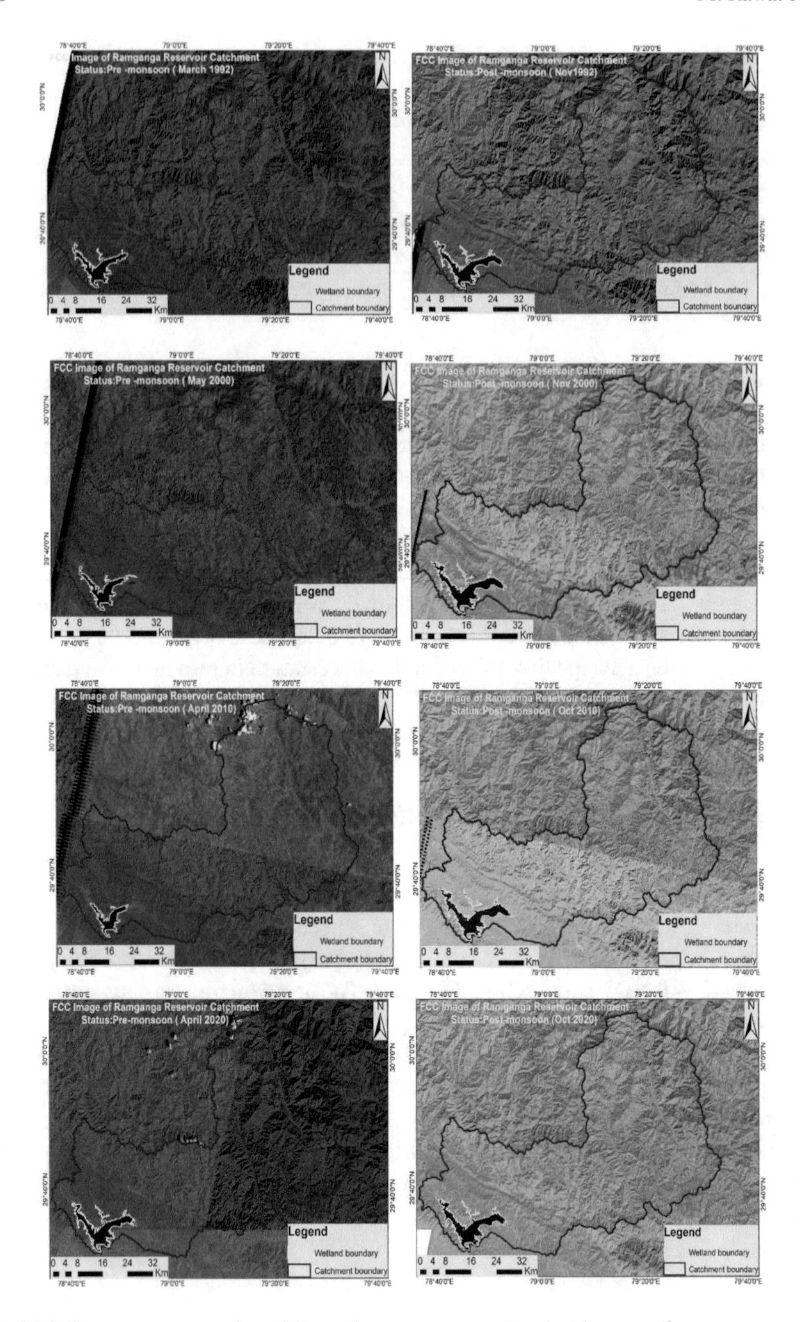

Fig. 21.2 Ramganga reservoir and its environs as seen on Landsat images of pre-monsoon and post-monsoon (1992–2020)

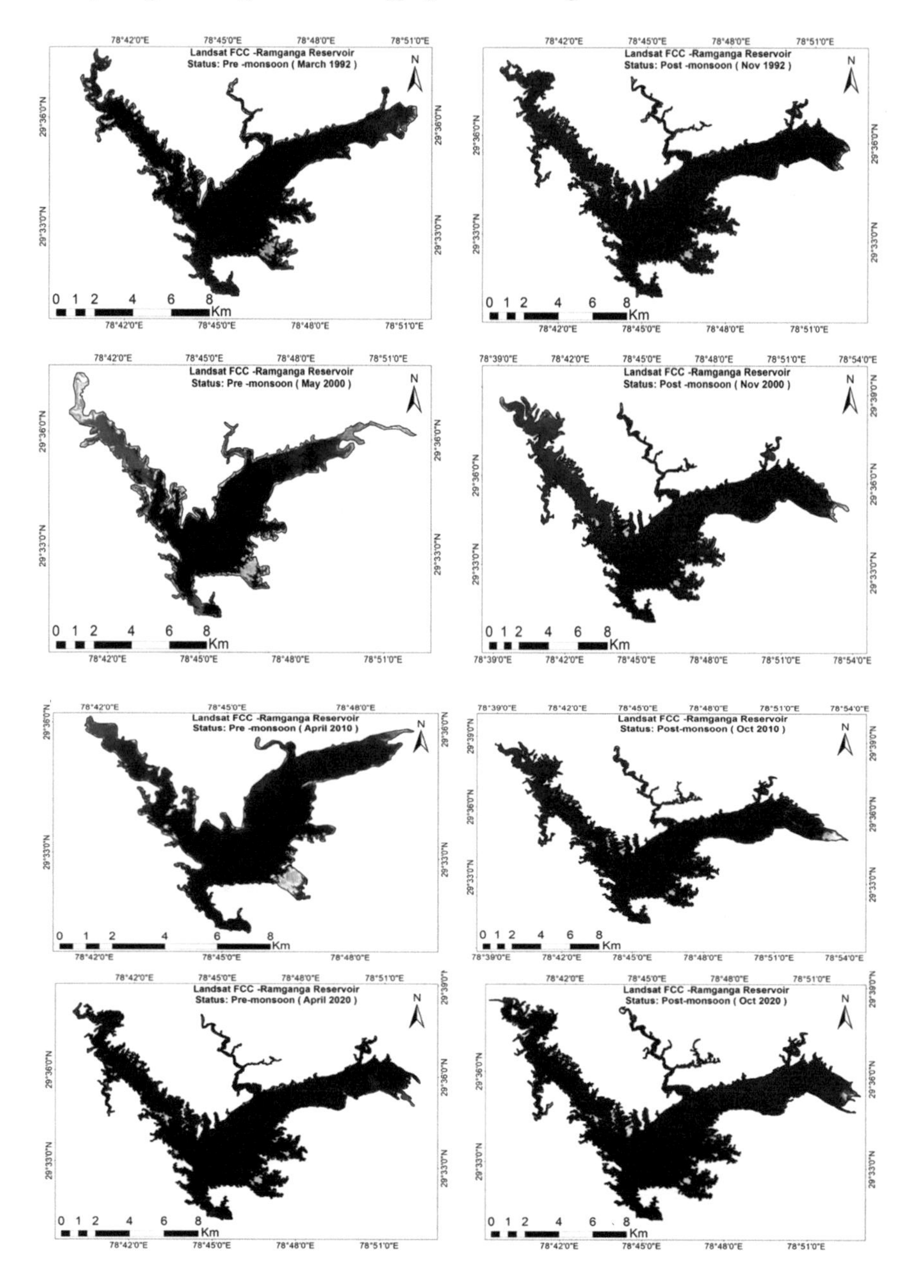

Fig. 21.3 FCC map of satellite imageries of Ramganga reservoir during pre- and post-monsoon periods dated 1992, 2000, 2010, and 2020

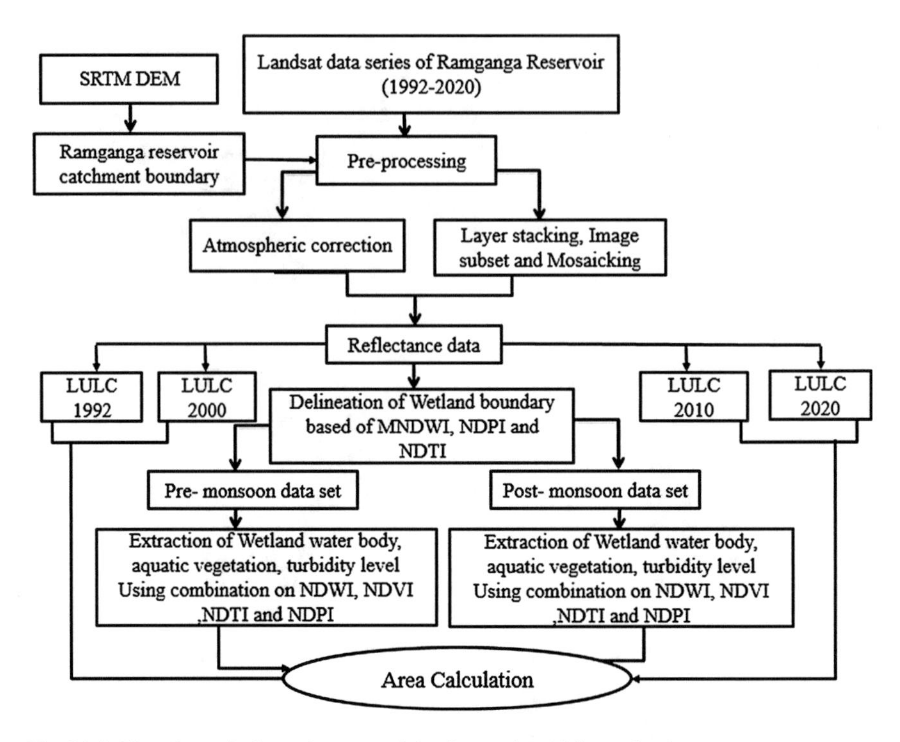

Fig. 21.4 Flowchart of schematic approach has been adopted for wetland

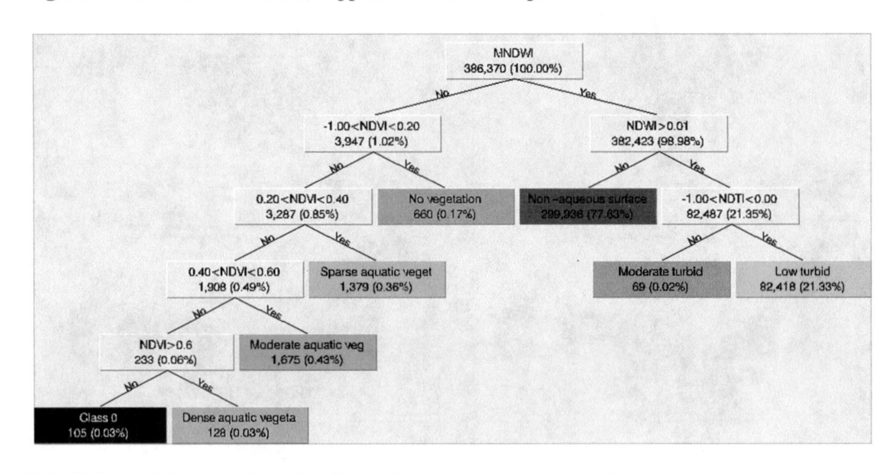

Fig. 21.5 Decision tree algorithm for wetland components estimation

21.2.3 *Change Detection of LULC in the Reservoir Catchment*

To develop the LULC map of the Ramganga reservoir catchment using Landsat images, a supervised classification technique has been used to define the different LULC classes. The total catchment area of Ramganga reservoir is 3134.66 km^2. For mapping the entire catchment area, six major LULC classes have been chosen: dense forest, open forest, cropland, waterbodies, built-up land, and barren land. LULC maps of the Ramganga catchment depicting land cover changes in each decade were prepared using classified images from 1992 to 2000, 2000 to 2010, and 2010 to 2020, respectively; a post-classification change detection technique was employed to analyze the changes.

21.2.4 *MNDWI for Wetland Boundary Mapping*

The MNDWI is a modification of the NDWI in which the near-infrared (NIR) band is being replaced by the middle infrared (MIR) band to improve surface water monitoring by noise suppression from built-up land cover. The MNDWI classification considers the wetland zone's swamp area based on soil moisture (Xu 2006). The wetland boundary was defined as the surface water-spread area with adjacent moist hydric soil area and swampy land. Xu (2006) explained the Modified Normalized Difference Water Index (MNDWI), which can be calculated using Eq. 21.1:

$$\mathrm{NDWI} = (\mathrm{Green} - \mathrm{MIR})/(\mathrm{Green} + \mathrm{MIR}) \tag{21.1}$$

The wetland boundary for pre-monsoon and post-monsoon was visually interpreted in the GIS environment using the MNDWI, NDPI, and NDTI combination from the stack of indices. After creating the wetland boundary, open water spread and vegetation extent of the wetland were interpreted using the combination of indices from the stack for both pre- and post-monsoon seasons.

21.2.5 *Mapping of Waterbody in the Reservoir Using NDWI*

Normalized Difference Water Index (NDWI) is the most frequently used method of mapping waterbodies in a wetland because it is easily distinguished in visible green and near-infrared (NIR) wavelengths region due to their strong absorptivity (Ashtekar et al. 2019; Yan et al. 2017; Bahera et al. 2012). The NDWI pixels value greater than zero shows the waterbodies in wetlands. The negative value of NDWI corresponds to agricultural land, marshy land, and other non-waterbody landscape.

The formula used for NDWI can be calculated using the following Eq. 21.2, which has been defined by McFeeters (1996) as:

$$\text{NDWI} = (\text{Green} - \text{NIR})/(\text{Green} + \text{NIR}) \quad (21.2)$$

The water-spread area in the wetland calculated through MNDWI is greater than the area calculated using NDWI. MNDWI also showed a positive value for weed-infested waterbodies and marshy land.

21.2.6 Aquatic Vegetation Mapping Using Vegetative Index

The presence of aquatic vegetation in a wetland is a good indicator of the trophic condition of the wetland, which affects the quality of water. Thus, the abundant aquatic vegetation known as eutrophication degrades water quality and threatens the ecological system of the wetlands. Consequently, regular monitoring of vegetation in wetlands is an important aspect. The concept of NDVI was used to evaluate the qualitative and quantitative statuses of aquatic vegetation in wetlands. The formula for calculating the Normalized Difference Vegetation Index (NDVI) has been defined as given below:

$$\text{NDVI} = (\text{NIR} - \text{Red})/(\text{NIR} + \text{Red}) \quad (21.3)$$

The NDVI time series pattern was thoroughly examined to illustrate the vegetation growth status in the reservoir ecosystem. The pixel values in NDVI images vary from -1 to $+1$, wherein in this study, NDVI values were classified into three intervals. The corresponding classes have been selected according to the earth observation system data and foundation of Remote Sensing Phenology shown in Table 21.2, in which low NDVI values (below 0.2) correspond to barren land, waterbodies, sand, or snow, sparse vegetation represents scrub, and grassland having NDVI values varies from 0.2 to 0.4, and moderate vegetation varies from 0.4 to 0.6. In contrast, a higher value of NDVI shows dense vegetation (above 0.6).

Table 21.2 NDVI classes according to the earth observation system data

NDVI values	Classes
−1 to 0.2	No vegetation
0.2 to 0.4	Sparse vegetation
0.4 to 0.6	Moderate vegetation
0.6 to 1	Dense vegetation

21.2.7 Turbidity Status Estimation in Reservoir

Satellite imagery data gives essential synoptic information on wetlands dynamics such as aquatic vegetation and turbidity. Water quality in wetlands must be sustained to support the wetland environment. The amount of turbidity level in the wetland is frequently reflected in the water quality. In this study, Normalized Difference Turbidity Index (NDTI) was used to measure the turbidity of the water in the Ramganga reservoir (NDTI) for pre- and post-monsoon seasons using Landsat data. Based on the pixel values, the NDTI classified image assigns qualitative water turbidity levels of the low, medium, and high. The NDTI pixel values range varies from -1 to $+1$. If the NDTI value increases, turbidity increases (Lacaux et al. 2007). The formula used for the NDTI was shown in Eq. 21.4:

$$\text{NDTI} = (\text{Red} - \text{Green})/(\text{Red} + \text{Green}) \tag{21.4}$$

Based on an assessment of the sensor data covering the Indian subcontinent, it has been observed that NDTI values ranging from -0.2 to 0.0 correspond to clear water, whereas values ranging from 0.0 to 0.2 are considered moderately turbid levels in waterbodies; however, while in case of highly turbidity level in waterbodies, the NDTI values are found to be greater than $+0.25$. In this study, two turbidity levels, low turbidity and moderate turbidity, were generated.

Bi-spectral indices-based classification methods are mainly used for qualitative assessment of the wetland elements in terms of extent of water-spread area, turbidity status, and aquatic vegetation concentration, and statistics exploration for pre-monsoon and post-monsoon interpreted data (water spread, vegetation extent) was calculated for all the wetlands.

21.3 Results and Discussion

21.3.1 LULC Status of Ramganga Reservoir Catchment

Multi-temporal Land use land cover classification encompassing six key major classes: dense forest, open forest, cropland, waterbodies, built-up land, and barren land of 1992, 2000, 2010, and 2020 is shown in Fig. 21.6. Table 21.3 shows the spatial variability pattern of LULC resultant from supervised classification (Figs. 21.7, 21.8 and 21.9).

For the period 1992–2020, major changes have been observed from built-up land to open forest. About 18.61 km^2 of built-up land was increased, while dense forests have been converted into open forest (agroforestry) areas, probably due to deforestation and forest fire. In the last three decades, we also observed that area covered by open forest was increased by 316.98 km^2, while dense forest showed a decrease by 181.77 km^2. On the other hand, cropland has also increased between 1992 and 2000

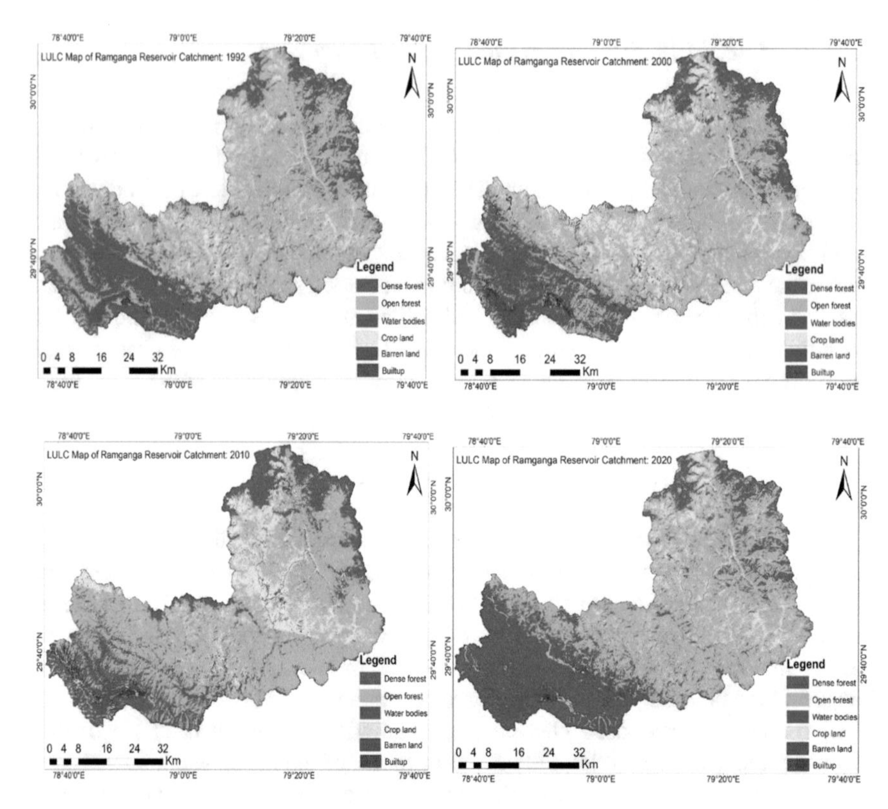

Fig. 21.6 Land use land cover change in different categories for the years 1992, 2000, 2010, and 2020 in Ramganga reservoir watershed

Table 21.3 Change in LULC classes in Ramganga reservoir catchment area

LULC classes	Area (km^2) 1992	Area (km^2) 2000	Area (km^2) 2010	Area (km^2) 2020
Dense forest	1290.43	1133.77	1119.21	1108.23
Open forest	1176.77	1209.37	1280.97	1493.75
Cropland	530.75	646.44	602.23	407.10
Waterbodies	69.70	55.63	36.22	69.16
Barren land	52.40	72.68	68.01	23.20
Built-up	14.61	16.77	28.02	33.22

from 530.75 to 646.44 km^2, then reduced from 602.23 to 407.10 km^2 between 2010 and 2020 because of conversion from cropland into fallow land. The cropland area was overall decreased during the last 28 years (Table 21.4).

The spatiotemporal variation of water-spread area, aquatic vegetation, and turbidity level in Ramganga reservoir was mapped in this study for the years 1992,

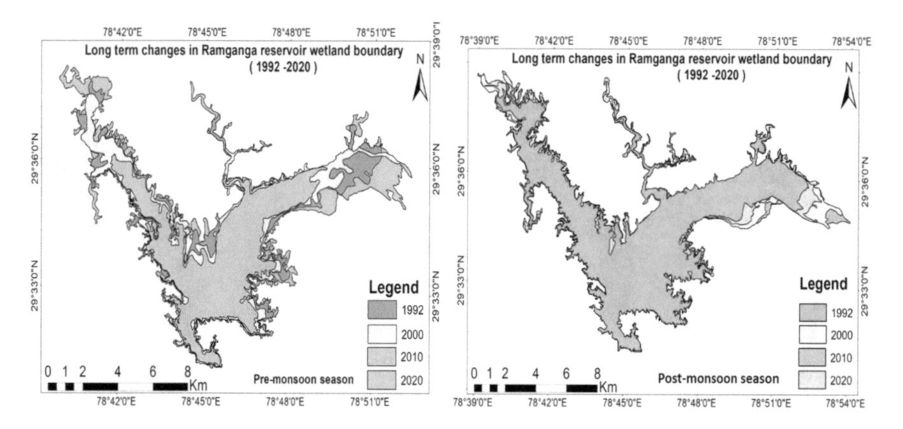

Fig. 21.7 Overlay map of the long-term seasonal changes of water spread in Ramganga reservoir

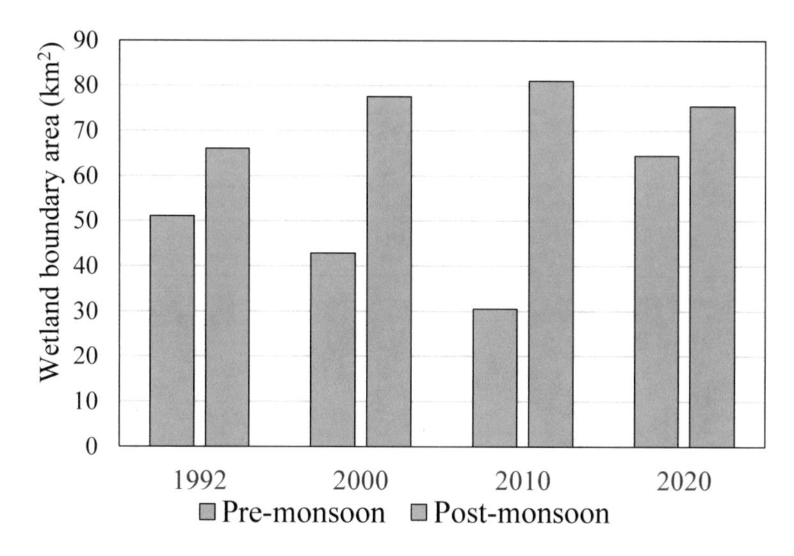

Fig. 21.8 Seasonal changes in reservoir water spread

2000, 2010, and 2020 to analyze the changing pattern over the last three decades. Figures 21.10 and 21.11, 21.12, 21.13 and 21.14 depict the classified NDWI and NDVI map of Ramganga reservoir for the years 1992, 2000, 2010, and 2020, respectively, whereas Tables 21.5 and 21.6 show the areas of NDWI and NDVI classes, respectively. The results of the analysis of NDVI images showed that the NDVI threshold values have classified the vegetation-covered area into four classes: NDVI values (below 0.2) correspond to the no-vegetation class, which is indicating barren land and open water surface of the lake, and NDVI values (range from 0.2 to 0.4) represent sparse aquatic vegetation which indicating mixture of small vegetation or unhealthy vegetation. Moderate NDVI values (0.40–0.60) represent moderate aquatic

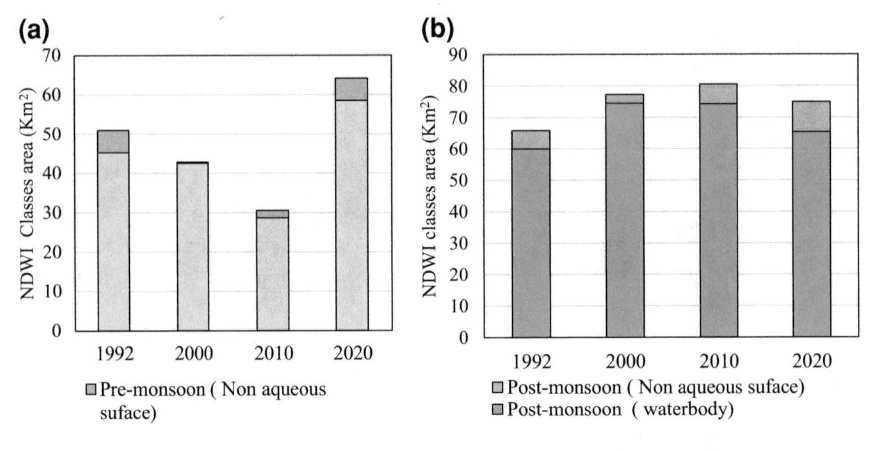

Fig. 21.9 **a** Bar charts depict NDWI area (pre-monsoon) and **b** NDWI area (post-monsoon)

Table 21.4 Decadal changes in Ramganga wetland boundary area

S. No.	Year	Wetland boundary (km^2)	
		Pre-monsoon	Post-monsoon
1	1992	51.12	66.11
2	2000	42.84	77.53
3	2010	30.53	80.97
4	2020	64.49	75.44

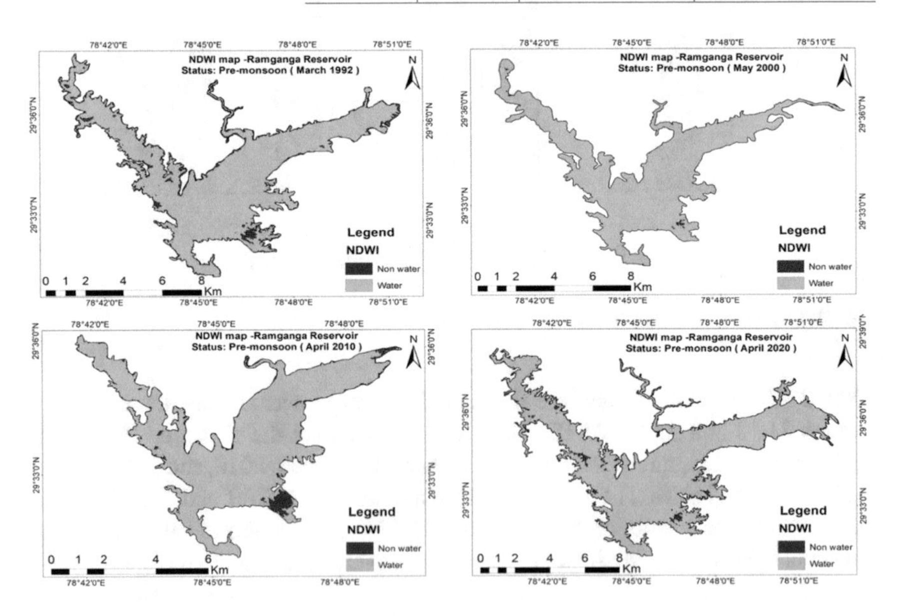

Fig. 21.10 Water-spread mapping of the Ramganga reservoir during the pre-monsoon season

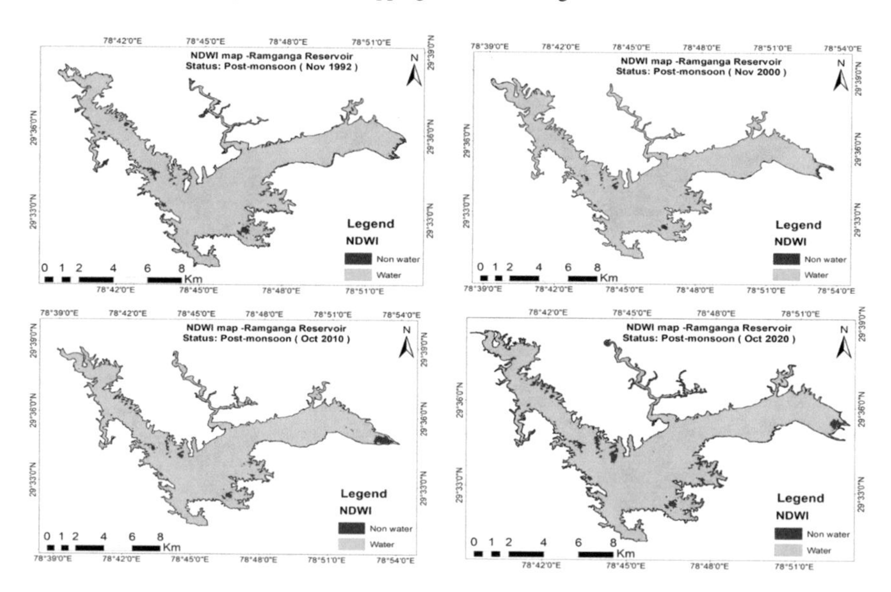

Fig. 21.11 Water-spread mapping of the Ramganga reservoir during the post-monsoon season

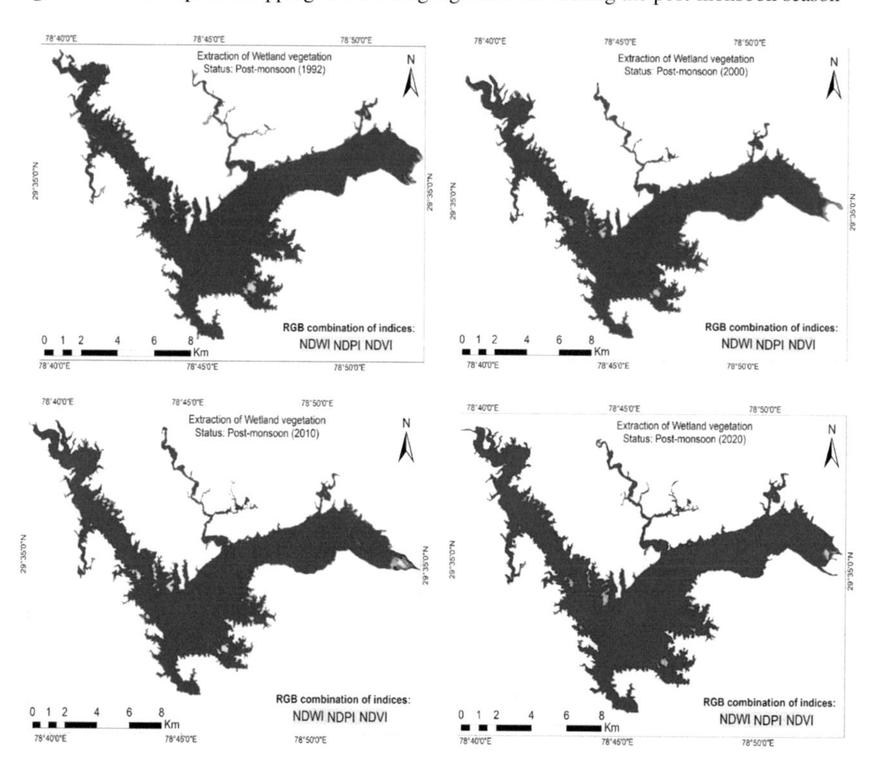

Fig. 21.12 RGB combination of indices display of wetland vegetation extraction

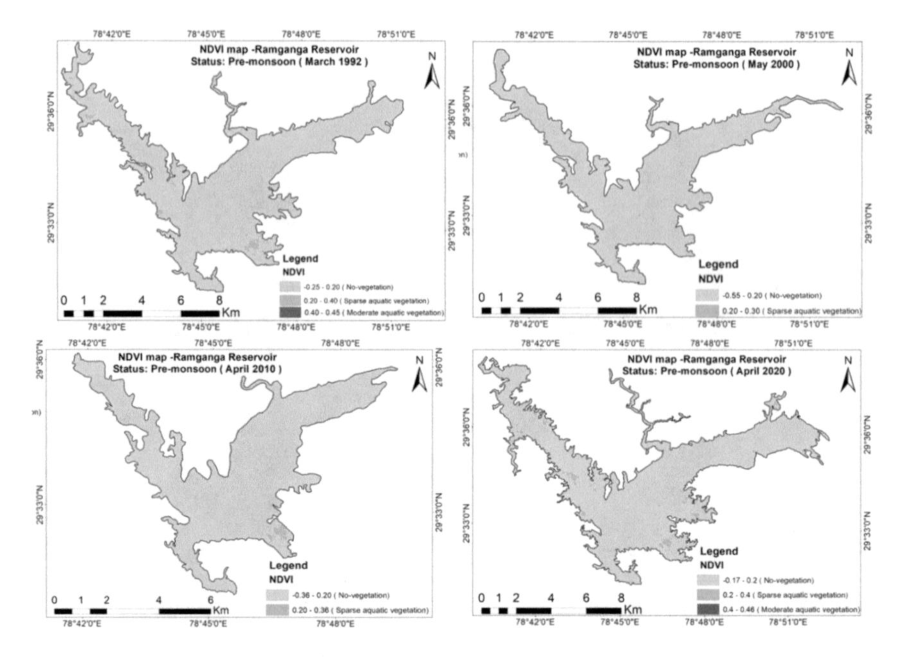

Fig. 21.13 Change detection map of NDVI for Ramganga reservoir during pre-monsoon period

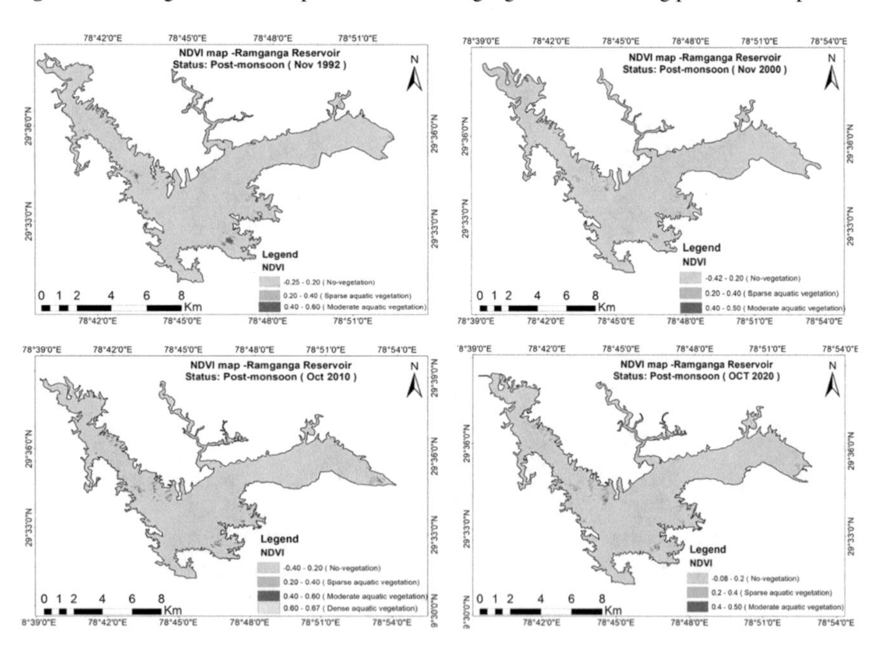

Fig. 21.14 Change detection map of NDVI for Ramganga reservoir during post-monsoon period

Table 21.5 Area of NDWI classes of the Ramganga reservoir

S. No.	Class name	1992 (km^2)		2000 (km^2)		2010 (km^2)		2020 (km^2)	
		Pre-monsoon	Post-monsoon	Pre-monsoon	Post-monsoon	Pre-monsoon	Post-monsoon	Pre-monsoon	Post-monsoon
1	Waterbody	45.319	59.987	42.555	74.559	28.697	74.340	58.536	65.460
2	Non-aqueous surface	5.734	5.882	0.243	2.780	1.835	6.230	5.661	9.546

Table 21.6 Area of NDVI classes of the Ramganga wetland

S. No.	Class name	1992 (km^2)		2000 (km^2)		2010 (km^2)		2020 (km^2)	
		Pre-monsoon	Post-monsoon	Pre-monsoon	Post-monsoon	Pre-monsoon	Post-monsoon	Pre-monsoon	Post-monsoon
1	No vegetation	50.1417	62.4555	42.795	75.6108	30.3732	75.7683	62.6265	70.8633
2	Sparse aquatic vegetation	0.9099	2.8242	0.0045	1.5669	0.1575	3.0519	1.5678	3.9528
3	Moderate aquatic vegetation	0.0018	0.5904	Nil	0.1629	Nil	1.6137	0.0018	0.1917
4	Dense aquatic vegetation	Nil	Nil	Nil	Nil	Nil	0.1422	Nil	Nil
5	Total aquatic vegetation	0.9117	3.4146	0.0045	1.7298	0.1575	4.8078	1.5696	4.1445

vegetation having healthy vegetated areas, while high value (greater than 0.6) indicates dense vegetation (Gandhi et al. 2015). Similarly, for assessment of water-spread area in the reservoir, NDWI threshold values have classified the area into two categories: aquatic and non-aquatic surface classes, which include vegetative and bare land classes of the NDVI.

The long-term decadal changes in the Ramganga reservoir wetland boundary area in the years 1992, 2000, 2010, and 2020 are shown in Fig. 21.7 and Table 21.4. The statistical findings are depicted in Fig. 21.8 and Table 21.4 and reveal that overall, it increased in the wetland surface area from 1992 to 2020 for both pre- and post-monsoon seasons, from 51.12 km^2 in 1990 to 64.49 km^2 in 2020 during the pre-monsoon season, and similarly, for post-monsoon season, 66.11 km^2 in 1992 increase up to 75.44 km^2 in 2020. The increased wetland surface area was caused by an increase in snowmelt water contribution from the Himalayan region, which increased the water inflow volume in the downstream rivers, which are hydrologically connected to the reservoir (Aggarwal et al. 2014). The results also showed that the minimum reservoir surface area was 30.53 km^2 in 2010 during the pre-monsoon period due to diverted reservoir water for other economic activities and lesser water inflow volume in the downstream river that period.

21.3.2 *Change Detection of Water-Spread Area*

NDWI is the most commonly remote sensing index to improve water characteristics and reduce soil and vegetation effects. It is advantageous for retrieving and monitoring of water-spread areas (Yang and He 2017). Many case studies have shown the efficacy of NDWI for monitoring ecological factors, such as soil moisture (Gu et al. 2008) and hydrological conditions (Maleki et al. 2016). Figures 21.10 and 21.11 show the water-spread mapping of the Ramganga reservoir for the pre- and post-monsoon periods, and Table 21.4 shows the estimated area of spread water in the reservoir. The water surface area of the reservoir in the pre-monsoon session was 45.319 km^2, 42.555 km^2, 28.697 km^2, and 58.536 km^2 for the years 1992, 2000, 2010, and 2020, respectively. According to the pre-monsoonal data shown in Table 21.5, the water-spread area was minimum in 2000 and maximum in 2020. The fluctuation in the water-spread area could be caused by an increase in water demand downstream of the reservoir (Doll et al. 2009).

Similarly, estimation of the reservoir's water-spread area during the post-monsoon season was 59.987 km^2, 74.559 km^2, 74.340 km^2, and 65.460 km^2 for years 1992, 2000, 2010, and 2020, respectively. The results show the water-spread area was minimum during 1992 and maximum during 2010. This temporal and spatial variation of water-spread area was due to increasing upstream river flow into the reservoir, and heavy rainfall enhanced the water volume in the reservoir in 2010.

The results show the water-spread area in 1992 was 59.987 km^2, which increased to 74.559 km^2 in 2000 but observed a decrease of 65.460 km^2 in the year 2020.

In 28 years (1992–2020), the wetland ponded area increased by 29.16% in the pre-monsoon period and 9.12% in the post-monsoon period. The increase of water-spread area in the reservoir throughout the years can be attributed to enhance rainfall patterns, as well as Himalayan snowmelt runoff contribution toward the downstream area, which contributes significantly to the increasing river volume, which is hydrologically connected with the reservoir (Adnan et al. 2017).

As illustrated in Figs. 21.9a, b, bar graphs were prepared to show the open water areas of the reservoir in km^2 through NDWI for both seasons throughout the study period (Figs. 21.10 and 21.11).

21.3.3 Status of Aquatic Vegetation in the Reservoir

The NDVI has been frequently performed to explore the relationship between spectrum heterogeneity and changes in vegetative growth rate. Table 21.6 shows three different aquatic vegetation classes retrieved from the NDVI image of the Ramganga reservoir. The aquatic vegetation grown in the Ramganga reservoir was classified into three classes: sparse aquatic vegetation, moderate aquatic vegetation, and dense aquatic vegetation. Due to the inaccessibility of the area and the non-availability of additional auxiliary data, it was not possible to verify different species of vegetation.

After analysis of NDVI images, it has been found the aquatic vegetation-covered area of the Ramganga reservoir ecosystem in the post-monsoon period was 3.41 km^2 in 1992 and increased to 4.14 km^2 in the year 2020; similarly, during the pre-monsoon period, aquatic vegetation-covered area also somewhat increased from 0.91 to 1.57 km^2 in last three decadal years. The waterlogged area was 62.4555 km^2 in the year 1992, further expanded to 70.8633 km^2 in 2020 during the post-monsoon period. The change of vegetation pattern can be seen in Table 21.6 and Figs. 21.13 and 21.14.

During the pre- and post-monsoon seasons of 1992, it can be observed that sparse aquatic vegetation-covered area increased from 0.99 to 2.82 km^2, whereas the moderate aquatic vegetation-covered area was also progressively increased from 0.0018 to 0.59 km^2 of the wetland area, respectively. After analyzing NDVI images, we found that the maximum vegetation area in the reservoir was covered under the sparse aquatic vegetation in both seasons. In the same year 1992, the waterlogged area has been increased from 50.14 to 62.45 km^2. Similarly, in the year 2020, during pre- and post-monsoon seasons, the area under sparse aquatic vegetation was increased from 1.5678 to 3.95 km^2, whereas consequences of this study revealed that the area covered by aquatic vegetation was significantly less than the ponded area in the reservoir and also found that there is no dense aquatic vegetation except of the year 2010, 0.14 km^2 in the post-monsoon season.

The above analysis clarifies that the wetland area covered by sparse aquatic vegetation is higher than moderate aquatic vegetation. The moderate aquatic vegetation is relatively less affected, while highly dense aquatic vegetation covers nearly insignificant areas. Changes in reservoir water levels can have a negative impact on aquatic vegetation in the vicinity of the reservoir (Vandel et al. 2016). Various plant species disintegrate during the post-monsoon season when the water level rises in the reservoir. This plant group mainly consists of shrubs and bushes. Figures 21.13 and 21.14 represent the temporal variation of NDVI maps of the study area during the pre- and post-monsoon seasons from 1992 to 2020, respectively.

The results obtained based on NDVI imagery analysis, presented in Table 21.6, show that the distribution of aquatic vegetation in wetland areas was higher during the post-monsoon period than the pre-monsoon period throughout the study time framework; the reason is waterlogging areas responsible for more vegetation growth after the end of monsoon period (Velmurugan et al. 2016). The estimated vegetation spread area of the reservoir in the post-monsoon session was 3.4146 km^2, 1.7298 km^2, 4.8078 km^2, and 4.1445 km^2 for the years 1992, 2000, 2010, and 2020, respectively, while the vegetation spread area in the pre-monsoon session was 0.911 km^2, 0.0045 km^2, 0.157 km^2, and 1.569 km^2 for the years 1992, 2000, 2010, and 2020, respectively.

Figure 21.12 shows the FCC of NDWI, NDPI, NDVI indices combination enhances wetland-specific vegetation and open water, thus facilitating easier delineation (Figs. 21.13 and 21.14).

21.3.4 Turbidity Status for Ramganga Reservoir Wetland

The Normalized Difference Turbidity Index (NDTI) was used to determine the turbidity of the water in the Ramganga reservoir. The turbidity levels of the Ramganga reservoir wetland were qualitatively described by two levels: low turbidity and moderate turbidity. The moderate turbidity levels, as well as the eutrophication rate, are found to be higher in the pre-monsoon season as compared to the post-monsoon season. The increased turbidity level during the pre-monsoon season is caused by silt and sediments delivered into the wetland along with runoff water (Bid and Siddique 2019). It should be noted that waterbodies in the reservoir shown in the satellite imagery were used to extract turbidity levels. In the pre-monsoon season, the calculated area under moderate turbidity level of the reservoir was 4.1238 km^2, 8.8272 km^2, 2.799 km^2, 0.0657 km^2 for the years 1992, 2000, 2010, and 2020, respectively; in contrast, during the post-monsoon season, the area under moderate turbidity level is observed to be less than 1 km^2 throughout the study period. Figure 21.15 shows the turbidity levels of the Ramganga reservoir for different years. Table 21.7 shows the area observed under different water turbidity levels from 1992 to 2020.

The result of the turbidity level given in Table 21.7 indicates that most of the water-spread areas in the Ramganga reservoir show low turbidity level or clear water, considering aquatic vegetation reflecting it. In addition to investigating spectral properties, the NDTI can also help us characterize the wetlands based on the turbidity level. Higher the threshold value of NDTI, the turbidity level will be higher.

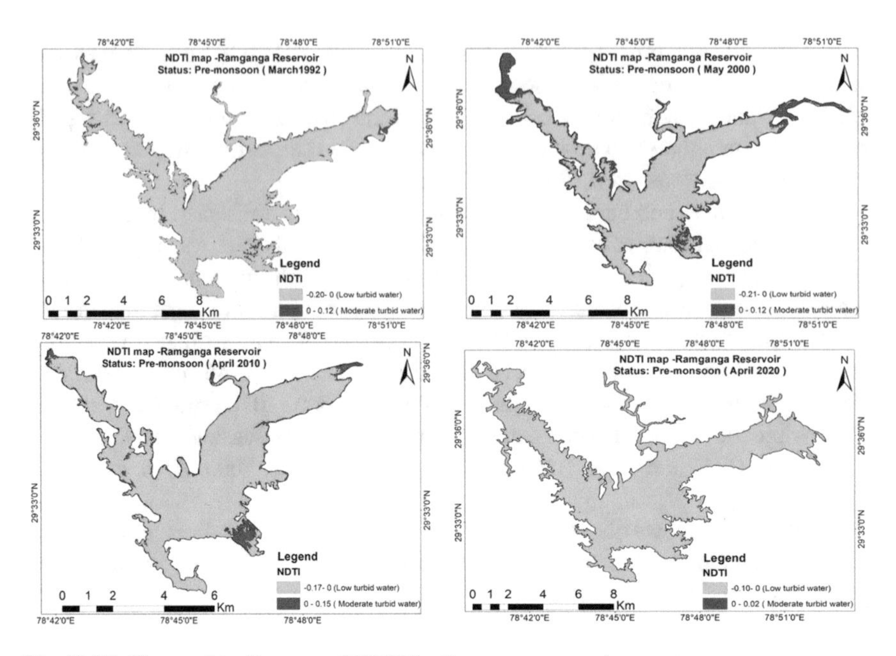

Fig. 21.15 Change detection map of NDTI for Ramganga reservoir

21.4 Conclusion

The present study used remote sensing and GIS techniques to analyze the spatiotemporal variations in wetland structural components such as water-spread area, vegetation, and water turbidity level in the Ramganga reservoir ecosystem, as well as to establish a fairly reliable wetland database and inventory over a decadal period from 1992 to 2020. In Ramganga reservoir, the area under sparse aquatic vegetation is higher than moderate aquatic vegetation in both the seasons and also found that water-spread area increased by 29.16% in pre-monsoon season and 9.12% in post-monsoon season throughout the study period 1992–2020. This increase in water-spread area of reservoir over the years can be ascribed to increased rainfall patterns and Himalayan snowmelt runoff contribution toward the downstream area, which is a major contributor in increasing river volume, which is directly connected with the reservoir. The massive presence of aquatic vegetation in the Ramganga reservoir ecosystem in the post-monsoon period was 3.41 km^2 in 1992 and increased to 4.14 km^2 in the year 2020; similarly, during the pre-monsoon period, the aquatic vegetation-covered area also somewhat increased from 0.91 to 1.57 km^2 in past three decades. As a result of the turbidity level of the Ramganga reservoir, most of the water spreads within a reservoir show low turbidity or clear water. Remote sensing techniques obtained from aquatic vegetation, turbidity level, and water spread have proven substantive indicators for evaluating wetland ecosystem health. These derived measures can help to monitor and prepare wetland inventory for the future purposes and planning. This study

Table 21.7 Area under different water turbidity levels from the year 1992 to 2020

S. No.	Class name	1992 (km^2)		2000 (km^2)		2010 (km^2)		2020 (km^2)	
		Pre-monsoon	Post-monsoon	Pre-monsoon	Post-monsoon	Pre-monsoon	Post-monsoon	Pre-monsoon	Post-monsoon
1	Low turbid	46.9296	65.6802	33.9723	77.1984	27.7326	80.0676	64.1322	75.006
2	Moderate turbid	4.1238	0.189	8.8272	0.1431	2.799	0.5085	0.0657	Nil

will assist policymakers in developing conservation and rehabilitation programs for these kinds of man-made wetlands. Nevertheless, a comprehensive perspective of the wetland is essential, which considers its correlations to other natural elements.

Acknowledgements We would like to express gratitude and sincere thanks to the Department of Water Resources Development and Management (WRD and M), IIT Roorkee, for providing a favorable environment and resources for this research and also highly thankful to Space Applications Centre, Indian Space Research Organisation, Ahmedabad, for valuable guidance toward this study.

References

Adnan M, Nabi G, Poomee MS, Ashraf A (2017) Snowmelt runoff prediction under changing climate in the Himalayan cryosphere: a case of Gilgit River Basin. Geosci Front 8(5):941–949
Aggarwal SP, Thakur PK, Nikam BR, Garg V (2014) Integrated approach for snowmelt run-off estimation using temperature index model, remote sensing and GIS. Curr Sci 106(3):397–407
Ahmed N (2016) Application of NDVI in vegetation monitoring using GIS and remote sensing in northern Ethiopian highlands. Abyssinia J Sci Technol 1(1):12–17
Alvarez-Cobelas M, Sanchez-Carrillo S, Cirujano S, Angeler DG (2008) Long-term changes in spatial patterns of emergent vegetation in a Mediterranean floodplain: natural versus anthropogenic constraints. Plant Ecol 194:257–271
Ashtekar AS, Mohammed-Aslam MA, Moosvi AR (2019) Utility of normalized difference water index and GIS for mapping surface water dynamics in sub-upper Krishna Basin. J Indian Soc Rem Sens. https://doi.org/10.1007/s12524-019-01013-6.
Bassi N, Kumar MD, Sharma A, Pardha-Saradhi P (2014) Status of wetlands in India: a review of extent, ecosystem benefits, threats and management strategies. J Hydrol Reg Stud 2:1–19
Behera MD, Chitale VS, Shaw A, Roy PS, Murthy MSR (2012) Wetland monitoring, serving as an index of land use change—a study in Samaspur Wetlands, Uttar Pradesh, India. J Indian Soc Remote Sens 40(2):287–297
Bhandari AK, Kumar A, Singh GK (2012) Feature extraction using normalized difference vegetation index (NDVI): a case study of Jabalpur city. Procedia Technol 6:612–621
Bid S, Siddique G (2019) Identification of seasonal variation of water turbidity using NDTI method in Panchet Hill Dam, India. Model Earth Syst Environ 5(4):1179–1200
Chopra G, Jakhar P (2016) Diversity and community composition of zooplankton in three wetlands of Fatehabad, Haryana. Curr World Environ 11(3):851
Doll P, Fiedler K, Zhang J (2009) Global-scale analysis of river flow alterations due to water withdrawals and reservoirs. Hydrol Earth Syst Sci 13(12):2413–2432
Du J, Song K, Yan B (2019) Impact of the Zhalong Wetland on neighboring land surface temperature based on remote sensing and GIS. Chin Geogra Sci 29(5):798–808
Gandhi GM, Parthiban S, Thummalu N, Christy A (2015) NDVI: vegetation change detection using remote sensing and gis-a case study of Vellore district. Procedia Comput Sci 57:1199–1210
Gu Y, Hunt E, Wardlow B, Basara JB, Brown JF, Verdin JP (2008) Evaluation of MODIS NDVI and NDWI for vegetation drought monitoring using Oklahoma Mesonet soil moisture data. Geophys Res Lett 35:L22401
Huising EJ (2002) Wetland monitoring in Uganda. Int Arch Photogramm Remote Sens Spat Inf Sci 36:127–135
Kingsford RT (2011) Conservation management of rivers and wetlands under climate change—a synthesis. Mar Freshw Res 62(3):217–222

Lacaux JP, Tourre YM, Vignolles C, Ndione JA, Lafaye M (2007) Classification of ponds from high-spatial resolution remote sensing: application to Rift Valley fever epidemics in Senegal. Remote Sens Environ 106(1):66–74

Maleki S, Soffianian AR, Koupaei SS, Saatchi S, Pourmanafi S, Sheikholeslam F (2016) Habitat mapping as a tool for water birds conservation planning in an arid zone wetland: the case study Hamun wetland. Ecol Eng 95:594–603

Mandal MH, Siddique G, Roy A (2018) Threats and opportunities of ecosystem services: a geographical study of Purbasthali oxbow Lake. J Geogr Environ Earth Sci Int 16(4):1–24

Manju G, Chowdary VM, Srivastava YK, Selvamani S, Jeyaram A, Adiga S (2005) Mapping and characterization of inland wetlands using remote sensing and GIS. J Indian Soc Remote Sens 33(1):51–61

McFeeters SK (1996) The use of the normalized difference water index (NDWI) in the delineation of open water features. Int J Remote Sens 17(7):1425–1432

Ray R, Mandal S, Dhara A (2012) Characterization and mapping of inland wetland: a case study on selected Bils on Nadia district. Int J Sci Res Publ 2(12):1–10

Shao Y, Lunetta RS, Wheeler B, Iiames JS, Campbell JB (2016) An evaluation of time-series smoothing algorithms for land-cover classifications using MODIS-NDVI multi-temporal data. Remote Sens Environ 174:258–265

Sivakumar R, Ghosh S (2016) Wetland spatial dynamics and mitigation study: an integrated remote sensing and GIS approach. Nat Hazards 80(2):975–995

Vandel E, Vaasma T, Koff T, Terasmaa J (2016) Impact of water-level changes to aquatic vegetation in small oligotrophic lakes from Estonia. Lakes Reserv Ponds 10(1):9–26

Velmurugan A, Swarnam TP, Ambast SK, Kumar N (2016) Managing waterlogging and soil salinity with a permanent raised bed and furrow system in coastal lowlands of humid tropics. Agric Water Manag 168:56–67

Wang Y, Sheng HF, He Y, Wu JY, Jiang YX, Tam NFY, Zhou HW (2012) Comparison of the levels of bacterial diversity in freshwater, intertidal wetland, and marine sediments by using millions of illumina tags. Appl Environ Microbiol 78(23):8264–8271

Xu H (2006) Modification of normalised difference water index (NDWI) to enhance open water features in remotely sensed imagery. Int J Remote Sens 27(14):3025–3033

Yang J, He Y (2017) Automated mapping of impervious surfaces in urban and suburban areas: linear spectral unmixing of high spatial resolution imagery. Int J Appl Earth Obs Geoinf 54:53–64

Yan D, Wang X, Zhu X, Huang C, Li W (2017) Analysis of the Use of NDWI green and NDWI red for Inland Water Mapping in the Yellow River Basin Using Landsat-8 OLI Imagery. Remote Sens Lett 8(10):996–1005

Chapter 22
Estimation of Water Quality Parameters Along the Indian Coast Using Satellite Observations

Chiranjivi Jayaram, Neethu Chacko, and V. M. Chowdary

Abstract Coastal regions across the world are the most densely populated areas that exert significant pressure on the water quality of the region. The rapid urbanization and industrialization add increasingly to the pollutants that alter the coastal water quality. Operational monitoring of water quality is exhaustive and cost-sensitive exercise that is imperative for sustainable management. With the advent of numerous high-resolution satellites and open data policies being followed by various space agencies, remote sensing of water quality has become robust and an important technique for researchers and managers. Chlorophyll-*a* concentration suspended sediment concentration and turbidity are some of the water quality parameters that could be derived from the satellite observations. The present study addresses the assessment of water quality using remote sensing for selected locations along the east coast (Hooghly estuary) and the west coast (Cochin backwaters). The impact of tropical cyclone 'Bulbul' on the water quality of the Hooghly estuary and the effect of COVID-19 induced lockdown during 2020 for the Cochin backwaters are considered as case studies to demonstrate the application of satellite remote sensing in the estimation of coastal water quality parameters.

22.1 Introduction

The physical, chemical, biological and thermal characteristics of water define the water quality of a region. In general, the water quality of a particular water body is influenced by the materials discharged into it (Ritchie et al. 2003). In light of the rapid urbanization and socio-economic development (~40% of the global population) especially along the coastal regions, effective monitoring of the water quality is imperative for productive management of the coastal ecosystems (Gholizadeh et al.

C. Jayaram (✉) · N. Chacko
Regional Remote Sensing Centre—East, NRSC/ISRO, New Town, Kolkata 700156, India

V. M. Chowdary
Regional Remote Sensing Centre—North, NRSC/ISRO, Sadiq Nagar, New Delhi, India

A. Pandey et al. (eds.), *Geospatial Technologies for Land and Water Resources Management*, Water Science and Technology Library 103,
https://doi.org/10.1007/978-3-030-90479-1_22

2016). The changes that occur due to the degradation of the water quality would have an adverse impact on the fisheries and other aquatic resources of freshwater, coastal and estuarine ecosystems across the world. Some of the major pollutants that have a significant effect on the coastal water quality are the toxic chemicals released from the industries, oil spills, domestic and industrial discharge into the rivers that eventually end up in the coastal oceans. The excessive nutrients due to the fertilizers and pesticides also influence the coastal water quality (Hafeez et al. 2019). Given the importance attached to the blue economy, it is imperative to have sustained and monitoring of the water quality for nurturing the aquatic resources.

The traditional method of water quality estimation involves water sample collection at different depths of the water column and subsequent laboratory analysis of these collected samples. A systematic field campaign requires trained manpower, approach to boats/coastal. The logistics involved in safely transporting the samples from the boat to the laboratory involve huge effort and cost. Though the in-situ water sampling and measurements are accurate, they are laborious, uneconomical and inconsistent in space and time (Cai et al. 2008). These limitations of the in-situ sampling could be mitigated using remote sensing as this technology provides the data on synoptic scale at varying spatio-temporal resolutions. The advancement in remote sensing technology together with the significant enhancement in the computational resources, to process the remote sensing data, has enabled water quality monitoring to a greater extent. Ocean colour remote sensing provides information pertaining to chlorophyll-*a* concentration, total suspended matter, coloured dissolved organic matter and other water leaving spectra at different wavelengths in the visible band of the electromagnetic spectrum. The spectral characteristics of the water body like absorption, scattering, reflection, etc., are governed by the chemical and hydrobiological properties (Gholizadeh et al. 2016). The reflected radiation from the water surface is detected by the sensors mounted on various satellite platforms that are used to derive the water quality parameters from the coastal and inland water bodies.

India has a long coastline with approximately greater than 7000 km spanning nine coastal states and two island groups, along both the east and west coasts. The east coast of India is dominated by the major east flowing rivers like the Ganges, Mahanadi, Godavari, Krishna and Cauvery that carry large quantities of polluting substances that are collected in the respective watersheds. Though the west-flowing rivers are less along the western coast of India compared to the east coast, minor rivulets that originate in the western ghats and the major rivers of Narmada and Tapi would have a profound impact on the coastal water quality. In the present study, the water quality of two coastal and estuarine regions viz., the Hooghly estuary and the Cochin backwaters along the east and west coast, respectively are analysed. The impact of cyclone 'Bubul' on the Hooghly estuarine waters and the influence of COVID-19 lockdown on the Cochin backwaters are considered case studies to demonstrate the utility of remote sensing observations in water quality measurements.

22.2 Study Region

22.2.1 The Hooghly Estuary

The study focussed on the northern coastal Bay of Bengal comprising of the Hooghly estuary and the adjoining coastal areas bounded by the longitudes 87° E–89° E and latitudes 21 N–22.5° N (Fig. 22.1). This area receives a huge amount of freshwater from the Hooghly River which enters into the Bay of Bengal through the estuary. Highest river discharge is exhibited in the summer monsoon season (3000 ± 1000 $m^3\ s^{-1}$) and minimum (1000 ± 80 $m^3\ s^{-1}$) during the pre-monsoon season (Mukhopadhyay et al. 2006; Ray et al. 2015). This region is also associated with the world's largest mangrove forest, the Sundarbans and forms a part of the Gangetic delta. This region is characterised by a complex network of interconnecting channels and creeks. The other adjoining mangrove regions of the Sundarbans also receive a discharge through small rivers and channels. Therefore, the discharges into the study region greatly influence its water quality. The entire estuarine system is relatively shallow (<6 m) and is subjected to strong meso- macro tides with amplitude ranging between 5 and 7 m. The lower part of the estuarine system often experiences rough

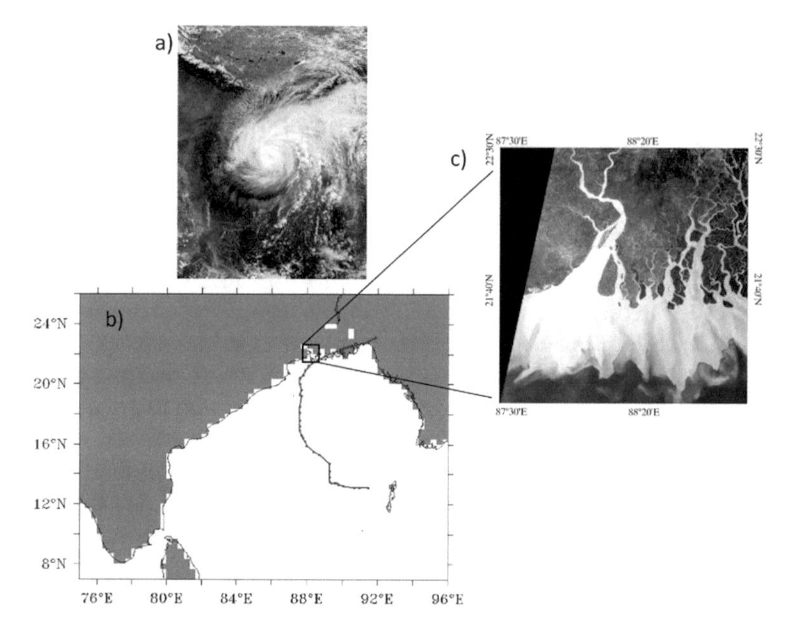

Fig. 22.1 Cyclone Bulbul and the study region. **a** True colour image of the cyclone 'Bulbul' just before it made landfall. **b** Track of the cyclone Bulbul overlaid on Bay of Bengal. The study region is marked in black rectangle. **c** The Landsat image of the study comprising Hooghly estuary and the adjoining coastal regions

weather conditions and cyclones that could strongly influence the estuarine water quality.

Bulbul was a strong and very damaging cyclone that achieved the very severe super cyclone status and struck the coast of India and Bangladesh in the year 2019. Bulbul originated as a low-pressure system over the north Andaman Sea on the 5th of November. It moved west-northwards and gained stronger winds in the next three days becoming very severe cyclone status on 8th November. On 9th November night, it made landfall on the coastal Bay of Bengal about 55 km east of Sagar Island in the Indian state of West Bengal close to the east of Sundarban Dhanchi forest with maximum sustained surface wind speed of 110–120 kmph gusting to 135 kmph. The cyclone caused heavy rains, storm surges and flash floods across the coastal region before its further weakening. The path of the cyclone Bulbul plotted using the best track data archived in the tropical cyclone database available online (http://www.rsmcnewdelhi.imd.gov.in/index.php) is shown in (Fig. 22.1).

22.2.2 Cochin Backwaters/Vembanad Lake System

The Cochin backwaters region (Lats. 9° 30′ N–10° 15′ N and Longs. 76° E–76° 30′ E) is a complex network of interconnected lagoons, swamps and canals that drain into a lake system called Vembanad Lake that extends up to 97 km in length and encompassing an area of over 300 km^2. This estuary, the third largest coastal wetland ecosystem in India, is a Ramsar site. This system has two openings to the Arabian Sea, one in the north known as the Azhikode inlet, which is narrow with a width of about 250 m and the other a more important one with a width of ~ 450 m, in the middle that acts as the entrance to the Cochin port (Vishnu et al. 2018). The Cochin estuarine region is subjected to heavy siltation, which necessitates continuous dredging to maintain channel depths suitable for navigation. Deepening of estuary by dredging increases estuary volume and alters the mixing processes and circulation patterns. Maximum load of suspended material occurs during the monsoon season, decreasing progressively during the post-monsoon periods. The nutrient load into the estuary is fed by the runoff of riverine and agricultural wastes as well as the discharge of municipal waste from the adjoining Cochin city (Shivaprasad et al. 2013a).

The Cochin backwater system is subjected to heavy pollution from the major cosmopolitan industrial city of Kochi that has a population of ~1.5 million and is a hub for more than 60% of the major chemical industrial units of the state of Kerala. Apart from the major port, the northern region of the estuary is also greatly affected by industrial activities in the Cochin shipyard, Kochi Refineries, fertilizer units and a host of medium and small-scale industrial clusters. The effluents that get drained into the estuary have a significant impact on the water quality. Similarly, in the southern region, the paddy fields and aquaculture alter the water quality of the backwaters. Indiscriminate use of fertilizers, pesticides and municipal waste generated in the Kochi city contributes about 104×10^3 m^3 of industrial waste and 260 m^3 of untreated domestic waste along with the agricultural wastes into the estuary per day (Thasneem et al. 2018). However, due to the rapid increase of COVID-19 cases in the state of Kerala since late February 2020, the industrial output and tourism activity around the lake had significantly reduced well ahead of the national lockdown enforced from 25th March 2020. Given the composition and levels of the effluents that are getting discharged into the backwater system, the lockdown substantially reduced the stress on the ecosystem (Yunus et al. 2020). This phenomenon is expected to lead to near-natural conditions in the backwaters (Saraswat and Saraswat 2020).

22.2.3 Data and Methods

Landsat-8 Operational Land Imager (OLI), Level-1C and Sentinel-2, Multi-Spectral Instrument (MSI), Level-1C, data sets were used in this study for extracting the water quality parameters. The level-1C product is orthorectified top-of-atmosphere reflectance data with sub-pixel multispectral registration. Landsat-8 has a temporal resolution of 16 days and a spatial resolution of 30 m in the visible region. To assess the impact of tropical cyclone induced rainfall on the Hooghly estuarine region, two Landsat-8 OLI level 1 image of 2019 covering the Hooghly estuary region during the last week of October to 1st fortnight of November for path = 139/row = 47 was downloaded from the USGS online archive Earth Explorer (https://earthexplorer.usgs.gov/). (i) The image on 29th October 2019 corresponds to the pre-cyclone conditions and (ii) the image on 14th November 2011 corresponds to the post-cyclone conditions. The impact of COVID-19 lockdown on the water quality of the Cochin backwaters was studied using Sentinel-2 imagery. Data for the period 2018–2020 are used in the present study. The entire study area is covered with two scenes of MSI images (T43PFL and T43PFM). Fifty-eight images were downloaded from the Copernicus Open Access Hub (https://scihub.copernicus.eu/) during the study period. A total of 38 images were used for the analysis after screening the data for the presence of cloud and sun glint. The details of the data are provided in Table 22.1.

The processing of the Landsat optical data is done on ENVI platform. The Landsat level 1 images downloaded are undergone system radiation correction and geometry

Table 22.1 Data availability for each month/year during the study period

Date of images	2018			2019				2020				
	Feb	Mar	Apr	Feb	Mar	Apr	May	Jan	Feb	Mar	Apr	May
	03rd	05th	04th	03rd	10th	04th	29th	5th	18th	19th	3rd	3rd
	13th	10th	09th	08th	15th	09th			28th	24th	28th	
	18th	15th	19th	18th	25th							
	28th	20th	24th									
			29th									

Dates that are underlined correspond to the screened-out images correction. The level images were processed in two steps: (1) calibration and (2) atmospheric correction. The Level 1 Landsat images were subjected to radiometric calibration for converting calibrated digital numbers into absolute units of at-sensor spectral radiance at each spectral band. The radiometrically calibrated images were then atmospherically corrected using the Fast Line-of-sight Atmospheric Analysis of Spectral Hypercubes (FLAASH) mode. The atmospheric correction eliminated the influence of atmospheric absorption and scattering. The FLAASH module facilitates automated atmospheric correction over water using NIR and SWIR channels. This is followed by elimination of cloud contaminated pixels using reflectance thresholding approach. After the cloud masking, land masking is done by using the shapefile of the Indian subcontinent. Atmospheric correction for Sentinel 2 was carried out using the ACOLITE software package (Version 20190326.0). ACOLITE was developed by the Royal Belgian Institute of Natural Science (RBINS) that employs dark spectrum fitting (DSF) method (Vanhellemont and Ruddick 2018) and (Vanhellemont 2019). The ACOLITE package computes the scattering due to aerosols using the Rayleigh-corrected reflectance for near infrared (NIR) bands in case of clear waters and shortwave infrared (SWIR) bands for turbid waters as the contribution of water towards reflectance is negligible in these bands (Martins et al. 2017). The glint correction includes the bands and pixels with the least estimate of aerosol optical thickness over the corresponding scene. This method allows the path reflectance that is insensitive to the sun glint (Vanhellemont 2019).

The reflectance bands are used to estimate TSM, turbidity and chlorophyll concentration employing the following algorithms. The algorithms used in this study are based on the single band turbidity and TSM algorithms which are functions of reflectance for the coastal waters. Total suspended matter (TSM) was derived following the retrieval algorithm developed by (Nechad et al. 2010)—water leaving reflectance of Red band ($\lambda = 665$ nm) was used to obtain the TSM from the expression (Eq. 22.1).

$$\text{TSM} = \frac{A\rho_w}{1 - \frac{\rho_w}{C}} \tag{22.1}$$

where, ρ_w is the water leaving reflectance (666 nm); A and C are the empirical coefficients obtained from (Nechad et al. 2010). The TSM thus retrieved has a root mean square error of < 10 mg/l.

Similarly, turbidity is derived from the following expression (Nechad et al. 2009):

$$\text{Turbidity} = \frac{A \cdot \rho_w}{1 - \frac{\rho_w}{C}} (\text{FNU}) \tag{22.2}$$

Chlorophyll-*a* was derived using the OC2 algorithm (Eq. 22.2) with the coefficients suggested by (Franz et al. 2015; Nazeer et al. 2020).

$$\text{Chlorphyll-a} = 10^{\left(0.1977 - 1.8117R + 1.9743*R^2 - 2.5635*R^3 - 0.7218*R^4\right)}. \tag{22.3}$$

where, R is the ratio of blue-green bands, which are 442 nm and 560 nm (S2A)/559 nm (S2B). The accuracy of OC2 algorithm for the Indian coastal waters is <0.73 mg m^{-3} (Poddar et al. 2019).

22.3 Variability of the Water Quality Parameters in the Hooghly Estuary

Estimates of the three important water quality parameters: TSM, turbidity and chlorophyll concentration are obtained from the atmospherically corrected surface reflectance. Figure 22.2 shows the TSM values before and after the cyclone. Fig. 22.2a forms the background conditions of the TSM based on which the impact of the cyclone Bulbul can be assessed. It can be seen that the pre-cyclone TSM concentration was lower in the entire coastal regions ranging from 20 to 30 g/m^3 in the mouth of the Hooghly estuary, increasing upstream with values reaching up to 35–40 g/m^3 in the upper estuary. The areas in the adjoining coastal regions and in the creeks of the Sundarbans, the TSM is much lower in the range 10–20 g/m^3. In the offshore regions, the TSM values are generally below 5 g/m^3. The concentration of TSM within the Hooghly estuary has a spatial distribution characteristic of high in the upstream and low in the downstream. Post-bulbul, it can be observed from Fig. 22.2b that the concentration of TSM increases in the study region. The elevation in the concentration of TSM is pronounced within the Hooghly estuary than elsewhere. The satellite observations reveal a large spatial extent of the higher TSM distribution, suggesting

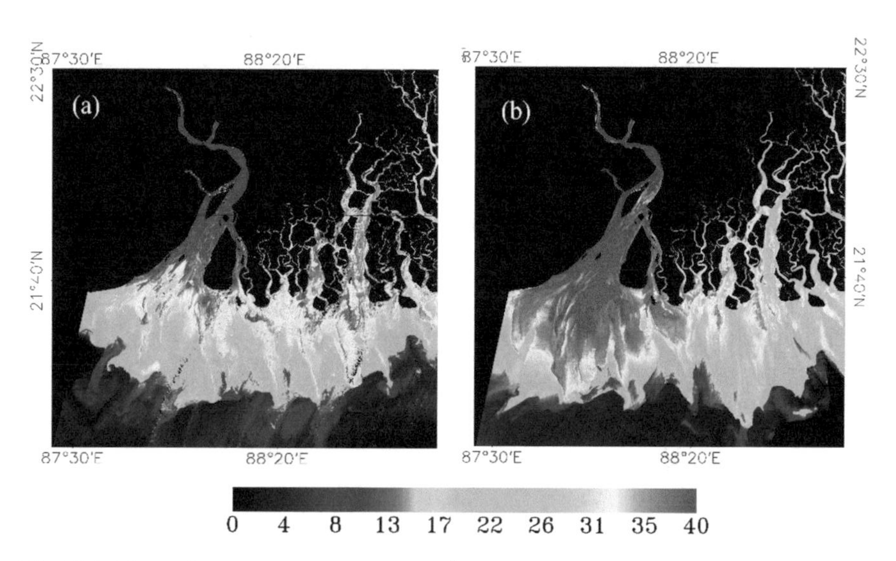

Fig. 22.2 Maps of Total suspended matter (g/m^3) estimated from Landsat-8: (a) before cyclone 'Bulbul' (29th October 2019) and (b) after cyclone 'Bulbul' (14th November 2019)

a strong impact of the cyclone on the TSM. The TSM concentration ranges between 40 and 45 g/m^3 in the entire Hooghly estuary towards the estuary mouth. The values of TSM concentration in the adjoining coastal regions also increased to 25–30 g/m^3. The highest change in TSM concentration occurred within the Hooghly estuary and its mouth where the tongues of higher TSM can be seen flushing out into the offshore waters.

The impact of the cyclone Bulbul on the water turbidity is shown in Fig. 22.3. Before the cyclone, the turbidity values in the coastal and offshore regions were low, while values increased in the Hooghly region. Within the Hooghly estuary, the turbidity was relatively higher particularly in the shallower northern upper estuary (Fig. 22.3a).

The plumes of turbidity high observed in the upper estuary range in concentration from 275 to 300 FNU which seem to be primarily related to the higher TSM concentration prevailed before the cyclone. The pre-cyclone turbidity in the regions outside the estuary and within the creeks of Sundarbans (70–80 FNU). Post-cyclone, the turbidity values are increased inside the estuary ranging from 300 to 400 FNU. The turbidity values in the offshore regions also exhibit a slight increase in turbidity after the passage of Bulbul. It can be observed that a significant increase in the turbidity happened inside the Hooghly estuary than elsewhere. The overall pattern in the variation of turbidity is similar to that of the TSM which points out the contribution of TSM in increasing turbidity in the dynamic Hooghly estuary.

Impact of cyclone Bulbul on the chlorophyll concentration in the northern coastal Bay of Bengal is shown in Fig. 22.4. Before Bulbul (Fig. 22.4a) made its landfall, the chlorophyll concentration was generally high within the Hooghly estuary and

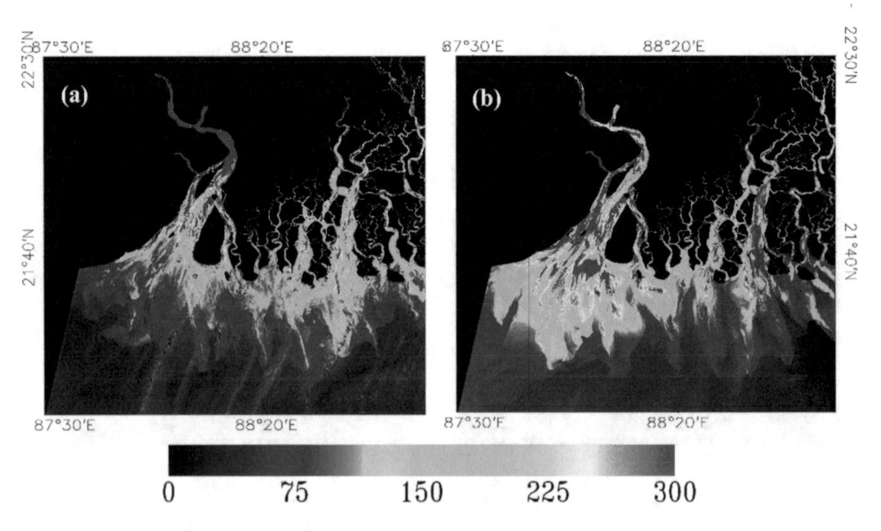

Fig. 22.3 Maps of turbidity (FNU) estimated from Landsat-8 **a** before Bulbul (29th October 2019) and **b** after Bulbul (14th November 2019)

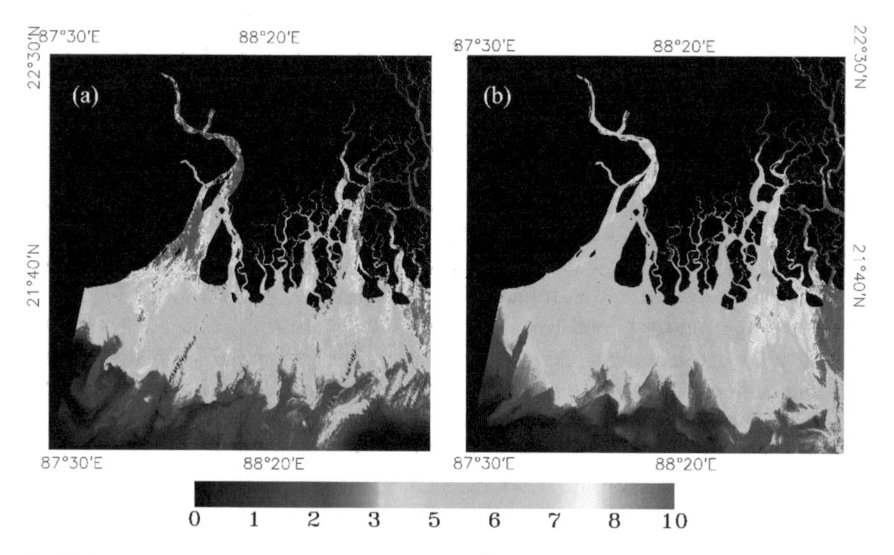

Fig. 22.4 Maps of chlorophyll concentration (mg/m^3) estimated from Landsat-8 **a** before Bulbul (29th October 2019) and **b** after Bulbul (14th November 2019)

in the creeks in the Sundarbans located in the eastern part of the study region (~7–10 mg/m^3). Filaments of chlorophyll-rich waters can be observed as emanating from the creeks and estuary towards the open ocean. The landfall of the Bulbul apparently caused an interestingly different pattern of chlorophyll concentration (Fig. 22.4b) than the other two parameters. In the Hooghly estuary and adjacent creeks and channels, the chlorophyll concentration is observed as decreasing after the cyclone Bulbul. The chlorophyll concentration decreased to ~6–7.5 mg/m^3 as compared to the pre-cyclone values >7–10 mg/m^3. Many studies have reported on the increase of chlorophyll concentration in the aftermath of cyclones (Subrahmanyam et al. 2002; Babin et al. 2004; Chacko 2017, 2019). But in this case, a clear large spatial scale reduction of chlorophyll concentration occurred except in the eastern part of the study region. A visual comparison of the variability of post-cyclone TSM, turbidity and chlorophyll concentration reveal that the decrease in the chlorophyll concentration happened over the regions where there is an increase in the concentration of TSM and turbidity.

The discharge and the resuspension induced by the physical forcing introduce a large amount of nutrients into the water column. Shelby et al. (2006) reported that the input of nutrients after major hurricanes was almost equal to the average annual input over several years. Williams et al. (2008) showed that total organic carbon, ammonium and other soluble reactive phosphorous concentrations increased two to five times after the passage of a series of hurricanes in Florida Bay. The increased input of nutrients into the region can have a profound impact in increasing the phytoplankton biomass by increasing the photosynthesis mechanism (Havens et al. 2012). However, the increased river discharge and mixing have a huge impact on the penetration of light into the water column. As a result of these, there will be an increase in the light attenuating substances in the water column which reduces the overall clarity of the water in the coastal systems. Light and nutrient availability are the major limiting factors for the chlorophyll concentration in the open ocean as well as coastal waters (Chacko 2017). Though nutrient input increases in the coastal waters which can support the algal growth, without sufficient light penetration, the production of biomass will be retarded.

Studies by Manna et al. (2010), Mitra et al. (2009) and Roshith et al. (2018) points that Hooghly estuary remains eutrophic throughout the year. This implies that the chlorophyll concentration remains high in the estuary as shown in Fig. 22.4a due to the high nutrient load from river discharges (Roshith et al. 2018). Havens et al. (2011) had examined that in eutrophic waters, an increase in the nutrient load due to flood water surge will not result in an increase in biomass as it would impact oligotrophic waters. In oligotrophic and open oceanic regions, however, the impacts of cyclones result in the entrainment of nutrients which are limited in the upper near surface waters, thereby increasing the chlorophyll concertation by new production (Gierach and Subrahmanyam 2008; Chacko 2017). Thus, in eutrophic waters like Hooghly estuary, light penetration plays a major role in determining the phytoplankton biomass. Similar Observations are made by Mallin et al. (2002) and Srichandan et al. (2015) indicating the importance of light limitation in inducing the changes of chlorophyll concentration in coastal environments where nutrients are abundant.

22.4 Impact of COVID-19 Lockdown on the Cochin Backwaters

*22.4.1 Average of TSM and Chlorophyll-*a *in the Study Region*

The three year mean of TSM was computed for the months of February—April as shown in Fig. 22.5. Based on the TSM concentration from the figure, it is evident that the backwater system could be categorised into the northern region and the southern region. The sediment concentration is higher (~40 g m^{-3}) in the northern part than in the southern part (~15 g m^{-3}) during all three months with higher values observed in April. The patches of high sediment concentration observed in the southern region arise from the Kuttanad paddy fields. There are two reasons for the relatively high concentration in the northern region: (1) One of the openings into the ocean is the bar mouth area or the Cochin inlet, where there is a constant interaction between the ocean waters and the backwaters. The tidal forces churn the subsurface thereby increasing the suspended sediments. (2) The presence of active port and Naval facilities in Cochin city requires regular dredging activities to maintain the channel depth for safe navigation. Dredging resuspends the sediments deposited on the seabed.

The three year mean of chlorophyll-*a* (Fig. 22.6) exhibited a gradual decrease

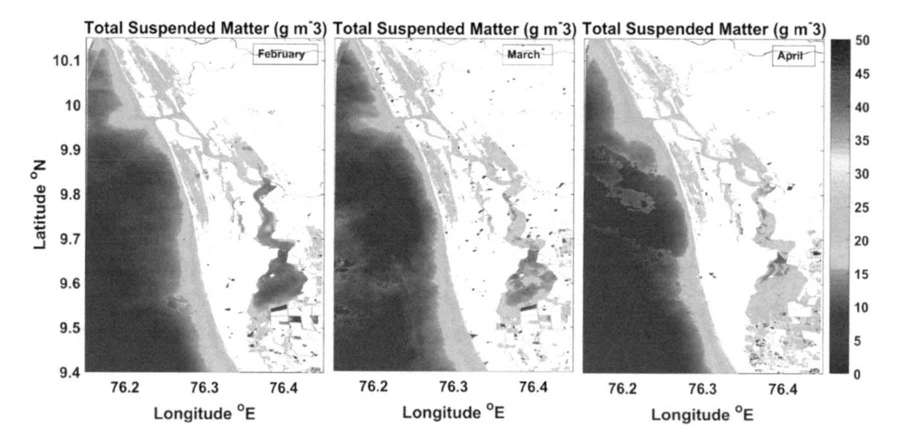

Fig. 22.5 Three year mean of TSM for the months of February, March and April

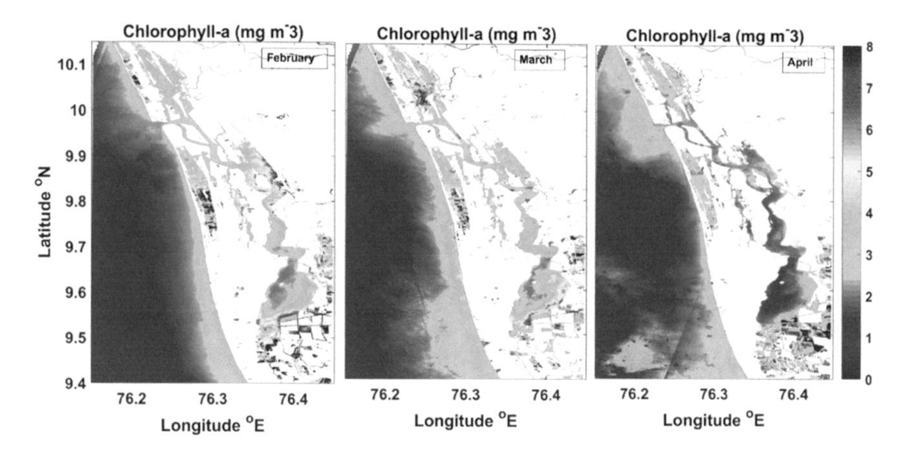

Fig. 22.6 Three year mean of Chlorophyll-*a* for the months of February, March and April

from February (high in the middle of the estuary with values <2 mg m^{-3}) to April (>5 mg m^{-3}) across the region except for the few patches in the southern region that is dominated by the paddy fields of Kuttanad region. The waters are generally oligotrophic due to little precipitation and river runoff during the pre-monsoon season that is evident in April when the entire coastal region and backwaters are observed to have relatively lower chlorophyll concentrations. The average surface chlorophyll-*a* of the backwater system during the pre-monsoon season is less than 3 mg m^{-3} going on *in-situ* observations of (Shivaprasad et al. 2013b) that are corroborated the satellite data.

22.4.2 TSM During the Lockdown Period

Monthly mean TSM and chlorophyll-*a* were mapped for the months of February—April 2020 with the February data coming from the pre-lockdown period, the March observations representing the initial/preparation phases of lockdown and April the data representing the peak of the lockdown (Fig. 22.7). The monthly values clearly depict the impact of reduced pollutant levels on the surface water quality of the entire Cochin backwater system, especially near the Cochin inlet and southern part of the study area.

To avoid the ambiguity due to the interannual variability, the monthly anomalies were computed for these three months as shown in Fig. 22.8. It is observed that the TSM concentration decreased by ~10 g m^{-3} during March and April near the

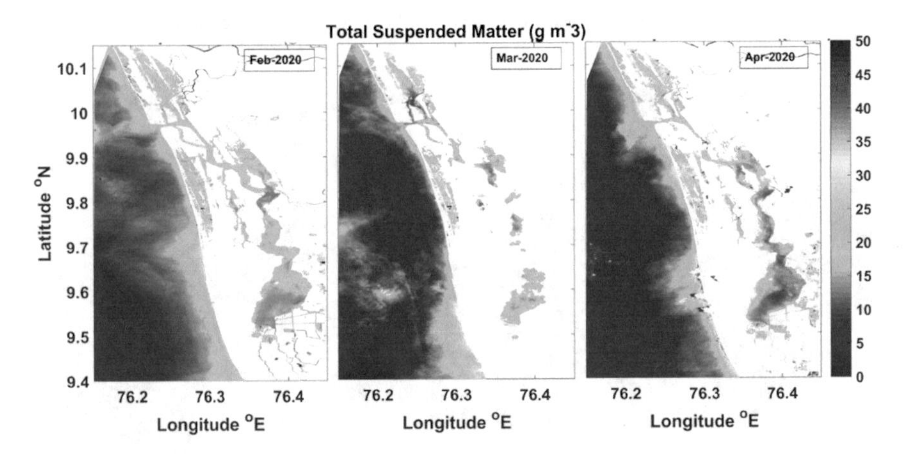

Fig. 22.7 Monthly mean of TSM for February, March and April 2020

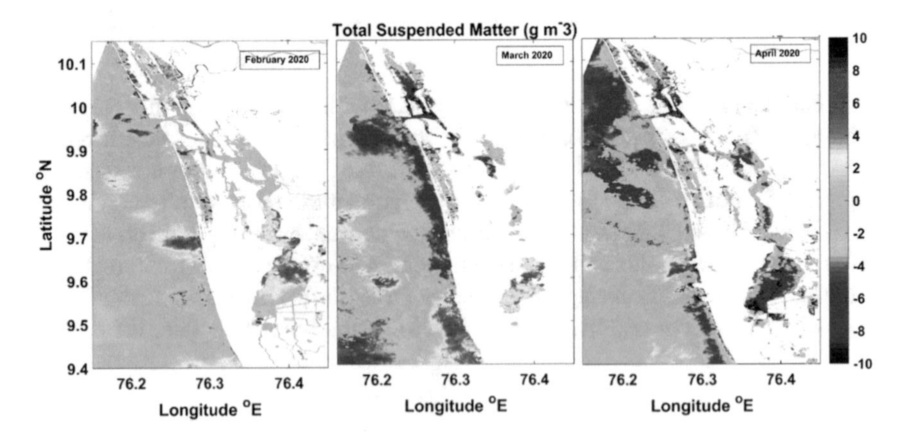

Fig. 22.8 Monthly anomaly of TSM for February, March and April 2020

Cochin inlet that could be attributed to the reduced ship traffic in March and complete lockdown in April. However, the southern region showed a positive anomaly in March, indicating the continuation of farming activities in the region prior to the lockdown period. The TSM in the entire study region showed negative anomalies in April when the lockdown impact was at its maximum. Positive or non-negative anomalies are observed at a few places very close to the land–water boundary that could be attributed to sewage discharge in the region. This infers the contribution of industrial discharge on the total suspended matter concentration of the estuarine region during pre-monsoon season when the runoff is at its minimum.

22.4.3 *Chlorophyll-*a *Concentration During the Lockdown Period*

Mean chlorophyll-*a* concentrations for February–April 2020 is shown in Fig. 22.9. Chlorophyll-*a* in the estuary varied between 1 and 8 mg m^{-3}. Chlorophyll-*a* was relatively high to the north of the Cochin inlet and at the southern fringes of the study region both in February and March with the highest concentration is observed in March to the north of Cochin inlet. The chlorophyll-*a* concentration in April is comparatively lower all along the estuary with values ranging from 1 to 3 mg m^{-3}. To ascertain the increase in the chlorophyll-*a* concentration, the monthly chlorophyll anomaly is mapped in Fig. 22.10.

The chlorophyll anomaly in February 2020 is negative towards the southern region of the estuary and positive in the northern part. During the month of March, anomalously high chlorophyll-*a* concentration is observed all over the backwaters with positive anomalies up to 2 mg m^{-3}. During the lockdown month of April, positive anomalies are observed in the Cochin inlet and the southern fringes of the estuary.

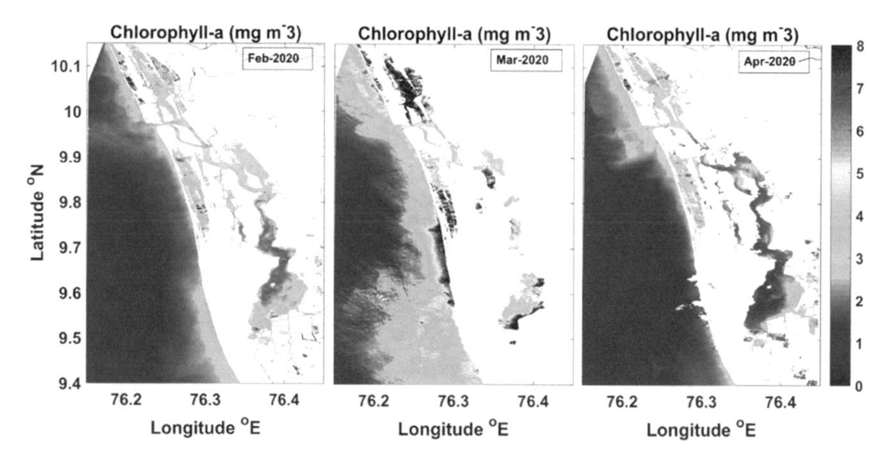

Fig. 22.9 Monthly mean of chlorophyll-*a* for February, March and April 2020

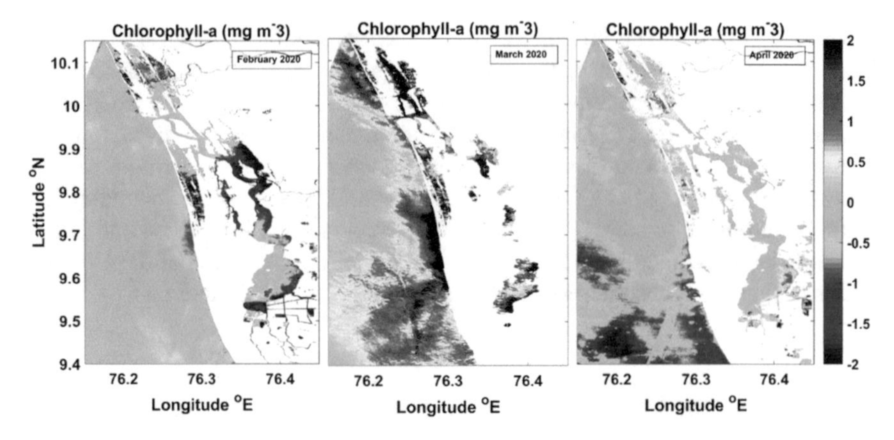

Fig. 22.10 Monthly anomaly of chlorophyll-*a* for February, March and April 2020

The positive anomalies in March and April could be attributed to the reduction in the TSM concentration during these two months thus allowing increased penetration of light that enhanced the productivity of the region. This is further supported by the lack of positive anomalies of chlorophyll during March, April of 2018 and 2019, respectively.

22.4.4 *Influence of Tides on TSM and Chlorophyll-*a *in the Cochin Backwaters*

The TSM concentration increases during the high tide as the bottom sediments get churned thereby resulting in their resuspension as evident from high concentration of TSM in the satellite imagery. In order to check the influence of tide on the suspended sediment distribution of the region, the hourly tide data near the bar mouth region was obtained from Cochin Port Trust. The time of satellite pass over the study region is observed to be between 0506 and 0508 h UTC that is approximately 1036-1038 IST. As per the available tide data, the low tide is observed between 1000 and 1100 h IST, for the dates (March 19th, April 3rd and 28th, May 3rd of 2020) on which the satellite data are available. Therefore, the influence of tide on the TSM concentration would be small and the observed TSM concentration and chlorophyll-*a* must result from lockdown.

22.5 Conclusions

The present study demonstrates the application of remote sensing technologies to assess the water quality along the Indian coast through two case studies viz., the effects of tropical cyclones on the water quality parameters of the northern coastal Bay of Bengal and the impact of COVID-19 lockdown on the Cochin backwaters. The parameters that were considered as proxies of water quality are the TSM, turbidity and chlorophyll.

High-resolution images from Landsat-8 were used to characterise the impact of cyclone 'Bulbul' on the water quality of the Hooghly estuary and the adjoining coastal areas comprising the Sundarbans. The observations of TSM, turbidity and chlorophyll concentration before and after Bulbul showed that the water quality of the coastal system deteriorated due to the cyclone. The strong rainfall associated with the cyclone induced high surface runoff into the study area which resulted in bringing large number of suspended solids and reduced the water clarity. The increase in TSM and the turbidity greatly deteriorated the light penetration of the water column. Decreased light penetration due to increased turbidity is implicated as the plausible reason which resulted in the observed decrease in chlorophyll concentration. This study shows that in the eutrophic estuarine systems like Hooghly, the algal growth is strongly dependent on light availability and any factor which can reduce the light availability can reduce the phytoplankton biomass in the region.

The impact of COVID-19 lockdown on the estuarine water quality is elucidated in Cochin backwaters as a case study. The TSM and chlorophyll-*a* concentration are considered in this study to assess the impact of the lockdown period of April 2020 on water quality. Satellite imagery from Sentinel-2/MSI is used to derive TSM and chlorophyll during February–April for the years 2018–2020. Monthly means and anomalies of TSM and chlorophyll are generated to determine the anomalous values during March and April 2020. The TSM is observed to fall by 10 g m^{-3} especially in the Cochin inlet region, where relatively high turbid waters prevail owing to the port activities and discharge of industrial and municipal wastes into the region. Th COVID-19 lockdown has led to remarkable improvement in water quality including an increase in chlorophyll due to deeper light penetration. This study emphasises the influence of anthropogenic activities on the coastal water quality and throws light on the need for pollution mitigation for sustainable development with cleaner coasts.

Acknowledgements Authors thank the USGS and ESA for making available the Landsat—8/OLI and Sentinel-2 A/B datasets. We are also thankful to the RBIN team for the development and distribution of the ACOLITE package. Authors thank the Head (Applications) and the General Manager, RRSC-East for the support.

References

Babin SM, Carton JA, Dickey JA, Wiggert JD (2004) Satellite evidence of hurricane-induced phytoplankton blooms in an oceanic desert. J Geophys Res 109:C03043. https://doi.org/10.1029/2003JC001988

Cai L, Liu P, Zhi C (2008) Discussion on remote sensing based on water quality monitoring methods. Geomat Spat Inf Technol 31:68–73. https://doi.org/10.3969/j.issn.1672-5867.2008.04.021

Chacko N (2017) Chlorophyll bloom in response to tropical cyclone Hudhud in the Bay of Bengal: Bio-Argo subsurface observations. Deep Sea Res 1(124), 66–72. https://doi.org/10.1016/j.dsr.2017.04.010

Chacko N (2019) Differential chlorophyll blooms induced by tropical cyclones and ther relation to cyclone characteristics and ocean pre-conditions in the Indian Ocean. J Earth Syst Sci 128:177. https://doi.org/10.1007/s12040-019-1207-5

Franz BA, Bailey SW, Kuring N, Werdell JP (2015) Ocean color measurements with the Operational Land Imager on Landsat-8: implementation and evaluation in SeaDAS. J Appl Remote Sens 9, 096070. https://doi.org/10.1117/1.JRS.9.096070

Gholizadeh MH, Melesse AM, Reddi L (2016) A comprehensive review on water quality parameters estimation using remote sensing techniques. Sensors 16:1298. https://doi.org/10.3390/s16081298

Gierach MM, Subrahmanyam B (2008) Biophysical responses of the upper ocean to major Gulf of Mexico hurricanes in 2005. J Geophys Res 113:C04029. https://doi.org/10.1029/2007JC004419

Hafeez S, Wong MS, Abbas S, Kwok CYT, Nichol J, Lee KH, Tang D, Pun L (2019) Detection and monitoring of marine pollution using remote sensing technologies. In: Fouzia HB (ed) Monitoring of marine pollution. IntechOpen. https://doi.org/10.5772/intechopen.76739

Havens KE, Beaver JR, Casamatta DA, East TL, James RT, Mccormick P, Phlips EJ, Rodusky AJ (2011) Hurricane effects on the planktonic food web of a large subtropical lake. J Plankton Res 33:1081–1094

Mallin MA, Posey MH, McIver MR, Parsons DC, Ensign SH, Alphin TD (2002) Impacts and recovery from multiple hurricanes in a Piedmont-coastal plain river system. Biosciences 52:999–1010

Manna S, Chaudhuri K, Bhattacharya S, Bhattacharya M (2010) Dynamics of Sundarban estuarine ecosystem: eutrophication induced threat to mangroves. Aquat Biosyst 6:8. https://doi.org/10.1186/1746-1448-6-8

Martins VS, Barbosa CCF, DeCaralho LAS, Jorge DSF, Lobo FDL, Novo EML (2017) Assessment of atmospheric correction methods for sentinel-2 MSI images applied to Amazon flood plain lakes. Remote Sens 9:322. https://doi.org/10.3390/rs9040322

Mitra A, Gangopadhayay V, Dube A, Schmidt ACK, Banerjee K (2009) Observed changes in water mass properties in the Indian Sundarbans (north-western Bay of Bengal) during 1980–2007. Curr Sci 97:1445–1452

Mukhopadhyay SK, Biswas H, De T, Jana TK (2006) Fluxes of nutrients from the tropical River Hooghly at the land-ocean boundary of Sundarbans, NE Coast of Bay of Bengal. J Mar Syst 62:9–21. https://doi.org/10.1016/j.jmarsys.2006.03.004

Nazeer M, Bilal M, Nichol JE, Wu W, Alsahli MMM, Shahzad MI et al (2020) First experiences with the Landsat-8 aquatic reflectance product: evaluation of the regional and ocean color algorithms in a coastal environment. Remote Sens 12:1938

Nechad B, Ruddick KG, Neukermans G (2009) Calibration and validation of a generic multi sensor algorithm for mapping of turbidity in coastal waters. In: Ch R Bostater SP Jr, Mertikas XN, Velez-Reyes M (eds) SPIE, remote sensing of the ocean, sea ice, and large water regions, vol 7473. Berlin, Germany, p 74730H

Nechad B, Ruddick K, Park Y (2010) Calibration and validation of a generic multi-sensor algorithm for mapping of total suspended matter in turbid waters. Remote Sens Environ 114:854–866. https://doi.org/10.1016/j.rse.2009.11.022

Poddar S, Chacko N, Swain D (2019) Estimation of chlorophyll-a in the northern coastal Bay of Bengal using Landsat-8 OLI and Sentinel-2 MSI sensors. Front Mar Sci 6:598. https://doi.org/10.3389/fmars.2019.00598

Ray R, Rixen T, Baum A, Malik A, Gleixner G, Jana TK (2015) Distribution, sources and biogeochemistry of organic matter in a mangrove dominated estuarine system (Indian Sundarabans) during the pre-monsoon. Estuarine Coastal Shelf Sci 167:404–413. https://doi.org/10.1016/j.ecss.2015.10.017

Ritchie JC, Zimba PV, Everitt JH (2003) Remote sensing techniques to assess water quality. Photogramm Eng Remote Sens 69:695–704

Roshith CM, Meena DK, Manna RK, Sahoo AK, Swain HS, Raman RK, Sengupta A, Das BK (2018) Phytoplankton community structure of the Gangetic (Hooghly-Matla) estuary: status and ecological implications in relation to eco-climatic variability. Flora 240:133–143. https://doi.org/10.1016/j.flora.2018.01.001
Saraswat R, Saraswat DA (2020) Research opportunities in pandemic lockdown. Science 368:594–595. https://doi.org/10.1126/science.abc3372
Shelby JD, Chescheir GM, Skaggs RW, Amatya DM (2006) Hydrologic and water quality response of forested and agricultural lands during the 1999 extreme weather conditions in Eastern North Carolina. Am Soc Agricult Eng 48:2179–2188
Shivaprasad A, Vinita J, Revichandran C, Reny PD, Deepak MP, Muraleedharan KR, Naveen Kumar KR (2013a) Seasonal stratification and property distributions in a tropical estuary (Cochin estuary, west coast, India). Hydrol Earth Syst Sci 17:187–199. https://doi.org/10.5194/hess-17-187-2013
Shivaprasad A, Vinita J, Revichandran C, Manoj NT, Srinivas K, Reny PD, Ashwini R, Muraleedharan KR (2013b) Influence of saltwater barrage on tides, salinity and chlorophyll-a in Cochin estuary, India. J Coastal Res 29:1382–1390. https://doi.org/10.2112/JCOASTRES-D-12-00067.1
Srichandan S, Kim JY, Kumar A, Mishra DR, Bhadury P, Muduli PR, Pattnaik AK, Rastogi G (2015) Interannual and cyclone-driven variability in phytoplankton communities of a tropical coastal lagoon. Mar Pollut Bull 101:39–52
Subrahmanyam B, Rao KH, Rao NS, Murty VSN, Sharp RJ (2002) Influence of a tropical cyclone on Chlorophyll-a concentration in the Arabian Sea. Geophys Res Lett 29:2065. https://doi.org/10.1029/2002GL015892
Thasneem TA, Nandan SB, Geetha PN (2018) Water quality status of Cochin estuary, India. Indian J Geo-Mar Sci 47:978–989
Vanhellemont Q, Ruddick K (2018) Atmospheric correction of metre-scale optical satellite data for inland and coastal water applications. Remote Sens Environ 216:586–597. https://doi.org/10.1016/j.rse.2018.07.015
Vanhellemont Q (2019) Adaptation of the dark spectrum fitting atmospheric correction for aquatic applications of the Landsat and Sentinel-2 archives. Remote Sens Environ 225:175–192. https://doi.org/10.1016/j.rse.2019.03.010
Vishnu PS, Shaju SS, Tiwari SP, Menon N, Nashad M, Joseph CA, Raman M, Hatha M, Prabhakaran MP, Mohandas A (2018) Seasonal variability in bio-optical properties along the coastal waters off Cochin. Int J Appl Earth Obs Geoinf 66:184–195. https://doi.org/10.1016/j.jag.2017.12.002
Williams CJ, Boyer JN, Jochem FJ (2008) Indirect hurricane effects on resource availability and microbial communities in a subtropical wetland–estuary transition zone. Estuaries Coasts 31:204–214
Yunus AP, Masago Y, Hijoka Y (2020) COVID-19 and surface water quality: improved lake water quality due to the lockdown. Sci Total Environ 731:13902. https://doi.org/10.1016/j.scitotenv.2020.139012

Chapter 23
Morphometric Characterization and Flash Flood Zonation of a Mountainous Catchment Using Weighted Sum Approach

Gagandeep Singh and Ashish Pandey

Abstract Uttarakhand is a hill state of North India with a unique and diverse topographic, morphologic, and climatic setting. The higher elevation areas have been experiencing repeated flash flood events caused by cloud bursts or heavy precipitation. This study demonstrates the potential of remotely sensed data and geographical information system for assessing the flash flood risk in the Alaknanda River Basin (ARB) located in the Northwestern Himalayan Region in India. Multiple sub-watersheds in this river basin experience flash flood events almost every year, causing massive property damages and life loss. SRTM 30 m DEM was processed and utilized for conducting morphometric analysis to prepare a flash flood risk prioritization map for the 7 sub-watersheds of ARB. The morphometric parameters considered for this purpose were sub-watershed area, perimeter, stream order, stream length, basin length, bifurcation ratio, drainage density, stream frequency, texture ratio, form factor, circularity ratio, and elongation ratio. The correlation matrix for all these morphometric parameters for each sub-watershed was analyzed to obtain a compound parameter value using weighted sum approach (WSA), which was used for performing flash flood zonation (low, medium, high, and very high). The analysis results indicate that out of the 7 sub-watersheds, SW-2 and SW-3 were categorized under the high-risk zone, and SW-5 was categorized under the very high-risk zone.

23.1 Introduction

Flash flood is one of the most devastating naturally occurring events capable of causing widespread destruction in a very short time. These events are usually a consequence of extreme rainfall events (EREs), which occur for a short period over

G. Singh (✉) · A. Pandey
Department of Water Resources Development and Management, Indian Institute of Technology Roorkee, Roorkee, Uttarakhand 247 667, India
e-mail: gsingh@wr.iitr.ac.in

A. Pandey
e-mail: ashish.pandey@wr.iitr.ac.in

A. Pandey et al. (eds.), *Geospatial Technologies for Land and Water Resources Management*, Water Science and Technology Library 103,
https://doi.org/10.1007/978-3-030-90479-1_23

a relatively smaller area resulting in excessive discharge, landslides, and mudflows, possessing a high damage-causing potential. Predicting and controlling such events have been very difficult due to the highly dynamic nature of the climate and their sudden occurrence. According to the United Nations World Water Development Report 2020 (Connor 2020), in this decade, events of floods around the world and EREs have increased by more than 50% and are now transpiring at four times higher rates than in 1980. Also, various researchers have highlighted a confirmed observation of increased frequency and intensity of EREs worldwide due to noteworthy climate change (Burt 2005; Joshi and Rajeevan 2006; Devrani et al. 2015; Kumar et al. 2018). According to the United Nations Office for Disaster Risk Reduction report published in 2015, between a span of 10 years from 1995 to 2004, an increment in the annual average of the number of flood events was noted from 127 to 171. Figure 23.1 shows the increasing global trend in flood-related natural disasters.

India is struck by frequent climatic hazards in recent years (Swain et al. 2021a; b, c; Kalura et al. 2021; Guptha et al. 2021; Singh and Pandey 2021a, b), among which floods have resulted in maximum detrimental impacts (Singh and Kumar 2017; Singh and Pandey 2021; Kumar 2021). In recent years, the Himalayan region has been enduring increased frequency of natural hazards, making it one of the world's highly vulnerable regions (Maikhuri et al. 2017; Himanshu et al. 2013; Dhami et al. 2018). The region experiences cloud burst induced heavy rains and flash flood events, especially in the monsoon season (Prasad and Pani 2017). Adding to the misery is the highly undulated topography of the area featuring steep slopes and narrow valleys, which triggers erosion and debris removal when a large quantity of water flows through the rivers and streams (Thakur et al. 2012; Ghosh et al. 2019). Such region-specific characteristics make a natural hazard extremely devastating. Every year, substantial damages are reported in terms of property, infrastructure, and lives of humans and cattle (Bhatt et al. 2014; Ghosh et al. 2019).

Flash floods in mountains can be attributed to two significant causes, viz. the atmospheric or meteorological condition and the drainage basin (Bisht et al. 2018).

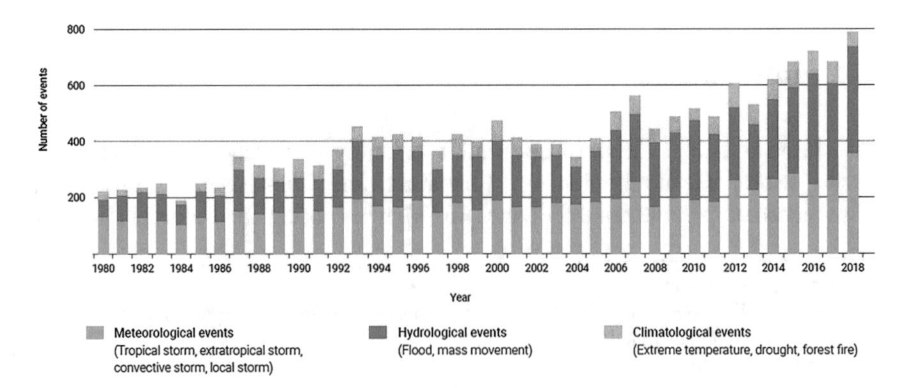

Fig. 23.1 Global trends in flood-related natural disasters. *Note* The figure has been imported from page 36 of the United Nations World Water Development Report 2020

The morphometric characteristics and behavior of a watershed reflect its hydrological response (Aher et al. 2014). Quantitative morphometric assessment of a watershed is a very effective method to interpret various aspects of its drainage network and compare multiple sub-watersheds developed in different geological and climatic settings. Geospatial technologies provide a highly effective and impactful means to conduct morphometric studies and consequent sub-watershed prioritization (Pandey et al. 2004; Aher et al. 2014; Bhatt and Ahmed 2014; Gajbhiye et al. 2014; Himanshu et al. 2013, 2015; Prasad and Pani 2017; Rai et al. 2020; Sharma et al. 2015). Several studies have been carried out on flash flood zonation based on morphometric characterization using parameters, viz. basin relief, circularity ratio, compactness number, drainage density, elongation ratio, form factor, overland flow, relief ratio, and stream frequency (Youssef and Hegab 2005; Angillieri 2008; Sreedevi et al. 2009; Thomas et al. 2012; Magesh et al. 2013). Researchers have also demonstrated the use of morphometric studies in proposing suitable flood preventive measures (Papatheodorou et al. 2014; Suresh et al. 2004; Yadav et al. 2014).

Uttarakhand is a hill state situated in India's Northwestern Himalayan Region which has a unique and diverse topographic, morphologic, and climatic setting. The higher elevation areas of the state have experienced repeated flash flood events caused by cloud bursts or heavy precipitation (Dimri et al. 2016). In view of the above, this study is based on the morphometric characterization and subsequent sub-watershed prioritization and flash flood zonation of the Alaknanda River Basin using the weighted sum approach (WSA). WSA method effectively makes use of ranking mechanism along with statistical correlation analysis for the sub-watershed prioritization.

23.2 Material and Methods

23.2.1 Study Area

The Alaknanda River Basin (ARB) is located between 78°33′ E and 80°15′ E longitude and 29°59′24″ N and 31°04′ 51″ N latitude in the mountain state of Uttarakhand in India. The river basin is spread across an area of 11,035.3 km^2 (Fig. 23.2). The study area features a significant altitudinal variation from 557 m to 7184 m (Fig. 23.2), justifying the existence of a diverse range of climates like subtropical, temperate, subalpine, and alpine. The magnitude and amount of rainfall in the catchment feature substantial variation depending upon the location of the place, viz. windward or leeward side of the high ridges (Joshi and Kumar 2006). Another essential characteristic of the study area is that 70 to 80% of the total rainfall experienced annually occurs during the monsoon season (Pandey and Prasad 2018).

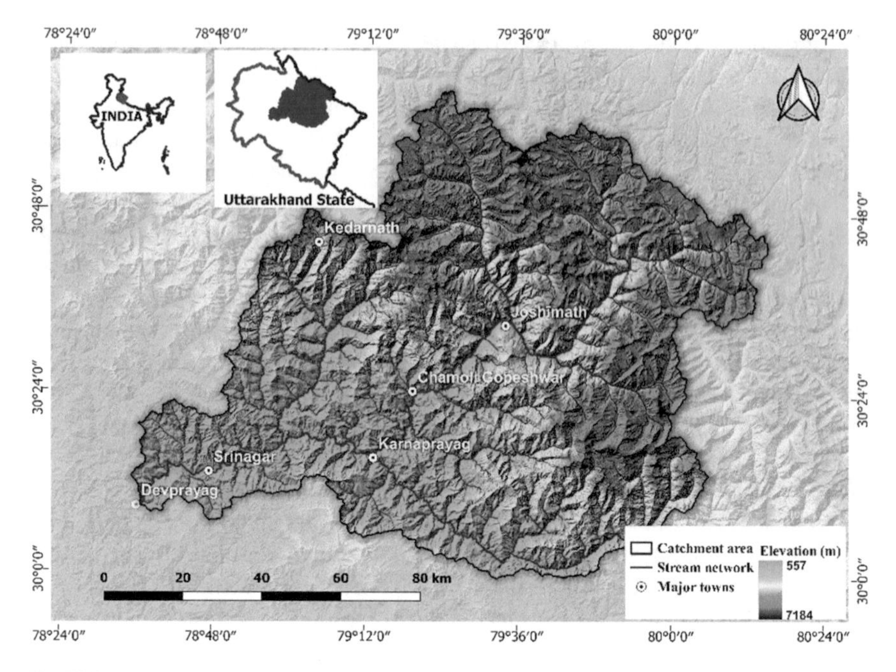

Fig. 23.2 Location map of the study area

23.2.2 Data Used

In this study, SRTM DEM data was used for the catchment delineation, derivation of drainage/stream network, and further division of the catchment area into seven sub-watersheds. The freely available SRTM DEM tiles with a spatial resolution of 30 m were downloaded in a Tagged Information File Format (TIFF) from https://dwtkns.com/srtm30m/, which requires NASA Earthdata login credentials. A total of four (Table 23.1) DEM tiles were downloaded to cover the entire Alaknanda River Basin.

Table 23.1 Details of satellite data downloaded

S. no	North grid ID	East grid ID
1	30	78
2	30	79
3	30	80
4	31	79

23.2.3 Tools and Techniques Used

In this study, ArcGIS 10.4 software package was used to process the downloaded DEM tiles for the catchment delineation and consequent generation of sub-watersheds. QGIS 3.10 software package was utilized to prepare the final maps. Devprayag (30°08′ 45″ N, 78°38′ 16″ E), the outlet of the catchment area, is the confluence of rivers Alaknanda and Bhagirathi situated in the Tehri district of Uttarakhand. Furthermore, the Arc Hydro tools developed by ESRI in ArcGIS 10.4 were used to derive the drainage network and associated entities. A threshold value of 10,000 cells was used to derive the streams in the catchment area. Using this value, it was possible to obtain a stream network that closely followed the actual streams. Besides, the Arc Hydro tools were also employed to compute various attributes for the morphometric characterization of the catchment area. Finally, the sub-watershed prioritization was conducted using a statistical method known as the weighted sum approach (WSA).

23.2.4 DEM Processing for Catchment Delineation, Sub-Watershed Extraction, and Stream Network Analysis

The four downloaded DEM tiles were mosaiced together to create a single raster, which was then reprojected to the UTM zone 43 N projection system. In this study, the catchment delineation was done using the Arc Hydro tools. Figure 23.3 shows the delineated sub-watersheds and stream ordering of the network.

The first step in the DEM processing was to identify sinks or depressions in the raw DEM that were filled up to generate a depressional DEM. Using the filled DEM, a flow direction raster was developed, one of the most critical aspects of the hydrological perspective. This D8 flow model was employed to obtain the output raster, which evaluates the possibility of flow from each pixel in eight different directions (Jenson and Domingue 1988). Using the flow direction raster, the flow accumulation raster was derived, representing the accumulated flow into every pixel. Each pixel represents the number of higher elevation pixels that contribute to the flow into it. To define the streams in the study area, a threshold value of 10,000 was used. As a result, a binary raster was generated with a pixel value of '1' for all the cells with an inflow accumulation value of 10,000 or more. All the remaining cells were assigned a value of '0'. The next step involved segmentation of the stream network, wherein a grid of stream segments was generated, and a unique grid code was assigned to each segment. Finally, the catchment was delineated with an outlet at Devprayag. Furthermore, the entire catchment was subdivided into seven sub-watersheds.

The entire drainage network was designated as a 5th-order catchment using Strahler's (1964) stream ordering scheme. Details of the number of various order

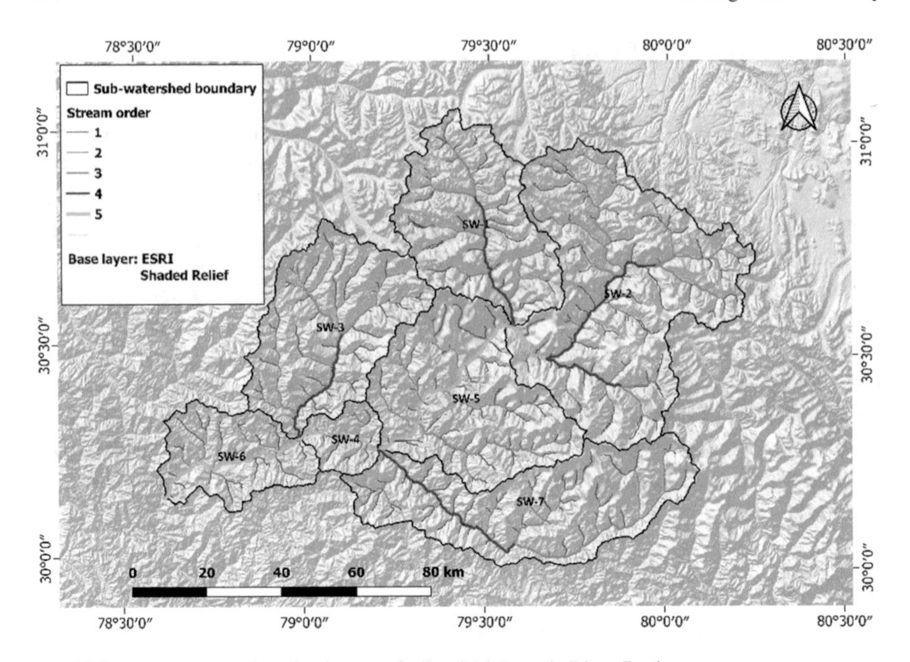

Fig. 23.3 Stream network and sub-watersheds of Alaknanda River Basin

(i.e., 1st order to 5th order) streams in each sub-watershed along with their respective lengths are presented in Table 23.2.

Table 23.2 Stream order and lengths of various sub-watersheds of ARB

Sub-watershed	SW-1	SW-2	SW-3	SW-4	SW-5	SW-6	SW-7
Number of different order streams							
1st order	104	196	95	21	116	51	113
2nd order	25	48	21	7	24	14	23
3rd order	7	12	6	1	6	2	4
4th order	2	3	2	0	1	1	1
5th order	1	1	1	0	0	0	0
Total	139	260	125	29	147	68	141
Order-wise stream length (km)							
1st order	253.36	489.11	231.70	24.73	326.00	139.25	278.23
2nd order	137.39	257.37	156.72	27.64	116.00	96.42	142.41
3rd order	62.85	140.39	76.97	22.40	112.53	38.58	68.71
4th order	28.00	66.83	36.87	0.00	41.16	13.66	92.33
5th order	28.76	49.58	8.04	0.00	0.00	0.00	0.00
Total	510.36	1003.28	510.31	74.77	595.68	287.91	581.68

23.2.5 Morphometric Analysis and Primary Priority Ranking of SWs

The morphometric analysis has been very widely used for sub-watershed prioritization (Anees et al. 2019; Bisht et al. 2018; Chauhan et al. 2016; Gajbhiye et al. 2014; Mahmood and Rahman 2019; Meraj et al. 2015; Prasad and Pani 2017). It is a comprehensive representation of the watershed geometry, topography, and stream network, which in turn provides an interpretation of the linear, areal, and relief aspects of the watershed (Aher et al. 2014). Each of these aspects plays a significant role in directly or indirectly influencing the surface runoff in the watershed (Aher et al. 2014) and can further be comprehended into vital information to prioritize the area concerning flash flood risk. Table 23.3 presents a list of morphometric parameters considered for the prioritization and flash flood risk zonation. All these parameters were calculated using well-accepted relations proposed by Strahler (1964), Horton (1945), Miller (1953), Hadley and Schumm (1961), Schumm (1956), Gravelius (1914), and Melton (1957).

In this study, ten parameters were considered for the morphometric characterization, namely drainage density (D_d), stream frequency (F_s), and texture ratio (T_r) representing the areal parameters, shape factor (S_f), form factor (F_f), circularity ratio

Table 23.3 Formulae used for computing the morphometric parameters

Morphometric parameter	Unit	Formula	References
Basin area (A)	km^2	Estimated in GIS	
Basin perimeter (P)	km	Estimated in GIS	
Stream order (u)	Dimensionless	Hierarchical rank	Strahler (1957)
Stream length (L_u)	km	Length of the stream	Horton (1945)
Basin length (L_b)	km	Estimated in GIS	
Drainage density (D_d)	km^{-1}	$D_d = L_u/A$	Horton (1932)
Stream frequency (F_s)	km^{-2}	$F_s = Nu/A$ Nu = Total no. of streams of all orders	Horton (1932)
Circulatory ratio (C_r)	Dimensionless	$C_r = 4\pi A/P^2$	Miller (1953)
Texture ratio (T_r)	Dimensionless	$T_r = S_1/P$ S_1 = Total no. of 1st order streams	Horton (1945)
Average slope (A_s)	Percent	Estimated in GIS	
Basin relief (R)	km	$R = H_{max} - H_{min}$	Hadley and Schumm (1961)
Relief ratio (R_h)	Dimensionless	$R_h = R/L_b$	Schumm (1956)
Elongation ratio (E_r)	Dimensionless	$E_r = 2(A/\pi)0.5/L_b$	Schumm (1956)
Form factor (F_f)	Dimensionless	$F_f = A/{L_b}^2$	Horton (1932)
Shape factor (S_f)	Dimensionless	$S_f = {L_b}^2/A$	Horton (1932)

(C_r), and elongation ratio (E_r) representing the shape parameters, average slope (A_s), basin relief (R), and relief ratio (R_h) depicting the relief parameters. The majority of these parameters were derived using linear parameters, namely basin area (A), basin perimeter (P), stream order (u), stream length (L_u), and basin length (L_b). All these parameters describe the unidimensional and multidimensional character of the basin in aerial, linear, relief, and shape aspects. Tables 1.4 and 1.5 present the values of all the above-listed parameters for each of the seven sub-watersheds of ARB.

For prioritizing sub-watersheds, a primary priority ranking (PPR) procedure was adopted, wherein the morphometric parameters were split into two groups based on their capability to influence (viz. direct or inverse) adversity of flash flood events. The first group of directly related parameters (Prasad and Pani 2017; Sujatha et al. 2015; Horton 1945) was average slope (A_s), basin relief (R), circularity ratio (C_r), drainage density (D_d), relief ratio (R_h), stream frequency (F_s), and texture ratio (T_r). The second group of inversely related parameters was elongation ratio (E_r), form factor (F_f), and shape factor (S_f). The study area is a part of the Himalayan region and experiences frequent occurrence of extreme rainfall events leading to cloud bursts. In such cases, an enormous volume of water suddenly starts flowing in the streams, which often exceeds the stream carrying capacity, causing massive destruction to human-made infrastructure and posing a threat to human lives. Higher the estimated values of morphometric parameters classified in this group, more noteworthy is the flash flood risk associated with it. Therefore, the primary priority ranking for parameters in the first group was assigned in such a way that for each parameter, the sub-watershed with the highest value was ranked 1, the next highest value was ranked 2, and likewise for the remaining sub-watersheds as presented in Table 23.7. Conversely, the primary priority ranking for parameters in the second group was assigned in a manner that for each parameter, the sub-watershed with the least value was ranked 1, the next lower value was ranked 2, and likewise for the remaining sub-watersheds as presented in Table 23.7.

23.2.6 Weighted Sum Approach (WSA) and Final Priority Ranking

WSA is a multi-criterion decision-making method employed to select the best among multiple alternatives. This method is implemented through an importance-based weight allotment to each parameter derived from a correlation matrix of the primarily ranked parameters. Finally, compound parameter values for each sub-watershed are calculated using the parameter rankings obtained through WSA. Figure 23.4 shows the complete workflow adopted for this study. The compound factors were calculated using the following equation.

$$\text{CF} = \text{PPR} \times W$$

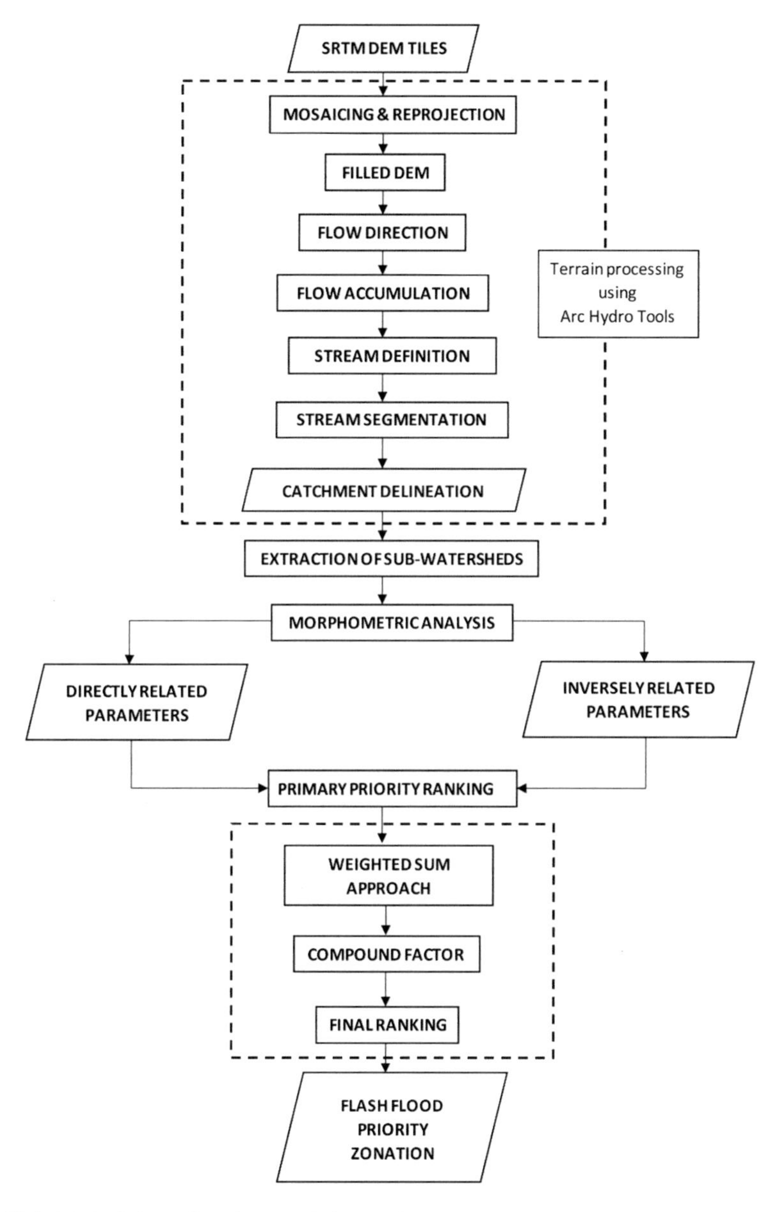

Fig. 23.4 Methodology flowchart depicting weighted sum approach used for sub-watershed prioritization and flash flood risk

where CF = compound factor; PPR = primary priority rank of the morphometric parameter; and W = weight of morphometric parameter, which is calculated using the following equation:

$$W = \frac{\text{sum of correlation coefficient}}{\text{grand total of correlations}}$$

23.3 Result and Discussion

The flash flood risk zonation was accomplished based on the morphometric characterization of the seven sub-watersheds of the Alaknanda River Basin.

23.3.1 *Stream Network Analysis of Watershed*

The stream network for the entire study area was processed, considering it as a 5th-order catchment. Basin area (A) is an essential linear aspect primarily because it directly attributes to the quantity of runoff generated. A larger area can be associated with a more significant number of streams, which leads to a more cumulative flow. The area of individual sub-watersheds in the basin varies from 294.61 km^2 (SW-4) to 3022.18 km^2 (SW-2), as presented in Table 23.4. Basin perimeter (P) for the study area in consideration varies from 103.76 km (SW-4) to 431.72 km (SW-2), as presented in Table 23.4. Stream length (L_u) specifics of each sub-watershed regarding the number of different order streams and stream lengths for each order are presented in Table 1.2. The stream network consists of 696, 162, 38, 10, and 3, 1st-, 2nd-, 3rd-, 4th-, and 5th-order streams. The highest number of streams was found in SW-2 (260) and the lowest number of streams in SW-4 (29). Basin length is referred to as the maximum length, measured from the farthest point on the catchment boundary to

Table 23.4 Area and perimeter of individual sub-watersheds

Sub-watershed	Perimeter (km)	Area (km^2)
SW-1	281.38	1538.50
SW-2	431.72	3022.18
SW-3	274.53	1641.23
SW-4	103.76	294.61
SW-5	287.77	1841.09
SW-6	192.49	807.90
SW-7	354.49	1889.80

Table 23.5 Morphometric parameters representing areal and shape aspects of individual sub-watersheds

Sub-watershed	Drainage density	Stream frequency	Circularity ratio	Texture ratio	Elongation ratio	Form factor	Shape factor
SW-1	0.33	0.09	0.24	0.37	0.76	0.45	2.21
SW-2	0.65	0.17	0.1	0.45	0.64	0.32	3.14
SW-3	0.33	0.08	0.26	0.35	0.77	0.46	2.17
SW-4	0.07	0.02	1.79	0.20	1.93	2.93	0.34
SW-5	0.39	0.10	0.23	0.40	0.79	0.49	2.06
SW-6	0.19	0.04	0.52	0.26	1.03	0.84	1.19
SW-7	0.38	0.09	0.15	0.32	0.7	0.39	2.57

the confluence point. The values of 'L_{bs}' in the basin range from s22.90 km (SW-4) to 69.39 km (SW-2) (Table 23.5).

The two relief parameters considered in this study were basin relief (R) and relief ratio (R_h). Basin relief is the maximum vertical distance between the outlet and the dividing point in a watershed (Aher et al. 2014). In this study, the highest relief was observed in SW-3 as presented in Table 23.6, indicating the availability of potential energy, capable of moving water along with the sediment, down the slope. The relief ratio is a measure of the overall steepness of the area. Low values of relief indicate a basin having flat to gentle slopes, and a high value indicates a hilly region and governs higher runoff potential. The values of 'R_h' for the Alaknanda River Basin (Table 23.6) are between 0.052 (SW-6) and 0.105 (SW-5).

Form factor (F_f) is a dimensionless ratio and an indicator of basin shape. It is the ratio of basin area to the square of basin length (Horton 1932) and can be effectively related to the occurrence of flood, the intensity of erosion, and sediment load transportation capacity in a watershed (Bisht et al. 2018). In this study area, the form factor values in the sub-watersheds range from 0.32 (SW-2) to 2.93 (SW-4). Circularity ratio (C_r) is calculated as the ratio of basin area to the area of a circle having the basin perimeter (Miller 1953). For a perfectly circular basin, the value of the circularity ratio is unity (Miller 1953). For ARB, 'C_r' ranges from 0.1 to 1.79; since all the values are less than 1 except for SW-4, it is evident that almost all the sub-watersheds

Table 23.6 Morphometric parameters representing relief aspects of individual sub-watersheds

Sub-watershed	Basin relief (m)	Relief ratio	Avg. slope (°)
SW-1	5571	0.096	26.602
SW-2	5583	0.080	26.969
SW-3	5999	0.104	27.693
SW-4	2318	0.101	26.618
SW-5	5889	0.105	28.752
SW-6	2215	0.052	25.778
SW-7	5792	0.092	28.581

have somewhat elongated shapes. Elongation ratio (E_r) is a measurement of basin dimensions or basin shape (Schumm 1956). According to a modified classification scheme, sub-watershed with E_r greater than 0.9 is categorized under circular shape, E_r values between 0.9 and 0.8 are categorized as oval-shaped, E_r values between 0.7 and 0.8 are categorized as less elongated, E_r values between 0.5 and 0.7 are categorized as elongated, and E_r values less than 0.5 are categorized as more elongated (Strahler 1964). The values of elongation ratio for all the sub-watersheds in the study area ranged from 0.64 to 1.93 (Table 23.5). Based on these values, it can be inferred that out of 7 sub-watersheds, two are circular, three are less elongated, and two are elongated.

The areal parameters considered were drainage density (D_d), stream frequency (F_s), and texture ratio (T_r). Factors like soil permeability, type of vegetation, and relief influence the drainage density (Aher et al. 2014). It also gives a characteristic description of infiltration capacity, runoff potential, and vegetation cover of the sub-watershed (Chorley and Kennday 1971; Macka 2001). Generally, drainage density is grouped into 4 categories (Bisht et al. 2018), viz. low (<2), medium (2–4), high (4–6), and very high (>6). All the sub-watersheds in the basin had low drainage density with values ranging from 0.07 to 0.65. Stream frequency is expressed as the ratio of the number of streams in a basin to the basin area. Melton (1957) documented that runoff processes are governed and influenced by two morphometric variables, namely drainage density and stream frequency. Table 23.5 shows the stream frequency of each sub-watershed. Texture ratio is controlled by a sole significant parameter of infiltration capacity (Aher et al. 2014). In ARB, the texture ratio ranges from 0.2 to 0.45. Higher relief values are attributed to steep slopes in the watershed, which renders them more prone to flash flooding. The greater the slope of a watershed, the more the capability of the same watershed to generate a rapid streamflow concentration, which plays a crucial role in causing flash floods (Mu et al. 2015). Table 23.5 shows the average slope values for each of the 7 sub-watersheds.

23.3.2 Prioritization Based on Morphometric Characterization Using Weighted Sum Approach

A correlation matrix of the 10 morphometric parameters presented in Table 23.8 was estimated using the primary priority ranks (Table 23.7) by employing 'data analysis' functionality in MS Excel. Weights for each of the morphometric parameters (Table 23.8) were computed based on their importance using the WSA model.

The computed parameter weights were used to formulate a model to estimate the ranks which were further used to calculate compound parameter values presented in Table 23.9 for individual sub-watershed prioritization. The equation for the developed model is:

$$\text{Prioritization} = (0.192 \times R) + (0.110 \times Rh)$$

Table 23.7 Primary priority ranking (PPR) of the morphometric parameters

Sub-watershed	R	R_h	D_d	F_s	C_r	T_r	A_s	E_r	F_f	S_f
SW-1	5	4	4	4	4	3	6	3	3	5
SW-2	4	6	1	1	7	1	4	1	1	7
SW-3	1	2	5	5	3	4	3	4	4	4
SW-4	6	3	7	7	1	7	5	7	7	1
SW-5	2	1	2	2	5	2	1	5	5	3
SW-6	7	7	6	6	2	6	7	6	6	2
SW-7	3	5	3	3	6	5	2	2	2	6

$$+ (0.185 \times Dd) + (0.185 \times Fs) + (0.169 \times Cr)$$
$$+ (0.181 \times Tr) + (0.183 \times As) - (0.134 \times Er)$$
$$- (0.134 \times Ff) - (0.134 \times Sf)$$

The final prioritization ranks (Table 23.10) were assigned in a manner that rank 1 was allotted to the sub-watershed having the least value of the compound factor, and the last rank was allocated to the sub-watershed having the highest value of the compound factor. The remaining sub-watersheds were ranked chronologically.

Table 23.10 also shows the priority categorization of the seven sub-watersheds based on the CF values. The sub-watersheds were grouped into four priority categories: (i) very high (0.88 and above) with area coverage of 16.68%; (ii) high (2.43–2.68) with area coverage of 42.26%; (iii) medium (3.18–3.72) with area coverage of 31.06%; and (iv) low (4.4–5.15) with area coverage of 9.99%. This categorization was done statistically using the Jenks natural breaks optimization technique in Microsoft Excel. Figure 23.5 shows the final priority map.

23.4 Conclusion

This study illustrates the potential application of remote sensing data and GIS for assessing the flash flood risk in the Alaknanda River Basin (ARB) in India. Geospatial technology was effectively for the morphometric characterization of 7 sub-watersheds of ARB using ten different parameters, namely drainage density (D_d), stream frequency (F_s), texture ratio (T_r), shape factor (S_f), form factor (F_f), circularity ratio (C_r), elongation ratio (E_r), average slope (A_s), basin relief (R), and relief ratio (R_h). The workflow involved processing and analysis of a digital elevation dataset in a GIS environment. Thereafter, the weighted sum approach (WSA) statistical method was implemented for the morphometric characterization and subsequent prioritization of all the seven sub-watersheds in the study area. This approach is based on the concept of a weighted average, wherein the overall compound factor value for every sub-watershed is equivalent to the products' total sum. The advantage of this

Table 23.8 Correlation matrix of the morphometric parameters

Correlation parameters	R	R_h	D_d	F_s	C_r	T_r	A_s	E_r	F_f	S_f
R	1	0.643	0.500	0.500	−0.464	0.500	0.857	0.393	0.393	−0.393
R_h	0.643	1	0.000	0.000	0.143	0.143	0.607	−0.286	−0.286	0.286
D_d	0.500	0.000	1	1.000	−0.964	0.893	0.571	0.786	0.786	−0.786
F_s	0.500	0.000	1.000	1	−0.964	0.893	0.571	0.786	0.786	−0.786
C_r	−0.464	0.143	−0.964	−0.964	1	−0.786	−0.536	−0.893	−0.893	0.893
T_r	0.500	0.143	0.893	0.893	−0.786	1	0.393	0.679	0.679	−0.679
A_s	0.857	0.607	0.571	0.571	−0.536	0.393	1	0.286	0.286	−0.286
E_r	0.393	−0.286	0.786	0.786	−0.893	0.679	0.286	1	1.000	−1.000
F_f	0.393	−0.286	0.786	0.786	−0.893	0.679	0.286	1.000	1	−1.000
S_f	−0.393	0.286	−0.786	−0.786	0.893	−0.679	−0.286	−1.000	−1.000	1
Sum	3.929	2.250	3.786	3.786	−3.464	3.714	3.750	2.750	2.750	−2.750
Weight = (sum/grand total)	0.192	0.110	0.185	0.185	−0.169	0.181	0.183	0.134	0.134	−0.134

Table 23.9 Calculation of compound parameter value

Sub-water-shed	$W*R$	$W*R_h$	$W*D_d$	$W*F_s$	$W*C_r$	$W*T_r$	$W*A_s$	$W*E_r$	$W*F_f$	$W*S_f$	+ve	−ve	CF
SW-1	0.96	0.44	0.74	0.74	0.68	0.54	1.10	0.40	0.40	0.67	5.19	1.48	3.72
SW-2	0.77	0.66	0.18	0.18	1.18	0.18	0.73	0.13	0.13	0.94	3.89	1.21	2.68
SW-3	0.19	0.22	0.92	0.92	0.51	0.72	0.55	0.54	0.54	0.54	4.04	1.61	2.43
SW-4	1.15	0.33	1.29	1.29	0.17	1.27	0.91	0.94	0.94	0.13	6.42	2.01	4.40
SW-5	0.38	0.11	0.37	0.37	0.84	0.36	0.18	0.67	0.67	0.40	2.62	1.74	0.88
SW-6	1.34	0.77	1.11	1.11	0.34	1.09	1.28	0.80	0.80	0.27	7.03	1.88	5.15
SW-7	0.57	0.55	0.55	0.55	1.01	0.91	0.37	0.27	0.27	0.80	4.52	1.34	3.18

Table 23.10 Final ranks and priority zones of sub-watersheds

Sub-watershed	Compound factor	Rank	Priority class	% area
SW-1	3.72	5	Medium	13.94
SW-2	2.68	3	High	27.39
SW-3	2.43	2	High	14.87
SW-4	4.40	6	Low	2.67
SW-5	0.88	1	Very high	16.68
SW-6	5.15	7	Low	7.32
SW-7	3.18	4	Medium	17.12

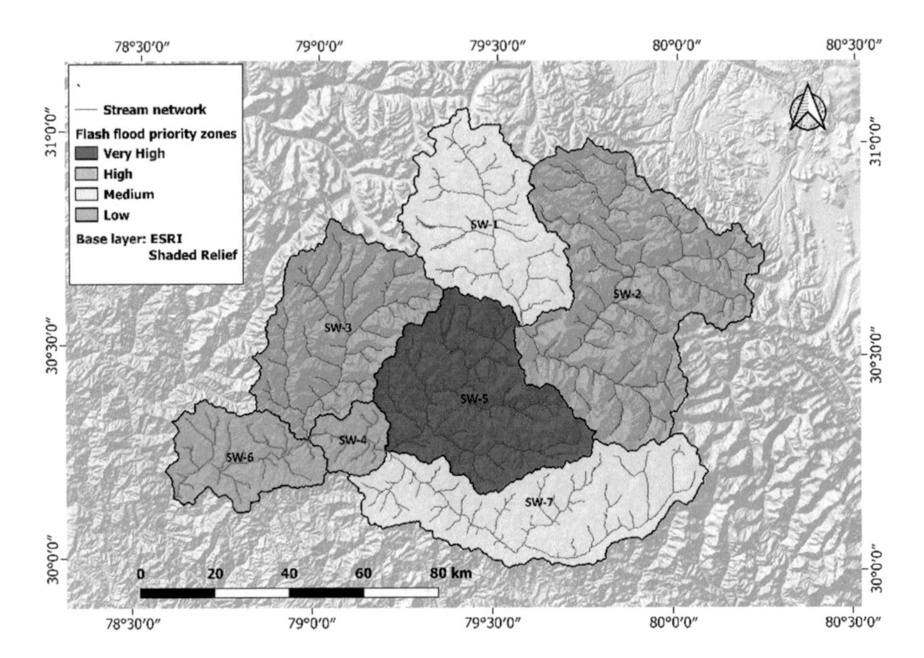

Fig. 23.5 Sub-watershed priority map based on morphometric analysis

method is that the transformation of raw data is linear in proportion, which means that the relative order of the standardized score magnitude remains equal (Aher et al. 2014; Malik et al. 2019). The morphometric characterization was also conducted by considering linear, areal, and relief aspects of the watershed, illustrating their correspondence with flash flood risk. The result revealed that SW-5 is categorized under very high-priority zone, SW-2 and SW-3 fall under high-priority zone, SW-1 and SW-7 are categorized under medium-priority zone, and SW-4 and SW-6 fall under low-priority zone. A map has been prepared as a final output, which can be utilized by the policymakers, water resources managers, and disaster management authorities to prioritize and plan for necessary conservation measures. Such studies give a strong indication that to minimize the loss of life and property, the expansion and growth of human activities in the sub-watersheds of high-priority zones need suitable flash flood preventive measures.

Acknowledgements We wish to express a deep sense of gratitude and sincere thanks to the Department of Water Resources Development and Management (WRD&M), IIT Roorkee, for providing a conducive environment and resources to conduct the research work.

References

Aher PD, Adinarayana J, Gorantiwar SD (2014) Quantification of morphometric characterization and prioritization for management planning in semi-arid tropics of India: a remote sensing and GIS approach. J Hydrol 511:850–860. https://doi.org/10.1016/j.jhydrol.2014.02.028

Anees MT, Abdullah K, Nawawi MNM, Nik Norulaini NNN, Ismail AZ, Syakir MI, Abdul Kadir MO (2019) Prioritization of flood vulnerability zones using remote sensing and GIS for hydrological modelling. Irrig Drain 68(2):176–190. https://doi.org/10.1002/ird.2293

Angillieri MYE (2008) Morphometric analysis of Colangüil river basin and flash flood hazard, San Juan, Argentina. Environ Geol 55(1):107–111

Bhatt GD, Sinha K, Deka PK, Kumar A (2014) Flood hazard and risk assessment in Chamoli District, Uttarakhand using satellite remote sensing and GIS techniques. Int J Innovative Res Sci Eng Technol 03(08):15348–15356. https://doi.org/10.15680/ijirset.2014.0308039

Bhatt S, Ahmed SA (2014) Morphometric analysis to determine floods in the Upper Krishna basin using Cartosat DEM. Geocarto Int 29(8):878–894. https://doi.org/10.1080/10106049.2013.868042

Bisht S, Chaudhry S, Sharma S, Soni S (2018) Assessment of flash flood vulnerability zonation through Geospatial technique in high altitude Himalayan watershed, Himachal Pradesh India. Remote Sens Appl Soc Environ 12(September):35–47. https://doi.org/10.1016/j.rsase.2018.09.001

Burt S (2005) Cloudburst upon Hendraburnick down: the Boscastle storm of 16 August 564 2004. Weather 60(8):219–227

Chauhan P, Chauniyal DD, Singh N, Tiwari RK (2016) Quantitative geo-morphometric and land cover-based micro-watershed prioritization in the Tons river basin of the lesser Himalaya. Environ Earth Sci 75(6). https://doi.org/10.1007/s12665-016-5342-x

Chorley RJ, Kennday BA (1971) Physical geography: a system approach. Longman Group Ltd., London, p 369

Connor R (2020) The United Nations world water development report 2020: water and climate change, vol 1. UNESCO, Paris

Devrani R, Singh V, Mudd SM, Sinclair HD (2015) Prediction of 585 flash flood hazard impact from Himalayan river profiles. Geophys Res Lett 42:5888–5894

Dhami B, Himanshu SK, Pandey A, Gautam AK (2018) Evaluation of the SWAT model for water balance study of a mountainous snowfed river basin of Nepal. Environ Earth Sci 77(1). https://doi.org/10.1007/s12665-017-7210-8

Dimri AP, Thayyen RJ, Kibler K, Stanton A, Jain SK, Tullos D, Singh VP (2016) A review of atmospheric and land surface processes with emphasis on flood generation in the Southern Himalayan rivers. Sci Total Environ 556:98–115. https://doi.org/10.1016/j.scitotenv.2016.02.206

Gajbhiye S, Mishra SK, Pandey A (2014) Prioritizing erosion-prone area through morphometric analysis: an RS and GIS perspective. Appl Water Sci 4(1):51–61. https://doi.org/10.1007/s13201-013-0129-7

Ghosh TK, Jakobsen F, Joshi M, Pareta K (2019) Extreme rainfall and vulnerability assessment: case study of Uttarakhand rivers. Nat Hazards 99(2):665–687. https://doi.org/10.1007/s11069-019-03765-3

Gravelius H (1914) Morphometry of Drainage Basins. Elsevier, Amsterdam

Guptha GC, Swain S, Al-Ansari N, Taloor AK, Dayal D (2021) Evaluation of an urban drainage system and its resilience using remote sensing and GIS. Remote Sens Appl: Soc Environ 23:100601. https://doi.org/10.1016/j.rsase.2021.100601

Hadley RF, Schumm SA, (1961) Sediment sources and drainage basin characteristics in upper Cheyenne River basin. Water Supply Paper 1531-B, U.S. Geological Survey, pp 137–196

Himanshu SK, Garg N, Rautela S, Anuja KM, Tiwari M (2013) Remote sensing and GIS applications in determination of geomorphological parameters and design flood for a Himalayan river basin, India. Int Res J Earth Sci 1(3):11–15

Himanshu SK, Pandey A, Palmate SS (2015) Derivation of nash model parameters from geomorphological instantaneous unit hydrograph for a Himalayan river using ASTER DEM. In: Proceedings of international conference on structural architectural and civil engineering, Dubai

Horton RE (1932) Drainage basin characteristics. Trans Am Geophys Unions 13:350–361

Horton RE (1945) Erosional development of streams and their drainage basins; hydro- physical approach to quantitative morphology. Geol Soc Am Bull 56:275–370

Jenson SK, Domingue JO (1988) Extracting topographic structure from digital elevation data for geographic information system analysis. Photogramm Eng Remote Sens 54(11):1593–1600

Joshi U, Rajeevan M (2006) Trends in precipitation extremes over India, National Climate Centre Research report. India Meteorological Department, Pune–411005

Joshi V, Kumar K (2006) Extreme rainfall events and associated natural hazards in Alaknanda valley, Indian Himalayan region. J Mt Sci 3(3):228–236

Kalura P, Pandey A, Chowdary VM, Raju PV (2021) Assessment of Hydrological Drought Vulnerability using Geospatial Techniques in the Tons River Basin India. J Indian Soc Remote Sens 49(11):2623–2637. https://doi.org/10.1007/s12524-021-01413-7

Kumar A, Gupta AK, Bhambri R, Verma A, Tiwari SK, Asthana AKL (2018) Assessment and review of hydrometeorological aspects for cloudburst and flash flood events in the third pole region (Indian Himalaya). Polar Sci 18:5–20

Kumar R (2021) Flood damage assessment in a part of the Ganga-Brahmaputra Plain Region, India. Adv Remote Sens Nat Resour Monit 389–404

Macka Z (2001) Determination of texture of topography from large scale contour maps. Geogr Vestn 73(2):53–62

Magesh NS, Jitheshlal KV, Chandrasekar N, Jini KV (2013) Geographical information system-based morphometric analysis of Bharathapuzha river basin, Kerala, India. Appl Water Sci 3(2):467–477

Mahmood S, Rahman AU (2019) Flash flood susceptibility modeling using geo-morphometric and hydrological approaches in Panjkora Basin, Eastern Hindu Kush, Pakistan. Environ Earth Sci 78(1):1–16. https://doi.org/10.1007/s12665-018-8041-y

Maikhuri RK, Nautiyal A, Jha NK, Rawat LS, Maletha A, Phondani PC, Bhatt GC (2017) Socio-ecological vulnerability: assessment and coping strategy to environmental disaster in Kedarnath valley, Uttarakhand, Indian Himalayan Region. Int J Disaster Risk Reduction 25:111–124

Malik A, Kumar A, Kandpal H (2019) Morphometric analysis and prioritization of sub-watersheds in a hilly watershed using weighted sum approach. Arab J Geosci 12(4):118

Melton MA (1957) An Analysis of the Relations among Elements of Climate. Surface Properties and Geomorphology (No. CU-TR-11). New York, Columbia Univ

Meraj G, Romshoo SA, Yousuf AR, Altaf S, Altaf F (2015) Assessing the influence of watershed characteristics on the flood vulnerability of Jhelum basin in Kashmir Himalaya. Nat Hazards 77(1):153–175. https://doi.org/10.1007/s11069-015-1605-1

Miller VC (1953) A quantitative geomorphic study of drainage basin characteristics in the clinch mountain area, Virgina and Tennessee. Technical Report (3), Department of Geology New York: Columbia University, pp 389–402

Mu W, Yu F, Li C, Xie Y, Tian J, Liu J, Zhao N (2015) Effects of rainfall intensity and slope gradient on runoff and soil moisture content on different growing stages of spring maize. Water 7(6):2990–3008
Pandey A, Chowdary VM, Mal BC (2004) Morphological analysis and watershed management using GIS. Hydrology J 27(3):71–84
Pandey BW, Prasad AS (2018) Slope vulnerability, mass wasting and hydrological hazards in Himalaya: a case study of Alaknanda Basin, Uttarakhand. Terræ Didatica 14(4):395–404
Papatheodorou KA, Tzanou EA, Ntouros KD (2014) Flash flood hazard prevention using morphometric and hydraulic models. An example implementation. In: International congress on "green infrastructure and sustainable socities/cities" GreInSus', vol 14, p 30
Prasad RN, Pani P (2017) Geo-hydrological analysis and sub watershed prioritization for flash flood risk using weighted sum model and Snyder's synthetic unit hydrograph. Model Earth Syst Environ 3(4):1491–1502. https://doi.org/10.1007/s40808-017-0354-4
Rai PK, Singh P, Mishra VN, Singh A, Sajan B, Shahi AP (2020) Geospatial approach for quantitative drainage morphometric analysis of varuna river basin, India. J Landscape Ecol (Czech Republic) 12(2):1–25. https://doi.org/10.2478/jlecol-2019-0007
Schumm SA (1956) The evolution of drainage systems and slopes in bad lands at Perth Amboy, New Jersey. Geol Soc Am Bull 67:597–646
Sharma SK, Gajbhiye S, Tignath S (2015) Application of principal component analysis in grouping geomorphic parameters of a watershed for hydrological modeling. Appl Water Sci 5(1):89–96. https://doi.org/10.1007/s13201-014-0170-1
Singh O, Kumar M (2017) Flood occurrences, damages, and management challenges in India: a geographical perspective. Arab J Geosci 10(5):102
Singh G, Pandey A (2021) Mapping Punjab flood using multi-temporal open-access synthetic aperture radar data in google earth engine. In Hydrological Extremes. Springer, Cham, pp 75–85
Singh G, Pandey A (2021a) Flash flood vulnerability assessment and zonation through an integrated approach in the Upper Ganga Basin of the Northwest Himalayan region in Uttarakhand. Int J Disaster Risk Reduction 66:102573. https://doi.org/10.1016/j.ijdrr.2021.102573
Singh G, Pandey A (2021b) Evaluation of classification algorithms for land use land cover mapping in the snow-fed Alaknanda River Basin of the Northwest Himalayan Region. Appl Geomatics. https://doi.org/10.1007/s12518-021-00401-3
Sreedevi PD, Owais SHHK, Khan HH, Ahmed S (2009) Morphometric analysis of a watershed of South India using SRTM data and GIS. J Geol Soc India 73(4):543–552
Strahler A (1964) Quantitative geomorphology of drainage basins and channel networks. In: Chow VT (ed) Handbook of applied hydrology. McGraw-Hill, New York
Strahler AN (1957) Quantitative analysis of watershed geomorphology. Trans Am Geophys Union 38:913–920
Sujatha ER, Selvakumar R, Rajasimman UAB, Victor RG (2015) Morphometric analysis of subwatershed in parts of Western Ghats, South India using ASTER DEM. Geomatics Nat Hazards Risk 6(4):326–341. https://doi.org/10.1080/19475705.2013.845114
Suresh M, Sudhakar S, Tiwari KN, Chowdary VM (2004) Prioritization of watersheds using morphometric parameters and assessment of surface water potential using remote sensing. J Indian Soc Remote Sens 32(3):249–259
Swain S, Mishra SK, Pandey A (2021a) A detailed assessment of meteorological drought characteristics using simplified rainfall index over Narmada River Basin, India. Environ Earth Sci 80(6):1–15
Swain S, Mishra SK, Pandey A (2021b) Assessing contributions of intensity-based rainfall classes to annual rainfall and wet days over Tehri Catchment, India. In: Advances in water resources and transportation engineering. Springer, Singapore, pp 113–121
Swain S, Mishra SK, Pandey A, Dayal D (2021c) Identification of meteorological extreme years over central division of Odisha using an index-based approach. In Hydrological extremes. Springer, Cham, pp 161–174

Thakur PK, Laha C, Aggarwal SP (2012) River bank erosion hazard study of river Ganga, upstream of Farakka barrage using remote sensing and GIS. Nat Hazards 61(3):967–987. https://doi.org/10.1007/s11069-011-9944-z

Thomas J, Joseph S, Thrivikramji KP, Abe G, Kannan N (2012) Morphometrical analysis of two tropical mountain river basins of contrasting environmental settings, the southern Western Ghats, India. Environ Earth Sci 66(8):2353–2366

Yadav SK, Singh SK, Gupta M, Srivastava PK (2014) Morphometric analysis of Upper Tons basin from Northern Foreland of Peninsular India using CARTOSAT satellite and GIS. Geocarto Int 29(8):895–914

Youssef AM, Hegab MA (2005) Using geographic information systems and statistics for developing a database management system of the flood hazard for Ras Gharib area, Eastern Desert, Egypt. In: The fourth international conference on the geology of Africa, vol 2. pp 1–15

Chapter 24
Flood Forecasting Using Simple and Ensemble Artificial Neural Networks

Bhabagrahi Sahoo, Trushnamayee Nanda, and Chandranath Chatterjee

Abstract With the increased flood havoc in many river basins worldwide, flood forecasting has been recognized as one of the feasible nonstructural measures of flood management. For accurate and reliable extreme flood forecasts, different artificial neural networks (ANNs)-based modelling approaches are developed in this study. For daily streamflow forecasting, a non-clustered feedforward backpropagation ANN (NCANN) model is developed using the daily observed streamflow training dataset. Further, in order to forecast high flow with ANN, two different types of models are developed, namely pre-classified ANN and post-processed ANN model. The pre-classified ANN models are basically the cluster-based ANN (CANN), and the post-processed ANN models are the ensemble ANNs (EANN). The high flow forecasting efficacy of the ANN models is compared for a common set of high flow regimes at one-, two- and three-day lead times in the flood-prone Mahanadi River basin in Eastern India. The results reveal that consideration of the training dataset consisting of various flow stratifications does not ensure a better forecasting of the high flow regime by NCANN model; rather the use of homogeneous set of training dataset gives improved forecasting during testing. Moreover, the high flow forecasting error is further reduced when the model ensembles are used. Seasonal flow assessment can be improved by using these developed models equally effectively.

24.1 Introduction

With the global climate change becoming a reality, flood forecasting is an important nonstructural measure for flood damage reduction and for minimizing flood-related deaths; hence, its implementation as an effective tool requires accurate short- and

B. Sahoo (✉)
School of Water Resources, Indian Institute of Technology Kharagpur, Kharagpur, India
e-mail: bsahoo@swr.iitkgp.ac.in

T. Nanda · C. Chatterjee
Agricultural and Food Engineering Department, Indian Institute of Technology Kharagpur, Kharagpur, India
e-mail: cchatterjee@agfe.iitkgp.ac.in

A. Pandey et al. (eds.), *Geospatial Technologies for Land and Water Resources Management*, Water Science and Technology Library 103,
https://doi.org/10.1007/978-3-030-90479-1_24

long-term forecasting. Generally, the hydrologic rules for streamflow generation vary in different seasons and for different hydrologic conditions. This may be attributed to the processes such as rainfall, baseflow, dam release or any intermediate lateral flow contribution into the river. The variation of streamflow processes results in nonlinear, seasonal and dynamic flow patterns over the input space for low-, medium- and high-magnitudes. Since river discharge during high flow periods is crucial for flood management, high flows should be predicted in advance with utmost accuracy.

The ability of artificial neural networks (ANNs) for nonlinear input–output mapping, high speed learning, pattern recognition and classification has made it an increasingly popular tool in hydrologic modelling, specifically for river flow modelling. The efficacy of the ANN has been widely tested in the fields of rainfall-runoff modelling (Hsu et al. 1995; Minns and Hall 1996; Shamseldin 1997; Dawson and Wilby 1998; Rajurkar et al. 2002, 2004; Abrahart and See 2007) and river flow forecasting (Campolo et al. 1999; Dawson and Wilby 1999; Imrie et al. 2000; Kim and Barros 2001; Kumar et al. 2004). Although ANN has been successfully implemented in streamflow time series prediction, it is recognized that this technique is highly sensitive to the quality, length and outlier characteristics of input data; hence, a number of ANN frameworks have evolved depending on the type of training data. The limitation of using the finite training dataset is that there is no certainty that the selected minimization algorithm can achieve the global minimum rather it may stop at the local minima. For river flow forecasting, several studies emphasized for the use of different kinds of hybrid ANN models such as the modular ANN, integrated ANN and bootstrapping and wavelet analysis (Zhang and Govindaraju 2000; Wang et al. 2006; Tiwari and Chatterjee 2010a, b; Tiwari and Chatterjee 2011; Huo et al. 2012; Nanda et al. 2016, 2019). Similarly, to deal with the complex time series flow pattern of river discharge, data clustering technique has been found satisfactory to avoid the ambiguity associated with the different patterns in the time series data (Ju et al. 2009). Wang et al. (2006) advocated three different techniques, namely cluster-based, threshold-based and seasonal-based to pre-classify the hydrological time series. Subsequently, separate ANN models are developed for discharge forecasting using each cluster of data and achieved improved performance as compared to a single ANN when faced with a complex flow data pattern. This performance could be attributed to more homogeneity in the sub-cluster datasets than the whole dataset. Hence, ANN using the raw data series may result in overall reliable forecast; however, it cannot simulate well for high flow periods which are important for flood prediction. Ju et al. (2009) found that backpropagation ANN based on grouping of data produced better streamflow prediction, especially the baseflow, as compared to the normal backpropagation ANN and the conceptual Xinanjiang model (Zhao 1992, e.g., Sahoo 2005). Similarly, to produce hybrid prediction, self-organizing map (SOM)-based decomposition of dataset into different number of clusters and development of separate clustered data specific ANN models was proposed by many researchers (Abrahart and See 2000; Zhang and Govindaraju 2000; Nanda et al. 2017). A hybrid network solution using SOM neural network for data clustering was advocated by Abrahart and See (2000). Zhang and Govindaraju (2000) demonstrated that for monthly discharge forecasting, the performance of the modular neural network (MNN) that divides the dataset into three classes of low, medium and high flows, with an expert ANN developed for each class, are better than that of a single ANN. For temperature prediction, Pal et al. (2003) combined the self-organizing feature map (SOFM) and the multilayer perceptions (MLP) network to generate hybrid ANN model in which SOFM was involved in partitioning the training data. Furthermore, the method of data clustering by fuzzy c-means (FCM) has been applied in a number of studies due to its simplicity. See and Openshaw (1999) used fuzzy logic to find subsets of river level dataset to train a series of neural network models for river level forecasting. Wang et al. (2006) developed cluster-based ANN as one of the hybrid models using the FCM technique for daily streamflow forecasting.

Post-processing of different model outputs helps each prediction model to produce an output of more reliable prediction. One of the widely used post-processing methods is by combining the outputs to get a weighted output. River flow forecast combination (ensemble) is a methodology that simultaneously utilizes the discharge forecasts of a number of different forecasting models to produce an aggregate discharge forecast which is generally superior to those of the individual models involved in the combination. The basis of this ensemble technique is that the limitation of a particular model in describing a process component is compensated by other model or models. For river flow forecasting, McLeod et al. (1987) introduced for the first time, the concept of combining the monthly river flows obtained from different time series models, which was further enhanced by Shamseldin et al. (1997) and Shamseldin and O'Connor (1999) using the daily discharge forecasts of different rainfall-runoff models. Recently, many researchers have studied the concept of ensemble prediction through multi-model combinations in river flow forecasting (See and Abrahart 2001; Abrahart and See 2002; Regonda et al. 2006). The concept of ensemble in ANN involves development and combination of member ANNs. Kim et al. (2006) used ANN as one of the methods for ensemble of rainfall-runoff models. Generally, the ANN ensembles have the capability for improved flood forecasting which is less sensitive to the selection of initial parameters as compared to a single ANN (Shu and Burn 2004). Ensemble of different ANN models combined with K-nearest neighbourhood (KNN) approach gives reasonably improved forecast in comparison with the individual MLP-ANN models (Agrahinejad et al. 2011).

Already, many researchers have explored the use of various frameworks of ANN such as ANN with data division, model hybridization and model ensemble. Researchers have proved that, for high flow forecasting in operational flood studies, ANN models based on data division perform better than that of the global ANN models developed without data division (Pal et al. 2003; Wang et al. 2006; Ju et al. 2009). Based on a clustering approach, Hu et al. (2001) proposed range-dependent neural network. Based on the initial study, See and Openshaw (1999) opined that, whenever a global feedforward ANN (FFANN) model is used for flow or water level forecasting, it specifically learns the low flow regime which accounts for about 95% of the total database (See et al. 1997). Hence, the FFANN testing result for the high flow regime shows very poor accuracy. Although many researchers have followed different methods of data decomposition in their ANN models, none of the studies have inter-compared the efficacy of data clustering by those methods, subsequently, their effect on the prediction accuracy. It would be difficult for a neural network with satisfactory validation accuracy for high flow events to simultaneously reproduce low flow events very well. This deficiency has been fulfilled by ensemble technique, which employs member models generated either from single model by varying the input data, training algorithm and structure (Shu and Burn 2004; Agrahinejad et al. 2011) or multi-model being contemporary to each other (Kim et al. 2006). The information contained in any poor model could be equally important in ensemble technique. In this regard to forecast the high flows, this could be an important approach to test the performance of ensemble by using a global model and a finite cluster-based model by the combination of different simple ANN models which has been undertaken in the current study.

In the light of the above discussion, this study has been carried out to forecast the high flows at one-, two- and three-day lead times in the Mahanadi River basin in Eastern India. To forecast the streamflows, feedforward type of ANNs trained with backpropagation algorithm is used. In Sect. 24.4, FFANN trained on non-clustered global streamflow time series (NCANN) is developed. In Sect. 24.5, development of different pre-classified and post-processed ANN models is discussed and is compared with the NCANN model for reproduction of daily high flows. The pre-classified ANN models, named as cluster-based ANNs (CANNs) are developed for high flow clusters using Kohonen self-organizing map (KSOM) and fuzzy c-means (FCM) clustering techniques. The post-processed ANN models are also developed as the ensemble ANNs (ANNEs) having two member models, viz., NCANN and CANN models that are varying on training pattern.

24.2 Study Area and Data Used

24.2.1 Description of Study Area

This study is conducted in the Mahanadi River basin, which is one of the major inter-state river basins in Eastern India. The area of this river basin is about 141,589 km^2 accounting for about 4.3% of the total geographical area of India. The basin lies within the longitudes of 80° 30′ to 86° 50′ East and latitudes of 19° 20′ to 23° 35′ North. About 52.9% (74,970 km^2) of the area of this basin is in the state of Chhattisgarh, 46.3% (65,600 km^2) is in the coastal state of Odisha and the remainder is in the states of Jharkhand and Maharashtra. The Mahanadi River originates from Chhattisgarh state having a total length from its origin to its outfall into the Bay of Bengal of about 851 km. The climate in Mahanadi River basin is tropical. This basin is mainly rainfed, and there is large variation and seasonal fluctuation of river discharge. This basin is highly vulnerable to floods which are affected by catastrophic flood disasters almost every year. A number of dams, irrigation projects and barrages are constructed at different reaches of the Mahanadi River, amongst which the most prominent is the Hirakud Dam constructed in the middle reaches. Many

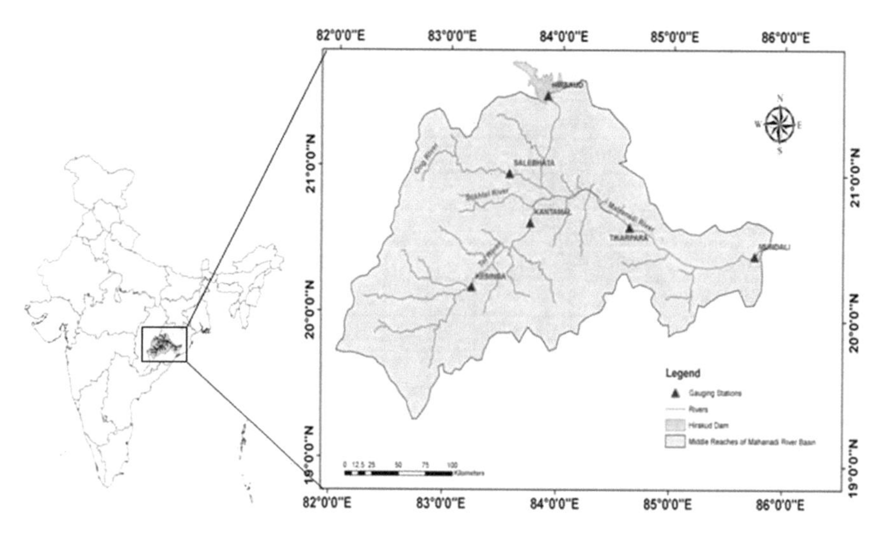

Fig. 24.1 Index map of the Mahanadi River basin in India showing location of different gauging stations

of the historical devastating high flows generally occur in the middle reaches due to synchronous dam releases at the upstream and significant lateral flow contribution along the river reaches downstream due to rainfall.

The middle reaches of the Mahanadi River, having a length of 338 km, comprise of the Hirakud Dam at the upstream and Mundali barrage at the downstream. This is located in Odisha between 82°–86° E and 19°–22° N, having a geographical area of 50,702.54 km^2 (Fig. 24.1). In the middle reaches of the Mahanadi River basin, although Hirakud Dam solves most of the critical issues of water resources planning and management, almost every year, flooding occurs in the delta region during monsoon season. The Mundali gauging station is at the mouth of the delta; hence, this middle reach has been selected for river flow forecasting in the current study. Daily discharge forecasting is carried out in this study for effective planning and management during flood period.

24.2.2 Data Used

Daily discharge time series data of five gauging stations from 2000 to 2010 were collected from the Central Water Commission (CWC), New Delhi; and Odisha Water Resources Department, Bhubaneswar. The releases from Hirakud Dam operation were collected from the Main Dam Division, Burla, Odisha. The locations of daily discharge gauging stations are shown in Fig. 24.1. The discharge datasets are analyzed for their mean, standard deviation, maximum and minimum and based on these

Table 24.1 Statistics of daily discharge data at Mundali gauging site

Year	Mean (m^3/s)	Standard deviation (m^3/s)	Maximum (m^3/s)	Minimum (m^3/s)
2000	825.09	823.20	4952.50	326.63
2001	2951.18	5940.21	36,698.45	291.49
2002	940.73	1738.90	16,589.46	289.96
2003	3069.46	6046.92	37,214.50	283.00
2004	1919.65	3017.92	21,681.73	297.15
2005	2104.58	4002.55	25,562.43	283.00
2006	2720.30	5488.75	36,084.23	318.96
2007	2301.60	3532.38	21,016.46	312.60
2008	2113.55	4880.95	44,750.45	261.68
2009	1665.96	3468.43	24,492.91	283.00
2010	1554.26	2801.58	19,527.00	283.00

statistics, the dataset is divided into training and testing datasets. The statistical analysis for discharge data at Mundali gauging site is shown in Table 24.1.

To incorporate the significant temporal effect of hydrologic variables in the model output, it is important to reconstruct the multidimensional input space temporally. This provides the identification of the input vector that can best represent the hydrological process avoiding loss of information (that may result if emphatic input variables are excluded). Wang et al. (2006) applied state space reconstruction using time delay coordinate method as proposed by Packard et al. (1980) and Takens (1981). The physical basis of such a reconstruction is that a nonlinear system is characterized by self-interaction, so that a time series of a single variable can carry the information about the dynamics of the entire multivariable system.

A statistical approach based on cross-correlation, autocorrelation and partial autocorrelation between the variables as proposed by Sudheer et al. (2002) was used (e.g., Tiwari and Chatterjee 2010a). The current study uses this method for a firm understanding of the hydrological system behaviour under consideration. This method is based on the heuristic assumption that the emphatic variables corresponding to different time lags can be identified through statistical analysis of the data series; which is useful to investigate the dependence between the variables. This method is applied herein to select the significant daily discharge inputs from the Hirakud release and from five gauging stations, viz., Salebhata, Kesinga, Kantamal, Tikarpara and Mundali.

The correlation functions of the daily discharges with different lag times of 0–10 days are obtained for selecting significant inputs (Fig. 24.2). To select significant input streamflow time series data from the Mundali gauging station, autocorrelation function (ACF) and partial autocorrelation function (PACF) were used. The ACF (Fig. 24.2a) shows a significant correlation at 93.5% confidence level for lag of zero to more than 10 days, whereas the PACF (Fig. 24.2b) shows a significant correlation up to three days lag after which it lies within the confidence limits. Correlation for

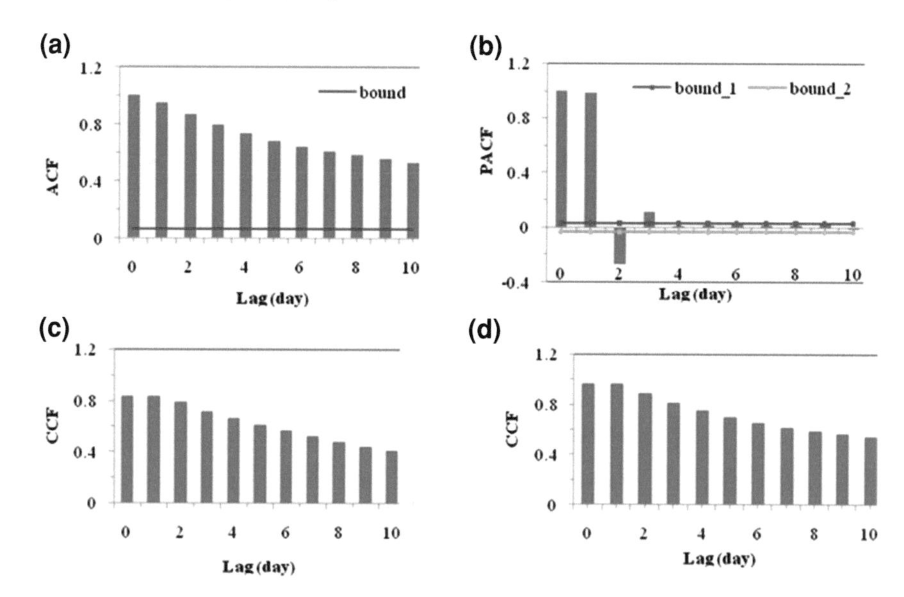

Fig. 24.2 Correlation statistics of discharge showing: **a** ACF at Mundali, **b** PACF at Mundali, **c** CCF between Hirakud and Mundali and **d** CCF between Tikarpara and Mundali stations

the discharge at Hirakud and Tikarpara site with that of Mundali is significant for lag up to three days as shown in Figs. 24.2c and 24.2d, respectively. The cross-correlation function (CCF) for the discharge time series at Mundali has a significant correlation with that both at Kantamal and Tikarpara up to three days lag. Similarly, significant lagged inputs for the discharge time series at the Salebhata and Kesinga gauging stations were obtained by this procedure of lag correlation analysis. In total, 21 discharge variables were found to be significant for input in the ANN modelling as shown in Table 24.2.

Table 24.2 Selection of input variables for ANN using correlation statistics (CCF, ACF and PACF) of daily discharge time series at different gauging stations

Stations	Input variables[a]
Hirakud	Q_{t-3}, Q_{t-2}, Q_{t-1}, Q_t
Salebhata	Q_{t-3}, Q_{t-2}, Q_{t-1}
Kesinga	Q_{t-2}, Q_{t-1}
Kantamal	Q_{t-3}, Q_{t-2}, Q_{t-1}, Q_t
Tikarpara	Q_{t-3}, Q_{t-2}, Q_{t-1}, Q_t
Mundali	Q_{t-3}, Q_{t-2}, Q_{t-1}, Q_t

[a]e.g., Q_{t-1} represents discharge at one-day lag time

24.3 Model Evaluation Criteria

To test the efficacy of the ANN models, five performance evaluation measures were used: (i) correlation coefficient (r), (ii) Nash–Sutcliffe efficiency (E_{NS}), (iii) root mean square error (RMSE), (iv) mean absolute error (MAE) and (v) threshold statistics (TS). Different correlation-based measures evaluate the linear relationship between the observed and predicted variables. However, there are several limitations associated with the correlation-based measures (Legates and Davis 1997; Legates and McCabe 1999).

The correlation coefficient (r) is expressed as

$$r = \frac{\sum_{i=1}^{n}\left[\left(O_i - \overline{O}\right)\left(P_i - \overline{P}\right)\right]}{\sqrt{\left[\left(\sum_{i=1}^{n}\left(O_i - \overline{O}\right)^2\right)\left(\sum_{i=1}^{n}\left(P_i - \overline{P}\right)^2\right)\right]}} \tag{24.1}$$

where O_i and P_i are the observed and predicted values, $\overline{O}$ is the mean of the observed values and n is the number of data points.

The Nash–Sutcliffe efficiency (E_{NS}), (Nash and Sutcliffe 1970; ASCE Task Committee 1993) which ranges from -∞ to 1, is expressed as

$$E_{NS} = \left[1 - \frac{\sum_{i=1}^{n}\left(O_i - P_i\right)^2}{\sum_{i=1}^{n}\left(O_i - \overline{O}\right)^2}\right] \tag{24.2}$$

E_{NS} provides a measure of the ability of a model to predict values that are different from the mean. Because of squared differences, E_{NS} is overly sensitive to high flow values.

The root mean square error (RMSE) is expressed as

$$\text{RMSE} = \sqrt{\frac{1}{n}\sum_{i=1}^{n}\left(O_i - P_i\right)^2} \tag{24.3}$$

The mean absolute error (MAE) is expressed as

$$\text{MAE} = \frac{1}{n}\sum_{i=1}^{n}\left|O_i - P_i\right| \tag{24.4}$$

It has been emphasized in the literature that the performance measures like summary statistics and relative error should also be used in addition to the efficiency measure and absolute measures like RMSE or MAE to properly evaluate a model (Legates and McCabe 1999). Therefore, in order to test the effectiveness of the model developed herein, it is important to use some other performance evaluation criteria such as the absolute relative error (ARE) and threshold statistics (TS) (Jain and Ormsbee 2002; Nayak et al. 2005). The distribution of the prediction errors over the input data space can be obtained by these two measures.

The absolute relative error, ARE is expressed as

$$\text{ARE} = \left| \frac{O_i - P_i}{O_i} \right| \tag{24.5}$$

The threshold statistics, TS_x for a level of $x\%$ is a measure of the consistency in forecasting errors of a particular model, expressed as

$$\text{TS}_x = \frac{Y_x}{n} \tag{24.6}$$

where Y_x is the number of predicted values (out of the total predicted) for which ARE is less than the $x\%$ threshold.

24.4 Streamflow Forecasting Using NCANN Models

24.4.1 Feedforward ANN Modelling Strategy

Artificial neural network, termed as multilayer perception, is a multilayer series of nodes or neurons, namely input layer, hidden layer and output layer. Amongst the artificial neural networks, FFANN trained using backpropagation algorithm is very often used in streamflow and rainfall-runoff time series modelling. In the present study, different FFANNs are used for forecasting discharges at Mundali gauging site in the middle reaches of the Mahanadi river basin with one- to three-day lead times. Significant lag effect of discharges at different gauging sites was found based on the results of lag correlation analysis (i.e., CCF, ACF and PACF).

The FFANN models using backpropagation training algorithm were developed using the MATLAB software (MATLAB 7.10.0 R2010a). ANNs were adjusted or trained so that a particular set of inputs lead to a particular target output. When the network was initialized, it generated a set of weights and biases for inputs to the hidden and output layers. The moment training started, a weighted sum of the inputs moved to the subsequent layer, got transformed through a transfer function and produced the output. During the course of training, the performance function was optimized for better matching of observed and target outputs. Backpropagation learning is a standard gradient descent algorithm which updates network weights and biases in the direction in which the performance function decreases most rapidly (the negative of the gradient). In each iteration of the algorithm, weights are updated as $x_{k+1} = x_k - \alpha_k g_k$, where x_k is the vector of current weights, g_k is the current gradient and α_k is the learning rate. In each iteration, the learning rate defines the magnitude of weight adjustment.

24.4.2 Modelling the Mahanadi River Floods Using NCANN

Normal FFANN models were first fitted to the daily flow time series without any data clustering procedure. The NCANN models were developed using significant set of input variables of discharges in which supervised backpropagation training algorithm was verified for multistep ahead daily discharge forecasting. In total, 21 number of significant discharge variables were identified as inputs to the model. Models were fitted using this set of daily discharge data of eight years (2003–2010) available at Hirakud and five gauging stations of Kesinga, Salebhata, Kantamal, Tikarpara and Mundali. Subsequently, one- to three-day ahead discharge forecasts were made at Mundali for the period 2000–2002.

The input data transformation was carried out using the following steps: (i) Significant set of discharges was first normalized as $Q = Q + 0.0001$ and $Q = \log Q$; (ii) then, it was standardized as $Q = (Q - \overline{Q}) \big/ \sigma_Q$ and (iii) finally, it was linearly scaled to the range [0, 1]. In the model structure, one hidden layer with two neural units was found to be the optimal by trial and error method. Hence, the model structure obtained was 21–2–1 for all the lead time forecasts. The network uses a logistic sigmoid transfer function in the hidden layer in the form of $n = 1/(1 + \exp(-a))$ for which the output varies from 0 to 1, where 'a' is input to and 'n' is output from the hidden layer as shown in Fig. 24.3. A linear transfer function was used for the output

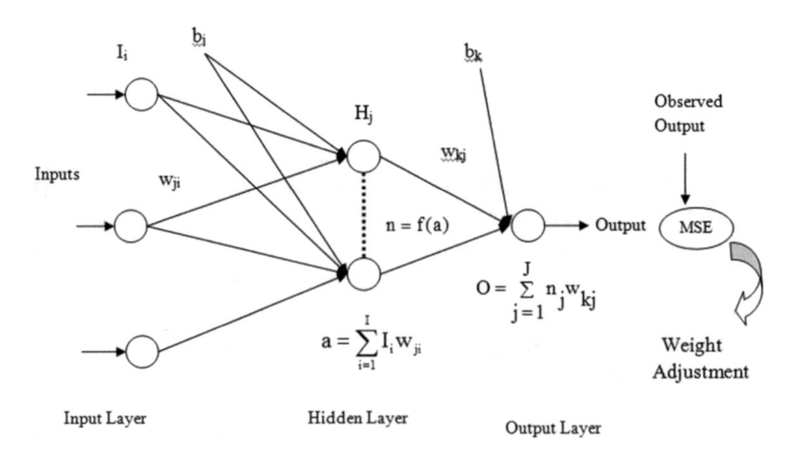

Fig. 24.3 Feedforward backpropagation artificial neural network (FFANN)

Table 24.3 Performance evaluation measures of the NCANN model for daily discharge forecasting using at one-day, two-day and three-day lead times

Criteria	Lead time (day)					
	1		2		3	
	Training	Testing	Training	Testing	Training	Testing
RMSE (m^3/s)	886.67	565.19	1690.90	1252.40	2402.90	1748.30
MAE (m^3/s)	253.96	166.38	487.91	336.68	743.58	490.21
E_{NS} (%)	95.82	97.72	84.79	88.80	69.28	78.18
r	0.98	0.99	0.92	0.95	0.84	0.89

layer. A typical feedforward ANN is shown in Fig. 24.3. In the backpropagation algorithm, the ANN weights and biases were updated using the Levenberg–Marquardt (LM) optimization in order to minimize the mean squared error between the forecast and observed values. After several iterations, it was convinced that 500 numbers of iterations are sufficient for the network to converge.

Table 24.3 shows the performance evaluation measures of the NCANN models during training and testing at one-day, two-day and three-day lead time streamflow forecasts. The reproduction of the observed daily discharge hydrographs forecasted at one-day, two-day and three-day lead times is shown in Figs. 24.4, 24.5 and 24.6, during testing of the NCANN models, respectively. It can be observed from Table 24.2 and Figs. 24.4, 24.5 and 24.6 that the performance of the NCANN model decreases with the increase in forecasting lead times. Figure 24.7 shows the scatter plots of the observed and predicted flows of the NCANN models for one- to three-day lead times. From both the hydrographs and scatter plots, it can be observed that most of the high flows are largely under-estimated at two-day and three-day lead time forecasts. This could be due to the dynamic flow patterns over the input space consisting of low-, medium- and high-magnitude flows, which was resulting in poor reproduction of individual flow patterns, especially for the high flows during testing. Hence, in order to improve the high flow forecasting accuracy, two different types of ANNs, namely pre-classified ANN and post-processed ANN models, are developed in the following sections.

24.5 Streamflow Forecasting Using Pre-Classified and Post-Processed ANN Models

24.5.1 Streamflow Forecasting Using Pre-Classified ANN Models

In the pre-classified ANN models, clustering was used for classification of data to obtain the high flow data series separately. Clustering is one of the most important

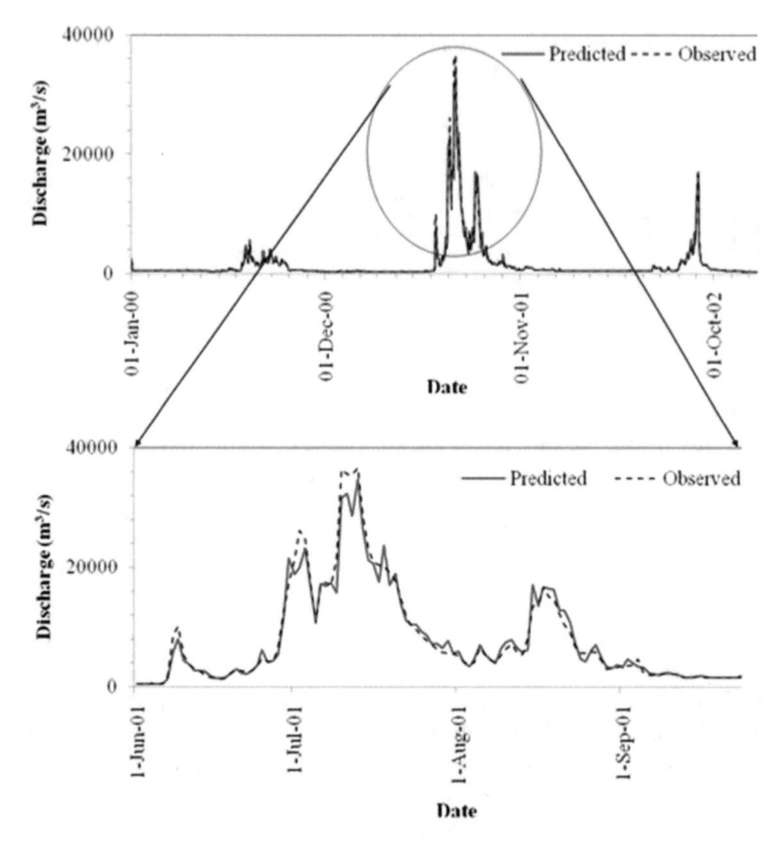

Fig. 24.4 Reproduction of the observed time series by the NCANN model at one-day lead time forecast during testing period

issues in pattern recognition in the input data (Bezdek 1981). In order to generate homogeneous series of flows, the discharge time series was divided into the 'low' and 'high' flows using two cluster analysis techniques: (1) ANN-based Kohonen self-organizing map (KSOM) and (2) fuzzy logic based fuzzy c-means (FCM).

24.5.1.1 KSOM Cluster Analysis

The Kohonen self-organizing map (KSOM) is one of the most widely used unsupervised ANN algorithms (Kohonen et al. 1996). It consists of a multidimensional input layer and a two-dimensional output layer whose units (nodes or neurons) accommodate different input data patterns. The KSOM map, also called the feature map, is setup using the training and testing datasets separately, and the obtained patterns in the maps are used for clustering the datasets. During training of KSOM, it involves clustering of the input patterns in such a way that similar data patterns belong to the same output neurons or any of its neighbours (Back et al. 1998). Hence, a KSOM

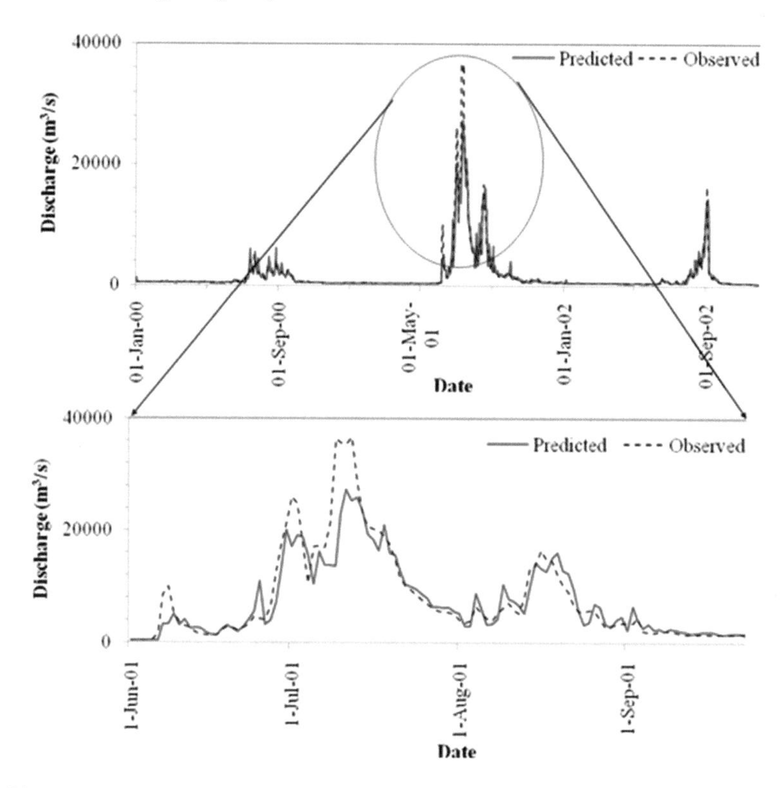

Fig. 24.5 Same as Fig. 24.4, but at two-day lead time forecast

feature map converts the high dimensional data into a two-dimensional display with separation boundaries amongst the different patterns in the high dimensional data (Kohonen et al. 1996; Zhang et al. 2009).

In clustering, the purpose is to partition the given dataset of r number of observations $\mathbf{X} = \{\mathbf{x}_1, \mathbf{x}_2, \ldots \mathbf{x}_i \ldots, \mathbf{x}_r\}$ into groups or clusters; in which each observation is given by $\mathbf{x}_i = [x_1, x_2, \ldots, x_n]^{tr}$, where n is the dimension of the input data pattern $\mathbf{x}$. The output layer consists of M neurons arranged in a two-dimensional grid of nodes. Each node or neuron j ($j = 1, 2, \ldots, m$) is represented by an n-dimensional weight or prototype vector. The weight vectors of the neurons of the Kohonen layer can be denoted by $\mathbf{W}_j = [w_1, w_2, \ldots, w_n]^{tr}$ where $\mathbf{W}_j$ is the weight or prototype vector of neuron j in the Kohonen layer. The weight vectors of the KSOM form a codebook of size equal to mapunits × dimension $[m \times n]$.

In this study, the KSOM feature maps were generated in MATLAB 10 using the SOM toolbox developed by Helsinki University of Technology, Finland (available at http://www.cis.hut.fi). Use of the KSOM toolbox has also been found in clustering dataset of wastewater treatment records, evapotranspiration and nonuniform set of rainfall data series (Garcia, and Gonzalez 2004; Adeloye et al. 2011; Mwale et al. 2012). All the default parameters of this SOM tool box were used as listed in

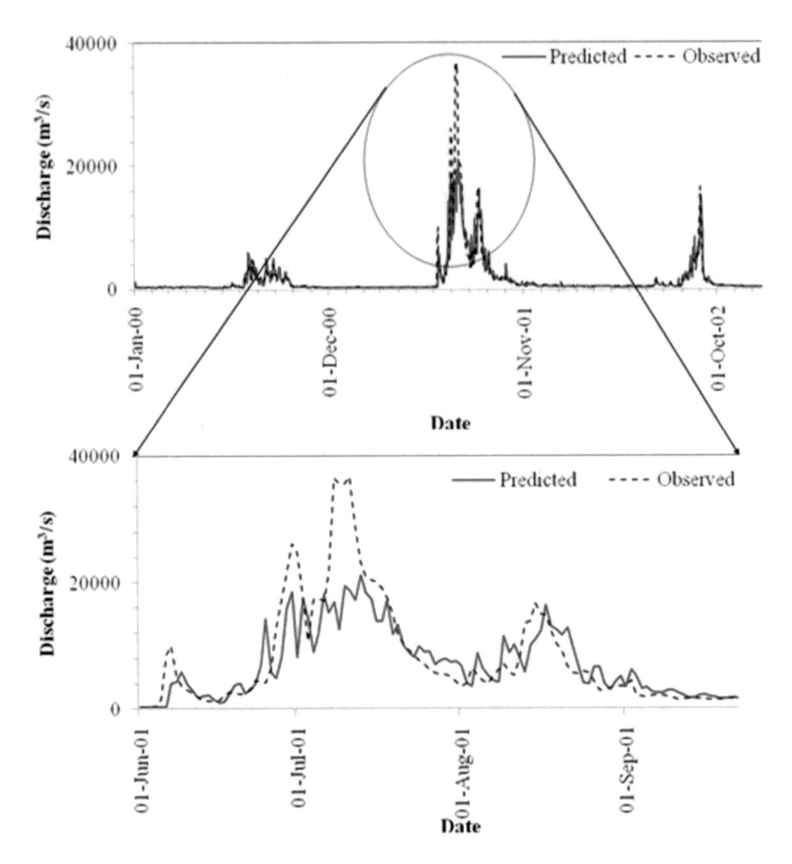

Fig. 24.6 Same as Fig. 24.4, but at three-day lead time forecast

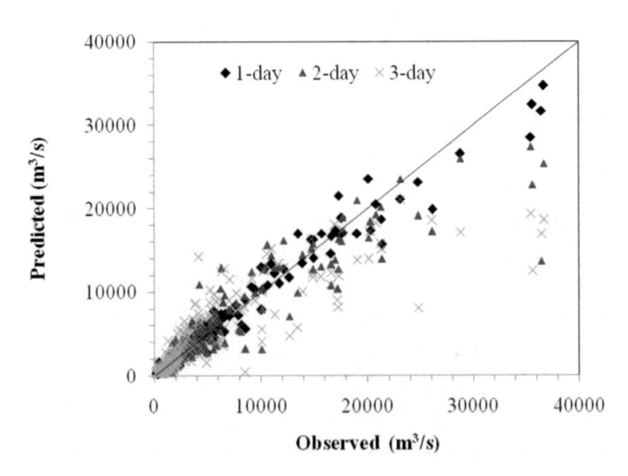

Fig. 24.7 Observed and forecasted discharges estimated using the NCANN model at one-day, two-day and three-day lead times

Table 24.4 Structural characteristics of developed KSOM in the MATLAB environment

Characteristics	Type
Data normalization	'var': $x_n = \frac{(x-\overline{x})}{\sigma_x}$
Map size	Training: 27 × 10 Validation: 21 × 8
Initialization and training	'Linear', 'Batch'
Lattice	'Hex'
Shape	'Sheet'
Neighbourhood function	'Gaussian'
Initial radius, final radius	4, 1
Alpha type	'inv'

Table 24.4 to generate Kohonen layer for the training and testing datasets separately. The Kohonen map was initialized with small random numbers, called weight or prototype vectors, having the same dimension as the input vector in the beginning of the training process. The prototype vectors generated were stored in the codebook of the map. During training, the SOM algorithm computes the similarity measures between the output neurons in terms of Euclidean distance between the input pattern and the weight vector as

$$d_j = \sum_{i=1}^{n} \left(x_i - w_{ji}\right)^2 \tag{24.7}$$

where d_j is the Euclidian distance for a data vector from the jth neuron.

A smaller Euclidean distance indicates higher similarity between the input pattern and the weight vector. Hence, the neuron having the smallest distance from the current input vector to all neurons of the Kohonen layer is selected and is called as the winning neuron or the best matching unit (BMU) of the current vector. After finding the BMU, the weight vectors of the SOM were updated so that the BMU is moved closer to the input vector in the input space. The topological neighbours of the BMU were treated similarly. This adoption procedure stretches the BMU and the neighbours towards the sample vector as shown in Fig. 24.8. The feature map achieves the distribution of weight vector that approximates the distribution of input datasets.

The distance from each neuron to its neighbouring neurons was calculated in the unified distance matrix (U-matrix). The U-matrix helps in identifying the clusters as it displays the distance between each pair of computed weight vectors. The resulting U-matrix, consisting of distance between neighbouring neurons and scaled between the maximum and minimum value, acts as an important visualization method as shown in Fig. 24.9. The U-matrices, as shown in Fig. 24.9 for training and testing, give the idea that there are two possible numbers of clusters in both the training and testing datasets. Figure 24.10 shows the KSOM produced for training and testing datasets indicating the separate clusters. The bar plots shown therein for discharges

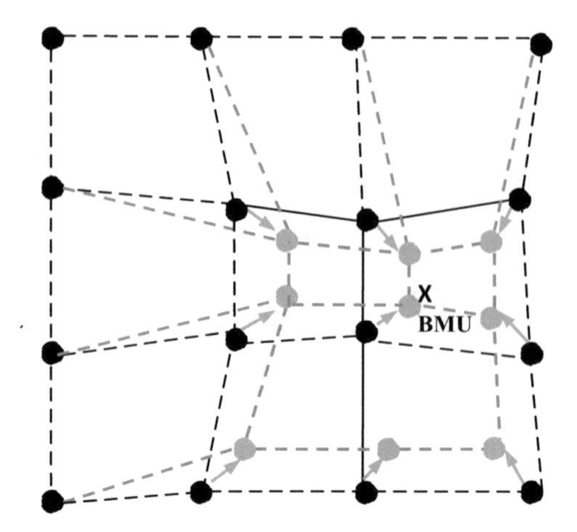

Fig. 24.8 Updating the BMU and its neighbour towards input vector **x** during clustering process

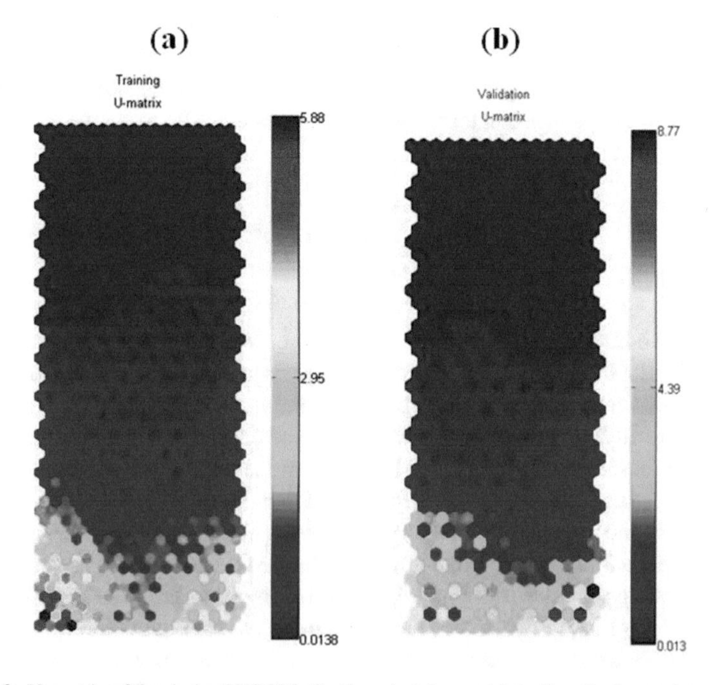

Fig. 24.9 U-matrix of the derived KSOMs for the **a** training and **b** testing discharge datasets at the Mundali gauging station

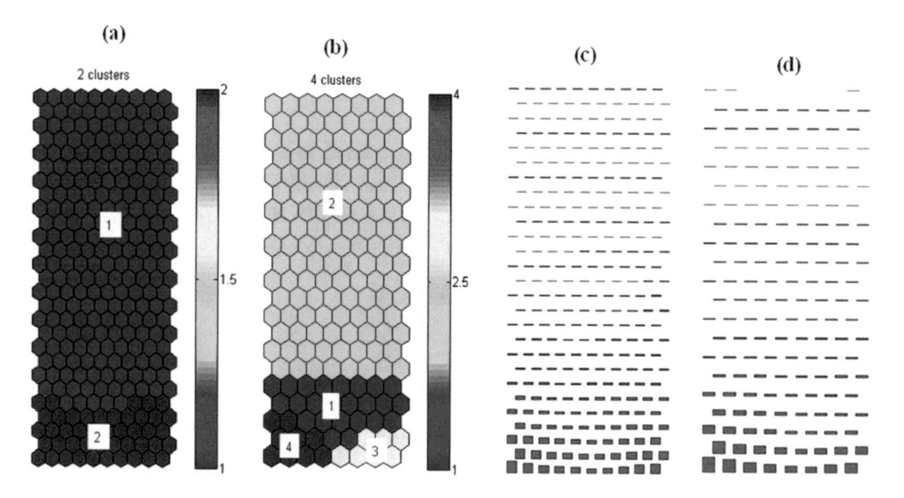

Fig. 24.10 KSOM showing clusters of **a** training and **b** testing datasets, and the corresponding bar plots for **c** training dataset and **d** testing datasets for the discharge data at Mundali

at Mundali gauging station distinguish the varied flow pattern in which the high flow clusters are illustrated. From Fig. 24.10a and c, it can be observed that cluster two consists of a group of high flows in training dataset. Similarly, from Fig. 24.10b and d, it can be observed that the clusters three and four in testing dataset together form the high flow clusters.

24.5.1.2 FCM Cluster Analysis

In KSOM clustering method, similar to any standard clustering methods, the cluster boundaries are well-demarcated based on the colour patterns. Hence, it is easy to define the belonging of a data point to a particular cluster. However, in hydrological datasets such as rainfall and runoff, some of the data points can be presented by more than one cluster by their partial memberships in each of the neighbouring clusters. This type of clustering or partitioning is referred to as fuzzy clustering. This concept of partial membership in clustering is the foundation of fuzzy clustering technique (Cios et al. 2007). In this study, the fuzzy c-means clustering method was also utilized to cluster the Mahanadi River discharge time series datasets into 'average' and 'extreme' flows as in Kentel (2009).

Suppose, a dataset of r vectors $\mathbf{X} = \{\mathbf{x}_1, \mathbf{x}_2, \ldots \mathbf{x}_i \ldots, \mathbf{x}_r\}$ is to be clustered into groups. In fuzzy c-means clustering, the dataset is partitioned into c-groups or clusters which are defined as a *priori*. The clusters are represented by cluster centre matrix $\mathbf{C} = \{\mathbf{c}_1, \mathbf{c}_2, \ldots, \mathbf{c}_c\}$ consisting of c cluster centres with each centre $\mathbf{c}_c = [c_1, c_2, \ldots c_n]^{tr}$. Each data point may belong to a cluster by certain membership value for that cluster. The membership values are together called as fuzzy membership matrix $\mathbf{U} = [u_{ij}]$, $j = 1, 2, \ldots, c$. The values of u_{ij} vary from 0 to 1, wherein the sum of the membership functions for each feature of data is equal to 1 expressed as

$$\sum_{j=1}^{c} u_{ji} = 1, \forall i = 1, \ldots, r \tag{24.8}$$

Subsequently, the membership values are allocated to every cluster in the centre matrix **C** for each data in **X.** In FCM algorithm, the cluster centres are updated over the iterations to minimize the distance between the centre and the data point. For an optimal partitioning, the objective function is to minimize the generalized least-square error function given by

$$F = \sum_{j=1}^{c} \sum_{i=1}^{r} u_{ji}^{g} \left\| c_j - x_i \right\|^2, \; 1 \le g < \infty \tag{24.9}$$

where g is a real number that governs the relative weight of membership grades to each centre.

The output arguments of this function are matrix of final cluster centres where each row provides the centre coordinates, fuzzy membership matrix 'U' and the values of the objective function during iterations. The solution of the optimization problem provides the partial membership value of each dataset in each cluster.

The separation of these high flows provides a homogeneous input dataset with minimum bias. The generated clusters for two groups of flows in training and testing datasets are presented in Fig. 24.11, wherein the data represent discharge at Mundali. The presence of various flow patterns in the datasets was observed; hence, the resulting two clusters formed herein represent the high and low flows. Table 24.5 shows the number and percentage of data points in the high flows of training and testing datasets after clustering by both the KSOM and FCM clustering methods. Furthermore, the high flows that are separated out during the course of all these analyzes are compared to find the flows that are common to both the high flow clusters, which are then used for comparing the model results. After the pattern classification, separate ANN models are built using only the high flow training data series and are named as cluster-based ANNs (CANNs). According to the KSOM and FCM cluster generating methods, the two types of CANN models developed herein are KSOM-CANN and FCM-CANN, respectively.

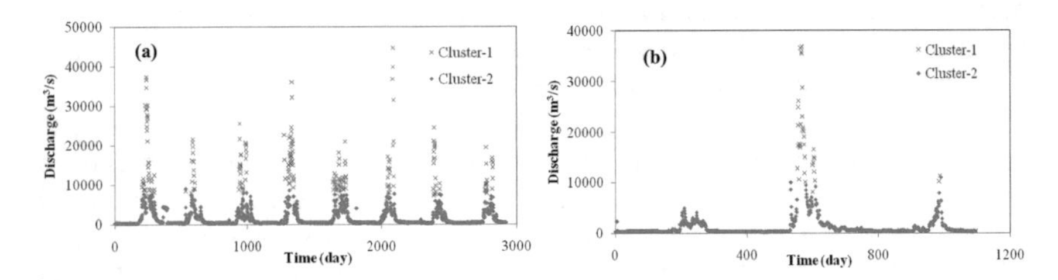

Fig. 24.11 Fuzzy c-means (FCM) clusters for **a** training and **b** testing datasets for two different discharge profiles at Mundali

Table 24.5 Features of KSOM and FCM clustering for training and testing datasets

Clustering method	High flow	Number of data points		Percentage of data points	
		Training	Testing	Training	Testing
KSOM	SM_1^*	249	39	86.46	13.54
FCM	F_1^*	211	37	85.08	14.92

SM_1^* = High flow of KSOM, F_1^* = High flow of FCM

24.5.2 *Streamflow Forecasting Using Post-Processed ANN Models*

The post-processed ANN models were developed based on the combination of two different ANN model forecasts. In this combination, the external inputs to the network for a particular time period are the discharge forecasts of the individual forecasting models for that time period. Each of the individual model forecasts is assigned to one neuron in the input layer. When the feedforward network is trained, the optimal weights are assigned to the individual inputs and the nonlinear relationship of the weighted sum in the hidden layer of the ANN produces the desired output in the output layer (Shamseldin et al. 2002; Kim et al. 2006).

In this study, the ensemble ANN (EANN) models were developed by combining the two ensemble members, that is, the NCANN and individual CANN models. The member models were combined to make two different ensemble ANNs, namely EANN-1 and EANN-2 models. For making the above ensembles, a Levenberg–Marquardt backpropagation neural network is trained, where the input layer receives all the external inputs (individual model results) to the network and the predicted output based on the nonlinear relationship of the two model results gives the combined forecast result.

24.5.2.1 Performance Evaluation of Developed Models for High Flow Forecasting

The high flow forecasting ANN models developed herein are the cluster-based ANN (CANN) models which are based on the separate input scenarios generated in the KSOM (KSOM-CANN) and FCM (FCM-CANN) methods. The CANN models are the feedforward type of backpropagation neural network models developed for cluster of high flows over the period 2003–2010 which were tested using the observed high flow data of 2000–2002. The structure of the models is same as the backpropagation type NCANN models i.e., 21–2–1 fitted to ungrouped dataset. Subsequently, each type of CANN models was replicated for high flow forecasting at one-day, two-day and three-day lead times.

Once the daily discharge time series was pre-classified and the observed high flows were clustered together in the two methods of clustering, the common high flows in both type of clusters were sorted out. First of all, the forecast results of those high flows were evaluated for the NCANN flood forecasting model and used as a basis for comparing the other two types of models viz., CANN and EANN. For a common testing dataset, the performance of the CANN models for high flow forecasting was compared with that of the NCANN as shown in Table 24.6. Figure 24.12 shows the high flow forecasting capability of the NCANN and CANN models at three different lead times of one-day, two-day and three-day. From Table 24.6 and Fig. 24.12, it can be seen that streamflow forecasting is better in the KSOM-CANN and FCM-CANN than that in the NCANN in terms of RMSE, MAE and E_{NS}. The KSOM-CANN produces better forecast results at two-day lead time; whereas, the FCM-CANN produces better results at three-day lead time forecast. Overall, both the clustering methods are able to identify the data patterns equally likely, which is resulting in better model performance. The performance evaluation measures of the ensemble ANN models (EANN) for high flow forecasting are presented in Table 24.7. The high flow forecast time series by the EANN-1 and EANN-2 models are shown in Figs. 24.13 and 24.14, respectively. From Table 24.7 and Figs. 24.13 and 24.14, it is clearly seen that there is significant increase in model efficiency and decrease in forecast error over the individual member models. The low efficient NCANN model significantly contributes to improve the high flow forecast results; resulting in better performance of the EANNs over CANN models. The uncertainty in the data between high and low flow regimes, which is present in the NCANN model, is handled by the ensemble of the two models.

The performances of the developed NCANN, CANNs and EANNs were also compared using threshold statistics. The high flow forecast errors for different models are shown in Table 24.8. The cumulative distribution of forecast errors for different models in high flow forecasting is represented in Fig. 24.15. The results illustrated

Table 24.6 Performance of different ANN models for high flow forecasting for a common testing dataset at one-day, two-day and three-day lead times

Approach	Criteria	Lead time (day)		
		1	2	3
NCANN	RMSE (m^3/s)	2756.20	6176.50	8611.50
	MAE (m^3/s)	2085.90	4260.00	6355.90
	E_{NS} (%)	41.69	31.98	−32.23
	r	0.95	0.70	0.43
KSOM_CANN	RMSE (m^3/s)	2584.30	5719.10	7578.80
	MAE (m^3/s)	1912.00	3808.60	5174.20
	E_{NS} (%)	88.09	41.69	−2.41
	r	0.95	0.66	0.45
FCM_CANN	RMSE (m^3/s)	2572.50	6075.70	6798.10
	MAE (m^3/s)	1933.70	4414.30	5500.70
	E_{NS} (%)	88.20	34.18	17.60
	R	0.94	0.63	0.53

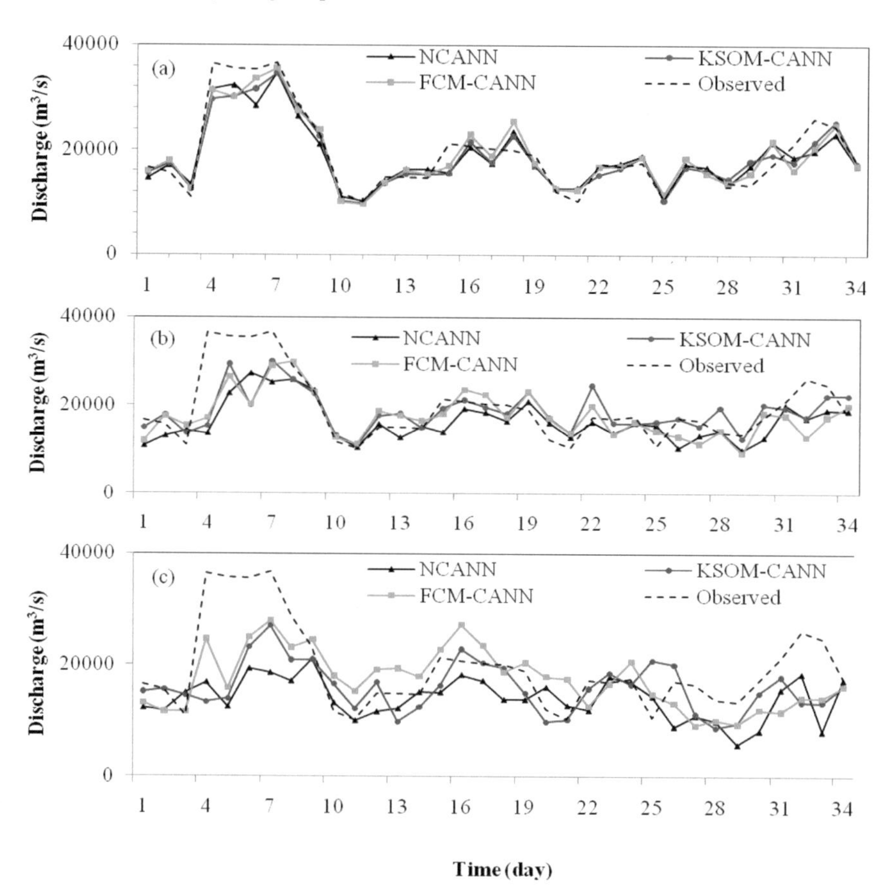

Fig. 24.12 High flow forecasting by the NCANN, KSOM-CANN and FCM-CANN at **a** one-day, **b** two-day and **c** three-day lead times

Table 24.7 Performance of ensemble ANN models for high flow forecasting at one-day, two-day and three-day lead times

Approach	Criteria	Lead time (day)		
		1	2	3
EANN-1	RMSE (m^3/s)	2003.40	4805.70	6152.10
	MAE (m^3/s)	1413.60	3272.70	4477.50
	E_{NS} (%)	92.84	58.82	32.51
	r	0.96	0.77	0.57
EANN-2	RMSE (m^3/s)	2054.10	4811.00	5537.00
	MAE (m^3/s)	1411.80	3454.60	4027.00
	E_{NS} (%)	92.48	58.73	45.33
	r	0.96	0.77	0.68

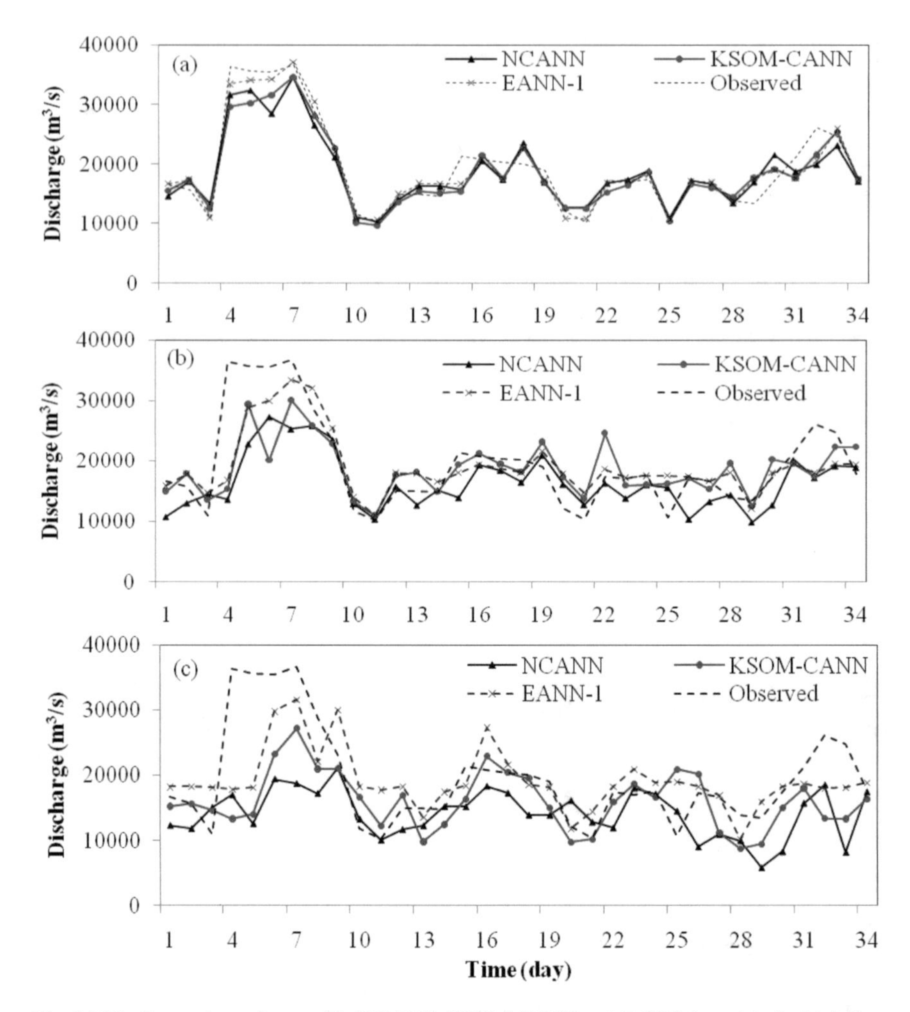

Fig. 24.13 Comparison of ensemble NCANN, KSOM-CANN and EANN-1 models for high flow forecasting at **a** one-day, **b** two-day and **c** three-day lead times

in Fig. 24.15 reveal that the high flow forecasts are improved in the order NCANN < CANN < EANN. Table 24.8 reveals that the KSOM-CANN model forecasts about 55.88% of high discharge within 20% absolute relative error which is an improvement of 17.64% over the 38.24% forecast by the NCANN model at three-day lead time. The EANN-1 model also forecasts about 91.18%, 73.53% and 55.88% of high discharge values within 20% absolute relative error at one-day, two-day and three-day lead times, respectively. The commonly used criterion by the Central Water Commission (CWC 1989), New Delhi, for evaluating the river flow forecasts in India allows ±20% variation between the predicted and observed discharges. The threshold statistics of the FCM-CANN show that it could not produce remarkable improvement, whereas its ensemble with the NCANN model i.e., EANN-2 reproduced the high flow better than

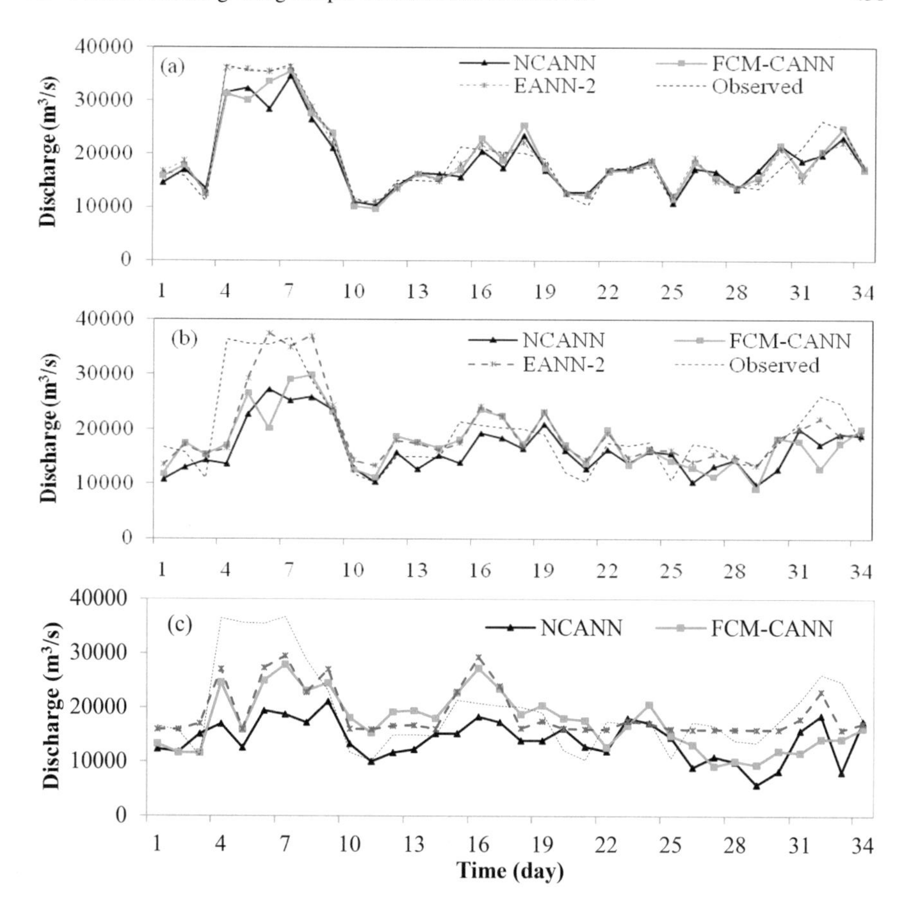

Fig. 24.14 Comparison of ensemble NCANN, FCM-CANN and EANN-2 models for high flow forecasting at **a** one-day, **b** two-day and **c** three-day lead times

the individual member models. Moreover, all the CANN models and their ensembles with NCANN produced more reliable high flow forecasts.

24.6 Summary and Conclusions

As a proof of the concept, artificial neural networks (ANNs)-based daily discharge forecasting were carried out at Mundali gauging station in the Mahanadi River basin using the historical daily discharge time series of 11 years collected from five gauging stations and the Hirakud Dam releases at the upstream of the Mahanadi River. The Levenberg–Marquardt backpropagation training was employed for developing various ANN forecasting models at one-, two- and three-day lead times for high flow forecasting at Mundali gauging station.

Table 24.8 Distribution of discharge forecast errors across different models and threshold levels at one-day, two-day and three-day lead times

Threshold statistics (%)	Lead time (days)		
	1	2	3
NCANN			
TS_5	35.29	14.71	11.76
TS_{10}	50.00	38.24	17.65
TS_{20}	85.29	64.71	38.24
TS_{25}	88.24	70.59	52.94
TS_{50}	100.00	91.18	82.35
KSOM-CANN			
TS_5	41.18	17.65	11.76
TS_{10}	64.71	38.24	26.47
TS_{20}	85.29	67.65	55.88
TS_{25}	91.18	73.53	61.76
TS_{50}	100.00	94.12	82.35
FCM-CANN			
TS_5	35.29	8.82	5.88
TS_{10}	67.65	26.47	20.59
TS_{20}	79.41	61.76	35.29
TS_{25}	85.29	76.47	55.88
TS_{50}	100.00	91.18	85.29
EANN-1			
TS_5	44.12	20.59	14.71
TS_{10}	76.47	41.18	35.29
TS_{20}	91.18	73.53	55.88
TS_{25}	97.06	79.41	61.76
TS_{50}	100.00	97.06	94.12
EANN-2			
TS_5	47.06	2.94	17.65
TS_{10}	76.47	35.29	41.18
TS_{20}	94.12	70.59	55.88
TS_{25}	94.12	85.29	67.65
TS_{50}	100.00	97.06	97.06

The results reveal that for daily discharge forecasting, the NCANN model could not forecast the observed hydrograph during the high flow periods at two- and three-day lead times. Therefore, it is summarized that a more complex ANN structure does not necessarily ensure an enhanced accuracy level in flow forecasting. Amongst the two clustering techniques (viz., KSOM and FCM) studied herein, though data clustering using the KSOM is a cumbersome job in comparison with FCM, it performs

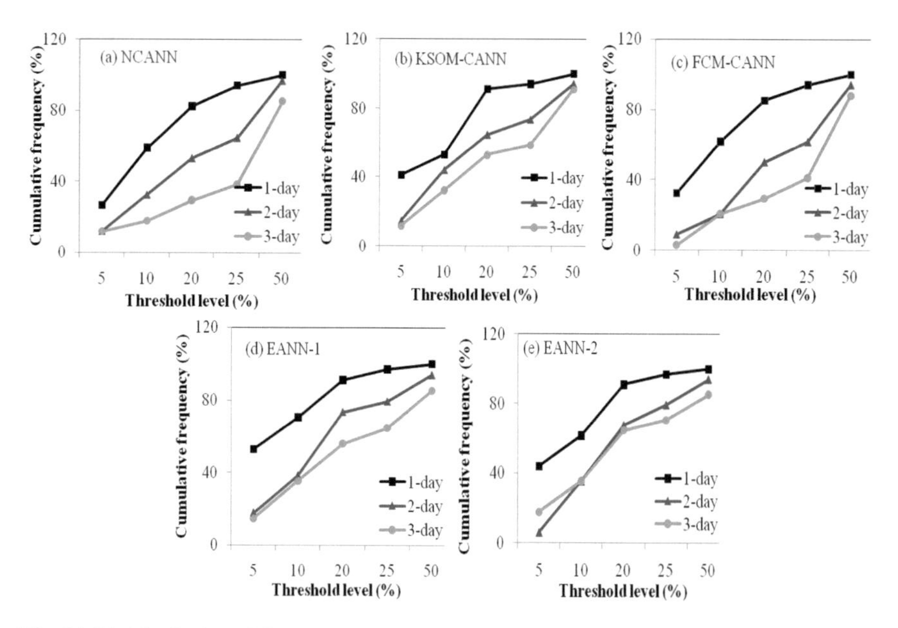

Fig. 24.15 Distribution of discharge forecast errors across different error thresholds of **a** NCANNs, **b** KSOM-CANN, **c** FCM-CANN, **d** EANN-1, and **e** EANN-2 at 1–3 day lead time forecasts

better in all the developed forecast models at one- to three-day lead time. The EANN-1, EANN-2 and KSOM-CANN forecast the high flows more closely, with more than 50% of forecasts having the absolute relative error within 20%. Therefore, these developed models can be effectively used in other world river basins for high flow forecasting.

Acknowledgements The streamflow data of the Mahanadi River used in this study were supplied by the Central Water Commission (CWC), Bhubaneswar, Odisha. The authors are thankful to the Hirakud Dam Circle (HDC), Burla, Odisha for providing the Hirakud Dam release data.

References

Abrahart RJ, See LM (2000) Comparing neural network and autoregressive moving average techniques for the provision of continuous river flow forecasts in two contrasting catchments. Hydrol Process 14:2157–2172

Abrahart RJ, See LM (2002) Multi-model data fusion for river flow forecasting: an evaluation of six alternative methods based on two contrasting catchments. Hydrol Earth Syst Sci 6(4):655–670

Abrahart RJ, See LM (2007) Neural network modeling of non-linear hydrological relationships. Hydrol Earth Syst Sci 11(5):1563–1579

Adeloye AJ, Rustum R, Kariyama ID (2011) Kohonen self-organizing map estimator for the reference crop evapotranspiration. Water Resour Res 47:W08523. https://doi.org/10.1029/2011WR010690

Agrahinejad S, Azmi M, Kholghi M (2011) Application of artificial neural network ensembles in probabilistic hydrological forecasting. J Hydrol 407:94–104
ASCE (1993) Criteria for evaluation of watershed models. J Irrig Drain Eng 119(3):429–442
Back B, Sere K, Hanna V (1998) Managing complexity in large database using self organizing map. Account Manag Inf Technol 8:191–210
Bezdek JC (1981) Pattern Recognition with Fuzzy Objective Function Algorithms. Plenum, NY
Campolo M, Andreussi P, Soldati A (1999) River flood forecasting with a neural network model. Water Resour Res 35(4):1191–1197
Central Water Commission (1989) Manual on flood forecasting. River Management Wing, New Delhi
Cios KJ, Pedrycz W, Swiniarski RW, Kurgan LA (2007) Data mining-a knowledge discovery approach. Springer, NY
Dawson CW, Wilby R (1998) An artificial neural network approach to rainfall runoff modeling. Hydrol Sci J 43(1):47–66
Dawson CW, Wilby R (1999) A comparison of artificial neural networks used for river flow river flow forecasting. Hydrol Earth Syst Sci 3(4):529–540
Garcia H, Gonzalez L (2004) Self-organizing map and clustering for wastewater treatment monitoring. Eng Appl Artif Intell 17(3):215–225
Hsu K, Gupta HV, Sorooshian S (1995) Artificial neural network modeling of the rainfall-runoff process. Water Resour Res 31(10):2517–2530
Hu TS, Lam KC, Ng ST (2001) River flow time series prediction with a range-dependent neural network. Hydrol Sci J 46(5):729–745
Huo Z, Feng S, Kang S, Huang G, Wanga F, Guo P (2012) Integrated neural networks for monthly river flow estimation in arid inland basin of Northwest China. J Hydrol 420–421:159–170
Imrie CE, Durucan S, Korre A (2000) River flow prediction using artificial neural networks: generalizing beyond the calibration range. J Hydrol 233:138–153
Jain A, Ormsbee LE (2002) Evaluation of short-term water demand forecast modeling techniques: conventional methods versus AI. J Am Water Works Assoc 94(7):64–72
Ju Q, Yu Z, Hao Z, Ouc G, Zhao J, Liu D (2009) Division-based rainfall-runoff simulations with BP neural networks and Xinanjiang model. Neurocomputing 72:2873–2883
Kentel E (2009) Estimation of river flow by artificial neural networks and identification of input vectors susceptible to producing unreliable flow estimates. J Hydrol 375:481–488
Kim G, Barros AP (2001) Quantitative flood forecasting using multisensor data and neural networks. J Hydrol 246:45–62
Kim Y, Jeong D, Ko IH (2006) Combining rainfall-runoff model outputs for improving ensemble streamflow prediction. J Hydrol Eng 11(6):578–588
Kohonen T, Oja E, Simula O, Visa A, Kangas J (1996) Engineering applications of the self-organizing map. Proc IEEE 84(10):1358–1384
Kumar DN, Raju KS, Sathish T (2004) River flow forecasting using recurrent neural networks. Water Resour Manage 18:143–161
Legates DR, Davis RE (1997) The continuing search for an anthropogenic climate change signal: limitations of correlation-based approaches. Geophys Res Lett 24:2319–2322
Legates DR, McCabe GJ (1999) Evaluating the use the goodness-of-fit measure in hydrologic and hydroclimatic model validation. Water Resour Res 35:233–241
McLeod AI, Noakes DJ, Hipel KW, Thompstone RM (1987) Combining hydrologic forecasts. Water Resour Plann Manag 113:29–41
Minns AW, Hall MJ (1996) Artificial neural networks as rainfall runoff models. Hydrol Sci J 41(3):399–417
Mwale FD, Adeloye AJ, Rustum R (2012). Infilling of missing rainfall and streamflow data: a self organizing map approach. In: BHS Eleventh national symposium, hydrology for a changing world, Dundeee. ISBN: 1903741181
Nanda T, Sahoo B, Beria H, Chatterjee C (2016) A wavelet-based non-linear autoregressive with exogenous inputs (WNARX) dynamic neural network model for real-time flood forecasting

using satellite-based rainfall products. J Hydrol 539:57–73. https://doi.org/10.1016/j.jhydrol.2016.05.014

Nanda T, Sahoo B, Chatterjee C (2017) Enhancing the applicability of Kohonen Self-Organizing Map (KSOM) estimator for gap-filling in hydrometeorological timeseries data. J Hydrol 549(6):133–147. https://doi.org/10.1016/j.jhydrol.2017.03.072

Nanda T, Sahoo B, Chatterjee C (2019) Enhancing real-time streamflow forecasts with wavelet-neural network based error-updating schemes and ECMWF meteorological predictions in variable infiltration capacity model. J Hydrol 575:890–910. https://doi.org/10.1016/j.jhydrol.2019.05.051

Nash JE, Sutcliffe JV (1970) River flow forcasting through conceptual models. I: J Hydrol 10:282–290

Nayak PC, Sudheer KP, Ramasastri KS (2005) Fuzzy computing based rainfall–runoff model for real time flood forecasting. Hydrol Process 19:955–968

Packard NH, Crutchfield JP, Farmer JD, Shaw RS (1980) Geometry from a time series. Phys Rev Lett 45(9):712–716

Pal NR, Pal S, Das J, Majumdar K (2003) SOFM-MLP: a hybrid neural network for atmospheric temperature prediction. IEEE Trans Geosci Remote Sens 41(12):2783–2791

Rajurkar MP, Kothyari UC, Chaube UC (2002) Artificial neural networks for daily rainfall-runoff modeling. Hydrologcal Sci 47(6):865–877

Rajurkar MP, Kothyari UC, Chaube UC (2004) Modeling of the daily rainfall runoff relationship with artificial neural network. J Hydrol 285:96–113

Regonda SK, Rajagopalan B, Clark ME (2006) A multimodel ensemble forecast framework: application to spring seasonal flows in the Gunnison River Basin. Water Resour Res 42:W09404. https://doi.org/10.1029/2005WR00453

Sahoo B (2005) The Xinanjiang model and its derivatives for modelling soil-moisture variability in the Land-Surface Schemes of the climate change models: an overview. In: Perumal M, Singhal DC, Arya DS, Srivastava DK, Goel NK, Mathur BS, Joshi H, Singh R, Nautiyal MD (eds) Proceedings of the international conference on hydrological perspectives for sustainable development (HYPESD-2005), vol I. IIT Roorkee, India, Allied Publishers Pvt. Ltd., New Delhi, ISBN: 81-7764-786-5, 23–25 Feb 2005, pp 518–532

See L, Abrahart RJ (2001) Multi-model data fusion for hydrological forecasting. Comput Geosci 27:987–994

See L, Corne S, Dougherty M, Openshaw S (1997) Some initial experiments with neural network models of flood forecasting on the River Ouse. In: Proceeding of second international conferance on geocomputation. Spatial Information Research Centre, University of Otago, Dunedin, New Zealand, pp 59–67

See L, Openshaw S (1999) Applying soft computing approaches to river level forecasting. Hydrol Sci J 44(5):763–778

Shamseldin AY (1997) Application of a neural network technique to rainfall runoff modeling. J Hydrol 199:272–294

Shamseldin AY, Elssidig NA, O'Connor KM (2002) Comparison of different forms of the multi-layer feed-forward neural network method used for river flow forecast combination. Hydrol Earth Syst Sci 6(4):671–684

Shamseldin AY, O'Connor KM (1999) A real-time combination Method for the outputs of different rainfall-runoff models. Hydrol Sci J 44:895–912

Shamseldin AY, O'Connor KM, Liang GC (1997) Methods for combining the outputs of different rainfall-runoff rodels. J Hydrol 197:203–229

Shu C, Burn DH (2004) Artificial neural network ensembles and their application in pooled flood frequency analysis. Water Resour Res 40:W09301. https://doi.org/10.1029/2003WR002816

Sudheer KP, Gosain AK, Ramasastri KS (2002) A data-driven algorithm for constructing artificial neural network rainfall-runoff models. Hydrol Process 16:1325–1330

Takens F (1981) Detecting strange attractors in turbulence. Lect Notes Math 898:366–381

Tiwari MK, Chatterjee C (2010a) Uncertainty assessment and ensemble flood forecasting using bootstrap based artificial neural networks (BANNs). J Hydrol 382(1–4):20–33

Tiwari MK, Chatterjee C (2010b) Development of an accurate and reliable hourly flood forecasting model using wavelet-bootstrap-ANN hybrid approach. J Hydrol 394(3–4):458–470
Tiwari MK, Chatterjee C (2011) A new wavelet-bootstrap-ANN hybrid model for daily discharge forecasting. J Hydroinf 13(3):500–519
Wang W, Van Gelder PHAJM, Vrijling JK, Ma J (2006) Forecasting daily streamflow using hybrid ANN models. J Hydrol 324:383–399
Zhang B, Govindaraju RS (2000) Prediction of watershed runoff using Bayesian concepts and modular neural networks. Water Resour Res 36(3):753–762
Zhang L, Scholz M, Mustafa A, Harrington R (2009) Application of the self-organizing map as a prediction tool for an integrated constructed wetland agro-ecosystem treating agricultural runoff. Bio-Resour Technol 100(2):559–565
Zhao RJ (1992) The Xinanjiang model applied in China. J Hydrol 135:371–381

Chapter 25
Application of Active Space-Borne Microwave Remote Sensing in Flood Hazard Management

C. M. Bhatt, Praveen K. Thakur, Dharmendra Singh, Prakash Chauhan, Ashish Pandey, and Arijit Roy

Abstract Globally, floods are attributed to be one of the leading natural hazards responsible for recurrent major economic losses, population affected, and mortality. The rapid assessment of flood hazard dynamics at regional scale during flood crisis is one of the few elements which is required by the agencies involved on ground for relief and rescue operations. Due to the weather-independent and day and night acquisition capability offered through microwave sensors, space-borne remote sensing for flood hazard management has undergone a paradigm change. Today, globally data from synthetic aperture radar (SAR) has emerged as invaluable source for monitoring flood hazard. From demonstrating the proof of concept in its initial launch campaigns, the SAR technology has matured to be competent enough to provide operational support for major flood disasters. In recent times, the continuously streaming of free SAR datasets from Sentinel-1 mission and together with emergence of advanced cloud-based computing and processing technologies like the Google Earth Engine (GEE), automated, and quasi-real-time flood mapping services have evolved. The future missions like the NISAR in conjunction with Sentinel-1 C-band data will help in providing more accurate and faster response during flood crisis and see application of SAR data grow multi-fold in coming years for flood hazard mitigation. This chapter attempts to provide a broad overview of the active microwave remote sensing for flood hazard studies. The first part of the chapter discusses about the interaction of the SAR signal for flooding in open, vegetated, and urbanized areas, followed by the role of the sensor parameters like the wavelength, polarization, and incidence angle on the backscattering of SAR signal. The latter half of the chapter discusses about the flood mapping techniques, SAR satellite mission contributing to flood hazard mapping, various applications of SAR derived flood hazard information, the Indian nationwide near-real-time (NRT) flood hazard mapping under ISRO DMS program, and the emergence of Web-based cloud computing techniques and open-source data policies revolutionizing the flood hazard mitigation activities.

C. M. Bhatt (✉) · P. K. Thakur · P. Chauhan · A. Roy
Indian Institute of Remote Sensing (IIRS), Dehradun, India
e-mail: cmbhatt@iirs.gov.in

D. Singh · A. Pandey
Indian Institute of Technology (IIT), Roorkee, India

A. Pandey et al. (eds.), *Geospatial Technologies for Land and Water Resources Management*, Water Science and Technology Library 103,
https://doi.org/10.1007/978-3-030-90479-1_25

25.1 Introduction

As per the EM-DAT database, the year 2020 witnessed an increase in terms of number of total recorded events and economic damages (US $151.6 billion) as compared to the previous two decades (2000–2019) due to natural hazards. The analysis of natural hazards category-wise quite clearly reveals that the hydrological hazards specially floods are one of the leading natural disasters causing major losses in terms of mortality, affected population, and economic damages. A further closer look at the EM-DAT database for the year 2020 reveals that the number of flood events and the deaths associated with the floods far exceed the annual average. An increase of 23% in case of flood events against the annual average of 102 events and 18% increase in flood mortality as compared to the annual average of 5,233 deaths was observed for the year 2020 (https://emdat.be/sites/default/files/adsr_2020.pdf).

There has been a tremendous improvement in the operational flood hazard activities since last few decades, especially after the emergence of satellite remote sensing technology. Remote sensing technology because of its synoptic and multi-temporal coverage of large, remote, and inaccessible areas aids in providing rapid information on the flood extent, duration, progression, and recession of flood waters. Though it is known floods cannot be completely stopped, but space technology can aid in providing accurate and timely spatial occurrence of flood hazard and allow planning of measures to minimize their impact through proper management (Smith 1997).

Conventional flood monitoring systems due to their limited spatial extent, damage to gauging instruments during major floods, involvement of high cost, and time has been replaced by advanced satellite remote sensing technology for faster emergency response and flood early warning activities. The other major limitation from conventional in-situ gauging instruments is that they measure only the water level, but not the spatial spread of the flood, which is needed for emergency response. Trans-boundary nature of many major rivers and problem of data sharing together with above-listed factors remote sensing techniques have gained popularity in flood hazard mapping and monitoring operations globally. Since the launch of the Earth Resources Technology Satellite, later renamed to Landsat in 1970's, satellite images have continuously aided the flood hazard management activities (McGinnis and Rango 1975). Satellite images have been able to provide cost-effective information for flood mapping and monitoring activities, especially required for the developing countries with poorly gauged basins (Smith 1997; Rao et al. 1998; Sanyal and Lu 2004). The space-based flood hazard mapping helps the community to understand the flood risk and make more informed decisions to manage risk (Skakun et al. 2014). However, the presences of cloud cover limited the effective utilization of data acquired from optical sensors operating in the visible, thermal, and microwave domain due to their shorter wavelengths.

Due to the weather-independent and day and night acquisition capability offered through microwave sensors, space-borne remote sensing for flood hazard management has undergone a paradigm change. Today, globally data from synthetic aperture radar (SAR) has emerged as invaluable source for monitoring flood hazard. From demonstrating the proof of concept in its initial launch campaigns, the SAR technology has matured to be competent enough to provide operational support for major flood disasters. In recent times, the continuously streaming of free SAR datasets and together with emergence of advanced cloud-based computing and processing technology many services providing automatic and quasi-real-time flood mapping have evolved (Notti et al. 2018; DeVries et al. 2020).

This chapter attempts to provide a broad overview of the backscattering of SAR signal for flooding observed in open, vegetated, and urbanized areas. The influence of sensor parameters like the wavelength, polarization, and incidence angle for detection of flooded areas was discussed. The later part discussed about the mapping techniques, SAR mission contributing to flood hazard mapping, the Indian nationwide NRT real-time flood hazard mapping program, and the emergence of Web-based cloud computing techniques and open source data policies revolutionizing the flood hazard mitigation activities.

25.2 Active Microwave Remote Sensing of Flood Inundation

Both optical sensors and passive and active microwave sensors support inundation mapping surface water through time. Optical sensors operate in the visible portion of the wavelength spectrum (~400–700 nm), whereas the microwave sensors operate at wavelengths of approximately 1 mm to 1 m. A major disadvantage in using sensors operating in the visible and/or infrared portion of the electromagnetic spectrum is their inability to penetrate the clouds. Microwave sensors because of their longer wavelengths have the ability to penetrate through cloud cover, including the rain, atmospheric haze, dust, and smoke, and therefore are not impacted by the atmospheric scattering which generally affects the shorter optical wavelengths. Figure 25.1 shows the comparison between acquisitions by optical (IRS AWiFS)

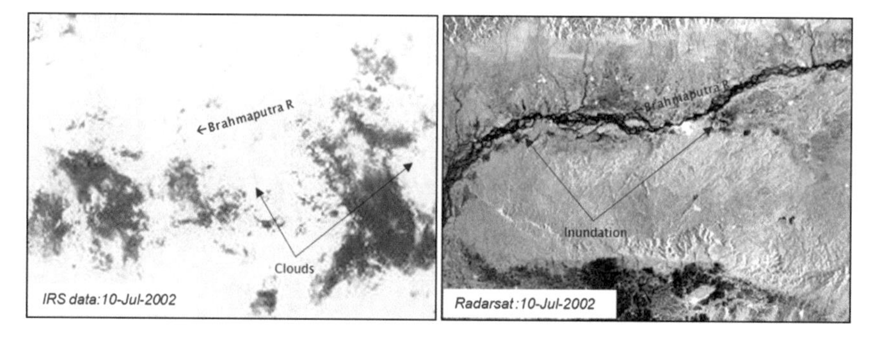

Fig. 25.1 Comparison between acquisitions by optical (IRS AWiFS) and microwave (RADARSAT-1) sensors over Brahmaputra, Assam, on same date (July 10, 2002). *Source* NRSC/ISRO-2002

and microwave (RADARSAT-1) sensors over Brahmaputra, Assam, on same date (July 10, 2002). The area being under intense clouds could not be viewed by optical sensors, whereas on same date the SAR sensors clearly depict the flood inundation along areas adjoining to River Brahmaputra. The capability of weather-independent image acquisition by microwave sensors has therefore seen corresponding increase in the demand to cater to the needs of supporting NRT flood mapping and monitoring activities.

25.2.1 Interaction Mechanism of SAR Signal

Active microwave sensor transmits a radar signal toward the earth's surface and detects the backscattered portion of the signal along with the time it takes for the pulse to return. The backscattered energy provides an idea of the target characteristics, whereas the time measurement gives the location of the target. Each pixel of a radar image represents the amount of backscattered energy reflected from that area on the ground. The pixel intensity values are often converted to a physical quantity called the backscattering coefficient ($\sigma°$) measured in decibel (dB) units. The amount and the characteristics of the backscattered energy returned to the sensor are a function of the physical and electrical properties of the target, along with the wavelength (λ), polarization, and incidence angle (θ) of the radar. Figure 25.2 shows the different backscattering mechanism of the incident radar signal in flood-inundated areas. Darker tone in the image represents low amount of backscatter, whereas the brighter tone represents higher backscatter. Bright features observed in a radar image indicate that a higher amount of the radar energy was reflected back to the sensor, whereas the dark tone features suggest that less energy was reflected back to the sensor. The amount of backscatter reflected from a surface is a function

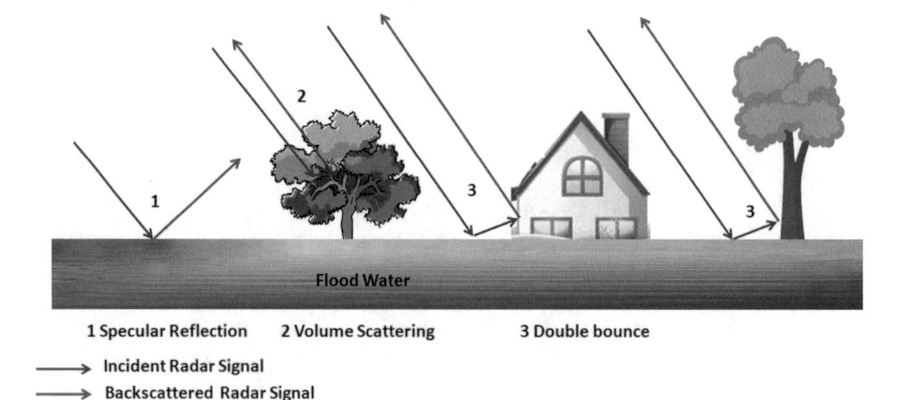

Fig. 25.2 Different backscattering mechanism of the incident radar signal. *Source* Author

of sensor parameters (wavelength, polarization, incidence angle of the electromagnetic waves emitted) as well the surface parameters (roughness, geometric shape, dielectric properties of the target, vegetation type, and phenology of plants).

25.2.2 Backscattering of SAR Signal in Open Flooded Areas

The detection of open-flooded areas where the incoming radar signal is not obscured by vegetation or buildings and the water surface is calm is relatively easier as compared to the identification of flooding in vegetated and urbanized areas. The smooth surface of water bodies behaves like a mirror for the incoming radar signal, reflecting most of the radar energy away from the sensor, described as specular reflection. Due to entire energy being reflected away, the radar antenna receives no backscattered energy, and therefore, water appears in dark tone in the SAR image. The adjacent land (non-flooded) appears in contrasting bright tone because of the diffused reflection (incident radar signal reflected at many angles) produced by the rough soil and vegetation, resulting in a high backscatter (Baldassarre et al. 2001, Ulaby and Dobson 1989) (Fig. 25.3). Flooded area can be reliably delineated if the water is calm having a smooth water surface with low backscattering values. However, under even light wind conditions, the ripples on water surface during the SAR data acquisition roughen the water surface, and the same water may exhibit high backscatter appearing brighter (Geudtner et al. 1996; Yang et al. 1999).The mapping of flooded areas may be overestimated in presence of sandy areas especially in arid regions. The sand surfaces are characterized by low backscatter response and act as water look-alike areas (Martinis et al. 2018). Therefore, NRT flood mapping operations based on microwave sensors requires careful exclusion of look-alikes to have reliable estimation of flooded areas.

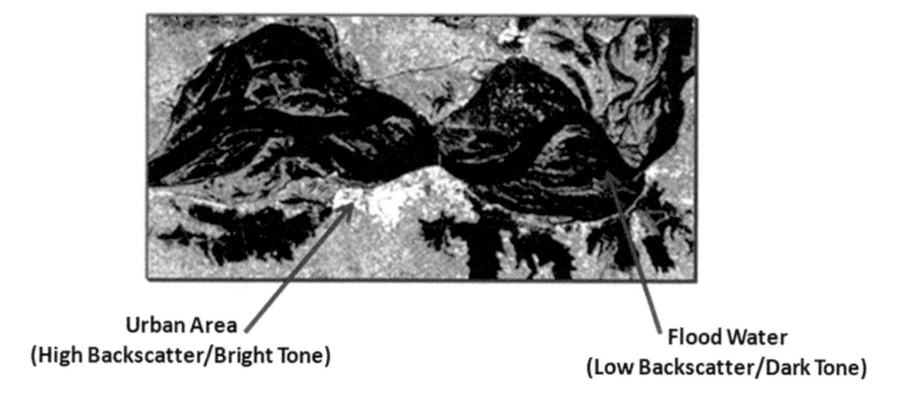

Fig. 25.3 Backscattering of radar signal for open flood waters. *Source* Sentinel-1

25.2.3 Backscattering of SAR Signal in Flooded Vegetated Areas

The main advantage of SAR technology, in case of flooded vegetation, is its deeper penetration through vegetation canopy (Tsyganskaya et al. 2018). The incoming radar signal is subjected to multiple scattering from canopy top, trunk, and stem to the floor of the inundated surface. The scattering of the signal is strongly dependent on the vegetation structure, density, and depth of submergence, along with the sensor-related factors such as wavelength, incidence angle, and polarization. It is therefore a full understanding of the scattering mechanism for interpreting and identifying flooded vegetation is needed. The backscattering from the surface, canopy (volume scattering), and multi-path interactions of canopy-ground and trunk-ground (double bounce) interactions all together constitute the return signal in case of flooded forests (Ormsby et al. 1985; Richards et al. 1987a; Wang et al. 1995). However, in several studies, the double-bounce scattering is pointed out as the prominent scattering mechanism between the plain surface of water and vertical vegetation components, viz the trunks and stems (Moser et al. 2016; Pulvirenti et al. 2013; Adeli et al. 2020). Due to the composite scattering happening in flooded vegetated areas, the backscatter return from these areas is higher (brighter tone) as compared to non-flooded conditions.

25.2.4 Backscattering of SAR Signal in Flooded Urban Areas

In the case of flooding in urban areas, the identification of flooded areas is not so straightforward. The SAR signal return in presence of building of varying heights, vegetated areas, and different road topologies presents a challenge in analyzing the complex backscatter return in accurate interpretation of flooded areas (Mason et al. 2021; Li et al. 2019; Pulverenti et al. 2015; Rykhus and Lu 2005). Many studies have demonstrated enhanced backscatter returns in flooded urban areas, due to the double-bounce scattering of the incoming SAR signal taking place between the flooded streets and adjacent buildings which can help in identifying the flooded areas (Li et al. 2019; Mason et al. 2021). The SAR signal traveling from sensor strikes over the water surface presence in front of building, which causes signal to be specularly reflected, which then strikes with the wall of adjacent building and is received back by the sensor (or vice versa). The double-bounce scattering is a function of the height of the building and the azimuth direction (i.e., aspect angle ϕ) including the floodwater level (Li et al. 2019). Due to the changes in level of standing water over a time period between observations, produces a decorrelation, which also helps to detect flooded areas in urban areas. The flood area estimation in urban areas is also hindered because of the side-looking geometry of the SAR, which causes the ground surface to not be detected by the SAR sensor due to the radar layover and shadow produced by the tall buildings and trees. Soergel et al. (2003) have observed that about 39% of the flooded urban area may not be visible due to the side viewing geometry of SAR instrument in urban areas. The area under shadow appears dark, having backscatter characteristics similar to water and therefore likely to be misclassified as flooded, even if the area under shadow is non-flooded. On the contrary, the layover area even being flooded may have high backscatter response (bright tone), resulting in underestimation of flooded area. In urban areas, the low backscatter return from smooth surfaces like the parking lots, tarmac roads, and airstrips has to be excluded from analysis to avoid overestimation of flooded areas (Shen et al. 2019).

25.3 Role of SAR Sensor Parameters in Detecting Flooded Areas

The amount of backscatter reflected from a surface is a function of sensor parameters, viz wavelength, polarization, and incidence angle of the electromagnetic waves emitted. Hence, careful understanding and selection of the sensor parameters are required for taking maximum advantage of SAR sensor system.

25.3.1 *Wavelength (λ)*

Radar signals may be transmitted at a range of wavelengths (Table). The radar return from flooded surface mainly depends on the wavelength used. The terrain features have a unique interaction in a particular wavelength, and hence, the return is also unique at each wavelength used. For an object to be detected, the size of the object should be of the same spatial magnitude as the incoming radar wavelength and larger. The longer wavelength radar signals therefore have the potential to penetrate through the forest canopy (Hess et al. 1990). It is well established from the understanding gained through several studies carried out globally over the years that the L-band (24 cm wavelength) and SAR missions are more well suited to find inundation beneath forested canopy (Adeli et al. 2020; Plank et al. 2017; Betbeder et al. 2015; Henderson and Lewis 2008; Hess et al. 1990; Richards et al. 1987a). In case of the shorter wavelengths (viz C-band and X-band), radar signal penetration is obstructed due to the presence of dense canopy, and therefore, the backscattering in such cases is prominently characterized by the volume scattering. The shorter wavelengths have been found to be useful in case of sparse forest canopy, at the initial growth stages and leaf-off conditions (Clement et al. 2018; Moser et al. 2016; Cohen et al. 2016; Zhang et al. 2016, Voormansik et al. 2014; Brisco et al. 2013b). High-resolution Terra SAR-X images have found wide application in studying the SAR signal response over flooded urban regions (Mason et al. 2009; Giustarini et al. 2012; Pradhan et al. 2016). However, the shorter wavelengths have also found to be impacted by the atmospheric effects. The degree of signal attenuation and scattering of microwave signal by rain drops is inversely proportional to the radar wavelength. C-band (~5.6 cm) has very small attenuation observed for low to moderate rain rates on the contrary to X-band (~3.1 cm) which shows more attenuation for same rain rate (Alpers et al. 2016). Many studies support the application of multi-frequency SAR especially while mapping of flooded vegetation (Bourgeau-Chavez et al. 2001).

25.3.2 Polarization

The radar polarization is an important parameter which affects the strength of the backscattered signal, and hence, the selection of the right kind of polarization for identifying the object is very critical. HH polarized signal is less scattered by open water as compared to HV or VV and hence is found to be more suitable in identification of open flooded areas (Pierdicca et al. 2013; Manjusree et al. 2012, Gstaiger et al. 2012; Henry et al. 2006). Henry et al. (2006) while investigating the multi-polarized Envisat Advanced Synthetic Aperture Radar (ASAR) data (HH and HV) and ERS-2 data (VV) for the 2002 Elbe river flooding observed that the HH polarization because of its wider histogram and higher radiometric dynamics help in better discrimination of flooded areas. VV polarized data is considered to be sensitive to surface roughness conditions. Studies have also pointed out that for open smooth water surfaces, co-polarized images give better distinction as compared to the cross polarization images (Martinis and Rieke 2015). In case of flooded vegetation, HH polarization is able to penetrate the vegetation canopy better than VV polarization because of the orientation of the SAR signals, and also double-bounce scattering (trunk-ground) is higher for HH compared to VV (Pierdicca et al. 2013; Wang et al. 1995). Studies have suggested the co-polarized data for identification of flooded vegetation because of its higher sensitivity to the double-bounce scattering (Hess et al. 1990). The application of multi-polarized SAR datasets for the identification of flooded areas is advocated to have more detailed and accurate information of the inundated status than single-polarized datasets (Henry et al. 2006; Manjusree et al. 2012). The dual-polarized SAR data allows going for polarization ratios, whereas the quad polarized data allows polarimetric decomposition, i.e., separating SAR signals into different scattering mechanisms (volume, single, specular, and double-bounce), and provides more enhanced information about the flooded vegetation (Souza-Filho et al. 2011).

25.3.3 Incidence Angle (θ)

Incidence angle is another important parameter which influences the backscattering component and therefore the image classification. The incidence angles are termed as steep (smaller incidence angles) and shallow (larger incidence angles). For mapping of flood in open areas, shallow incidence angle enhances the distinction between the flood water surface and the adjacent ground surface because of stronger specular reflection from smooth water surface (Henry et al. 2006). Solbø and Solheim (2005) using RADARSAT SAR data have observed that the contrast between the land and water increases with the shallow (45°) incidence angle as compared to the steep (23°) incidence angle. In case of flooded vegetated areas, steep incidence angles are preferred due to higher penetration capabilities (Wang et al. 1995; Hess et al. 1990; Richards et al. 1987a). The radar signals for steep incidence angles have a shorter travel path, increased transmissivity, and enhanced energy interaction between the flooded surface and tree trunks, whereas the shallow incidence angles have higher volume scattering. Moderate incidence angle have been recommended in case of sparse vegetation cover.

25.4 Mapping of Flooded Areas

The flood hazard information extraction using the SAR datasets involves image preprocessing, classification of flood, followed by post-processing refinement of flood hazard layer, and then finally integration with spatial database to generate inundation statistics and mitigation measures. Figure 25.4 shows the flowchart procedure followed for processing of Sentinel-1 (pre- and post-flood), Level-1 Ground Range Detected (GRD) data using Sentinel Application Platform (SNAP) software. The preprocessing steps basically consist of applying of orbit file, thermal and border noise removal, radiometric calibration, speckle filtering, followed by Range–Doppler terrain correction. The orbit state vectors are updated using SNAP, by providing an accurate satellite position and velocity information for each SAR scene. Thermal noise removal minimizes the noise effects in the inter-sub-swath texture, in particular, normalizing the backscatter signal within the scene. The border noise removal includes the removal of low intensity noise and invalid data on scene edges. Radiometric calibration converts the radar reflectivity (digital number) to physical units (radar backscatter). The radiometrically calibrated SAR images are then filtered to minimize the noise caused by the interferences of coherent echoes from individual scatterers within one pixel. Range–Doppler ortho-rectification correction is then implemented to account for geometric distortions caused by topography using a digital elevation model (DEM). Finally, the unit-less backscatter coefficient product is converted to dB using a logarithmic transformation.

After the initial processing of the data, classification of SAR data is done to delineate the flooded areas. Various techniques have been used for identifying the flooded areas. SAR-intensity-based histogram thresholding technique is one of the most widely used method for identification of flooded areas mapping (Brivio et al. 2002; Schumann et al. 2009). Change detection techniques like the rationing and differencing are also used for mapping of flooded regions (Martinez and Toan 2007; Herrera-Cruz et al. 2009; Schlaffer et al. 2016) along with the amplitude as well as the coherence change detection (Nico et al. 2000; Buck and Monni 2000). Supervised and classification techniques are detecting the flooded areas based on the statistical properties of the training datasets (Gan et al. 2012). Region growing algorithms

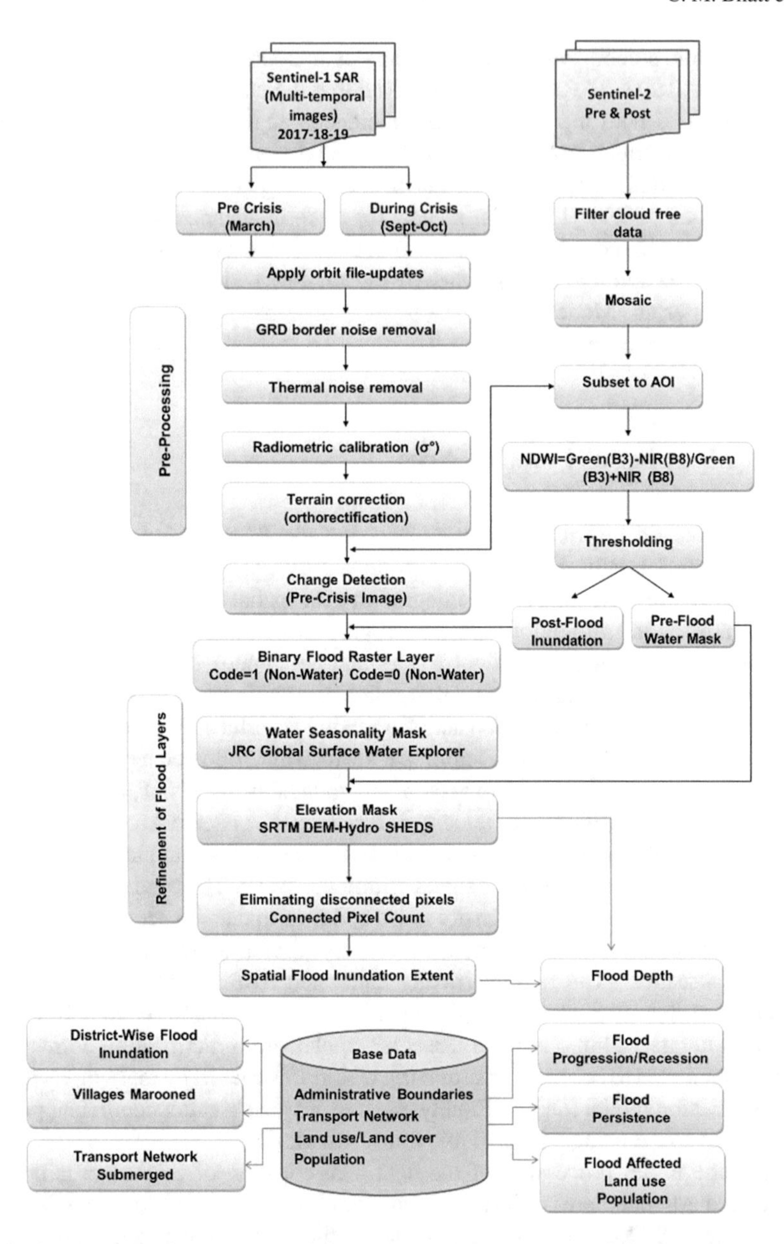

Fig. 25.4 Flowchart of steps executed for preprocessing, extraction of flood-inundated areas, refinement of flood hazard layer, validation, and deriving value-added products

using a seed pixel or group of seed pixels which expand according to their statistical properties until the defined conditions are attained have also been used (Mason et al. 2009). For the high-resolution SAR images, image segmentation techniques which group together the homogeneous pixels characterized by their spectral values, shape, size, texture, their neighborhood relations, and other properties have been employed for identification of flooded urban areas (Hess 2003; Pulvirenti et al. 2011a). The coherence data in conjunction with intensity is also utilized for flood detection in urban areas (Li et al. 2019; Pulvirenti et al. 2016).

The flood hazard layer delineated through above-mentioned techniques are then further refined during the post-processing step by applying the pre-flood water mask. This mask incorporates the pre-flood existing permanent water features like the river channel, lakes, waterlogged, and marshy areas, which are extracted from the satellite data acquired before the flood season (Fig. 25.5). The elevation mask is also applied to remove any misclassification of the shadow areas. The final flood hazard layer is then further refined by removing the stray and isolated flood pixels using clump and sieve (GIS-based operations) to improve the homogeneity of the extracted flooded areas. Finally, through image binarization, the pixels are classified into flooded and non-flooded classes. The flood layer is then converted to vector form and integrated with different administrative layers to generate flood inundation map derive the inundation statistics.

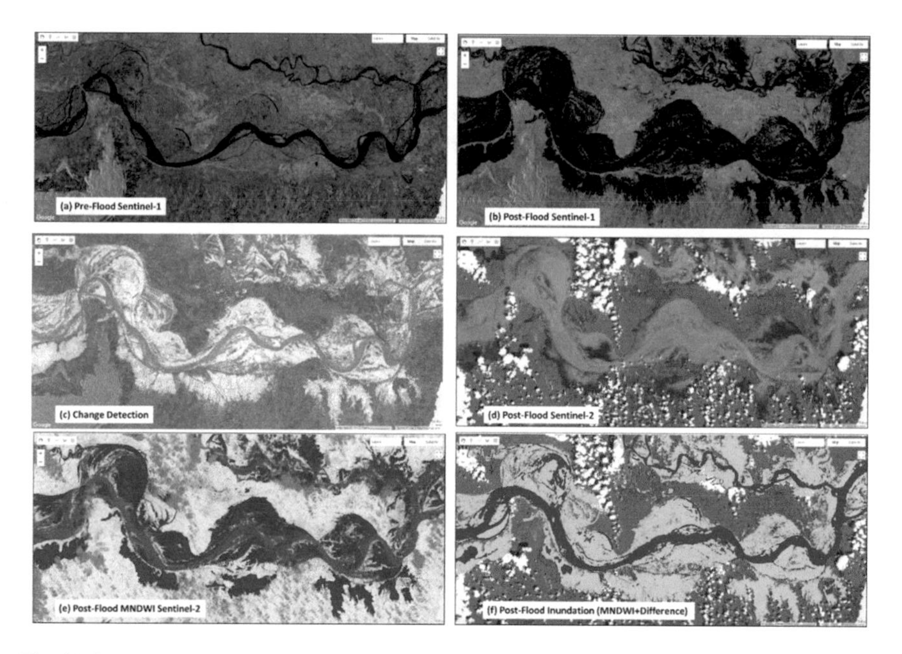

Fig. 25.5 Pre-flood Sentinel-1 SAR image (**a**) post-flood Sentinel-1 SAR image (**b**) change detection image (**c**) Sentinel-2 post-flood image (**d**) flood hazard layer classification using MNDWI (**e**) and SAR and MNDWI derived flood hazard layer superimposed over Sentinel-2 images for validation. *Source* Bhatt et al. (2021)

25.5 Application of SAR-Derived Flood Hazard Information

The flood hazard layer delineated using above-mentioned techniques becomes an important input in immediate flood response and long-term flood mitigation measures. During the flood crisis response, SAR data provides invaluable response through rapid mode data analysis to ground authorities dealing with relief and rescue operations (Bhatt et al. 2010, 2013, 2016).

Bhatt et al. (2010) have demonstrated the successful operational use of temporal SAR datasets for flood mapping, monitoring, and damage assessment during the August 18, 2008, Kosi floods in Bihar. During this event, about 200 flood maps (between August and October, 2008) derived from 30 satellite images were provided in NRT to the state agencies. The maps were used by state agencies successfully to identify the villages marooned and transport network submerged, crop area submerged, flood persistence, flood progression/recession, dropping of food packets, and evacuation of people during the ongoing flood crisis to name a few.

Apart from the immediate response activities, the flood hazard information derived from SAR images input together with other space-based derived geospatial inputs (topography, precipitation, and slope, aspect of the terrain, drainage network, and land cover) help in taking up long-term non-structural flood mitigation measures which finds application in implementing long-term non-structural flood mitigation measures. The main applications where these inputs are widely being used are for flood hazard zonation and flood risk assessment (Manjusree et al. 2015; Skakun et al. 2014). Manjusree et al. (2015) have analyzed about more than 120 SAR images acquired for different flood magnitudes during between 1998 and 2010 (13 years period) in conjunction with the hydrological data, for assessing the frequency of inundation, severity of flood hazard and cropped land under flood hazard. A district-wise flood hazard zonation map atlas is prepared, which indicates the flood hazard severity in terms of the frequency of inundation. The area of flood frequency under five categories very high (flooded 11–13 times), high (8–10 times), moderate (5–7 times), low (3–4), and very low (<2times). Such maps are an important source for policy makers for undertaking long-term sustainable flood mitigation activities.

Space-based flood inundation information is also used for calibration and validation of hydraulic models, in absence of ground data from well distributed sites. This in turn helps to improve hydrologic prediction and flood management strategies especially for ungauged catchments (e.g., Horritt 2000). Dasgupta et al. (2020) have investigated the reliability of SAR-derived flood inundation maps for assessing the hydrodynamic model-derived results for a stretch of Dhanua River, part of Mahanadi River Basin in Odisha. They have used RADARSAT-2 SAR images of ScanSAR narrow mode acquired on the September 4, 2003, and RISAT-1 Medium Resolution ScanSAR (MRS) of August 9, 2014 in HV mode for mapping of flooded areas. The SAR images were processed using visual interpretation, histogram thresholding, and texture-based classification techniques to derive the probabilistic flood inundation spatial extent. Using the MIKE-FLOOD, 1D2D fully hydrodynamic model, the channel and floodplain roughness parameters were calibrated for the 2003 flood event, through binary pattern matching objective functions. The best-fit parameter identified through model calibration was then validated by simulating the flood event which occurred during 2014 derived from an independent SAR image analysis. Through their study, they have found good agreement between the flood inundation extents derived through modeling and SAR data analysis. The modeled and observed extents showed an R^2 value of 0.938 and a RMSE (root mean squared error) of 0.278 pixels for the validation indicating the potential of SAR data to support flood model calibration.

Thakur et al. (2012) have demonstrated the role of multi-temporal RADARSAT-1 SAR images for 2003 flood event and geospatial tools in identifying the structural elements at risk for part of Kendrapara, Odisha. SAR images were used to derive flood maps, high-resolution Cartosat-1 images for mapping building footprints, and flood depth which was derived from Cartosat-1 DEM. The space input along with field-based data collected were integrated using geospatial tools and risk assessment. The study concluded that the damage to various structures (houses) depended on the houses type of construction material used for the house, depth of floodwater depth, and the duration of flooding. Figure 25.6 shows the flowchart for the approach adopted for the study.

Surface soil moisture estimation is another important parameter where the active microwave remote sensing is contributing (Tripathi and Tewari 2020; Singh et al.

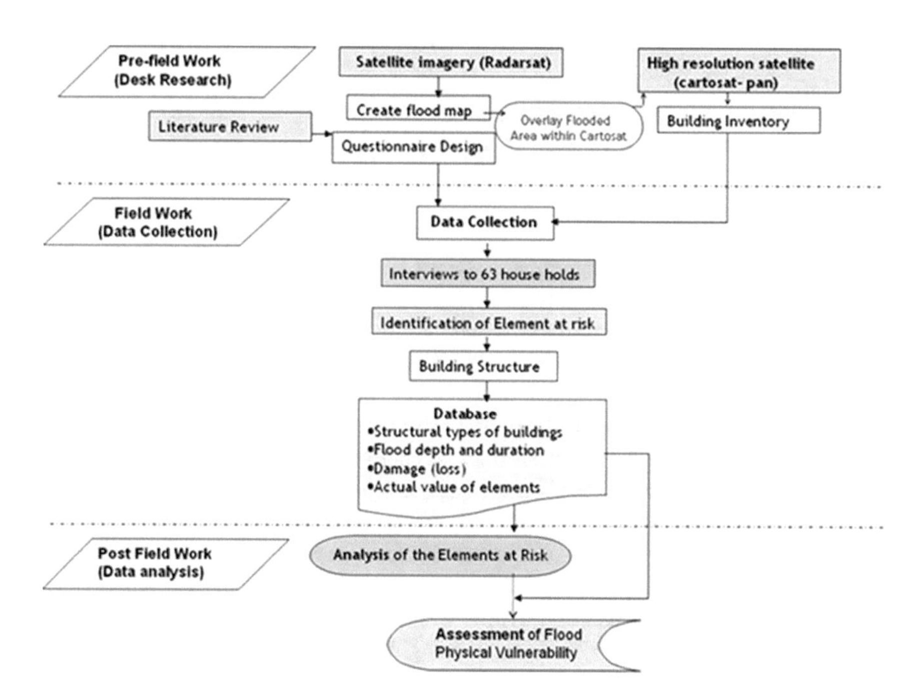

Fig. 25.6 Flowchart for flood-induced structural vulnerability assessment (Thakur et al. 2012)

2020; Prakash et al. 2011). Soil moisture estimation is very important as it directly governs the flood hydrology and therefore becomes an important input for many flood hazard studies. Prakash et al. (2011) have developed an algorithm to retrieve the soil moisture over vegetated areas by fusion of Advanced Land Observation Satellite—Phased Array type L-band Synthetic Aperture Radar (ALOS-PALSAR), a SAR data, and Moderate-resolution Imaging Spectroradiometer (MODIS), an optical data. Utilizing the polarimetric capability of SAR data, the land cover was classified, through MODIS normalized difference vegetation index (NDVI) was derived. The backscattering coefficient from PALSAR data was normalized to develop an empirical relationship to provide the backscattering coefficient over bare soil in HH- and VV-polarizations. Dubois model was then solved with co-polarization ratio approach to get the volumetric soil moisture estimates. The flowchart of the methodology for the present study is shown in Fig. 25.7. The results from the study were validated with the ground data and showed a good match with the observed data. The approach developed had the potential to be used for soil surface moisture estimation with minimal a priori information and applicable for various other sensors like RADARSAT and RISAT.

The space-based SAR images have also been used for a wide number of applications like the flood shelter suitability mapping (Uddin et al. 2021), surveillance of water and vector borne diseases (Kalluri et al. 2007; Kaya et al. 2004), flood risk insurance (Bach et al. 2005; Amarnath et al. 2019), crop loss assessment (Rahman and Di 2020; Shrestha et al. 2018), flood damage survey and assessment (Pulvirenti et al. 2014), and developing of flood inundation libraries (Bhatt et al. 2017a, b).

25.6 SAR Remote Sensing Satellites Supporting Flood Mapping

Though there had been experimental missions like the Seasat (L-band) and spaceborne imaging radar (SIR)-C/X-band SAR, but the major thrust and development toward flood hazard mapping activities were witnessed from the launch of the European Remote Sensing (ERS) series of satellites ERS-1, ERS-2, than Envisat, and later with launch of the Canadian satellite RADARSAT-1, all operating in C-band. The experimental missions in initial period were launched as technology demonstration. In the past, most of the missions operated between 1990 and 2005 have largely focused on the development of C-band sensors. These satellites helped to establish the relevance of SAR sensors for flood hazard mapping and monitoring. Table 25.1 gives the details on the major SAR satellite missions in past, operation, and planned for future which has supported flood hazard activities.

With improvements in technology, the next generations (2005 onward) of SAR satellites were equipped with multi-mode imaging and full polarimetric data acquisition capabilities. The fully polarimetric (transmit and receive energy in all four

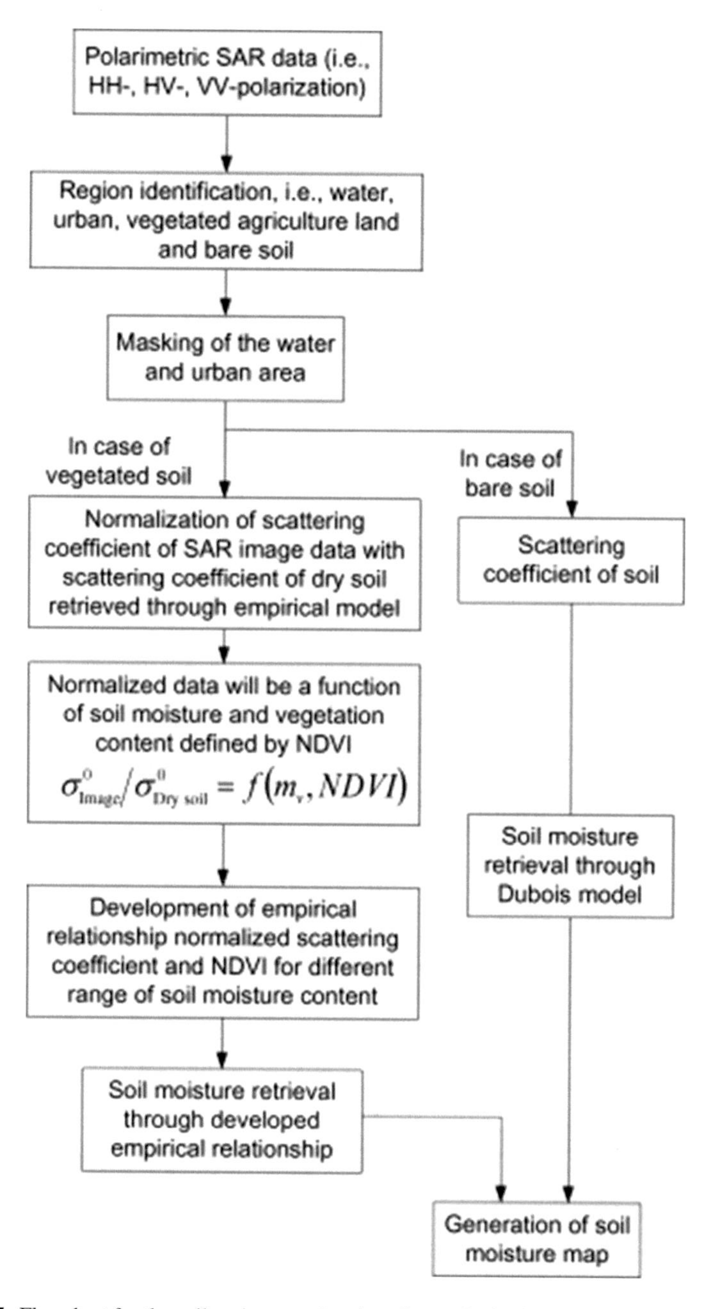

Fig. 25.7 Flowchart for the soil moisture estimation. *Source* Prakash et al. (2011)

Table 25.1 SAR satellites (non-operational, operational, and planned) for flood hazard mapping

Satellite	Band	Polarization	Incidence angle	Repeat Cycle (days)	Swath (km)	Agency	Launch date
Not in operation							
ERS-1	C	VV	21–26°	35	100	ESA	1991
JERS-1	L	HH	32–38°	44	75	JAXA	1992
ERS-2	C	VV	21–26°	35	100	ESA	1995
RADARSAT-1	C	HH	20–60°	24	45–510	CSA	1995
ENVISAT	C	Alternating polarization	15–45°	35	Up to 405	ESA	2002
ALOS	L	Quad polarized	8–60°	46	40–350	JAXA	2006
RISAT-1	C	Quad polarized	12–55°		10–225	25	2012
Kompsat-5	X	Variable polarization	20–55°	28	5–100	KARI	2013
RADARSAT-2	C	Quad polarized	20–60°	24	10–500	CSA	2008
TerraSAR-X	X	Dual polarized	20–45°	28	5–100	DLR	2007
COSMO-SkyMed	X	Quad polarized	25–50°	16	10–200	ASI	2007
RISAT-1	C	Quad polarized	12–55°		10–225	25	2012
Operational							
ALOS-2	L	Compact	8–70°	14	25–350	JAXA	2014
Sentinel-1A	C	Dual polarized	20–46°	12	20–400	ESA	2014
PAZ	X	Dual polarized	15–60°	11	10–100		2015
Sentinel-1B	C	Dual polarized	20–46°	12	20–400	ESA	2016
SAOCOM	L	Quad polarized	20–50°	16	20–350	CONAE/ASI	2017
ICEYE satellite constellation	X	VV	15–35°	21–2	30–100	ICEYE	2018
Capella constellation	X	HH	25–40°	1 h	10–50	Capella	2018
RADARSAT constellation	C	Compact		4	14–500	CSA	2019

(continued)

Table 25.1 (continued)

Satellite	Band	Polarization	Incidence angle	Repeat Cycle (days)	Swath (km)	Agency	Launch date
Planned							
Kompsat-6	L	Quad polarized	20–60°	11	5–100	KARI	2021
NISAR	L&S	Quad polarized	33–47°	12	>240 km	NASA/ISRO	2023
Tandem-L	L	Quad polarized	24–77°	16	350 km	DLR	2022

planes, i.e., HH, VV, HV, and VH) data acquisition capability by the advanced satellite systems equipped the researchers with the capability to map the individual scattering mechanisms taking place for flooded vegetation. Polarimetric data is important to researchers as it has the capability in providing physical information about the target (i.e., shape, orientation, symmetry, and non-symmetry) and therefore useful in land cover classification where a priori information is not available (Mishr and Singh 2013). Development of systems with compact polarimetric (CP) SAR was another major advancement which radar remote sensing seen during this generation included in RISAT-1, ALOS-2, and RADARSAT constellation mission (RCM). There were also more missions in this phase acquiring data in X-band (Terra SAR-X, TanDEM-X, and COSMO-SkyMed) and L-band (Advanced Land Observing Satellite (ALOS-1&2), PAZ, SAOCOM-1A&B), with high-resolution imaging capability. The interesting development was from European Space Agency (ESA) by launching constellation of Sentinel-1 missions in C-band with improved spatial resolution and high revisit frequency and allowing open access.

In the context of India, the nationwide flood mapping and monitoring activity also received significant boost due to the launch of its own Radar Imaging Satellite (RISAT-1) in 2012 operating in C-band 5.35 GHz (Bhatt et al. 2017a, b). RISAT-I being the first SAR satellite of ISRO had capability to acquire data in wide variety of spatial resolutions (1–50 m) with variable swath widths (30–240 km), multiple imaging modes (spotlight/strip map/scan SAR), in single or dual polarization or in quad polarization modes (including the circular polarization) and can be operated over any incidence angle ranging from 12 to 55°. RISAT-1 in fact emerged as the main workhorse for flood hazard mapping activities of ISRO and complemented the operational information from IRS optical satellites. It played a major role in providing flood dynamics information during the Cyclone Phailin, Odisha in 2013, Jhelum floods, Jammu and Kashmir in 2014 (Bhatt et al. 2016), and Ganga floods, Bihar in 2016. Figure 25.8 shows the flood progression through SAR acquisitions from RISAT-1 at 600 h, 1800 h on October 13, 2013, and then again next morning by RADARSAT-1 at 600 h on October 14, 2013, post-Cyclone Phailin landfall. The progression of flooding along Budhabalanga River upstream of Balasore on October

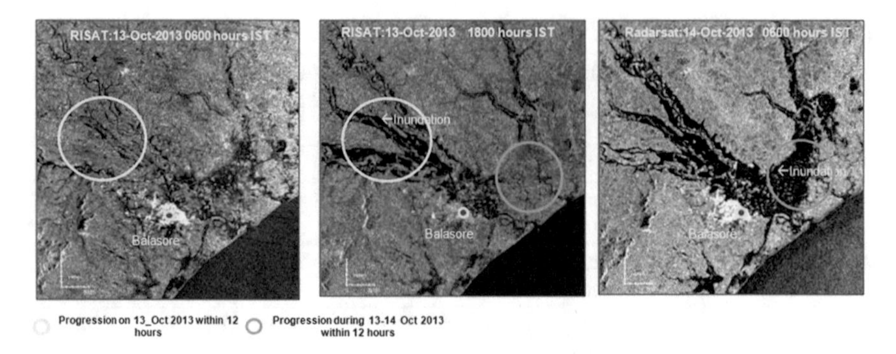

Fig. 25.8 Multi-temporal SAR images showing flood progression between Oct 13 and 14, 2013, in Balasore district, Odisha due to cyclone Phailin © NRSC/ISRO 2013

13, 2013, afternoon is indicated by yellow circle, whereas the fresh inundation on 600 h on October 14, 2013, along Subarnarekha River is indicated by blue circle.

The concept of small constellation of SAR microsatellites as a potential tool for rapid response for flood monitoring was also realized with the launch of first ICEYE-X1 satellite by the Finland's ICEYE commercial space company in the year 2018. ICEYE constellation provides data imaging in X-band with resolution varying between 0.3 m and 15 m at different angle imaging multiple times a day. Till the year 2021, beginning ICEYE constellation had grown up to ten satellites which is expected to grow further. Capella constellation was launched by San-Francisco-based company in 2018 and plans to add 36 satellites in group to have revisit of less than an hour. Several other commercial space companies like the as iQPS (Japan), Synspective (Japan), PredaSAR (Florida), and Umbra Lab (California) to name a few have planned to develop such microsatellites constellation especially keeping up flood hazard into mind.

The future upcoming joint mission (planned to launch in 2023) between NASA and ISRO called as NISAR is a very interesting mission from hydrological disaster perspective especially flood hazard. This mission because of longer wavelength L-band (24 cm), polarimetric SAR imaging capability and free access to data is further bound to enhance the operational flood activities and aid flood disaster response activities in a big way. L-band utility for flooded vegetation a lot of literature is available, but the S-band SAR data which has not much been focused will further enhance for deriving the forest structural information, especially for temperate mixed forests, secondary forests, and savanna woodlands of the tropics (Ningthoujam et al. 2017) in perspective of flooding. The integrated L- and S-band flood hazard mapping will provide more accurate and updated information.

25.7 Near-Real-Time Flood Mapping and Monitoring Under ISRO DMS Program

Taking the advantage of SAR-based sensors which provide time and weather-independent data acquisitions, complemented with optical datasets today India has its own near-real-time operation flood hazard mapping and monitoring activity under ISRO DMS program. India is one of the very few countries to have its own dedicated necessary expertise and infrastructure for mapping and monitoring of flood hazard on near-real-time (NRT) basis in the country. The establishment of Decision Support Centre, at National Remote Sensing Centre (NRSC), was the effort of country's progress from demonstrating the potential to operationalization and then subsequently institutionalization of the flood hazard mapping program through its own constellation of satellites (Nair 2004). DSC keeps a continuous close watch on the flood situation in the country based on various proxy indicators (river gauge data, precipitation, cloud persistence, news media, etc.) to program SAR datasets. Rapid assessment of data is carried out using established automated procedures for preparing various flood-based products like flood inundation maps, damage statistics, flood progression, flood recession, and flood persistence to national and state agencies. Figure 25.9 shows the NRT flood mapping carried out by DSC between 2017 and 2020 for the major flood events and the number of maps disseminated to state relief agencies.

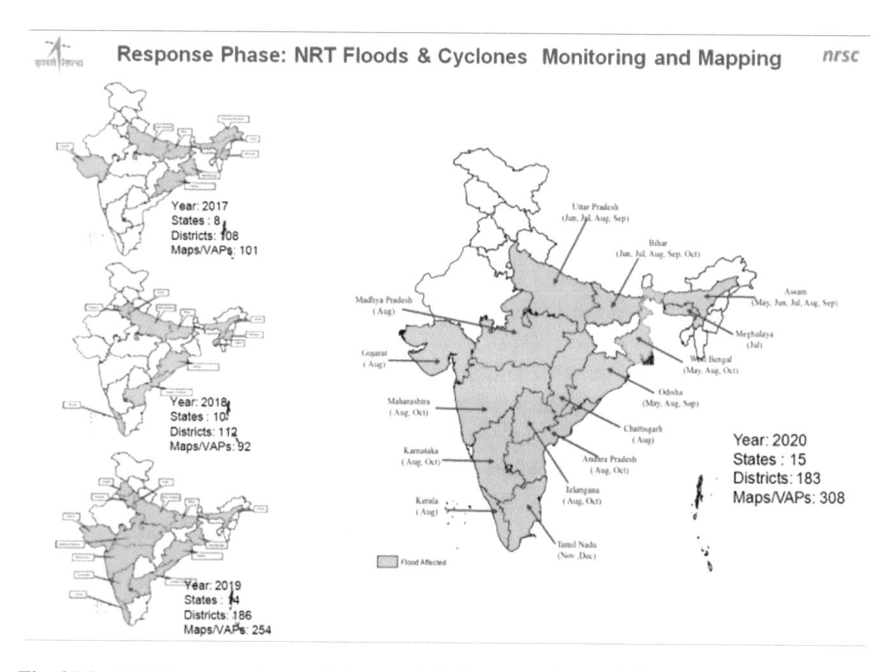

Fig. 25.9 NRT flood mapping carried out by DSC between 2017 and 2020. *Source* DMSG/NRSC

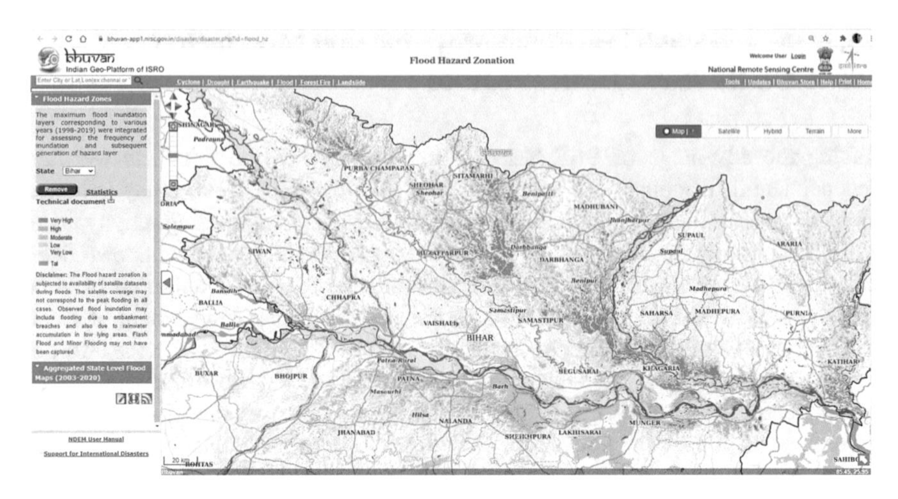

Fig. 25.10 Flood hazard zonation map for Bihar state. *Source* www.bhuvan.nrsc.gov.in

The flood hazard information generated for more than two decades under the program mainly from the SAR datasets is also utilized for planning long-term flood mitigation measures. The flood hazard zonation atlases and GIS-enabled maps generated integrating flood inundation frequency and flood waves data for the states of Assam, Bihar, and Odisha are examples of this effort. The first version of the Bihar hazard zonation atlas based on historical satellite data acquired between 1998 and 2010 was released in 2013. This flood hazard atlas was a major milestone in making the state of Bihar prepare for mitigating the flood hazards (Fig. 25.10) (Manjusree et al. 2015). The second version of the atlas updated with recent flood hazard information derived from satellite images acquired between 2011 and 2019 years made available to district authorities will help the state in moving toward a more safer and resilient Bihar (https://bhuvan.nrsc.gov.in/pdf/Flood-Hazard-Atlas-Bihar.pdf).

The flood hazard zonation maps can be very useful to mitigate flood hazard problem. Flood hazard zonation derived from space data have been translated to ground application by Assam State Disaster Management Agency (ASDMA) by bringing out suggestive application of flood hazard atlas for district administration and lying departments. Few recommendations incorporated under the suggestive measures for the district administration to give emphasis to high and very high flood hazard villages, while planning developmental activities, deploy rescue and relief teams for quick evacuation, propagate use of flood tolerant crops, identify shelters, prepositioning of mobile medical teams, raise hand pump level above high flood level, and maintain sufficient stock of food commodities.

25.8 Open Source SAR Data, Tools, and Cloud Computing Platforms for Flood Hazard Analysis

For flood hazard mapping, SAR data is possibility the best source from satellite perspective to detect flood dynamics under cloud cover. However, the high cost of the SAR datasets has always remained an issue for regional-scale operational flood hazard mapping applications. The announcement of free, full, and open data (https://scihub.copernicus.eu/) policy of Sentinel-1A (launched 2014) and Sentinel-1B (launched 2016) by ESA (European Space Agency) under the Copernicus Earth observation program has seen the application of SAR data growing many fold over the last few years (Berezowski et al. 2020). Sentinel-1 mission predefined conflict-free observation scenario and repeat cycle of six days have seen increase in development of automated processes for supporting operational flood hazard (Uddin et al. 2019; Cao et al. 2019; Twele et al. 2016). The Sentinel Application Platform (SNAP), an open source platform with a series of executable tools and Application Programming Interfaces (APIs) from ESA provided in open domain, has helped to maximize the benefits of the SAR data especially for flood hazard mitigation. Recently, the Canadian Space Agency (CSA) also announced the archival RADARSAT-1 SAR images for free access to researchers, industry, and the public at no cost (https://www.eodms-sgdot.nrcan-rncan.gc.ca/index-en.html). The upcoming NISAR (L- and S-band) mission announcement of free access to data and products (https://www.sac.gov.in/nisar/NisarMission.html) can be a major milestone in radar application especially for flooded vegetation.

Despite the free data availability, the rapid processing services critical for a flood response could not be executed particularly in most of the developing countries due to inadequate systems to support the downloading of huge volume of data, tools for processing the data, and storage space. The preloaded radiometrically calibrated and terrain corrected Sentinel-1 images along with a wide number of geospatial datasets and parallel processing capability offered by Google Earth Engine (GEE) have overcome most of the limitations (Lal et al. 2020; Bhatt et al. 2021; Singha et al. 2020; Uddin et al. 2019). Figure 25.11 shows the GEE interface with Sentinel-1 SAR data over Bihar and corresponding flood inundation delineated using GEE script. Today with emergence of cloud-based processing platforms like GEE, SAR data application has been able to find wide proliferation and application for humanitarian response and mitigation. DeVries et al. (2020) have generated a number of flood maps for the recent flood events following Hurricane Harvey in late August 2017 in few minutes, using the GEE's web-based java script application programming interface (API), which else could have taken several days. Singha et al. (2020) have analyzed the Sentinel-1 SAR images using and the GEE platform for Bangladesh between 2014 and 2018 to analyze the spatiotemporal pattern of flood-affected paddy rice fields. Bhatt et al. (2021) have processed the multi-temporal Sentinel-1A and Sentinel-1B datasets for assessing the impact of late September, 2019, floods in Bihar (Figure and Figure) using a customized GEE script run using GEE code editor (https://code.earthengine.google.com/).

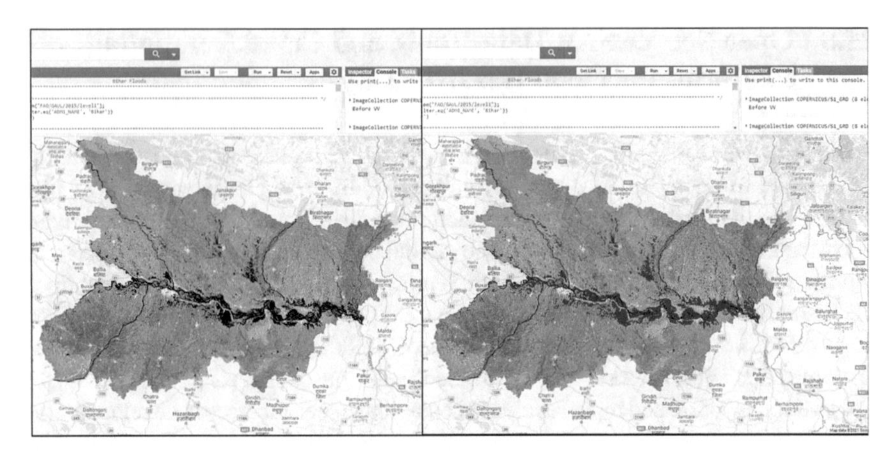

Fig. 25.11 GEE interface with Sentinel-1 SAR data over Bihar and corresponding flood inundation delineated using GEE script

25.9 Conclusions

The review of literature suggests that today the implementation of flood hazard mapping for open flooded areas using SAR data has well matured and has reached a stage of operational flood mapping. There has been considerable advancement in the understanding the mechanism of flooded vegetation. The future L-band missions and their free access will help in integrating them in operational chain mode to improve the accuracy of flood mapping and minimize the misclassification. There have been a very few selected prototype studies for assessing the floods in urban areas till now. Therefore, there exists a lot of scope to investigate, enhance the SAR response in urban areas to develop operational mechanism of response. These studies though have given important insights into the backscattering mechanisms but still more research is needed in terms of sensor and ground parameters to actually develop an operational mechanism for mapping of flooded urban areas. This is also important from the perspective that urban flooding has emerged as a major challenge in last few years than compared to rural flooding and has seen a multi-fold increase in frequency of occurrence and severity globally.

The Sentinel-1 missions and software processing tools like SAR availability on public domain has further strengthened the flood hazard mapping and humanitarian response activities. These activities are further likely to get boosted with free access of data from upcoming NISAR mission. The future open access NISAR L-band data in conjunction with Sentinel-1 C-band data will help in complementing the information on flooded vegetated areas and provide faster response time during crisis. The open-source cloud-based platforms together with multiple other spatial datasets including Sentienl-1 SAR data available through GEE have made seen flood application studies growing multi-fold since last few years. In Indian context, the upcoming SAR data from RISAT-1A and 1B will also boost flood mapping applications.

The launch of SAR microconstellations from number of players will enable SAR data democratization. These constellations will help in faster and accurate NRT mapping and monitoring of floods, detailed flood damage assessment, flood depth estimation, better flood insurance index-based instruments for processing claims of who have suffered flood damages. Therefore, active SAR applications for flood hazard in particular are bound to see many improvements and mitigate vulnerabilities due to emerging climate changes.

Acknowledgements The authors sincerely acknowledge the studies cited in the present chapter to highlight the SAR potential. All the sources of data used from various sources are also duly acknowledged.

References

Alpers W, Zhang B, Mouche A, Zeng K, Chan PW (2016) Rain footprints on C-band synthetic aperture radar images of the ocean-Revisited. Remote Sens Environ 187:169–185

Adeli S, Salehi B, Mahdianpari M, Quackenbush LJ, Brisco B, Tamiminia H, Shaw S (2020) Wetland monitoring using SAR data: a meta-analysis and comprehensive review. Remote Sens 12(14):2190

Amarnath G, Ghosh S, Alahacoon N, Ravan SK, Taneja PK, Dave N, Srivastava SK (2019) Insurance as an agricultural disaster risk management tool: evidence and lessons learned from South Asia

Bhatt CM, Rao GS, Diwakar PG, Dadhwal VK (2017a) Development of flood inundation extent libraries over a range of potential flood levels: a practical framework for quick flood response. Geomat Nat Hazards Risk 8(2):384–401

Bhatt CM, Rao GS, Farooq M, Manjusree P, Shukla A, Sharma SVSP, Kulkarni SS et al (2017b) Satellite-based assessment of the catastrophic Jhelum floods of September 2014, Jammu & Kashmir, India. Geomat Nat Hazards Risk 8(2):309–327

Bhatt CM, Gupta A, Roy A, Dalal P, Chauhan P (2021) Geospatial analysis of September, 2019 floods in the lower Gangetic plains of Bihar using multi-temporal satellites and river gauge data. Geomat Nat Haz Risk 12(1):84–102

Bhatt CM, Rao GS, Begum A, Manjusree P, Sharma SVSP, Prasanna L, Bhanumurthy V (2013) Satellite images for extraction of flood disaster footprints and assessing the disaster impact: Brahmaputra floods of June–July 2012, Assam, India. Curr Sci 1692–1700

Betbeder J, Rapinel S, Corgne S, Pottier E, Hubert-Moy L (2015) TerraSAR-X dual-pol time-series for mapping of Wetland vegetation. ISPRS J Photogrammetry Remote Sens 107:90–98. https://doi.org/10.1016/j.isprsjprs.2015.05.001

Bach H, Appel F, Fellah K, de Fraipont P (2005). Application of flood monitoring from satellite for insurances. In Proceedings 2005 IEEE international geoscience and remote sensing symposium. IGARSS'05. IEEE, vol 1, July 2005, p 4

Berezowski T, Bieliński T, Osowicki J (2020) Flooding extent mapping for synthetic aperture radar time series using river gauge observations. IEEE J Sel Top Appl Earth Observations Remote Sens 13:2626–2638

Brisco B, Kun L, Tedford B, Charbonneau F, Yun S, Murnaghan K (2013b) Compact polarimetry assessment for rice and Wetland mapping. Int J Remote Sens 34(6):1949–1964. https://doi.org/10.1080/01431161.2012.730156

Brivio PA, Colombo R, Maggi M, Tomasoni R (2002) Integration of remote sensing data and GIS for accurate mapping of flooded areas. Int J Remote Sens 23(3):429–441. https://doi.org/10.1080/01431160010014729

Bourgeau-Chavez LL, Kasischke ES, Brunzell SM, Mudd JP, Smith KB, Frick AL (2001) Analysis of space-borne SAR data for wetland mapping in Virginia Riparian ecosystems. Int J Remote Sens 22(18):3665–3687. https://doi.org/10.1080/01431160010029174

Buck C, Monni S (2000) Application of SAR interferometry to flood damage assessment. In SAR workshop: CEOS committee on earth observation satellites, vol 450. March 2000, p 473

Cao H, Zhang H, Wang C, Zhang B (2019) Operational flood detection using Sentinel-1 SAR data over large areas. Water 11(4):786

Clement MA, Kilsby CG, Moore P (2018) Multi-temporal synthetic aperture radar flood mapping using change detection. J Flood Risk Manage 11(2):152–168

Cohen J, Riihimäki H, Pulliainen J, Lemmetyinen J, Heilimo J (2016) Implications of boreal forest stand characteristics for X-band SAR flood mapping accuracy. Remote Sens Environ 186:47–63

Dasgupta A, Thakur PK, Gupta PK (2020) Potential of SAR-derived flood maps for hydrodynamic model calibration in data scarce regions. J Hydrol Eng 25(9):05020028

DeVries B, Huang C, Armston J, Huang W, Jones JW, Lang MW (2020) Rapid and robust monitoring of flood events using Sentinel-1 and Landsat data on the google earth engine. Remote Sens Environ 240:111664

Di Baldassarre G, Schumann G, Brandimarte L, Bates P (2001) Timely low resolution SAR imagery to support floodplain modeling: a case study review. Sur Geophys 32(3):255–269

Gan TY, Zunic F, Kuo CC, Strobl T (2012) Flood mapping of Danube River at Romania using single and multi-date ERS2-SAR images. Int J Appl Earth Obs Geoinf 18:69–81

Geudtner D, Winter R, Vachon PW (1996) Flood monitoring using ERS-1 SAR interferometry coherence maps. In IGARSS'96. 1996 International Geoscience and Remote Sensing Symposium, vol 2. IEEE, pp 966–968

Giustarini L, Hostache R, Matgen P, Schumann GJP, Bates PD, Mason DC (2012) A change detection approach to flood mapping in urban areas using TerraSAR-X. IEEE Trans Geosci Remote Sens 51(4):2417–2430

Henry JB, Chastanet P, Fellah K, Desnos YL (2006) Envisat multi-polarized ASAR data for flood mapping. Int J of Remote Sens 2(10):1921–1929. https://doi.org/10.1080/01431160500486724

Henderson FM, Lewis AJ (2008) Radar detection of wetland ecosystems: a review. Int J Remote Sens 29:5809–5835

Hess LL, Melack JM, Simonett DS (1990) Radar detection of flooding beneath the forest canopy: a review. Int J Remote Sens 11:1313–1325

Horritt MS (2000) Calibration of a two-dimensional finite element flood flow model using satellite radar imagery. Water Resour Res 36(11):3279–3291

Kalluri S, Gilruth P, Rogers D, Szczur M (2007) Surveillance of arthropod vector-borne infectious diseases using remote sensing techniques: a review. PLoS pathogens 3(10):e116

Kaya S, Sokol J, Pultz TJ (2004) Monitoring environmental indicators of vector-borne disease from space: a new opportunity for RADARSAT-2. Can J Remote Sens 30(3):560–565

Li Y, Martinis S, Wieland M, Schlaffer S, Natsuaki R (2019) Urban flood mapping using SAR intensity and interferometric coherence via Bayesian network fusion. Remote Sens 11(19):2231

Lal P, Prakash A, Kumar A (2020) Google earth engine for concurrent flood monitoring in the lower basin of Indo-Gangetic-Brahmaputra plains. Nat Hazards 104(2):1947–1952

Martinis S, Rieke C (2015) Backscatter analysis using multi-temporal and multi-frequency SAR data in the context of flood mapping at River Saale Germany. Remote Sens 7(6):7732–7752

Martinis S, Plank S, Ćwik K (2018) The use of Sentinel-1 time-series data to improve flood monitoring in Arid areas. Remote Sens 10(4):583

Martinez J, Le Toan T (2007) Mapping of flood dynamics and spatial distribution of vegetation in the amazon floodplain using multitemporal SAR data. Remote Sens Environ 108(3):209–223. https://doi.org/10.1016/j.rse.2006.11.012

Mason DC, Speck R, Devereux B, Schumann GJP, Neal JC, Bates PD (2009) Flood detection in urban areas using TerraSAR-X. IEEE Trans Geosci Remote Sens 48(2):882–894

Mason DC, Dance SL, Cloke HL (2021) Floodwater detection in urban areas using Sentinel-1 and WorldDEM data. J Appl Remote Sens 15(3):032003

Manjusree P, Kumar LP, Bhatt CM, Rao GS, Bhanumurthy V (2012) Optimization of threshold ranges for rapid flood inundation mapping by evaluating backscatter profiles of high incidence angle SAR images. Int J Disaster Risk Sci 3(2):113–122

Manjusree P, Bhatt CM, Begum A, Rao GS, Bhanumurthy V (2015) A decadal historical satellite data analysis for flood hazard evaluation: a case study of Bihar (North India). Singap J Trop Geogr 36(3):308–323

McGinnis DF, Rango A (1975) Earth resources satellite systems for flood monitoring. Geophys Res Lett 2:132–135

Mishra P, Singh D (2013) A statistical-measure-based adaptive land cover classification algorithm by efficient utilization of polarimetric SAR observables. IEEE Trans Geosci Remote Sens 52(5):2889–2900

Moser L, Schmitt A, Wendleder A, Roth A (2016) Monitoring of the Lac Bam Wetland extent using dual-polarized X-band SAR data. Remote Sens 8(4):302. https://doi.org/10.3390/rs8040302

Notti D, Giordan D, Caló F, Pepe A, Zucca F, Galve JP (2018) Potential and limitations of open satellite data for flood mapping. Remote Sens 10(11):1673

Nair MG (2004) Space for disaster management: Indian perspectives. 55th international astronautical congress, 4–8 Octo, Vancouver, Canada

Ningthoujam RK, Balzter H, Tansey K, Feldpausch TR, Mitchard ET, Wani AA, Joshi PK (2017) Relationships of S-band radar backscatter and forest aboveground biomass in different forest types. Remote Sens 9(11):1116

Ormsby JP, Blanchard BJ, Blanchard AJ (1985) Detection of lowland flooding using active microwave systems

Prakash R, Singh D, Pathak NP (2011) A fusion approach to retrieve soil moisture with SAR and optical data. IEEE J Sel Topics Appl Earth Observations Remote Sens 5(1):196–206

Pradhan B, Tehrany MS, Jebur MN (2016) A new semiautomated detection mapping of flood extent from TerraSAR-X satellite image using rule-based classification and taguchi optimization techniques. IEEE Trans Geosci Remote Sens 54(7):4331–4342

Plank S, Jüssi M, Martinis S, Twele A (2017) Mapping of flooded vegetation by means of polarimetric Sentinel-1 and ALOS-2/PALSAR-2 imagery. Int J Remote Sens 38:3831–3850

Pulvirenti L, Chini M, Pierdicca N, Boni G (2016) Use of SAR data for detecting floodwater in urban and agricultural areas: the role of the interferometric coherence. IEEE Trans Geosci Remote Sens 54(3):1532–1544. https://doi.org/10.1109/TGRS.2015.2482001

Pulvirenti L, Chini M, Pierdicca N, Guerriero L, Ferrazzoli P (2011a) Flood monitoring using multitemporal COSMO-SkyMed data. Image segmentation and signature interpretation. Remote Sens of Environ 115 (4): 990–1002. https://doi.org/10.1016/j.rse.2010.12.002

Pulvirenti L, Pierdicca N, Chini M, Guerriero L (2013) Monitoring flood evolution in vegetated areas using COSMO-SkyMed data: the Tuscany 2009 case study. IEEE J Sel Top Appl Earth Observations Remote Sens 6(4):1807–1816. https://doi.org/10.1109/JSTARS.2012.2219509

Pulvirenti L, Pierdicca N, Boni G, Fiorini M, Rudari R (2014) Flood damage assessment through multitemporal COSMO-SkyMed data and hydrodynamic models: the Albania 2010 case study. IEEE J Sel Top Appl Earth Observations Remote Sens 7(7):2848–2855

Pierdicca N, Pulvirenti L, Chini M, Guerriero L, Candela L (2013) Observing floods from space: experience gained from COSMO-SkyMed observations. Acta Astronaut 84:122–133. https://doi.org/10.1016/j.actaastro.2012.10.034

Rao DP, Bhanumurthy V, Rao GS, Manjusri P (1998) Remote sensing and GIS in flood management in India. Mem Geol Soci of India 41:195–218

Rahman MS, Di L (2020) A systematic review on case studies of remote-sensing-based flood crop loss assessment. Agriculture 10(4):131

Rykhus R, Lu Z (2005) Hurricane Katrina flooding and oil slicks mapped with satellite imagery. Science and the storms: the USGS response to the Hurricanes of, pp 49–52

Richards JA, Sun G, Simonett DS (1987a) L-band radar backscatter modeling of forest stands. IEEE Trans Geosci Remote Sens GE-25, 487–498

Sanyal J, Lu XX (2004) Application of remote sensing in flood management with special reference to monsoon Asia: a review. Nat Hazards 33(2):283–301
Schumann G, Bates PD, Horritt MS, Matgen P, Pappenberger F (2009) Progress in integration of remote sensing–derived flood extent and stage data and hydraulic models. Rev Geophys 47:RG4001
Schlaffer S, Chini M, Dettmering D, Wagner W (2016) Mapping Wetlands in Zambia using seasonal backscatter signatures derived from ENVISAT ASAR time series. Remote Sens 8:5. https://doi.org/10.3390/rs8050402
Shen X, Wang D, Mao K, Anagnostou E, Hong Y (2019) Inundation extent mapping by synthetic aperture radar: a review. Remote Sens 11(7):879
Solbø S, Solheim I (2005) Towards operational flood mapping with satellite SAR. In Envisat & ERS Symposium, vol 572. April 2005
Singh A, Gaurav K, Meena GK, Kumar S (2020) Estimation of soil moisture applying modified Dubois model to Sentinel-1; A regional study from central India. Remote Sens 12(14):2266
Singha M, Dong J, Sarmah S, You N, Zhou Y, Zhang G, Doughty R, Xiao X (2020) Identifying floods and flood-affected paddy rice fields in Bangladesh based on Sentinel-1 imagery and google earth engine. ISPRS J Photogramm Remote Sens 166:278–293
Smith LC (1997) Satellite remote sensing of river inundation area, stage, and discharge: a review. Hydrol Process 11(10):1427–1439
Shrestha BB, Sawano H, Ohara M, Yamazaki Y, Tokunaga Y (2018) Methodology for agricultural flood damage assessment. In: Recent advances in flood risk management. IntechOpen
Skakun S, Kussul N, Shelestov A, Kussul O (2014) Flood hazard and flood risk assessment using a time series of satellite images: a case study in Namibia. Risk Anal 34(8):1521–1537
Soergel U, Thoennessen U, Stilla U (2003) Visibility analysis of man-made objects in SAR images. In 2003 2nd GRSS/ISPRS joint workshop on remote sensing and data fusion over urban areas. IEEE, pp 120–124
Souza-Filho PWM, Paradella WR, Rodrigues SWP, Costa FR, Mura JC, Gonçalves FD (2011) Discrimination of coastal wetland environments in the Amazon region based on multi-polarized L-band airborne synthetic aperture radar imagery. Estuar Coast Shelf Sci 95:88–98
Thakur PK, Maiti S, Kingma NC, Prasad VH, Aggarwal SP, Bhardwaj A (2012) Estimation of structural vulnerability for flooding using geospatial tools in the rural area of Orissa India. Nat Hazards 61(2):501–520
Tsyganskaya V, Martinis S, Marzahn P, Ludwig R (2018) SAR-based detection of flooded vegetation–a review of characteristics and approaches. Int J Remote Sens 39(8):2255–2293
Tripathi A, Tiwari RK (2020) Synergetic utilization of Sentinel-1 SAR and Sentinel-2 optical remote sensing data for surface soil moisture estimation for Rupnagar, Punjab, India. Geocarto Int 1–22
Twele A, Cao W, Plank S, Martinis S (2016) Sentinel-1-based flood mapping: a fully automated processing chain. Int J Remote Sens 37(13):2990–3004
Uddin K, Matin MA, Meyer FJ (2019) Operational flood mapping using multi-temporal sentinel-1 SAR images: a case study from Bangladesh. Remote Sens 11(13):1581
Ulaby FT, Dobson MC (1989) Handbook of radar scattering statistics for terrain. Artech House, Norwood, MA
Voormansik K, Praks J, Antropov O, Jagomagi J, Zalite K (2014) Flood mapping with TerraSAR-X in forested regions in Estonia. IEEE J Sel Topics Appl Earth Observations Remote Sens 7(2):562–577. https://doi.org/10.1109/JSTARS.2013.2283340
Wang Y, Hess LL, Filoso S, Melack JM (1995) Understanding the radar backscattering from flooded and nonflooded Amazonian forests. Results from canopy backscatter modeling. Remote Sens Environ 54(3):324–332. https://doi.org/10.1016/0034-4257(95)00140-9
Zhang M, Zhen L, Tian B, Zhou J, Tang P (2016) The backscattering characteristics of wetland vegetation and water-level changes detection using multi-mode SAR. A case study. Int J Appl Earth Observation Geoinf 45:1–13. https://doi.org/10.1016/j.jag.2015.10.001

Chapter 26
Role of Geospatial Technology in Hydrological and Hydrodynamic Modeling-With Focus on Floods Studies

Praveen K. Thakur, Pratiman Patel, Vaibhav Garg, Adrija Roy, Pankaj Dhote, C. M. Bhatt, Bhaskar R. Nikam, Arpit Chouksey, and S. P. Aggarwal

Abstract Assessment of surface water with higher accuracy is very critical in the present changing environment. Such assessment requires understanding of each and every hydrological process involved in hydrological cycle. The land characteristics along with climate variables make it a daunting task. With the advent of geospatial technology, both land surface and climate parameters may be studied with higher accuracy. Some of the hydrological parameters such as precipitation, soil moisture, evapotranspiration, and water level can now directly be retrieved using remote sensing data. However, other hydrological components such as rainfall-runoff, snowmelt-runoff, peak discharge, or flood hydrograph need modeling approach. Most of the surface and climate inputs required for hydrological and hydrodynamic (H&H) modeling nowadays can be quantified using the geospatial datasets. It makes

---

P. K. Thakur (✉) · P. Patel · V. Garg · A. Roy · P. Dhote · C. M. Bhatt · B. R. Nikam · A. Chouksey · S. P. Aggarwal
Water Resources Department, Indian Institute of Remote Sensing, ISRO, Dehradun, Uttarakhand, India
e-mail: praveen@iirs.gov.in

V. Garg
e-mail: vaibhav@iirs.gov.in

P. Dhote
e-mail: pdh@iirs.gov.in

C. M. Bhatt
e-mail: cmbhatt@iirs.gov.in

B. R. Nikam
e-mail: bhaskarnikam@iirs.gov.in

A. Chouksey
e-mail: arpit@iirs.gov.in

S. P. Aggarwal
e-mail: spaggarwal2010@gmail.com

P. Patel · A. Roy
Department of Civil Engineering, Indian Institute of Technology, Bombay, Mumbai, India

A. Pandey et al. (eds.), *Geospatial Technologies for Land and Water Resources Management*, Water Science and Technology Library 103,
https://doi.org/10.1007/978-3-030-90479-1_26

modeling more realistic, and hydrological response of large basins can be studied. The H&H models are utilized for studying the hydrological extremes such as flood and droughts. Floods are one of the most naturally re-occurring hazards, which significantly impact the long-term sustainable use of land and water resources of a geographical region. The chapter focuses the use of geospatial technology for H&H modeling with relevant case studies on flood modeling. Further, the improvements in modeling outputs may be done by assimilating the geospatial inputs in near-real time. Moreover, these geospatial inputs are being updated and improved in spatial–temporal domain with the advancement in geospatial technology.

26.1 Introduction

The exponential growth of open Earth observation (EO) data, open-source, and freely available softwares for image processing, geographical information system (GIS) analysis, and hydrological modeling have led to a significant increase in the use of these geospatial data and tools for improved management of land and water resources. In water resources management and hydrology, the physical and climatic characteristics of a given area or a river basin are geographical in nature, i.e., they vary both in space and time. The physical characteristics such as soil type, Land Use Land Cover (LULC), drainage networks, elevation, slope, and geomorphology vary predominantly in the space domain. The climatic and weather characteristics such as rainfall, temperature, humidity, solar radiation, energy fluxes, soil moisture, and river discharge vary both in space and time domains (Thakur 2013).

These spatial–temporal variations can be well-captured using remote sensing (RS) data (Schultz 1988; Schultz and Engman 2000; see Fig. 26.1), ground-based observations from weather stations, soil, geomorphology, and topographical maps. The

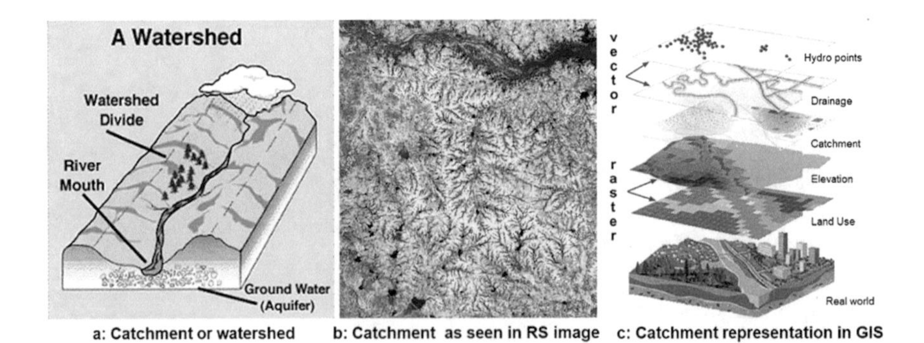

Fig. 26.1 Representation of watershed or a river basin **a:** Schematic of a watershed, **b:** A typical watershed and a river system as seen in RS image from Landsat-7, and **c:** Representation of real-world in GIS in the form of various raster and vector data. Source of these figures: see Web references at the end of chapter)

magnitude of changes in hydrological variables such as rainfall and river flow for a given watershed or river reach (Fig. 26.1) is significantly higher for the extreme hydrological case of floods (Agarwal et al. 2009). The intense and long-duration flood events can also cause the inundation of large land and water areas, causing heavy damages to these natural resources and artificial infrastructure (Agarwal et al. 2009; Bhatt et al. 2010, 2017; Rahman and Thakur 2017). Similarly, to represent and model any normal or extreme hydrological event, it is essential to create, store and retrieve all the required geographical data related to the catchment and climate, to a standard projection and coordinate system, in a suitable GIS platform (Meijerink et al. 1994, see Fig. 26.1). Therefore, RS, GIS, and global navigation satellite system (GNSS) play an important role in providing base data and suitable platforms for creating, analyzing, and modeling such geospatial data and hydrological variables for a given catchment (Sharma and Thakur 2007; Arya et al. 2010; Kushwaha et al. 2010; Pandey et al. 2011; Thakur et al. 2013, 2018).

Once all the catchment's physical and climate-related data is integrated into GIS, the movement and dynamics of surface water over land, subsurface, and river channels can be simulated using a set of mathematical equations, which represent the hydrological and open channel flow process, and popularly called a hydrological and hydrodynamic model (Maidment and Djokic 2000; Meijerink and Mannaerts 2000). The basin-level computer-aided simulation of various hydrological processes such as surface runoff, infiltration, evapotranspiration, interflow, soil moisture, and river flow, commonly known as hydrological modeling, where main input is precipitation and output is river flow or discharge (Q) hydrograph (Singh 1995). The simulation of water flow or any fluid flow in general, once it enters into a natural or artificial channel or conduit, from surface runoff or any other sources of inflow, and associated flow routing at various locations along or across river channel or floodplain, is known as hydrodynamic modeling. The main inputs for hydrodynamic models are inflow hydrograph, boundary conditions and channel geometry, and output as water level and discharge. The RS-, GIS-, and GNSS-based hydrometric data are essential for hydrological and hydrodynamic models (Maidment and Djokic 2000).

This chapter aims to provide an easy-to-understand text and information that can be implemented and utilized with various geospatial data and hydrological and hydrodynamic models to analyze multiple water resources management problems, particularly on flood studies. The chapter is organized into the following seven sections. Section 26.1 introduces the subject and importance of geospatial technology for effective water and land resources management with emphasis on hydrological models. Section 26.2 briefly introduces the present status of geospatial data and associated technology for supporting hydrological studies. Section 26.3 provides a brief introduction to hydrological and hydrodynamic models. Sections 26.4 and 26.5 provide a detailed description of the role of geospatial data in hydrological and hydrodynamic modeling, respectively. Section 26.6 gives a brief idea about upcoming geospatial technological advances and their integration with hydro models. Section 26.7 concludes and summarizes the chapter.

26.2 Geospatial Data and Technological Support for Hydrological Studies

The combined use of remote sensing images, GNSS data, and spatial and non-spatial data in a GIS environment is commonly known as geospatial technology or techniques (Thakur et al. 2012). Table 26.1 summarizes the main component of geospatial technology and its corresponding use in hydrological studies. The primary datasets, which are most commonly used in many hydrological studies, are remote sensing image-based LULC map, topographical information in the form of digital elevation models (DEM), RS, soil physiography and ground survey-based soil maps, RS image-based drainage, and geomorphological maps, EO satellites-based hydrological variables (García et al. 2016; Thakur et al. 2017), and GNSS-based location of various hydro-meteorological stations. The significant progress in remote sensing sensors and imaging capabilities in different parts of the electromagnetic (EM) spectrum during the last 30 years has led to the development of a new generation of RS satellites. These include EO satellites working in the active microwave (synthetic aperture radars (SAR), altimeters, scatterometers, atmospheric sounders, precipitation radars, and LiDARs), multi-frequency or high-resolution (HR) single-frequency passive microwave radiometers, thermal infrared (TIR) sensors, high spatial–temporal resolution multispectral, and hyperspectral data in visible, infrared and short, long-wave infrared region of EM spectrum (VIS, IR, SWIR, LWIR), and HR stereo imaging satellites. This technological progress has enabled the quantification of various hydrological cycle variables such as precipitation, soil moisture (SM), evapotranspiration (ET), water level (WL) (Thakur et al. 2021), interception, and change in terrestrial water storage (ΔTWS), which can be utilized for hydrological studies (Thakur et al. 2017; Montanari and Sideris 2018). Various national space agencies, hydro-meteorological organizations, private space, and weather companies provide these geospatial datasets used extensively for hydrological modeling (Thakur et al. 2017; Sheffield et al. 2018; Jiang and Wang 2019), covering both data-rich and data-poor parts of the world.

26.3 Hydrological and Hydrodynamic (H&H) Modeling

The flow or movement and tracking of water in all three phases (i.e., vapor, liquid, and solid form) via the atmosphere, land surface, vegetation, subsurface, man-made structures, and river channels is called a hydrological cycle (Fig. 26.2). A river basin or a watershed is a closed system, defined as a natural land surface draining all its surface water into a common outlet (see Fig. 26.1a), which is the most unit for the quantitative study of hydrological cycle components. The hydrological and hydrodynamic (H&H) models simulate water movement through natural and human-made water systems, utilizing various RS-based inputs and in a GIS-enabled environment (Maidment and Djokic 2000). The use of geospatial technology, data, and

Table 26.1 Component of geospatial technology and their hydrological relevance*

Geospatial data	Data type	Hydrological relevance
Remote sensing images (VIS–NIR and SAR)	Raster	LULC, soil physiographic units, geomorphology, water, snow/ice, and drainage, rainfall, used with hydro models
Remote sensing images (in TIR)	Raster	Rainfall, land surface temperature estimation, and energy balance studies
Remote sensing images (passive microwave)	Raster	Hydrological variables: precipitation, soil moisture, snow depth (SD), and snow water equivalent (SWE)
Remote sensing images (Precipitation radars, space, and ground-based)	Raster	Precipitation and its use in hydrological models, including flood forecasting models
Remote sensing non-imaging data (Altimeter/LiDAR)	High-density points	Water level and bathymetry, and its use in hydrodynamic models
Spatial and non-spatial data (hydromet. data)	vector	Geolocation of Rain gauges, discharge sites, and weather stations
RS-, GNSS-, and topo sheets-based DEM	Vector and Raster	Catchment characterization, river and floodplain topography, gauge locations
High-resolution RS data	Raster	Water infrastructure (dams, barrages, canals, culverts, etc.) maps
Gravity-based RS data	Raster	Change in terrestrial water and ice storage at regional scale
GNNS and IRNSS	Vector	Radio occultation and reflectometry for hydrological variables (WL, SM, SD)
GIS software	Tools and Software	GIS environment for integration of various vector and raster data in common project system (GRASS GIS, QGIS, ArcGIS, GeoHMS, GeoRAS, Mike Hydro, ILWIS)
Hydro model with GIS interface	Tools and Hydro Models	Hydrological and hydrodynamic models integrated and linked into the GIS environment (RiverGIS, HEC-RAS, QSWAT, QEPANET, QGEP-SWMM, ArcSWAT, Mike-11, 21, Basin, Visual Modflow, Flo-2D)

*QGIS: Quantum GIS; ILWIS: Integrated Land and Water Information System; SWMM: Storm Water Management Model; SWAT: Soil and Water Assessment Tool; EPANET: Environmental Protection Agency's (EPA) Net. HMS: Hydrological modeling system; RAS: River Analysis System

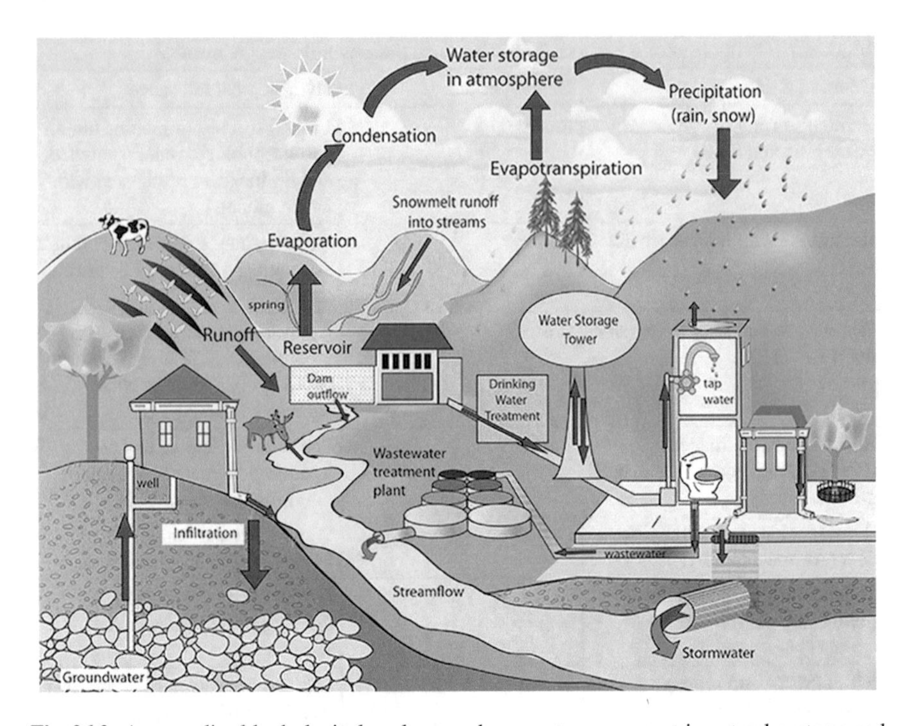

Fig. 26.2 A generalized hydrological cycle map shows water movement in natural systems and man-made water infrastructure for storage, redistribution, and utilization. Source of image https://www.caryinstitute.org/eco-inquiry/teaching-materials/water-watersheds

modeling tools for various H&H modeling studies is highlighted in studies such as Paiva et al. (2011), Patel (2015), Sindhu and Durga Rao (2017), Amir et al. (2018), Roy (2018), and Kalogeropoulos et al. (2020). These studies have extensively utilized the RS derived precipitation, DEM, LULC, and river topographical inputs for H&H modeling studies.

26.3.1 Hydrological Modeling

"A hydrologic model can be defined as a mathematical representation of the flow of water and its constituents on some part of the land surface or subsurface environment." (Singh 1995). Most of the hydrological models attempt to represent the real-world hydrological processes using simplified concepts and mathematical equations. These models can be made as physical models, electrical analog, or mathematical models (Refsgaard 1990). The hydrological complexities of a given area, space and time scale of hydrological simulations, and representation of hydrological processes in simulation model govern the type or class of hydrological model (Chow et al. 1988; Schulze 1998; Plate 2009) (Fig. 26.3).

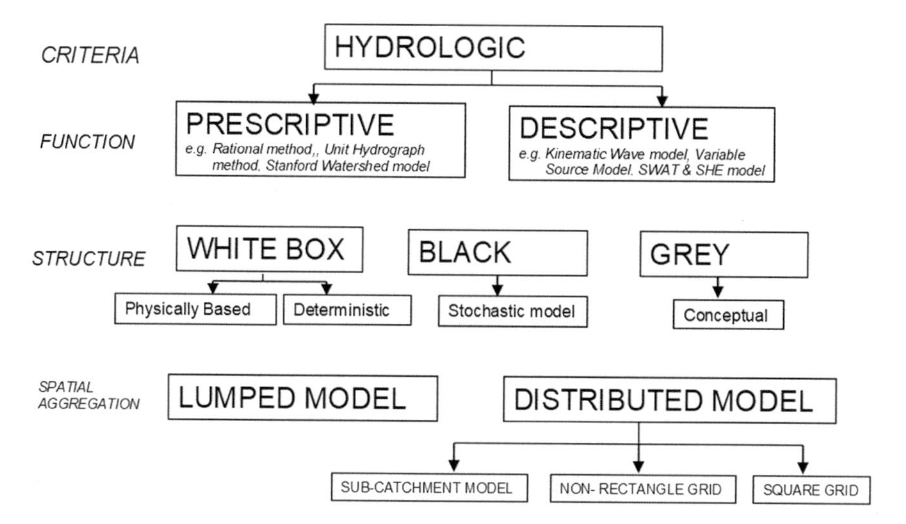

Fig. 26.3 Classification of hydrological models. Schulze (1998)

In case of flood studies, the strategy or choice of the model needs to consider the topography, geomorphology, and geographical location of a given watershed (Plate 2009). The hilly area is more prone to rapid and flash floods, with short concentration time. In contrast, flat regions downstream of dams or below hills are susceptible to large-scale and long-duration floods (Plate 2009). Similarly, highly urbanized catchments are affected by waterlogging and short-duration floods caused by high-intensity rainfall and inadequate drainage system. This necessity requires different types of geospatial data and hydrological models (Thakur et al. 2018).

26.3.2 Hydrodynamic Modeling

The surface or subsurface water comes out of the river catchment or watershed from a hydrograph. It mainly flows in a channel. The scientific study of water or any fluid in such natural or artificial channels is dealt with in hydrodynamic (HD) modeling. In general, hydrodynamic models include all kind of fluid flow, ranging from marine, surface waters, costal/tidal water flows, estuaries, and pressurized water/fluid flow in closed conduits or pipes. But in the specific case of surface hydrology, HD modeling refers to the quantification of water levels and discharge at various sections of river channel or canal or reservoir or floodplain at different times, with hydrological inputs at boundary conditions, using one- or two-dimensional HD models (Chow 1959; Thakur et al. 2006, 2016). Other parameters, which HD models can simulate, are sediment and water quality parameters (Davis et al. 2009).

26.4 Role of Geospatial Data and Tools in Hydrological Modeling

Geospatial datasets and technology are gaining importance in hydrological simulation concerning the study of flood and drought patterns and forecasting the hydrological extremes of a watershed. Geospatial technology can provide data time series for weather parameters, LULC, and soil information, and other numerous auxiliary data like soil moisture, snow cover, snowmelt, elevation, drainage lines, slope, etc., which are essential inputs needed hydrological simulation models (see Table 26.1). Moreover, apart from catering for inputs in hydrological models, RS products can support in validating the output parameters from such simulations, e.g., river flow hydrographs generated from satellite altimeter-based inputs (Tarpanelli et al. 2013; Demaria and Capdevila, 2016; Ghosh et al. 2017; Dhote et al. 2020). In fact, SCS-CN method-based surface runoff estimation was one the first applications of Landsat data in hydrological modeling (Slack and Welch 1980; Sharma and Singh 1992; Chakraborti, 1993; Verma et al. 2017).

GIS interface provides effective maneuvers to delineate drainage structures and gain insights about watershed morphometry by utilizing the DEM (Fleming and Doan 2013). DEM can be processed in a GIS-based interface to derive significant parameters like slope, aspect, drainage density, stream length and frequency, and parameters defining the shape of the watershed, which are necessary for watershed prioritization study and flood susceptibility estimation. After the basin is delineated and selected, the meteorological inputs can be obtained from weather stations or RS observed/reanalysis data without spatio-temporally continuous observed weather data. Figure 26.4 shows an integration of various RS-GIS data in the hydrological model.

Performing hydrological simulations require extensive information for soil type and texture and land use map for generating discharge for the watershed, which can be created directly from RS datasets or updated and processed using geospatial technologies in the GIS platform. Visualization of the output generated from the hydrological simulation again requires geospatial technologies. Flood inundation, water level, snow, glacier mapping, melt timing, etc., can most effectively be interpreted using RS data in any GIS-based architecture. Thus, geospatial data and technologies are of utmost importance in hydrological modeling to produce better results efficiently and are equally important for hydrologically modeling gauged and ungauged watersheds.

Depending on the space and time scale of hydrological models, they can be used to estimate annual, monthly, and daily to sub-daily water balance at global, regional, national, river basin, or watershed scale (Singh 2018). The use of Thronthwaite and Mather (1955, 1957) climatic water balance models to estimate the watershed or national level water budget at annual scale (Singh et al. 2004; NRSC 2013, Durga Rao et al. 2014; CWC 2019); fully distributed physical-based regional hydrological model for national level hydrological modeling (Agarwal et al. 2013); quantifying impacts of LULC change on hydrological components (Agarwal et al. 2012; Garg et al. 2017,

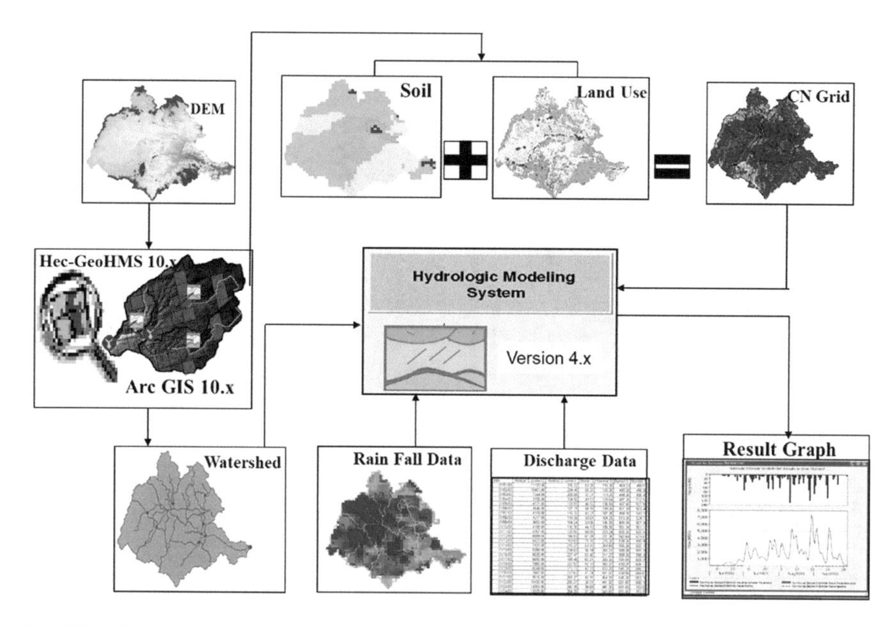

Fig. 26.4 Pictorial flowchart showing various geospatial and meteorological data integration in GIS environment for aiding the hydrological simulation. Modified from Mishra and Mohanty (2007)

2019a, b; Nikam et al. 2018) and drought studies (Mishra et al. 2014) are some of the studies which have utilized geospatial and climate data for hydrological modeling. Many studies have highlighted the use of hydrological models and geospatial data for flood studies (Bekhira et al. 2018; Zhuohang et al. 2019; Wijayarathne and Coulibaly 2020), and this topic forms the following sub-section case study of this chapter.

26.4.1 A Case Study on Hydrological Modeling of Beas River Basin

In an attempt to utilize geospatial data in hydrological models for deriving components for snowmelt and rainfall-runoff, hydrological models had been set up in the Beas basin in the Northwestern Himalayan (NWH) region of India (Fig. 26.5). In this study, an effort has been made to contribute to flood modeling and hydrological studies in the Beas river basin and provide guidance on selecting model calibration parameters judiciously. This method of hydrological analysis can be extended to a certain extent to other NWH basins with some modifications for simulating streamflow characteristics with calibrated hydrological models.

A semi-distributed precipitation-runoff model, HEC-HMS was used, which is efficient for event-based simulations (Feldman 2000; Scharffenberg et al. 2018). Here, in our case, initially, the HEC-HMS setup was calibrated for major flood

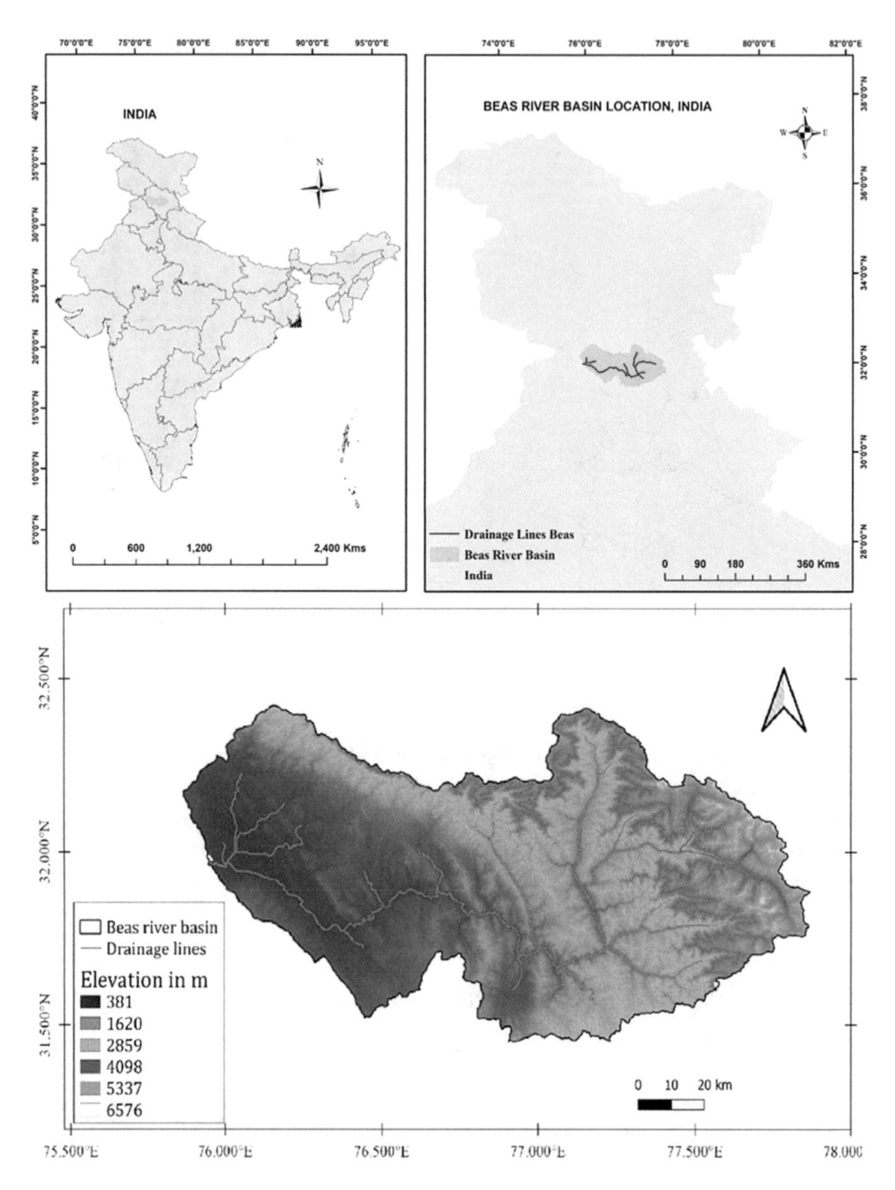

Fig. 26.5 Location and DEM map of Beas River Basin

events of 2014 and validated for flood events of 2015. The RS-based inputs were used for LULC, DEM, and rainfall. However, a different model parameterization scheme was found to be suitable for the different time steps. A time step of 3 h was most suitable for modeling single storm events at a sub-daily time scale. The selected methods for this case were soil conservation service, curve number, SCS-CN loss

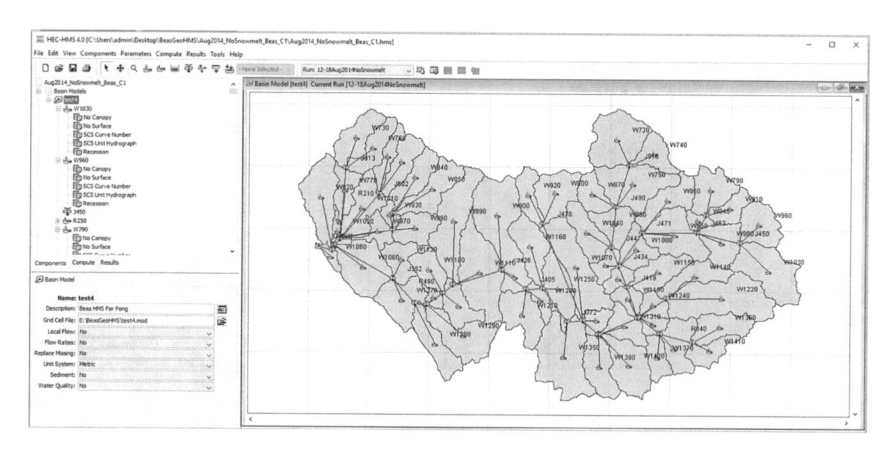

Fig. 26.6 Sub-basins and reaches as elements in final HEC-HMS model setup

method, the SCS Unit Hydrograph transform method, and the recession base flow method for all sub-basins (Roy 2018).

The input parameters of HEC-HMS were taken from different geospatial data sources: ASTER GDEM (30 m resolution), LULC, and soil maps from ISRO and USGS maps (Roy et al. 2016) and NBSS-LUP soil texture map, respectively, the meteorological inputs were obtained from forecast data and historical reanalysis data and satellite information. The HEC-HMS model has been shown in Fig. 26.6 with all sub-basins, drainage network, and various components of the HMS model for the Beas river basin (Roy 2018). In the current situation, two extreme rainfall events were used to calibrate and verify the flood events using the HMS model at two river flow gauging stations of Bhakra and Beas Management Board (BBMB), i.e., Nadaun and Sujanpur Tira.

The HEC-HMS model setup for the Beas basin was tested for extreme rainfall events, one occurred on 13–16 August in 2014, and another was 5–9 August in 2015. The comparison of simulated and observed discharge at Nadaun and Sujanpur Tira is plotted in Fig. 26.7. Hydrographs for the events 2014 and 2015 are generated. It was found that the N–S efficiency coefficient value for the calibration period is 0.601 and 0.650 for Sujanpur Tira and Nadaun, respectively. For the validation period, the N–S efficiency is found to be 0.418 for Sujanpur Tira and 0.423 for Nadaun, with R^2 ranging from 0.85 to 0.95 for two sites, showing high accuracy of HMS-based flood simulations (Roy 2018).

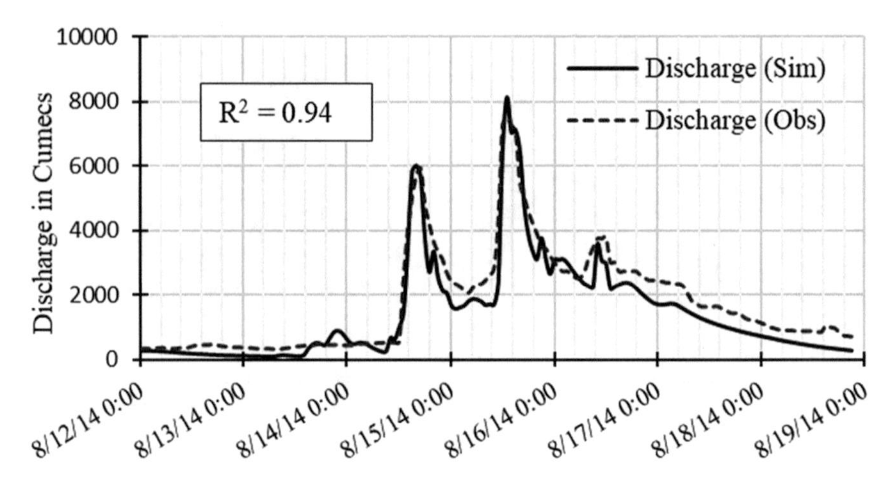

Fig. 26.7 Observed and HMS Simulated flood hydrograph for a major flood event during 13–17 August 2014 at Nadaun river gauging site (Roy 2018)

26.5 Role of Geospatial Data and Tools in Hydrodynamic Modeling

The hydrodynamic models are tools to study the motion of fluids. In terms of floods, hydrodynamic models primarily represent the spatial extent of flood inundation, flood depth, and velocity of the particles (Chow 1959). The dimensions of the flow representation they represent, such as 1-dimensional, 2-dimensional, and 3-dimensional, are generally used to classify these hydrodynamic models. 1D hydrodynamic models represent flow only in main longitudinal direction only at a particular cross-section (Pramanik et al. 2010), while 2D models represent flow properties at 2D grids, i.e., in both X–Y direction, which is more realistic from the flood point of view. Lastly, 3D models, which represent vertical variations in in flow properties, in addition to X–Y changes, are rarely used in flood modeling. Examples for the 1D HD model include getting a flow and WL at a particular cross-section of a river channel; in 2D models, flood inundations, flood depth, and velocities are available at each grid point within the computational domain. Here, overtopping is considered. Lastly, 3D flow calculates the forces exerted by the flow over a physical structure such as a dam or reservoir. A commonly used list of flood modeling software includes HEC-RAS, TUFLOW, LISFLOOD, and MIKE suites of software (Teng et al. 2017), and most of these models have a strong GIS interface, which can be used to read, write, and analyze geospatial data. GIS-enabled HD modeling approach is given in Fig. 26.8 (Thakur et al. 2016), which can be utilized for any river system.

Hydrodynamic models involve datasets for cross-sections, bathymetry, land use/land cover, hydraulic structures, inflow and boundary conditions, and meteorological data. These datasets can be obtained either by field surveys that are generally expensive or using geospatial techniques. For example, in the case of non-availability of

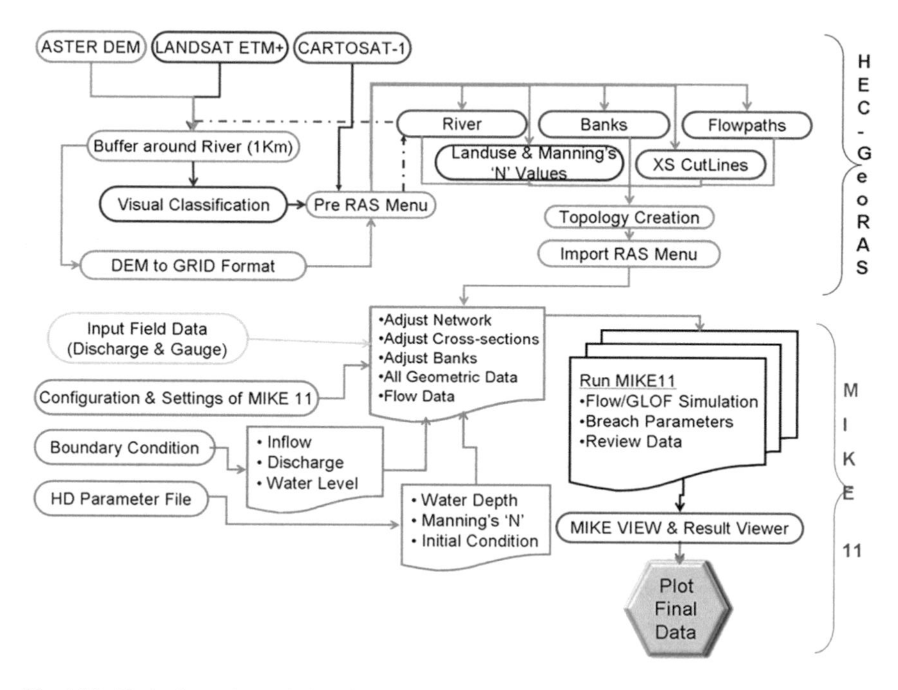

Fig. 26.8 Hydrodynamic modeling flow chart, highlighting the integration of various geospatial data in 1D HD models, such as Mike-11. *Source* Thakur et al. (2016)

the field data for cross-sections, DTM is generally used as a replacement (Thakur and Sumangala 2006; Yan et al. 2015). Also, bathymetric information can be extracted out using the altimeter products (Adnan and Atkinson 2012). The land use/land cover can be derived from satellite imagery such as Resourcesat, Landsat, and Sentinel satellites. Several geospatial algorithms are available from minimum distance to machine learning, such as random forest and deep neural networks. If someone does not want to generate their datasets, several pre-tested global land cover products are available (Pandey et al. 2021). The inflow and boundary conditions, if not viable from the observation, then hydrologic simulations outputs can be utilized (Grimaldi et al. 2013). One of the essential meteorological parameters used is rainfall, which can be either provided by the rainfall stations or in the form of gridded datasets such as interpolated gridded products, reanalysis, satellite datasets, and forecasted rainfall products (Hu et al. 2019).

Some examples of hydrodynamic modeling studies include developing a flood forecasting system where the hydrological model is run using forecasted rainfall, and then, the output is used to provide the flood inundation using the hydrodynamic model (Patel 2015). Preparing the flood risk maps for a 100-year return period of rainfall for a given region (Patro et al. 2009). Using geospatial techniques such as microwave radar sensors for near-real-time flood monitoring by updating the flood inundation maps (Schumann and Moller 2015). Reanalyzing the natural hazards such as glacial lake outbursts, flash floods (Rao et al. 2014), and urban floods (Rangari et al. 2019). Understanding the potential inundation areas after a dam or reservoir breach (Yi 2011). The overall methodology of the 1D HD model is shown in Fig. 26.8, where the DEM is used to generate the river, banks, flow paths, and cross-section lines. The Manning's N value is calculated based on the land use/land cover. The overall generated network of river, flow paths, and cross-sections are adjusted to avoid pitfalls. Next, the initial and boundary are provided based on the observations or simulations. Lastly, the simulations were run based on the hydrodynamic parameters.

26.5.1 A Case Study on Hydrodynamic Modeling

The aim of this case study is to generate flood inundation maps using combination of 1D and 2D HD models. The study area is a part of Tirthan River, Himachal Pradesh, India, for a total length of 9 km. The narrow channels characterize the reach and steep slopes making it difficult do the field survey. Therefore, this study shows the potential of geospatial datasets in places where it is difficult to perform field surveys. The overall setup of the MIKE FLOOD which is 1D–2D coupled model is shown in Fig. 26.9. The 1D model is run (shown in solid black line), while the 2D model runs in the area shown in the grids. The two HD models are coupled to save computational time. The results of the simulations suggest that the HD model performs well (Fig. 26.10) and can be used for further modeling studies, such as flood forecasting in the region, generating flood risk maps, and flood inundation maps (Dhote et al. 2019). This study has concluded that that 1D approach gives efficient

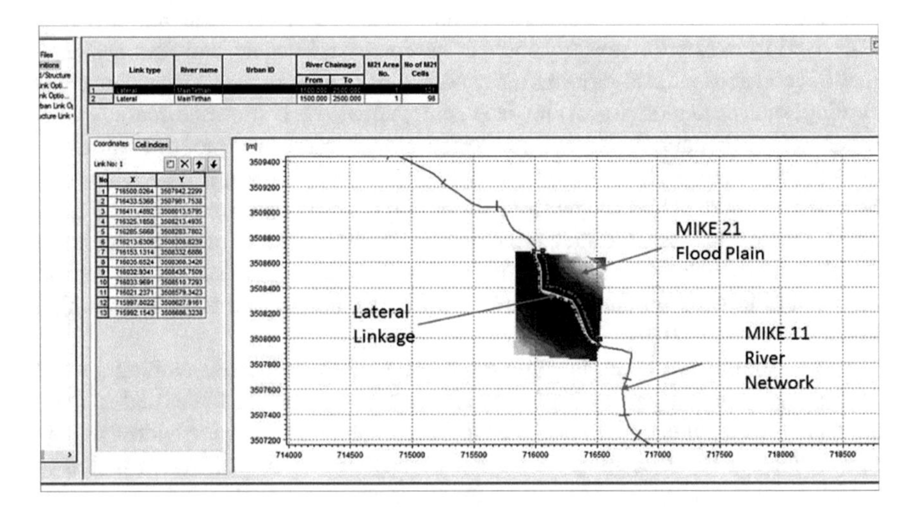

Fig. 26.9 MIKE FLOOD setup for the Tirthan Basin. *Source* Dhote et al. (2019)

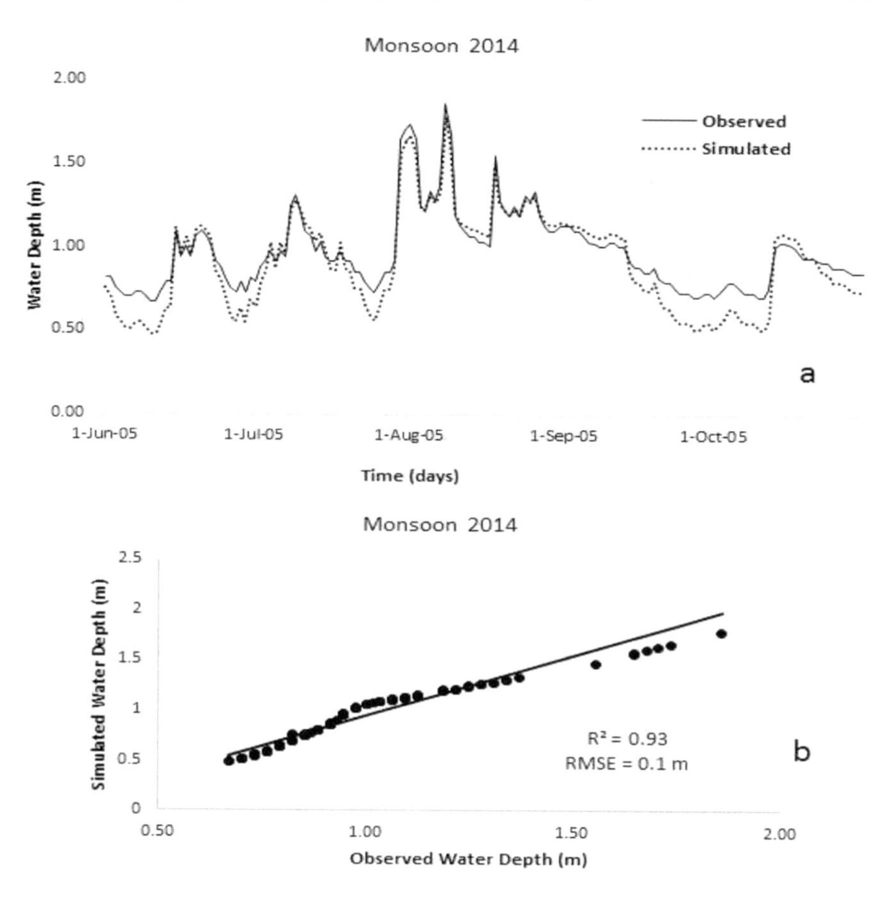

Fig. 26.10 Comparison of observed and 1D HD model simulated water depth at Larji for validation in Tirthan River, Himachal Pradesh, India. *Source* Dhote et al. (2019)

results in the mountainous terrain of Tirthan River while river channel-floodplain characteristics restrict the application of 1D–2D coupled model.

26.6 Advances and Future Directions for Integration of Geospatial Technology in Hydrological/hydrodynamic Modeling

The traditional methods or hydrological models utilizes the geospatial data in a standalone closed loop system, i.e., using various thematic layers to characterize the land surface (LULC, soil, and DEM) and time series of RS-derived meteorological inputs (precipitation, snow cover) as inputs for H&H models, without any feedback from actual observations or model outputs. However, the various land surface and hydro-meteorological parameters vary daily to sub-daily space time scales (e.g., soil moisture, SWE, water level), which standard H&H models cannot incorporate in static systems.

In this regard, data assimilation (DA) techniques that incorporate the satellite-based inputs into H&H models in a more direct and forward loop feedback provide the best strategy for improving the simulation accuracy of these models. Moradkhani (2008) provides a comprehensive review of various techniques used in hydrologic remote sensing and land surface DA. Griessinger et al. (2016) showed that DA of snow cover improved the hydrological simulations of an alpine catchment. DeChant and Moradkhani (2011) heightened the impact of DA in improving the characterization of initial conditions for ensemble streamflow prediction.

Additionally, the recent constellation of small satellite such as Planet, ICEYE, and SPIRE have capability to provide daily to demand-based sub-daily data in optical, microwave region radio occultation/GNSS reflectometry domains. In future, more such constellations are planned from various national space agencies and private satellite companies covering various hydrological variables (such as precipitation, SM, and WL). Few of the future dedicated important satellite missions such as NASA-ISRO Synthetic Aperture Radar (NISAR), Surface Water and Ocean Topography (SWOT), Global Change Observing Mission (GCOM), ISRO GISAT, and Copernicus Polar Ice and Snow Topography Altimeter (CRISTAL) will provide vital geospatial inputs for hydrological studies.

26.7 Conclusions

The importance of remote sensing-derived geospatial inputs for various hydrological studies is highlighted. It can be seen that a large number of sensors and their data products are available in public domain with global coverage for around last 50 years. The water resources and hydrological applications were the first applications of the satellite data from the historic geostationary and polar orbiting satellites, such as TIROS and Landsat series. During the last 30 years, the RS has matured to provide quantitative estimates on hydrological parameters and improved inputs for H&H models. The main conclusions of the chapter are listed below:

- Geospatial data and technology are indispensable to undertake any water resources management or hydrological assessment including use of H&H models.
- The major geospatial inputs for hydrological models are, LULC, soil, DEM, and hydrological variables. Most of the hydrological models now work with a dedicated GIS interface.
- The main geospatial inputs for HD models are, river morphology using medium–high-resolution satellite data; DEM derived river cross-sections; satellite-based water levels and rating curves at virtual gauges; and RS-based flood inundation.
- It has been noted that the use of geospatial data has improved the accuracy of H&H models in both gauged and ungauged basins.

- The recent advances and future sensors would provide high spatial and temporal geospatial data enabling improvements in land surface characterization and data integration for H&H models, enhancing their modeling accuracy.

Acknowledgements The authors are thankful to the Director, Indian Institute of Remote Sensing, Dehradun and Department of Civil Engineering, Interdisciplinary Programme (IDP) in Climate Studies and remote sensing facilities at Civil Engineering Department, IIT Bombay, for providing the required facilities.

References

Web References

Adnan NA, Atkinson PM (2012) Remote sensing of river bathymetry for use in hydraulic model prediction of flood inundation. In: 2012 IEEE 8th international colloquium on signal processing and its applications, pp 159–163

Aggarwal SP, Thakur PK, Dadhwal VK (2009) Remote sensing and GIS applications in flood management. J Hydrol Res Dev Theme Flood Manage 24:145–158

Aggarwal SP, Garg V, Gupta PK, Nikam BR, Thakur PK, Roy PS (2013) Runoff potential assessment over Indian landmass: a macro-scale hydrological modeling approach. Curr Sci 104(7):950–959

Aggarwal SP, Garg V, Gupta PK, Nikam BR, Thakur PK (2012) Climate and LULC change scenarios to study its impact on hydrological regime. Int Arch Photogramm Remote Sens Spat Inf Sci XXXIX-B8:147–152. https://doi.org/10.5194/isprsarchives-XXXIX-B8-147-2012

Amir MSII, Khan MMK, Rasul MG, Sharma RH, Akram F (2018) Hydrologic and hydrodynamic modelling of extreme flood events to assess the impact of climate change in a large basin with limited data. J Flood Risk Manage 11:S147–S157. https://doi.org/10.1111/jfr3.12189

Arya VS, Kumar R, Hooda RS (2010) Evaluation and geo-database creation of watersheds in Siwaliks, Haryana. Curr Sci 98(9):1219–1223

Bekhira A, Habi M, Morsli B (2018) Hydrological modeling of floods in the Wadi Bechar watershed and evaluation of the climate impact in arid zones (southwest of Algeria). Appl Water Sci 8:185. https://doi.org/10.1007/s13201-018-0834-3

Bhatt CM, Rao GS, Farooq M, Manjusree P, Shukla A, Sharma SVSP, Kulkarni SS, Begum A, Bhanumurthy V, Diwakar PG, Dadhwal VK (2017) Satellite-based assessment of the catastrophic Jhelum floods of September 2014, Jammu & Kashmir, India, Geomatics. Nat Hazards Risk 8(2):309–327

Bhatt CM, Srinivasa Rao G, Manjushree PV, Bhanumurthy V (2010) Space based disaster management of 2008 Kosi floods, North Bihar, India. J Indian Soc Remote Sens 38:99–108

Chakraborti AK (1993) Strategies for watershed management planning using remote sensing techniques. J Indian Soc Remote Sens 21(2):87–97. https://doi.org/10.1007/BF02996346

Chow VT (1959) Open channel hydraulics. MacGraw-Hill Book Co., Inc., New York, NY, p 206

Chow VT, Maidment DR, Mays LW (1988) Applied hydrology. McGraw-Hill, New York, NY

Central Water Commission, CWC (2019) Reassessment of water availability in India using space inputs. A project report by Basin Planning & Management Organisation (BPMO), CWC New Delhi and NRSC Hyderabad, No: CWC/BPMO/BP/WRA/June 2019/Main Report, 120 pages

Davies S, Mirfenderesk H, Tomlinson R, Szylkarski S (2009) Hydrodynamic, water quality and sediment transport modeling of estuarine and coastal waters on the gold coast Australia. J Coastal Res 56:937–941

DeChant CM, Moradkhani H (2011) Improving the characterization of initial condition for ensemble streamflow prediction using data assimilation. Hydrol Earth Syst Sci 15:3399–3410

Demaria MCE, Capdevila SA (2016) Validation of streamflow outputs from models using remote sensing inputs. In: García L, Rodríguez D, Wijnen M, Pakulski I (eds) Chapter 10 in earth observation for water resources management: current use and future opportunities for the water sector, Pages, 185–192, https://doi.org/10.1596/978-1-4648-0475-5_ch10

Dhote PR, Aggarwal SP, Thakur PK, Garg V (2019) Flood inundation prediction for extreme flood events: a case study of Tirthan River, North West Himalaya. Himalayan Geol 40(2):128–140

Dhote PR, Thakur PK, Domeneghetti A, Chouksey A, Garg V, Aggarwal SP, Chauhan P (2020) The use of SARAL/AltiKa altimeter measurements for multi-site hydrodynamic model validation and rating curves estimation: an application to Brahmaputra River, a COSPAR publication. Adv Space Res 68(2):691–702. https://doi.org/10.1016/j.asr.2020.05.012

Durga Rao KHV, Venkateshwar Rao V, Dadhwal VK (2014) Improvement to the Thornthwaite method to study the discharge at a basin scale using temporal remote sensing data. Int J Water Resour Manage 28:1567–1578

Feldman AD (2000) Hydrologic modeling system, HEC-HMS, Technical Reference Manual. US Army Corps of Engineers, Hydrologic Engineering Center, HEC, 609 Second St., Davis, CA, CPD-74B, 145 pages

Fleming MJ, Doan JH (2013) HEC-GeoHMS - Geospatial Hydrologic Modeling Extension, User's Manual, Version 10.1, US Army Corps of Engineers, USACE, Institute for Water Resources, Hydrologic Engineering Center (HEC), 609 Second Street, Davis, CA 95616, 193 pages

García L, Rodríguez D, Wijnen M, Pakulski I (2016) Earth observation for water resources management: current use and future opportunities for the water sector, Editors, 220 pages, https://doi.org/10.1596/978-1-4648-0475-5

Garg V, Sambare RS, Thakur PK, Dhote PR, Nikam BR, Aggarwal SP (2019a) Improving stream flow estimation by incorporating time delay approach in soft computing models. ISH J Hydraulic Eng. https://doi.org/10.1080/09715010.2019.1676171

Garg V, Nikam BR, Thakur PK, Gupta PK, Aggarwal SP, Srivastav SK (2019b) Human-Induced Land Use Land Cover Change and its Impact on Hydrology. HydroResearch 1:48–56

Garg V, Aggarwal SP, Gupta PK, Nikam BR, Thakur PK. Srivastav SK, Senthil Kumar A (2017) Assessment of land use land cover change impact on hydrological regime of a Basin. Environ Earth Sci 76.https://doi.org/10.1007/s12665-017-6976-z

Ghosh S, Thakur PK, Sharma R, Nandy S, Garg V, Amarnath G, Bhattacharyya S (2017) The potential applications of satellite altimetry with SARAL/AltiKa for Indian inland waters. Proc Natl Acad Sci, India, Sect A 87(4):661–677

Griessinger N, Seibert J, Magnusson J, Jonas T (2016) Assessing the benefit of snow data assimilation for runoff modeling in Alpine catchments. Hydrol Earth Syst Sci 20:3895–3905

Grimaldi S, Petroselli A, Arcangeletti E, Nardi F (2013) Flood mapping in ungauged basins using fully continuous hydrologic–hydraulic modeling. J Hydrol 487:39–47

Hu Q, Li Z, Wang L, Huang Y, Wang Y, Li L (2019) Rainfall spatial estimations: a review from spatial interpolation to multi-source data merging. Water 11(3):579

Jiang D, Wang K (2019) The role of satellite-based remote sensing in improving simulated streamflow: a review. MDPI Water 11(1615):1–29. https://doi.org/10.3390/w11081615

Kalogeropoulos K, Stathopoulos N, Psarogiannis A, Pissias E, Louka P, Petropoulos GP, Chalkias C (2020) An integrated GIS-hydro modeling methodology for surface runoff exploitation via small-scale reservoirs. Water 12(3182):1–18. https://doi.org/10.3390/w12113182

Kushwaha SPS, Mukhopadhyay S, Hariprasad V, Kumar S (2010) Sustainable development planning in Pathri Rao sub-watershed using geospatial techniques. Curr Sci 98(11):1479–1486

Maidment D, Djokic D (2000) Hydrologic and hydraulic modeling support with geographic information systems. ESRI Press, Redland, USA, Environmental Systems Research Institute, p 216

Meijerink AMJ, Mannaerts CMM (2000) Introduction to and general aspects of water management with the aid of remote sensing. In: Schultz GA, Engman ET (eds) Remote sensing in hydrology and water management. Springer, Berlin. https://doi.org/10.1007/978-3-642-59583-7_15

Meijerink AMJ, De Brouwer JAM, Mannaerts CMM, Valenzuela CR (1994) Introduction to the use of geographical information systems for practical hydrology. UNESCO—ITC Publication N° 23, International Institute of Aerospace Surveys & Earth Sciences. IHP-IV M 2.3, ITC, Enschede, the Netherlands

Mishra RN, Mohanty T (2007) Hydrological simulation of Mahanadi river basin. Post Graduate Diploma (PGD) report of Indian Institute of Remote Sensing (IIRS) and International Institute for Geoinformation Science and Earth Observation (ITC), a Joint PGD in Geohazards, 134 pages

Mishra V, Shah R, Thrasher B (2014) Soil moisture droughts under the retrospective and projected climate in India. J Hydrometeorol 15(6):2267–2292

Montanari A, Sideris M (2018) Satellite remote sensing of hydrological change. In: Beer T, Li J, Alverson K (eds), Global change and future earth: the geoscience perspective (Special Publications of the International Union of Geodesy and Geophysics, pp. 57–71). Cambridge University Press, Cambridge. https://doi.org/10.1017/9781316761489.008

Moradkhani H (2008) Hydrologic remote sensing and land surface data assimilation. Sensors 8:2986–3004

Nikam BR, Garg V, Jaya K, Gupta PK, Srivastav SK, Thakur PK, Aggarwal SP (2018) Analyzing Future Water Availability and Hydrological Extremes in Krishna Basin under Changing Climatic Conditions. Arab J Geosci 11(19):581, 1–16. https://doi.org/10.1007/s12517-018-3936-1

National Remote Sensing Centre, NRSC (2013) Assessment of water resources at basin scale using space inputs—A pilot study in Godavari and Brahmani-Baitarani basins, NRSC-TR369

Paiva RCD, Collischonn W, Tucci CEM (2011) Large scale hydrologic and hydrodynamic modeling using limited data and a GIS based approach. J Hydrol 406(3–4):170–181. https://doi.org/10.1016/j.jhydrol.2011.06.007

Pandey A, Chowdary VM, Mal BC, Dabral PP. (2011) Remote sensing and GIS for identification of suitable sites for soil and water conservation structures. Land Degrad. Develop 22:359–372. https://doi.org/10.1002/ldr.1012

Pandey PC, Koutsias N, Petropoulos GP, Srivastava PK, Ben Dor E (2021) Land use/land cover in view of earth observation: data sources, input dimensions, and classifiers—a review of the state of the art. Geocarto Int 36(9):957–988

Patel P (2015) Flood simulation using weather forecasting and hydrological models. A Master of Technology (M.Tech)—Remote Sensing and GIS, project report of Water Resources Department, WRD of IIRS Dehradun. 68 pages

Patro S, Chatterjee C, Mohanty S, Singh R, Raghuwanshi NS (2009) Flood inundation modeling using MIKE FLOOD and remote sensing data. J Indian Soc Remote Sens 37(1):107–118

Plate EJ (2009) HESS opinions: classification of hydrological models for flood management. Hydrol Earth Syst Sci 13:1939–1951. www.hydrol-earth-syst-sci.net/13/1939/2009/

Pramanik N, Panda RK, Sen D (2010) One dimensional hydrodynamic modeling of river flow using DEM extracted river cross-sections. Water Resour Manage 24:835–852. https://doi.org/10.1007/s11269-009-9474-6

Rahman MR, Thakur PK (2017) Detecting, mapping and analysing of flood water propagation using synthetic aperture radar (SAR) satellite data and GIS: A case study from the Kendrapara District of Orissa State of India. Egypt J Remote Sens Space Sci 21(S1):S37–S41

Rangari VA, Umamahesh NV, Bhatt CM (2019) Assessment of inundation risk in urban floods using HEC RAS 2D. Model Earth Syst Environ 5(4):1839–1851

Rao KD, Rao VV, Dadhwal VK, Diwakar PG (2014) Kedarnath flash floods: a hydrological and hydraulic simulation study. Curr Sci 598–603

Refsgaard JC (1990) Terminology, modelling protocol and classification of hydrological model codes. In: Abbott MB, Refsgaard JC (eds) Distributed hydrological modelling. Water science and technology library, vol 22. Springer, Dordrecht. https://doi.org/10.1007/978-94-009-0257-2_2.

Roy PS, Meiyappan P, Joshi PK, Kale MP, Srivastav VK, Srivasatava SK, Behera MD, Roy A, Sharma Y, Ramachandran RM, Bhavani P, Jain AK, Krishnamurthy YVN (2016) Decadal land use and land cover classifications across India, 1985, 1995, 2005. ORNL DAAC, Oak Ridge, Tennessee, USA. https://doi.org/10.3334/ORNLDAAC/1336

Roy A (2018) Flood early warning system development in North Western Himalaya. A Master of Technology (M.Tech)—Remote Sensing and GIS, project report of Water Resources Department, WRD of IIRS Dehradun. 138 pages

Scharffenberg B, Bartles M, Brauer T, Fleming M, Karlovits G (2018) Hydrologic Modeling System, HEC-HMS, User's Manual. US Army Corps of Engineers, Hydrologic Engineering Center, HEC, 609 Second St., Davis, CA, CPD-74A, 640 pages

Schultz GA (1988) Remote sensing in hydrology. J Hydrol 100:239–265

Schultz GA, Engman ET (2000) Remote sensing in hydrology and water management. Springer, Berlin Heidelberg, XX, 483.https://doi.org/10.1007/978-3-642-59583-7

Schulze RE (1998) Hydrological modelling: concepts and practice. International Institute of Infrastructural, Hydraulic and Environmental Engineering, Delft, Netherlands, p 134

Schumann GJP, Moller DK (2015) Microwave remote sensing of flood inundation. Phys Chem Earth, Parts a/b/c 83:84–95

Sharma KD, Singh S (1992) Runoff estimation using landsat thematic mapper data and the SCS model. Hydrol Sci J 37(1):39–52. https://doi.org/10.1080/02626669209492560

Sharma AK, Thakur PK (2007) Quantitative assessment of sustainability of proposed watershed development plans for Kharod watershed, Western India. J Indian Soc Remote Sens (JISRS) 35(3):231–241

Sheffield J, Wood EF, Pan M, Beck H, Coccia G, Serrat-Capdevila A, Verbist K (2018) Satellite remote sensing for water resources management: potential for supporting sustainable development in data-poor regions. Water Resour Res 54:9724–9758. https://doi.org/10.1029/2017WR022437

Sindhu K, Durga Rao KHV (2017) Hydrological and hydrodynamic modeling for flood damage mitigation in Brahmani–Baitarani River Basin, India. Geocarto Int 32(9):1004–1016.https://doi.org/10.1080/10106049.2016.1178818

Singh VP (ed) (1995) Computer models of watershed hydrology. Water Resources Publications, Littleton, Revised Edition, 1130 pages

Singh VP (2018) Hydrologic modeling: progress and future directions. Geosci Lett 5:15. https://doi.org/10.1186/s40562-018-0113-z

Singh RK, Prasad VH, Bhatt CM (2004) Remote sensing and GIS approach for assessment of the water balance of a watershed/Evaluation par télédétection et SIG du bilan hydrologique d'un bassin versant. Hydrol Sci J 49(1):131–141. https://doi.org/10.1623/hysj.49.1.131.53997

Slack RB, Welch R (1980) Soil conservation service runoff curve number estimates from Landsat data. Bull Wat Resour 16:887–893

Tarpanelli A, Barbetta S, Brocca L, Moramarco T (2013) River discharge estimation by using altimetry data and simplified flood routing modeling. Remote Sens 5:4145–4162

Teng J, Jakeman AJ, Vaze J, Croke BF, Dutta D, Kim S (2017) Flood inundation modelling: a review of methods, recent advances and uncertainty analysis. Environ Model Softw 90:201–216

Thakur PK, Sumangala A (2006) Flood inundation mapping and 1-D hydrodynamic modeling using remote sensing and GIS techniques. Published in ISPRS Orange Book Publications during ISPRS/ISRS commission IV symposium on: "Geospatial Database for Sustainable Development" at GOA, India, from 27–30, 2006, 1–6, http://www.isprs.org/proceedings/XXXVI/part4/WG-VIII-2-9.pdf

Thakur PK (2013) Catchment geo-data base creation-using geo-HMS and flood hydrograph estimation using HEC-HMS. Lecture notes for short course on Flood risk mapping, modeling and assessment using space technology, July, 22–26, 2013, CSSTEAP, Dehradun, India, Internal publications of Water Resources Department (WRD) of Indian Institute of Remote Sensing (IIRS), Dehradun, 19 pages

Thakur JK, Singh SK, Ramanathan A, Prasad MBK, Gossel W (eds) (2012) Geospatial techniques for managing environmental resources. Springer, XII, 296 pages. ISBN: 978-94-007-1858-6

Thakur PK, Aggarwal S, Aggarwal SP, Jain SK (2016) One-dimensional hydrodynamic modeling of GLOF and impact on hydropower projects in Dhauliganga River using remote sensing and GIS applications. Springer's Natural Hazards 83:1057–1075. https://doi.org/10.1007/s11069-016-2363-4

Thakur PK, Nikam BR, Garg V, Aggarwal SP, Chouksey A, Dhote PR, Ghosh S (2017) Hydrological parameters estimation using remote sensing and GIS for Indian Region—a review. Proc Natl Acad Sci India Sect A 87(4):641–659

Thakur PK, Aggarwal SP, Nikam BR, Garg V, Chouksey A, Dhote PR (2018) Training, education, research and capacity building needs and future requirements in applications of geospatial technology for water resources management. Int Arch Photogramm Remote Sens Spatial Inf Sci XLII-5:29–36. https://doi.org/10.5194/isprs-archives-XLII-5-29-2018

Thakur PK, Garg V, Kalura P, Agrawal B, Sharma V, Mohapatra M, Kalia M, Aggarwal SP, Calmant S, Ghosh S, Dhote PR, Sharma R, Chauhan P (2021) Water level status of Indian reservoirs: a synoptic view from altimeter observations. A COSPAR publication. Adv Space Res 68(2):619–640. https://doi.org/10.1016/j.asr.2020.06.015

Thornthwaite CW, Mather RJ (1955) The water balance. Publications in Climatology 8, 1–86. DIT, Laboratory of Climatology, Centerton, New Jersey, USA

Thronthwaite CW, Mather JR (1957) Instructions and tables for computing potential evapotranspiration and water balance, Laboratory of Climatology, Publication No. 10, Centerton, NJ

Verma S, Verma RK, Mishra SK, Singh A, Jayaraj GK (2017) A revisit of NRCS-CN inspired models coupled with RS and GIS for runoff estimation. Hydrol Sci J 62(12):1891–1930. https://doi.org/10.1080/02626667.2017.1334166

Wijayarathne DB, Coulibaly P (2020) Identification of hydrological models for operational flood forecasting in St John's, Newfoundland, Canada. J Hydrol Regional Stud 27(100646):1–16. https://doi.org/10.1016/j.ejrh.2019.100646

Yan K, Di Baldassarre G, Solomatine DP, Schumann GJP (2015) A review of low-cost space-borne data for flood modelling: topography, flood extent and water level. Hydrol Process 29(15):3368–3387

Yi X (2011) A dam break analysis using HEC-RAS. J Water Resour Protect

Zhuohang X, Shi K, Chenchen W, Wang L, Lei Y (2019) Applicability of hydrological models for flash flood simulation in small catchments of hilly area in China. Open Geosci 11(1):1168–1181

https://serc.carleton.edu/eyesinthesky2/week5/intro_gis.html. Source of figure 27.1c. Last accessed on 27 July 2021

https://snre.arizona.edu/research/watershed-management. Source of figure 27.1a. Last accessed on 27 July 2021

Chapter 27
Delineation of Frequently Flooded Areas Using Remote Sensing: A Case Study in Part of Indo-Gangetic Basin

Vinod K. Sharma, Rohit K. Azad, V. M. Chowdary, and C. S. Jha

Abstract Remote sensing is a useful tool for flood monitoring and damage assessment. Unlike traditional survey methods, it provides cost-effective solution with wider coverage and frequent revisit cycle. In general, coarser sensors provide high repetitivity with lower spatial resolution, whereas constellation of finer spatial resolution sensors can be useful in continuous flood monitoring. Different methods and techniques are used for delineating the flood extent and damage assessment based on the type of sensors. Hence, it is necessary to carry out a detailed and in-depth review of remote sensing technologies and approaches available for processing and analysing satellite data for flood response studies. In the present study, automated procedures were used for generation of flood layers and flood persistence maps at Gram panchayat level in the part of Indo-Gangetic plains, Uttar Pradesh. Further, attempt was made to plan the measures that can be useful in relief operations based on the detailed analysis of persistence maps. Methods based on thresholding were improvised by applying unsupervised classification and online Geo-processing platform of Google Earth Engine. Historical flood events for the period 2010–2020 were generated over part of Indo-Gangetic basin and integrated with administrative layers for identifying the villages vulnerable to floods. Accuracy of flood maps were improved by applying the conditioning factors to remove misclassification of flood extents. Particularly, villages located in Ghazipur, Allahabad, Ballia, Gorakhpur, Bahraich and Balrampur districts of UP state indicated high Flood Vulnerability Index (FVI) values. FVI computed at village level using the historical flood events can be of great help for identifying and planning relief shelter locations in the study area. Remote sensing and GIS technologies were successfully envisaged in the identification and planning of relief shelters for the most vulnerable villages.

V. K. Sharma (✉) · V. M. Chowdary
Regional Remote Sensing Center—North, National Remote Sensing Center, New Delhi 110049, India
e-mail: vinod_sharma@nrsc.gov.in

R. K. Azad
Center for Disaster Management, Jamia Millia Islamia University, New Delhi 110025, India

C. S. Jha
National Remote Sensing Center, Hyderabad 500037, India

A. Pandey et al. (eds.), *Geospatial Technologies for Land and Water Resources Management*, Water Science and Technology Library 103,
https://doi.org/10.1007/978-3-030-90479-1_27

27.1 Introduction

Water that temporarily submerges land is defined as flood (Patel and Srivastava 2013). High rainfall, river breach, tsunami and typhoon are the major causes of flood; however, other phenomena like glacier melting could be another source causing flood-like situation. The flood may submerge land from a few days to months and is considered as most frequent and destructive type of natural disaster across the world. Major consequences of flood disaster include losses to agricultural crops, human and animal lives, damages to infrastructure and communication networks (Brivio et al. 2002; Jonkman and Kelman 2005; Diakakis et al. 2016). Soil loss and subsequent transport of sediment loads and pollutants during floods is also another major recurring problem. Delineation of flood extent using traditional survey methods is a time-consuming process and involves lot of human resources. Delineation of flood areas through traditional survey mapping methods is a major challenge due to dynamic changes in surface water extent. Thus, mapping water surface changes continuously is not possible, thereby resulting in non-availability of flood extent frequently. Availability of remote sensing data proves to be an alternative method to traditional survey due to its frequent revisit capability, broad coverage and low cost. Capabilities of remote sensing data for flood mapping and management is widely recognised globally. Optical and microwave sensors are widely used for mapping the flood extent and damage assessment. Although both optical and microwave sensors have strong water recognition capabilities, availability of cloud-free optical data during monsoon season for flood mapping is difficult. Optical sensors cannot penetrate through clouds due to shorter wavelengths and can only be used during pre- and post-flood seasons (Sheng et al. 2001). Thus, usage of optical satellite data alone for flood assessment and monitoring is limited due to cloud availability. Although, NOAA/AVHRR (National Oceanic and Atmospheric Administration/Advance Advanced Very High-Resolution Radiometer) provides satellite data from its optical sensors, but due its high temporal resolution, it contributed significantly in flood monitoring process. Collection of temporal images from NOAA optical sensors provides cloud-free image over flood-affected area. Microwave data on the other hand has significantly contributed in flood mapping. Several researchers used longer wavelength microwave data for mapping flood extent due to its cloud penetration property and found to be useful for flood mapping during monsoon times (Hess et al. 1995; Kiage et al. 2005; Henry et al. 2006). Radarsat-1, Radarsat-2, SIR-C, ASAR, RISAT and Sentinel 1 are the commonly used longer wavelength microwave satellites providing radar data for flood mapping. Flood forecasting is of great importance for reducing the impact of flood and flood preparedness. Flood forecasting based on WRF model and multi-sensor precipitation estimates are useful for flood modelling (Yucel et al. 2015). Understanding the capabilities and limitations of different sensors, technologies and methods available for flood response studies necessitates to carry out a detailed survey showcasing the available data and tools. Recent advances in technologies enabled satellite data providers to share the satellite data as service. Online processing platforms are providing API's (application programmable interface), thereby facilitating geoprocessing of satellite data. Tremendous potential exists where remote sensing satellite data can be analysed in an automated manner to generate the flood layers (utilising both satellite data as service and analysing satellite data physically). This concept can be applied to historical datasets for generating flood layers over the past flood events. Analysis of historical flood layers will help in identifying the frequently flooded areas, for which relief management measures can be planned. Thus, the present study focused on the utilisation of automated procedures not only for generation of flood layers but also for generation of flood persistence map at Gram panchayat level in the part of Indo-Gangetic plains, Uttar Pradesh. Further, attempt was made to plan the measures that can be useful in relief operations based on the detailed analysis of persistence maps. The strengths and limitations of remote sensing technologies and various methods used for flood mapping and also for managing flood disasters are discussed in subsequent sections.

27.2 Role of Remote Sensing Satellite Data for Flood Mapping

Both optical and microwave data have their advantages and limitations in carrying out the flood mapping activities. Microwave remote sensing is all-weather and invaluable for flood monitoring, whereas multispectral optical images-based flood analysis can be managed with simple image processing (Schumann and Moller 2015).

27.2.1 Optical Remote Sensing Satellite Data for Flood Mapping

Optical remote sensing sensors have potential to detect naturally reflected energy or the energy emitted by Earth's surface in visible and infrared spectral bands. High-resolution optical sensors are proven to be good means for flood mapping activity, particularly for capturing cloud-free flood events with low tree canopy (Klemas 2015). Indian Remote Sensing Satellites (IRS), Landsat, NOAA/AVHRR, MODIS (Moderate Resolution Imaging Spectroradiometer) and S2 (Sentinel-2) are the major optical remote sensing data source for flood mapping activities. Temporal resolution is one of the important factors responsible for flood monitoring. In general, optical sensors can be categorised into coarse, moderate and high-resolution satellite data based on spatial resolution. Analysis of flood situation continuously on ground helps to identify whether the flood extent is increasing or decreasing, thereby enabling the decision-makers to plan for flood relief measures.

(a) Coarse spatial resolution satellite data

Moderate Resolution Imaging Spectroradiometer (MODIS) provides global coverage daily twice with different spectral band combinations. Spatial resolution (250 m) of band 1 and 2 are useful for flood mapping activities (Fayne et al. 2017; Lin et al. 2019). Researchers employed high temporal resolution MODIS data for establishing automated flood monitoring system for Southeast Asia (Ahamed and Bolten 2017). However, incapability of cloud penetration restricts its usage in flood mapping irrespective of high temporal resolution as most floods occurs with a heavy cloud coverage. It has been successfully used for change analyses of surface water and flood duration estimation (Rao et al. 2018; Lin et al. 2017).

Unlike MODIS, usually optical sensors are having low temporal resolution, making it non-suitable for flood change detection studies. Flood monitoring with NOAA/AVHRR sensor provides images over the study area at regular interval of time; however, the spatial resolution is not fine. Studies that involve large-scale flood events, which affect the complete states like Bihar and Assam, moderate spatial resolution NOAA/AVHRR data is quite useful. Researchers have successfully used AVHRR data for dynamic flood monitoring (Sheng et al. 2001). Temporal AVHRR images over the study area were used to identify and analyse the flood extent and damage. Various feature classification techniques that are useful for identifying the flood extent are briefly explained below.

Suomi NPP-VIIRS launched in 2011 is a wide swath multispectral sensor (Suomi National Polar-orbiting Partnership-Visible Infrared Imaging Radiometer Suite) and is an alternative to MODIS and AVHRR for flood mapping studies (Huang et al. 2015; Li et al. 2018a, b). Data from operational satellites, NPP-VIIRS, Joint Polar Satellite System (JPSS) series and Advanced Baseline Imager (ABI) onboard GOES-R series have been further used for mapping, monitoring and prediction of floods due to ice jam and snowmelt (Goldberg et al. 2020). Medium Resolution Imaging Spectrometer (MERIS) datasets were used for flood detection by many researchers (Yesou et al. 2007; Andreoli et al. 2007). As a substitute to MERIS, Sentinel-3 OLCI (Ocean and Land Colour Instrument) provides global coverage every two days using 21 visible and infrared bands at 300-m resolution. Sentinel 3 data have potential for different applications that include river flow monitoring and damage assessment to urban infrastructure exposed to flood (Pourghasemi et al. 2020; Tarpanelli et al. 2020).

Sensors mapping specific regions, rather than providing global coverage, can help in providing high temporal images over the region, which can be useful in dynamic flood monitoring. Indian Space Research Organisation (ISRO) is in the process of launching Geo-stationary satellites with 24-h orbital period over Indian region. It will be providing a range of swath ranging between 1000 × 1000-km, 3000 × 3000-km and 6000 × 6000-km in spectral bands namely multi-spectral visible and near-infrared (6 bands) with a resolution of 42 m, hyperspectral visible and near-infrared (158 bands) with a resolution of 318 m and hyperspectral short wave-infrared (256 bands) with a resolution of 191 m (Source: https://www.isro.gov.in/). It will be helpful for quick monitoring of natural disasters, episodic events and any short-term events. Integration of temporal images over a study area will be helpful in obtaining cloud-free images.

(b) Moderate spatial resolution satellite data

IRS 1A LISS-I false colour composite satellite images are used for flood mapping in the Punjab state during July, 1993. Flood-affected categories were identified using variation in photo elements like tone, texture etc. (Sharma et al. 1996). Simple density slicing, Tasseled Cap Transformation and water-specific index techniques were applied to IRS LISS-III datasets of 1999 over Bihar state for flood delineation. Different flood delineation techniques were compared, where Normalised Difference Water Index (NDWI) approach found to be best suited (Jain et al. 2005). IRS LISS-III and LISS-IV optical data observed to be a good source for validation of the flood classification, interpretation and for analysis of Rockslide Triggered Flood Event (Pandey et al. 2021).

Landsat optical data with a global coverage at periodic intervals facilitates and is advantageous for flood mapping events due to its availability in open-source domain (Jain et al. 2005; Zhou et al. 2002). Combination of Landsat and IRS can be used for increasing the data coverage needed for flood monitoring situation. Various feature classification techniques that include supervised classification, unsupervised classification, delta-cue change detection, normalised difference water index (NDWI) and modified NDWI can be applied to IRS or Landsat images for identification of flood water pixels. Researchers successfully used supervised machine learning techniques for coastal flood mapping, rapid flood mapping and flood frequency studies (Wang et al. 2002; Qi et al. 2009; Ireland et al. 2015; DeVries et al. 2020).

Systeme Probatoire d'Observation dela Tarre (SPOT) satellite series that provides optical data with a spatial resolution of 10-m was used for mapping the flood extent, flood risk mapping and integrated with other datasets including open street maps (OSM) for mapping poorly gauged areas (Blasco et al. 1992; Stancalie et al. 2012; Bozza et al. 2016).

(iii) Fine spatial resolution satellite data

Wide range of sensors namely Cartosat, IKONOS, QuickBird, WorldView and RapidEye provide satellite images at meter and sub-meter spatial resolutions. Particularly, these are highly useful for monitoring small water bodies. Natural disasters like flood impacts a larger area, thereby mapping of larger area extents with high-resolution satellite data is a challenge. Monitoring flood situations demand high temporal data, and acquisition of sub-meter data at regular intervals over a large study area need constellation of satellites, which is costly and may not be possible practically. However, the products generated using high-resolution satellite data like CartoDEM generated from Cartosat stereo data can be beneficial for flood modelling (Jena et al. 2016). Small localised floods can be mapped using the high-resolution satellite data (Malinowski et al. 2015). Usually, high/fine-resolution optical data have limited usage in flood mapping activities. Detailed comparison of optical remote sensors is given in Table 27.1.

27.2.1.1 Image Classification Techniques for Flood Delineation

(a) Supervised and unsupervised classification techniques for flood delineation

Table 27.1 Optical sensors for flood mapping activities

Spatial resolution group	Sensor	No. of bands	Spatial resolution (m)	Temporal resolution (days)	Data availability since
Coarse	NOAA/AVHRR	5	1100	0.5	1978 Onwards
	MODIS	36	250–1000	0.5	1999 Onwards
	Suomi NPP-VIIRS	22	375–750	0.5	2012 Onwards
	MERIS	15	300	3	2002–2012
	Sentinel-3 OLCI	21	300	2	2016 Onwards
Moderate	IRS	4	5.8–70	5–24	1988 Onwards
	Landsat	4–9	15–80	16	1972
	SPOT	4–5	2.5–20	26	1986
	ASTER	14	15–90	16	1999
	Sentinel-2 MSI	13	10–60	5	2015
High resolution	Cartosat	4	0.25–2.5	5	2005 Onwards
	IKONOS	5	1–4	1–3	1999 Onwards
	QuickBird	5	0.61–2.24	2–3	2001 Onwards
	WorldView	4–17	0.31–2.40	1–4	2007 Onwards
	RapidEye	5	5	1–6	2008 Onwards

Supervised and unsupervised classification techniques are important for classifying the remote sensing satellite images. Supervised learning facilitates model to predict future outcomes once the model is trained using ground truth/past data. Main objective of supervised learning is to develop a function capable of predicting outputs by providing new input. It can be used to solve the regression (continues outputs) and classification problems (categorical outputs) (Sen et al. 2020). Supervised classification can be used for both structured and unstructured datasets. Supervised classifiers were successfully used in the extraction of flooded areas during Mediterranean floods using Landsat TM imagery (Ireland et al. 2015). User can select the sample water pixels as a specific spectral signature class, and further, the pixels in image were classified using classifiers like maximum likelihood. However, output of the classification depends on user experience in choosing the training samples i.e. input water pixels.

Unlike supervised classification, unsupervised classification uses automated analysis for grouping pixels. It performs grouping of spectrally similar pixels without user provided sample pixels (for training). In this approach, user specifies the number of classes only and classification is carried out automatically, gathering information on spectral signatures. Unsupervised classification techniques were successfully used to identify the agricultural flood damage and also for assessing the sedimentation process during floods (Pantaleoni et al. 2007; Abbaszadeh et al. 2019). Classification algorithm like K-means can facilitate identification of different earth features and will be labelled post classification using the expert knowledge.

(b) Change detection using pre- and during flood images

Detection and analysis of changes between pre- and post-flood event satellite images over study area can be used to identify and demarcate the flood extent polygon. Extraction and comparison of water layers using two images (pre and during) can be helpful to know the change, and this change can be considered as a flood extent layer. The approach is similar to removal of permanent water bodies (computed from pre-event image) from the satellite image acquired during flood event. However, comparison and extraction of flood polygons manually is time-consuming and not a feasible solution for covering large areas affected due to floods.

(iii) Satellite data derived Index-based flood classification

Normalised Difference Water Index (NDWI) makes use of reflected near-infrared radiation and visible green light to enhance the presence of open water features while eliminating the presence of soil and terrestrial vegetation features. It is computed by using equation given below (McFeeters 1996).

$$\mathrm{NDWI} = \frac{\mathrm{Green} - \mathrm{NIR}}{\mathrm{Green} + \mathrm{NIR}}$$

where Green is a spectral band that encompasses reflected green light and NIR represents reflected near-infrared radiation. Differencing water indices from pre-and post-flood optical images is successfully used for rapid flood inundation mapping, monitoring and damage assessment (Amarnath 2014; Memon et al. 2015; Sivanpillai et al. 2021). Before and after flood event satellite images are also quite useful for identifying the waterlogged areas. LISS-III (Linear imaging self-scanning sensor) optical sensor data at 23-m spatial resolution was used for delineation of waterlogged areas for the year 2002–2003 (Chowdary et al. 2008). Optical data characteristics were used to delineate the waterlogged areas. Spectral properties of waterlogged areas can be easily picked by visible and infrared domain of optical sensors. The standing water areas appear as dark blue to black depending upon the depth of water, while the wet areas appear as dark grey to light grey in colour/tone on the imagery. Chowdary et al. (2008) successfully delineated surface waterlogged areas for both pre- and post-monsoon seasons in the Bihar state (Fig. 27.1).

Normalised Difference Water Index (NDWI) computed using near-infrared (NIR) and green wavelengths was modified by Xu (2006) by substituting the middle infrared

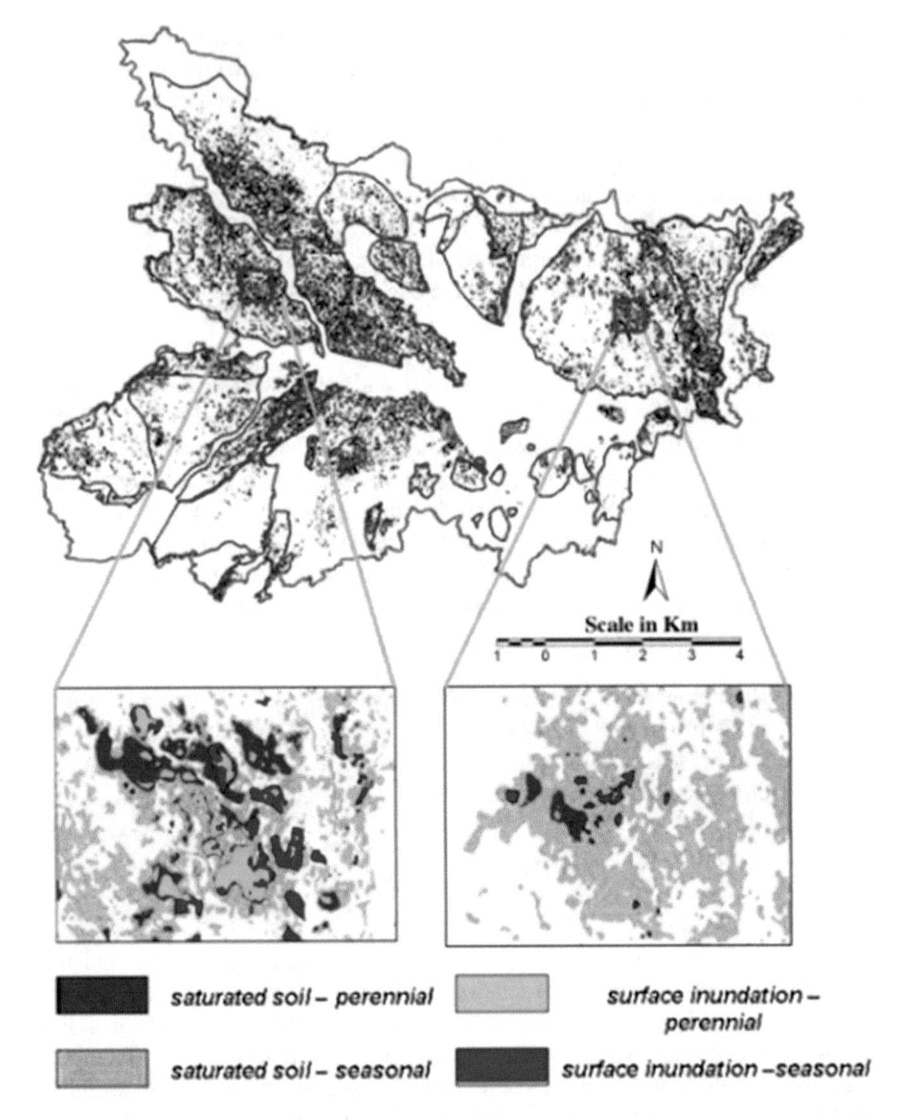

Fig. 27.1 Global trends in flood-related natural disasters; waterlogged areas obtained using analysis of pre-and post-monsoon satellite data. *Source* Chowdary et al. (2008)

band for NIR band and named it as MNDWI (Modified NDWI). Pal and Pani (2016) carried out hydrology studies where river flow dynamics due to heavy rains causing flood-like situation on Ganga River using MNDWI, satellite-based discharge data and Gumbel's flood frequency analysis. NDWI and MNDWI were further replaced with a new index called as Desert Flood Index (DFI) for studying river flooding that includes both desert and vegetation areas (Baig et al. 2013). Comparison of

MNDWI- and DFI-derived flood extents indicated better performance of MNDWI in delineating water body with enhanced contrast with background features.

The above-mentioned methods are used for identification of surface water and subsequently identifying the flood extent by differentiating the extent of pre-event water bodies/permanent water bodies and flood extent during events.

(iv) Flood mapping and prediction using machine learning-based models

Keum et al. (2020) integrated numerical analysis model with machine learning model to develop a classification-based real-time flood prediction model for urban areas. They used Latin hypercube sampling (LHS) and probabilistic neural network (PNN) classification techniques for higher-precision flood range prediction. Bivariate statistical models (frequency ratio and information value) and supervised statistical learning models (boosted regression trees and classification and regression trees) were used for mapping flood proneness in an arid region of southern Iraq (Al-Abadi and Al-Najar 2020). Different models were compared in this study and found that the machine learning models are optimal in mapping flood proneness in the study area, followed by the multivariate and bivariate models.

27.3 Microwave Remote Sensing Satellite Data for Flood Mapping

Limited availability of cloud-free data restricted the usage of optical data for carrying out the flood mapping activity. However, microwave sensors with longer wavelength radiations made them capable for flood extent mapping due to its cloud penetration capability, all-weather capability and day night capability. In some of the applications, Cheng et al. (2019) reported that the features can be classified more clearly as compared to optical sensors. Both passive and active microwave sensors can be used for inundation mapping, where active microwave sensors have better capability to capture the earth information any time. Capabilities of microwave/radar sensors for flood mapping is highlighted in subsequent sections.

Radar Imaging Satellite (RISAT-1) launched by ISRO, is a multi-mode Synthetic Aperture Radar (SAR) system supporting stripmap and scanSAR modes with dual polarisation operation modes. RISAT-1 operates with C-band SAR sensor with a spatial resolution of 3–50-m and swath ranging from 25 to 223-km. Its ability to penetrate the clouds and also daytime and night-time imaging capabilities make it suitable for flood mapping activities (Misra et al. 2013). Changes in the backscattering coefficients during non-flood (dry period) and during flood event were utilised by Bhatt et al. 2020 for determining the flood extents in the Northern India during the year 2014. Potential of Risat-1 SAR data was further utilised for assessment of embankment breach (Sivasankar et al. 2019). Urban flood mapping using SAR data is a challenge due to backscattering effects caused by high rise buildings. Urban areas exhibit high backscattering values in horizontal co-polarisation (HH) while low values in vertical co-polarisation (VV) due to geometric and dielectric properties.

Integration of multiple polarisations can enhance the accuracy of flood mapping. Vanama et al. (2020) combined different change detection methods for identification of permanent water bodies and flooded areas using Risat-1 SAR data. It was observed during this study that HH polarisation is highly sensitive and useful in identifying the permanent water bodies and flood-affected areas over the study area as compared to other polarisation combinations. RISAT-1 SAR data proves to be a good source for identifying the flood extent.

Canadian Space Agency under Radarsat programme launched pair of remote sensing satellites Radarsat-1 (Nov 1995–March 2013) and Radarsat-2 (Dec 2007 onwards). Radarsat-1 operates in C-band (5.3 GHz) and HH polarisation with near-range and far-range incidence angles of 20° and 49.42°, respectively. The resolution of the system is 50-m. Radarsat-2 (RS2) operates in C-band (5.405 GHz), with ScanSAR wide & narrow, wide, fine, standard and extended high & low modes with spatial resolution ranging from 1 to 100 m. ScanSAR wide mode is generally used for flood mapping activities. Radarsat-1 (RS1) data was widely used for flood monitoring and flood mapping caused by hurricanes (Kiage et al. 2005; Kuehn et al. 2002). Constellation of RS1 and RS2 provide regular coverages needed for flood monitoring. Despite its all-weather capabilities, speckle noise needs to be minimised to obtain satisfactory classification results. Unlike RS1 (HH), RS2 has different polarimetric modes including HH, VV, (HV, VH), (HH, VV) and (HH, VV, HV, VH) which make it advantageous to map urban floods also. RS2 has been successfully used in different applications that include flood and flooded vegetation monitoring, wetland mapping and monitoring, soil moisture estimation, river ice estimation and monitoring (Brisco et al. 2008).

Advanced Microwave Scanning Radiometer Earth Observing System (AMSR-E) by National Space Development Agency of Japan (NASDA) provides SAR data with temporal resolution up to two days and spectral resolution ranging from 5 to 60-km. AMSR-E, X band data was successfully used for flood monitoring in Brahamputra basin and also for coastal flood mapping by minimising the noise sensitivity (Singh et al. 2013; Zheng et al. 2016). Although Ka band is available in this sensor, it could not be used efficiently for flood mapping due to its high noise sensitivity.

TerraSAR-X1 is a German SAR satellite mission working in X-band providing high-resolution spotlight, stripmap, ScanSAR and wide ScanSAR polarisation modes with spatial resolutions ranges from 25-cm to 40-m. It was used in rapid flood mapping and integrated with other optical and SAR data for regular coverage over the study area (Herrera-Cruz et al. 2009).

The SAR data illustrated above is not free for usage hence needs advanced programming for data acquisition. European Space Agency (ESA) through Sentinels Data Hub is providing basic pre-processed ground range detected (GRD) Sentinel 1 SAR products, which are free for usage and can be accessed as a service in online processing environment like Google Earth Engine and also can be downloaded using FTP (file transfer protocol) server.

Sentinel-1 mission is a pair of two polar-orbiting satellites, capable of imaging using C-band SAR that enables acquiring imagery in day and night regardless of weather. Sentinel-1 data products are available in single polarisation (HH or VV) and in dual polarisation (VV + VH or HH + HV). The level-1 GRD products can be used in full resolution, high resolution and medium resolution, where in strip mode, it provides 5 × 5-m spatial resolution; in interferometric wide mode, it provides 5 × 20-m and in extra-wide mode, it provides 20 × 40 m spatial resolution. It is designed to address medium- to high-resolution applications through a main mode of operation that features both wide swath (250-km) and high geometric (5-m × 20-m) and radiometric resolution. Sentinel-1 data was successfully used in Operational Flood Mapping in Bangladesh, where the temporal SAR data has been used for damage assessment and also for hazard response (Uddin et al. 2019). It has potential to be used in automated flood extent mapping and in unsupervised flood mapping process (Li et al. 2018a, b; Amitrano et al. 2018).

27.3.1 SAR Inundation Mapping Techniques

Study and analysis of surface roughness is the basic principle in inundation detection using SAR data. Low roughness of flood water produces ideal reflective scattering, which can easily be detected and can be used for delineation of flood extent. Inundation mapping using SAR data is little complex as compared to optical data, as identification and removal of water-like surfaces, speckle noise and geometric corrections can induce false-positives and over detection of flood extent. Techniques usually used for deriving flood extent are simple visual interpretation, image histogram thresholding, automatic classification algorithms, image texture algorithm and multi-temporal change detection methods (Schumann and Moller 2015) and are briefly discussed below.

(a) Visual inspection and editing of generated flood layers

Visual delineation and subsequent editing can generate flood layers with less errors depending on the person's expertise involved in mapping exercise. Manual interpretation involves large man power and is time-consuming. It is not suitable for rapid flood mapping and also for handling multiple flood events. Manual editing can be done to remove the false flood polygons once the flood layer is generated.

(b) Image histogram thresholding

Efforts to determine a threshold where pixels in satellite data below a particular threshold can be identified as water based on specular reflective properties of water was attempted by researchers (Yamada 2001; Matgen et al. 2011). However, a single threshold value cannot be hold good for a larger area due to heterogeneity of the environment. However, for smaller areas or with multiple threshold values, it can provide good results. Generation of flood layers manually by testing set of threshold values is a time-consuming process, while automation of flood mapping steps can save time and resources. An event-driven flood management framework was proposed and implemented that is capable of collecting heterogeneous distributed flood-related information for analysing and alerting probable flood events for generation of automatic flood extent maps (Sharma et al. 2016). Attempts towards generation of threshold ranges for rapid flood inundation mapping by evaluating backscatter profiles of high incidence angle SAR images were attempted (Manjusree et al. 2012), but threshold globalisation is complex. Other flood delineation techniques include segmentation, supervised and unsupervised classification methods.

(iii) Segmentation-based approach for flood delineation

Image segmentation techniques combine and group the homogeneous pixels into patches. It sub-divides the data into disjoint regions that are uniform with respect to homogeneity criteria such as spectral or textural features. Unlike pixel-based application, it allows addition of texture, object geometry and contextual information integration, thereby improving the classification accuracy. Segmentation approach was successfully used by Martinis et al. (2015) for flood extent mapping, semi-automatic and automatic flood rapid mapping, and global water mask generation.

(iv) Supervised and unsupervised classification

Classification methods are used for extracting water layers, where supervised classification demands a rich set of training datasets using which the likely flood polygons can be classified, whereas unsupervised classification classifies the pixels automatically. Generation of rich labelled datasets for supervised classification is a time-consuming process; here, unsupervised classification can play an important role. To overcome the dependency of thresholding and labelled training datasets, Sharma et al. (2017) used ISO clustering algorithm (unsupervised algorithm) for extraction of flood extent. It generates different feature classes and segregate each feature in respective classes by iterative process. The overall process was automated using the Python-based scripts and the flood extent map generated using ISO clustering algorithm using RS2 satellite data is shown in Fig. 27.2. Classification techniques may have false flood polygons and may need a quality check before finalising the flood layer.

(e) Change detection technique for flood delineation

Flood delineation using change detection approach involves comparison of backscattering intensity images acquired during pre- and during flood to detect changes in pixels that caused flood-like situation over a study area (Bazi et al. 2005). Implementation of change detection approach usually involves two images, where permanent water bodies can be demarcated using pre-disaster event image and further can be differenced with water layers extracted from post images for flood layer generation.

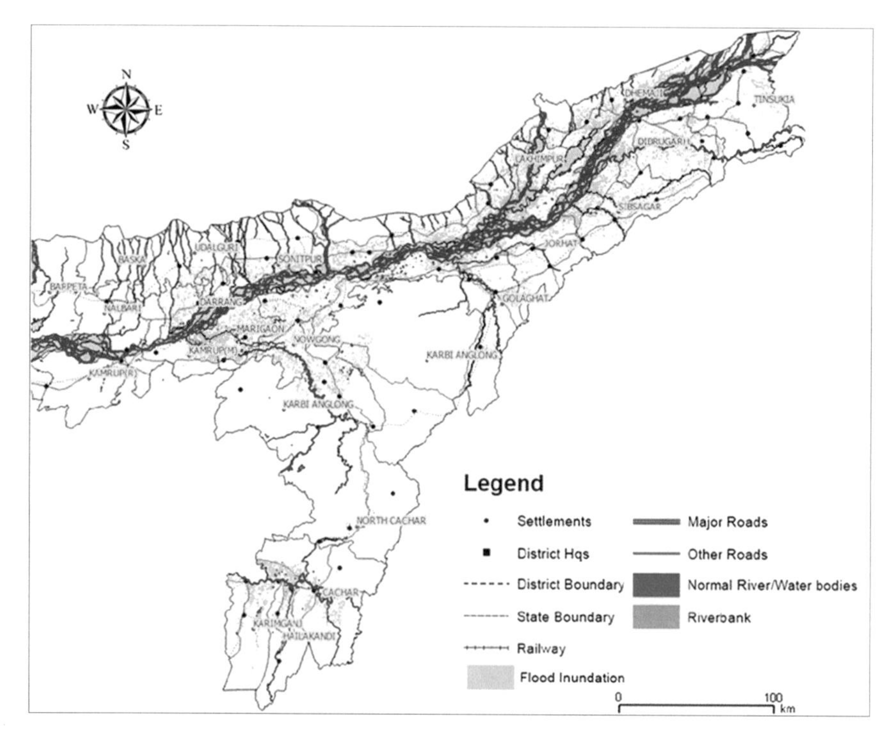

Fig. 27.2 Flood inundation map in Assam state as on 24 July 2016 derived using automated procedure. *Source* Sharma et al. (2017)

27.4 Materials and Methods

27.4.1 Study Area

Uttar Pradesh, part of Indo Gangetic plains lies in north–central part of India and geographically situated between 23°52′ and 31°28′ North latitude and 77°3′–84°39′ East longitude, respectively. It is the fourth largest and most densely populated state with well-distributed network of many rivers and tributaries. Ganga, Yamuna, Ghagra and Gomti are the major rivers flowing through the Uttar Pradesh state causing flood-like situation during the monsoon season (From June to September). Majority of the population lives along the banks of rivers, which get affected due to flood every year. Location map of the study area is shown in Fig. 27.3.

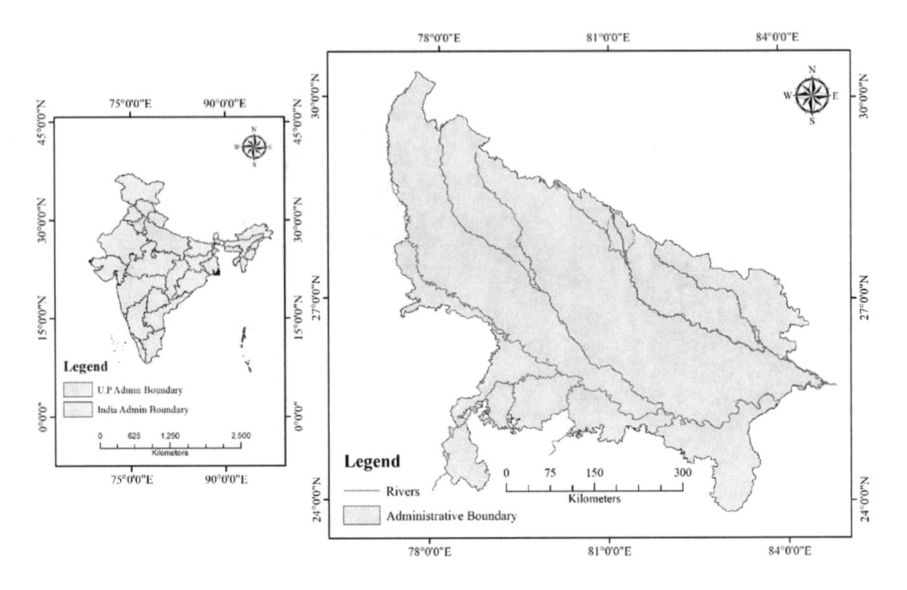

Fig. 27.3 Location map of the study area

27.4.2 Data Used

In this study, C-band SAR data of RISAT-1, RADARSAT-2 and Sentinel-1 were used for mapping the flood events for the period 2010–2020. Combination of offline and online satellite data processing tools were used for flood delineation. Flood events during the period 2010–2016 were mapped by using offline satellite data processing using semi-automated procedures. For the analysis of 2010–2016 flood events, flood products available on disaster management services Web page of Bhuvan geo-portal was considered as base. For the events, where satellite data was not available, flood layers from Bhuvan geo-portal was digitized and used for the study (https://bhuvan-app1.nrsc.gov.in/disaster/disaster.php?id=flood). For the period 2017–2020, freely available Sentinel-1 SAR data was used and yearly flood layer information is provided in Table 27.2. Sentinel data products are available online in the Copernicus Data and Information Access Service (DIAS) cloud environments. The Google Earth Engine (GEE) hosts pre-processed Sentinel-1 Ground Range Detected (GRD) data acquired in different modes for processing and usage. GEE (https://earthengine.google.com/) online geoprocessing platform was used in this study for delineation of flood layers by customizing GEE Web APIs.

SRTM DEM data was used as input in HAND (Height Above Nearest Drainage) for identifying the low-lying areas. The freely available SRTM DEM tiles with a spatial resolution of 30-m were downloaded from https://dwtkns.com/srtm30m. Administrative layers and thematic layers that include road, rail and Land Use Land Cover data were obtained from SIS-DP project (Space-Based Information Support for Decentralised Planning (https://bhuvan-panchayat3.nrsc.gov.in/).

Table 27.2 Major flood events mapped at yearly scale based on satellite data availability

S. No	Year	Major flood mapped using available satellite data
1	2010	2
2	2011–2012	No major flood event
3	2013	7
4	2014	No major flood event
5	2015	6
6	2016	11
7	2017	19
8	2018	15
9	2019	13
10	2020	19

27.5 Methodology

Flood layers were generated by analysis of SAR satellite data available physically and as service. Detailed methodology used for generation of flood layers by integration of offline and online satellite data processing is shown in Fig. 27.4. The satellite data was downloaded from data providers FTP and pre-processed using GIS and Erdas softwares. Autosync functionality of Erdas imagine was successfully employed to geo-reference the input satellite images. A rich library of master images in HH (Horizontal–Horizontal) and HV (Horizontal–Vertical) were analysed

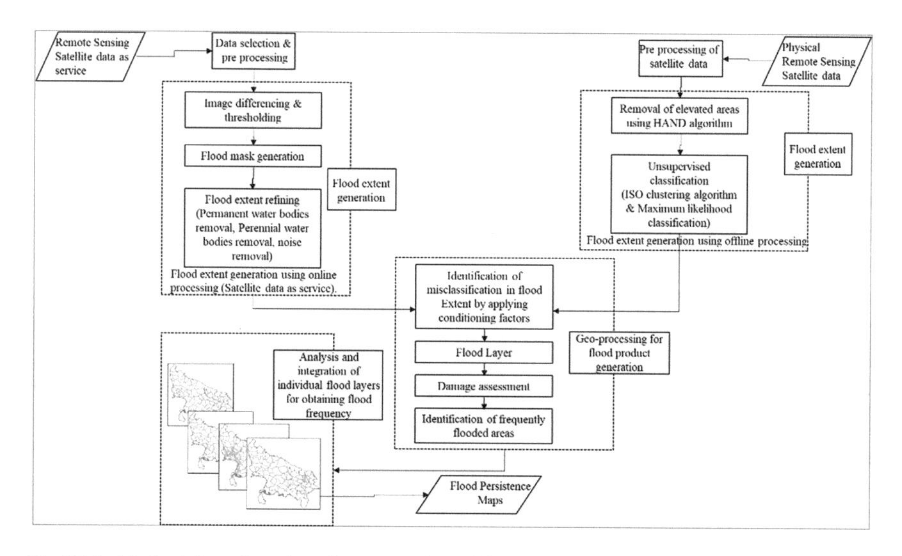

Fig. 27.4 Methodology flowchart for generation of flood layers and flood persistence map

in master data generation and further used for geo-referencing. ISO clustering unsupervised algorithm with five classes were programmed using ArcPython for flood delineation. Outputs of the clustering algorithm were further refined to remove the noise and isolated water pixels by applying maximum likelihood classification and water mask derived from HAND tool. False flood polygons as a result of misclassification were identified and removed by applying conditioning factors that include altitude, slope, aspect and distance from river and land use/land cover map. Freely available Sentinel-1 GRD data as service on GEE was accessed and analysed using an automatic script in Python code. Python is a powerful scripting language and was used in this study. The script accepts the study area as shape file or user can specify the bounding box over which the data can be searched and processed. User has to provide a date range for obtaining the pre-satellite data (date range must be greater than 6, Sentinel-1 revisit time). Data over the user-specified area acquired in date range will be searched and available for visualisation purpose. Similarly, post-disaster date range over same study area can be given as input, where choice can be made to choose the required polarisation (HH/HV) and mode (ascending/descending). Once pre- and post-images are selected, differencing of the images will be performed with a threshold value. Threshold value of 1.25 shows less false positive for the study area.

Additional parameters needed for accessing the Sentinel-1 GRD products include polarisation, orbit properties and AOI mode and were set for initiating the pre-processing of the satellite data. As the study area is large and covered in multiple scenes, options of mosaic and individual tile processing using looping were attempted. Smoothing radius of 50 to 55 was used to reduce the speckles. Once the pre-processing is completed, differencing of the pre- and post-images was done using user-specified threshold value. State mask containing permanent water bodies, perennial waterbodies, elevation areas and river bank, generated using post-flood LISS-IV data was applied on the difference image to remove the permanent water bodies. Isolated flood pixels were also removed by applying filters, resulting in generation of flood extent layers. The flood extent layer still may contain the false polygons, which were minimised using conditioning factors. Further, LULC, Gram panchayat boundaries (GP), rail and road layers were used to compute the damage caused to infrastructure and crop. GPs inundated, road and rail submerged, agriculture under water were computed using an automated model.

The above-mentioned procedure is repeated for each flood event in order to obtain the flood layers for historical flood events. The flood layers were integrated with GP boundaries to identify the flood-affected GPs. Persistency and frequently flooded GP's information were extracted by analysing monthly and annual flood layers. Based on annual flood layer, information about frequently flooded GPs were obtained, and relief measures for them were planned. The process for assessing flood persistency was automated using an ArcGIS model. Frequently flooded areas at GP level for the period 2010–2020 were identified in this study. Further, feasible relief shelter locations were identified by integration of asset/point of interest (POI)/health facility data with the flood persistence map using GIS operations. The resultant products can be used for planning rescue and relief operations in the study area by decision-makers during flood periods.

27.6 Results and Discussion

27.6.1 Flood Layer Generation Using Automated Procedures

Individual flood layers were generated based on the availability of satellite coverage over the study area, while flood persistence maps were generated by integrating these individual flood layers. Administrative layers such as village, GP, taluk, district and state layers were integrated with flood layers so as to identify the extent of flooding at different spatial scales. Individual flood layers were integrated at monthly and yearly scale for generation of monthly and yearly flood layers, while yearly flood layers were integrated to determine the flood persistence at village level. Flood extent as on 27 July 2020 was derived from Sentinel-1 satellite data using GEE platform. Extent of flood damage village scale is shown in Fig. 27.5 and annual commutative flood layer in Fig. 27.6.

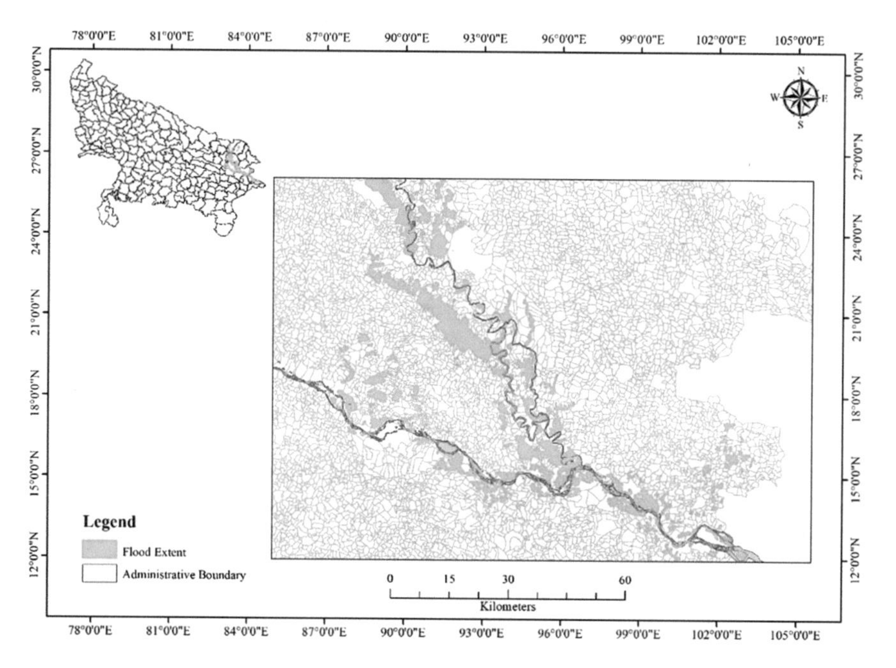

Fig. 27.5 Flood extent as on 27 July 2020 over Uttar Pradesh state, derived from Sentinel-1 satellite data using GEE, showing villages inundated

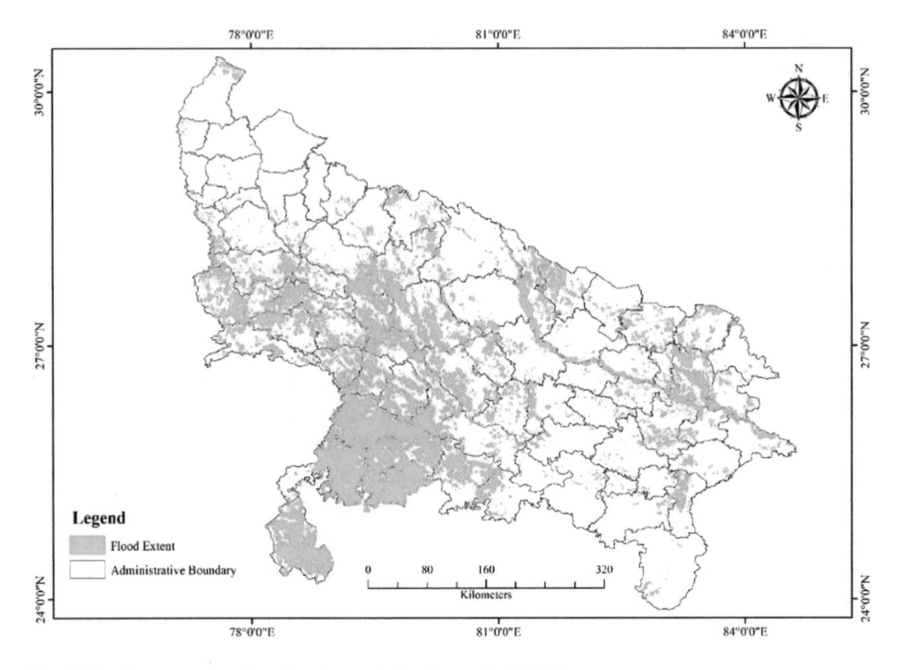

Fig. 27.6 Commutative flooding (annual flood layer) of 2020

Damage assessment to infrastructure (railway track, road) and crop under water can be derived at various time scales using daily, monthly and annual flood layers (Fig. 27.7).

Damage statistics showcasing the area inundated, calculated using annual flood layer can be generated. Gram panchayats with maximum flood inundation in 2017 is shown in Table 27.3.

27.6.2 Analysis of Flood Persistence

The generated flood layers were used to identify the villages that are frequently affected by flood, i.e. proneness to floods. Decadal commutative flood layer was generated by combining all the flood layers with village information. Flood frequency count was computed for each village based on its unique village ID and categorised for computing flood vulnerably index (FVI) at village level. Occurrence of flood in a village is categorised into three classes namely low (0–5 times), medium (5–10 times) and high (>10 times). Spatial distribution of villages categorised into different vulnerability classes in Uttar Pradesh state based on flood occurrence was shown in Fig. 27.8. Particularly, villages located in Ghazipur, Allahabad, Ballia, Gorakhpur, Bahraich and Balrampur districts of UP state indicated high FVI values.

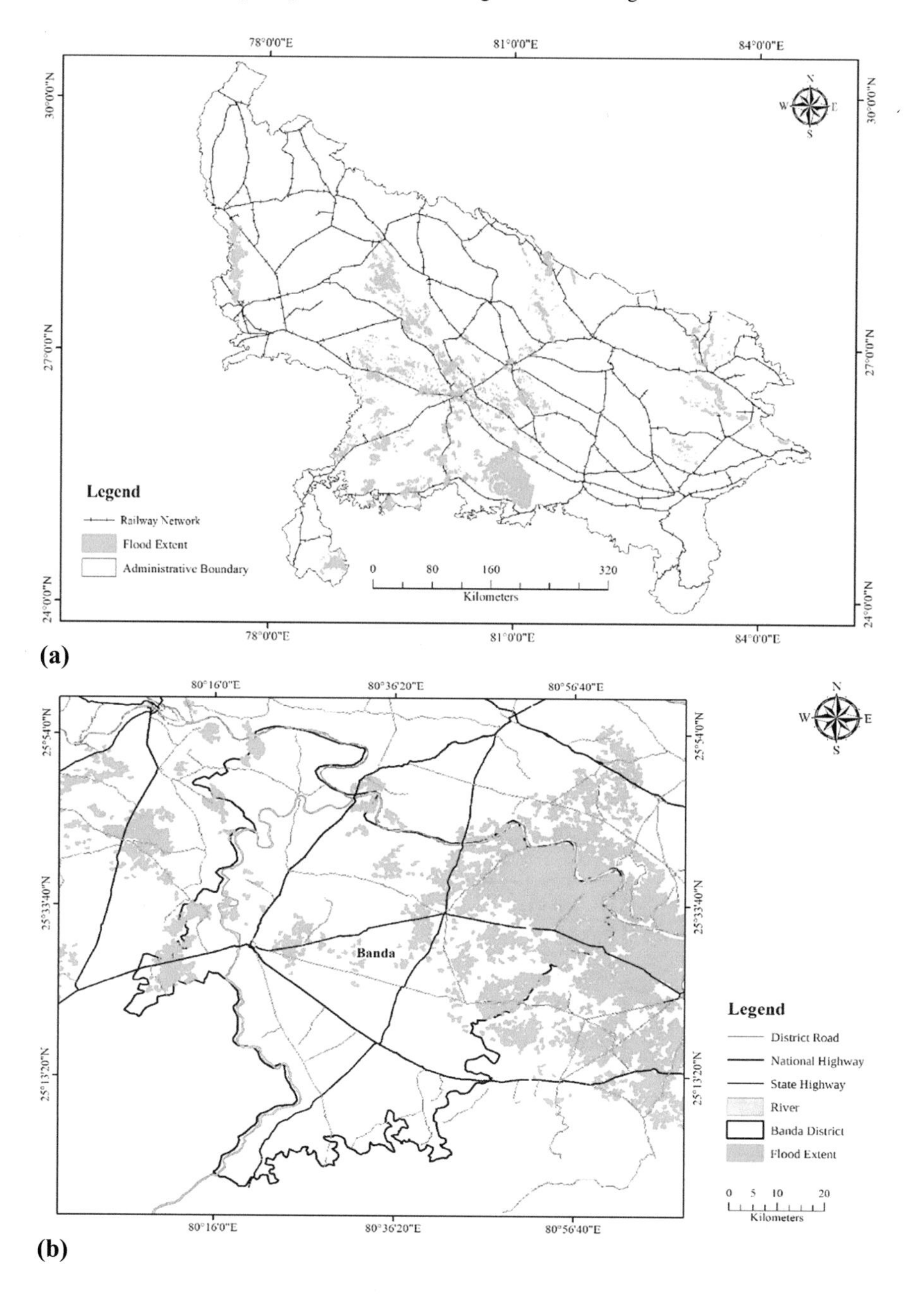

Fig. 27.7 2020 Flood damage assessment **a** Submerged rail network during August 2017. **b** Submerged Road network of Banda district, UP during August 2017. **c** Submerged crops during August 2017

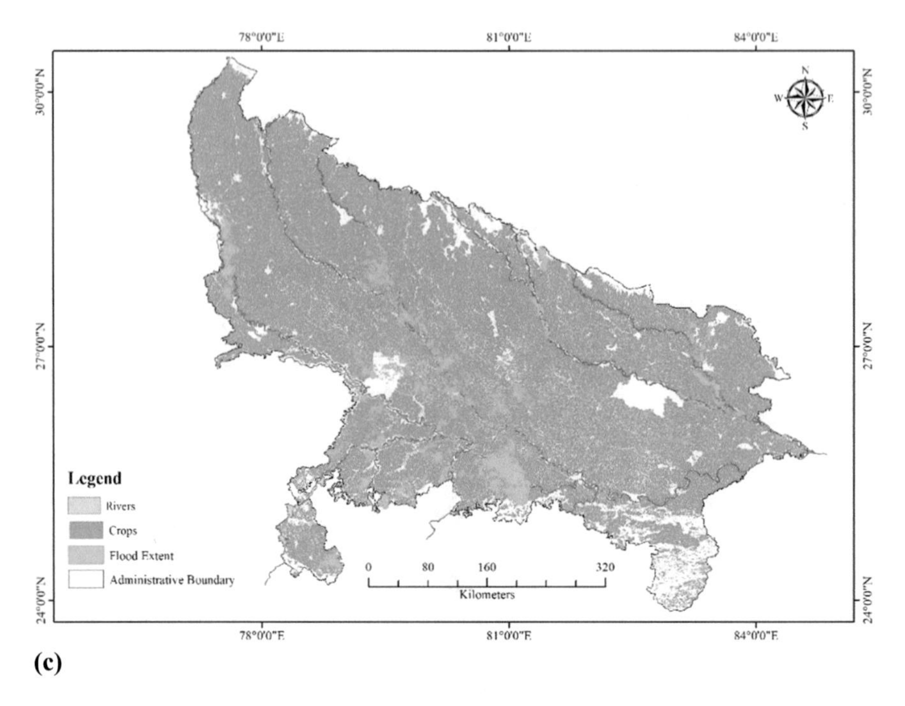

(c)

Fig. 27.7 (continued)

Table 27.3 Damage statistics (flood-affected area) calculated using 2017 annual flood layer

S. No	Village name	Gram Panchayat name	District name	GP flood effected area (Sq. Km.)
1	Kharela dehat	Charkhari	Mahoba	32.15
2	Rewai	Charkhari	Mahoba	29.72
3	Supa	Charkhari	Mahoba	29.68
4	Kulpahar Gramin	Jaitpur	Mahoba	23.77
5	Islampur	Sarila	Hamirpur	23.45
6	Chureh Kesarua	Manikur	Chitrakoot	22.72
7	Gudha	Charkhari	Mahoba	21.10
8	Bhatewara kalan	Charkhari	Mahoba	20.39
9	Durkhuroo	Bamaur (part)	Jhansi	19.33
10	Rukama Bujurg	Manikur	Chitrakoot	18.62

FVI computed at village level using the historical flood events can be of great help for identifying and planning relief shelter locations in the study area.

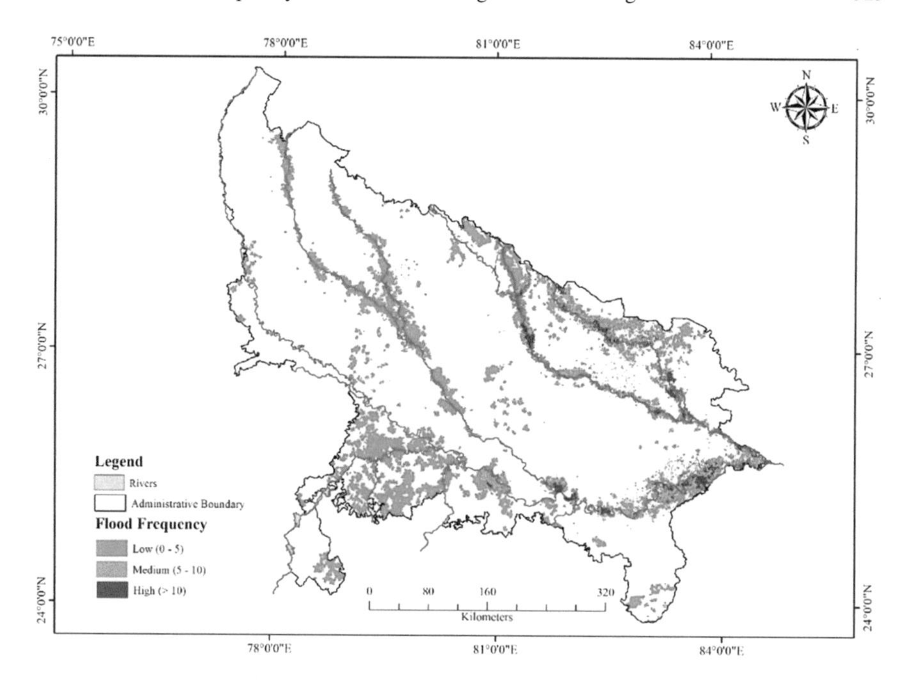

Fig. 27.8 Spatial distribution of villages of Uttar Pradesh state vulnerable to flood based on flood occurrence

27.6.3 Relief Shelter Location Identification and Planning for Persistent Flooded Villages

Minimisation of losses to human lives during floods is possible by evacuating people to safer locations situated at higher elevated areas. Even the relief operations that involve transfer of equipments like boats, food and clothes have to be planned. Identification of highly vulnerable villages in advance can help in the efficient management of resources during the disaster times. The first requirement for identifying the relief shelter location is the information on the availability facilities at higher elevations, which can be converted to relief shelters/safe houses for relocating people from flooded areas. Govt. schools, colleges and buildings are the best choice for choosing the place for relief shelter, which further integrated with DEM to identify the facilities located at higher elevations. Relief shelter locations identified for Jangl Bahadur Ali village of Gorakhpur district, UP is shown in Fig. 27.9. Various GIS operations can be performed to generate maps that can identify shortest path to reach hospital at the time of emergencies and disasters.

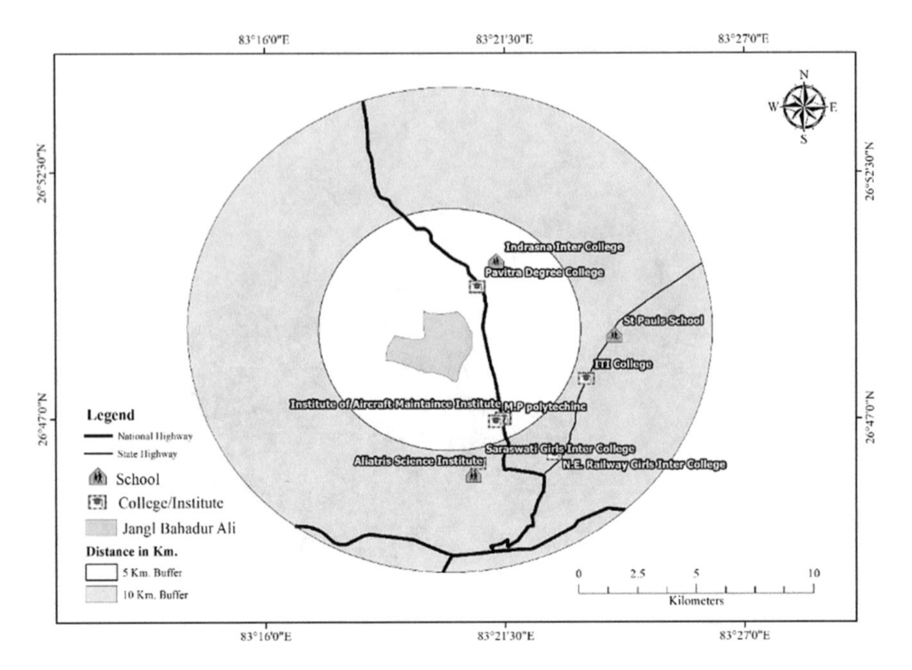

Fig. 27.9 Relief shelter location in the vicinity of Jangl Bahadur Ali village, Gorakhpur, UP

27.7 Conclusions

Popular sensors that provide satellite data for flood mapping activities along with various classification techniques available for flood mapping activities were discussed in this study. Integration of online and offline approaches for satellite data (physical data and as a service) to generation of flood layers was also presented. Flood layers generated during the procedure will be used to identify the frequently flooded areas using flood persistence maps so that it can be integrated as an input for relief management activities. This study illustrated the potential of remote sensing data and GIS for assessing the flood vulnerability in the Uttar Pradesh state of India. Geospatial technologies together with online services were effectively utilised to automate the flood mapping process. Flood situation for the period 2010–2020 was assessed based on the availability of satellite datasets. Offline processing of satellite data using supervised classification is implemented in semi-automated way to generate flood extent for the period 2010–2016, whereas online data services and Python scripts were used on GEE platform to generate flood inundation layers during the period 2017–2020. Further, accuracy of flood maps was improved by applying the conditioning factors such as altitude, slope, aspect and distance from river and land use/cover map to remove misclassification of flood extents. Administrative boundaries of villages were integrated with flood layer to identify the persistently flood-affected villages in the study area. Flood vulnerability index for villages of Uttar Pradesh state is derived based on decadal flood layers and the villages were categorised into low, medium and high FVI. GIS operations were executed to identify relief shelter locations in most vulnerable villages and also identified path connectivity to nearest hospitals with a user-specified radius. Gap analysis in terms insufficient health resources near high FVI villages can be attempted and wherever adequate resources are not there, construction of new resources can be planned. Such studies provide a strong indication that the expansion and growth of human activities in the high FVI areas have to be minimised in order to minimise the losses to life and property.

Acknowledgements We wish to express a deep sense of gratitude and sincere thanks to the Disaster Management Support Group (DMSG), Bhuvan Web Services Group and NRSC for providing satellite data and resources to conduct this study.

References

Abbaszadeh M, Mahdavi R, Rezai M (2019). Assessment of sedimentation process in flood water spreading system using IRS (P5) and supervised classification algorithms (case study: Dahandar plain, Minab city, south of Iran). Remote Sens Appl Soc Environ 16:100269

Ahamed A, Bolten JD (2017) A MODIS-based automated flood monitoring system for southeast Asia. Int J Appl Earth Obs Geoinf 61:104–117

Al-Abadi AM, Al-Najar NA (2020) Comparative assessment of bivariate, multivariate and machine learning models for mapping flood proneness. Nat Hazards 100(2):461–491

Amarnath G (2014) An algorithm for rapid flood inundation mapping from optical data using a reflectance differencing technique. J Flood Risk Manage 7(3):239–250

Amitrano D, Di Martino G, Iodice A, Riccio D, Ruello G (2018) Unsupervised rapid flood mapping using Sentinel-1 GRD SAR images. IEEE Trans Geosci Remote Sens 56(6):3290–3299

Baig MHA, Zhang L, Wang S, Jiang G, Lu S, Tong Q (2013) Comparison of MNDWI and DFI for water mapping in flooding season. In: 2013 IEEE international geoscience and remote sensing symposium-IGARSS. IEEE, pp 2876–2879

Bazi Y, Bruzzone L, Melgani F (2005) An unsupervised approach based on the generalized Gaussian model to automatic change detection in multitemporal SAR images. IEEE Trans Geosci Remote Sens 43(4):874–887

Bhatt CM, Rao GS, Jangam S (2020) Detection of urban flood inundation using RISAT-1 SAR images: a case study of Srinagar, Jammu and Kashmir (North India) floods of September 2014. Model Earth Syst Environ 6(1):429–438

Blasco F, Bellan MF, Chaudhury MU (1992) Estimating the extent of floods in Bangladesh using SPOT data. Remote Sens Environ 39(3):167–178

Bozza A, Durand A, Confortola G, Soncini A, Allenbach B, Bocchiola D (2016) Potential of remote sensing and open street data for flood mapping in poorly gauged areas: a case study in Gonaives, Haiti. Appl Geomatics 8(2):117–131

Brisco B, Touzi R, van der Sanden JJ, Charbonneau F, Pultz TJ, D'Iorio M (2008) Water resource applications with RADARSAT-2–a preview. Int J Digit Earth 1(1):130–147

Brivio PA, Colombo R, Maggi M, Tomasoni R (2002) Integration of remote sensing data and GIS for accurate mapping of flooded areas. Int J Remote Sens 23(3):429–441

Cheng J, Yin Q, Hong W (2019) Classification capability analysis of Polarimetric features obtained by decision tree. In: 2019 6th Asia-pacific conference on synthetic aperture radar (APSAR). IEEE, pp 1–5

Chowdary VM et al (2008) Assessment of surface and sub-surface waterlogged areas in irrigation command areas of Bihar state using remote sensing and GIS. Agric Water Manag 95(7):754–766

DeVries B, Huang C, Armston J, Huang W, Jones JW, Lang MW (2020) Rapid and robust monitoring of flood events using Sentinel-1 and Landsat data on the Google Earth Engine. Remote Sens Environ 240:111664

Diakakis M, Deligiannakis G, Katsetsiadou K, Lekkas E, Melaki M, Antoniadis Z (2016) Mapping and classification of direct effects of the flood of October 2014 in Athens. Bull Geol Soc Greece 50(2):681–690

Fayne JV, Bolten JD, Doyle CS, Fuhrmann S, Rice MT, Houser PR, Lakshmi V (2017) Flood mapping in the lower Mekong River Basin using daily MODIS observations. Int J Remote Sens 38(6):1737–1757

Goldberg MD, Li S, Lindsey DT, Sjoberg W, Zhou L, Sun D (2020) Mapping, monitoring, and prediction of floods due to ice jam and snowmelt with operational weather satellites. Remote Sens 12(11):1865

Henry JB, Chastanet P, Fellah K, Desnos YL (2006) Envisat multi-polarized ASAR data for flood mapping. Int J Remote Sens 27(10):1921–1929

Hess LL, Melack JM, Filoso S, Wang Y (1995) Delineation of inundated area and vegetation along the Amazon floodplain with the SIR-C synthetic aperture radar. IEEE Trans Geosci Remote Sens 33(4):896–904

Herrera-Cruz V, Koudogbo F, Herrera V (2009) TerraSAR-X rapid mapping for flood events. In: Proceedings of the international society for photogrammetry and remote sensing (earth imaging for geospatial information), Hannover, Germany, pp 170–175

Huang C, Chen Y, Wu J, Li L, Liu R (2015) An evaluation of Suomi NPP-VIIRS data for surface water detection. Remote Sens Lett 6(2):155–164

Ireland G, Volpi M, Petropoulos GP (2015) Examining the capability of supervised machine learning classifiers in extracting flooded areas from Landsat TM imagery: a case study from a Mediterranean flood. Remote Sens 7(3):3372–3399

Jain SK, Singh RD, Jain MK, Lohani AK (2005) Delineation of flood-prone areas using remote sensing techniques. Water Resour Manage 19(4):333–347

Jena PP, Panigrahi B, Chatterjee C (2016) Assessment of Cartosat-1 DEM for modeling floods in data scarce regions. Water Resour Manage 30(3):1293–1309

Jonkman SN, Kelman I (2005) An analysis of the causes and circumstances of flood disaster deaths. Disasters 29(1):75–97

Keum HJ, Han KY, Kim HI (2020) Real-time flood disaster prediction system by applying machine learning technique. KSCE J Civ Eng 24(9):2835–2848

Kiage LM, Walker ND, Balasubramanian S, Babin A, Barras J (2005) Applications of Radarsat-1 synthetic aperture radar imagery to assess hurricane-related flooding of coastal Louisiana. Int J Remote Sens 26(24):5359–5380

Klemas V (2015) Remote sensing of floods and flood-prone areas: an overview. J Coastal Res 31(4):1005–1013

Kuehn S, Benz U, Hurley J (2002) Efficient flood monitoring based on RADARSAT-1 images data and information fusion with object-oriented technology. In: IEEE international geoscience and remote sensing symposium, vol 5. IEEE, pp 2862–2864

Li Y, Martinis S, Plank S, Ludwig R (2018a) An automatic change detection approach for rapid flood mapping in Sentinel-1 SAR data. Int J Appl Earth Obs Geoinf 73:123–135

Li S, Sun D, Goldberg MD, Sjoberg B, Santek D, Hoffman JP, DeWeese M, Restrepo P, Lindsey S, Holloway E (2018b) Automatic near real-time flood detection using Suomi-NPP/VIIRS data. Remote Sens Environ 204:672–689

Lin L, Di L, Yu EG, Tang J, Shrestha R, Rahman MS, Kang L, Sun Z, Zhang C, Hu L, Yang Z (2017) Extract flood duration from Dartmouth flood observatory flood product. In: 2017 6th international conference on agro-geoinformatics. IEEE, pp 1–4

Lin L, Di L, Tang J, Yu E, Zhang C, Rahman M et al (2019) Improvement and validation of NASA/MODIS NRT global flood mapping. Remote Sens 11(2):205

Malinowski R, Groom G, Schwanghart W, Heckrath G (2015) Detection and delineation of localized flooding from WorldView-2 multispectral data. Remote Sens 7(11):14853–14875

Manjusree P, Kumar LP, Bhatt CM, Rao GS, Bhanumurthy V (2012) Optimization of threshold ranges for rapid flood inundation mapping by evaluating backscatter profiles of high incidence angle SAR images. Int J Disaster Risk Sci 3(2):113–122

Martinis S, Kuenzer C, Wendleder A, Huth J, Twele A, Roth A, Dech S (2015) Comparing four operational SAR-based water and flood detection approaches. Int J Remote Sens 36(13):3519–3543

Matgen P, Hostache R, Schumann G, Pfister L, Hoffmann L, Savenije HHG (2011) Towards an automated SAR-based flood monitoring system: lessons learned from two case studies. Phys Chem Earth Parts a/b/c 36(7–8):241–252

McFeeters SK (1996) The use of the Normalized Difference Water Index (NDWI) in the delineation of open water features. Int J Remote Sens 17(7):1425–1432

Memon AA, Muhammad S, Rahman S, Haq M (2015) Flood monitoring and damage assessment using water indices: a case study of Pakistan flood-2012. Egypt J Remote Sens Space Sci 18(1):99–106

Misra T, Rana SS, Desai NM, Dave DB, Rajeevjyoti, Arora RK, Rao CVN, Bakori BV, Neelakantan R, Vachchani JG (2013) Synthetic Aperture Radar payload on-board RISAT-1: configuration, technology and performance. Curr Sci 446–461

Pandey P, Chauhan P, Bhatt CM, Thakur PK, Kannaujia S, Dhote PR, Roy A, Kumar S, Chopra S, Bhardwaj A, Aggrawal SP (2021) Cause and process mechanism of rockslide triggered flood event in Rishiganga and Dhauliganga river valleys, Chamoli, Uttarakhand, India using satellite remote sensing and in situ observations. J Indian Soc Remote Sens 49(5):1011–1024

Patel DP, Srivastava PK (2013) Flood hazards mitigation analysis using remote sensing and GIS: correspondence with town planning scheme. Water Resour Manage 27(7):2353–2368

Pal R, Pani P (2016) Seasonality, barrage (Farakka) regulated hydrology and flood scenarios of the Ganga River: a study based on MNDWI and simple Gumbel model. Model Earth Syst Environ 2(2):57

Pantaleoni E, Engel BA, Johannsen CJ (2007) Identifying agricultural flood damage using Landsat imagery. Precis Agric 8(1):27–36

Pourghasemi HR, Amiri M, Edalat M, Ahrari AH, Panahi M, Sadhasivam N, Lee S (2020) Assessment of urban infrastructures exposed to flood using susceptibility map and google earth engine. IEEE J Selected Topics Appl Earth Observ Remote Sens 14:1923–1937

Qi S, Brown DG, Tian Q, Jiang L, Zhao T, Bergen KM (2009) Inundation extent and flood frequency mapping using LANDSAT imagery and digital elevation models. Gisci Remote Sens 46(1):101–127

Rao P, Jiang W, Hou Y, Chen Z, Jia K (2018) Dynamic change analysis of surface water in the Yangtze River Basin based on MODIS products. Remote Sens 10(7):1025

Schumann GJP, Moller DK (2015) Microwave remote sensing of flood inundation. Phys Chem Earth, Parts a/b/c 83:84–95

Sen PC, Hajra M, Ghosh M (2020) Supervised classification algorithms in machine learning: a survey and review. In: Emerging technology in modelling and graphics. Springer, Singapore, pp 99–111

Sharma PK, Chopra R, Verma VK, Thomas A (1996) Technical note flood management using remote sensing technology: the Punjab (India) experience. Int J Remote Sens 17(17):3511–3521

Sharma VK, Rao GS, Amminedu E, Nagamani PV, Shukla A, Rao KRM, Bhanumurthy V (2016) Event-driven flood management: design and computational modules. Geo-Spatial Inf Sci 19(1):39–55

Sharma VK, Rao GS, Bhatt CM, Shukla AK, Mishra AK, Bhanumurthy V (2017) Automatic procedures analyzing remote sensing data to minimize flood response time: a step towards National flood mapping service. Spat Inf Res 25(5):657–663

Sheng Y, Gong P, Xiao Q (2001) Quantitative dynamic flood monitoring with NOAA AVHRR. Int J Remote Sens 22(9):1709–1724

Sivanpillai R, Jacobs KM, Mattilio CM, Piskorski EV (2021) Rapid flood inundation mapping by differencing water indices from pre-and post-flood Landsat images. Front Earth Sci 15(1):1–11

Singh Y, Ferrazzoli P, Rahmoune R (2013) Flood monitoring using microwave passive remote sensing (AMSR-E) in part of the Brahmaputra basin, India. Int J Remote Sens 34(14):4967–4985

Sivasankar T, Das R, Borah SB, Raju PLN (2019) Insight to the potentials of Sentinel-1 SAR data for embankment breach assessment. In: Proceedings of international conference on remote sensing for disaster management. Springer, Cham, pp 33–41

Stancalie G, Craciunescu V, Mihailescu D (2012) Contribution of satellite data to flood risk mapping in Romania. In: 2012 IEEE international geoscience and remote sensing symposium. IEEE, pp 899–902

Tarpanelli A, Iodice F, Brocca L, Restano M, Benveniste J (2020) River flow monitoring by Sentinel-3 OLCI and MODIS: comparison and combination. Remote Sens 12(23):3867

Uddin K, Matin MA, Meyer FJ (2019) Operational flood mapping using multi-temporal sentinel-1 SAR images: a case study from Bangladesh. Remote Sens 11(13):1581

Vanama VSK, Shitole S, Rao YS (2020) Urban flood mapping with C-band RISAT-1 SAR Images: 2016 flood event of Bangalore city, India. In: 2020 international conference on convergence to digital world-Quo Vadis (ICCDW). IEEE, pp 1–4

Wang Y, Colby JD, Mulcahy KA (2002) An efficient method for mapping flood extent in a coastal floodplain using Landsat TM and DEM data. Int J Remote Sens 23(18):3681–3696

Xu H (2006) Modification of normalised difference water index (NDWI) to enhance open water features in remotely sensed imagery. Int J Remote Sens 27(14):3025–3033

Yamada Y (2001) Detection of flood-inundated area and relation between the area and micro-geomorphology using SAR and GIS. In IGARSS 2001. In: Scanning the present and resolving the future. Proceedings. IEEE 2001 international geoscience and remote sensing symposium (Cat. No. 01CH37217), vol 7. IEEE, pp 3282–3284

Yesou H, Andreoli R, Fellah K, Tholey N, Clandillon S, Batiston S, Allenbach B, Meyer C, Bestault C, de Fraipont P (2007) Large plain flood mapping and monitoring based on EO data. In: 2007 IEEE international geoscience and remote sensing symposium. IEEE, pp 1155–1158

Yucel I, Onen A, Yilmaz KK, Gochis DJ (2015) Calibration and evaluation of a flood forecasting system: Utility of numerical weather prediction model, data assimilation and satellite-based rainfall. J Hydrol 523:49–66

Zheng W, Sun D, Li S (2016) Coastal flood monitoring based on AMSR-E data. In: 2016 IEEE international geoscience and remote sensing symposium (IGARSS). IEEE, pp 4399–4401

Zhou X, Dandan L, Huiming Y, Honggen C, Leping S, Guojing Y, Qingbiao H, Brown L, Malone JB (2002) Use of landsat TM satellite surveillance data to measure the impact of the 1998 flood on snail intermediate host dispersal in the lower Yangtze River Basin. Acta Trop 82(2):199–205

Chapter 28
Drought Characterisation and Impact Assessment on Basin Hydrology—A Geospatial Approach

Bhaskar R. Nikam, Satyajeet Sahoo, Vaibhav Garg, Abhishek Dhanodia, Praveen K. Thakur, Arpit Chouksey, and S. P. Aggarwal

Abstract Characterisation and quantification of drought are important considerations in the planning and management of water resources. Present study examines the impacts of meteorological droughts on various hydrological aspects of the Godavari River basin by characterising and quantifying the droughts using a geospatial approach. Standardised indicators such as Standardised Precipitation Index (SPI) and Standardised Stream Flow Index (SSI) were used to identify the meteorological and hydrological droughts, respectively at different accumulation periods (1 month to 24 months) for the years 1951–2018. The meteorological drought index was used to characterise the droughts in the Godavari River basin, considering their severity, spatial extent and duration at various time scales. The long-term hydrological regime of the basin (1951–2018) was simulated using the Variable Infiltration Capacity (VIC) hydrological model. The time periods (of each accumulation period) with normal rainfall were identified and segregated. The mean (average) discharge (both observed and modelled) of each accumulation period (1 month to 24 months) was estimated from discharge data of these normal meteorological periods. The percentage deviation of observed and simulated discharge values from the 'mean' values during each drought period for each station were obtained and their relation with the various category of meteorological drought was quantified. The hydrological drought index (SSI) was computed by applying the appropriate probability distribution (concluded from the KS test) for both observed and simulated discharge values for the Polavaram discharge station representing the whole basin. Majorly for all timescales, log-normal was observed as the optimum probability distribution for both the observed and modelled hydrological data. Hydrological drought events were then analysed in accordance with the meteorological droughts in order to examine their interrelationships. The results showed a linear relationship when we compared

B. R. Nikam (✉) · S. Sahoo · V. Garg · P. K. Thakur · A. Chouksey · S. P. Aggarwal
Water Resources Department, Indian Institute of Remote Sensing, ISRO, Dehradun, Uttarakhand, India
e-mail: bhaskarnikam@iirs.gov.in

A. Dhanodia
Agriculture and Soils Department, Indian Institute of Remote Sensing, ISRO, Dehradun, Uttarakhand, India

A. Pandey et al. (eds.), *Geospatial Technologies for Land and Water Resources Management*, Water Science and Technology Library 103,
https://doi.org/10.1007/978-3-030-90479-1_28

the meteorological droughts with percentage deviation in the modelled discharge. However, similar relation doesn't exist in the case observed discharge values. This might be due to the fact that every basin will have its anthropogenic resilience towards droughts. The observed discharge data from the Godavari basin does not describe the natural flow regime of the river, which is why the behaviour of hydrological droughts in the basin is anthropogenically controlled. The numerical impact of each category of meteorological drought on the hydrological regime of the Godavari basin and its sub-basins was quantified successfully for the virgin flow conditions.

28.1 Introduction

Drought is one of nature's most dangerous and underrated disasters (Wilhite et al. 2000). It occurs in all climatic zones, from high- to low-rainfall locations, albeit the incidence and severity may vary. This subtle natural hazard has the potential to create catastrophic water shortages, crop production losses, economic losses and negative social repercussions, costing billions of dollars annually throughout the world (Narasimhan and Srinivasan 2005). Droughts are fuzzier than other water-related crises since defining their start and progression is challenging due to the numerous elements that influence them. A drought begins with a lack of precipitation across a broad area and can last for a month or several years. This precipitation deficit propagates through the different parts of the hydrological cycle, resulting in various forms of drought. In general, droughts can be classified as meteorological, hydrological, agricultural and socio-economical (Wilhite and Glantz 2009). Long periods of below-normal precipitations are linked to meteorological drought. Due to decreased precipitation in the basin, hydrological drought arises when streamflow, groundwater levels, or reservoir storage are less than the normal long-term average. Drought in agriculture refers to a situation in which the amount of water available in the soil is insufficient for plant development and survival. This soil moisture shortage can arise for a variety of causes, including climatic and hydrological factors. When demand for an economic good exceeds supply especially due to a weather-related water shortage, socio-economic drought occurs. The insufficiency of a water resources system to meet demand is linked to socio-economic drought. Impacts resulting from the propagation of meteorological drought to agricultural drought, hydrological drought and further socio-economic drought are filled with complexity.

Drought has a comparatively large impact on humanity compared to other natural disasters (Wilhite et al. 2000). According to Obasi (1994), about 2.8 billion people were affected by various disasters throughout the world between 1949 and 1991, with drought impacting the majority of them (around 51%). Between 1871 and 2002, India saw 22 significant drought years, with the 1987 drought being amongst the worst throughout the century, with a 19% decrease in the total precipitation. Around 60% of farmland and 280 million people were affected (Ministry of Agriculture and Farmer's Welfare, India 2016). Drought frequency has fluctuated over time; between 1899 and 1920, there were seven drought years; between 1941 and 1965, droughts were less prevalent, with just three drought years in total. After then, ten out of the 21 years between 1965 and 1987 drought occurred, with the ENSO effect being cited for their increased frequency.

The future drought risks are larger in India because of irregular monsoon rainfall, diminishing groundwater and the pressure of food demand from ever-increasing population. Drought situations occur nearly every year owing to rising demand for water in the agricultural and industrial sectors as a result of population expansion and climate change all over the world. Droughts are expected to become more catastrophic as climate forecasts indicate that droughts will become more intense in many areas (Downing 1993; IPCC 2012). Also, various regional studies (Chen and Sun 2017; Kumar et al. 2013; Lee et al. 2018; Nam et al. 2015; Spinoni et al. 2018; Thilakarathne and Sridhar 2017) conducted in several regions of the world have consistently indicated an increase in drought occurrences as climatic conditions change.

Drought characterisation and impact assessment are important considerations in water resource planning and management. Where geospatial approach is beneficial because it offers an analytical tool for analysing drought in a better way. Several studies have been done earlier for identification and characterisation of drought for the Indian region and other parts of the world (Aswathi et al. 2018; Barker et al. 2016; Chouhan et al. 2017; Das et al. 2020; Edossa et al. 2010; Hisdal and Tallaksen 2003; Padhee et al. 2014, 2017; Khatiwada and Pandey 2019; Mishra 2020; Nikam et al. 2020; Thomas et al. 2015; Tigkas et al. 2012) using standardised indices such as Standardised Precipitation Index (SPI) and Standardised Streamflow Index (SSI).

Various traits of drought propagation have indicated a lag between meteorological, soil moisture (Van Loon and Van Lanen 2012; Padhee et al. 2014, 2017) and hydrological deficits. Many times hydrological droughts last for longer time periods than meteorological drought and agricultural (or soil moisture) droughts (Hisdal and Tallaksen 2003) and that the meteorological droughts are attenuated when storage is high at the time of its onset (Van Loon and Van Lanen 2012; Wang et al. 2021). It has been observed that not all meteorological droughts result in a hydrological drought (Deshpande et al. 2019; Nanda et al. 2021; Wang et al. 2021). Although it is unsurprising that drought characteristics (such as inception, persistence, intensity/severity, location and spatial extent) change throughout its development, it is difficult to illustrate how droughts propagate in real-world scenarios from a time series of simultaneously observed precipitation, soil moisture, groundwater levels and streamflow (Peters et al. 2003).

Drought propagation is not only influenced by various linked meteorological/climatic factors, basin characteristics and land–atmosphere interactions. Also, human-induced factors including land cover change, reservoir/storage control, irrigation/domestic water supply and water extraction from rivers or streams must all be taken into account (Van Loon and Laaha 2015). According to Peters et al. (2005) on a regional scale, where the climate is thought to be fairly consistent, the geology, spatial extent, slope and the groundwater system are also important variables in determining the length of a hydrological drought as it progresses from meteorological drought conditions.

Over the last decade, the propagation of meteorological drought to another form of droughts has been a significant topic in the hydrological research community (Barker et al. 2016; Nanda et al. 2021; Van Loon and Laaha 2015; Wang et al. 2021) but with only one peer-reviewed study (Bhardwaj et al. 2020) available for Indian region. Detailed investigation examining—How the meteorological drought induced by the climatic anomalies develops and propagates considering the anthropogenic interventions is required for Indian river basins. With this background, in the present study, the characterisation of meteorological droughts that occurred in a river basin is done using geospatial techniques. Propagation (or interrelation) of meteorological drought to the hydrological droughts was analysed with the help of long-term observed and simulated discharge data.

28.2 Study Area

Godavari basin (Fig. 28.1) covers 0.312 million km^2 area in central and southern India, which is about 9.5% of the total geographical area of the country. Godavari (the second largest river basin in India) is a perennial river that originates at an altitude

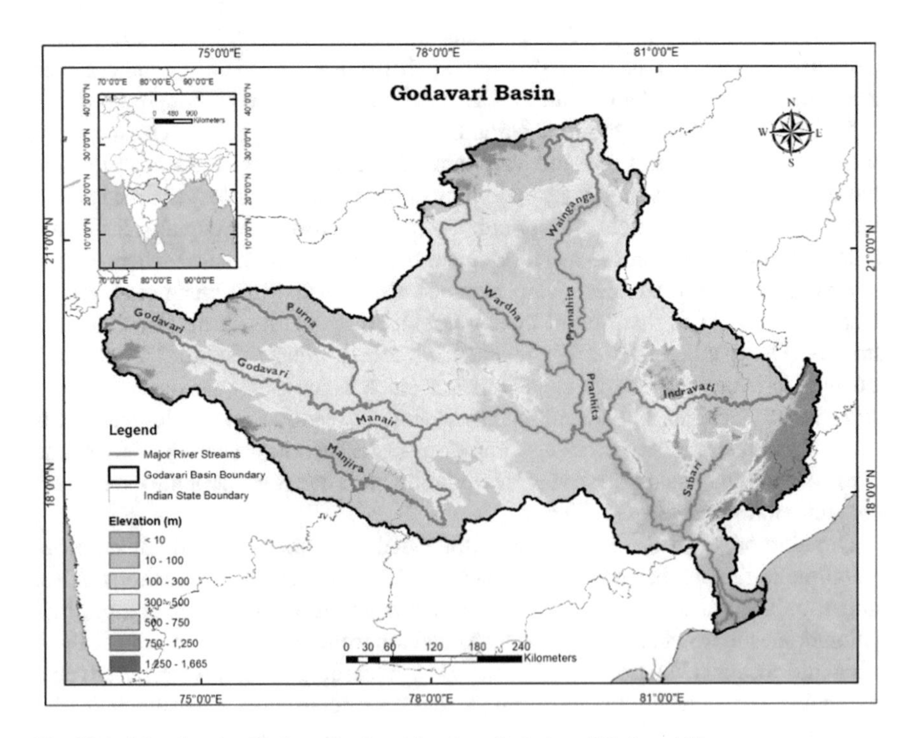

Fig. 28.1 Map showing Godavari basin and major tributaries of Godavari River

of 1068 m near Trimbakeshwar in Nasik District of Maharashtra and submerges into the Bay of Bengal after flowing 1465 km at Antarvedi (Andhra Pradesh).

Godavari basin is spatially extended over states of Maharashtra (48.7%), Telangana (19.87%), Chhattisgarh (10.69%), Madhya Pradesh (10.17%), Odisha (5.7%) Andhra Pradesh (3.53%), small areas of Karnataka state (1.41%) and union territory of Puducherry (0.01%). Satamala, Balaghat and Mahadeo Hills surround the northern part of the basin, whereas the Ajanta range surrounds the north-west region, the Sahyadri range in the Western Ghats surrounds the west the Eastern Ghats surrounds the south and southeast part. Dandakaranya range covers the eastern part of the Godavari basin along with the Eastern Ghats. The majority of the basin's interior part lies on the Deccan Plateau, with elevation in the range of 300–600 m sloping eastward (CWC & NRSC 2014; GRMB 2019). The entire population of the basin is estimated to be around 60.48 million (According to the 2001 census). There are over 33 cities, 58,072 settlements and 441 towns. The region's population is concentrated in plains, lowlands and isolated basins where good soil cover and water are available. The basin's least inhabited areas are Dantewada, Kanker, Bastar districts and the Sabari and Indravati plains. The major part of the Godavari basin is under agricultural land use (59.57% of the basin). Waterbodies account for 2.06% of the total basin area, with forests accounting for 29.78%. The basin's north-eastern region is notably densely forested, with deciduous forest constituting 26.16% of its total area.

28.2.1 River and Sub-basin System

The Godavari River originates in the Eastern Ghats and flows into the lowlands. The length of the river is 1465 km. The Godavari Central Delta lies between two river branches, namely the Gautami and Vasishta, divided at Dhawaleswaram. Pranhita, Indravati, Purna, Manjira, Manair and Sabri/Kolab rivers are the prime tributaries of the Godavari River. Among all, the largest tributary of the Godavari River is Pranhita. It is joined by the Penganga, the Wainganga and the Wardha River, which ultimately merges with the Godavari along with the Manjira River. The Godavari basin is demarcated into eight sub-basins as shown in Table 28.1. The Godavari middle sub-basin is the source of the Dudhana, Kapra, Sindphana and Siddha rivers. The Godavari upper sub-basin is the source of the Kadva, Pravara, Shivna, Mula and Godavari rivers. The Wardha sub-basin originates the rivers Penganga, Kayadhu, Arunavati, Aran, Wardha, Vena and Yashoda.

28.2.2 Rainfall and Climate

The Godavari basin has a tropical climate with yearly precipitation varying from 1000 to 3000 mm in the heavy rainfall zone of Western Ghats and some parts of

Table 28.1 Sub-basins of the Godavari basin (CWC & NRSC 2014)

S. No.	Name of sub-basin	Area (km^2)
1	Wardha	46,242.09
2	Weinganga	49,695.4
3	Godavari Upper	44,492.93
4	Godavari Middle	36,290.47
5	Godavari Upper	21,443.23
6	Indravati	38,306.1
7	Manjra	29,472.88
8	Pranhita and others	36,119.6

the Western Ghats' eastern belt receive less than 600 mm of yearly rainfall. Moving towards the East coast in the Godavari basin, annual rainfall gradually increases to 900 mm. During the Southwest Monsoon, the Godavari basin receives the majority of its rainfall. From July through September, more than 85% of the rainfall occurs. January and February with rainfall less than 15 mm are the driest months for the basin. In majority of the basin, during the months of March–April–May, the precipitation varies from 20 to 50 mm approximately (GRMB 2019). Due to climate variability, almost 20.6% of the overall Godavari basin (64.5 thousand km^2) has been prone to periodic droughts for decades, including several districts in Maharashtra's Marathwada area.

28.2.3 Water Resources

The average annual water resource potential of the Godavari basin is around 110.54 BCM, including 76.30 BCM surface water resources that can be utilised (GRMB 2019). There are around 870 Reservoirs in the Godavari basin (CWC & NRSC 2014). Jayakwadi, Sriramsagar, Balimela, Yeldari, Singur, Pench, Isapur, Upper Kolab, Indrawati and Nizam Sagar are among the major reservoirs in the basin. The majority of the water bodies are between 0 and 25 ha in size. However, 29 large water bodies are covering more than 25 km^2 area. According to the data collected from various sources (CWC & NRSC 2014; GRMB 2019), the combined annual storage capacity of all the major water resources projects in the basin is 45.60 BCM which is about 41.25% of the average annual water resource potential of the Godavari basin.

28.3 Datasets Used

For estimation of meteorological drought indices (SPI) and for generating the meteorological forcing for the hydrological model, India Meteorological Department (IMD) daily gridded rainfall data of spatial resolution (0.25° × 0.25°) and daily gridded temperature data of spatial resolution (1° × 1°) for the period (1951–2019) were used. IMD gridded rainfall data is developed using precipitation data of 6995 rain gauge stations across India (Pai et al. 2014). IMD gridded temperature data is developed by implying datasets of 395 quality controlled weather stations across the country (Srivastava et al. 2009). Interpolation of the observed point temperature data was made at 1° × 1° of spatial resolution using an angular distance weighting algorithm (Shepard 1968).

The Multi-Error-Removed Improved-Terrain Digital Elevation Model" (MERIT DEM) was used as input data to represent terrain characteristics of the basin in the Variable Infiltration Capacity (VIC) hydrological model. MERIT DEM was developed by (Yamazaki et al. 2017) after eliminating multiple error components from the existing space borne DEMs ("SRTM3 v2.1" and "AW3D-30m v1") with the resolution of 3 arc-second (i.e. ~90 m at the equator).

Version 1.2 of Harmonised World Soil Database (30 arcsec or 0.05° × 0.05° spatial resolution) provided by the Food and Agriculture Organization (FAO) is used to generate a soil parameter file as an input for VIC model. HSWD provides the physical and chemical characteristics for standardised soil parameter layers of topsoil (0–30 cm) and sub-soil profiles (30–100 cm). Land use land cover data of spatial resolution of 100 × 100 m generated at decadal time intervals for 1985–1995, 1995–2005 under the ISRO Geosphere-Biosphere Programme (Roy et al. 2016) was used in the present study for characterising the land cover in the hydrological model. The observed discharge data of Godavari River and its selected tributaries was obtained from India Water Resource Information System (India-WRIS) portal for all the available years till 2018 on a monthly time scale. Observed discharged data was used to calibrate and validate the hydrological model, calculate hydrological drought indices and analyse the propagation of meteorological drought to hydrological drought.

28.4 Methodology

This section describes the steps followed and methods used, as shown in Fig. 28.2, to achieve the objectives of the present study. The meteorological data were used to estimate meteorological drought index, in the present case, it was Standardised Drought Index (SPI). The characterisation of meteorological drought was done using a geospatial approach, this gave us the flexibility to include the location and extent of drought as additional characteristics of drought, which otherwise goes unattended. The long-term hydrological behaviour of the basin was simulated using well calibrated and validated hydrological model (VIC) parameterised using gridded meteorological data

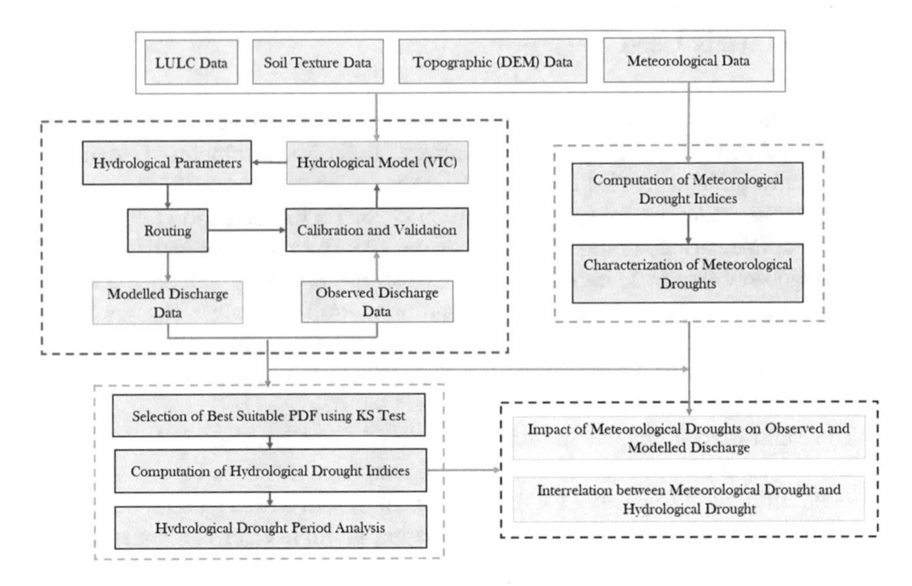

Fig. 28.2 Methodology flowchart for the study

from IMD, terrain data derived using MERIT DEM, land cover data generated from IGBP-LULC maps, soil data obtained FAO HSWD map. The hydrological drought index was calculated using both observed and simulated discharge data. The impact of meteorological droughts on river discharge in ideal case (simulated discharge) and actual case (observed discharge) was quantified. The interrelation between meteorological droughts and hydrological droughts was analysed using the observed and simulated discharge values. The detailed description of each of these steps is given in subsequent sub-sections.

28.4.1 Computation of Meteorological Drought Indices

Based on the literature survey, the Standardised Precipitation Index (SPI) a standardised meteorological drought index, proposed by Mckee et al. (1993), that only depends on the precipitation data, was selected as the best suitable meteorological drought index for the study. Guttman (1999), Stagge et al. (2015) and Thom (1966) found out that Gamma distribution fits best for climatological time series data of precipitation, so IMD gridded datasets of precipitation were fitted to a Gamma distribution and then shape (α) and scale (β) parameters were estimated to compute cumulative distribution function (CDF).

Probability density function $f(x)$ defining Gamma distribution is given by Eq. (28.1):

$$f(x) = \left(\frac{1}{\beta^{\alpha}\Gamma(\alpha)}\right)x^{\alpha-1}e^{-\left(\frac{x}{\beta}\right)} \tag{28.1}$$

where, $\Gamma(\alpha)$ represents Gamma function, also (α and β) > 0.

$\Gamma(\alpha)$ is defined by the Eq. (28.2):

$$\Gamma(\alpha) = \int_{0}^{\infty} x^{\alpha-1}e^{-x}\mathrm{d}x \tag{28.2}$$

To estimate required fitting parameters α and β, the maximum likelihood method from (Thom 1966) is similar (Mckee et al. 1993).

$$\alpha = \frac{1}{4A}\left(1 + \sqrt{1 + \frac{4A}{3}}\right) \tag{28.3}$$

$$\beta = \frac{x}{\alpha} \tag{28.4}$$

$$A = \ln\overline{x} - \frac{\sum \ln(x)}{n} \tag{28.5}$$

where n = number of observations (precipitation).

Cumulative probability distribution function $G(x)$ is given by the Eq. (28.6).

$$G(x) = \int_{0}^{x} f(x)\mathrm{d}x = \frac{1}{\beta^{\alpha}\Gamma(\alpha)}\int_{0}^{x} x^{\alpha-1}e^{-\left(\frac{x}{\beta}\right)} \tag{28.6}$$

Since for $x = 0$ (zero precipitation events) Gamma function $\Gamma(\alpha)$ is undefined. Therefore, the cumulative probability distribution function (CPDF) becomes:

$$H(x) = q + (1 - q)G(x) \tag{28.7}$$

Here, q = probability of zero precipitation.

Then, obtained cumulative probability distribution is transformed to a standardised normal distribution to make the mean SPI for the desired period zero and variance equals one. For cases with many grid points or stations method described by (Abramowitz 1965) is used as an alternative technique to approximately convert cumulative probability to a standardised z variable defined as Eqs. (28.8) (28.9) and (28.10).

$$z = \pm\left(t - \frac{(c_0 + c_1t + c_2t^2)}{(1 + d_1t + d_2t^2 + d_3t^3)}\right) \tag{28.8}$$

For (+) sign $0 < H(x) < 0.5$,

$$t = \sqrt{\ln\left(\frac{1}{(H(x))^2}\right)} \tag{28.9}$$

For (−) sign $0.5 < H(x) < 1$,

$$t = \sqrt{\ln\left(\frac{1}{(1 - H(x))^2}\right)} \tag{28.10}$$

where c_0, c_1, c_2, d_1, d_2 and d_3 are constants with the following values:

$c_0 = 2.515517$, $c_1 = 0.802853$, $c_2 = 0.010328$, $d_1 = 1.432788$, $d_2 = 0.189269$ and $d_3 = 0.001308$.

Figure 28.3 shows the example for the computed SPI values at 1-month timescale for August 1972.

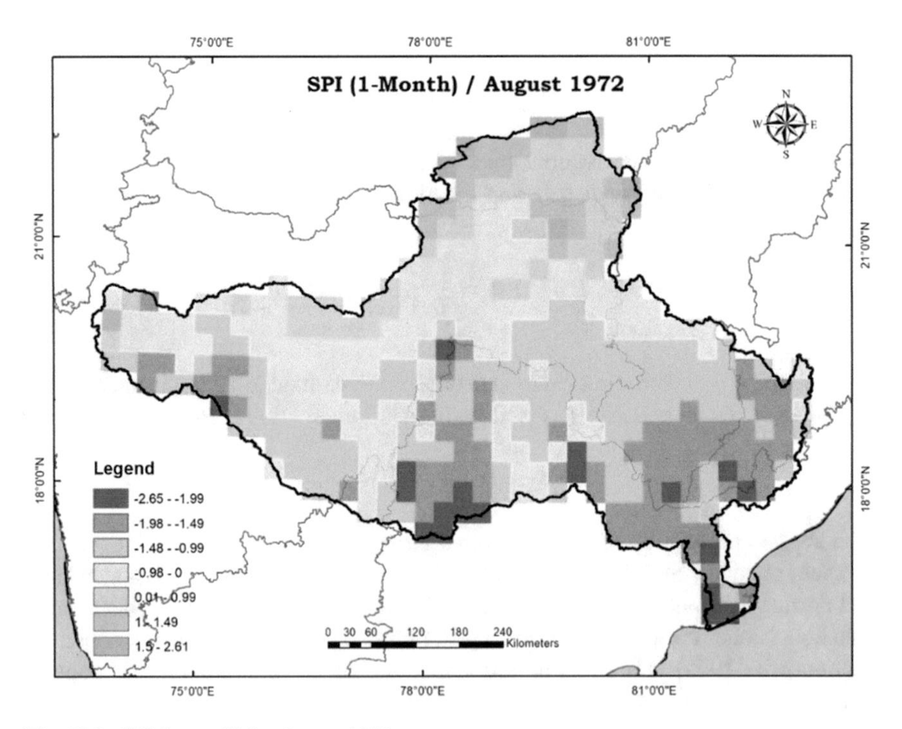

Fig. 28.3 SPI (1-month) for August 1972

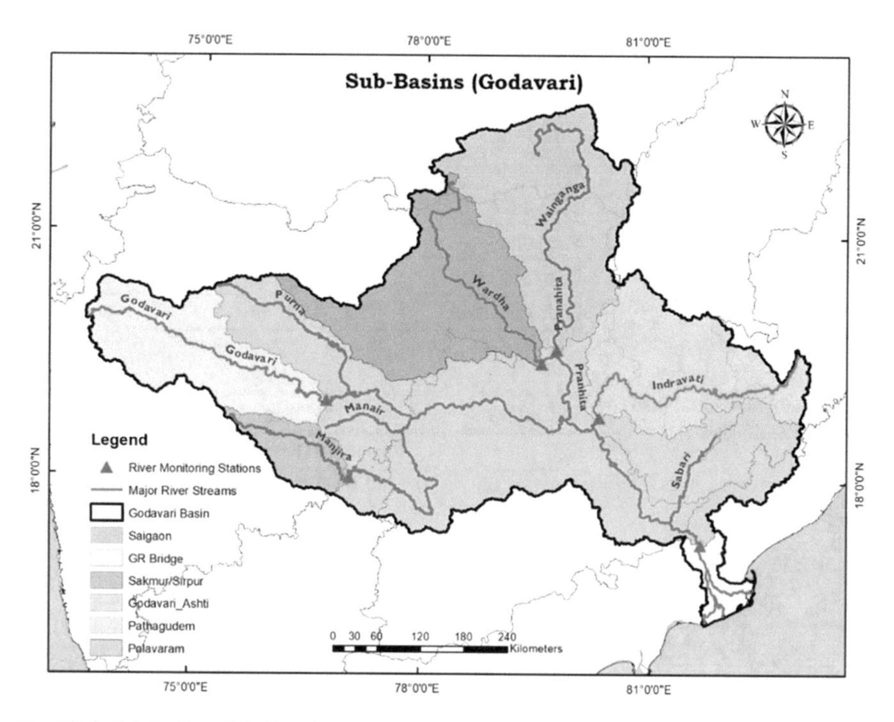

Fig. 28.4 Sub-basins of Godavari

28.4.2 Delineation Sub-basins

To add the location and extend as geospatial characteristics of meteorological droughts sub-basin of six outlets, which were river discharge monitoring stations, were delineated using HEC-GeoHMS extension in ArcGIS as shown in Fig. 28.4. These outlets were selected due to their isolated behaviour with each other in terms of contribution to the discharge to the station except Polavaram. Polavaram was selected in order to represent the characteristics of entire Godavari basin. Total number of precipitation grids (0.25° × 0.25° resolution) covered by each sub-basin were extracted using a python script for further characterisation of meteorological drought. Table 28.2 provides the pixel count for each sub-basin.

28.4.3 Characterisation of Meteorological Droughts

Characterisation of meteorological droughts is an important step in the analysis of their impact. As discussed by various researchers, the impact of a meteorological drought event depends on its characteristics such as onset, duration, severity, spatial

Table 28.2 Total grid count for each basin

Name of sub-basin	Total grid count
Ashti	73
G.R Bridge	44
Pathagudem	53
Polavaram	423
Saigaon	14
Sakmur/Sirpur	63
Godavari basin	440

Table 28.3 Meteorological drought classification

SPI values	Class
>2	Extreme wet
1.50 to 1.99	Severe wet
1.00 to 1.49	Moderately wet
0 to 0.99	Mild wet
0 to −0.99	Mild drought
−1.0 to −1.49	Moderate drought
−1.5 to −1.99	Severe drought
−2>	Extreme drought

extent and location in the basin when considering the hydrological drought. Meteorological drought events that occurred over the Godavari basin were categorised into eight intensity/severity classes according to the classification scheme (Mckee et al. 1993) given in Table 28.3.

Generally, the onset of a drought event is considered when the SPI value is below −1.0 (in this case only moderate, severe and extreme types of events are called drought events). The computed monthly raster files of SPI values were classified into the above-stated classification scheme. To analyse and characterise the monthly drought events based on their spatial variability, the total (%) area under drought (Eq. 28.11) as well as the percentage area under each class (severity) of drought out of the total area under drought (Eq. 28.12) was estimated for all sub-basins (as described in results section).

$$\text{Total \% area under drought} = \frac{\text{No. of grid or pixel under drought}}{\text{No. of grid or pixel under each sub-basin}} \times 100 \tag{28.11}$$

$$\text{\% area under each drought class} = \frac{\text{No. of grid or pixel under a particular drought class}}{\text{No. of grid or pixel under drought}} \times 100 \tag{28.12}$$

Table 28.4 Type-I categorisation based on spatial extend of drought event

% Area under drought	Drought category
>10	0
10 < 30	1
30 < 50	2
50 < 70	3
70 < 90	4
90 < 100	5

Using these values, meteorological droughts are categorised as type-I and type-II classification. There were total '6' and '13' categories under type-I and type-II characterisation scheme, respectively (as shown in Tables 28.4 and 28.5).

All the drought events in each sub-basin were characterised for their onset using monthly SPI analysis (of all the accumulation periods). The location of drought was taken into account by analysing each sub-basin separately. The severity and spatial extent and their complex were incorporated in the drought characterisation by classifying each drought event using the scheme mentioned in Tables 28.4 and 28.5. This helped us in analysing or identifying each drought based on its onset (in which month it begins), spatial extend (Type-I categorisation), overall severity (Type-II categorisation).

Table 28.5 Type-II categorisation of drought event based on composite of different severity classes in the total drought extend

Sub-category of drought	% Area under moderate drought	% Area under severe drought	% Area under extreme drought
a	≧0 & ≦35	≧0 & ≦35	≧0 & ≦35
b	≧0 & ≦35	≧0 & ≦35	>35 & ≦70
c	≧ 0 & ≦35	≧ 0 & ≦35	>70 & ≦100
d	≧ 0 & ≦35	>35 & ≦70	≧0 & ≦35
e	≧0 &≦35	>35 & ≦70	>35 & ≦70
f	≧0 & ≦35	>35 & ≦70	>70 & ≦100
g	≧0 & ≦35	>70 &≦ 100	≧0 & ≦35
h	≧0 & ≦35	>70 & ≦100	>35 & ≦70
i	>35 & ≦70	≧0 & ≦35	≧0 & ≦35
j	>35 & ≦70	≧0 & ≦35	>35 & ≦70
k	>35 & ≦70	>35 & ≦70	≧0 & ≦35
l	>35 & ≦70	>35 & ≦70	>35 & ≦70
m	>70 & ≦100	≧0 &≦ 35	≧0 & ≦35

28.4.4 Hydrological Modelling of Godavari Basin

Availability of long-term hydrological observations is a limiting factor in hydro-meteorological studies in developing countries (Garg et al. 2019; Nikam et al. 2018). One can overcome this limitation using hydrological modelling. In the present study, though long-term hydrological data for all the six outlets (of each sub-basin) was available, we were faced with a different challenge of high degree of anthropogenic control over the flow regime of the basin. As discussed in the Sect. 28.2.3 'Water Resources' the man-made reservoirs in the Godavari basin have a total capacity to store around 41.25% of the average annual water resource potential of the basin. In such a scenario it is difficult to get natural/virgin flow regime of the basin/sub-basin. To understand and quantify the natural response of the basins towards meteorological fluctuations the hydrological regime of the basin is simulated using a macro-scaled, semi-distributed, physic based hydrological model, i.e. Variable Infiltration Capacity (VIC) model (Liang 1994). VIC hydrological model has been used earlier for analysing the impact of land use land cover change (Aggarwal et al. 2012; Garg et al. 2017; Garg et al. 2019), climate change (Aggarwal et al. 2016; Garg et al. 2013, Sharma et al. 2020) on hydrology of river basins. It is also being used to simulate hydrological processes for assessment, characterisation and evolution of droughts (Bhardwaj et al. 2020; Mishra et al. 2010; Niu et al. 2015; Shamshirband et al. 2020).

VIC, a grid-based land surface representation model simulates land surface-atmosphere fluxes of moisture and energy. Each grid cell is divided into various elevation and land cover (vegetation) classes, resulting in a sub-grid representation of vegetation and elevation. The Penman–Monteith equation is used to compute evapotranspiration. Three soil layers are often used to divide the soil column. Variable infiltration curve (Zhao et al. 1980) is used to calculate surface runoff in the topsoil layer and a non-linear Arno recession curve (Todini 1996) is utilised in simulating the baseflow amount released from the lowermost soil layer. VIC uses a routing model (Lohmann et al. 1998) that employs the unit hydrograph concept inside the grid cells and St. Venant's equations to route river flow (surface runoff and baseflow) through the stream channel to the basin outlet. The observed discharge data of six stations are used in the present study to calibrate the VIC model. The validation of model behaviour is done utilizing remain part of observed data. The values of coefficient of determination (R^2) are used to quantify the closeness of model behaviour with actual conditions. The long-term simulated discharge of six stations derived through implementation of well calibrated and validated VIC model was further used for understanding and quantifying the nexus between meteorological and hydrological droughts under natural/virgin flow condition of the basin.

28.4.5 Hydrological Drought Indices Computation

The hydrological drought identification and characterisation were done using Standardised Streamflow Index (SSI). The SSI is estimated using both observed as well and simulated discharge data. SSI is estimated for different accumulation periods, e.g. 1-month, 2-months, 6-months, 12-months and 24-months to match the accumulation period of hydrological drought index with the accumulation period of meteorological drought index. Various accumulation periods are also selected to help us in better identifying both short-term and long-term hydrological drought events. The method used for SSI calculation is similar to SPI calculation as they have similar concepts and theoretical backgrounds. The only difference between SPI and SSI calculations is the discharge values used instead of precipitation values, along with the use of a suitable probability distribution function in place of Gamma distribution as used for SPI. Studies across the globe have highlighted the lack of consensus on which probability distribution is best suitable in the case of hydrological drought index calculation. When compared with SPI calculation, there is no such widely accepted probability distribution that can be used to best fit the streamflow data (Vicente-Serrano et al. 2012). Due to this reason, in the present study, all the popular probability distributions were tested with the observed and modelled discharge data using Kolmogorov-Smirnov (KS) Test. Using the KS test, the best fitting probability distributions are selected for the discharge data of various accumulation periods. The selected probability distribution is used for the estimation of SSI of the various accumulation period.

28.4.5.1 Kolmogorov–Smirnov One-Sample Test

Non-parametric goodness of fit test is used to compare the probability distribution of the sample with a reference probability distribution (one-sample test) or between the probability distribution of the two samples (two-sample test). K-S test is applied to check whether a sample belongs to a population of any specific probability distribution or not by quantifying the distance between their distribution function.

Hypothesis defined:

H_o (Null Hypothesis): sample probability distribution and reference distribution are from the same distribution (assumes that there is no difference between the sample F_o and reference distribution function F_r).

H_a (Alternative Hypothesis): sample probability distribution and reference distribution are from different distributions.

For a collection of random samples (x_1, x_2, x_3, … x_n) the observed or empirical probability distribution of the sample F_o is defined as Eq. (28.13):

$$F_o(x_i) = \frac{k}{n} \tag{28.13}$$

Table 28.6 K-S test statistic D(critical) values for large no. of samples

α	0.10	0.05	0.025	0.01	0.005	0.001
$c(\alpha)$	1.22	1.36	1.48	1.63	1.73	1.95

where n is the total no. of observations in the sample, k is the no. of observation less than x_i and k is ordered in the.

The test statistic (D) for the K-S test is defined as the largest vertical distance between F_o and F_r. Where D is formulated as Eq. (28.14):

$$D = \max_{1 \le i \le n} \left(F_r(x_i) - F_o(x_i), F_o(x_i) - F_r(x_i) + \frac{1}{n} \right) \tag{28.14}$$

If the value of calculated test statistic (D) is less than critical values (D_α) then the null hypothesis is accepted else rejected.

For the level of significance (α) and n number of sample D_α is expressed as Eq. (28.15)

$$D_\alpha = \frac{c(\alpha)}{\sqrt{n}} \tag{28.15}$$

where $c(\alpha)$ is the coefficient depending upon the level of confidence α as shown in Table 28.6.

28.4.6 Identification and Characterisation of Hydrological Droughts

After the computation of the hydrological drought index (SSI) using both the observed and modelled discharge data, the SSI values were classified into eight categories using the similar classification scheme used for SPI as described in Table 28.3. The hydrological drought events were then identified using SSI values at each accumulation period only for the drought categories of "Moderate," "Severe," and "Extreme." The mean, median and maximum values for the duration and severity of the identified drought events were also obtained. Drought periods were determined using the respective category's threshold value (i.e. −1.0 for moderate drought, −1.5 for severe drought and −2.0 for extreme drought). The end of the drought period was defined when the succeeding monthly SSI values returned to normal (i.e. SSI ≥ 0). The results obtained were analysed considering the basin's natural and controlled flow regimes.

28.4.7 *Impact of Meteorological Drought on Hydrological Regime*

The impact of meteorological droughts on the observed and simulated discharge was assessed numerically and graphically. For each sub-basin, the percentage deviation of observed and simulated monthly discharge values from their 'mean' values was determined, as well as their relationship with the various categories of meteorological droughts was drafted using boxplots. This aided in the investigation of the change in discharge levels in each sub-basin in response to various meteorological drought classes and sub-classes. To deduce the connection between meteorological and hydrological droughts in order to analyse and quantify the influence of meteorological drought on the hydrological regime of the Godavari River basin, Polavaram sub-basin was selected as study area. Polavaram sub-basin is the most downstream of the Godavari basin and represents the majority of the basin. Monthly SSI values for both actual and simulated discharge, as well as the corresponding various classes/sub-classes of meteorological droughts, were compared for this purpose.

28.5 Results and Discussion

The drought intensity classified maps were sub-categorised using type-I and type-II categorisation scheme (as shown in Tables 28.4 and 28.5). The number of months under the type-I and type-II droughts for Polavaram catchment are shown in Tables 28.7 and 28.8, respectively.

Droughts under the category of four and above (drought area greater than 70%) can be seen to have occurred only in the months of June, July, August, September and October. Out of 60 years under analysis about 56% times, the basin was under Drought category '0 m' (i.e. total % area under drought $\leq$10 with the majority of drought affected area under moderate category of drought). Most numbers of drought events under the type-I category above '0' (i.e. greater than 10% area under drought) occurred in the month of March with a total '55' then in April with '44' such events.

Table 28.7 Number of droughts under each type-I categorisation for Polavaram sub-basin

SPI category	Annual	Jan	Feb	Mar	Apr	May	Jun	Jul	Aug	Sep	Oct	Nov	Dec
0	500	63	65	14	27	32	36	37	36	37	39	46	68
1	232	6	3	43	33	26	20	24	22	19	12	23	1
2	71	0	1	11	9	11	10	3	8	8	10	0	0
3	15	0	0	1	0	0	2	2	2	3	5	0	0
4	8	0	0	0	0	0	1	2	1	2	2	0	0
5	2	0	0	0	0	0	0	1	0	0	1	0	0

Table 28.8 Number of droughts under each type-II categorisation for Polavaram sub-basin

SPI Category	Annual	Jan	Feb	Mar	Apr	May	Jun	Jul	Aug	Sep	Oct	Nov	Dec
0a	43	0	0	0	1	2	3	2	4	7	10	4	10
0b	4	0	1	0	0	0	1	1	0	1	0	0	0
0c	1	0	0	0	0	0	0	1	0	0	0	0	0
0d	4	0	0	3	0	0	0	1	0	0	0	0	0
0g	3	0	0	0	0	0	0	1	1	0	1	0	0
0i	35	0	1	6	3	3	4	3	5	9	1	0	0
0j	2	0	0	1	0	0	0	0	0	1	0	0	0
0k	12	0	1	3	0	0	1	0	3	0	4	0	0
0m	396	63	62	1	23	27	27	28	23	19	23	42	58
1a	1	0	0	1	0	0	0	0	0	0	0	0	0
1b	1	0	0	1	0	0	0	0	0	0	0	0	0
1d	4	0	0	3	0	0	0	0	0	0	0	1	0
1e	1	0	0	1	0	0	0	0	0	0	0	0	0
1i	76	1	0	14	8	2	15	9	12	6	8	0	1
1j	2	0	0	0	0	0	1	0	1	0	0	0	0
1k	28	0	0	22	0	0	0	1	1	3	1	0	0
1m	0	5	3	1	25	24	4	14	8	10	3	22	0
2a	1	0	0	0	0	0	0	0	1	0	0	0	0
2b	1	0	0	0	0	0	1	0	0	0	0	0	0
2d	3	0	1	0	0	0	0	0	1	0	1	0	0
2i	31	0	0	5	2	3	5	3	4	7	2	0	0
2k	20	0	0	6	1	1	2	0	2	1	7	0	0
2m	15	0	0	0	6	7	2	0	0	0	0	0	0
3b	1	0	0	0	0	0	0	0	0	1	0	0	0
3d	1	0	0	0	0	0	0	0	0	0	1	0	0
3i	8	0	0	1	0	0	1	1	2	1	2	0	0
3k	5	0	0	0	0	0	1	1	0	1	2	0	0
4b	3	0	0	0	0	0	0	1	1	1	0	0	0
4d	1	0	0	0	0	0	0	0	0	0	1	0	0
4e	2	0	0	0	0	0	0	1	0	0	1	0	0
4i	1	0	0	0	0	0	1	0	0	0	0	0	0
4k	1	0	0	0	0	0	0	0	0	1	0	0	0
5b	1	0	0	0	0	0	0	1	0	0	0	0	0
5d	1	0	0	0	0	0	0	0	0	0	1	0	0

Most numbers of drought events under the type-I category above ‘2’ (i.e. greater than 50% area under drought) occurred in the month of October, with total ‘8’ such events. Most numbers of the drought event occurred under the type-II category was for the ‘m’ type (i.e. 70 to 100 of the total area under drought are moderate drought events). It is clear from these results that, more drought events occur of less severe categories with the lowest percentage area under drought and the frequency of occurrences drops when the drought severity or spatial extend of drought increases. It must be taken into account that the results of these analyses may change if done on another sub-basin of the Godavari basin. However, to avoid repetition of results this analysis is only done on Polavaram sub-basin.

28.5.1 Observed Rainfall-Discharge Analysis

To select years where rainfall appears to have more control on observed discharge, the analysis of basin average rainfall and discharge is done. The observed discharge data were compared with observed annual average (spatial) rainfall (IMD) for each discharge station and corresponding delineated sub-basins. It was observed that in some years, the rainfall and discharge don’t follow a good correlation when compared

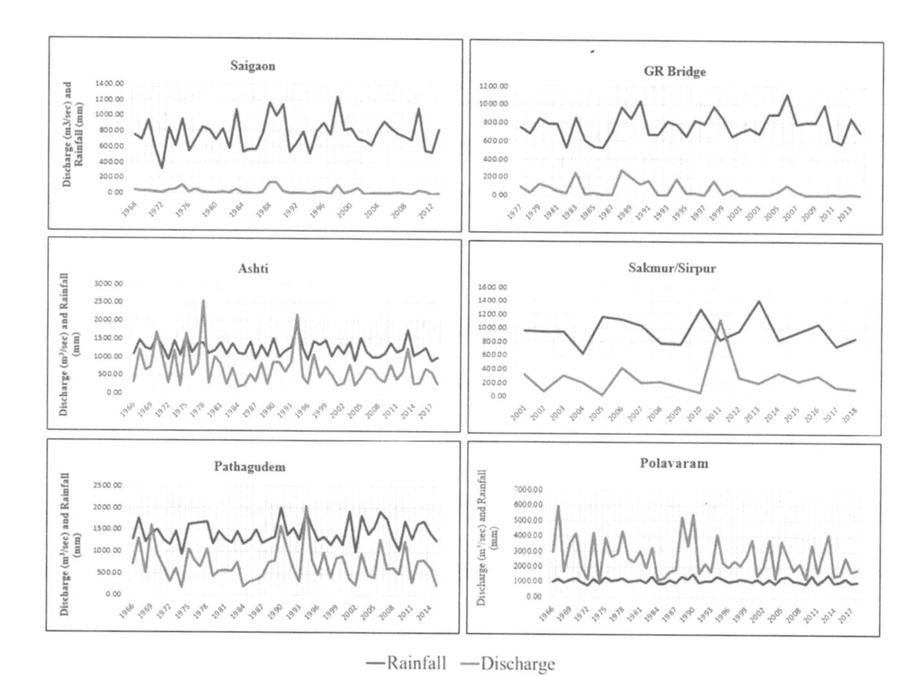

Fig. 28.5 Line graph of observed annual average (spatial) rainfall and discharge at all six discharge stations

with the rest majority of years. Figure 28.5 describes the rainfall-runoff relationship in the form of a line graph.

The non-linearity of the relationship between the average rainfall and discharge can be grouped in following four categories:

(i) Discharge showing decreasing trend with the increased rainfall values. (Example—1998, 2001 and 2006 at Polavaram)
(ii) Discharge showing an increasing trend with the decreased rainfall values. (Example—1987, 2000 and 2002 at Polavaram)
(iii) Lower increase of discharge in proportion to the rainfall. (Example—1983 at Polavaram)
(iv) Higher increase of discharge in proportion to the rainfall. (Example—1967 at Polavaram)

The reason for such behaviours in the Godavari basin could be due to the basin's very high amount of storage capacity, as discussed in the Sect. 28.2 Study Area. Around 41% of the average annual water potential is stored in hundreds of reservoirs present in the basin. The large storage potential of the water resources projects in the basin alters the natural flow regime of the rivers. To avoid erroneous calibration of the model, the altered river flow data must be eliminated from the calibration and validation process of the hydrological model. For each sub-basins, certain years with the above-stated behaviour were selected out of all data to remove the non-linearity and storage effect from the relationship between annual average rainfall and discharge values. During the calibration and validation of the hydrological model to show the efficiency of the model in accurately simulating discharge values when compared to the observed values, these years were excluded for each sub-basins respectively, such as for Ashti (16 years out of 53), for G.R. Bridge (15 years out of 38), for Saigaon (8 years out of 46), for Sakmur/Sirpur (9 years out of 18), for Pathagudem (17 years out of 50) and for Polavaram (13 years out of 53).

28.5.2 Hydrological Modelling (VIC) Results

The VIC model was run for the period of 1951–2018 at daily time-step in water balance mode. Total active grids used to run the model were 440 out of 1040 grids ($0.25° \times 0.25°$ resolution) covering the extent of the study area. The output fluxes generated for each active grid cell from the hydrological module of the model were used in the routing model. The model was calibrated using soil parameters to improve the model's accuracy and the model was then run again using the amended soil parameter file. The calibration procedure assumes that there may be inaccuracies in the soil parameter file that need to be corrected, resulting in a better match between observed and simulated discharge (Garg et al. 2017; Nikam et al. 2018). The calibration and validation time period was selected based on the average annual rainfall-discharge analysis discussed in an earlier section. Table 28.9 provides detailed information about the time period selected for calibration and validation for each sub-basin, along with the coefficient of determination between observed data and modelled data. Line

Table 28.9 Results of calibration and validation of the VIC model

Calibration			Validation		
Polavaram					
Years	Period	R^2	Years	Period	R^2
20	1966-1990	*0.83*	20	1991-2018	*0.82*
Ashti					
Years	Period	R^2	Years	Period	R^2
19	1966-1992	*0.78*	18	1993-2018	*0.84*
GR Bridge					
Years	Period	R^2	Years	Period	R^2
12	1977-1994	*0.85*	11	1995-2014	*0.74*
Pathagudem					
Years	Period	R^2	Years	Period	R^2
17	1966-1988	*0.77*	16	1989-2015	*0.82*
Sakmur/Sirpur					
Years	Period	R^2	Years	Period	R^2
5	2001-2009	*0.87*	4	2009-2018	*0.95*
Saigaon					
Years	Period	R^2	Years	Period	R^2
19	1968-1992	*0.87*	19	1993-2013	*0.89*

hydrograph of the observed and simulated annual discharge for Polavaram is shown in Fig. 28.6. The long-term simulated discharge data was further used in estimation of the impact of each meteorological drought on hydrology and nexus between different meteorological droughts and natural flow regimes of the basin.

28.5.3 *Impact of Meteorological Droughts on Observed and Modelled Discharge*

Observed discharge values at any station are controlled or influenced by the anthropogenic activities performed upstream of the basin (such as storage, diversion, irrigation supply, public water supply, industrial water supply, etc.). Modelled/simulated runoff and baseflow values routed to a station in the VIC model don't consider any of the man-made intrusions during its simulation. It considers that the stream will flow freely throughout the basin though it is not the actual scenario of the basin. The simulated discharge values can be used to highlight the impact of various anthropogenic interventions on the natural hydrological regime of a basin in relation to the meteorological extremes or droughts.

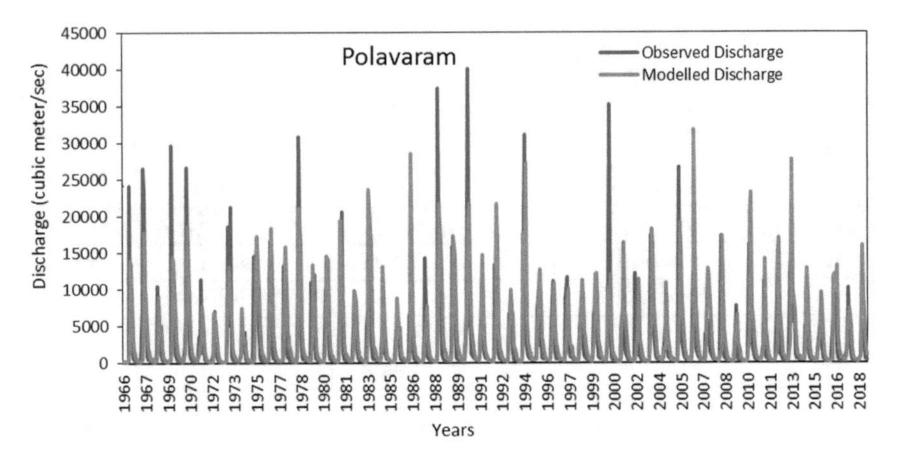

Fig. 28.6 Line graph between simulated and observed annual discharge at Polavaram during the calibration Phase

The percentage deviation of observed and simulated monthly discharge values from its 'mean' values for each station was calculated (only the years with normal annual rainfall-discharge relationship as discussed earlier were used to derive mean from observed monthly discharge data) and their relation with the various category of the meteorological droughts was schematised using boxplots, as shown in Fig. 28.7.

A boxplot is a standardised method of showing numerical data distributions using the quartiles ('minimum', 'first quartile (Q1)', 'median', 'third quartile (Q3)' and 'maximum'). These five elements constitute a boxplot along with outliers represented by the dots. Whiskers or the boxes (blue coloured) represents the interquartile range of the data with a median line (green coloured) in it. The impact of each category

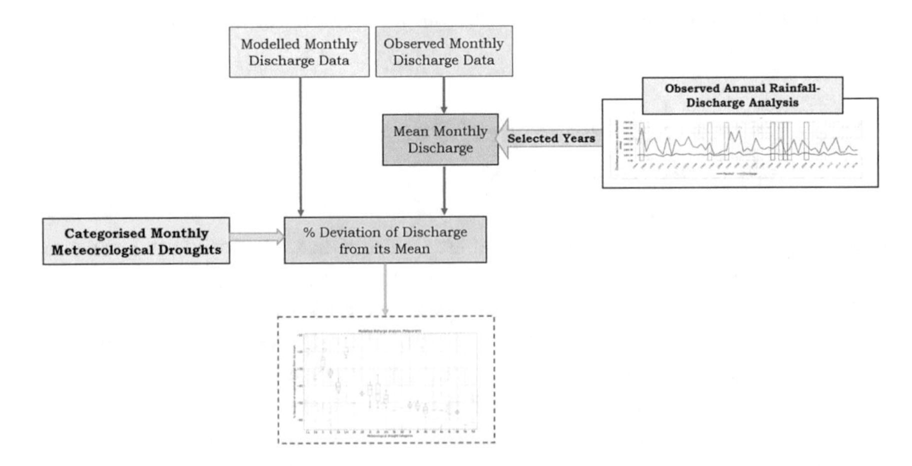

Fig. 28.7 Flowchart for relating the meteorological drought categories with the percentage deviation of observed discharge from its mean

of meteorological drought on the river discharge at each outlet was quantified by estimating the deviation of river flow, both observed and simulated. The hydrological behaviour of each sub-basin vis-à-vis different categories of meteorological drought is represented in Figs. 28.8 and 28.9. These figures indicate the minimum, median and maximum change (reduction) in the river discharge (observed and simulated) when the contributing sub-basin is exposed to each of the type-II categories of meteorological drought.

From the observed discharge analysis in relation to the meteorological drought categories for all sub-basin, we can lay down the following points:

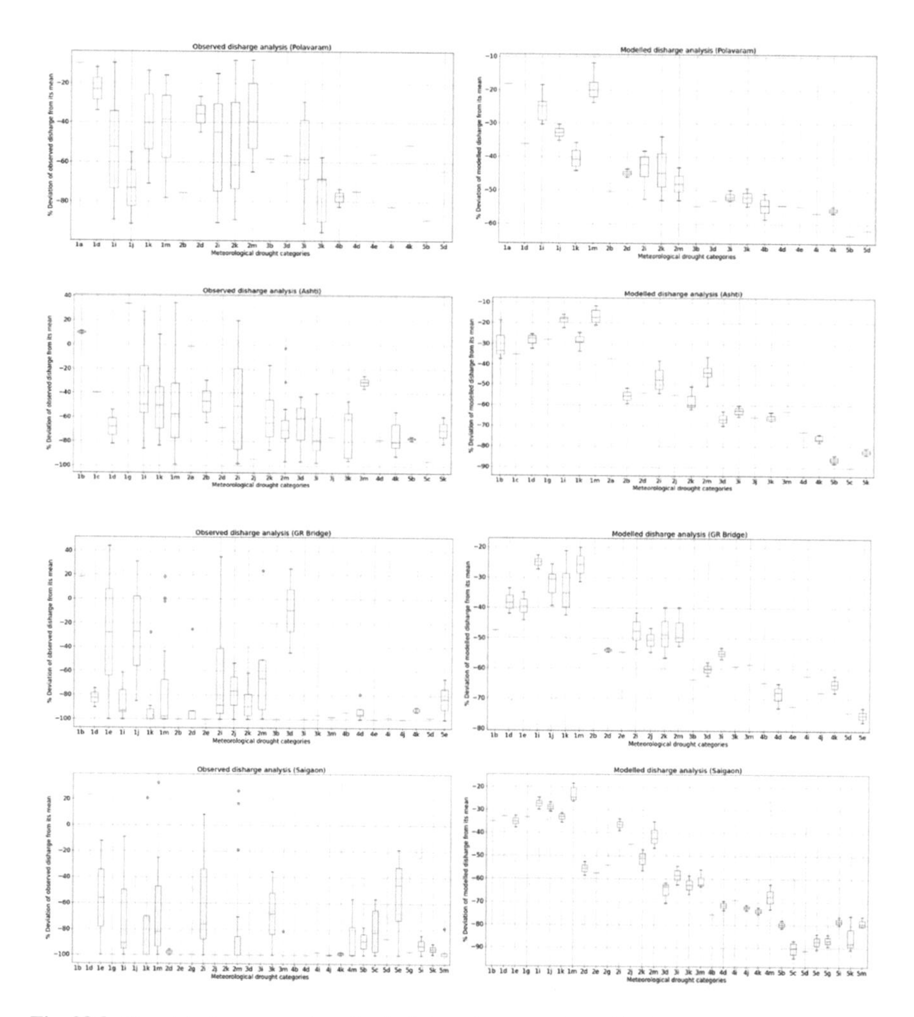

Fig. 28.8 Boxplot showing meteorological drought categories with the percentage deviation of observed and simulated discharge from its mean at Polavaram, Ashti, G.R. Bridge and Saigaon

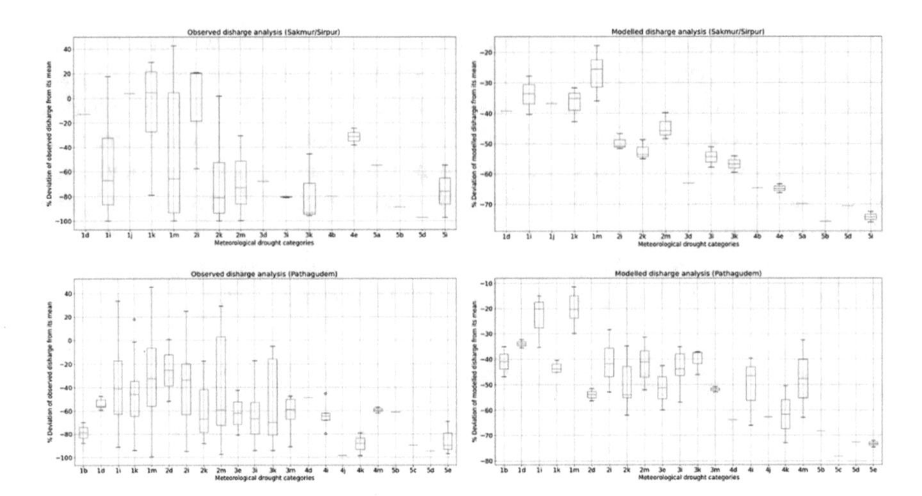

Fig. 28.9 Boxplot showing meteorological drought categories with the percentage deviation of observed and simulated discharge from its mean at Sakmur/Sirpur and Pathagudem

- Any relationship or trend was not observed between the percentage deviations in discharge values in relation to the various meteorological drought categories as a whole. This might be due to the dominant anthropogenic control over the river discharge in the Godavari basin.
- There are some instances where we are getting positive deviation in discharge values even during meteorological drought periods for all sub-basins except for Polavaram. These periods also highlights the prominent anthropogenic control over the river flow in the Godavari Basin. The water released in the river from all the major reservoirs during large-scale drought (meteorological) forces the observed discharge data towards positive deviation.
- For meteorological categories above 3 (i.e. more than 50% sub-basin area under meteorological drought), we could only see the negative deviation in discharge for all sub-basin. This shows the susceptibility of Godavari basins towards large-scale (spatial) meteorological droughts. The transformation of many meteorological droughts that have occurred on a smaller spatial scale must have been controlled by the reservoir release operations in the basin. However, meteorological droughts covering 50% or more area of sub-basin tend to have a greater impact on the natural flow regime of the rivers. The deviation from mean discharge value varies even in these cases also due to the emergency water release in the river to avert the transformation of meteorological disaster into a hydrological, agriculture and socio-economic disaster.
- For all sub-basins except Polavaram (especially Saigaon and G.R Bridge), −100% deviation was observed for lower as well as higher meteorological drought categories because of the presence of zero discharge values in the observed data, most probably due to the reservoir operations controlling the flow regime.

- For meteorological droughts of higher severity with four and above (% area under drought above 50 percent), a higher negative percentage deviation was observed for all sub-basins. The reason for this could be that at meteorological droughts events of higher severity, the storage factor tends to reduce over the streamflow.

From the simulated discharge analysis in relation to the meteorological drought categories, we can lay down the following points:

- A linear and direct relationship was observed between the percentage deviations of discharge values with the scale (both in terms of area and severity) of the various meteorological drought events.
- With the increase in the severity of meteorological drought events, the percentage deviation of discharge values also increases in a negative direction generally for all sub-basins.
- Among all sub-basins, Polavaram shows a lesser reduction in discharge values as we move towards higher meteorological drought categories. Mostly due to the reason that the other five sub-basins are isolated in nature when it comes to discharge contribution to the outlet station. Whereas, Polavaram at the lower part of the Godavari basin gets discharge contribution from all the upstream basin areas. This again goes to prove the fact that the hydrological response of a basin towards a meteorological drought becomes more and more complex with the increase in the spatial extent of the meteorological drought or basin.

28.5.4 Best Probability Distribution Function for Hydrological Drought Index Estimation

Using a single probability distribution function for different discharge stations and accumulation periods requires a high degree of spatial homogeneity throughout the basin, which is unlikely to be found in reality with varying physical characteristics, climatic and hydrological regimes. Therefore, different distributions (Normal, Log-Normal, Gamma and Weibull) were evaluated to determine the best probability distribution to fits the observed and modelled discharge data. Kolmogorov–Smirnov (KS) test was performed for six different accumulation periods/time scales (1, 3, 6, 9, 12 and 24-month) to measure the goodness of fit for various distributions at chosen confidence level (99%). KS test results for both observed and simulated discharge data of all six stations can be seen in Tables 28.10 and 28.11, respectively.

From Table 28.10, it was observed that for the 1 and 3-month accumulation discharge (observed) series, Normal and Log-Normal distribution fits better. For discharge (observed) series of higher time scales (9, 12 and 24-month), Log-Normal fits best for the majority of the outlets except for the 6-month scale in which Weibull distribution also fits better along with Log-Normal and Normal distribution. Also, the Normal, Log-Normal, Gamma and Weibull distribution are acceptable to fit data for flow at Polavaram (representing the whole basin) at a 24-month timescale. From Table 28.11, it was observed that the log-normal distribution is the best fit for almost

Table 28.10 KS test results for observed discharge data of all the six stations

Accumulation Periods	1 Month				3 Month				6 Month				9 Month				12 Month				24 Month			
Discharge Stations	Normal	Log-Normal	Gamma	Weibull	Normal	Log-Normal	Gamma	Weibull	Normal	Log-Normal	Gamma	Weibull	Normal	Log-Normal	Gamma	Weibull	Normal	Log-Normal	Gamma	Weibull	Normal	Log-Normal	Gamma	Weibull
Saigaon																								
GR Bridge																								
Sakmur/Sirpur																								
Ashti																								
Pathagudem																								
Polavaram																								

Table 28.11 KS test results for simulated discharge data of six stations

Accumulation Periods	1 Month				3 Month				6 Month				9 Month				12 Month				24 Month			
Discharge Stations	Normal	Log-Normal	Gamma	Weibull	Normal	Log-Normal	Gamma	Weibull	Normal	Log-Normal	Gamma	Weibull	Normal	Log-Normal	Gamma	Weibull	Normal	Log-Normal	Gamma	Weibull	Normal	Log-Normal	Gamma	Weibull
Saigaon																								
GR Bridge																								
Sakmur/Sirpur																								
Ashti																								
Pathagudem																								
Polavaram																								

Distribution acceptable to fit data at confidence level 99%

Optimum distribution fitting to data

all timescales and at all stations similar to (Nalbantis and Tsakiris 2008; Zaidman et al. 2002). The Normal, Log-Normal, Gamma and Weibull distributions are suitable to fit data at all stations (except for Saigaon) at higher timescales of 12 and 24 months. Hydrological drought indices at Polavaram for different time scales from modelled discharge are estimated based on the interpretation of KS test results (i.e. Optimum distribution fitting to data).

28.5.5 Hydrological Drought Period Analysis

Standardised Streamflow Index (SSI) was calculated from discharge (both observed and modelled) data of Polavaram station for accumulation period of 1, 3, 6, 9, 12 and 24-month using the suitable probability distribution function based on the KS test results. Hydrological drought index values (SSI) were classified into eight classes similar to SPI using Table 28.3. Then the hydrological drought events were identified using SSI values at each timescale for 'Moderate', 'Severe' and 'Extreme' drought categories. Duration and severity of the identified drought events were also estimated with their mean, median and maximum values.

Drought periods were defined using the threshold value of the corresponding category (i.e. −1.0 for moderate drought, −1.5 for severe drought and −2.0 for extreme drought). The end of the drought period was considered when the subsequent monthly SSI values attain normal condition (i.e. above zero value). This analysis was performed for both observed and simulated discharge-based SSI values for each timescale, whose results are represented in a tabulated form as shown in Tables 28.12 and 28.13, respectively.

From the hydrological drought period analysis of observed discharge data following inferences were made:

- Number of hydrological droughts occurred decreases with increasing severity and accumulation period. Out of all hydrological droughts occurred historically more

Table 28.12 Hydrological drought characteristics calculated for Polavaram station using SSI values from observed discharge

SSI Category	SSI Accumulation Period (months)	Number of Events	Duration			Severity		
			Mean	Median	Max	Mean	Median	Max
Moderate Drought (-1.5 < - 0.1)	1	37	6.63	4.00	19.00	-5.87	-3.50	-17.80
	3	24	7.58	7.00	21.00	-7.86	-7.04	-26.78
	6	19	8.63	8.00	18.00	-9.63	-9.88	-24.75
	9	15	9.87	11.00	22.00	-12.00	-11.19	-30.04
	12	11	15.18	13.00	46.00	-17.07	-15.72	-54.09
	24	5	21.00	17.00	35.00	-28.25	-21.67	-64.62
Severe Drought (-2.0 < - 1.5)	1	17	4.94	3.00	13.00	-6.01	-4.78	-17.44
	3	11	9.00	7.00	21.00	-11.15	-8.31	-26.78
	6	10	8.20	7.50	18.00	-10.96	-10.22	-24.75
	9	8	10.50	11.50	21.00	-14.47	-14.93	-28.64
	12	7	17.00	13.00	46.00	-20.04	-19.05	-54.09
	24	3	20.33	14.00	34.00	-30.17	-21.59	-63.24
Extreme Drought (<-2.0)	1	-	-	-	-	-	-	-
	3	2	9.50	9.50	13.00	-15.19	-15.19	-21.54
	6	4	7.25	6.50	13.00	-12.50	-12.01	-20.73
	9	3	8.33	10.00	13.00	-14.10	-18.16	-19.95
	12	3	16.67	13.00	35.00	-18.86	-19.05	-35.39
	24	2	18.50	18.50	34.00	-33.95	-33.95	-63.24

Table 28.13 Hydrological drought characteristics calculated for Polavaram station using SSI values from modelled discharge

SSI Category	SSI Accumulation Period (months)	Number of Events	Duration			Severity		
			Mean	Median	Max	Mean	Median	Max
Moderate Drought (-1.5 < -0.1)	1	26	7.77	7.50	22.00	-8.16	-8.04	-20.58
	3	17	10.82	11.00	20.00	-10.27	-9.61	-19.46
	6	16	10.94	11.00	17.00	-10.57	-9.30	-18.25
	9	16	10.31	11.00	23.00	-12.62	-11.05	-36.83
	12	10	13.20	12.50	26.00	-17.93	-18.59	-40.74
	24	6	26.50	24.00	47.00	-28.66	-22.77	-63.55
Severe Drought (-2.0 < -1.5)	1	15	8.38	10.00	17.00	-9.94	-9.97	-20.58
	3	13	10.00	10.00	11.00	-11.81	-12.32	-12.75
	6	12	11.00	10.50	17.00	-13.16	-13.18	-17.69
	9	8	10.75	11.50	23.00	-16.22	-17.93	-36.83
	12	7	13.57	12.00	24.00	-20.33	-17.68	-37.97
	24	4	27.00	24.50	46.00	-32.60	-28.20	-62.18
Extreme Drought (<-2.0)	1	6	11.00	11.00	14.00	-13.91	-14.63	-19.30
	3	5	6.50	5.50	13.00	-11.05	-9.02	-21.54
	6	4	8.25	8.00	11.00	-11.77	-13.46	-16.79
	9	3	9.67	9.00	11.00	-13.17	-16.48	-17.39
	12	3	11.67	12.00	12.00	-13.59	-17.32	-21.00
	24	2	39.50	39.50	46.00	-48.41	-48.41	-62.18

than 50% were of moderate intensity, around 28% of severe intensity and 0–20% were of extreme intensity.

- The number of hydrological droughts occurred at the station reduces with an increasing accumulation period, which indicates that the hydrological drought does not persist for long duration at Polavaram station. Only 10–20% of hydrological drought events were observed in the accumulation period of more than a year. However, the number of events of hydrological drought due to accumulation of deficit over 24 months decreases with increasing intensity/severity of the drought, which means historically very few (only 02) severe hydrological droughts of extreme intensity of accumulation period of 24 months have occurred at Polavaram.

- The frequency of occurrences drops, the duration lengthens and the severity worsens as the accumulation period increases except for the extreme drought events at 1 and 3-month accumulation periods.
- No extreme drought events were observed at the 1-month accumulation period; this might be due to immense storage capacity of the basin which can be utilised to overcome short duration extreme hydrological drought events.

From the hydrological drought period analysis of simulated discharge data following inferences were made:

- Similar to the hydrological droughts detected using analysis of observed discharge data in case of simulated discharge data also indicates higher (more than 50%) occurrence of moderate intensity hydrological droughts in case of all the accumulation periods.
- The occurrence of hydrological droughts reduces with an increase in both accumulation period and drought intensity. Less than 10% of total events are caused by accumulation period longer than a year.
- In case of hydrological droughts detected using simulated discharge data, the number of drought events of moderate intensity is less in comparison to a number of events of same intensity detected using observed discharge data. However, a number of drought events of sever and extreme intensity are detected more in number in case of simulated data as compared with a number of events of same intensity detected during analysis using observed discharge data. This observation strongly hints towards the obvious anthropogenic behaviour of the basin. In the highly controlled basin, water from upstream reservoirs is released in case of severe and extreme hydrological drought conditions to avoid damage to the natural and man-made ecosystems. However, during moderate drought periods, no such concessions are given to the downstream to protect the water based economy of the upstream projects.
- For 1 and 3-month accumulation periods, more extreme drought events are observed in case of simulated discharge data.

28.5.6 Interrelation Between Meteorological Drought and Hydrological Drought

To analyse and quantify the impact of meteorological drought on the hydrological regime of the Godavari River basin, we tried to derive the interrelation between the meteorological and hydrological drought. For that, monthly SSI values derived from both observed and simulated discharge were compared along with the coinciding meteorological droughts as shown in Table 28.15 using the colour coding for different categories of meteorological drought (type-I categorisation) as shown in Table 28.14.

From the above-stated analysis following observations were made:

Table 28.14 Colour code for SPI type-I categories

Colour Gradient	SPI Category
	0
	1
	2
	3
	4
	5

- The occurrence of hydrological drought events sometimes corresponds with the incidence of meteorological drought, although not all of the meteorological drought events get migrated into hydrological droughts, especially in case of meteorological droughts occurring over small spatial extend. This can be seen during the March and April months of the year 1971, when class-1 meteorological drought does not migrate into hydrological drought in both observed and simulated discharge, due to smaller areas (10–30% of total basin area) under meteorological drought.
- In many instances, the meteorological drought propagates to hydrological drought with an inconsistent lag time depending on the subsequent drought developments. This also indicates the impact of the storage effect on the hydrological regime of the basin.
- In some cases, even after a meteorological drought event has ended, the hydrological drought event continues to last for prolonged periods of time, or we can say that when a meteorological drought occurs, the influence on river discharge might take months to manifest.
- In various cases, the SSI values from observed discharge data either overestimates or underestimate the incidence of hydrological droughts.

28.6 Conclusion

Meteorological droughts that occurred in the Godavari basin were identified and characterised using drought onset time, intensity, location of occurrence and spatial extent. The addition of location of occurrence and spatial extent in drought characterisation was only possible in a geospatial environment. The severity of drought, its spatial extent and their complex were incorporated in the drought characterisation by classifying each drought event using the scheme considering total area under each drought and percent contribution of each intensity/severity class in the total affected area. It was observed that more than 55% times the basin was under meteorological drought with less than 10% of the basin area being affected and within the affected area, the majority was under moderate drought. Most numbers of drought events occurring on the less than 10% of basin area occurred in the month of March with a

Table 28.15 Monthly SSI values [both observed (Obs.) and simulated (Mod.)] for the period 1970–1990 along with coinciding meteorological drought categories

Years	Type	1	2	3	4	5	6	7	8	9	10	11	12
1970	Obs.	-0.13	-0.69	-0.69	-0.79	-0.33	0.02	-0.32	1.86	1.49	-0.19	0.03	-0.27
	Mod.	0.77	0.66	1.00	0.68	0.28	1.23	0.13	0.09	1.18	0.67	0.50	0.64
1971	Obs.	-0.21	-0.76	-0.63	0.17	0.11	0.37	-0.85	0.09	-0.90	0.15	-0.47	-0.56
	Mod.	0.69	0.68	0.99	1.03	1.60	0.85	-0.85	-2.60	-0.68	-0.55	-0.92	-1.15
1972	Obs.	-0.58	-1.08	-1.30	-1.08	-1.38	-0.35	-1.05	-0.41	-0.51	-0.48	-0.54	-0.90
	Mod.	-1.28	-1.22	-1.62	-1.10	-1.54	-1.99	0.71	-2.53	-1.24	-1.39	-1.41	-1.51
1973	Obs.	-1.49	-1.89	-1.58	-1.25	-1.11	-0.66	-0.74	0.93	0.04	6.51	0.40	0.58
	Mod.	-2.11	-1.98	-2.13	-2.15	-2.11	-0.74	1.14	-0.54	0.43	0.63	2.12	1.27
1974	Obs.	0.27	0.09	-0.36	0.04	-0.36	-0.54	-1.06	-0.92	-0.88	0.78	-0.09	-0.75
	Mod.	1.31	1.25	1.04	0.98	0.66	-0.05	-1.70	-0.91	-2.61	-1.12	-0.22	-1.09
1975	Obs.	-0.92	-0.45	-1.10	-1.00	-0.74	3.68	-0.44	-0.36	1.88	0.77	0.18	0.11
	Mod.	-1.13	-0.83	-0.27	-1.23	-1.26	-0.39	0.80	0.23	1.02	1.41	2.02	1.72
1976	Obs.	0.56	-0.53	-0.35	-0.19	-0.56	-0.65	0.14	0.70	-0.73	-0.41	0.09	-0.51
	Mod.	1.68	1.56	1.20	1.35	0.76	-0.37	1.46	0.18	-1.14	-0.63	-0.23	0.14
1977	Obs.	-0.55	-0.85	-0.78	-0.28	-0.10	-0.57	0.54	0.23	-0.78	-0.40	6.49	0.03
	Mod.	0.03	-0.01	0.02	0.50	0.75	-0.59	0.30	-0.09	0.85	-0.50	-0.19	1.32
1978	Obs.	-0.54	-0.30	-0.48	-0.74	-0.38	-0.14	0.42	2.34	-0.19	-0.33	-0.40	0.46
	Mod.	0.45	0.59	0.32	-0.02	-0.13	1.11	1.13	1.08	0.93	-0.26	-0.06	0.19
1979	Obs.	0.72	3.25	0.62	0.24	0.64	2.60	-0.93	-1.01	1.82	-0.41	-0.44	0.11
	Mod.	0.03	1.09	0.09	-0.19	0.28	-1.01	-0.16	0.21	-2.19	-0.59	-0.94	-0.52
1980	Obs.	-0.63	-1.20	-0.59	-0.73	-0.75	0.53	0.11	-0.32	-0.39	-0.36	-0.49	-0.42
	Mod.	-1.10	-1.23	-1.25	-1.03	-1.12	0.77	-0.02	0.36	0.58	-0.28	-0.64	-0.12
1981	Obs.	0.34	-0.05	-0.29	-0.03	-0.45	-0.46	-0.37	-0.68	3.98	-0.22	-0.23	0.05
	Mod.	0.47	-0.11	0.32	-0.07	0.15	-0.95	-0.99	0.92	-0.75	-0.59	-0.43	0.54
1982	Obs.	0.78	0.66	-0.28	-0.32	0.00	-0.19	-0.54	-0.13	-0.59	-0.07	0.05	-0.27
	Mod.	0.73	0.98	0.54	0.57	0.35	-0.70	-1.20	-0.40	-0.53	-0.69	-0.31	-0.77
1983	Obs.	0.21	0.77	-0.48	-0.56	-0.50	0.82	-0.28	-0.42	2.43	0.18	0.33	0.90
	Mod.	-0.73	-0.56	-0.77	-0.94	-0.57	-1.30	-0.13	1.30	1.16	2.55	1.61	1.70
1984	Obs.	1.49	1.32	0.95	1.44	1.11	-0.55	-0.69	-0.78	-0.92	-0.29	-0.33	0.27
	Mod.	2.00	1.92	1.47	1.59	0.51	0.40	-0.43	0.17	-1.58	-1.54	-1.98	-2.09
1985	Obs.	0.08	-0.20	-0.57	-0.67	-0.64	-0.10	-0.59	-0.89	-0.56	-0.29	-0.48	-0.53
	Mod.	-1.43	-1.78	-2.05	-1.09	-1.52	-0.45	-0.35	-0.59	-1.78	-0.30	-0.71	-0.82
1986	Obs.	-0.52	0.67	-0.67	-0.33	-0.25	1.15	-0.26	-0.87	-0.89	-0.48	-0.57	-0.44
	Mod.	-0.52	0.24	-0.46	-1.18	-0.53	-0.24	0.72	1.65	-1.37	-1.39	-1.11	-1.08
1987	Obs.	0.20	-0.74	-0.22	-0.73	-0.41	-0.59	-0.71	0.42	-1.04	-0.44	0.12	-1.02
	Mod.	-0.25	-0.98	-0.43	-1.15	-0.38	-0.64	-0.99	-1.96	-0.66	-1.13	-0.58	-0.77
1988	Obs.	-0.93	-0.59	-1.16	-1.11	-1.03	-0.24	3.63	0.34	0.58	-0.27	-0.12	-0.02
	Mod.	-1.19	-1.21	-1.19	-1.34	-0.74	-0.44	1.31	1.13	0.90	1.93	1.24	1.42
1989	Obs.	-0.25	-0.57	-0.06	-0.16	-0.33	-0.07	0.86	0.28	0.40	-0.39	-0.30	-0.80
	Mod.	1.51	1.41	1.48	1.46	0.52	0.19	0.49	0.69	0.67	0.32	0.18	0.28
1990	Obs.	-0.58	-0.23	-0.30	-0.28	3.04	0.16	-0.25	3.42	-0.01	1.49	1.17	2.27
	Mod.	0.71	0.47	0.56	0.22	4.71	2.41	0.62	0.95	1.49	1.86	1.86	1.55

total '55' then in April with '44' such events. It is clear from the results that, more drought events have occured of less severe categories with the lowest percentage area under drought and the frequency of occurrences drops when the drought severity or spatial extend increases. It must be taken into account that the results of these analyses may change if done on another sub-basin of the Godavari basin.

Hydrological droughts were identified, characterised and analysed for their inter-relations for Godavari River basin using SSI for various accumulation periods. Log-normal was found as the best probability distribution to fit both the observed and modelled hydrological data for almost all timescales. More hydrological drought events have occurred of less severe drought categories with smaller duration at shorter accumulation periods. The occurrence of hydrological drought events drops when the duration increases and the severity worsens as the accumulation period extends.

For higher accumulation periods, the hydrological drought characteristic remains the same, with slight variation for both observed and modelled discharge-based drought periods. For modelled discharge values, a linear relationship is followed when we compare meteorological droughts with the percentage deviation of discharge from its mean. But not necessarily the same linear relation exists in actual condition for the reason that every basin will have its resilience towards droughts and also we have not considered diversion and water storage in the basin.

Observed discharge data of the Godavari basin doesn't represent the natural flow regime due to which behaviour of hydrological drought is anthropogenically controlled. In a natural basin (without any man-made interventions or storage), almost all significant metrological drought events will migrate to hydrological drought events. But for a controlled basin to respond in terms of hydrological drought would require an even higher, larger and a severe amount of meteorological drought. Generally, it's not necessary all the time that anthropogenic storage and control structure would have a negative impact on the natural and hydrological processes. From the drought point of view, the more the basin will have storage, the more it will have resilience against the hydrological drought.

In case of hydrological droughts detected using simulated discharge data, the number of drought events of moderate intensity is less in comparison to a number of events of same intensity detected using observed discharge data. However, a number of drought events of severe and extreme intensity are detected more in number in case of simulated data as compared with number of events of same intensity detected during analysis using observed discharge data. This observation strongly hints towards the obvious anthropogenic behaviour of the basin. In the highly controlled basin, water from upstream reservoirs is released in case of severe and extreme hydrological drought conditions to avoid damage to the natural and man-made ecosystems. However, during moderate drought periods, no such concessions are given to the downstream to protect the water based economy of the upstream projects.

The behaviour of the basin towards the propagation of meteorological drought to hydrological drought is very complex due to the reasons discussed above, for which a detailed investigation is required. More attention is needed towards the spatial and temporal development of hydrological droughts as a response to meteorological droughts and the related soil moisture droughts (i.e. Drought Propagation). There is little that can be done to avert the shortfall of precipitation, but drought monitoring and early warning systems (M&EWs) can help to reduce its socio-economic vulnerability by giving authorities more time to prepare solutions and avert a crisis scenario.

References

Abramowitz M (1965) Handbook of mathematical functions with formulas, graphs, and mathematical tables. Dover Publications, Dover. https://ci.nii.ac.jp/naid/10011544844

Aggarwal SP, Garg V, Gupta PK, Nikam BR, Thakur PK (2012) Climate and LULC change scenarios to study its impact on hydrological regime. In: The international archives of the photogrammetry, remote sensing and spatial information sciences, vol XXXIX-B8, pp 147–152. https://doi.org/10.5194/ISPRSARCHIVES-XXXIX-B8-147-2012

Aggarwal SP, Thakur PK, Garg V, Nikam BR, Chouksey A, Dhote P, Bhattacharya T (2016) Water resources status and availability assessment in current and future climate change scenarios for beas river basin of North Western Himalaya. In: The international archives of the photogrammetry, remote sensing and spatial information sciences (ISPRS), XLI-B8 2016, XXIII ISPRS Congress, 12–19 July 2016, Prague, Czech Republic, pp 1389–1396. https://doi.org/10.5194/isprs-archives-XLI-B8-1389-2016

Aswathi PV, Nikam BR, Chouksey A, Aggarwal SP (2018) Assessment and monitoring of agricultural droughts in Maharashtra using meteorological and remote sensing based indices. In: ISPRS annals photogrammetry, remote sensing and spatial information science, vol IV-5, pp 253–264. https://doi.org/10.5194/isprs-annals-IV-5-253-2018.

Barker LJ, Hannaford J, Chiverton A, Svensson C (2016) From meteorological to hydrological drought using standardised indicators. Hydrol Earth Syst Sci 20(6):2483–2505. https://doi.org/10.5194/hess-20-2483-2016

Bhardwaj K, Shah D, Aadhar S, Mishra V (2020) Propagation of meteorological to hydrological droughts in India. J Geophys Res: Atmosp 125(22):e2020JD033455. https://doi.org/10.1029/2020JD033455

Chen H, Sun J (2017) Characterizing present and future drought changes over eastern China. Int J Climatol 37:138–156. https://doi.org/10.1002/JOC.4987

Chouhan H, Garg V, Nikam BR, Chouksey A, Aggarwal SP (2017) Assessment and characterization of meteorological drought using standardized precipitation index in the Upper Luni River Basin, Rajasthan. Int J Emerg Technol 8(1):265–271

CWC, & NRSC (2014) Godavari basin report. www.india-wris.nrsc.gov.in

Das J, Gayen A, Saha P, Bhattacharya SK (2020) Meteorological drought analysis using standardized precipitation index over Luni River Basin in Rajasthan, India. SN Appl Sci 2:9, 2(9), 1–17. https://doi.org/10.1007/S42452-020-03321-W

Deshpande S, Nikam BR, Garg V, Mohapatra M (2019) Assessment of meteorological droughts and their impact on the hydrological regime of Godavari River Basin. In: Best paper award 'HYDRO 2019 INTERNATIONAL', 24th International conference on hydraulics, water resources, coastal & environmental engineering, December 18–20, 2019, At Osmania University, Hyderabad, India

Downing TE (1993) Natural hazards, E. A. Bryant. Cambridge University Press, Cambridge, 1991. No. of pages: 294 + xvii. Price: (h/b) £40, US$79.50 (ISBN 0 521 37295 X); (p/b) £14.95, US$29.95 (ISBN 0 521 37889). Int J Climatol 13(3):344–346. https://doi.org/10.1002/joc.3370130310

Edossa DC, Babel MS, Gupta AD (2010) Drought analysis in the Awash River Basin, Ethiopia. Water Resour Manage 24(7):1441–1460. https://doi.org/10.1007/s11269-009-9508-0

Field CB, Barros V, Stocker TF, Qin D, Dokken DJ, Ebi KL, Mastrandrea MD, Mach KJ, Plattner G-K, Allen SK, Tignor M, Midgley PM (2012) Managing the risks of extreme events and disasters to advance climate change adaptation. https://www.ipcc.ch/report/managing-the-risks-of-extreme-events-and-disasters-to-advance-climate-change-adaptation/

Garg V, Aggarwal SP, Nikam BR, Thakur PK (2013) Hypothetical scenario based impact assessment of climate change on runoff potential of a Basin. ISH J Hydraul Eng 19(3):244–249. https://doi.org/10.1080/09715010.2013.804673

Garg V, Aggarwal SP, Gupta PK, Nikam BR, Thakur PK, Srivastav SK, Senthil Kumar A (2017) Assessment of land use land cover change impact on hydrological regime of a basin. Environ Earth Sci 76(18). https://doi.org/10.1007/s12665-017-6976-z

Garg V, Nikam BR, Thakur PK, Gupta PK, Aggarwal SP, Srivastav SK (2019) Human-induced land use land cover change and its impact on hydrology. HydroResearch 1:48–56. https://doi.org/10.1016/j.hydres.2019.06.001

GRMB (2019) Annual report. https://grmb.gov.in/uploadedFiles/publications/AR2018-19.pdf

Guttman NB (1999) Accepting the standardized precipitation index: a calculation algorithm. J Am Water Resour Assoc 35(2):311–322. https://doi.org/10.1111/j.1752-1688.1999.tb03592.x

Hisdal H, Tallaksen LM (2003) Estimation of regional meteorological and hydrological drought characteristics: A case study for Denmark. J Hydrol 281(3):230–247. https://doi.org/10.1016/S0022-1694(03)00233-6

Khatiwada KR, Pandey VP (2019) Characterization of hydro-meteorological drought in Nepal Himalaya: a case of Karnali River Basin. Weather Clim Extremes 26:100239. https://doi.org/10.1016/J.WACE.2019.100239

Kumar RODN, Kumar RODN, Sharma MAA, Mehrotra R (2013) Assessing severe drought and wet events over India in a future climate using a nested bias-correction approach. J Hydrol Eng 760–772. http://citeseerx.ist.psu.edu/viewdoc/summary?doi=10.1.1.728.5046

Lee JH, Park SY, Kim JS, Sur C, Chen J (2018) Extreme drought hotspot analysis for adaptation to a changing climate: assessment of applicability to the five major river basins of the Korean Peninsula. Int J Climatol 38(10):4025–4032. https://doi.org/10.1002/JOC.5532

Liang X (1994) A two-layer variable infiltration capacity land surface representation for general circulation models [Elsevier B.V.]. In: Water resources series. https://doi.org/10.1016/0921-8181(95)00046-1

Lohmann D, Raschke E, Lettenmaier DP (1998) Regional scale hydrology: I. Formulation of the VIC-2L model coupled to a routing model. Hydrol Sci J 43(1):131–141. https://doi.org/10.1080/02626669809492107

Mckee TB, Doesken NJ, Kleist J (1993) The relationship of drought frequency and duration to time scales. In: Eighth conference on applied climatology

Ministry of Agriculture and Farmer's Welfare (India) (2016) Manual for drought management. In: Agricultural water management. https://agricoop.nic.in/sites/default/files/ManualDrought2016.pdf

Mishra V (2020) Long-term (1870–2018) drought reconstruction in context of surface water security in India. J Hydrol 580. https://doi.org/10.1016/J.JHYDROL.2019.124228

Mishra V, Cherkauer KA, Shukla S (2010) Assessment of drought due to historic climate variability and projected future climate change in the Midwestern United States. J Hydrometeorol 11(1):46–68. https://doi.org/10.1175/2009JHM1156.1

Nalbantis I, Tsakiris G (2008) Assessment of hydrological drought revisited. Water Resour Manage 23:5, 23(5), 881–897. https://doi.org/10.1007/S11269-008-9305-1

Nam WH, Hayes MJ, Svoboda MD, Tadesse T, Wilhite DA (2015) Drought hazard assessment in the context of climate change for South Korea. Agric Water Manag 160:106–117. https://doi.org/10.1016/J.AGWAT.2015.06.029

Nanda A, Nikam BR, Garg V, Aggarwal SP (2021) Analysing the response of meteorological droughts on stream flow in the Pennar River Basin. In: Proceedings of 'HYDRO 2020 INTERNATIONAL', 25th International conference on hydraulics, water resources, coastal & environmental engineering, March 26–28, 2021, at National Institute of Technology Rourkela, Odisha, India

Narasimhan B, Srinivasan R (2005) Development and evaluation of Soil Moisture Deficit Index (SMDI) and Evapotranspiration Deficit Index (ETDI) for agricultural drought monitoring. Agric for Meteorol 133(1–4):69–88. https://doi.org/10.1016/J.AGRFORMET.2005.07.012

Nikam BR, Aggarwal SP, Thakur PK, Garg V, Roy S, Chouksey A, Dhote PR, Chauhan P (2020) Assessment of early season agricultural drought using remote sensing. In: International archives photogrammetry, remote sensing spatial information science, XLIII-B3-2020, pp 1691–1695. https://doi.org/10.5194/isprs-archives-XLIII-B3-2020-1691-2020

Nikam BR, Garg V, Jeyaprakash K, Gupta PK, Srivastav SK, Thakur PK, Aggarwal SP (2018) Analyzing future water availability and hydrological extremes in the Krishna basin under changing climatic conditions. Arab J Geosci 11(19). https://doi.org/10.1007/s12517-018-3936-1

Niu J, Chen J, Sun L (2015) Exploration of drought evolution using numerical simulations over the Xijiang (West River) basin in South China. J Hydrol 526:68–77. https://doi.org/10.1016/J.JHYDROL.2014.11.029

Obasi GOP (1994) WMO's role in the international decade for natural disaster reduction. Bull Am Meteor Soc 75(9):1655–1661. https://doi.org/10.1175/1520-0477(1994)075%3c1655:writid%3e2.0.co;2

Padhee SK, Nikam BR, Aggarwal SP, Garg V (2014) Integrating Effective Drought Index (EDI) and remote sensing derived parameters for agricultural drought assessment and prediction in Bundelkhand Region of India. In: The international archives of the photogrammetry, remote sensing and spatial information sciences (ISPRS), Xl-8, pp 89–100. https://doi.org/10.5194/isprsarchives-XL-8-89-2014

Padhee S, Nikam BR, Dutta S, Aggarwal SP (2017) Using satellite based soil moisture to detect and monitor spatiotemporal traces of agricultural drought over Bundelkhand region of India. Geosci Remote Sens 54(2):144–166. https://doi.org/10.1080/15481603.2017.1286725

Pai DS, Sridhar L, Rajeevan M, Sreejith OP, Satbhai NS, Mukhopadyay B (2014) Development of a new high spatial resolution (0.25° × 0.25°) long period (1901–2010) daily gridded rainfall data set over India and its comparison with existing data sets over the region (vol 65, Issue 1)

Peters E, Torfs PJJF, van Lanen HAJ, Bier G (2003) Propagation of drought through groundwater: a new approach using linear reservoir theory. Hydrol Process 17(15):3023–3040. https://doi.org/10.1002/HYP.1274

Peters E, Van Lanen HAJ, Torfs PJJF, Bier G (2005) Drought in groundwater—drought distribution and performance indicators. J Hydrol 306(1–4):302–317. https://doi.org/10.1016/J.JHYDROL.2004.09.014

Shamshirband S, Hashemi S, Salimi H, Samadianfard S, Asadi E, Shadkani S, Kargar K, Mosavi A, Nabipour N, Chau K-W (2020) Predicting standardized streamflow index for hydrological drought using machine learning models 14(1):339–350. https://doi.org/10.1080/19942060.2020.1715844, Http://Www.Tandfonline.Com/Action/AuthorSubmission?JournalCode=tcfm20&page=instructions

Sharma V, Nikam BR, Thakur PK, Garg V, Aggarwal SP, Srivastav SK, Chauhan P (2020) Estimation of hydro-meteorological extremes in Beas Basin over historic, present and future scenario. In: International archives photogrammetry, remote sensing spatial information science, XLIII-B5-2020, pp 139–147. https://doi.org/10.5194/isprs-archives-XLIII-B5-2020-139-2020

Shepard D (1968) A two-dimensional interpolation function for irregularly-spaced data. In: Proceedings of the 1968 23rd ACM national conference, ACM 1968, pp 517–524. https://doi.org/10.1145/800186.810616

Spinoni J, Vogt JV, Naumann G, Barbosa P, Dosio A (2018) Will drought events become more frequent and severe in Europe? Int J Climatol 38(4):1718–1736. https://doi.org/10.1002/JOC.5291

Srivastava AK, Rajeevan M, Kshirsagar SR (2009) Development of a high resolution daily gridded temperature data set (1969–2005) for the Indian region. Atmos Sci Let https://doi.org/10.1002/asl.232

Stagge JH, Tallaksen LM, Gudmundsson L, Van Loon AF, Stahl K (2015) Candidate distributions for climatological drought indices (SPI and SPEI). Int J Climatol 35(13):4027–4040. https://doi.org/10.1002/joc.4267

Thilakarathne M, Sridhar V (2017) Characterization of future drought conditions in the Lower Mekong River Basin. Weather Clim Extremes 17:47–58. https://doi.org/10.1016/J.WACE.2017.07.004

Thom HCS (1966) Some methods of climatological analysis. WMO

Thomas T, Jaiswal RK, Nayak PC, Ghosh NC (2015) Comprehensive evaluation of the changing drought characteristics in Bundelkhand region of Central India. Meteorol Atmos Phys 127(2):163–182. https://doi.org/10.1007/S00703-014-0361-1

Tigkas D, Vangelis H, Tsakiris G (2012) Drought and climatic change impact on streamflow in small watersheds. Sci Total Environ 440:33–41. https://doi.org/10.1016/j.scitotenv.2012.08.035

Todini E (1996) The ARNO rainfall-runoff model. J Hydrol 175(1–4):339–382. https://doi.org/10.1016/S0022-1694(96)80016-3

Van Loon AF, Laaha G (2015) Hydrological drought severity explained by climate and catchment characteristics. J Hydrol 526:3–14. https://doi.org/10.1016/J.JHYDROL.2014.10.059

Van Loon AF, Van Lanen HAJ (2012) A process-based typology of hydrological drought. Hydrol Earth Syst Sci 16(7):1915–1946. https://doi.org/10.5194/HESS-16-1915-2012

Vicente-Serrano SM, Beguería S, Lorenzo-Lacruz J, Camarero JJ, López-Moreno JI, Azorin-Molina C, Revuelto J, Morán-Tejeda E, Sanchez-Lorenzo A (2012) Performance of drought indices for ecological, agricultural, and hydrological applications. Earth Interact 16(10):1–27. https://doi.org/10.1175/2012EI000434.1

Wang J, Wang W, Cheng H, Wang H, Zhu Y (2021) Propagation from meteorological to hydrological drought and its influencing factors in the Huaihe River Basin. Water 13(14):1985. https://doi.org/10.3390/W13141985

Wilhite DA, Glantz MH (2009) Understanding: the drought phenomenon: the role of definitions. 10(3):111–120. https://doi.org/10.1080/02508068508686328

Wilhite D, Hayes M, Knutson C, Smith K (2000) Planning for drought: moving from crisis to risk management. Drought Mitigation Center Faculty Publications. https://digitalcommons.unl.edu/droughtfacpub/33

Yamazaki D, Ikeshima D, Tawatari R, Yamaguchi T, O'Loughlin F, Neal JC, Sampson CC, Kanae S, Bates PD (2017) A high-accuracy map of global terrain elevations. Geophys Res Lett 44(11):5844–5853. https://doi.org/10.1002/2017GL072874

Zaidman MD, Rees HG, Young AR (2002) Spatio-temporal development of streamflow droughts in north-west Europe. Hydrol Earth Syst Sci 6(4):733–751. https://doi.org/10.5194/HESS-6-733-2002

Zhao RJ, Zuang YL, Fang LR, Liu XR, Zhang QS (1980) The Xinanjiang model. In: Hydrological forecasting, proceedings of the Oxford symposium, vol 129, pp 351–356. https://ci.nii.ac.jp/naid/10015695212

Chapter 29
Evaluation of Multiple Satellite Precipitation Gridded Products for Standard Precipitation Index Based Drought Assessment at Different Time Scales

Neeti Neeti, V. M. Chowdary, C. S. Jha, S. R. Chowdhury, and R. C. Srivastava

Abstract Standard Precipitation Index (SPI) computed at multiple time scales is considered as a key indicator for short-term agricultural to long-term hydrological drought monitoring. SPI computed at multiple timescales namely 1, 3–6 and 12 months represent meteorological, agricultural and hydrological droughts, respectively. Traditionally, precipitation based SPI drought index is being computed using raingauge station data, which is often limited by sparse and uneven distribution of raingauge stations. However, uncertainty in aerial estimation of rainfall and data scarcity regions can be overcome by envisaging various quasi-global satellite derived precipitation products such as TRMM multi-satellite Precipitation Analysis (TMPA), Climate Hazards Group Infrared Precipitation with Stations data (CHIRPS), Climate Research Unit (CRU) data for drought monitoring. Hence in this study, characterization of spatially varying drought occurrences and its severity at different time scales was carried out using various precipitation gridded products (TRMM, CRU and CHIRPS) for part of Indo-Gangetic Plain, India. Further, drought severities assessed by these gridded products were relatively evaluated with reference to rain gauge based IMD gridded product. The spatial pattern of TRMM and IMD based SPI for all the time scales were observed to be similar for both wet and dry years. The spatial pattern of low and high number of drought events is mostly similar for CHIRPS, TRMM and IMD. Overall, it was observed that spatial pattern of drought frequency identified through CRU based SPI was completely distinct compared to

N. Neeti (✉)
TERI School of Advanced Studies, New Delhi, India
e-mail: neeti@terisas.ac.in

V. M. Chowdary
Regional Remote Sensing Center-North, National Remote Sensing Centre, New Delhi, India

C. S. Jha
National Remote Sensing Centre, Hyderabad, India

S. R. Chowdhury · R. C. Srivastava
Dr. Rajendra Prasad Central Agricultural University, Pusa, Samastipur, Bihar, India

A. Pandey et al. (eds.), *Geospatial Technologies for Land and Water Resources Management*, Water Science and Technology Library 103,
https://doi.org/10.1007/978-3-030-90479-1_29

other datasets. In general, CHIRPS appears to have overestimated the drought area and frequency compared to IMD, while TRMM data exhibited a similar pattern as that of IMD product which could be due to the fact that CHIRPS underestimates the rainfall.

29.1 Introduction

Drought is a natural, slow and progressive hydro-meteorological phenomenon. Although there is no proper definition of drought, it can be understood as a decrease in water resource availability relative to normal conditions. Indian Meteorological Department (IMD) reported that the drought situation arises when the rainfall is less than 25% of the long-term mean annual rainfall (Wankhede et al. 2014). It is considered to be one of the most devastating natural hazards affecting both environmental health and socio-economic security of people (Nagarajan 2010). Globally, 30% of land portion of southern hemisphere and 19% land portion of northern hemisphere witnessed significant drying conditions during the period 1980–2012 (Damberg and AghaKouchak 2014). In India, frequent drought situation has affected more than 50 million people (Dutta et al. 2015) and can be attributed to the deficient and uneven rainfall distribution.

The long-term analysis of causative and responsive factors is necessary for understanding drought dynamics (Bhuiyan et al. 2006). In literature, different types of droughts are defined on the basis of deficiency in rainfall, soil water, surface and sub-surface water availability (Mishra and Singh 2010). Deficiency in rainfall for an extended period of time leads to meteorological drought. Deficiency in soil water required for plant and forage growth leads to agricultural drought. Deficiency in availability of surface and sub-surface water from normal at a particular instant of time leads to hydrological drought. The collective effects of the different types of drought influence environmental, economic and socio-economic conditions of the region.

Traditionally drought monitoring is carried out using in-situ data from the rainfall monitoring stations (Rhee and Carbone 2011). The main problem with this type of approach is the unavailability of near real-time data. With the availability of a plethora of Earth Observation (EO) time series data, it has become possible to monitor drought at multiple spatial and temporal scales. Several drought indices have been developed over the years as drought indicators using these EO time series. Among the several indices recognized for drought assessment, World Meteorological Organization (WMO) recommended Standard Precipitation Index (SPI) as a key indicator for drought monitoring (Hayes et al. 2011). SPI at multiple timescales reflect short-term (SPI-1) meteorological, medium-term agricultural (SPI-3 and SPI-6) and long-term (SPI-12) hydrological drought conditions.

Precipitation is the key component for various drought indices and has been captured traditionally by raingauge stations. However, spatially sparse and uneven distribution of raingauge stations increases uncertainty in aerial estimation of rainfall. Various quasi-global satellite products such as TRMM multi-satellite Precipitation Analysis (TMPA), Precipitation Estimation from Remotely Sensed Information using Artificial Neural Networks (PERSIANN) (Pan et al. 2010a, b; Sahoo et al. 2015; Ashouri et al. 2016), Climate Hazards Group Infrared Precipitation with Stations data (CHIRPS) (Funk et al. 2015), Climate Research Unit (CRU) data (Harris et al. 2020; Kalisa et al. 2020) have been used for drought assessment. Therefore, it is important to assess their utility and agreement among them in drought assessment.

Recent studies suggest that drought events have shifted to the agricultural important regions of the country including Indo-Gangetic Plain (e.g. Mallya et al. 2016; Kishore et al. 2015). Thus, this study focuses on evaluation of satellite based gridded precipitation products namely TRMM, CRU and CHIRPS for SPI based drought assessment for multiple time scales in part of Indo-Gangetic Plain, India. This study investigated the potential of various precipitation gridded products for characterization of spatially varying drought occurrences and its severity at multiple time scales representing meteorological (SPI-1), agricultural (SPI-3, SPI-6) and hydrological droughts (SPI-12). Further, drought severity assessed by TRMM, CRU and IMD rainfall products were relatively evaluated with respect to gauge based IMD gridded products.

29.2 Study Area

The Indian state Bihar is situated in the Indo-Gangetic Plain and is situated between 83° 19.83′ to 88° 17.67′ E longitude and 24° 20.17′ to 27° 31.25′ N latitude (Fig. 29.1). Geographical area of the study region is nearly 94,163 km^2 that encompasses 38 administrative districts. The study region falls under subtropical climate zone with an annual rainfall of 1200 mm (Chowdary et al. 2008). The southern part of the study region is usually affected by drought while northern part is flood affected. The study region has experienced moderate to severe droughts during 2009–2013, where this period was considered as the driest period until now. In the year 2009, 26 districts of the state faced drought while in the year 2013, 33 districts were affected (Goswami et al. 2018).

29.3 Data Used

Drought assessment was carried out using four gridded precipitation products namely Tropical Rainfall Measuring Mission (TRMM), Climate Research Unit (CRU), Climate Hazards Group Infrared Precipitation with Station (CHIRPS) and Indian Meteorological Department (IMD) during the study period 1998–2018.

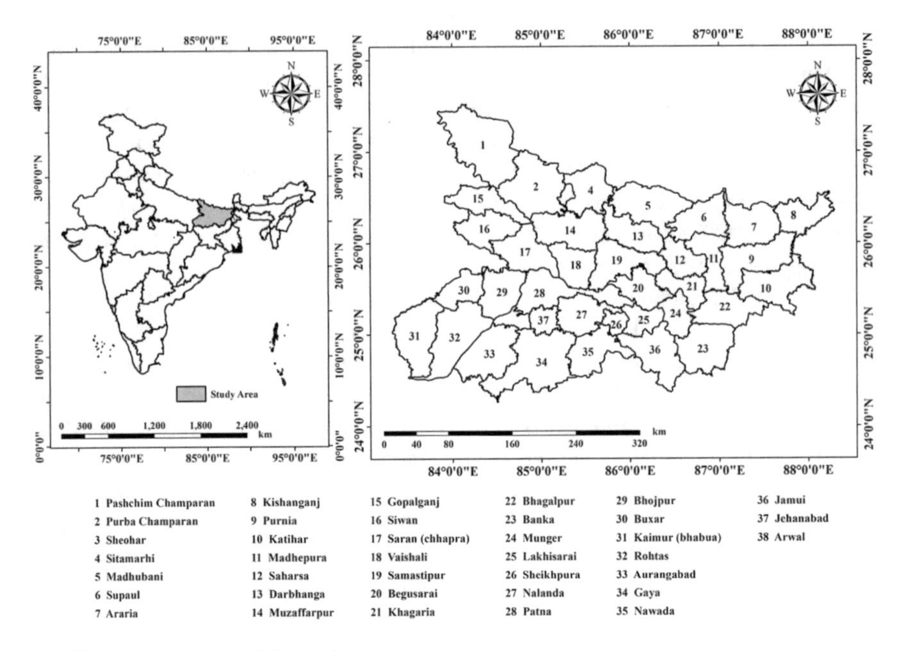

Fig. 29.1 Location map of the study area

29.3.1 TRMM 3B43 Data

The TRMM level 3 product 3B43, a joint mission between the space agencies of the United States and Japan was envisaged in this study. There are four instruments onboard TRMM for precipitation measurements namely precipitation radar (PR), TRMM microwave imager (TMI) and visible and infrared radiometer system (VIRS) (Kummerow et al. 1998; Almazroui 2011). This study uses the level 3 product 3B43, which is the result of integration of the estimates produced by TRMM and other products (3B42) with the gridded rain gauge data from the National Oceanic and Atmospheric Administration Climate Prediction Center and the global rain gauge product produced by the Global Precipitation Climatology Centre (GPCC) (Naumann et al. 2012). The level 3 product 3B43 of TRMM includes the precipitation rate (mm/h) for each month at $0.25° \times 0.25°$ spatial resolution with a spatial coverage from 50° S to 50° N (TRMM 2011).

29.3.2 CRU Data

The CRU TS v4.05 dataset comprises fourth version of the gridded product by Climate Research Unit (CRU) of the University of East Angelia (Harris et al. 2020). It includes a number of variables such as rainfall, temperature, PET among others.

The gridded rainfall data is available globally excluding Antarctica and is interpolated with the monthly data retrieved from the archives of World Meteorological Organization (Harris et al. 2014, 2020). The spatial resolution of this product is 0.5° × 0.5° and is available for the period 1901–2018.

29.3.3 CHIRPS Data

Climate Hazards Group Infrared Precipitation with Station data (CHIRPS) is publicly available long-term re-analysis precipitation near-global (50° N–50° S) dataset with a spatial resolution of 0.05° × 0.05° and available from 1981 to present (Funk et al. 2015). This product was developed especially to focus on drought monitoring and forecasting and other land surface modelling activity (Funk et al. 2015). It is a combined product generated using global precipitation climatology, geostationary TIR satellite estimates and in-situ rain gauge measurements (Funk et al. 2015).

29.3.4 IMD Data

Indian Meteorological Department rainfall gridded product with a spatial resolution of 0.25° is generated based on the daily rainfall data collected from 6955 rain gauge stations. The Inverse Distance Weighted (IDW) interpolation technique was employed to generate the gridded precipitation data (Pai et al. 2014). The dataset was used as a reference for comparing other satellite precipitation data products in this study.

29.4 Methodology

The overall drought assessment methodology includes two major steps namely (i) disaggregation of coarse resolution gridded products and (ii) SPI computations at multiple timescales. The TRMM, CRU and IMD rainfall products were disaggregated to 0.05° to match with the resolution of CHIRPS dataset for relative evaluation among gridded precipitation products.

29.4.1 Drought Assessment Using SPI Computed at Multiple Time Scales

Multi-scale SPI was computed using 20 years of monthly TRMM, CHIRPS, CRU and IMD gridded products. SPI computation involves application of Gamma distribution function to long-term precipitation time series at different time scales such as 1 month, 3 months, 9 months, 12 months. In this study, SPI-1, SPI-3, SPI-6 and SPI-12 were computed for drought assessment and relatively evaluated different rainfall products with respect to gauge based IMD products. In SPI-1 individual month, precipitation is compared to the historical record for that month. In SPI-3, precipitations for three months are compared to historical records for the aggregated precipitation over the same months. Similarly, SPI-6 and SPI 12 used 6 months and 12 months rainfall, respectively for comparison. Since there is zero precipitation at a number of observations and two-parameter Gamma function is undefined for zero value, a mixed distribution function is applied and modified cumulative distribution function after incorporation of zero as defined in McKee et al. (1993). The details of the mathematics behind SPI calculation can be found in McKee et al. (1993).

The standard drought severity classes as defined by Svobada et al. (2002) were used for drought categorization in this study (Table 29.1).

The drought frequency for different severity levels was computed by quantifying the number of times that the SPI value was below the threshold as defined in Table 29.1.

29.5 Results and Discussion

Drought indicator SPI was computed for different time scales using CHIRPS, TRMM and CRU precipitation data and was used for drought monitoring in the study area for the period 1998–2019. The frequency of the drought events for different severity levels was computed at pixel level for characterization of meteorological (SPI-1), agricultural (SPI-3 and SPI-6) and hydrological (SPI-12) droughts in the study region.

Table 29.1 Drought severity categorization (D-scale) based on SPI values (Svoboda et al. 2002)

Drought classes	SPI
Exceptional drought (D4)	≤ -2.0
Extreme drought (D3)	Between −1.6 and −1.9
Severe drought (D2)	Between −1.3 and −1.5
Moderate drought (D1)	Between −0.8 and −1.2
Abnormally dry (D0)	Between −0.5 and −0.7

29.5.1 Meteorological Drought Assessment Using SPI-1

The SPI computed at a 1 monthly scale was used for assessment of meteorological drought. SPI-1 was computed for 2007 and 2013 that corresponds to wet and dry years, respectively (Fig. 29.2). CHIRPS and TRMM datasets indicated wet north and northwest and dry east and southeast districts, which is consistent with IMD product. However, CRU indicated a completely different pattern with north south divide between wet and dry years, respectively. In the dry year 2013 CHIRPS, TRMM and IMD datasets indicated dry north and central districts and wet southeast and eastern districts contrary to the year 2007. However, CRU data has a completely different pattern in 2013 as well.

Meteorological drought events with different severity levels were represented in Fig. 29.3. Maximum number of drought events with D1 severity level was identified in all the datasets that are consistent with IMD data. However, maximum number of abnormally dry and moderately dry conditions was identified in CHIRPS dataset compared to other datasets. CHIRPS overestimated the number of drought events, while other gridded products are almost at par with IMD. However, in terms of spatial pattern of the number of drought events, CHIRPS and TRMM match with IMD. This is consistent with the findings of other researchers (e.g. Zhao and Ma 2019). The northwest, southwest and eastern districts experienced more drought events in the three datasets. With an increase in the severity level, reduction in the number of drought events can be seen across all datasets consistent with IMD dataset.

The proportion of study regions that experienced different drought severity levels is given in Fig. 29.4. The results indicated that the total area under drought is much higher in SPI-1 computed using CHIRPS dataset compared to CRU and TRMM datasets in most of the time steps. The percentage of area under drought computed using TRMM data is observed to be much closer to IMD compared to the other two. CHIRPS and CRU based SPI computations indicated that a much larger area experienced drought for most of the time steps. However, all the datasets indicated that areas experienced drought at higher severity levels were nearly the same for most of the time steps.

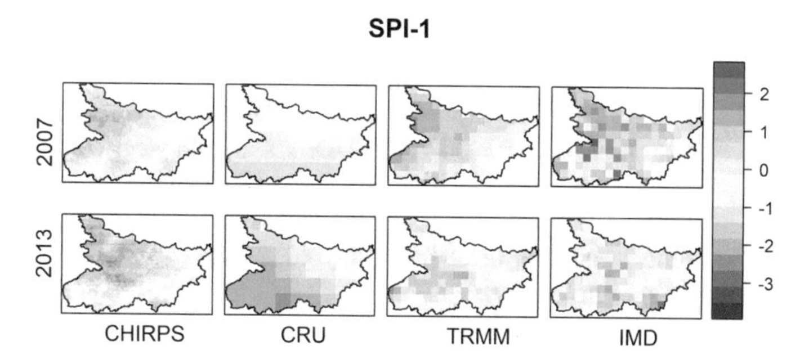

Fig. 29.2 Spatial pattern of SPI-1 across four datasets for monsoon month of August for the year 2007 (wet) and 2013 (dry)

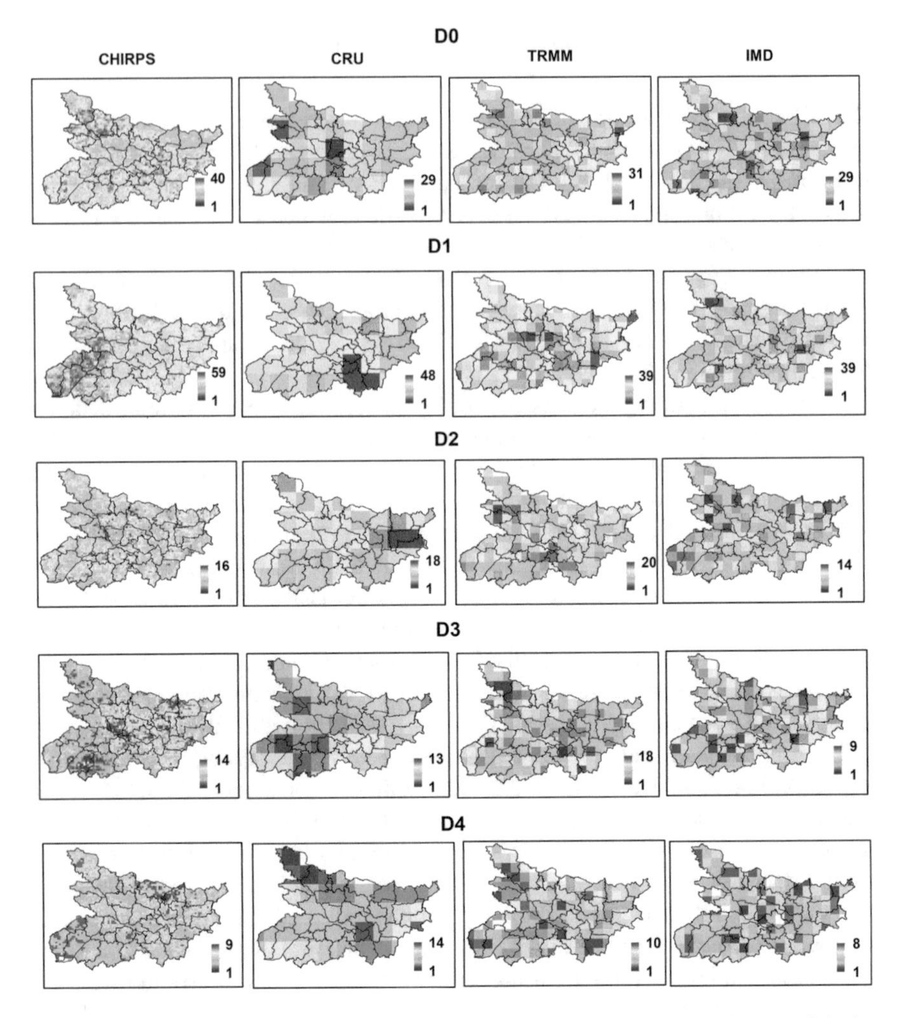

Fig. 29.3 Drought frequency for different severity levels across state from four precipitation datasets

Analysis of drought using IMD data indicated that maximum area under drought was observed to be 23% (April 2016), 44% (2000 October), 21% (May 2005), 21% (May 2012) and 29% (July 2013) for D0, D1, D2, D3 and D4. respectively. For most of these time steps, all the gridded products underestimated the total area under drought except for CRU and CHIRPS for D3 and D4 categories, respectively. CRU data estimated 68% of the area under drought in May 2012 whereas CHIRPS estimated 78% area under drought in May 2012.

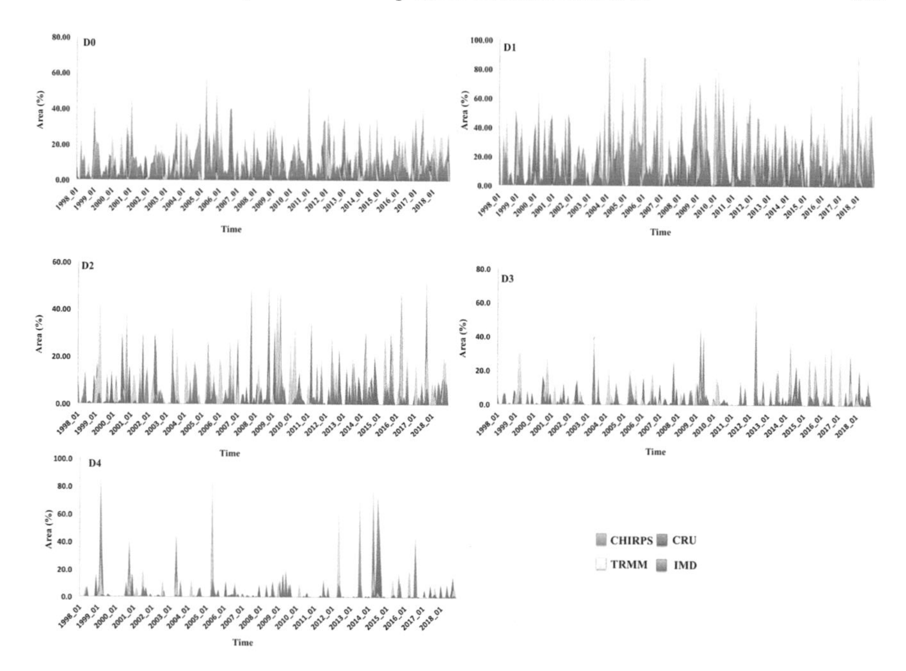

Fig. 29.4 Percentage of state area under different severity drought between 1998 and 2018 on the basis of SPI-1

29.5.2 *Agricultural Drought (SPI-3 and SPI-6) Assessment*

The SPI computed at 3-month and 6-month scales are considered to be representative of agricultural drought. The spatial pattern of SPI-3 for a wet (2007) and a dry year (2013) for the month of August was represented in Fig. 29.5. The spatial pattern of SPI-3 for TRMM data is much similar to IMD data compared to CRU data. SPI-3 computed using CHIRPS data also has spatial patterns similar to IMD data except

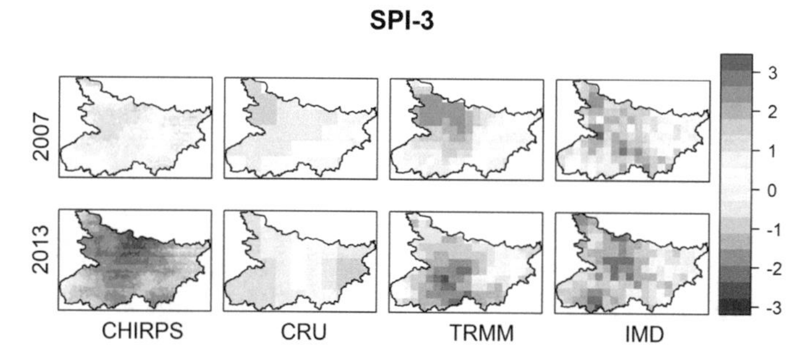

Fig. 29.5 Spatial pattern of SPI-3 across four datasets for monsoon month of August for the year 2007 (wet) and 2013 (dry)

for northwest districts in 2007 and northeast districts in 2013. SPI values computed based on CHIRPS and IMD products in the year 2007 were observed to be negative and positive in the northwest districts, respectively. Similarly negative SPI value can be seen in the northeast district in CHIRPS, while IMD based SPI indicated positive for the same region.

The spatial patterns of SPI-3 based drought frequency (SPI-3) computed using four gridded products were represented in Fig. 29.6. Similar to meteorological drought, maximum number of agricultural drought events were detected by CHIRPS data especially for mild and moderate severity levels (Fig. 29.6). With an increase in severity level, the number of droughts identified through the four datasets was found

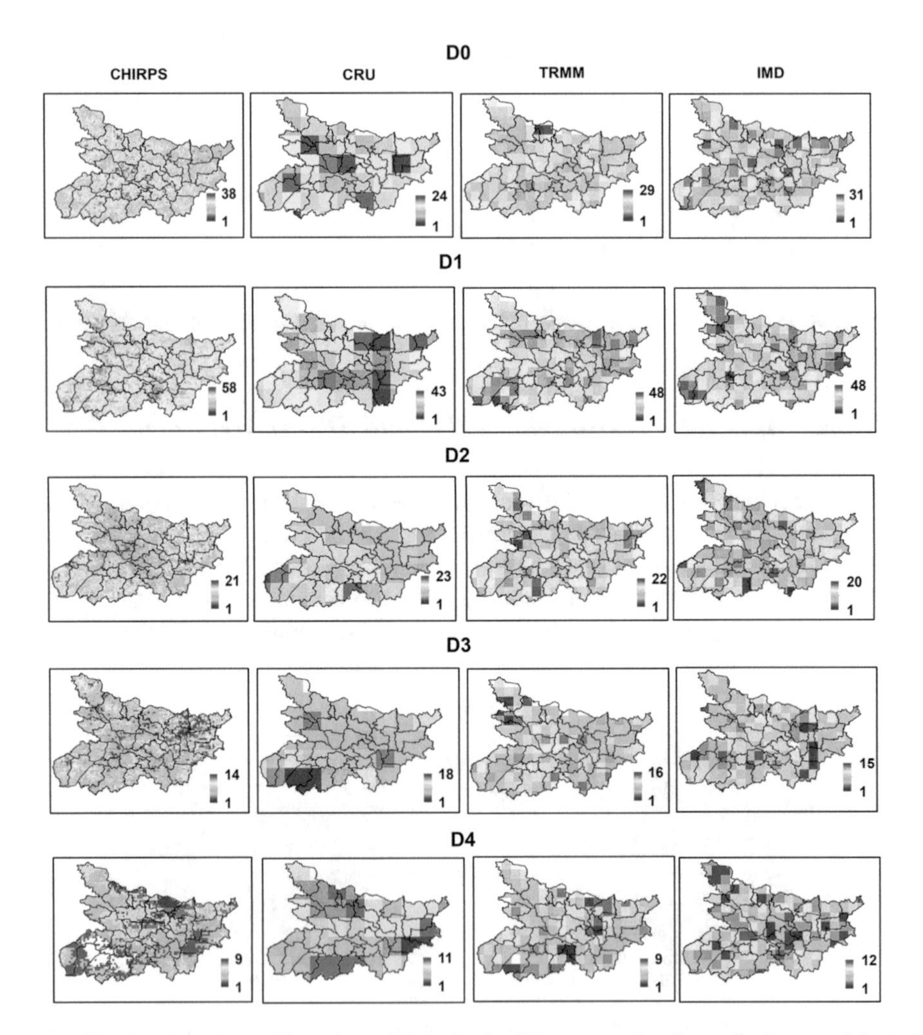

Fig. 29.6 Spatial pattern of frequency of droughts for different severity class using four precipitation datasets for SPI-3

to be quite similar. Spatial patterns for CHIRPS and CRU are found to be more similar especially for D0 and D1 drought severity levels. But, none of the satellite data had a pattern similar to IMD gridded dataset across the entire study region. For example, CHIRPS, CRU and IMD indicated large number of drought events with D0 severity levels in the northwest and central districts. However, in the northeast district, CHIRPS data has less droughts but IMD data indicated a large number of drought events.

The total area under drought for different severity classes is given in Fig. 29.7. In contrast to SPI-1, all the datasets indicated less area under mild drought conditions compared to IMD for most of the time steps. But, consistency can be observed in temporal pattern of the area under drought among all the datasets. Similar to meteorological drought, CHIRPS and CRU data based analysis indicated a larger area under drought in a given year compared to TRMM and IMD data (Fig. 29.7). However, all the four datasets identified approximately the same proportion of area under drought under D3 and D4 severity across the four datasets.

The spatial pattern of SPI-6 for wet year (2007) and dry year (2013) is represented in Fig. 29.8. It is evident from the figure that the spatial pattern of TRMM based SPI-6 is mostly comparable to IMD with wet northwest and central districts and dry north east districts. The spatial pattern of CHIRPS based SPI- is also similar to IMD except for northwest districts of the state in the year 2007. Similar to SPI-3, CHIRPS and IMD based SPI-6 indicated distinct patterns in the northeast districts for a dry year

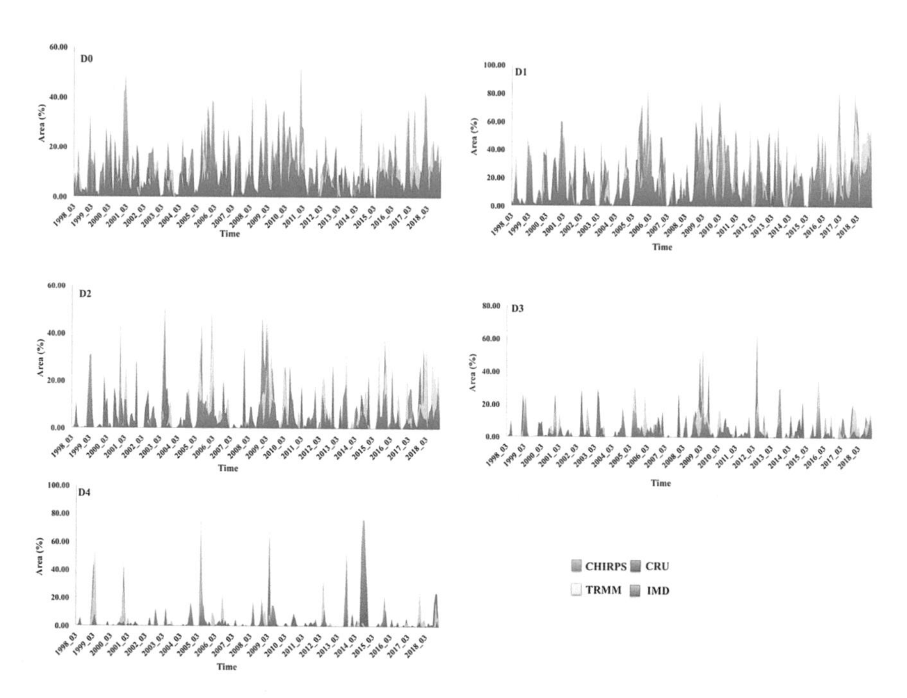

Fig. 29.7 Percentage of state area under different severity drought between 1998 and 2018 on the basis of SPI-3

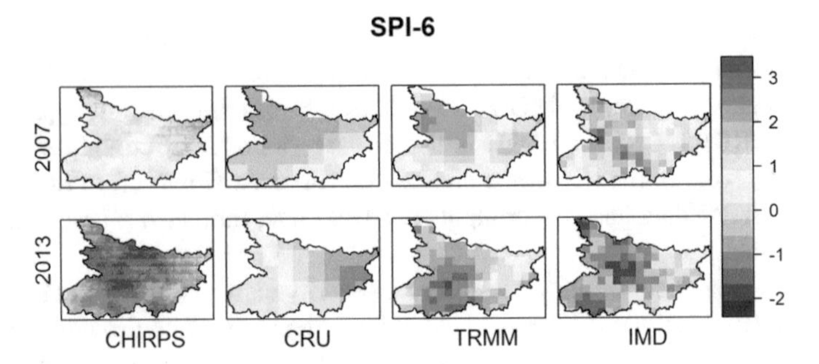

Fig. 29.8 Spatial pattern of SPI-6 across four datasets for monsoon month of August for the year 2007 (wet) and 2013 (dry)

(2013). The spatial pattern of CRU based SPI-6 is completely distinct from the three datasets for both 2007 and 2013 years.

The spatial pattern of SPI-6 based drought frequency is represented in Fig. 29.9. Overall it was observed that the three products could identify an almost comparable number of droughts across different severity levels at larger time scales, compared to smaller time scales (i.e. SPI-1 and SPI-3). It was also observed that the spatial patterns of high and low number of drought events are quite similar for CHIRPS and IMD among all severity categories. The spatial similarity between any two products was not consistent across the entire study region except for D1 severity for which CHIRPS and CRU were observed to be more similar compared to TRMM. Consistent with SPI-1 and SPI-3, geographical pattern of a number of agricultural drought events for CRU data is different compared to other datasets.

The percentage of area under different drought severity categories computed based on SPI-6 is given in Fig. 29.10. Among all the severity levels, CRU based SPI-6 indicated maximum area under drought for most of the time steps compared to CHIRPS, TRMM and IMD based SPI-6 computations. The maximum area under D0 drought category as per IMD is in August 2018, where approximately 23% study region experienced drought conditions. Contrary to IMD based estimates, CHIRPS, CRU and TRMM based drought assessments indicated 10%, 6% and 0.1% area of study region under drought, respectively for the same time step, which signifies underestimation. The maximum difference between CHIRPS and IMD based drought patterns was observed in June 2004. CHIRPS and IMD based drought assessment indicated nearly 37% and 6% of study regions experienced drought, respectively. The difference in area estimates between CRU (48%) and IMD data (8%) was observed to be maximum for January 2004. The maximum area difference between TRMM and IMD was observed to be nearly 38% more in case of TRMM for April 2004. With the increase in the severity level, the area estimates for all the datasets indicated a similar pattern except for exceptional drought conditions. CRU indicated a much larger area under D4 drought severity category in the March and April months of 2009 in comparison to IMD and other datasets. Similarly, CRU based area estimates

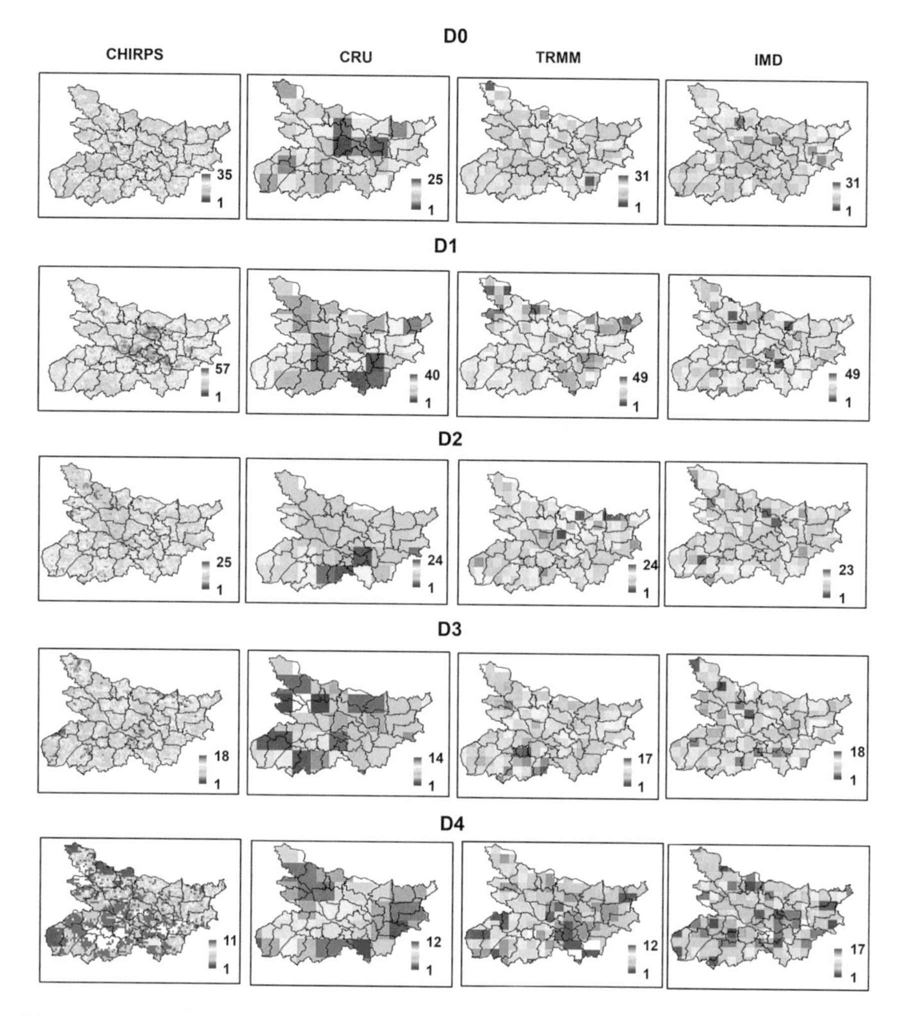

Fig. 29.9 Spatial pattern of frequency of droughts for different severity classes using four precipitation datasets for SPI-6

for drought between September and December 2018 was also much larger compared to other datasets.

29.5.3 Hydrological Drought Assessment Using SPI-12

The SPI-12 was computed to assess hydrological drought in the study region. The SPI-12 values computed for both wet (2007) and dry (2013) years were spatially represented in Fig. 29.11. Similar to the pattern seen for other SPI's, CHIRPS and

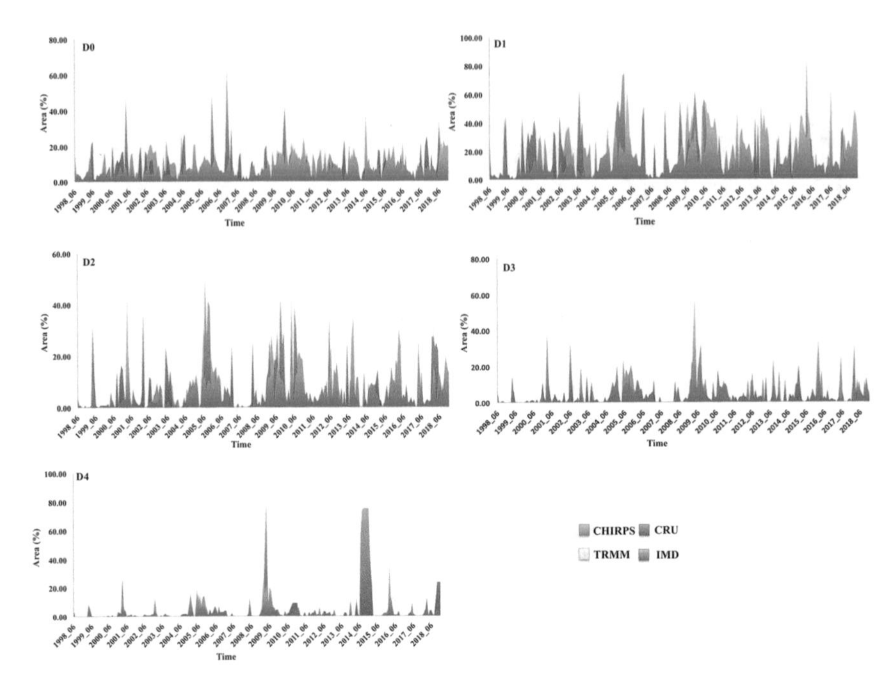

Fig. 29.10 Percentage of state area under different severity drought between 1998 and 2018 on the basis of SPI-6

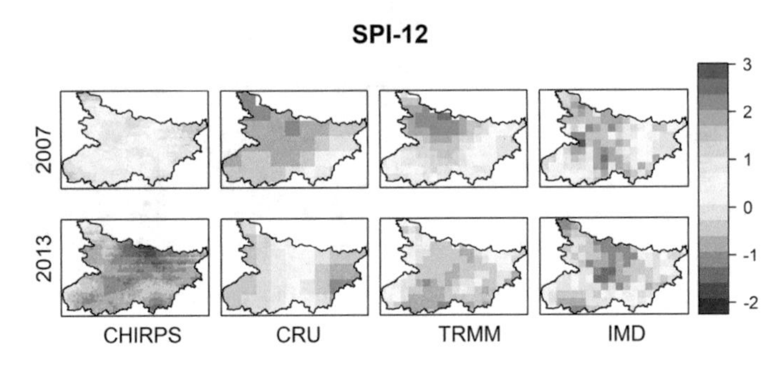

Fig. 29.11 Spatial patterns of SPI-12 across four datasets for monsoon month of August for the year 2007 (wet) and 2013 (dry)

TRMM based SPI-12 also indicated a similar pattern as represented by IMD.

Drought frequency of longer time scale was observed to be higher among all the three datasets relative to 3- and 6-month time scales (Fig. 29.12). CHIRPS data based SPI-12 identified drought frequency as high compared to other datasets, especially in mild and moderate severity drought levels. The geographical pattern of drought frequency calculated using TRMM is more similar to IMD compared to

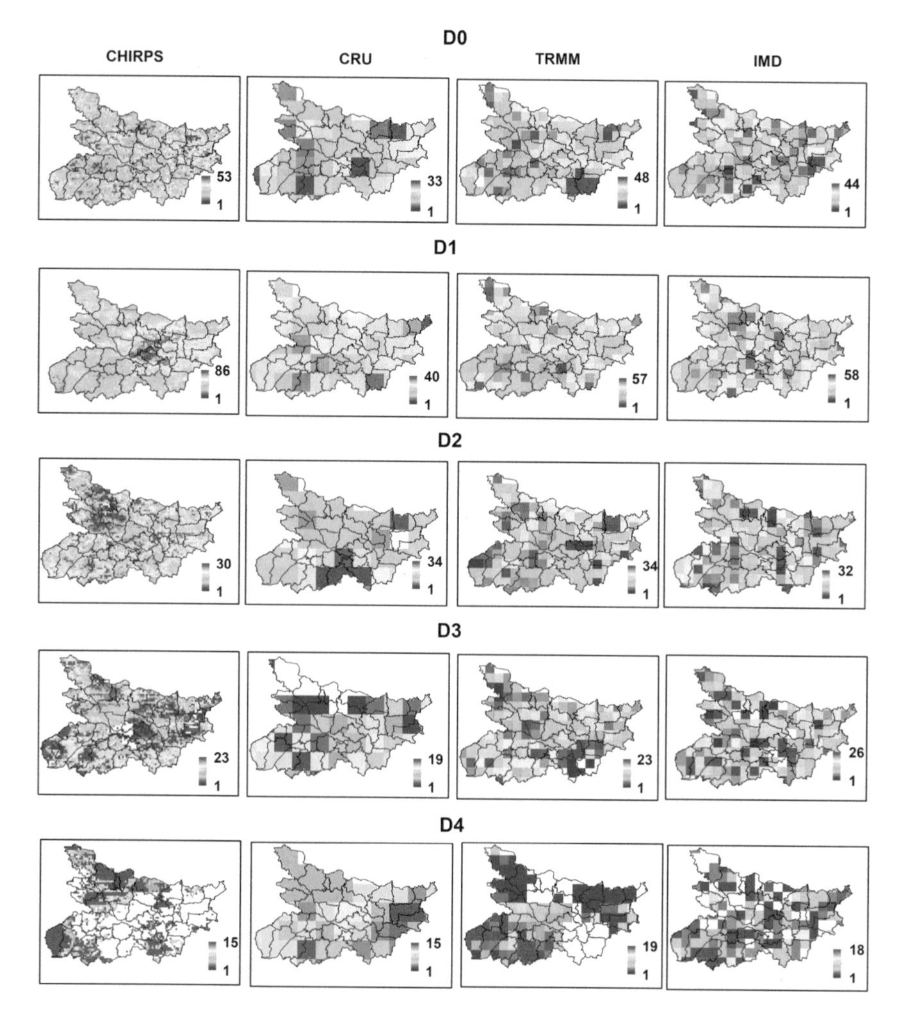

Fig. 29.12 Spatial pattern of frequency of droughts for different severity class using four precipitation datasets for SPI-12

other datasets. The similarity between spatial patterns of TRMM and IMD can be seen consistently across all severity classes.

Compared to SPI-1, SPI-3 and SPI-6 based drought assessment, the percentage of area under drought in SPI-12 across three datasets was found to be very similar (Fig. 29.13). IMD based SPI analysis at multiple time scales indicated that the maximum area under drought in D0, D1, D2, D3 and D4 severity categories were 15% (March 2016), 48% (Aug 2010), 19% (May 2010), 12% (Feb 2011) and 17% (July 2015), respectively. However, all the datasets underestimated the area under D0 drought. CHIRPS and CRU overestimated DI category while TRMM underestimated the total area under drought. TRMM underestimated D1 severity level, while the other

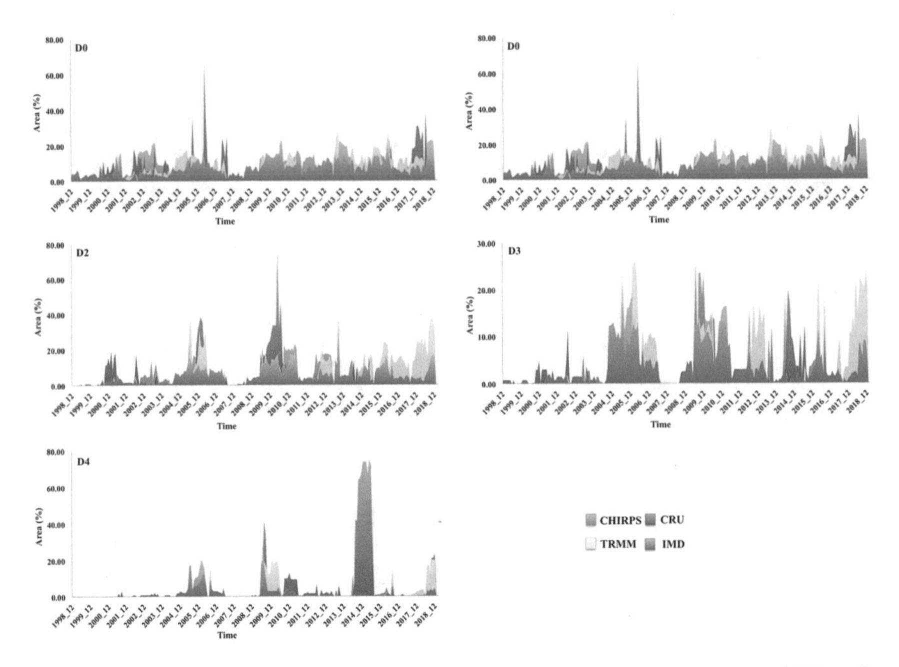

Fig. 29.13 Percentage of state area under different severity drought between 1998 and 2018 on the basis of SPI-12

two datasets overestimated the area under drought. CRU estimated 78% of the study region under D2 category whereas the other two datasets slightly underestimated the area. CHIRPS overestimated D3 category, while other datasets underestimated the total area under drought. For, Overall, all the datasets underestimated the D4 severity category in comparison to IMD dataset.

The comparison of three satellite derived gridded products with reference to IMD indicated that TRMM and CHIRPS are more closely related to IMD in identifying frequency, area and geographical pattern of droughts compared to CRU dataset. This is consistent with the findings of other researchers, where Zhao and Ma (2019) indicated similarity between TRMM and CHIRPS for global scale drought assessment. This could be due to the fact that CHIRPS is calibrated with 3B42 product of TRMM data (Funk et al. 2015). There has been limited number of studies on the evaluation of satellite derived gridded products for drought assessment over the Indian subcontinent. Pandey et al. (2020) are one of the few studies carried over Central India on comparison of different satellite products with reference to IMD data for drought assessment. They also noted a similarity between CHIRPS and IMD. However, the authors noted the frequency of droughts detected by CHIRPS to be very similar to the IMD based analysis that is contrary to the results observed in this study. It observed that CHIRPS always detected more droughts compared to any other dataset including IMD. This difference in result could be due to the fact that Pandey et al. (2020) has compared the result at the study area scale instead of pixel level. Analysis in this study found that the output of TRMM based analysis was more consistent with the study carried out using IMD data. This is consistent with the findings of study by Pandey and Srivastava (2019) for Bundelkhand region. This difference between TRMM and CHIRPS regarding identifying a number of drought events could be due to the fact that CHIRPS underestimates while TRMM overestimates the rainfall as found by other researchers (e.g. Usman and Nichol 2020). It was also observed that CRU based drought assessment resulted in showing a much larger area under drought compared to other datasets. This could be due to the fact that CRU based rainfall estimates are lower than the ground level observations (Shi et al. 2020). Further, Nkunzimana et al. (2020) reported that CRU overestimated the rainfall in less wet areas whereas underestimated in higher rainfall regions, which could be the reason. This may be one of the reasons for CRU based analysis had completely different spatial patterns compared to the other products.

29.6 Conclusions

The drought assessment was carried out for Bihar state, a North Indian state using multiple different precipitation gridded products namely CHIRPS, CRU and TRMM products and were relatively evaluated with reference to IMD gridded products. CHIRPS dataset has the finest while CRU has the coarsest spatial resolution. The SPI index was computed at multiple time scales namely 1-, 3-, 6- and 12 months representing meteorological, agricultural and hydrological droughts, respectively. The spatial pattern of TRMM and IMD based SPI for all the time scales was observed to be similar for both wet and dry years. The SPI-pattern for CHIRPS data was also observed to be similar to IMD during both 2007 and 2013 years. However, CHIRPS overestimated drought frequency relative to IMD based estimate. The spatial pattern of a low and high number of drought events is mostly similar for CHIRPS, TRMM and IMD. Overall, it was observed that the spatial pattern of drought frequency identified through CRU based SPI was completely distinct compared to other datasets. This study suggests that one needs to be very careful about the selection of a precipitation gridded product for drought monitoring. This study indicated that a drastically different drought level might be identified when different gridded products were envisaged for drought monitoring. In general, CHIRPS appears to have overestimated the drought area and frequency compared to IMD, while TRMM data exhibited a similar pattern as that of IMD product which could be due to the fact that CHIRPS product underestimates the rainfall.

References

Almazroui M (2011) Calibration of TRMM rainfall climatology over Saudi Arabia during 1998–2009. Atmos Res 99(3–4):400–414

Ashouri H, Sorooshian S, Hsu KL, Bosilovich MG, Lee J, Wehner MF, Collow A (2016) Evaluation of NASA's MERRA precipitation product in reproducing the observed trend and distribution of extreme precipitation events in the United States. J Hydrometeorol 17(2):693–711

Bhuiyan C, Singh RP, Kogan FN (2006) Monitoring drought dynamics in the Aravalli region (India) using different indices based on ground and remote sensing data. Int J Appl Earth Obs Geoinf 8(4):289–302

Chowdary VM, Chandran RV, Neeti N, Bothale RV, Srivastava YK, Ingle P, Singh R (2008) Assessment of surface and sub-surface waterlogged areas in irrigation command areas of Bihar state using remote sensing and GIS. Agric Water Manag 95(7):754–766

Damberg L, AghaKouchak A (2014) Global trends and patterns of drought from space. Theoret Appl Climatol 117(3):441–448

Dutta D, Kundu A, Patel NR, Saha SK, Siddiqui AR (2015) Assessment of agricultural drought in Rajasthan (India) using remote sensing derived vegetation condition index (VCI) and Standardized Precipitation Index (SPI). Egypt J Remote Sens Space Sci 18(1):53–63

Funk C, Peterson P, Landsfeld M, Pedreros D, Verdin J, Shukla S, Husak G, Rowland J, Harrison L, Hoell A, Michaelsen J (2015) The climate hazards infrared precipitation with stations—a new environmental record for monitoring extremes. Sci Data 2:150066

Goswami RK, Maiti S, Garai S, Jha SK, Bhakat M, Chandel BS, Kadian KS (2018) Coping mechanisms adopted by the livestock dependents of drought prone districts of Bihar India. Indian J Anim Sci 88(3):356–364

Harris I, Ones PD, Osborn TJ, Lister DH (2014) Updated high-resolution grids of monthly climatic observations—the CRU TS3.10 dataset. Int J Climatol 34:623–642

Harris I, Osborn TJ, Jones P, Lister D (2020) Version 4 of the CRU TS monthly high-resolution gridded multivariate climate dataset. Sci Data 7(1):1–18

Hayes M, Svoboda M, Wall N, Widhalm M (2011) The Lincoln declaration on drought indices: universal meteorological drought index recommended. Bull Am Meteor Soc 92(4):485–488

Kalisa W, Zhang J, Igbawua T, Ujoh F, Ebohon OJ, Namugize JN, Yao F (2020) Spatio-temporal analysis of drought and return periods over the East African region using standardized precipitation index from 1920 to 2016. Agric Water Manag 237:106195

Kishore A, Joshi PK, Pandey D (2015) Drought, distress, and a conditional cash transfer programme to mitigate the impact of drought in Bihar, India. Water Int 40(3):417–431

Kummerow C, Barnes W, Kozu T, Shiue J, Simpson J (1998) The tropical rainfall measuring mission (TRMM) sensor package. J Atmos Oceanic Technol 15(3):809–817

Mallya G, Mishra V, Niyogi D, Tripathi S, Govindaraju RS (2016) Trends and variability of droughts over the Indian monsoon region. Weather Clim Extremes 12:43–68

McKee TB, Doesken NJ, Kleist J (1993) The relationship of drought frequency and duration to time scales. In: Proceedings of the 8th conference on applied climatology, vol 17. No. 22, pp. 179–183

Mishra AK, Singh VP (2010) A review of drought concepts. J Hydrol 391(1–2):202–216

Nagarajan R (2010) Drought assessment, 1st edn. Springer, Netherlands

Naumann G, Barbosa P, Carrao H, Singleton A, Vogt J (2012) Monitoring drought conditions and their uncertainties in Africa using TRMM data. J Appl Meteorol Climatol 51(10):1867–1874

Nkunzimana A, Bi S, Alriah MAA, Zhi T, Kur NAD (2020) Comparative analysis of the performance of satellite based rainfall products over various topographical unities in Central East Africa: case of Burundi. Earth Space Sci 7(5):e2019EA000834

Pai D, Sridhar L, Rajeevan M, Sreejith O, Satbhai N, Mukhopadhyay B (2014) Development of a new high spatial resolution (0.25× 0.25) long period (1901–2010) daily gridded rainfall data set over India and its comparison with existing data sets over the region. Mausam 65(1):1–18

Pan M, Li H, Wood E (2010a) Assessing the skill of satellite based precipitation estimates in hydrologic applications. Water Resour Res 46(9). https://doi.org/10.1029/2009WR008290

Pan M, Li H, Wood E (2010b) Assessing the skill of satellite-based precipitation estimates in hydrologic applications. Water Resour Res 46(9)
Pandey V, Srivastava PK (2019) Evaluation of satellite precipitation data for drought monitoring in Bundelkhand Region, India. In: IGARSS 2019–2019 IEEE international geoscience and remote sensing symposium, IEEE, pp 9910–9913
Pandey V, Srivastava PK, Mall RK, Munoz-Arriola F, Han D (2020) Multi-satellite precipitation products for meteorological drought assessment and forecasting in Central India. Geocarto Int 1–20
Rhee J, Carbone GJ (2011) Estimating drought conditions for regions with limited precipitation data. J Appl Meteorol Climatol 50(3):548–559
Sahoo AK, Sheffield J, Pan M, Wood EF (2015) Evaluation of the tropical rainfall measuring mission multi-satellite precipitation analysis (TMPA) for assessment of large-scale meteorological drought. Remote Sens Environ 159:181–19
Shi H, Li T, Wei J (2017) Evaluation of the gridded CRU TS precipitation dataset with the point raingauge records over the three-river headwaters region. J Hydrol 548:322–332
Svoboda M, LeComte D, Hayes M, Heim R, Gleason K, Angel J, Miskus D (2002) The drought monitor. Bull Am Meteor Soc 83(8):1181–1190
TRMM (Tropical Rainfall Measuring Mission) (2011) M (TMPA/3B43) Rainfall Estimate L3 1 month 0.25 degree x 0.25 degree V7, Greenbelt, MD, Goddard Earth Sciences Data and Information Services Center (GES DISC), Accessed: [Data Access Date]. https://doi.org/10.5067/TRMM/TMPA/MONTH/7
Usman M, Nichol JE (2020) A spatio-temporal analysis of rainfall and drought monitoring in the Tharparkar region of Pakistan. Remote Sens 12(3):580
Wankhede S, Gandhi N, Armstrong L (2014) Role of ICTs in improving drought scenario management in India. In: Proceedings of the 9th Asian federation for information technologies in agriculture, (AFITA), Perth, pp 521–530
Zhao H, Ma Y (2019) Evaluating the drought-monitoring utility of four satellite-based quantitative precipitation estimation products at global scale. Remote Sens 11(17):2010

Chapter 30
Tropical Cyclones and Coastal Vulnerability: Assessment and Mitigation

Debadatta Swain

Abstract Tropical cyclone (TC) landfalls are among the most damaging natural disasters. The North Indian Ocean (NIO) experiences ~12% of all cyclones every year. TC damage is primarily due to high wind gusts, rainfall, storm surges, waves and coastal flooding which pose serious risks to life, property and coastal ecosystems. Extreme wave activities, vegetation loss due to gale winds and saltwater intrusion during coastal inundation cause coastal erosion and turn agricultural land infertile over extended periods of time. The rate of TC devastation also depends on coastal Land Use and Land Cover (LULC: vegetation density, barren lands, agricultural fields, etc.). TCs in turn change the LULC and soil characteristics, thus modulating the land surface properties. The extent of physical and social vulnerability due to TCs are directly associated with population density, coastal infrastructure and TC frequency and intensity. Improved forecast and advanced preparedness are crucial to reduction in TC related fatalities with early risk assessment being key to disaster mitigation. The coastal vulnerability and impact of land-falling TCs in the NIO were analyzed. Assessment frameworks, observational tools and mitigation strategies were reviewed and critical factors for better disaster preparedness and mitigation of TC impacts in the coastal regions were identified.

30.1 Introduction

Natural disasters have affected the earth since its very beginning. These are events with high destructive potential resulting from natural processes of the earth system. Natural disasters are in the form of earthquakes, landslides, volcanoes, dust storms, wildfires, droughts, famines, heavy precipitation, flooding, heat waves, cold waves, thunderstorms & lightning, Tsunamis, tropical cyclones (TC), typhoons, tornadoes and hurricanes. On average, natural disasters of various types were responsible for loss of nearly 60,000 precious human lives per year during the last decade accounting

D. Swain (✉)
School of Earth, Ocean and Climate Sciences, Indian Institute of Technology Bhubaneswar, Argul, Jatni, Odisha 752050, India

A. Pandey et al. (eds.), *Geospatial Technologies for Land and Water Resources Management*, Water Science and Technology Library 103,
https://doi.org/10.1007/978-3-030-90479-1_30

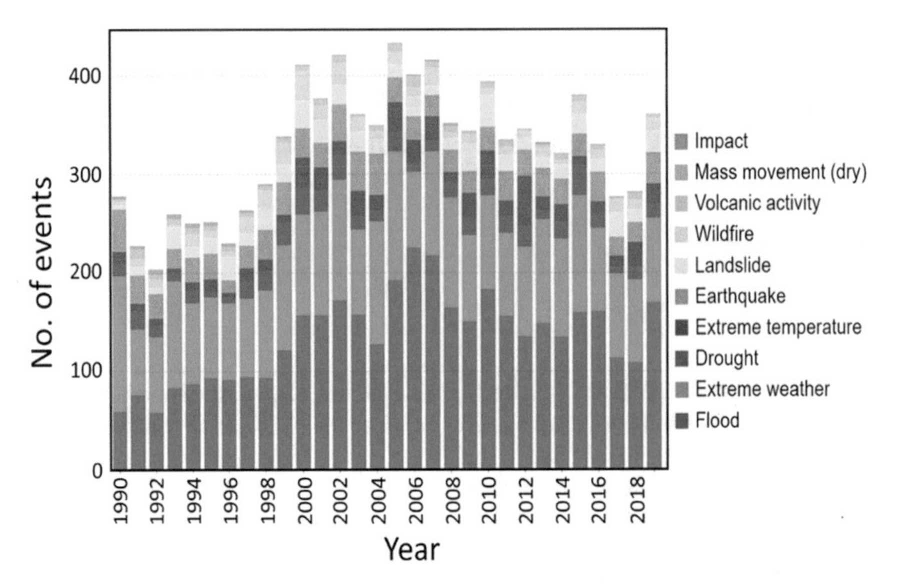

Fig. 30.1 The annual reported number of natural disasters globally during 1990–2019 along with the impact. Reproduced in part from Ritchie and Roser (2014). *Data Source* Emergency Events Database (2020)

for ~0.1% of deaths globally (Ritchie and Roser 2014). Figure 30.1 presents the annual reported number of natural disasters globally during the period 1990–2019. Among the various reported disasters, TCs, tornadoes, typhoons and hurricanes are among the most common and regular natural disasters on earth. While all of them are associated with violently rotating columns of air, TCs, typhoons and hurricanes form over the oceans and tornadoes originate over the land. The impact of TCs could also significantly alter the regional climate in long term.

Disaster preparedness is the key to facing any natural disaster. It is not possible to prevent the occurrence of extreme weather events which very often escalate into natural calamities. However, the consequences of such disasters could be minimized through advanced preparedness as witnessed with TC hazards in recent times owing to precise forecast quality and intelligent preparedness services. This has resulted in a significant reduction in TC related fatalities. Certain steps of disaster preparedness need to commence well in advance of any disaster, even in normal times. For example, building of infrastructure in terms of TC shelters, flood water channelling system, sewerage water disposal and setting up of communication and transport links are vital to TC preparedness and post landfall rescue and relief operations. Mitigation strategies are also long-term processes that include adaptive planning and resilience methodologies.

Following the introduction, the primary study area, the north Indian Ocean (NIO), is discussed followed by analyses of the lifecycle of TCs. An assessment of the destruction potential of land-falling TCs is then presented. Thereafter, a review of the existing works on coastal vulnerability assessment is carried out and various tools to evaluate damage due to TCs are explored. Finally, solutions for enabling better preparedness and mitigation strategies to contain the damaging impacts of TCs in the coastal regions are reviewed and suggestions are made based on the outcome of the analysis.

30.2 The North Indian Ocean

The NIO (0° N to 30° N and 30° E to 100° E) is one of the unique regions of the global oceans which has been of great interest to oceanographers and atmospheric scientists world over (Fig. 30.2). This region experiences two primary cyclone seasons every year (the transition periods between the winter and summer monsoons). The NIO can be broadly divided into two major water bodies based on their geographic location with the Indian peninsula almost bisecting it into the Arabian Sea (AS) in the west and the Bay of Bengal (BoB) in the east. To be more specific, the Indian Ocean borders the southern tip of India. Both the AS and BoB are unique and form an integral part of the atmospheric and oceanographic interactive feedback mechanisms over the NIO. In addition, they contribute extensively to the global circulation process in the world oceans and indirectly are responsible for climatic conditions the world over as the various ocean systems and weather systems are mutually interlinked. Both the basins are unique in their oceanographic and atmospheric dynamics as well. Further, the Bay is more influenced by cyclones compared to the AS due to its unique setting.

30.2.1 Regional Dynamics

The AS and the BoB have very distinct oceanographic and meteorological properties leading to interesting circulation patterns. Though the two water bodies occupy the same latitudinal range, a remarkably wide difference in characteristics is observed. The AS is unique in its topographic location making it more sensitive to the effects of atmospheric forcings. The BoB on the other hand is dominated by fresh water forcing in the form of fresh water inflows from rivers as well as precipitation. The

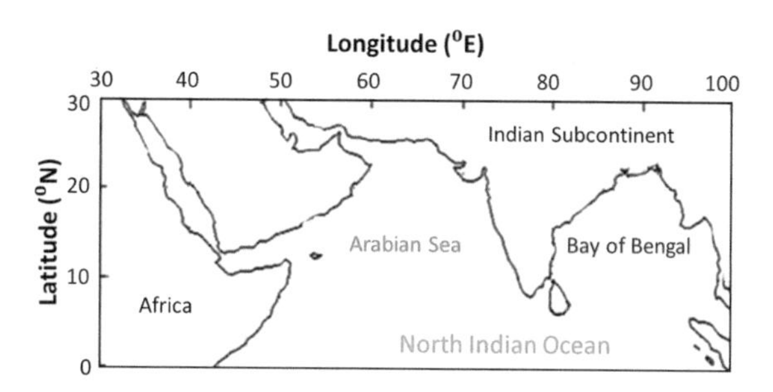

Fig. 30.2 Schematic of the region of interest (The North Indian Ocean)

NIO is forced by wind systems, solar heating owing to the lower latitudes, immense evaporation, precipitation and huge river run-off particularly in the BoB. The AS is connected to the warm, highly saline waters of the Persian Gulf and the Red sea. The coupled air-sea interaction between the ocean and the atmosphere combine to produce strong intra-seasonal and inter-annual variability in circulations and rainfall of the South-Asian summer monsoon. The major part of the west coast of India adjoining the AS undergoes intense upwelling associated with the southwest monsoon while the upwelling in the east coast adjoining the BoB is weak and is associated with the northeast monsoon. Specific climatic features are also influenced by the monsoon circulation dominating large areas of southern and eastern Asia (Rohli and Vega 2018). The BoB is warmer compared to the AS throughout the year, though the spatial variation of sea surface temperature (SST) is larger in the AS with a bimodal oscillation (Düing and Leetmaa 1980). It is well known that a warmer ocean is one of the essential conditions for cyclogenesis and SST remains greater than 28 °C in the BoB for the major part of the year. SST and sea surface winds (SSW) further are important parameters influencing the physical processes in the ocean, atmosphere and biological processes in the sea including the exchange of heat energy between the ocean and atmosphere. Their variability is linked not only to the local weather phenomena like depressions, cyclones, but also the long-term changes of the global environment.

30.2.2 Coastal Coverage

The AS expands over a total area of ~3,862,000 km^2 with a maximum depth of 4652 m (Verlaan et al. 2019). The BoB expands over an area of 2,600,000 km^2 with an average depth greater than 2600 m and a maximum depth of 4694 m (Verlaan et al. 2009). Many rivers drain into both the basins, but the Bay is mostly fed by several large perennial rivers, thus introducing unique subsurface dynamics in the Bay. The total coastline of India extends over 7500 km, distributed along nine coastal states and four union territories. It is dotted by several ports and a large number of major coastal towns and cities including three metros, Kolkata, Chennai and Mumbai. An area of over 40,230 km^2 in India is in the form of coastal wetlands with the near-shore marine ecosystem consisting of mangroves, salt marshes, vegetated wetlands, coral reefs, rocky coasts, sandy beaches, backwaters, lagoons, estuaries, deltas and tidal mudflats among others (Singh 2003).

Several islands also dot both the basins and the coastlines harbouring unique flora and fauna. However, marked differences are observed between the eastern and western coastlines of India with the east coast harbouring almost all the continental islands, whereas the major island formations on the west coast are oceanic atolls (Venkataraman and Wafar 2005). The east coast of India, bounded by the BoB is generally shelved with lagoons, beaches, delta and marshes, but with a narrow continental shelf (Venkataraman and Wafar 2005). The west coast bounded by the AS on the other hand is exposed with heavy surf, rocky shores and headlands with the shelf width varying from ~340 km in the north to less than ~60 km in the south (Sanil Kumar et al. 2006). Unfortunately, the Bay is affected by several cyclones every year, many of which are quite intense and destructive. The AS also experiences TCs but with much lesser frequency and varying intensities. Nevertheless, TCs in the region are known to cause havoc with short and long-term changes that have affected the life and economy of the region.

30.2.3 Demography and Economic Significance

The people habiting the coastal regions of the NIO are largely dependent on marine resources with farming and fisheries being the main occupation, though the city habitants are engaged in a variety of non-agricultural occupations. The people all along the Indian coasts are largely dependent on rice and seafood as their staple. India and other South-Asian countries in fact share the major chunk of low elevation coastal zones by population. Given the vast coastline, about 2.6% of the total area of India can be classified as a low elevation coastal zone which was inhabited by 6.1% (640 lakh) of its total population in 2000 (Neumann et al. 2015).

The Indian Ocean region is of major commercial significance with important economic trade routes running through this region. The Indian Ocean is the pathway to an estimated 40% of the world's oil supply and 64% of oil trade. The region is rich in natural resources like a variety of seafood (2546 of the total 21,730 species of fish found globally are present in India), offshore oil & gas deposits, gas hydrates, mineral resources both near the coasts and as nodules on the sea bed, various rare earth elements and precious and semi-precious gems.

30.3 Tropical Cyclones

TC landfalls are known to be dangerous and devastating natural disasters with huge effects on life, property and ecosystem throughout the world and thus are multi-hazard weather phenomena (Anthes 2016; Jangir et al. 2020). A TC is a large atmospheric system comprising a rapid inward circulation of air masses with a low-pressure centre and accompanied by stormy destructive weather. TCs are marked by an anti-clockwise circulation in the northern hemisphere and a clockwise flow in the southern hemisphere. The name TC refers to cyclones in the Indian and South Pacific Oceans. A hurricane is a TC in the Atlantic and north-eastern Pacific Oceans, while the TCs in the north-western Pacific Ocean region are known as typhoons. The NIO region (north of the equator) experiences about 7–12% of the total numbers of cyclones in the entire global oceans in a year, out of which 2–3 are quite intense every year.

30.3.1 Phases of a Tropical Cyclone

The life cycle of a TC is marked by its origin phase known as "cyclogenesis" and ends with "cyclolysis" or landfall when the TC hits the land and thus undergoes a very rapid dissipation. Broadly the phases can be divided into four, namely formation or genesis, intensification or immature stage, mature stage and finally the decay of TC. The origin or birth of a cyclone can be traced to the formation of low-pressure system over the warm ocean waters showing signs of weak circulation. This stage is known as a tropical disturbance. With suitable conditions, as the winds pick up, the system develops into a well-marked circulation which is then referred to as a tropical depression. With further intensification and increasing wind speeds, the system is now referred to as a tropical storm and when the maximum wind speed within the system exceeds a certain threshold, the storm is then classified as a TC. In most parts of the world, the low-pressure system is assigned a name once it matures into the category of a tropical storm or cyclonic storm. Further intensification leads it to be categorized based on various scales till the TC reaches the landfall, dissipating its utmost energy at the landfall location and entering the cyclolysis phase.

30.3.2 Cyclogenesis and Intensification Changes

TCs form only over the warm ocean waters close to the equatorial regions but at sufficient distance to be supported by a non-negligible Coriolis force. TC genesis marked by the development of a warm-core and associated circulation begins with significant convection assisted by the following favourable conditions (Gray 1968, 1979), even though the exact mechanisms of origin and development of a TC are still not well understood:

- Sufficiently warm sea with SST >26.5 °C and throughout a sufficient depth (at least of the order of 50–70 m from the sea surface)
- Existence of atmospheric instability or upper level air divergence
- Relatively large amount of warm and moist air in the lower to middle levels of the troposphere
- Presence of sufficient Coriolis force to support formation of a low-pressure centre
- A pre-existing low-level disturbance with sufficient vorticity and convergence
- Low or minimal vertical wind shear (<10 m/s) between the surface and the upper troposphere.

It is well observed that TCs originate over the tropical oceans with sufficiently high SSTs and equally warm subsurface, but not in the regions where cold ocean currents lower the surface and subsurface temperatures below the threshold level (e.g. eastern sides of the tropical oceans, South Atlantic Ocean). TCs derive their energy from the warm ocean waters and moist air in the form of latent heat. Since sufficient Coriolis force is necessary to support and sustain the low-pressure centre, TCs do not form very near the equator even though the ocean is sufficiently warm. Existence of an upper level divergence is essential to maintain the surface convergence providing the necessary lifting mechanism to the cyclone. Minimal vertical wind shear between the lower and upper troposphere is necessary for cyclogenesis as this would sustain the latent heat lifted to develop the core. In fact, large values of vertical wind shear could easily prevent genesis and even weaken or destroy already formed TCs. Additionally, pre-existing low-level disturbances such as easterly waves and mild atmospheric vortices in the Inter-Tropical Convergence Zone (ITCZ) could easily develop into TCs given the existence of abundant warm and moist air inducing instability through the formation of intense columns of latent heat. It is prudent to note that the above conditions are some of the necessary and favourable conditions but not the only ones for TC genesis. Rather a multitude of factors, but never operating in isolation, determine the origin and development of TCs. Several works have highlighted the role of other complex factors such as meso-scale convective complexes as well in TC genesis (Velasco and Fritsch 1987; Chen and Frank 1993; Emanuel 1993).

The genesis phase of TCs is followed by the intensification phase which is based on the available thermal energy of the underlying ocean waters and existence of the above suitable conditions. The intensity changes are denoted by the sustained winds, existence of rain bands and drops in the central pressure. The eye which is clearly developed too undergoes rapid transitions. This is also a stage of rapid changes to the circulation features, cyclone translational motion and propagation away from the place of origin. TC track usually leads to a landward direction assisted by the prevailing near-shore conditions and on land features, though some of the TCs also end up in the ocean waters itself.

TC on hitting the land (landfall) usually weakens or enters the lysis phase wherein the cyclone rapidly dissipates. This is due to the non-availability of energy over land, unlike the underlying warm ocean waters. However, this phase of the TCs is associated with severe rains within a short period of time. The typical lifespan of a TC ranges from a few days to about a week, with the exception of very few which have lasted close to a month. It is interesting to note that a TC can re-intensify if it re-enters a marine region post landfall (re-curving back to the ocean or even crossing across to basins, such as cyclones “Gaja”, “Ockhi”, “Vardah” in the NIO). However, most often the land-falling cyclones move far inland, shedding their moisture as rain and causing lots of wind damage before dissipating completely.

30.3.2.1 Measurement of Cyclone Intensity and Storm Classification

TCs are generally classified utilizing cyclone intensity scales based on the maximum sustained winds and the geographic location in terms of the classification scales being officially designated by the meteorological agencies monitoring them. The Dvorak technique (Dvorak 1975; Velden et al. 2006) is one of the most popular systems used to estimate TC intensity based on visible and infrared satellite images. Several versions of the Dvorak technique like the Objective Dvorak Technique (ODT) (Velden et al. 2006), the advanced ODT (Olander and Velden 2015), the Dvorak version involving only visible satellite imagery, the infrared imagery version of Dvorak, synthetic satellite imagery (Manion et al. 2015), pressure-wind relationships (e.g., Kraft 1961), or empirical reduction of flight level reconnaissance wind measurements to produce surface level estimates (Franklin et al. 2003), the Hebert-Poteat technique (Hebert and Poteat 1975) for subtropical cyclones and the Miller and Lander extratropical transition technique (Miller and Lander 1997) are some of the other cyclone intensity measurement techniques in use for classification of TCs.

The various stages of a TC in the NIO as classified by the India Meteorological Department (IMD), the nodal agency for meteorological observations, seismology and weather forecasting in India are presented in Table 30.1.

A slightly different wind scaled based classification, the Saffir-Simpson Hurricane Scale (Saffir 1973; Simpson 1974) is followed for hurricanes in the North Atlantic and central and eastern North Pacific Oceans (Table 30.2). The Saffir-Simpson Hurricane Wind Scale [and its adjusted version as in use currently (PIS 2012)] classifies the hurricanes into five categories based only on the hurricane's maximum sustained wind speed and not any of the other potentially deadly hazards such as storm surge, rainfall flooding and tornadoes.

Table 30.1 Classification of cyclonic disturbances over the North Indian Ocean

Disturbance type	Associated maximum sustained wind in km/h (knots)
Low-pressure area	Below 31 (below 17)
Depression	31–49 (17–27)
Deep depression	50–61 (28–33)
Cyclonic storm	62–88 (34–47)
Severe cyclonic storm	89–117 (48–63)
Very severe cyclonic storm	118–167 (64–90)
Extremely severe cyclonic storm	168–221 (91–119)
Super cyclonic storm	222 and above (120 and above)

Source India Meteorological Department (2021a)

Table 30.2 The Saffir-Simpson Hurricane wind scale based classification

Hurricane Category	Sustained Winds in km/h (knots)
I	119–153 (64–82)
II	154–177 (83–95)
III	178–208 (96–112)
IV	209–251 (113–136)
V	252 or higher (137 or higher)

Tropical Cyclone Classification in Global Oceans.

A summary of the TC classification in different basins as per the World Meteorological Organization (WMO) is provided in Table 30.3 (World Meteorological Organization 2019, 2021a, b).

30.3.3 Tropical Cyclones and Land Surface Processes

The rate of devastation by TCs depends upon the environmental factors as well, such as vegetation cover (e.g. mangrove density, forests), barren lands, agricultural fields, all of which constitute the Land Use and Land Cover (LULC). TCs can also modulate the LULC of the landfall location and nearby areas by changing the landform depending on the destructive potential of the cyclone. The LULC change in turn can, directly and indirectly, affect the local meteorology of the region post-cyclone landfall. For example, LULC change in terms of vegetation loss leads to a rise in maximum and minimum temperature along with the Land Surface Temperature (LST), soil temperature and soil moisture. Increased LST could easily boost convectional heating and thundershowers. For example, significant LULC changes were reported for TC "Monica" (category-5) and TC "Larry" (category-4) both hitting northern Australia in 2006 (Turton 2008). Similar observations were also made for Hurricane "Katrina" hitting the USA in August 2009 (Rodgers et al. 2009), Hurricane "Maria" (category-5) in Dominica (Hu and Smith 2018) and many such in the Indian subcontinent as well. TCs are thus capable of modulating the land surface processes of the region they strike over a period of time.

30.3.4 Impact of Cyclones in the North Indian Ocean

The damage by TCs in the NIO region has been quite significant, though it has been decreasing in recent years due to proactive disaster preparedness and mitigation strategies. Although mortality rates due to TCs could be reduced, infrastructure damage is difficult to contain. At best, preventive measures and long-term mitigation strategies are only sufficient to reduce the damage, but not substantially, unless they are significant enough. A historical perspective of the damages (life and economic) caused by all TCs (from the depression stage) in the NIO are shown in Fig. 30.3 (India Meteorological Department 2021a,b). From the figure, it is observed that deaths due to TCs in the NIO region have decreased considerably over the years from more than 130,000 in the years 1991 and 2008 (the highest in last 30 years) to less than 1000 in recent years. However, total damages have been continuously on the rise with the highest of 15.8 billion USD during 2020. Cyclone "Amphan" in the BoB, the 2nd

Table 30.3 Classification of tropical cyclones based on the maximum sustained wind speed criteria for different basins

km/h	*knot*	Beaufort[d]	Arabian Sea & Bay of Bengal[a]	South-west Indian Ocean[b]	North-west Pacific[a]	North Atlantic & North-east Pacific[c]	South-west Pacific & South-east Indian Ocean[b]
Average wind speed			**km/h** ***knot***	**km/h** ***knot***		**km/h** ***knot*** **mi/h**	**km/h** ***knot***
			Low pressure area	Zone of distributed weather			Tropical Disturbance
31 50	*17* *27*	- 6	Depression	Tropical disturbance	Tropical depression	Tropical depression	Tropical Low / depression
51 62	*28* *33*	7	Deep depression	Tropical depression			
63 - 88	*34* *40* *47*	8 - 9	Cyclonic storm	Moderate Tropical storm	Tropical storm	- *34* 39 Tropical storm ***(Name is assigned)*** - - 73	Tropical cyclone (Gale) / Cat-1
89 - 117	*48* *52* *63*	10 - 11	Severe cyclonic storm	Severe Tropical storm	Severe Tropical storm		Tropical cyclone (Storm) / Cat-2
118	*64*	12	- *64* Very severe cyclonic storm 166 *89*	- *64* Tropical cyclone 165 *89*	Typhoon	- *64* - Hurricane Cat-1 153 *82* 95	- *64* Severe Tropical cyclone (Hurricane)/Cat-3 159 *85*
						154 *83* 96 Hurricane Cat-2 177 *95* 110	
						178 *96* 111 Hurricane Cat-3 208 *112* 129	
			167 *90* Extremely severe cyclonic storm 221 *119*	166 *90* Intense Tropical cyclone 212 *115*		209 *113* 130 Hurricane Cat-4 251 *136* 156	160 *86* Tropical cyclone / Cat-4 199 *107*
			222 *120* Super cyclonic storm	213 *116* Very intense Tropical cyclone		252 *137* 157 Hurricane Cat-5	>200 *>107* Tropical cyclone / Cat-5

[a]10-min (recording) 3-min (non-recording) average wind speed; [b]10-min average wind speed; [c]1-min average wind speed; [d]Beaufort (Beaufort wind force scale) classification: an empirical measure relating the wind speed to observed sea condition (Huler, 2004); km/h: km per hour; mi/h: miles per hour.

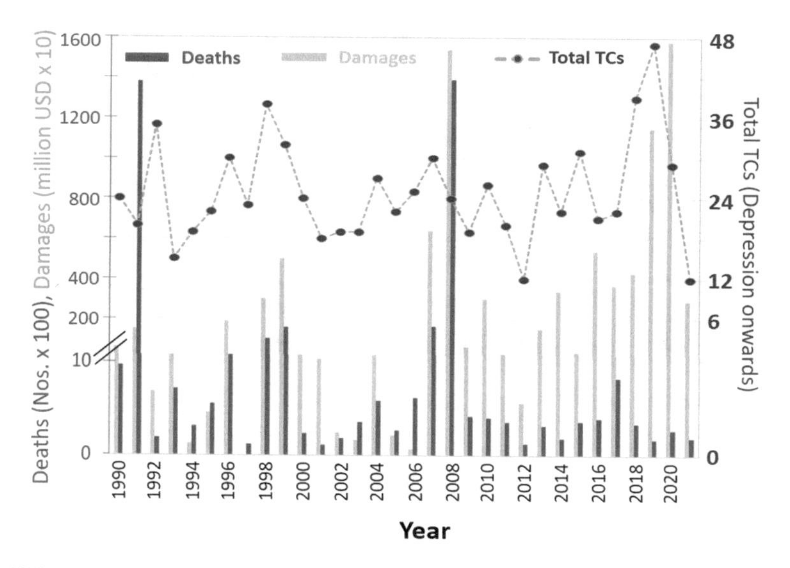

Fig. 30.3 Estimated damages by TCs in the NIO during 1990 and 2021. Damage data not available for the year 1997. *Data Source* IMD, JTWC

super cyclone in the Bay after the one in 1999 turned out to be the costliest TCs ever recorded in the region so far. The cyclone season of 2020 also marked the costliest on record in the entire NIO (India Meteorological Department 2021b).

30.3.4.1 Recent Trends

The cyclone frequency, as well as intensity, are on a significant increase in recent decades. Balaguru et al. (2014) showed a clear increasing trend in the intensity of major post-monsoon TCs in the BoB during the period 1981–2010 which were attributed to TC favourable changes in the environmental parameters. Similar increases in hurricane intensification were also observed in the central and eastern tropical Atlantic Oceans and north Pacific (Balaguru et al. 2018; Bhatia et al. 2019; Song et al. 2020). Additionally, Singh et al. (2000) based on their study of 122 years (1877–1998) of TCs in the AS and BoB concluded a significant increase in the frequency of TCs over the Bay during November to May (the main cyclone months), but decrease in TC frequency between June–September (the south-west monsoons) and no significant trend in TC frequency in the AS. Inundation of a residential area caused post the landfall of cyclone "Yaas" in May 2021 is shown in Fig. 30.4.

Fig. 30.4 Inundation post the recent cyclone Yaas (23–28 May 2021)

30.4 Damages Caused by Tropical Cyclones

The damages by cyclones in the open ocean are not very severe owing to the least presence of human lives or infrastructure. Further, in most parts of the ocean, the cyclone interaction is in the form of latent heat exchange and picking of moisture. The presence of vast ocean waters and its fluidity allows the cyclone energy to be relatively distributed. Therefore, the open ocean damages are primarily to any floating objects like mooring/buoys which get displaced, ships carrying men and material and oil and communication installations. Also, islands in the area under the influence of a passing TC could get affected in the form of damage to natural resources and life if habituated. However, communication systems like underwater cables, etc. remain unaffected.

TC damages are quite pronounced on land and near coast regions starting with the landfall location inwards. On hitting land, the TCs undergo a rapid dissipation of energy with high wind gusts, heavy rainfall, lightning and thunderstorms, storm surges and high waves contributing to the maximum damage. Destruction of natural vegetation, infrastructure and flooding of the coastal areas are the first visible signs of TC damage often accompanied by loss of life. The severity of the damage depends on factors ranging from the intensity of the TC at the time of landfall, density of population and infrastructure development, amount of coastal vegetation and type of land cover. On a longer time scale with frequent TCs hitting the same region (e.g. eastern coast of India), more long-term impacts are seen ranging from shoreline changes, modification of land cover, saltwater intrusion, infertility of agricultural lands to damage to the coastal ecosystem and near-shore environment. The post-cyclone landfall damage assessment forms an integral part of the TC disaster relief and rescue process. The preliminary damage assessment provides the first and most essential level of information to the relief agencies as well as to the public to outline the destructive level of a TC landfall. The damage assessment is carried out based on several factors and this forms an important input to any future cyclone related disaster mitigation approach.

30.4.1 Indian Scenario

The extreme weather events have been on a continuous rise in the Indian subcontinent. Ray et al. (2021) analyzed 50 years' of IMD data spanning from 1970–2019 for mortality rates of different extreme weather events like floods, TCs, heat waves, cold waves, lightning, etc. at different levels and observed that TCs accounted for 28.6% of mortality in India, second to 46.1% due to floods. Although the mortality rate owing to TCs has reduced by 94% in the past 20 years (Ray et al. 2021), it has not been possible to achieve a zero mortality number. They have further identified the states of Odisha, Andhra Pradesh, Assam, Bihar, Kerala and Maharashtra in India to have the maximum mortality rates due to extreme weather events, emphasizing the need to prioritize these states in developing disaster management and mitigation action plans.

30.4.2 Destructive Potential of Cyclones

The destructive potential of cyclones originates from their strong winds, wind gust and rainfall which assumes a disproportionate limit on striking land. Scales used for classification of cyclones are also indicative of their destructive potential. Some of the cyclone destructive potential assessment metrics in the form of indices widely being used are listed in Table 30.4. Saffir-Simpson Hurricane Wind Scale and Dvorak technique based classifications are also sometimes considered for destructive potential estimation of cyclones. The destructive potential based metrics attempt utilizing

Table 30.4 List of popular cyclone risk and hazard assessment indices

Cyclone/Hurricane hazard index	References
Accumulated Cyclone Energy (ACE)	Bell et al. (2000)
Carvill Hurricane Index (CHI)	Kantha (2006)
Hurricane Destruction Potential index (HDP)	Gray (1988)
Hurricane Hazard Index (HHI)	Kantha (2006)
Hurricane Intensity Index (HII)	Kantha (2006)
Hurricane Risk index (HRi)	Hernández et al. (2018)
Hurricane Severity Index (HSI)	Hebert et al. (2008)
Hurricane Surge Index (HSI)	Kantha (2006)
Integrated Kinetic Energy Index (IKE)	Powell and Reinhold (2007); Kantha (2008); Misra et al. (2013)
Maina Hurricane Index (MHI)	Maina (2010)
Power Dissipation Index (PDI)	Emanuel (2005)
Willis Hurricane Index (WHI)	Owens and Holland (2009)

dynamic variables like variations in damage type, real-time forecasts and adaptability to impacts of climate change and variability as well as information from the local cyclone observation centres. TC hazard assessment however has seen significant improvements in the form of TC intensity potential computations contributing to more reliable estimates of TC damage probability (Zeng et al. 2007; Tang and Emanuel 2012; Lin et al. 2013; Balaguru et al. 2015; Glazer and Ali 2020).

30.4.3 Coastal Vulnerability

The high vulnerability of coasts and low lying areas to extreme weather events, such as storms is one of the major challenges globally. Annually, about 12,00,00,000 human lives are exposed to TC hazards throughout the world and 250,000 lives were lost between 1980 and 2000 (Nicholls et al. 2007). Intensification of TCs accompanied by large waves, storm surges and inundation has been a major problem in recent years. Coastal vulnerability is further exacerbated by accelerated sea-level rise, coastal erosion, saltwater intrusion, rapidly changing rainfall patterns and runoff threatening the coastal wetland ecosystem. Further, SST rise leading to coral bleaching and ocean acidification has witnessed a fragile coastal ecosystem, especially for islands. A fast changing climate coupled with extreme weather events and accelerated anthropogenic activity has emerged as an adaptation challenge to humanity at large.

30.4.3.1 Short-Term Impacts

The major damage by TCs is to the landfall areas. When the cyclone strikes land, the first reaction is to dissipate the energy. The most devastation by the cyclones is in the form of high wind gusts, heavy rainfall, storm surges, high waves and flooding of the coastal areas. The flooding is due to the high walls of water pushed into the land by the TC which is also known as storm surge. If the landfall time of a cyclone coincides with the high tide duration at that location, flooding and reverse flooding are at its peak. Even though the landfall of cyclones are associated with rapid decay of the cyclone intensity due to land interference and unavailability of sufficient moisture which may last only few hours, heavy winds accompanied by heavy rains could have disastrous consequences. Further, as the coastal areas are densely populated, this could and does cause havoc in the inland socio-economic conditions as well. Winds, gales and high wind gusts, heavy rainfall, high waves and storm surges, lightning and thunderstorms are the factors associated with TCs for causing short-term damages which could be of very serious consequences. Such damages are also invigorated by environmental factors existing at and near the landfall locations of TCs which includes barren lands, thin coastal vegetation, or the presence of manmade infrastructure. The short-term impact of TCs may thus be summarized as:

- Storm Surge and Coastal Inundation
- Infrastructure/Property Damage
- Damage to Life (human & livestock)
- Saltwater Intrusion
- Land Cover Change (vegetation-infertile agricultural lands, build up, coastal ecosystem, etc.)
- Disruption of public utilities and health care.

30.4.3.2 Long-Term Impacts

TCs over the years are responsible for leaving permanent traces of destruction in the form of shoreline changes and coastal erosion which is largely due to the associated storm surge and extreme wave activities, coastal inundation and loss of vegetation due to gale winds and rainfall. Further, heavy rainfall during the cyclone landfall causes terrestrial flooding drowning life and property closer to the coasts. The inundated water sometimes sweeps out the sediments in addition to saltwater intrusion which leaves the agricultural land infertile over a period of time (Anthes 2016). In fact, the same short-term factors are responsible for long-term damages as well but when repeated over large durations.

The most pronounced long-term changes have been coastal erosion and thus shoreline deformation. Saltwater intrusion on a regular basis renders the coastal fresh groundwater aquifers saline over a period of time making it unfit for consumption. Though coastal erosion could reverse owing to accretion by rivers or by artificial means, degradation of freshwater aquifers is almost an irreversible process. Further, regions exposed to frequent TCs on a regular basis are very like to sustain changes on a climate scale. Change in their land cover (LULC) can result in alteration in land surface properties including LST through modification of soil cover and other soil properties. This could in turn vary the regional maximum and minimum temperature, precipitation and other meteorological parameters thus modifying the regional climate itself. In fact, significant changes in regional meteorology have been reported due to TC activity in different parts of the world (Visher 1925; Cooper and Orford 1998; Hickey 2011). Degradation of coastal ecosystems, especially wetlands and coral reefs, has serious implications for the well-being of societies dependent on the coastal ecosystems for goods and services (Nicholls et al. 2007).

Shoreline Changes

The physical damage of land-falling TCs is one of the most distinguishing characters. Though the short-term damages are temporary but sometimes serious (e.g. loss of life and property), the long-term impacts that are often a result of frequent exposure to TCs are indelible in the form of physical changes. Shoreline deformation is one of such major impacts which also leads to other vulnerabilities in due courses, like coastal flooding, inundation due to sea-level rise, saline water intrusion and a weak coastline vulnerable to further erosional losses. BoB has been the location of frequent cyclogenesis with the eastern coast of India facing the wrath of many land-falling

Fig. 30.5 Shoreline deformation map of Puri beach adjoining the Bay of Bengal within a two year period (Feb. 2006–Feb. 2008)

TCs. Consequently, it has been subjected to continuous shoreline deformation. The shoreline deformation owing to frequent TCs with landfall close to the religious town of Puri in Odisha, India during 2006 and 2008 is shown in Fig. 30.5 as an example. The coastal erosion and accretion are mapped onto Google Earth maps for easy representation. The amount of solids removed (erosion) or solids deposited (accretion) as obtained from inter-comparison of satellite imagery are mentioned in the Figure. More than 25 TCs and depressions were witnessed in the BoB during this period with as many as 5 TCs hitting the Odisha coast. Significant coastal deformation is observed primarily as a consequence of TC activity.

30.4.4 Coastal Vulnerability Assessment

Two broad types of vulnerability assessment frameworks are found to be used widely (i) natural-hazard based approach relying on physical hazards and mainly utilized by physical scientists (also referred to as the outcome vulnerability) and (ii) based on the underlying characteristics of populations and the urban systems that predispose them from a shock or stress (also referred as contextual vulnerability), primarily used by social scientists (Füssel 2010; Bukvic et al. 2020). The Coastal Vulnerability Index (CVI) is one of the widely used coastal vulnerability assessment indices based on metrics of physical exposure (Gornitz et al.1994), while the Social Vulnerability Index (SoVI) is one that relies on contextual vulnerability (Cutter et al. 2003; Hung et al. 2016). Several other vulnerability indices were also developed based on a varied range of relevant factors (Table 30.5).

Table 30.5 List of few widely used coastal vulnerability assessment indices

Coastal vulnerability indices	References
Coastal Vulnerability Index (CVI)	Gornitz et al. (1994), Thieler and Hammer-Klose (1999)
Computed Coastal area Flood Vulnerability Index (CCFVI)	Dinh et al. (2012)
Dynamic Interactive Vulnerability Assessment (DIVA) Model/Tool	DINAS-COAST Consortium Staff (2006)
Place Vulnerability Index (PVI) [combination of CVI & SoVI]	Boruff et al. (2005)
TC Mortality Risk Index (TMRI)	Peduzzi et al. (2012
Vulnerability Extreme Index (VEI)	Kleinosky et al. (2007)
Vulnerability Index for Human Life (VIH)	Dube et al. (2006)

30.4.4.1 Physical Vulnerability

The major coastal zones facing TC impacts have wide variability in their physical and demographic characteristics. Vulnerability of the coastal zone hence depends on both of these factors. The degree of physical and social vulnerability associated with natural hazards such as TCs varies and has varying consequences depending on the physical characteristics of the coast under TC impact and the population density along the coastal belts (Sahoo and Bhaskaran 2018). The level of physical vulnerability further contributes to the extent of socio-economic damage. The physical vulnerability of a region to a TC is primarily dependent on physical properties like the coastal structure, topography, inland landforms, sea level, meteorological and environmental conditions. These factors are crucial to the extent of post landfall destruction caused by a TC like inundation and storm surge, loss of natural resources, LULC change, landform deformation, shoreline changes (accretion and erosion) and change in the regional meteorology which on a longer time scale lead to permanent damages like regional climate change.

30.4.4.2 Socio-Economic Vulnerability

Vulnerability is associated with the susceptibility of a geographical area (such as coastal zone) to an external event such as TC, with the susceptibility factor conditioned by the past, current and future population, settlement pattern and rate of change of the area, together with its aggregate and per capita level of economic wealth (Turner et al. 1996). The population density and concentration is representative of the socio-economic conditions of a region (Nallathiga 2010). The coastal vulnerability in terms of damage is significantly aggravated by the coastal population concentration as well as their occupation as these control the socio-economic vulnerability (e.g. poverty, infrastructure development and information accessibility). Considering the Indian coastline as an example, it is home to diverse human races and cultures with varied economic activities (industries, infrastructure, trade and commerce, major and minor port activities), population density, natural ecosystems, physiography, land use and geomorphology (Nallathiga 2010). They have listed 11 major ports and 130 minor ones along the entire Indian coastline providing vast logistical support for international trade and commerce. Further, the coast is lined with major towns and cities supporting a huge coastal population. The population density of India with specific distribution across a 10 m low elevation coastal zone as in 2000 is shown in Fig. 30.6 (Center for International Earth Science Information Network 2007; Neumann et al. 2015). The authors further projected this to increase nearly three-fold by the year 2060 (2160 lakhs which would be 10.3% of the total population) from a 2000 baseline population of 640 lakhs (or 6.1% of the total population of India). Similarly, considering the south-east Asian region which is bounded by the NIO, the population in the low elevation coastal zone amounted to 31.1% of the total population in this region in 2000 as per the same study.

To sum up, all types of coastal vulnerability assessment fall broadly into four categories (Sudha Rani et al. 2015):

1. Index-based methods: classification of the coastlines into uniform entities possessing similar features
2. Indicator-based approach: vulnerability indicator based on variables that influence and change the vulnerability of the risk element
3. Decision support system (DSS) based: vulnerability indicator relying on community or dynamic vulnerability assessment tools such as Geographic Information System (GIS) based DSS
4. Methods based on dynamic computer models: sector and integrated assessment models for analyzing and mapping coastal vulnerability risks.

Further, a predictive coastal vulnerability assessment also depends on the assessment of the rise in coastal population and economic development with time. The growth of coastal populations are expected to be more rapid and projected to continue to grow much faster than inland areas (Neumann et al. 2015; Merkens et al. 2016; Bukvic et al. 2020). It is hence important to understand that formulations for assessing the destructive potential of TCs restricted to meteorological and/or oceanographic inputs alone are not sufficient. Moreover, these do not account for devastation in terms of socio-economic loss to the affected area. At the same time, destruction in the affected areas when quantified in terms of mortality and economic loss, cannot easily be associated with cyclones of a given size and intensity (Powell and Reinhold 2007). Instead, the degree of physical and social vulnerability is directly linked with the population density, resilience of the coastal population and the preparedness measures too (Sahoo and Bhaskaran 2017, 2018). Peduzzi et al. (2012) have postulated that despite the projected reduction in TC frequency, projected increases in both demographic pressure and TC intensity over the subsequent two decades could be expected to greatly increase the number of people exposed per year and exacerbate disaster risk, despite potential progression in development and governance.

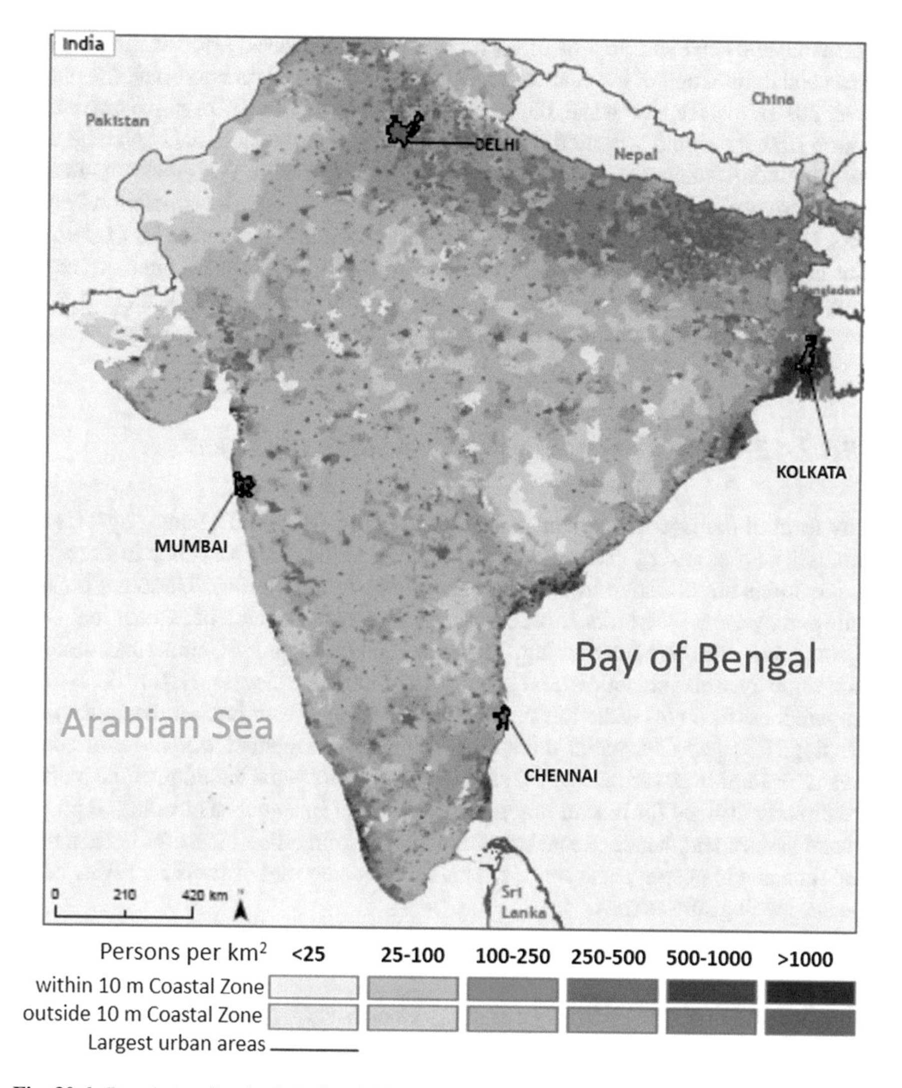

Fig. 30.6 Population density in India within and outside a 10 m low elevation coastal zone. Reproduced in part from PreventionWeb 2021. *Data Source* Center for International Earth Science Information Network (2007), McGranahan et al. (2007)

30.5 Tropical Cyclone Hazard Preparedness and Mitigation

Any step of hazard preparedness including long-term mitigation requires development of accurate forecasting capability of the impending hazard (TC in the present case), the socio-economic hazard assessment as well as public awareness and outreach. Hence, preparedness for TC hazards and their mitigation is quite complex and a multi-dimensional effort. TC disaster preparedness and mitigation to a large extent depends on identification of vulnerable coasts. However, prioritization of such areas and delineation of evacuation zones during TC hazards are complex (Kennedy et al. 2012). In such a scenario, the successful execution of a TC or any hazard mitigation strategy would largely depend on the cultural and educational awareness of the population irrespective of the technical improvements in TC forecast and early warning or policy formulation, adaptation being the key. Lower the population awareness, lower would be their ability to adopt prevention and mitigation strategies and thus higher the damage potential. Mortality risk thus depends on TC intensity, exposure, levels of poverty and governance (Peduzzi et al. 2012) and hence all of these are critical considerations in any TC hazard preparedness and mitigation strategy.

30.5.1 Tools in Tropical Cyclone Hazard Assessment

Any form of damage assessment requires accurate field data. Damages by TCs too can only be assessed by collection of field observations. Near real-time in situ*ocean* observations are effective in TC track and intensity monitoring. These are closely followed by modelling tools, most of which are directly dependent on near real-time observations for their functioning. These models consisting of numerical weather prediction models, statistical and empirical models are necessary for TC hazard preparedness through prediction and forecast based early warning. Similarly, Remote Sensing (RS) and Geospatial tools form the third component which are effective means for remote assessment of TC affected areas. Geospatial techniques and models are directly utilized for hazard mapping and disaster risk analysis leading to proper hazard assessment, hence execution of appropriate mitigation measures both in near real time and in post-event scenarios. These three categories of tools are reviewed in the succeeding subsections.

30.5.1.1 In Situ Observations

The sources of in situ observations in the oceans primarily consist of moorings and buoys deployed in the ocean. Several oceans and coastal observation networks are active globally providing real-time in situ observations such as the Research Moored Array for African-Asian-Australian Monsoon Analysis and Prediction (RAMA) mooring network (McPhaden et al. 1998), Ocean Moored Buoy Network in NIO (OMNI) (Venkatesan et al. 2013) and National Data Buoy Center (NDBC) Hurricane Data Buoys (Evans et al. 2003; Bouchard et al. 2014) to list a few. Argo profiling floats and even unmanned/autonomous aerial, underwater and surface vehicles are other tools being used for TC monitoring (Chen et al. 2021). Extensive spatiotemporal meteorological and oceanographic information are regularly collected from these offshore moored surface buoys which have led to a better understanding of ocean dynamics, atmosphere–ocean interactions as well as improvement of weather and ocean state forecast (McPhaden et al. 1998; Venkatesan et al. 2014). Further, coastal high frequency (HF) Radars, wave rider buoys, coastal tide gauges and coastal current meter arrays allow for real-time observations of high wave activity, surface currents and other ocean surface dynamics over longer distances from the shore during the cyclonic period are also regularly utilized for effective early warning on TC hazards (Jena et al. 2015). These data sets also constitute the backbone for near real-time model forecasts and potential hazard zone identification. Similarly, real-time information on environmental conditions is relayed through actual observations by meteorological stations and on-ground personnel or crowd sourced during TC events. All of these are then effectively used for disaster relief and rescue planning and execution.

30.5.1.2 Modelling and Simulations

Several numerical, statistical, empirical, semi-empirical models and soft-computing techniques are in use for the prediction of TC genesis, intensification, tracks and landfalls. Some of the dynamical climate models were successfully used in simulating the trends in global TCs and hurricanes and correctly predicting the anthropogenic contributions to TC intensification changes (Bhatia et al. 2019). Models like Hurricane Weather Research and Forecasting (HWRF) model are in regular use to forecast the track and intensity of TCs (Bender et al. 1993; Gopalakrishnan et al. 2011; Zhang et al. 2014). TC track and intensity are also simulated using the WRF model, upper ocean circulation during TC events using the Regional Ocean Modelling System (ROMS) model and other features utilizing coupled modelling systems (Bender et al. 1993; Chan et al. 2001; Prakash and Pant 2017).

Wave models such as the Wave Action Model (WAM) (Group 1988) and WAVEWATCH III (The WAVEWATCH III Development Group 2019) are often used to simulate the growth, propagation and decay of ocean waves based on the prevailing winds. Apart from these, several hydrodynamic, coupled hydrologic-hydrodynamic and dynamic flood models are used to model the impact of TC landfalls in terms of regional sediment transport and coastal inundation (Barnard et al. 2014; Chen and Liu 2014; Ikeuchi et al. 2017; Li et al. 2021). Almost all the dynamical models are however dependent on data obtained from satellite observations, buoys and other remote sensing means or other model simulations for their operation.

In addition to the above and existing statistical (Neumann and Randrianarison 1976; Neumann and Mandal 1978) and deterministic methods (Mohanty and Gupta 1997), approaches based on hybrid genetic algorithms, Artificial Neural Network techniques and other soft-computing techniques are regularly utilized for TC parameter estimation and intensity, category and track forecasts (Ali et al. 2007; Sharma et al. 2013; Karthick and Malathi 2020).

30.5.1.3 Remote Sensing and Geospatial Tools

Extreme events like TCs make it impossible to be in field during the event or immediately after the event. Since field observations are critical to any hazard assessment and mitigation, remotely collected data sets serve as an alternative and many a time the only option. Consequently, RS techniques have found wide application in TC studies. As seen in earlier sections, RS tools are the only option for studying the storm characteristics, their passage as well as intensity. A varied range of RS platforms is utilized for TC tracking and post landfall damage assessment. Satellite based microwave and Infrared sensors provide synoptic observations of a cyclone spread, water vapour and rainfall (Katsaros et al. 2002; Xiang et al. 2019). Altimeters measures the actual pressure drops associated with the TCs as well as wind distribution (Young and Vinoth 2013; Tamizi and Young 2020), while scatterometers have been widely used to measure the prevailing winds, the primary input for estimation of TC intensity (Quilfen et al. 1998; Jayaram et al. 2014; Mahesh Kumar et al. 2015; Sasamal and Jena 2016; Ribal and Young 2020).

Similarly, GIS-based approaches are helpful in mapping coastal inundation, flood levels and social vulnerability post a TC landfall. The information stored on various communication and transport networks is an asset for policy planners as well as rescue and relief teams. In other words, GIS is an important component of a DSS being widely used for immediate cyclone hazard assessment and planning. It can be well utilized to assess prior potential landform damage or post likely losses as an adaptable modelling framework. In recent times, hybrid approaches combining conventional and spatial data sets are in use for very successful disaster management. Further, with the continuous changing coastlines, RS & GIS-based solutions are the most effective means for their frequent analysis (Sudha Rani et al. 2015; Islam et al. 2016). Multi-Hazard GIS-based modelling methods like GIS-based flood simulation models, are quite popular and have also been integrated for hazard early warning and strategy development (Taramelli et al. 2010). With the availability of huge data sets from a plethora of satellite sensors over the years, optimal utilization of RS and GIS has emerged as key to early assessment and situation analysis of any cyclone related damages and their mitigation.

30.5.2 Disaster Preparedness

The most important aspect of disaster mitigation starts with disaster preparedness. As with various other disasters, though TCs are unavoidable, a better advanced preparedness results in lesser damage to life in particular by any land-falling TC. Disaster preparedness can be categorized into two, one is to be able to make a better forecast of the damage potential of the approaching cyclones well in advance, the second is to be physically prepared to face any eventuality to be ushered in by the approaching event. The first one is dependent on a few skilled personnel like atmospheric and ocean scientists or cyclone experts, while the second calls for more direct participation of the society as well as the Government bodies. A brief overview of a few aspects of disaster preparedness is presented.

30.5.2.1 Scientific Preparedness

The foremost challenge in TC disaster preparedness and mitigation is the accurate forecast of TC intensity, track and landfall locations. Accuracy in TC brings to the fore uncertainties inherent in meteorological forecasts as well as in inundation models. Most conventional TC intensity estimates are limited in terms of accurate forecasting when TCs intensify to maximum sustained winds between 105 km/h and 160 km/h. At these intensities, the TCs can have their centre of circulations obscured by cloudiness within visible and infrared satellite imagery, making a diagnosis of their intensity very challenging (Wimmers and Velden 2012). Winds within TCs could also be estimated by tracking cyclone features using rapid scan geostationary satellite imagery, whose pictures are taken minutes apart rather than at longer time durations as with polar orbiting satellites (Rodgers et al. 1979; Cintineo et al. 2020). This also provides an opportunity for utilizing fused multi-satellite and multi-sensor observations for better TC now-casts and forecasts.

Modelling of TC intensity and track has witnessed significant improvement over the years. Models have transitioned from simple statistical ones to complex numerical weather prediction models with a multitude of forcings, parameterization schemes and assimilation techniques. The use of latest computational approaches and methodologies like machine learning, deep learning, data analytics have significantly improved TC early warning and prediction capabilities as well and would continue to do so in the coming years (Rozoff et al. 2011; Matsuoka et al. 2018; Wimmers et al. 2019).

30.5.2.2 Physical Preparedness

One aspect of disaster preparedness is the pre-emptive efforts for safety of lives and livestock. Physical preparedness involves building and maintenance of TC resistant infrastructure and coastal development to slow down the impact of an approaching TC. Providing shelter to the local masses in case they need to be vacated in view of the impending danger from an approaching TC is critical. A wide practice in the Indian peninsula has been the construction of cyclone shelters at vantage locations in the cyclone vulnerable areas along the coasts. In fact, one of the key goals of the National Cyclone Risk Mitigation Project of India has been to ensure accessibility of every household within 10 km from seashore to a cyclone shelter not exceeding a distance of 2.5 km (Nirupama 2013). These TC shelters are specific purpose constructions taking into account the possible cyclone impacts for temporary evacuation of the vulnerable masses. They are usually built deeper in the land located at a sufficient height to be safe from inundation or surges with prime considerations including vulnerability of the region, evacuation cost and conveniences to evacuees as well as easy accessibility to rescue and relief teams. These shelters are multi-purpose in nature to support the community during TCs and doubling up as community centres, schools, drought relief centres, isolation centres or self-help group offices during normal times. Some of these are also made state-of-the-art to be safe as well as provide shelter from other natural hazards like rainfall induced floods, Tsunamis and earthquakes among others. In addition, housing constructions in the TC prone areas are encouraged to be flood proof constructions as a preliminary step. In long run, these facilities contribute to disaster mitigation measures themselves.

30.5.2.3 Community Preparedness

Although physical exposure is a cause for significant vulnerability for both human populations and natural systems, human hotspots are a consequence of a lack of adaptive capacity, which is largely dependent upon development status. Investment in preparation to face the TCs and a sound post-event plan to deal with the possible exigencies due to TCs are extremely significant. Community level participation is quite essential to deal with the threat from any TC. This involves participatory contribution at the local level through proper information dissemination, resource sharing and active support to conserve coastal vegetation, maintaining health and hygiene as well as sharing of traditional knowledge with respect to cyclone early warning and mitigation strategies. It is pertinent to mention that in recent years, the Government in various countries prone to cyclone related disasters has been quite active in minimizing damage to life and property. To enable their success, community level engagement is necessary as a part of TC disaster preparedness and mitigation.

30.5.3 Hazard Mitigation

Mitigation strategies are crucial to face natural disasters and with minimal casualties. The success of mitigation strategies being adopted by policy makers and people alike has emerged as a primary reason for reduction in TC related fatalities in recent times even though TCs have increased in frequency and intensity over the years (Fig. 30.3). Mitigation involves both adaptation strategies as well as active pre-emptive measures for disaster preparedness. A few of the TC related adaptation strategies being followed in the NIO region are summarized in Table 30.6.

Natural vegetation significantly reduces the damage potentiation of land-falling TCs. Consequently, shorelines harbouring coastal vegetation are less prone to undergo cyclone related coastal damages. Globally most coastlines are lined with mangroves with mangrove resources found to spread over in approximately 117 countries, covering an area of 1,90,000 to 2,40,000 km^2, sustaining the highest level of productivity among natural ecosystems and performing several ecosystem services (Upadhyay et al. 2002). It has been widely observed that mangroves have effectively acted as barriers to multiple disasters. They are significant in reducing the

Table 30.6 A list of adaptation strategies obtained from literature

Concern	Adaptation strategy	References
Sea-level rise	Dune afforestation, restoration & management of mangroves, beach nourishment and construction of seawalls and groins	Dwarakish et al. (2009), Kumar et al. (2012)
Shoreline erosion	Beach filling & nourishment	Dwarakish et al. (2009), Mujabar and Chandrasekhar (2011)
Extreme events	Protecting mangroves and dense vegetation	Das (2009)
Saltwater intrusion	Use of strategically located recharge wells planned pumping and strategy development	Datta et al. (2009), Selvakumar et al. (2012)

Source Sudha Rani et al. (2015)

impact of land-falling cyclones considerably including preservation of the coastline. Mangroves have been effective in reducing storm surge water levels by slowing the flow of water and reducing surface waves. Numerical model studies also suggest mangroves to be more effective at reducing the water levels of fast moving surges than those of slow moving surges (Zhang et al. 2012). However, they are also exploited worldwide leading to the destruction of about half of the mangroves in recent times. Disaster mitigation strategies hence need to accommodate comprehensive management plans with specific thrust on mangrove ecosystems and spreading of environmental awareness on the importance of their restoration as an effective means for TC hazard mitigation along the coasts (Upadhyay et al. 2002). Similarly, coral reefs too contribute to naturally protecting the coasts from TC damage through a reduction in the impact of large waves heading for the coastline, thus limiting coastal erosion (Cuttler et al. 2018).

30.5.3.1 Policy and Planning

TCs have been identified as one of the major short-term challenges which are fast becoming a long-term challenge under the impact of climate change. Climate change has been identified to intensify TCs, accelerating sea-level rise and increasing coastal flooding as well. Since the coastal areas including the river deltas are densely populated but lie at low elevations, these regions are highly vulnerable to flooding. Edmonds et al. (2020) have estimated that 3,390 people inhabited the river deltas globally in 2017, 89% of which lived in the same latitudinal zone as most TC activity. They further estimated 41% of the global population to be exposed to TCs flooding the deltas directly, with 92% population of this 41% residing in developing or least-developed economies. Further, a significant percentage of this number are in the coastal regions which cannot naturally mitigate flooding through sediment deposition and hence calling for reframing strategies to cover the people who would be impacted disproportionately, particularly in developing and least-developed economies (Edmonds et al. 2020). Turner et al. (1996) had attributed an interrelated set of planning failures including information, economic market and policy intervention failures to the increasing stress on the coastal zones. They further stressed the evolution of integrated coastal zone management for the co-evolution of natural and human systems with a pre-emptive emphasis on adequate information acquisition, economically efficient resource pricing and proactive coastal planning.

It is necessary for adoption of policies that enhance social and economic equity, reduce poverty, increase consumption efficiencies, decrease the discharge of wastes, improve environmental management and increase the quality of life of vulnerable and other marginal coastal groups, thus leading to advance sustainable development and hence strengthening of adaptive capacity and coping mechanisms (Nicholls et al. 2007). Several key steps namely, mainstreaming the building of resilience and reduction of vulnerability (Agrawala and van Aalst 2005; McFadden et al. 2007), public participation, open data exchange, exploration scenarios for future adaptation policies and practices (Poumadère et al. 2005), study of ecological and socio-economic responses (Parson et al. 2003), generation of useful, usable and actionable information leading to science and policy gap reduction (Hay and Mimura 2006), linkages between comprehensive coastal management plans (Contreras-Espinosa and Warner 2004), strengthening institutions and enhancing regional cooperation and coordination (Bettencourt et al. 2005), coordination among oceans related agencies (West 2003) and short-term training for practitioners at all levels of management (Smith 2002) have been recognized as significant to TC disaster reduction planning and mitigation (Nicholls et al. 2007). This is also the means to develop active resilience to damages by TCs which is quite essential in addition to TC disaster mitigation.

30.5.3.2 Technology Intervention

The use of technology has proven crucial to TC disaster preparedness planning and mitigation. Various facets of technology intervention involve effective communication networks for conveying TC related warnings and messages through mobile phones and satellite radios, positioning systems for fishing boats, coastal wireless warning displays as well as remote assistance during cyclone events.

Advances and use of super computing technology in TC track and intensity modelling has led to better forecast capability thus improving TC hazard early warning. GIS-based modelling has enabled easy and fast disaster assessment. Besides the conventional use of technology, rather totally new attempts have been made towards eliminating the cause of damage itself as a means of TC damage control. It is well recognized that TCs dissipate when they are unable to extract sufficient energy from warm ocean water. Hence, it could be possible to either weaken TCs or completely dissipate them over the ocean itself artificially, thus preventing their landfall. Such attempts have gained attention in some parts of the world in which storms and TCs were artificially dissipated through the use of chemicals and explosives. Projects STORMFURY, CIRRUS and BATON were few such attempts to weaken TCs by flying aircraft into them and seeding with silver iodide (Havens 1952; Black et al. 1972; Willoughby et al. 1985). However, some of them were reported to have resulted in the formation of multiple hurricane eye walls rather than completely dissipating the main one. Hence, the success of such attempts is yet to be established.

30.5.3.3 International Cooperation

Global efforts with international cooperation are key to tackling TC hazards on an international level. This also involves formulation and implementation of globally recognized mitigation strategies. Modalities for TC/Hurricane management and adaptation strategies are regularly developed under the aegis of the World Meteorological Organization (WMO) with the regional bodies contributing to it (World Meteorological Organization 2019, 2020a, b, 2021a, b). The TC Programme under WMO enables easy accessibility to TC forecasters to various sources of conventional and specialized data and products, including numerical weather predictions and RS observations, as well as forecasting tools on the development, movement, intensification and wind distribution of TCs. The programme is instrumental in establishing national and regionally coordinated systems to ensure minimal loss of life and damage by TCs and functioning on national and regional levels through cooperative action.

In India, National Disaster Management Authority (NDMA) is the nodal agency to deal with TC disaster and mitigation along with its state level bodies. IMD is the nodal agency for providing TC forecasts and early warning as are the Joint Typhoon Warning Center (JTWC) National Hurricane Center and Central Pacific Hurricane Center for the Pacific and parts of Atlantic Oceans. The above organizations also represent India in the international for dealing with natural disasters including those of TCs. Thus, a comprehensive and cooperative approach forms the basis of international cooperation for TC hazard assessment and mitigation.

30.6 Summary

TCs or hurricanes have been among the worst natural hazards faced by mankind which have been intensifying with every passing year. Many scientists have attributed this to the impact of changing climate scenarios. NIO is one of the worst TC affected regions in the world due to several reasons including the large population density in the coastal regions, still developing economic conditions and rapidly depleting natural resources. However, over the years, it has also emerged as one of the regions most successful in facing TC hazards with least loss of lives. This has been partly due to the exhaustive mitigation strategies and policies being adopted by the regional Governments, inter-governmental cooperation, focused execution of disaster preparedness and the rapid awareness of the masses. This chapter attempted to review the various facets of TC damage, regional demography, resources, the damage assessment measures or metrics and disaster mitigation and sustainable policies being implemented in this part of the world in context of global efforts to contain the damage from TCs. Notable traces of TC hazards were analyzed in the form of short-term and long-term changes ranging from instantaneous damages to more visible long-term impacts as shoreline deformation and coastal erosion. It could be established that the extent of physical and social vulnerability due to TCs along the coastal regions bore a direct association with the population density as well as frequency and intensity of the hazards. TC events are not avoidable and so are the landfalls or a coast's physical vulnerability. Though some damages may be reversible with time, every effort should be made to safeguard the region from damages of the permanent kind through early assessment, preparedness and adequate situation analysis and management. Overview of the recent advances in TC forecast, strategies for disaster preparedness and long-term mitigation efforts both by the Government and the regional masses is the key to reduction in fatalities during TC hazards. The study revealed that maintaining and increasing coastal vegetation, avoiding excessive exploitation of coastal landforms as well as minimizing the factors inducing climate changes need to be particularly addressed for the coastal regions in the NIO as a part of better preparedness and mitigation strategy to contain the impacts of TCs in the region.

Acknowledgements The various organizations and data centres like IMD, ISRO, NOAA, JTWC, EM-DAT, CIESIN-Columbia University and several others are gratefully acknowledged for free accessibility to the various data sets forming part of the work. All literature cited in the text has been duly referenced. The author also thankfully acknowledges IIT Bhubaneswar for facilitating the completion of the work.

References

Agrawala S,van Aalst M (2005) Bridging the gap between climate change and development, In: Agrawala S (ed) Bridge over troubled waters: Linking climate change and development, Organisation for Economic Co-operation and Development (OECD), Paris, p 154. https://doi.org/10.1787/9789264012769-en

Ali MM, Kishtawal CM, Jain S (2007) Predicting cyclone tracks in the north Indian Ocean: An artificial neural network approach. Geophys Res Lett 34(L04603):1–5

Anthes R (ed) (2016), Tropical cyclones: Their evolution, structure and effects. Meteorological monograms, American Meteorological Society, vol 19. Springer, 208

Balaguru K, Taraphdar S, Leung LR, Foltz GR (2014) Increase in the intensity of post-monsoon Bay of Bengal tropical cyclones. Geophys Res Lett 41:3594–3601

Balaguru K, Foltz GR, Leung LR, D'Asaro E, Emanuel KA, Liu H, Zedler SE (2015) Dynamic Potential Intensity: An improved representation of the ocean's impact on tropical cyclones. Geophys Res Lett 42:1–8

Balaguru K, Foltz GR, Leung LR (2018) Increasing magnitude of hurricane rapid intensification in the central and eastern tropical Atlantic. Geophys Res Lett 45:1–10

Barnard PL, van Ormondt M, Erikson LH, Eshleman J, Hapke C, Ruggiero P, Adams PN, Foxgrover AC (2014) Development of the Coastal Storm Modeling System (CoSMoS) for predicting the impact of storms on high-energy, active-margin coasts. Nat Hazards 74:1095–1125
Bell GD, Halpert MS, Schnell RC, Higgins RW, Lawrimore J, Kousky VE, Tinker R, Thiaw W, Chelliah M, Artusa A (2000) Climate assessment for 1999. Bull Amer Meteor Soc 81(6):S1–S50
Bender MA, Ginis I, Kurihara Y (1993) Numerical simulations of tropical cyclone-ocean interaction with a high-resolution coupled model. J Gophys Res-Atmos 98(D12):23245–23263
Bettencourt S, Croad R, Freeman P, Hay J, Jones R, King P, Lal P, Mearns A, Miller G, Paswarayi-Riddihough I, Simpson A, Teuatabo N, Trotz U, Van Aalst (2005) Not if but when: Adapting to natural hazards in the Pacific Islands region, A policy note. Pacific Islands Country Management Unit, East Asia and Pacific Region, World Bank, p 46
Bhatia KT, Vecchi GA, Knutson TR, Murakami H, Kossin J, Dixon KW, Whitlock CE (2019) Recent increases in tropical cyclone intensification rates. Nat Commun 10(635):1–9
Black P, Senn H, Courtright C (1972) Airborne radar observations of eye configuration changes, bright band distribution, and precipitation tilt during the 1969 multiple seeding experiments in hurricane Debbie. Mon Weather Rev 100(3):208–217
Boruff BJ, Emrich C (2005) Cutter SL (2005) Erosion hazard vulnerability of US coastal counties. J Coastal Res 215:932–942
Bouchard, RH, Elliott J, Pounder D, Kern K (2014) Sea-change in ocean observations on moored buoys from the National Data Buoy Center (NDBC). In: Abstracts of the 2014 AGU Fall Meeting, San Francisco, 15–19 December 2014
Bukvic A, Rohat G, Apotsos A, de Sherbinin A (2020) A systematic review of coastal vulnerability mapping. Sustainability 12(2822):1–26
Center for International Earth Science Information Network (2007) Columbia University. http://sedac.ciesin.columbia.edu/gpw/lecz.jsp, Accessed 21 Jun 2021
Chan JCL, Duan Y, Shay LK (2001) Tropical cyclone intensity change from a simple ocean–atmosphere coupled model. J Atmos Sci 58(2):154–172
Chen SA, Frank WM (1993) A numerical study of the genesis of extratropical convective mesovortices. Part I: Evolution and dynamics. J Atmos Sci 50:2401–2426
Chen WB, Liu WC (2014) Modeling flood inundation induced by river flow and storm surges over a river basin. Water 6(10):3182–3199
Chen H, Li J, He W, Ma S, We Y, Pan J, Zhao Y, Zhang X, Hu S (2021) IAP's solar-powered unmanned surface vehicle actively passes through the center of typhoon Sinlaku (2020). Adv Atmos Sci 38:538–545
Cintineo JL, Pavolonis MJ, Sieglaff JM, Wimmers A, Brunner J, Bellon W (2020) A deep-learning model for automated detection of intense midlatitude convection using geostationary satellite images. Weather Forecast 35(6):2567–2588
Contreras-Espinosa F, Warner BG (2004) Ecosystem characteristics and management considerations for coastal wetlands in Mexico. Hydrobiologia 511:233–245
Cooper JAG, Orford JD (1998) Hurricanes as agents of mesoscale coastal changes in western Britain and Ireland. J Coastal Res 26:71–77
Cutter SL, Boru BJ, Shirley WL (2003) Social vulnerability to environmental hazards. Soc Sci Quart 84:242–261
Cuttler MVW, Hansen JE, Lowe RJ, Drost EJF (2018) Response of a fringing reef coastline to the direct impact of a tropical cyclone. Limnol Oceanogr Lett 3(2):31–38
Das S (2009) Addressing coastal vulnerability at the village level: The role of socio-economic and physical factors. Working paper series No. E/295/2009, p 30
Datta B, Vennalakanti H, Dhar A (2009) Modeling and control of saltwater intrusion in a coastal acquifier of Andhra Pradesh. India. J Hydroenviron Res 3(3):148–159
DINAS-COAST Consortium Staff (2006) DIVA 1.0. Potsdam Institute for Climate Impact Research, Potsdam, Germany, CD-ROM

Dinh Q, Balica S, Popescu I, Jonosk A (2012) @@ Climate change impact on flood hazard, vulnerability and risk of the Long Xuyen Quadrangle in the Mekong Delta. Int J River Basin Manag 10(1):103–120

Dube SK, Mazumder T, Das A (2006) An approach to vulnerability assessment for tropical cyclones: a case study of a coastal district in West Bengal. Inst Town Panners India J 3:15–27

Düing W, Leetmaa A (1980) Arabian Sea cooling, a preliminary heat budget. J Phys Oceanogr 10:307–312

Dvorak YE (1975) Tropical cyclone intensity analysis and forecasting from satellite imagery. Mon Weather Rev 103:420–430

Dwarakish GS, Vinay SA, Natesan U (2009) Coastal vulnerability assessment of the future sea level rise in Udupi coastal zone of Karnataka state, west coast of India. J Ocean Coast Manag 52:467–478

Edmonds DA, Caldwell RL, Brondizio ES, Sacha MO, Siani ES (2020) Coastal flooding will disproportionately impact people on river deltas. Nat Commun 11:4741

Emanuel K (1993) The physics of tropical cyclogenesis over the Eastern Pacific. In: Lighthill J, Zhemin Z, Holland GJ, Emanuel K (eds) Tropical cyclone disasters. Peking University Press, Beijing, pp 136–142

Emanuel K (2005) Increasing destructiveness of tropical cyclones over the past 30 years. Nature 436(7051):686–688

Emergency Events Database (2020) OFDA/CRED international disaster database. Université catholique de Louvain, Brussels, Belgium. https://www.emdat.be. Accessed 23 June 2021

Evans D, Conrad CL, Paul FM (2003) Handbook of automated data quality control checks and procedures of the National Data Buoy Center. NOAA/National Data Buoy Center Technical Document 03-02, NDBC, Stennis Space Center, Mississippi, p 44

Franklin JL, Black ML, Valde K (2003) GPS dropwindsonde wind profiles in hurricanes and their operational implications. Weather Forecast 18:32–44

Füssel H-M (2010) Review and quantitative analysis of indices of climate change exposure, adaptive capacity, sensitivity, and impacts. The World Bank, Washington, DC, USA. https://openknowledge.worldbank.org/handle/10986/9193. Accessed 23 June 2021

Glazer RH, Ali MM (2020) An improved potential intensity estimate for Bay of Bengal tropical cyclones. Nat Hazards 104:2635–2644

Gopalakrishnan SG, Marks F Jr, Zhang X, Bao J, Yeh K, Atlas R (2011) The Experimental HWRF system: a study on the influence of horizontal resolution on the structure and intensity changes in tropical cyclones using an idealized framework. Mon Weather Rev 139(6):1762–1784

Gornitz VM, Daniels RC, White TW, Birdwell KR (1994) The development of a coastal risk assessment database: vulnerability to sea-level rise in the US Southeast. J Coastal Res 12:327–338

Gray WM (1968) A global view of the origin of tropical disturbances and storms. Mon Weather Rev 96:669–700

Gray WM (1979) Hurricanes: Their formation, structure and likely role in the tropical circulation. In: Shaw DB (ed) Meteorology over tropical oceans. Royal Meteorological Society, Bracknell, Berkshire, pp 155–218

Gray WM (1988) Forecast of Atlantic seasonal hurricane activity for 1988. Colorado State University, Fort Collins, Colorado, pp 13–14

Group TW (1988) The WAM model—a third generation ocean wave prediction model. J Phys Oceanogr 18(12):1775–1810

Havens BS (1952) History of project cirrus. General Electric Company, Research Laboratory, RL-756, Schenectady, New York, p 103

Hay JE, Mimura N (2006) Supporting climate change vulnerability and adaptation assessments in the Asia-Pacific Region–an example of sustainability science. Sustain Sci 1:23–25

Hebert PH, Poteat KO (1975) A satellite classification technique for subtropical cyclones. NOAA Tech Memo NWS SR-83, p 25

Hebert C, Weinzapfel B, Chambers M (2008) The Hurricane Severity Index–a destructive potential rating system for tropical cyclones. In: Poster presented at the 28th conference on hurricanes and tropical meteorology, American Meteorological Society, Orlando, FL, 28 Apr–2 May 2008
Hernández ML, Carreño ML, Castillo L (2018) Methodologies and tools of risk management: Hurricane Risk index (HRi). Int J Disast Risk Re 31:926–937
Hickey KR (2011) The impact of Hurricanes on the weather of Western Europe. In: Lupo AR (ed) Recent hurricane research–climate, dynamics and societal impacts. IntechOpen. https://doi.org/10.5772/14487
Hu T, Smith RB (2018) The impact of hurricane Maria on the vegetation of Dominica and Puerto Rico using multispectral remote sensing. Remote Sens 10(827):1–20
Huler S (2004) Defining the wind: the Beaufort scale, and how a nineteenth-century admiral turned science into poetry. Crown Publishers, New York, p 290
Hung L-S, Wang C, Yarnal B (2016) Vulnerability of families and households to natural hazards: a case study of storm surge flooding in Sarasota County, Florida. Appl Geogr 76:184–197
Ikeuchi H, Hirabayashi Y, Yamazaki D, Muis S, Ward PJ, Winsemius HC, Verlaan M, Kanae S (2017) Compound simulation of fluvial floods and storm surges in a global coupled river-coast flood model: model development and its application to 2007 Cyclone Sidr in Bangladesh. J Adv Model Earth Sy 9(4):1847–1862
India Meteorological Department (2021a) Cyclone warning in India: standard operation procedure. India Meteorological Dept, MoES, Govt of India, p 220. https://mausam.imd.gov.in/imd_latest/contents/pdf/cyclone_sop.pdf. Accessed 15 June 2021
India Meteorological Department (2021b) Annual frequency of cyclonic disturbances (Maximum sustained wind speeds of 17 knots or more), Cyclones (34 knots or more) and Severe Cyclones (48 knots or more) over the Bay of Bengal (BOB), Arabian Sea (AS) and land surface of India. India Meteorological Department. http://www.imd.gov.in/section/nhac/dynamic/ANNUAL_FREQ_CYCLONIC_DISTURBANCES.pdf. Accessed 29 June 2021
Islam MA, Mitra D, Dewan A, Akhter SH (2016) Coastal multi-hazard vulnerability assessment along the Ganges deltaic coast of Bangladesh—a geospatial approach. Ocean Coast Manage 127:1–15
Jangir B, Swain D, Ghose SK, Goyal R, Udaya Bhaskar TVS (2020) Inter-comparison of model, satellite and in situ tropical cyclone heat potential in the North Indian Ocean. Nat Hazards 102(2):557–574
Jayaram Ch, Udaya Bhaskar TVS, Swain D, Rama Rao EP, Bansal S, Dutta D, Rao KH (2014) Daily composite wind fields from Oceansat-2 scatterometer. Remote Sens Lett 5(3):258–267
Jena BK, Sivakholundu KM, John M (2015) Surface current and wave measurement during cyclone Phailin by high frequency radars along the Indian coast. Curr Sci 108(3):405–409
Kantha L (2006) Time to replace the Saffir-Simpson Hurricane scale? Eos Trans AGU 87(1):3–6
Kantha L (2008) Tropical cyclone destruction potential by integrated kinetic energy. Bull Am Meteor Soc 88(11):219–221
Karthick S, Malathi D (2020) A performance evaluation of hybrid genetic algorithm approach for forecasting tropical cyclone categories. Int J Electr Eng Educ. https://doi.org/10.1177/0020720919891076
Katsaros KB, Vachon PW, Liu WT, Black PG (2002) Microwave remote sensing of tropical cyclones from space. J Oceanogr 58:137–151
Kennedy AB, Westerink JJ, Smith JM, Hope ME, Hartman M, Taflanidis AA, Tanaka S, Westerink H, Cheung KF, Smith T, Hamann M, Minamide M, Ota A, Dawson C (2012) Tropical cyclone inundation potential on the Hawaiian Islands of Oahu and Kauai. Ocean Model 52–53:54–68
Kleinosky LR, Yarnal B, Fisher A (2007) Vulnerability of Hampton Roads, virginia to storm-surge flooding and sea-level rise. Nat Hazards 40:43–70
Kraft RH (1961) The hurricane's central pressure and highest wind. Mariners Weather Log 5:157
Kumar VS, Dora GU, Philip CS (2012) Nearshore currents along the Karnataka coast, west coast of India. Int J Ocean Clim Syst 3(1):71–84

Li Z, Chen M, Gao S, Luo X, Gourley JJ, Kirstetter P, Yang T, Kolar R, McGovern A, Wen Y, Rao B, Yami T, Hong Y (2021) CREST-iMAP v1.0: a fully coupled hydrologic-hydraulic modeling framework dedicated to flood inundation mapping and prediction. Environ Model Softw 141(105051)

Lin I-I, Black P, Price JF, Yang C–Y, Chen SS, Lien C–C, Harr P, Chi N–H, Wu C–C, D'Asaro EA (2013) An ocean coupling potential intensity index for tropical cyclones. Geophys Res Lett 40:1878–1882

Mahesh Kumar U, Sasamal SK, Swain D, Narendra Reddy N, Ramanjappa T (2015) Intercomparison of geophysical parameters from SARAL/AltiKa and Jason-2 altimeters. IEEE J Sel Top Appl Earth Obs Remote Sens 8(10):4863–4870

Maina SN (2010) Developing a hurricane damage index. https://doi.org/10.5065/j0t6-vr11. Accessed 22 June 2021

Manion A, Evans C, Olander TL, Velden CS, Grasso LD (2015) An evaluation of advanced Dvorak technique–derived tropical cyclone intensity estimates during extratropical transition using synthetic satellite imagery. Weather Forecast 30(4):984–1009

Matsuoka D, Nakano M, Sugiyama D, Uchida S (2018) Deep learning approach for detecting tropical cyclones and their precursors in the simulation by a cloud-resolving global nonhydrostatic atmospheric model. Prog Earth Planet Sci 5(80):1–16

McFadden L, Nicholls RJ, Penning-Rowsell E (2007) Managing coastal vulnerability. Elsevier, Oxford, p 282

McGranahan, G, Balk D, Anderson B (2007) Low elevation coastal zone (LECZ) urban-rural population estimates. Global Rural-Urban Mapping Project (GRUMP), Alpha Version. NASA Socioeconomic Data and Applications Center (SEDAC), Palisades, NY. https://doi.org/10.7927/H4TM782G. Accessed 21 June 2021

McPhaden MJ, Busalacchi AJ, Cheney R, Donguy JR, Gage KS, Halpern D, Ji M, Julian P, Meyers G, Mitchum GT, Niiler PP, Picaut J, Reynolds RW, Smith N, Takeuchi K (1998) The tropical ocean global atmosphere (TOGA) observing system: a decade of progress. J Gophys Res-Oceans 103(C7):14169–14260

Merkens J-L, Reimann L, Hinkel J, Vafeidis AT (2016) Gridded population projections for the coastal zone under the Shared Socioeconomic Pathways. Glob Planet Change 145:57–66

Miller DW, Lander MA (1997) Intensity estimation of tropical cyclones during extratropical transition. JTWC Rep JTWC/SATOPS/TN-97/002, p 9

Misra V, DiNapoli S, Powell M (2013) The track integrated kinetic energy of Atlantic tropical cyclones. Mon Weather Rev 141(7):2383–2389

Mohanty UC, Gupta A (1997) Deterministic methods for prediction of tropical cyclone tracks. Mausam 48(2):257–272

Mujabar PS, Chandrasekhar N (2011) Coastal erosion hazard and vulnerability assessment for southern coastal Tamil Nadu of India using remote sensing and GIS. Nat Hazards 69:1295–1314

Nallathiga R (2010) Analysing the physical, demographic and vulnerability profile of Indian coastal zone. Asia Pac Bus Rev 6(1):90–104

Neumann B, Vafeidis AT, Zimmermann J, Nicholls RJ (2015) Future coastal population growth and exposure to sea-level rise and coastal flooding—a global assessment. PLoS One 10(3):e0118571

Neumann CJ, Mandal GS (1978) Statistical prediction of tropical storm motion over the Bay of Bengal and Arabian Sea. Indian J Meteorol Hydrol Geophys (mausam) 29(3):487–500

Neumann CJ, Randrianarison EA (1976) Statistical prediction of tropical cyclone motion over the southwest Indian Ocean. Mon Weather Rev 104:76–85

Nicholls RJ, Wong PP, Burkett VR, Codignotto JO, Hay JE, McLean RF, Ragoonaden S, Woodroffe CD (2007) Coastal systems and low-lying areas. In: Parry ML, Canziani OF, Palutikof JP, van der Linden PJ, Hanson CE (eds) Climate change 2007: impacts, adaptation and vulnerability. Cambridge University Press, Cambridge, UK, pp 315–356

Nirupama N (2013) Vertical evacuation during cyclones: suitable for developing countries. Nat Hazards 69:1137–1142

Olander TL, Velden CS (2015) ADT–Advanced Dvorak Technique Users' Guide (McIDAS Ver. 8.2.1). Cooperative Institute for Meteorological Satellite Studies Rep, University of Wisconsin, Madison, p 49
Owens B, Holland GJ (2009) The Willis Hurricane Index. In: Paper presented at the 2nd international summit on hurricanes and climate change, Corfu, Greece, 31 May–5 June 2009
Parson EA, Corell RW, Barron EJ, Burkett V, Janetos A, Joyce L, Karl TR, MacCracken MC, Melillo J, Morgan MG, Schimel DS, Wilbanks T (2003) Understanding climatic impacts, vulnerabilities, and adaptation in the United States: Building a capacity for assessment. Clim Change 57(1):9–42
Peduzzi P, Chatenoux B, Dao H, De Bono A, Herold C, Kossin J, Mouton F, Nordbeck O (2012) Global trends in tropical cyclone risk. Nat Clim Change 2:289–294
PIS (2012) Public information statement, NOAA. https://www.nhc.noaa.gov/news/20120301_pis_sshws.php. Accessed 25 June 2021
Poumadère M, Mays C, Pfeifle G, Vafeidis AT (2005) Worst case scenario and stakeholder group decision: A 5–6 meter sea level rise in the Rhone Delta, France. Working Paper FNU-76, Hamburg University and Centre for Marine and Atmospheric Science, Hamburg, p 30. https://epub.sub.uni-hamburg.de/epub/volltexte/2012/16505/pdf/waiscamarguewp_FNU76.pdf. Accessed 21 June 2021
Powell MD, Reinhold TA (2007) Tropical cyclone destructive potential by integrated kinetic energy. Bull Amer Meteor Soc 88(4):513–526
Prakash KR, Pant V (2017) Upper oceanic response to tropical cyclone Phailin in the Bay of Bengal using a coupled atmosphere-ocean model. Ocean Dyn 67:51–64
PreventionWeb (2021) PreventionWeb: the knowledge platform for disaster risk reduction. http://preventionweb.net/go/7689. Accessed 21 June 2021
Quilfen Y, Chapron B, Elfouhaily T, Katsaros K, Tournadre J (1998) Observation of tropical cyclones by high-resolution scatterometry. J Gophys Res-Oceans 103(C4):7767–7786
Ray K, Giri RK, Ray SS, Dimri AP, Rajeevan M (2021) An assessment of long-term changes in mortalities due to extreme weather events in India: a study of 50 years' data, 1970–2019. Weather Clim Extremes 32(100315):1–10
Ribal A, Young IR (2020) Calibration and cross-validation of global ocean wind speed based on scatterometer observations. J Atmos Ocean Technol 37:279–297
Ritchie H, Roser M (2014) Natural disasters. Available on OurWorldInData.org. https://ourworldindata.org/natural-disaster. Accessed 23 June 2021
Rodgers E, Gentry RC, Shenk W, Oliver V (1979) The Bbnefits of using short-interval satellite images to derive winds for tropical cyclones. Mon Weather Rev 107(5):575–584
Rodgers JC III, Murrah AW, Cooke WH (2009) The impact of Hurricane Katrina on the coastal vegetation of the Weeks Bay Reserve Alabama from NDVI Data. Estuar Coast 32(3):496–507
Rohli RV, Vega AJ (2018) Climatology, 4th edn. Jones & Bartlett Learning, Sudbury, Massachusetts, Burlington, p 418
Rozoff CM, Kossin J, Velden C, Wimmers A, Kieper M, Kaplan J, Knaff J, DeMaria M (2011) Improvements in the statistical prediction of tropical cyclone rapid intensification. In: Extended abstracts of the 65th interdepartmental hurricane conference, US Department of Commerce, National Oceanic and Atmospheric Administration (NOAA), Miami, FL, 28 Feb–3 Mar 2011
Saffir HS (1973) Hurricane wind and storm surge. Military Eng 423:4–5
Sanil Kumar V, Pathak KC, Pednekar P, Raju NSN, Gowthaman R (2006) Coastal processes along the Indian coastline. Curr Sci 91(4):530–536
Sasamal S, Jena B (2016) Monitoring of ocean environmental changes under influence of cyclonic system: utilization of SARAL and contemporary satellite observations. Remote Sens Appl Soc Environ 4:204–210
Sahoo B, Bhaskaran PK (2017) Coastal vulnerability index and its projection for Odisha coast, east coast of India. Int J Environ Ecol Eng 11(6):529–533
Sahoo B, Bhaskaran PK (2018) Multi-hazard risk assessment of coastal vulnerability from tropical cyclones-A GIS based approach for the Odisha coast. J Environ Manage 206(2018):1166–1178

Selvakumar S, Chandrasekhar N, Magesh NS (2012) Preliminary investigation of groundwater quality along the coastal aquifers of southern Tamil Nadu using GIS techniques. Bonfring Int J Ind Eng Manag Sci 2(1):46–52
Sharma N, Ali MM, Knaff JA, Purna Chand C (2013) A soft-computing cyclone intensity prediction scheme for the Western North Pacific Ocean. Atmos Sci Lett 14:187–192
Simpson RH (1974) The hurricane disaster potential scale. Weatherwise 27(169):169–186
Singh HS (2003) Marine protected areas in India. Indian J Mar Sci 32(3):226–233
Singh O, Ali Khan T, Rahman M (2000) Changes in the frequency of tropical cyclones over the North Indian Ocean. Meteorol Atmos Phys 75:11–20
Smith HD (2002) The role of the social sciences in capacity building in ocean and coastal management. Ocean Coast Manage 45:573–582
Song J, Duan Y, Klotzbach PJ (2020) Increasing trend in rapid intensification magnitude of tropical cyclones over the western North Pacific. Environ Res Lett 15(8:084043):1–10
Sudha Rani NNV, Satyanarayana ANV, Bhaskaran PK (2015) Coastal vulnerability assessment studies over India: a review. Nat Hazards 77:405–428
Tamizi A, Young IR (2020) The spatial distribution of ocean waves in tropical cyclones. J Phys Oceanogr 50(8):2123–2139
Tang B, Emanuel K (2012) A ventilation index for tropical cyclones. Bull Am Meteor Soc 93:1901–1912
Taramelli A, Melelli L, Pasqui M, Sorichetta A (2010) Modelling hurricane element at risk in potentially affected areas by GIS system. Geomat Nat Haz Risk 1(4):349–373
The WAVEWATCH III Development Group (WW3DG) (2019) User manual and system documentation of WAVEWATCH III version 6.07. NOAA/NWS/NCEP/MMAB Tech Note 333, College Park, MD, USA, p 465
Thieler ER, Hammar-Klose ES (1999) National assessment of coastal vulnerability to sea-level rise: preliminary results for the U.S. Atlantic coast. U.S. Geological Survey Open-File Rep 99-593, U.S. Geological Survey, Woods Hole, MA, USA. https://pubs.usgs.gov/of/1999/of99-593. Accessed 12 June 2021
Turner RK, Subak S, Adger WN (1996) Pressures, trends, and impacts in coastal zones: interactions between socioeconomic and natural systems. Environ Manage 20:159–173
Turton S (2008) Editorial: cyclones Larry and Monica: ecological effects of two major disturbance events. Austral Ecol 33(4):365–367
Upadhyay VP, Ranjan R, Singh JS (2002) Human–mangrove conflicts: the way out. Curr Sci 83(11):1328–1336
Velasco I, Fritsch JM (1987) Mesoscale convective complexes in the Americas. J Geophys Res 92:9561–9613
Velden C, Harper B, Wells F, Beven II JL, Zehr R, Olander T, Mayfield M, Guard C, Lander M, Edson R, Avila L, Burton A, Turk M, Kikuchi A, Christian A, Caroff P, McCrone P (2006) The Dvorak tropical cyclone intensity estimation technique: a satellite-based method that has endured for over 30 years. Bull Amer Meteor Soc 87(9):1195–1214
Venkataraman K, Wafar M (2005) Coastal and marine biodiversity of India. Ind J Mar Sci 34(1):57–75
Venkatesan R, Shamji VR, Latha G, Mathew S, Rao RR, Arul Muthiah M, Atmanand MA (2013) New in-situ ocean subsurface time series measurements from OMNI buoy network in the Bay of Bengal. Curr Sci 104(9):1166–1177
Venkatesan R, Mathew S, Vimala J, Latha G, Muthiah MA, Ramasundaram S, Sundar R, Lavanya R, Atmanand MA (2014) Signatures of very severe cyclonic storm Phailin in met–ocean parameters observed by moored buoy network in the Bay of Bengal. Curr Sci 107(4):589–595
Verlaan PA, Morgan JR, Balakrishna S (2009) Bay of Bengal. In: Encyclopedia Britannica https://www.britannica.com/place/Bay-of-Bengal. Accessed 28 June 2021
Verlaan PA, Aleem AA, Morgan JR (2019) Arabian Sea. In: Encyclopedia Britannica. https://www.britannica.com/place/Arabian-Sea. Accessed 28 June 2021

Visher S (1925) Effects of tropical cyclones upon the weather of Mid-Latitudes. Geogr Rev 15(1):106–114

West MB (2003) Improving science applications to coastal management. Mar Policy 27:291–293

Willoughby HE, Jorgensen DP, Black RA, Rosenthal SL (1985) Project STORMFURY: a scientific chronicle 1962–1983. Bull Amer Meteor Soc 66(5):505–514

Wimmers A, Velden C, Cossuth JH (2019) Using deep learning to estimate tropical cyclone intensity from satellite passive microwave imagery. Mon Weather Rev 147(6):2261–2282

Wimmers A, Velden C (2012) Advances in objective tropical cyclone center fixing using Multi-spectral satellite imagery. In: Poster presented at the 30th conference on hurricanes and tropical meteorology, American Meteorological Society, Ponte Vedra Beach, FL, 15–20 Apr 2012

World Meteorological Organisation (2021a) Regional association IV–Hurricane operational plan for North America, Central America and the Caribbean. Tropical Cyclone Programme Rep TCP-30, WMO No. 1163, Geneva, Switzerland

World Meteorological Organisation (2021b) Typhoon committee operational manual: meteorological component. Tropical Cyclone Programme Rep TCP-23, WMO TD No. 196, Geneva, Switzerland, p 133

World Meteorological Organisation (2019) Regional association I–Tropical cyclone operational plan for the South-West Indian Ocean. Tropical Cyclone Programme Rep TCP-12, WMO No. 1178, TCP-12, Geneva, Switzerland

Xiang K, Yang X, Zhang M, Ziwei L, Kong F (2019) Objective estimation of tropical cyclone intensity from active and passive microwave remote sensing observations in the Northwestern Pacific Ocean. Remote Sens 11(627):1–18

Young IR, Vinoth J (2013) An 'extended fetch' model for the spatial distribution of tropical cyclone wind-waves as observed by altimeter. Ocean Eng 70:14–24

Zeng Z, Wang Y, Wu C-C (2007) Environmental dynamical control of tropical cyclone intensity–an observational study. Mon Weather Rev 135:38–59

Zhang KQ, Liu H, Li Y, Hongzhou X, Jian S, Rhome J, Smith TJ III (2012) The role of mangroves in attenuating storm surges. Estuar Coast Shelf Sci 102:11–23

Zhang Z, Tallapragada V, Kieu C, Trahan S, Wang W (2014) HWRF based ensemble prediction system using perturbations from GEFS and stochastic convective trigger function. Trop Cyclone Res Rev 3(3):145–161

Printed in the United States
by Baker & Taylor Publisher Services